T0250373

Combinatorics of Spreads and Parallelisms

PURE AND APPLIED MATHEMATICS

A Program of Monographs, Textbooks, and Lecture Notes

MONOGRAPHS AND TEXTBOOKS IN PURE AND APPLIED MATHEMATICS

Recent Titles

Combinatorics of Spreads and Parallelisms

Norman L. Johnson

University of Iowa
Iowa City, U.S.A.

CRC Press
Taylor & Francis Group
Boca Raton   London   New York

CRC Press is an imprint of the
Taylor & Francis Group, an **informa** business

A CHAPMAN & HALL BOOK

CRC Press
Taylor & Francis Group
6000 Broken Sound Parkway NW, Suite 300
Boca Raton, FL 33487-2742

© 2010 by Taylor and Francis Group, LLC
CRC Press is an imprint of Taylor & Francis Group, an Informa business

No claim to original U.S. Government works

Printed in the United States of America on acid-free paper
10 9 8 7 6 5 4 3 2 1

International Standard Book Number: 978-1-4398-1946-3 (Hardback)

Library of Congress Cataloging-in-Publication Data

Johnson, Norman L. (Norman Lloyd), 1939-
    Combinatorics of spreads and parallelisms / Norman L. Johnson.
        p. cm. -- (Pure and applied mathematics)
    Includes bibliographical references and index.
    ISBN 978-1-4398-1946-3 (hardcover : alk. paper)
        1. Vector spaces. 2. Algebraic spaces. 3. Geometry, Projective. 4. Incidence algebras.
    I. Title. II. Series.

QA186.J64 2010
512'.52--dc22   2010018110

Visit the Taylor & Francis Web site at
http://www.taylorandfrancis.com

and the CRC Press Web site at
http://www.crcpress.com

I dedicate this text to my wife
Bonnie Lynn Hemenover
and to my children
Catherine Elizabeth Johnson
Garret Norman Johnson
Scott Hamilton Hemenover

Contents

Preface

It may be said that a substantial part of the theory of incidence geometry is devoted to questions about 'covers' of the set of points or lines (or 'blocks') by various means. It has been fashionable recently to consider 'flocks' of quadric sets, which are covers of elliptic, hyperbolic quadrics or quadratic cones by maximal sets of mutually disjoint conics. All of these covers are, in turn, equivalent to certain partitions of a four-dimensional vector space by sets of mutually disjoint two-dimensional vector subspaces, which is one example of a 'spread' of a vector space.

Indeed, vector space 'spreads' in this context are equivalent to 'projective spreads' of a three-dimensional projective space. In this context, projective spreads are sets of mutually skew lines that form a disjoint cover of the set of points. These objects are of considerable interest in that they construct very interesting examples of projective or affine planes. Going a bit further with the ideas of covers, a 'packing' or 'parallelism' of the three-dimensional projective space is a disjoint cover of the set of lines by a set of spreads. In a very real sense, this is perhaps the most fundamental of the covering objects in incidence geometry and at the same time one that has, until recently, very few 'interesting' or 'nice' examples.

Many geometers are drawn to the subject of incidence geometry by the elegant beauty and simplicity of the examples. Of course, there now are a number of fascinating examples of 'parallelisms,' and the point of the text is to show how to construct most if not all of the more interesting ones.

It seems that spreads are also very much of interest in the more general sense, that of a set of mutually disjoint vector subspaces of the same dimension of a given finite dimensional vector space. Not much is really known of spreads that have nice group properties, but what is known shall be essentially given in this text. More generally, arbitrary partitions of vectors by subspaces not necessarily of the same dimension are beginning to have applications. Recently, general spreads of vector spaces have been used to study and construct 'subgeometry partitions of projective spaces.' Here there are some very basic questions, such as what might be the variety of the subgeometries involved in a partition of a projective space?

DEFINITION 1. *Let **P** be a projective space. A 'parallelism' of **P** is an equivalence relation on the set $\mathcal{L}$ of lines such that the Euclidean parallel postulate holds.*

Let ℓ be a line and denote by $[\ell]$, the equivalence class containing ℓ. Then let P be any point of **P** that is not incident with ℓ. Then there exists a unique line ℓ_P, incident with P and 'parallel' to ℓ. Hence, there is a partition of the points of **P** by any equivalence class.

DEFINITION 2. *A partition of the points of a projective space **P** by a set of disjoint lines is said to be a 'line spread' of **P**.*

Just as we have pointed out previously:

REMARK 1. *Since partitions of the line set of a projective space and equivalence relations are equivalent, we see that a partition of the lines of a projective space by spreads is equivalent to a parallelism of a projective space.*

Why are parallelisms of projective spaces of interest to geometers? Suppose the spreads of a parallelism produce 'field' affine planes (or projective planes). We shall get to the exact meaning of this in due course, but merely think of the affine Euclidean plane over the field of real numbers $\mathcal{R}$, where we insert an arbitrary field for the field $\mathcal{R}$. Since field planes are so very easy to construct, it would seem likely that there are many parallelisms consisting of field spreads. However, it turns out that such parallelisms are quite hard to come by. For reasons that will come later, we call such parallelisms 'regular' and the effort to find regular parallelisms does pay off in the construction of new and quite amazing affine planes.

In graduate school in 1964 at Washington State University, I was studying algebraic geometry when my advisor thought that I should broaden my reading and suggested that I take a course in finite geometry from T.G. Ostrom, who was charting some very new directions in the construction of affine and projective planes. The short version of this story is that I never went back to algebraic geometry and became fascinated by the ideas of 'derivation' that Ostrom was developing. Derivation is a construction method by which a renaming of certain subplanes (called 'Baer subplanes') as 'new' lines together with certain of the 'old' lines sometimes can produce a totally different affine plane from a given one. Projectively, a projective Baer subplane is a subplane that touches all points and lines, in the sense that every point is incident with some line of the subplane and every line is incident with some point of the subplane. The derivation process is perhaps the most important construction procedure in the construction of affine and projective planes. The reader interested in more details on derivation is directed to the author's text Subplane Covered Nets, [114].

Actually, this text forms the fourth volume of a series of works on finite geometry and a word on where the present volume fits is appropriate. In the text Subplane Covered Nets, [114], there is a complete theory of derivation. The analysis of a derivable affine plane focuses more properly on the associated net that contains the Baer subplanes that are used in the process and which are redefined as lines in the derived affine plane. Considered in this manner, a derivable net is a net that is covered by subplanes in the sense that given any two affine points P and Q of the net that are collinear, there is a unique subplane $\pi_{P,Q}$ of the net that contains P and Q and which shares the set of parallel classes of the net as the line at infinity of the corresponding projective plane. The text then concerns the analogous theory of nets that are covered by subplanes as in the previous manner. In this text, there is the introduction of the direct product of affine planes of the same order (or cardinality), which becomes a net containing the two affine subplanes. When the two affine subplanes are Pappian, the constructed net is a derivable net, and in this case, we identify the affine subplanes. If the Pappian subplanes correspond to quadratic field extensions of a given field K, then repeated application of these ideas produces a set of derivable nets that share a regulus net coordinatized by K. If L is a component of this regulus net and is a four-dimensional K-space, then any Pappian K-spread of L induces a direct product derivable net sharing the regulus net. Considering L as a projective space $PG(3, K)$, then any Pappian parallelism of $PG(3, K)$ induces a set of derivable nets, the union of which produces a spread in $PG(7, K)$ for a translation plane admitting $SL(2, K)$.

Thus, the text Subplane Covered Nets, provides ideas for the analysis of parallelisms of projective spaces. In this work, we shall repeat parts of the theory of direct products that applies to parallelisms, so as to keep this text as self-contained as possible.

The study of parallelisms in $PG(3, q)$ involves, of course, a set of $q^2 + q + 1$ spreads in $PG(3, q)$ that construct translation planes. In the work Foundations of Translation Planes, [21], a complete background is given not of only of translation planes of "dimension two," which arise from spreads in $PG(3, q)$, but also of general translation planes. The reader is directed to the foundations text for general information on translation planes, but again, we shall include enough of the theory here so that the reader is not necessarily dependent on this previous text.

Suppose that a parallelism in $PG(3, q)$ admits a collineation group that leaves one spread invariant and acts transitively on the remaining $q^2 + q$ spreads. Then we shall show that the fixed spread defines a Desarguesian translation plane and the remaining spreads define translation planes of order q^2 with associated spreads in $PG(3, q)$ that admit Baer groups of order q (a group of order q that fixes a Baer subplane pointwise). Such translation planes actually correspond to flocks of quadratic cones and so certain of the theory of flocks of quadratic sets shall be developed in this text.

Considering that a parallelism is defined as a covering of the line set by a spread, a natural generalization would be to consider analogous geometries as follows: If a spread is a covering of the point set by mutually disjoint 'blocks,' then a covering of the block set by spreads can be called a parallelism. When we consider parallelisms of quadric sets in $PG(3, K)$, for K a field, we would used the term 'flock' instead of spread so that a parallelism of a quadric set is a covering of the conics by a set of mutually disjoint flocks. Generalizations include parallelisms of sharply k-transitive geometries, for $k \geq 2$.

In general, there is a tremendous variety of connections with the theory of translation planes and other point-line geometries, all of which are explicated in the text Handbook of Finite Translation Planes, [138]. What has not been done is to give a comprehensive and "up to date" treatment of parallelisms of projective spaces and other geometries, and, of course, the intent here is to do just that. We have mentioned translation planes and the vector space partitions that produce them but there are other spreads that do not produce translation planes but may produce Sperner Spaces. One important class of geometric structures associated with Sperner Spaces is that of a subgeometry partition of a projective space. More generally, a subgeometry partition of a projective space need not produce or correspond to a Sperner space or a translation plane but always gives rise to a general partition of a vector space. Even though there are some examples of partitions of vector spaces known, there is almost no theory that has developed from such covers. Here we also provide a census of most of the known examples of partitions and what can be said theoretically of associated point-line geometries. Since there are many new subgeometry partitions constructed, we then provide a complete description of all known examples.

Therefore, it is felt that this fourth text Combinatorics of Spreads and Parallelisms, is a strong complement to the previous three texts

Subplane Covered Nets, Foundations of Translation Planes, and Handbook of Finite Translation Planes.

Acknowledgments. It seems that perhaps an indication of how I became interested in the general study of parallelisms might be appropriate, and for which I feel I need to acknowledge a number of people.

In 1985, my son Scott and I spent the spring semester and summer in Italy, where I met many Italian geometers and many still creating wonderful mathematics. In particular, in Lecce, I am indebted to Giuseppe (Pino) Micelli and Rosanna Marinosci for their wonderful hospitality to Scott, my wife Bonnie (who joined Scott and me later in the spring semester) and of course, to me. Also, many thanks are due to Gabriella Murciano, who was instrumental in introducing Scott to a wonderful group of young people.

While I was working mostly with Mauro Biliotti of Lecce, we persuaded Vik Jha to spend a few weeks with us and about that time we began looking at the work of Mike Walker that connected regular parallelisms with direct products of Desarguesian affine planes. I was also privileged to be invited to visit Naples and to work with Guglielmo Lunardon, who had independently determined the connection with regular parallelisms and direct products. And, this is where my real interest in 'parallelisms' was brought to light, so to speak.

My wife Bonnie had joined Scott and myself for a month in Palermo while I was working with Claudio Bartolone and most of that time we persuaded Vik Jha to come to Palermo as well. Sometime during that visit and the previous period in Lecce, Vik and I proved that regular parallelisms could be realized from translation planes of order q^4 that admit $SL(2, q) \times C_{1+q+q^2}$, with the idea of applying the general theory of Lunardon and Walker. However, there were only two examples of regular parallelisms known at that time, something that has since dramatically changed. For that great trip, I am very much indebted to Claudio Bartolone.

When Bonnie, Scott, and I finally arrived in Naples, after a very, very long night ship from Palermo, we found Guglielmo at the dock waving wildly to us. Bonnie and Scott went off to bed and Guglielmo and I went to talk—and we talked continuously for a week about 'parallelisms,' and more 'parallelisms.' Guglielmo and Vik really did hook me on this subject that year, and I very much appreciate both of their influences and help on my work, as well as on the general area.

I might add that I returned to Italy many times, and many times, I have worked with Mauro Biliotti and Vik Jha on various aspects of parallelisms and related geometries. I gladly acknowledge my debt to

Mauro Biliotti and Vik Jha for twenty-five years or so of great collaboration and lasting friendship (and many wonderful joint articles and two great books Foundations of Translation Planes and Handbook of Finite Translation Planes).

On another continent, later in 1985, I began collaborating with Rolando Pomareda of Santiago, Chile, also many times on parallelisms and returned to Santiago a number of times. I thank Sylvia and Rolando for their gracious hospitality and to Rolando for his help and insights with our joint work and for his friendship over these twenty-five years.

Parts of this text also appear in certain of the previous three works on finite geometry, predominantly in the Subplanes text [114], and I am grateful to Taylor and Francis for allowing the inclusion of this material here. Certain of the material found in this text follows work previously published and which has been suitably modified for the treatment given here. Much of the work, however, is quite new, and a set of open problems is given at the end of the text that should be accessible to most researchers in finite incidence geometry.

When thinking of this time and how it is that I actually became a mathematician interested in finite geometry, it is also impossible for me not to mention the influence my mother Catherine Elizabeth Lamb (Gabriel) has had on my career in mathematics. My mother was a talented mathematician in spirit and ability, but my grandfather was rather old-fashioned and was not willing to support a university education for his children. So, I feel that I was the recipient from my mother of whatever ability I may have in mathematics.

Of course, there is another Catherine Elizabeth, with very similar talents—my daughter, and yes, she was named after her grandmother. Her very strong analytical abilities are now being applied in the law as she is a disability rights attorney, graduated the University of Iowa, with a J.D. and a Master's in Education and now is specializing in the important area of defending the rights of children.

My son Garret Norman also inherited his grandmother's mathematical abilities and has opted rather for the field of engineering, holding a Bachelor's and Master's Degree in Electrical Engineering from the University of Florida he has become a genuine leader in his field.

All of these years later, I might add that the son who accompanied me to Italy, Scott Hamilton, pursued a Ph.D. from the University of Nebraska and is now a Professor of Psychology at Western Illinois University and is a very active researcher in a variety of exciting new directions in psychology.

My three children have been wonderful sources of support and encouragement for my work, for which I am most thankful. Considering support, I am happy to acknowledge support from the University of Iowa who awarded me a Career Development Award in 2009 in order to complete this text. I am indebted to Doug Slauson for help with the technical aspects of the preparation and with the electronic version of the diagram for the cover of the book, which, by the way, is encrypted in a more or less obvious manner. At the end of the text, all will be revealed.

I am glad to thank Andre Barnett for her copious editing of this text.

Finally, I gratefully acknowledge my wife Bonnie Lynn Hemenover, to whom I owe much, if not everything, and without whom none of the various books on geometry would have been written.

Norman L. Johnson
University of Iowa,
Iowa City, IA.
2010

Part 1

Partitions of Vector Spaces

In this part, we consider a variety of partitions of vector spaces, which are more general than the spreads that construct translation planes. This material is very general, and there are many new areas of research presented. We also focus on the so-called focal-spreads, which are partitions of a vector space of dimension $t + k$ by subspaces of two different dimensions, t and k, where there is a unique t-subspace. It will turn out that such focal-spreads can become the building blocks for general partitions.

CHAPTER 1

Quasi-subgeometry Partitions

0.1. Partitions of Vector Spaces. A 't-spread' of a finite dimensional vector space over a skewfield K is a partition of the non-zero vectors by vector subspaces of the same dimension t. Apart from their intrinsic interest, these partitions are important geometrically, which we shall see in due course.

We shall assume, for the most part, that the reader is familiar with basic aspects of affine and projective planes and their collineation groups. In our previous texts on finite geometry, the main emphasis has been on the type of projective and affine planes called 'translation planes.' Of course, the main reason for this is that all known finite projective planes are intrinsically related to finite vector spaces and for this reason are connected somehow to translation planes. In the following, we note the connection between translation planes and vector spaces.

We start at the beginning with the fundamental definitions.

DEFINITION 3. *An 'affine plane' is a triple $(\mathcal{P}, \mathcal{L}, \mathcal{I})$, of sets of 'points $\mathcal{P}$,' 'lines $\mathcal{L}$,' and 'incidence $\mathcal{I} \subseteq \mathcal{P} \times \mathcal{L}$,' with the following properties:*

(1) Given two distinct points P and Q, there is a unique line ℓ incident with P and Q, and which we shall denote by PQ.

(2) Given a line ℓ and a point P not incident with ℓ, there is a unique line m that is not incident with any points (is 'disjoint from') with ℓ and is incident with P. The line m is said to be 'parallel' to ℓ.

(3) There are at least four points no three of which are 'collinear' (are not incident with a common line).

REMARK 2. *It is easy to show that 'parallelism' is an equivalence relation on the set of lines of an affine plane. The equivalence classes are called 'parallel classes.' So, given two lines from distinct parallel classes, there is a unique point incident with the lines, and each point is incident with a unique line of each parallel class.*

DEFINITION 4. *A 'projective plane' is a triple $(\mathcal{P}, \mathcal{L}, \mathcal{I})$, of sets of 'points $\mathcal{P}$,' 'lines $\mathcal{L}$,' and 'incidence $\mathcal{I}$,' with the following properties:*

(1) Given two distinct points P and Q, there is a unique line ℓ incident with P and Q, and which we shall denote by PQ.

(2) Given two distinct lines p and q, there is a unique point L incident with p and q, and which we shall denote by $p \cap q$.

(3) There are at least four points no three of which are 'collinear' (are not incident with a common line).

REMARK 3. *It is well known and straightforward to check that given a projective plane $\pi^{\,\prime}$, it is possible to construct an affine plane by removing a line ℓ and the subset of the points incident with ℓ, $[\ell]$, and where incidence of the new structure is inherited from the projective plane. Hence, if $(\mathcal{P}, \mathcal{L}, \mathcal{I})$ is a triple that defines a projective plane, then $(\mathcal{P} - [\ell], \mathcal{L} - \{\ell\}, \mathcal{I} - \{(P, \ell); P \in [\ell]\})$ is an affine plane π.*

The affine plane is called the 'affine restriction of the projective plane by ℓ.' The line ℓ then defines the set of parallel classes of the affine plane and is sometimes called the 'line at infinity.'

If π is an affine plane, for each parallel class α, formally adjoin a 'point' to each line of α and let the union of the parallel classes be a formal line ℓ_∞, again called the 'line at infinity' of π. With the natural extension of the point, line, and incidence sets, an associated projective plane π^+ is constructed, called the 'projective completion of the affine plane π.'

We shall also be interested in point-line geometries with line parallelism with a slightly weaker set of axioms than that of an affine plane.

DEFINITION 5. *A 'Sperner Space' is a triple $(\mathcal{P}, \mathcal{L}, \mathcal{I})$, of sets of 'points $\mathcal{P}$,' 'lines $\mathcal{L}$,' and 'incidence $\mathcal{I} \subseteq \mathcal{P} \times \mathcal{L}$,' with the following properties:*

(1) Given two distinct points P and Q, there is a unique line ℓ incident with P and Q, and which we shall denote by PQ.

(2) There is an equivalence relation on the line set called 'parallelism.'

(3) Given a line ℓ and a point P not incident with ℓ, there is a unique line m that is parallel with ℓ and is incident with P.

(4) For any two distinct lines ℓ and m, the cardinal number of the set of points incident with ℓ, $[\ell]$, is equal to the cardinal number of the set of points incident with m, $[m]$.

REMARK 4. *Let π be an affine plane and α and β be distinct parallel classes and let lines $\ell \in \alpha$ and $m \in \beta$. Then clearly m and ℓ are incident with a unique common point, for otherwise m would be parallel to ℓ.*

More generally, if S is a Sperner space the lines from distinct parallel classes need not share a common incident point. For example, an affine space is a Sperner space.

DEFINITION 6. *A 'translation' of an affine plane (respectively, of a Sperner Space) is a collineation that leaves invariant all parallel classes and fixes each line of some parallel class. The group generated by the set of translations is called the 'translation group.'*

DEFINITION 7. *A 'translation plane' (respectively, a 'translation Sperner Space') is an affine plane (respectively, Sperner Space) that admits a translation group that acts transitively on the points (i.e., 'affine points').*

It turns out that a translation plane or a translation Sperner Space has an underlying vector space whose vectors are the points of the structure. We consider a construction that turns out to be canonical.

THEOREM 1. *Let V be a vector space of dimension kt over a skew-field K, whose non-zero vectors are partitioned by a set of mutually disjoint t-dimensional subspaces. We define such a partition to be a 't-spread.' Define 'points' to be vectors, define a 't-component' to be any t-subspace of the t-spread, and define 'lines' to be vector translates of t-components, with incidence inherited from the vector space.*

(1) Then the point-line geometry constructed is a translation Sperner Space.

(2) If the dimension of the vector space is $2t$, then the point-line geometry is a translation plane.

PROOF. Let P and Q be distinct points. Translate Q to the zero vector 0 by a translation τ_Q. Then $\tau_Q P$ is incident with a unique t-component ℓ_t. P and Q are incident with a unique line $\tau_Q^{-1}\ell_t$. Two lines are said to be parallel if and only if one is an image of the other by a translation. The translations act on the points and hence induce mappings on the corresponding points. Given any component ℓ_t, there is a translation subgroup that fixes ℓ_t and acts transitively on the points incident with ℓ_t. The parallel classes of lines are, then the set of orbits of the components under the vector space translation group (which, of course, is transitive on the points). Note that every vector space translation is a 'translation' in the sense of Definition 6, since every group element τ_Q^{-1} that maps 0 to Q will fix the unique t-space $0Q$ that contains 0 and Q and hence must fix each element of the orbit of $0Q$. It now follows directly that a translation Sperner Space is obtained. Now assume that the dimension of the vector space is $2t$. Let i and j

be distinct lines and let ℓ_t and m_t be the two distinct components such that i is in the translation orbit of ℓ_t and j is in the translation orbit of m_t. Then $\ell_t \oplus m_t = V$, so that if T_{ℓ_t} and T_{m_t} are the translation subgroups that fix ℓ_t and m_t, respectively, then the full translation group $T = T_{\ell_t} T_{m_t}$. Note that there exists an element τ_{m_t} of T_{m_t} that maps ℓ_t to i and an element τ_{ℓ_t} that maps m_t to j. Then $\tau = \tau_{\ell_t} \tau_{m_t}$ maps (ℓ_t, m_t) to (i, j), as the group is Abelian. Hence, $i \cap j = \tau 0$. Therefore, the point-line geometry is a translation plane. $\qquad \square$

REMARK 5. *The construction in the preceding can be made much more general. For example, we illustrate this in the finite case. Define a 'partition' of a finite vector space over a field K isomorphic to $GF(q)$, for $q = p^r$, a prime power, to be a set of vector subspaces whose union is a exact cover of the non-zero vectors of V. Then a point-line geometry $\mathcal{P}$ may be defined by taking the 'points' as vectors and 'lines' as translates of the partition subspaces ('components' of the partition), and where two lines are said to be 'parallel' if and only if they belong to the same orbit under the translation group of the vector space. In this setting, lines do not have to be incident with the same number of points.*

DEFINITION 8. *Let $\mathcal{P}$ be a point-line geometry such that any two distinct points are incident with a unique line, and which admits an equivalence relation on the line set satisfying the Euclidean parallel axiom. A translation is defined as a collineation that fixes all parallel classes and fixes one parallel class linewise. In the finite case, if there are at least three parallel classes and if there is a set $\{s_1, .., s_k\}$ such that each line of a parallel class is incident with exactly s_i points and each s_j is the number of points of some line, we shall say that the point-line geometry is of type $(s_1, .., s_k)$, for $s_1 > s_2 > ... > s_k$.*

The translation group is the group of translations (each element is a translation) and if the translation group is transitive on the points of $\mathcal{P}$ then we shall say that we have a 'translation geometry.'

In the partition case, we then say that we have a 'translation geometry of type $(s_1, .., s_k)$.'

We shall have the occasion to study translation geometries of various types. Later in this text, we shall be considering the theory of 'focal-planes' and, more generally, 'double-planes.'

DEFINITION 9. *A 'focal-plane' is a point-line geometry of type $\{s_1, s_2\}$, there is exactly one parallel class each of whose lines are incident with s_1 points and all remaining lines are incident with s_2 points.*

A 'translation focal-plane' is a translation geometry whose underlying point-line geometry is a focal-plane.

A 'double-plane' is a point-line geometry of type $\{s_1, s_2\}$ and a translation double-plane is a translation geometry whose underlying point-line geometry is a double-plane.

In the following, we shall give constructions of focal-spreads and of more general double spreads, where the 'spread' (or perhaps generalized spread or partition) is the underlying set of vector subspaces of a translation geometry that corresponds to a vector space.

0.2. Beutelspacher's Construction. Let V_{t+k} be a vector space of dimension $t + k$ over $GF(q)$ for $t > k$ and let L be a subspace of dimension t. Let V_{2t} be a vector space of dimension $2t$ containing V_{t+k} ($t > k$ required here) and let S_t be a t-spread containing L. There are always at least Desarguesian t-spreads with this property. Let M_t be a component of the spread S_t not equal L. Then $M_t \cap V_{t+k}$ is a subspace of V_{t+k} of dimension at least k. But, since M_t is disjoint from L, the dimension is precisely k and we then obtain a focal-spread with focus L. This construction may be found in Beutelspacher [**14**], Lemma 2, page 205. The corresponding focal-spread is said to be a 'k-cut of a t-spread,' and we shall adopt this terminology and use the notation $F = S_t \backslash V_{t+k}$ for the focal-spread F.

The similar formal definition is

DEFINITION 10. *A partition of a finite-dimensional vector space of dimension $t + k$ by a partial Sperner k-spread and a subspace of dimension $t \neq k$ shall be called a 'focal-spread of type (t, k).' The unique subspace of dimension t of the partition shall be called the 'focus' of the focal-spread.*

We define a 'planar extension' of a focal-spread as a t-spread such that the focal-spread is of type (t, k) and arises from the t-spread as a k-cut.

The 'kernel' of a focal-spread of dimension $t + k$, over $GF(q)$, with focus of dimension t, shall be defined as the endomorphism ring of the vector space that leaves the focus and each k-component invariant.

Now we turn to a similar construction of more general double-spreads.

0.3. Double-Spreads Constructed from t-Spreads. Of course, the idea of cutting or slicing a spread could conceivably lead to other interesting partitions. For example, from a t-spread of a $2t$-dimensional vector space, we could simply select a $2t - 1$ dimensional subspace

and form the possible intersections to construct a partition, leading
ultimately to a focal-spread of type $(t, t - 1)$. In general, it is very
difficult to achieve any sort of uniformity with such slicing with explicitly determined subspaces. However, one type of cut that does produce
interesting partitions arises as follows: Let $\mathcal{P}_t$ be a t-spread of a st-
dimensional vector space over $GF(q)$, where $k > 1$. Let $\mathcal{U}$ denote any
vector subspace of dimension $st - 1$. Consider any component of $\mathcal{P}_t$, it
follows easily that $\mathcal{P}_t \cap \mathcal{U}$ has dimension either t or $t - 1$. Let a denote
the number of components of intersection of dimension $t - 1$ so that
$(q^{st} - 1)/(q^t - 1) - a$ components then have dimension t. Therefore, by
intersections, we clearly have a cover of $\mathcal{U}$ so that

$$q^{st-1} - 1 = a(q^{t-1} - 1) + ((q^{st} - 1)/(q^t - 1) - a)(q^t - 1).$$

This implies that

$$aq^{t-1}(q - 1) = q^{st-1}(q - 1),$$

so that $a = q^{t(s-1)}$.

Therefore, we have constructed a double-spread (a partition of a
vector space with exactly two cardinalities of components) of type
$(t, t - 1)$ with $(q^{(s-1)t} - 1)/(q^t - 1)$ subspaces of dimension t and $q^{t(s-1)}$
subspaces of dimension $t - 1$.

DEFINITION 11. *The notation becomes also problematic, since we
use only the dimensions and say that the double-spread is of type $(t, t -
1)$. However, in this case, there are $\frac{q^{t(s-1)} - 1}{q^t - 1}$ subspaces of dimension t
and $q^{t(s-1)}$ subspaces of dimension $t - 1$. The double-spread of this type
is said to be a '$(t-1)$-cut' of a t-spread for an st-dimensional subspace.*

REMARK 6. *Note that $t + k = st - 1$ if and only if $t(s - 1) = k + 1$
and for $k < t$. So, $t(s - 1) \le t - 1 + 1$, implying that $s \le 2$. So,
when $s = 2$, we see that we have a focal-spread of type $(t, t-1)$. Hence,
the previous construction provides a focal-spread of type (t, k), exactly
when $s = 2$ and otherwise, a double-spread that is not a focal-spread is
constructed.*

REMARK 7. *Of course, given any double-spread of type $(t, t-1)$, in
a vector space of dimension $st - 1$, the natural question is whether it
arises from a t-spread of a st-dimensional vector space, as a $(t-1)$-cut.*

We shall develop the theory of focal-spreads more completely in a
later chapter and further give several other constructions of double-
spreads and triple-spreads.

0.4. Main Theorem on Finite Translation Geometries. The following material also is valid for arbitrary translation planes and the reader is referred to the Foundations text [**21**] to see this. The proof of the theorem stated below for finite translation geometries is more generally valid for arbitrary translation planes, with appropriate changes.

THEOREM 2. *Every finite translation geometry corresponds to a finite vector space over a field K. The field K may be identified with the endomorphism ring of the associated translation group that leaves invariant each parallel class.*

We shall give the proof as a series of lemmas, parts of which are left to the reader to complete. In the following, we assume the hypothesis of Theorem 2. We first point out some trivial facts.

REMARK 8. *(1) Each point is incident with a unique line of each parallel class.*

(2) A non-identity translation σ fixes a unique parallel class linewise.

(3) A non-identity translation is fixed-point-free so that the group T of translations is fixed-point-free.

PROOF. Part (1) follows immediately from the definitions. Let σ be a translation. Then σ fixes all parallel classes and fixes one parallel class α linewise. Suppose that σ fixes another parallel class $\beta \neq \alpha$ linewise. Choose any point P, so there is a line ℓ_α of α and a line ℓ_β of β both incident with P. Since σ fixes both of the indicated lines, it follows that σ fixes P, so that σ fixes all points and hence is the identity mapping. If σ fixes an affine point P, then σ fixes all lines incident with P as it fixes all parallel classes. Choose any point Q distinct from P and form the line QP. Let m be the line of α incident with Q. Since σ fixes QP and m, it follows that it also fixes Q. Hence, σ fixes all points and so is the identity translation. If τ is in the translation group it then fixes all parallel classes. If τ fixes an affine point P, then τ fixes all lines incident with P. □

DEFINITION 12. *If σ is a translation of a translation geometry, the unique parallel class linewise fixed by σ is called the 'center' of σ. If α is a parallel class, let $T(\alpha)$ denote the group of translations with center α.*

The first two parts of the following lemma are immediate.

LEMMA 1. *Let T denote the group of translations of a translation geometry.*

(1) Then $T(\alpha)$ is a normal subgroup for each parallel class α.

(2) $T = \cup_\alpha T(\alpha)$, a partition of T.

(3) T is Abelian.

PROOF. (3): Let g and h be elements of $T(\alpha)$ and $T(\beta)$, for $\alpha \neq \beta$. Then the commutator $ghg^{-1}h^{-1}$ is in the intersection of both translation subgroups, which are disjoint. Hence, elements of $T(\alpha)$ and elements of $T(\beta)$ commute. Let g and k be elements of $T(\alpha)$ and h an element of $T(\beta)$. Then $(gk)h = g(kh) = g(hk) = (gh)k$, and $kh \in T(\gamma)$, for a parallel class γ distinct from α or β, implying that $g(kh) = (kh)g = k(hg) = (hg)k$, using the same principle. Hence, $(gh)k = (hg)k$, so that $gh = hg$. This completes the proof. □

LEMMA 2. *Let K denote the set of endomorphisms of T that fix each $T(\alpha)$. Then K is a field and T is a vector space over K. Identifying the point set with T, then the translation geometry becomes a vector space over K and the lines incident with 0 are vector subspaces over K so that we have a partition of a vector space.*

PROOF. Let σ be a non-zero endomorphism. If σ is not injective, then since T is Abelian, we may assume that $\sigma(z) = 1$, for some non-identity element z of a partition element, say $T(\alpha)$. For u in $T(\beta)$, for $\alpha \neq \beta$, then $zu \in T(\gamma)$, for γ distinct from α or β. So, $\sigma(zu) = \sigma(z)\sigma(u) = \sigma(u)$, but $\sigma(zu) \in T(\gamma)$ and $\sigma(u) \in T(\beta)$ so both are in the intersection $T(\beta) \cap T(\gamma) = \langle 1 \rangle$. Therefore, $T(\beta) \subseteq Kernel\sigma$, which also implies that $\cup_{\delta \neq \alpha} T(\delta) \subseteq Kernel\sigma$. Now repeat the argument for a kernel element w in $T(\beta)$, to obtain $T(\alpha) \subseteq Kernel\sigma$, so that σ is the identity endomorphism. Since the translation geometry is finite, it follows that K is a field. Then considering the natural endomorphism ring of T, T becomes a K-vector space. Now identifying the points of the translation geometry with the elements of T and using additive notation, choose the point 0 (the 'origin') then it follows immediately that lines through the origin may be identified with the groups $T(\alpha)$, which are clearly K-vector subspaces. This completes the proof of the theorem. □

1. Collineation Groups of Translation Geometries

In this section, we show that the full collineation group of a translation geometry with underlying vector space V over the field K (or skewfield if the translation geometry is a translation plane) is a semidirect product of a subgroup of $\Gamma L(T, K)$ by the translation subgroup T. The proof given is essentially the same as that in Lüneburg [**167**] (1.10), determining isomorphisms of finite translation planes.

If G is a collineation group of a translation geometry π (finite with field kernel K or an arbitrary translation plane with skewfield kernel

K), let T denote the translation group and let G_0 denote the subgroup of G that fixes the zero vector 0. Clearly, T is a normal subgroup so that for $g \in G$, then if $g0 \neq 0$, there is a translation τ so that $\tau g0 = 0$, so that $\tau g \in G_0$, which implies that $G \subseteq G_0 T$. Now if h is a collineation that fixes 0, h will permute the lines incident with 0 (components), which means that h permutes the parallel classes. We wish to prove that $h \in \Gamma L(T, K)$, that is, h is a semi-linear group element, where T is a K-vector space. For $\tau_P \in T$, where $\tau_P(0) = P$, we, of course, know that $h\tau_P h^{-1}$ is a translation τ_Q, where $\tau_Q(0) = Q$, for points P and Q of π. Since h leaves 0 invariant, we note that $\tau_Q = \tau_{hPh^{-1}} = h\tau_P h^{-1} = \tau_{hP}$.

Now consider $\tau_{h(P+Q)h^{-1}} = \tau_{h(P+Q)}$:

$$
\begin{aligned}
\tau_{h(P+Q)} &= \tau_{h(P+Q)h^{-1}} = h\tau_{P+Q}h^{-1} = h\tau_P\tau_Q h^{-1} = h\tau_P h^{-1} h\tau_Q h^{-1} \\
&= \tau_{hPh^{-1}}\tau_{hQh^{-1}} = \tau_{hP}\tau_{hQ} \\
&= \tau_{hPh^{-1}+hQh^{-1}}.
\end{aligned}
$$

Therefore,

$$
h(P+Q) = hP + hQ.
$$

It remains to show that if k is any element in the kernel K of T then $h(kP) = k^\rho h(P)$, for all points P, where ρ is an automorphism of K. It is clear that K may be considered a normal subgroup of the collineation group that fixes 0. Let $hkh^{-1} = k^\rho$. First, $h(kP) = hkh^{-1}(hP) = k^\rho h(P)$, and for $k, k' \in K$, we have that

$$
\begin{aligned}
h(k+k')P &= (k+k')^\rho hP = h(kP + k'P) = \\
h(kP) + h(k'P) &= k^\rho h(P) + k'^\rho h(P) \\
&= (k^\rho + k'^\rho)(hP),
\end{aligned}
$$

$$
\text{implying that } (k+k')^\rho = (k^\rho + k'^\rho).
$$

Similarly,

$$
\begin{aligned}
h(kk'P) &= (kk')^\rho h(P) = h(k(k'P) = k^\rho h(k'P) \\
&= k^\rho k'^\rho h(P),
\end{aligned}
$$

$$
\text{implying that } (kk')^\rho = k^\rho k'^\rho.
$$

THEOREM 3. *In any translation geometry (or translation plane) with ambient vector space V over a field (or skewfield K), the full collineation group is a semi-direct product of the subgroup of $\Gamma L(V, K)$ by the translation subgroup T.*

The subcollineation group that fixes the zero vector shall be called the 'translation complement.' The subgroup of the translation complement of $GL(V, K)$ is called the 'linear translation complement.'

1.1. Collineation Groups of Translation Planes. We now specialize to finite translation planes. We shall be mostly interested in finite translation planes whose underlying vector space is 4-dimensional over a field K isomorphic to $GF(q)$, where $q = p^r$, for p a prime and r a positive integer. We shall refer to the parallel classes of a translation plane as the points on the line at infinity.

DEFINITION 13. *If V is a vector space of dimension z over a field L, the 'projective geometry $PG(z-1, L)$' is the lattice of vector subspaces of V. A 1-dimensional subspace is said to be a 'point,' a 2-dimensional vector subspace a 'line,' and a 3-dimensional vector space a 'plane.'*

An 'affine space $AG(z, L)$,' is the geometry of translates of the subspaces of a vector space.

DEFINITION 14. *A collineation σ of a (an affine) translation plane is a 'central collineation' provided σ fixes a line ℓ of the plane pointwise. The collineation is an 'elation' if σ fixes all lines parallel to ℓ, and we say the 'center' of σ is incident with ℓ (on the line at infinity) and ℓ is the 'axis' of σ. The collineation is a 'homology' if σ fixes all lines of a parallel class β not containing ℓ. If the line of β incident with 0 is denoted by 0β, we shall say that ℓ is the 'axis,' 0β is the 'coaxis' and β is the 'center' of σ.*

DEFINITION 15. *If π^+ is a finite projective plane of order q^2 ($q^2 + 1$ is the number of points per line), a projective subplane of order q is said to be a 'Baer' subplane π_0^+. Any associated affine restriction π of π^+ is also said to have order q^2. If the Baer subplane of π^+ contains the line of restriction, then the corresponding affine subplane of order q is a subplane of π.*

A 'Baer collineation group' of a finite affine plane is a collineation group that fixes a Baer subplane pointwise.

REMARK 9. *For finite translation planes of order q^2, it is known that any Baer collineation group has order dividing $q(q-1)$. For more details, the reader is directed to Biliotti, Jha, Johnson [21].*

NOTATION 1. *In this text, when considering translation planes of order q^2, constructed or equipped with a 2-spread over $GF(q)$, we may always represent vectors in the form (x_1, x_2, y_1, y_2), for all $x_i, y_i \in GF(q)$, $i = 1, 2$, and further let $x = (x_1, x_2)$, $y = (y_1, y_2)$. Then there is a set of q^2 2×2 matrices over $GF(q)$, including the zero matrix, so that the components of the 2-spread are as follows:*

$$x = 0, y = 0, y = x \begin{bmatrix} g(t, u) & f(t, u) \\ t & u \end{bmatrix}; u, t \in GF(q),$$

and g and f are functions from $GF(q) \times GF(q)$ such that the matrices and their differences are either non-singular or the zero matrix.

Conversely, any set of q^2 2×2 matrices, including the zero matrix with the above properties, determines a translation plane of order q^2.

This representation shall become extremely important when we consider flocks of quadratic cones.

The translation complement of any translation plane of this type is a subgroup of $\Gamma L(4, q)$.

CHAPTER 2

Finite Focal-Spreads

In the previous chapter, we briefly discussed focal-spreads and the Beutelspacher's construction. Although there are a number of examples of such partitions, there has been no theory developed about general partitions. Part of our treatment in this text is modeled from the article by Jha and Johnson [77], and the reader is directed to this article for additional details.

Although it is certainly possible to consider focal-spreads over arbitrary fields, all of the material that is presented in this text is for finite focal-spreads. The reader is directed to the open problem chapter 40 for more information.

In this chapter, we begin to build a theory of partitions based on focal-spreads and their obvious connections to translation planes. For general focal-spreads, however, nothing is known. For example, it is not known whether collineations that fix components pointwise or homologies are elations or whether an involution that does not fix a component pointwise becomes a Baer collineation or a kernel involution. In the 'Handbook of Finite Translation Planes' [138], it is mentioned that there are not very many known partitions of vector spaces, in the sense that the partitions are not refinements of Sperner t-spreads. Furthermore, the existence of such partitions has been established by Beutelspacher [14], who proves if the dimension of the vector space is n and it is required to find a partition, where the dimensions of the subspaces are $\{t_1, t_2, .., t_k\}$, where $t_i < t_{i+1}$, then if $\gcd\{t_1, t_2, .., t_k\} = d$ and $n > 2t_1([t_k/(d\cdot k)]+t_2+...+t_k)$, a partition may be constructed with various subspaces of dimension t_i. However, these partitions are basically constructed using focal-spreads and general spreads. We give some of the constructions later in the section on 'towers of focal-spreads.'

The main point of this chapter is to use of ideas and theory of finite translation planes in order to develop a certain theory of focal-spreads and, since these seem to be important building blocks of general spreads, such work might prove useful to the general theory of partitions.

Given a k-cut, of course, there is a corresponding spread for a translation plane that produces it. It is a central and potentially important question to ask if every focal-spread is a k-cut; if it can 'extended' to a spread for a translation plane.

Therefore, here we give the basics of focal-spreads and use some theory from translation planes for their study. We also show that the existence of focal-spreads of type $(k + 1, k)$ leads to a construction of designs of type $2 - (q^{k+1}, q, 1)$ and to other double-spreads or triple-spreads (partitions where there are subspaces in the partition of either two or three different dimensions, respectively).

The formal definition of a focal-spread has been given in Definition 10.

1. Towers of Focal-Spreads

As we begin our discussions on t-spreads, we have briefly discussed focal-spreads and found that they are readily constructed, but we might also just as well ask the obvious question: How hard is it to find general partitions of vector spaces? That is, partitions of the non-zero vectors by a set of subspaces of dimensions t_i, for $i = 1, 2, .., k$.

In Beutelspacher [16], and more generally in Heden [65], there are a variety of results of the existence of partitions of type $T = \{t_1, t_2, .., t_k\}$, in vector spaces of dimension n over $GF(q)$, where the partition admits exactly subspaces over $GF(q)$ of dimension t_i, for $i = 1, 2, .., n$, for $t_1 < t_2 < ...t_k$. (The reader is cautioned that our previous notation of partitions has the inequalities in the opposite order.) Many of these results rely basically on focal-spreads of type (t, k), in the sense that a variety of focal-spreads or t-spreads are used in the construction. We give here one construction that provides a partition of type T in any vector space of dimension $n = \sum_{i=1}^{k} \alpha_i t_i$ over $GF(q)$, showing that again general partitions are readily available.

First, we note the following lemma.

LEMMA 3. *In an st-dimensional vector space over $GF(q)$, there is always a t-spread over $GF(q)$.*

PROOF. Let the vectors be represented by $GF(q^{st})$. Take the cosets of $GF(q^t)^*$ in $GF(q^{st})^*$. Note that $GF(q^t)^* g \cup \{0\}$ is a $GF(q)$-subspace. Since the cosets partition $GF(q^{st})^*$, representing the non-zero vectors, the lemma is proved. □

LEMMA 4. *Let V be a $t_1 + t_2$-dimensional vector space over $GF(h)$, for h a prime power. If $t_2 > t_1$, then there is focal-spread of type (t_2, t_1).*

PROOF. Form a $2t_2$-dimensional vector space over $GF(h)$ and construct a t_2-spread. Since $t_2 > t_1$, there is a proper focal-spread of type (t_2, t_1) constructed as a t_1-cut.   □

THEOREM 4. *A partition of type $\{t_1, t_2, .., t_k\}$ exists in any vector space of dimension $\sum_{i=1}^k \alpha_i t_i$ over $GF(q)$.*

PROOF. First assume that $\alpha_i = 1$, for all $i = 1, 2, .., k$. Construct a (t_2, t_1)-focal-spread in a $t_1 + t_2$-dimensional vector space over $GF(q)$. Now there are three cases (1) $t_1 + t_2 > t_3$, (2) $t_1 + t_2 = t_3$, (3) $t_1 + t_2 < t_3$. In case (1), in a $t_1 + t_2 + t_3$-dimensional vector space, construct a $(t_1 + t_2, t_3)$-focal-spread. We consider the direct sum of a vector space of dimension $t_1 + t_2$ with a vector space of dimension t_3 and make the obvious identifications. On the unique focus, we identify our original (t_2, t_1)-focal-spread. We then would have subspaces of dimensions exactly t_1, t_2, t_3. In case (2), if $t_1 + t_2 = t_3$, in a $2t_3$-dimensional subspace, we always have a t_3-spread. In any proper subset of the components, we may consider forming (t_2, t_1)-focal-spreads, again providing the required subspaces of dimensions t_1, t_2, t_3. Finally, in case (3), we construct a $(t_3, t_2 + t_1)$-focal-spread. Then on all $(t_2 + t_1)$-components, we construct (t_2, t_1)-focal-spreads. Clearly, this sort of argument may be continued and completes the proof of the theorem, in the case when all $\alpha_i = 1$.

Now more generally, we claim that in a vector space of dimension $\alpha_1 t_1 + \alpha_2 t_2$, there is a partition of type $\{t_1, t_2\}$. Since $t_2 > t_1$, form a focal-spread of type $(\alpha_1 t_2 + \alpha_2 t_2, t_1)$. Then on the unique focus of dimension $\alpha_1 t_1 + \alpha_2 t_2$ construct another focal-spread of type $((\alpha_1 - t)t_1 + \alpha_2 t_2, t_1)$. If we continue with the same line of argument, we end up with a focal-spread with focus of dimension $\alpha_2 t_2$, which always contains a t_2-spread by Lemma 3. Now consider a vector space of dimension $\alpha_1 t_1 + \alpha_2 t_2 + \alpha_3 t_3 = z + \alpha_3 t_3$ and form a sequence of (e, t_3)-focal spreads until we obtain a focus (or vector space) of dimension $\alpha_1 t_1 + \alpha_2 t_2 + t_3$. We consider the cases (1) $\alpha_1 t_1 + \alpha_2 t_2 > t_3$, (2) $\alpha_1 t_1 + \alpha_2 t_2 = t_3$, and (3) $\alpha_1 t_1 + \alpha_2 t_2 < t_3$.

In case (1), we may form a $(\alpha_1 t_1 + \alpha_2 t_2, t_3)$-focal-spread. Then, on the focus of dimension $\alpha_1 t_1 + \alpha_2 t_2$, by the previous argument, we may find a $\{t_1, t_2\}$-partition. In case (2), we have a $2t_3$-dimensional vector space, which allows a t_3-partition. Take any component of dimension $t_3 = \alpha_1 t_1 + \alpha_2 t_2$ and again find a $\{t_1, t_2\}$-partition. In case (3), we form a series of (r, t_i)-focal-spreads on the corresponding foci, for $i = 1, 2$, possible since $t_3 > t_i$, for $i = 1, 2$. This leaves us with a focus of dimension t_3. Clearly, this argument is inductive, thus completing the proof of the theorem. Alternatively, in this last case (3), form a

$(t_3, \alpha_1 t_1 + \alpha_2 t_2)$-focal-spread and on each subspace of dimension $\alpha_1 t_1 + \alpha_2 t_2$ form $\{t_1, t_2\}$-partitions. $\qquad\qquad\qquad\qquad\qquad\qquad\qquad\square$

DEFINITION 16. *A partition of a vector space V of dimension n over $GF(q)$ shall be called a 'tower of general focal-spreads' if and only if there is a sequence of vector subspaces $F_1 \subseteq F_2 \subseteq F_3, .., \subseteq F_k = V$, such that F_j is a vector space of dimension $\sum_{i=1}^{j} \alpha_i t_i$ admitting a partition by a set of focal-spreads of type (e, t_z), for $z = 1, 2, .., j$, or by t_z-spreads, for α_i a positive integer, $i = 1, 2, .., j$.*

We note that built into the construction of towers of general focal-spreads, there is a certain variation. For example, consider the construction of a $\{4, 5, 9\}$-partition in a 18-dimensional vector space over $GF(q)$. We begin with the construction of a $(5, 4)$-focal spread in a 9-dimensional vector space and form the direct sum of a 9-dimensional and a second 9-dimensional vector space. In this 18-dimensional vector space, we take any 9-spread. Now take any proper subset of 9-dimensional components and form $(5, 4)$-focal-spreads. Since we may do this construction for any proper subset, we always end up with a partition of type $\{4, 5, 9\}$ that arises from a tower of focal-spreads. This complexity makes the problem listed above potentially quite challenging.

So, we are interested in what can be said about the general theory of focal-spreads, as focal-spreads and t-spreads seem to form basic building blocks of arbitrary partitions, about which we know essentially nothing beyond existence. We shall see in the next chapters that there are a number of ways of constructing focal-spreads that do not appear to arise from the Beutelspacher cut-method.

2. Focal-Spreads and Coordinatization

It is tempting to define a focal-spread simply as a partition by subspaces of two different dimensions, of which there is a unique subspace of one of the dimensions. For example, if the dimension of the ambient vector space is n and the subspaces have dimensions t and k, we call such partitions of 'type $(n; (t, k))$' and if $n = t + k$, then of type (t, k), over a finite field $GF(q)$. In fact, it is possible to construct such partitions as we have previously seen and which, we shall also do in due course. In the beginning, we insist that the dimension is $t+k$, for focal-spreads of type (t, k). We first consider how to represent focal-spreads using matrices, in a manner analogous to that of the representation of translation planes.

2.1. Matrix Representation of Focal-Spreads. As for planar spreads, it is usually helpful to express these results in terms of matrices. The reader might wish to work through this material carefully as the ideas closely parallel analogous matrix coordinatization for translation planes. For addition background, the Foundations text can be consulted [**21**].

Let B be a focal-spread of dimension $t + k$ over $GF(q)$ with focus L of dimension t. Fix any k-component M. We may choose a basis so that the vectors have the form $(x_1, x_2, .., x_k, y_1, y_2, .., y_t)$. Let $x = (x_1, x_2, .., x_k)$ and $y = (y_1, y_2, .., y_t)$, where the focus L has equation $x = 0 = (0, 0, 0, .., 0)$ (k-zeros) and the fixed k-component M has equation $y = 0 = (0, 0, 0, .., 0)$ (t-zeros). We note that $q^{t+k} - q^t = q^t(q^k - 1)$, which implies that there are exactly q^t k-subspaces in the focal-spread. We refer to this as the 'partial Sperner k-spread'. Take any k-component N distinct from $y = 0$. There are k basis vectors over $GF(q)$, which we represent as follows: $y = xZ_{k,t}$, where $Z_{k,t}$ is a $k \times t$ matrix over $GF(q)$, whose k rows are a basis for the k-component. It is clear that we obtain a set of q^t k-components, which we also represent as follows: Row 1 shall be given by $[u_1, u_2, .., u_t]$, as the u_i vary independently over $GF(q)$. Then the rows $2, .., k$ have entries that are (turn out to be) linear functions of the u_i. It then now follows directly that the $k \times t$ matrices in the focal-spread have rank k and the difference of any two distinct matrices associated with k-components also has rank k.

The following is the analogous result of translation planes:

THEOREM 5. *Let V_{t+k} be a $t + k$-dimensional vector space over $GF(q)$ and let S be a set of $q^t - 1$ $k \times t$ matrices of rank k such that the difference of any two distinct matrices also has rank k. Then there is an associated focal-spread constructed as*

$$x = 0, y = 0, y = xM; \ M \in S,$$

where x is a k-vector and y is a t-vector over $GF(q)$, where the focus is $x = 0$.

In particular, it is possible to choose one k-space to have 1's in the (i, i), position and 0's elsewhere in the $k \times t$ matrices. We denote this matrix by $I_{k \times t}$.

Conversely, any focal-spread has such a representation.

The reader is directed to the open problem chapter 40 for more discussion on how a generalization for arbitrary partitions might be phrased using a matrix approach.

There is, of course, a coordinate-free approach given as follows:

DEFINITION 17. *Let* $V = V_t \oplus V_k$ *be a finite vector space of rank* $t+k$ *expressed as a direct sum of subspaces* V_t *and* V_k, *having dimensions* t *and* k, *respectively. Then a '(t,k)-spread set on* V,' *based on axes* (V_k, V_t), *is a collection* $\mathcal{S}$ *of linear maps from* V_k *to* V_t, *such that*
  (1) $0 \in \mathcal{S}$;
  (2) the nonzero maps in $\mathcal{S}$ *are injective;*
  (3) the difference between any two members of $\mathcal{S}$ *is injective or zero;*
  (4) $\mathcal{S}$ *is transitive in the sense that for any pair of non-zero vectors*

The following remarks are then left to the reader to verify.

REMARK 10. *Every* (t, k)-*spread set* $\mathcal{S}$ *as above yields a focal-spread of type* (t, k), *with component set* $\{M_S : S \in \mathcal{S}\} \cup \{V_t\}$, *where* $M_S :=$ $\{(x, xS) : x \in V_k\}$.
  Every focal-spread of type (t, k) *may be coordinatized by a* (t, k)-*spread set, and any focal-spread of type* (t, k) *arises by coordinatization by a* (t, k)-*spread set, which is uniquely determined by the focus and any other component chosen as 'basis.'*

3. k-Cuts and Inherited Groups

In this section, if a focal-spread is a k-cut of a translation plane, we ask what sorts of collineations of the translation plane become collineations of the focal-spread? More generally, given an abstract focal-spread, how are central collineations defined? We begin with the simplest type of collineation of a focal-spread, a 'homology.'

DEFINITION 18. *In a focal-spread of dimension* $t + k$ *over* $GF(q)$ *and focus* L *of dimension* t, *every collineation is assumed to be an element of* $\Gamma L(t + k, q)$ *that leaves invariant the focus* L *and permutes the components of the partial Sperner* k-*space* $\mathcal{S}$.
  A 'homology' h *is a collineation of* $GL(t + k, q)$ *with the following properties:*
  (1) h *leaves invariant the focus* L *and another* k-*subspace of* $\mathcal{S}$,
  (2) h *fixes one of the two fixed components pointwise and acts fixed-point-free on another fixed component (in the case that a* k-*component is fixed pointwise, we assume that the group acts fixed-point-free on the focus).*

To show why there is a potential problem with dealing with collineations that one might wish to call a homology, consider the focal-spread of type $(2, 1)$ over $GF(q)$, represented as

$$x = 0, y = x[t, u], \text{ for all } t, u \in GF(q),$$

where the focus is $x = 0$, a 1-vector. Let $B = \begin{bmatrix} 1 & 1 \\ 0 & 1 \end{bmatrix}$, and consider the collineation $r_B : (x, y) \rightarrow (x, yB)$. We note that $(0, 1) \begin{bmatrix} 1 & 1 \\ 0 & 1 \end{bmatrix} = (0, 1)$, so that r_B is not semi-regular on the focus, and since $y = x[0, 1]$ is fixed by r_B, the collineation is also not semi-regular on the components other than the putative coaxis $x = 0$.

In terms of homologies that are not affine collineations that fix the line at infinity pointwise (the axis) and fix an affine point (the center), the situation is much more manageable. This is because a finite focal-spread is a finite translation geometry, and Theorem 3 shows that the subgroup that fixes each component (k-component and focus) is the cyclic subgroup of a field; the kernel is a field.

We now intend to show that a non-identity homology (see Definition 18) fixes exactly two components. The pointwise fixed subspace is called the 'axis' of h and the fixed subspace is the 'coaxis' of h. Thus, the focus is either the axis or coaxis of any homology.

PROPOSITION 1. *A non-identity homology of a focal-spread fixes exactly two components and permutes the remaining components semi-regularly.*

PROOF. We represent the focal-spread in the form

$$x = 0, \ y = 0, \ y = xM; \ M \text{ is a } k \times t \text{ matrix in set } \mathcal{M}$$

for x a k-vector and y a t-vector, where $x = 0$ is the focus, and $y = 0$ is a k-space, and where we choose $I_{k \times t} \in \mathcal{M}$. Assume that $y = xM_1$ is fixed pointwise by a collineation h. Choose a basis by the mapping $(x, y) \rightarrow (x, -xM_1 + y)$, to change the form to

$$x = 0, \ y = 0, \ y = x(M_1 - M); \ M \text{ is a } k \times t \text{ matrix in set } \mathcal{M}.$$

Now choose any of these matrices $M_1 - M_2$, of rank k, and column-reduce to $I_{k \times t}$. The other matrices $M_1 - M$ reduce to rank k matrices, whose differences are also of rank k. Hence, if $y = xM_1$ is fixed pointwise by h, we may assume without loss of generality that $y = 0$ is fixed pointwise by h. Therefore, any collineation in $GL(t + k, q)$ that fixes a component pointwise may be represented in either the form $l_A : (x, y) \rightarrow (xA, y)$, where A is a non-singular $k \times k$ matrix or $r_B : (x, y) \rightarrow (x, yB)$, where B is a non-singular $t \times t$ matrix. First, assume that r_B is a homology, so that B acts fixed-point-free on the focus. Assume that r_B fixes $x = 0, y = 0$, and $y = xM$. So, $xMB = xM$. Choose then any $x_0M = z_0$, for z_0 a t-vector, so that $z_0B = z_0$. However, this is contrary to the action of B. Since the previous argument is also valid for any power $B^j \neq I_t$, then $\langle r_B \rangle$ fixes exactly two components and acts semi-regularly on the remaining components. Now assume that l_A is a non-identity collineation that fixes the focus pointwise. Assume that l_A fixes $x = 0, y = 0$ and $y = xN$, so that $A^{-1}N = N$. Column reduce to $A^{-1}I_{k \times t} = I_{k \times t}$, and then realize that this forces $A^{-1} = I_k$, a contradiction. This proves (1) and (2). $\quad\square$

Now we prove that, contrary to when a k-component is fixed pointwise, if a collineation fixes the focus pointwise, it must be a homology, provided it fixes another component.

PROPOSITION 2. *If a collineation h fixes the focus pointwise and fixes another component then h is a homology.*

PROOF. Assume that h fixes a k-component $y = xM$ and fix the focus pointwise. Change representation so that the fixed k-component is $y = 0$. Then represent h as $l_A : (x, y) \rightarrow (xA, y)$, where A is a non-singular $k \times k$ matrix. The proof of the previous result shows that if A is not I_k, then l_A is semi-regular on the remaining components. Assume that $x_0 A = x_0$. Then $(x_0, x_0 I_{k \times t})$ is on both $y = xI_{k \times t}$ and $y = xA^{-1}I_{k \times t}$, a contradiction. Hence, h is a homology. $\quad\square$

We now define the concept of an 'elation' of a focal-spread.

DEFINITION 19. *An 'elation' e of a focal-spread of type (t, k) over $GF(q)$ is a collineation of $GL(t+k, q)$ with the following two properties: (1) e fixes the focus L pointwise, and (2) if V is the associated vector space of dimension $t + k$ over $GF(q)$, e fixes V/L pointwise.*

Note that a collineation e in $GL(t+k, q)$ that fixes the focus $x = 0$ pointwise may be represented in the form $e_{A,C} : (x, y) \rightarrow (xA, xC + y)$, where A is a non-singular $k \times k$ matrix, where C is a $k \times t$ matrix of rank k. Without additional assumptions, there is no reason to assume that A is necessarily I. But, under the assumption of Definition 19, condition (2), a collineation $e_{A,C}$ is an elation if and only if $A = I$, since on $V/(x = 0)$, $e_{A,C}$ acts on $(z, w) + (x = 0)$ as $(zA, zC + w) + (x = 0) = (z, w) + (x = 0)$ if and only if $(zA - z, zC) \in (x = 0)$, so $A = I$. We note then that when $A = I$, the collineation is additive and semi-regular or order p, because $e_{I,C}^j : (x, y) \rightarrow (x, jxC + y)$ (for $q = p^r$, p a prime) on $\mathcal{M} \cup \{0\}$. Furthermore, any group E of elations is an elementary Abelian p-group that acts semi-regularly on the partial Sperner spread.

In terms of inherited groups, the following is immediate:

LEMMA 5. *Let $V = V(2t, q)$ and let S be a t-spread on V and assume that G is a collineation group of S. Let W be a subspace of* *dimension* $t+k$ *containing a component* L *and assume that both* W *and* L *are* G *invariant. Then* G *acts faithfully as a group of collineations of the* k-cut $F = S \setminus W$.

So, obviously, any kernel homology group of the translation plane of order q^t that leaves k-subspaces invariant inherits as a collineation group of any k-cut. Therefore, we obtain the following corollary involving the inheritance of central collineations.

COROLLARY 1. *(1) Any affine homology group with axis* $y = 0$, *and coaxis* $x = 0$, *of the* t-spread *inherits as a collineation group of a* k-cut *focal-spread with focus* $x = 0$.
(2) Any affine elation group with axis $x = 0$ *of the* t-spread *inherits as a collineation group of any* k-cut *focal-spread with focus* $x = 0$.

PROOF. Consider the axis $y = 0$ of an affine homology group. Take the subspace V_{t+k} generated by any k-subspace of $y = 0$ and $x = 0$. Then the affine homology group will leave V_{t+k} invariant. This proves (1). Let E be an elation group with axis $x = 0$. An elation group acts trivially on the quotient space $V_{2t}/(x = 0)$ and hence will leave V_{t+k} invariant. This proves (2).   □

We note that planar extensions of focal-spreads do not have to be unique, as one extension might admit a particular collineation group that the other does not. We need, therefore, a definition of what it might be for a planar extension to admit a given collineation group.

DEFINITION 20. *Let* F *be a focal-spread of type* (t, k) *with collineation group* G *in* $\Gamma L(t+k, q)$ *and let* V *be a* $2t$-*dimensional vector space over* $GF(q)$. *Let* W *be a* $t + k$-*dimensional subspace of* V *containing the focus* L *of* F. *Then we shall say that* F *has a 'planar extension with group* G' *if and only if there is a* t-spread S *of* V *and a group* G' *in* $\Gamma L(2t, q)$, *such that there is a* $GF(q)$-*linear monomorphism* $\phi : F \to V$ *and a group isomorphism* $f : G \to G'$ *such that* $\phi(L)$ *lies in a unique spread component* L' *of* S, *where* $\phi(L^\alpha) \subseteq L'^{f(\alpha)}$, *for* $\alpha \in G$.
When $G = \langle 1 \rangle$, *we simply say that* F *has a 'planar extension.'*

Now we show at least one situation where a focal-spread necessarily is a k-cut. As a guide to the argument that we give, we formulate a matrix-based method to construct k-cuts from t-spreads (corresponding to translation planes of order q^t with kernel containing $GF(q)$). The situation that we then consider will show almost immediately that the focal-spread in question can arise as a k-cut.

3.1. Matrix Representation of k-Cuts.

THEOREM 6. *Let π be a translation plane of order q^t, and kernel containing $GF(q)$. Represent points of the $2t$-dimensional $GF(q)$-vector space as*

$$(x_1, .., x_t, y_1, .., y_t), \text{ where } x_i, y_i \in GF(q), \text{ for } i = 1, 2, .., t.$$

Assume that we have a matrix t-spread set

$$x = 0, y = 0, y = xM, \ M \in \mathcal{M},$$

where $\mathcal{M}$ is a set of non-singular $t \times t$ matrices, where the differences of distinct matrices are also non-singular, and where

$$x = (x_1, .., x_t), y = (y_1, .., y_t).$$

Let V_{t+k} be the $t + k$-dimensional subspace of vectors

$$(x_1, .., x_k, 0, .., 0, y_1, y_2, .., y_t)$$

$$\text{for all } x_j, y_i \ \in \ GF(q), \text{ for } j = 1, 2, .., k, i = 1, 2, .., t.$$

If we form the k-cut, $V_{t+k} \cap Z$, where Z is one of the components of the matrix t-spread, we have the focal-spread:

$$V_{t+k} \cap (x \ = \ 0) = (x = 0), \text{(note the two uses of } x = 0\text{)}$$
$$V_{t+k} \cap (y \ = \ 0) = \{(x_1, .., x_k, 0, 0, .., 0);$$
$$x_i \ \in \ GF(q), i = 1, 2, .., k\},$$
$$V_{t+k} \cap (y \ = \ xM) = \{(x_1, .., x_k, 0, .., 0), (x_1, .., x_k, 0, .., 0)M\},$$
$$\forall M \ \in \ \mathcal{M}.$$

Now if $I_{k \times t}$ is the $k \times t$ matrix with 1 in the (i, i) positions, for $i = 1, 2, .., k$ and zeros in all other positions, then

$$(x_1, .., x_k, 0, ...0)M = (x_1, .., x_k)I_{k \times t}M.$$

Now suppress the $t = k$ zeros in x and now use x to represent $(x_1, .., x_k)$. Then the focal spread of type (t, k) over $GF(q)$ is

$$x = 0, y = 0, y = xI_{k \times t}M, \text{ for } M \in \mathcal{M},$$

where x is a k-vector and y is a t-vector.

Two major questions regarding focal-spreads and homology or elation groups arise. If a focal-spread admits a collineation group fixing a line pointwise and acting regularly on the remaining non-fixed components, is the focal-spread a k-cut of a spread of a translation plane? If the group is a putative homology group, then we may prove that this is always the case. We begin with a fundamental proposition.

PROPOSITION 3. *A collineation h of a focal-spread that fixes a k-component pointwise and is semi-regular on the remaining components is a homology.*

PROOF. Represent the focal-spread in the form

$$x = 0, \ y = 0, \ y = xM; \ M \text{ is a } k \times t \text{ matrix in set } \mathcal{M}$$

for x a k-vector and y a t-vector, where $x = 0$ is the focus, and $y = 0$ is a k-space. Assume furthermore that $I_{k \times t} \in \mathcal{M}$. By Proposition 2, we may choose a representation so that the collineation is of the form $\sigma_B : (x, y) \to (x, yB)$, where B is a non-singular $t \times t$ matrix. Assume that some σ_B does not act fixed-point-free on the focus $x = 0$. Let $y_0 B = y_0$, where y_0 is a non-zero t-vector (that is, $(0, y_0)$ is fixed by σ_B). There exists a non-zero k-vector x_0 and a matrix M of $\mathcal{M}$ so that $x_0 M = y_0$. Hence, $(x_0, x_0 M B) = (x_0, y_0 B) = (x_0, y_0) = (x_0, x_0 M)$, so that (x_0, y_0) is a vector on $y = xMB$ and $y = xM$, so this means that σ_B leaves $y = xM$, invariant, a contradiction to our assumptions. Hence, h is a homology group and as such acts sharply transitive on the non-zero vectors of the focus $x = 0$. □

3.2. Nearfield Focal-Extension Theorem.

THEOREM 7. *Let $\mathcal{F}$ be a focal-spread of type (t, k) over $GF(q)$. Assume that there is an affine group G of order $q^t - 1$ fixing the focus, fixing a k-component pointwise and acting transitively on the remaining k-spaces of the partial Sperner k-spread.*

(1) Then G is a homology group (every non-identity element of G is an affine homology with the same axis and coaxis).

(2) There is a nearfield plane π of order q^t so that $\mathcal{F}$ is a k-cut of π. Hence, there is a planar extension with group G.

PROOF. By the previous proposition, since G is semi-regular, it follows that G is an affine homology group. In the context of the matrix cut procedure 3.1, recalling that we may always assume that $y = xI_{k \times t}$ is a k-component, we begin by first considering x and y t-vectors and form the associated $2t$-dimensional vector space over $GF(q)$ with vectors $(x_1, .., x_t, y_1, .., y_t) = (x, y)$. Let $\mathcal{C}$ denote the group $\{B; \sigma_B = \begin{bmatrix} I_k & 0 \\ 0 & B \end{bmatrix} \in G\}$. Form the putative t-spread:

$$x = 0, y = 0, y = xB, \ B \in \mathcal{C}.$$

Now we claim that this is a t-spread. To see this, we note that if $y = xB$ and $y = xD$, for $B, D \in \mathcal{C}$, share a vector $(x_0, x_0 B) = (x_0, x_0 D)$, for $x_0 \neq 0$, then $x_0 BD^{-1} = x_0$. However, $\mathcal{C}$ is fixed-point-free, as noted above. Hence, we obtain a t-spread and by the matrix cut procedure 3.1, it follows immediately that the focal-spread is a k-cut. This completes the proof of the theorem. □

We shall return to these ideas in Chapter 39, when we consider doubly transitive focal-spreads. In this setting, it is possible to produce elation groups as well as homology groups.

4. Spread-Theoretic Dual of a Semifield

We interrupt our discussion of focal-spreads to prepare for the ideas of an 'additive' focal-spread, which would be the natural analogue of a semifield spread. Although all of the material in this chapter can be given by various other methods, we treat the concepts here strictly from a spread-theoretic viewpoint. This chapter is modified from work of Jha and Johnson [**81**].

Given an affine translation plane, complete to the projective extension and dualize the projective plane. If the translation plane is a semifield plane and if the line at infinity becomes the parallel class (∞), then restricting to the affine plane from the dual semifield plane, another semifield plane is constructed. Normally all of this is done using the coordinate semifield, as taking the reverse or opposite multiplication, a coordinate semifield for the dual 'affine' semifield plane may be constructed. Specifically, given a finite semifield $(S, +, \circ)$, then the 'dual semifield' $(S, +, *)$ may be defined by the multiplication $a * b = b \circ a$.

It is also true that for finite translation planes, a duality of the associated projective space provides what we call a 'dual spread', which is always a spread in the finite case. From a matrix spread set, if

$$x = 0, y = x, y = xM, \text{ for } M \in S_{Mat}$$

is a matrix spread set for a semifield plane, then

$$x = 0, y = x, y = xM^t, \text{ for } M \in S_{Mat},$$

where M^t denotes the transpose of the matrix M, also gives a semifield (see, e.g., Johnson [**133**]). In general, there are six semifields associated with a given one, and in this chapter, we develop a notion of the dual of a semifield strictly from the matrix spreadset viewpoint. This material explicates the original results of Kantor [**156**] but given completely in terms of the spreads and gives a spread description of the six semifields arising from a given semifield.

5. The Dual Semifield Plane

We begin with the dual semifield. Of course, any finite semifield spread of order p^n, for p a prime, may be written in the form

$$x = 0, \ y = xM,$$

for M in an additive set $\mathcal{S}$ of $n \times n$ matrices, including the zero matrix and such that all non-zero matrices are non-singular, and where x and y are n-vectors (considered as row vectors). The reader should note that we are not asserting anything about the kernel of the associated translation plane, merely that we may always consider an arbitrary semifield spread in this form. It will turn out that by working over the prime field the ideas we shall be considering are more easily formulated.

The set of non-zero elements of $\mathcal{S}$ is a sharply transitive set acting on the set of all n-vectors. Fix a row vector w_0. Then given any non-zero n-vector w, there is a unique matrix M_w that maps w_0 to w. Hence, there is a bijection between the set of all n-vectors and the elements of $\mathcal{S}$ that maps the zero vector to the zero matrix. More generally, we have a (pre)semifield (i.e., lacks an identity) defined by any bijection $\phi : V \to \mathcal{S}$ (which maps zero to zero) in which multiplication has form $x \circ w = x M_w$, where M_w denotes the matrix $\phi(w)$.

The projection mappings A_i that map each matrix $M_w \in \mathcal{S}$ onto its i-th row are additive and hence linear, so we consider each A_i as a matrix such that the map $w \mapsto A_i w$ coincides with the i-th row of M_w. Hence, the spreadset $\mathcal{S}$ coincides with the set of matrices: $\{M_w \in \mathcal{S}\} =$

$$\left\{ \quad y = x \begin{bmatrix} wA_1 \\ wA_2 \\ wA_3 \\ \cdot \\ \cdot \\ \cdot \\ wA_n \end{bmatrix} \cdot \right\} \quad \text{Moreover, since the non-}$$

for all n-vectors w (rows) over $GF(p)$

zero M_w are required to be non-singular, it follows that the additive group generated by the matrices A_i, $\mathcal{A} = \langle A_1, A_2, \ldots A_n \rangle$ consists entirely of non-singular elements (and zero), hence $\mathcal{A}$ is a spreadset. Actually, this is the spreadset of the plane π', dual to π, coordinatized by the pre-semifield with multiplication $w \odot x := x \circ w$. Letting

$x = (x_1, x_2, .., x_n)$, we have

$$w \odot x \quad : \quad = x \circ w = x \begin{bmatrix} wA_1 \\ wA_2 \\ wA_3 \\ \cdot \\ \cdot \\ \cdot \\ wA_n \end{bmatrix}$$

$$= \sum_{i=1}^{n} x_i w A_i = w\left(\sum_{i=1}^{n} x_i A_i\right).$$

This means that the spread

$$x = 0, y = x\left(\sum_{i=1}^{n} \alpha_i A_i\right), \text{ for all } \alpha_i \in GF(p),$$

is the semifield spread given by the dual presemifield.

There is a similar partition of a spreadset $\mathcal{S}$ into its column vectors. If a semifield is represented in the form

(2.1) $x = 0, \ y = x[C_1 w^t, C_2 w^t, .., C_n w^t],$

where x and w are row vectors, then the semifield spread for the dual semifield is

(2.2) $x = 0, \ y = x[C_1^t w^t, C_2^t w^t, .., C_n^t w^t].$

Let $\odot$ be the dual of $\circ$:

$$w \odot x := x \circ w$$
$$= x \left[C_1 w^t, C_2 w^t, \ldots, C_n w^t\right]$$
$$= \left[x C_1 w^t, x C_2 w^t, \ldots, x C_n w^t\right],$$
$$= \left[w C_1^t x^t, w C_2^t x^t, \ldots, w C_n^t x^t\right],$$
(since $x^t C_i y$ are scalars, they may be transposed)
$$= w \left[C_1^t x^t, C_2^t x^t, \ldots, C_n^t x^t\right].$$

Notice that $[C_1^t x^t, C_2^t x^t, \ldots, C_n^t x^t]$ is the slope map of some element of the spreadset generated by $\{C_1^t, \ldots, C_n^t\}$, i.e. in the transposed spread of the initial spreadset $\mathcal{S}$. This provides the basic sketch of the following result, which makes the ideas of 'dual' clear spread theoretically.

THEOREM 8. *(1) Let π denote a semifield spread of order p^n written in the form*

$$x = 0, \; y = x \begin{bmatrix} wA_1 \\ wA_2 \\ wA_3 \\ . \\ . \\ . \\ wA_n \end{bmatrix}, \; \text{for all } n\text{-vectors } x \text{ over } GF(p),$$

Then the following is the semifield spread corresponding to the dual pre-semifield of π:

$$x = 0, y = x \left(\sum_{i=1}^{n} \alpha_i A_i \right), \; \text{for all } \alpha_i \in GF(p).$$

(2) Let the semifield spread be written in the form

(2.3) $$x = 0, \; y = x[C_1 w^t, C_2 w^t, .., C_n w^t],$$

then the semifield spread for the dual pre-semifield is:

(2.4) $$x = 0, \; y = x[C_1^t w^t, C_2^t w^t, .., C_n^t w^t].$$

6. The Six Associated Semifields

From the previous section, we recall that every semifield spread of order p^n may be written in the form:

(2.5) $$(*) : x = 0, y = x[C_1 w^t, C_2 w^t, .., C_n w^t],$$

for all n-vectors w, where $C_1, C_2, .., C_n$ are non-singular $n \times n$ matrices over $GF(p)$. The transposed semifield spread then may be represented as

$$x = 0, y = x \begin{bmatrix} wC_1 \\ wC_2 \\ wC_3 \\ . \\ . \\ . \\ wC_n \end{bmatrix}; w \text{ a } n\text{-vector.}$$

We have shown in Theorem 8 that the dual of the semifield spread may be given by

$$y = x \left(\sum_{i=1}^{n} \alpha_i C_i \right), \; \text{for all } \alpha_i \in GF(p).$$

These steps clearly also reverse. Hence, we have the following description of the transpose and dual of a semifield spread (here dual refers to the dual of the semifield).

THEOREM 9. *Let S be a semifield spread of order p^n. If D and E are semifields let D^t, E^d denote the spreads corresponding to the transpose and dual for the semifields D and E, respectively. Then there are non-singular matrices C_i, for $i = 1, 2, .., n$, such that we have the following representations for the various spreads.*

$$S \quad : \quad x = 0, y = x[C_1 w^t, C_2 w^t, .., C_n w^t]$$

$\Longleftrightarrow \quad transpose$

$$S^t \quad : \quad x = 0, y = x \begin{bmatrix} wC_1^t \\ wC_2^t \\ wC_3^t \\ . \\ . \\ . \\ wC_n^t \end{bmatrix}$$

$\Leftrightarrow \quad dualize$

$$S^{td} \quad : \quad x = 0, y = x \left(\sum_{i=1}^{n} \alpha_i C_i^t \right), \text{ for all } \alpha_i \in GF(p),$$

$\Longleftrightarrow \quad transpose$

$$S^{tdt} \quad : \quad x = 0, y = x \left(\sum_{i=1}^{n} \alpha_i C_i \right), \text{ for all } \alpha_i \in GF(p),$$

$\Longleftrightarrow \quad dualize$

$$S^{tdtd} \quad : \quad x = 0, y = x \begin{bmatrix} wC_1 \\ wC_2 \\ wC_3 \\ . \\ . \\ . \\ wC_n \end{bmatrix}$$

$\Longleftrightarrow \quad transpose$

$$S^{tdtdt} \quad : \quad x = 0, y = x[C_1^t w^t, C_2^t w^t, .., C_n^t w^t]$$

$\Longleftrightarrow \quad dualize$

$$S^{tdtdtd} \quad = \quad S : x = 0, y = x[C_1 w^t, C_2 w^t, .., C_n w^t].$$

Finally, we consider the connections between two representations of the same spread:

(2.6) $$x = 0, \quad y = x[C_1 w^t, C_2 w^t, .., C_n w^t],$$

and

$$x = 0, \ y = x \begin{bmatrix} wA_1 \\ wA_2 \\ wA_3 \\ . \\ . \\ . \\ wA_n \end{bmatrix}, \text{ for all } n\text{-vectors } x \text{ over } GF(p).$$

We have noted that

$$x = 0, y = x \left(\sum_{i=1}^{n} \alpha_i A_i \right), \text{ for all } \alpha_i \in GF(p),$$

represents the spread of the dual pre-semifield and we know that

$$x = 0, y = x \left(\sum_{i=1}^{n} \alpha_i C_i \right), \text{ for all } \alpha_i \in GF(p),$$

is also a spread, and it follows immediately that this is the transposed-dual-transposed spread of our spread.

6.1. The 'Transfer.' Here we simply note the transfer from the two possible representations of a semifield spread.

THEOREM 10. *If the semifield spread is represented both by*

(2.7) $$x = 0, \quad y = x[C_1 w^t, C_2 w^t, .., C_n w^t],$$

and by

$$x = 0, \ y = x \begin{bmatrix} wA_1 \\ wA_2 \\ wA_3 \\ . \\ . \\ . \\ wA_n \end{bmatrix}, \text{ for all } n\text{-vectors } x \text{ over } GF(p),$$

Then the ordered set of columns of A_j is the ordered set of transposes of the j-th rows of the C_i for $i = 1, 2, .., n$. Similarly, the ordered set of rows of C_j is the ordered set of transposes of the j-th columns of the A_i, for $i = 1, 2, .., n$.

PROOF. If $A_j = [D_{j1}, D_{j2}, .., D_{jn}]$, where the D_{ji} are the columns

of A_j, for $i = 1, 2, .., n$, and $C_j = \begin{bmatrix} E_{1j} \\ E_{2j} \\ . \\ E_{nj} \end{bmatrix}$, where E_{ij} are the rows of

C_j, for $i = 1, 2, .., n$. Then we obtain

$$[wD_{ij}] = [E_{ij}w^t]$$

so $D_{ij} = E_{ij}^t$. Then $A_j = [E_{j1}^t, E_{j2}^t, .., E_{jn}^t]$, so the set of columns of
A_j is the set of transposes of the j-th rows of the C_i for $i = 1, 2, .., n$.

Similarly, $C_j = \begin{bmatrix} D_{1j}^t \\ D_{2j}^t \\ . \\ D_{nj}^t \end{bmatrix}$, and we have that the set of rows of C_j is the

set of transposes of the j-th columns of the A_i, for $i = 1, 2, .., n$. □

7. Symplectic Spreads

In Kantor [**156**], there is a beautiful and fundamental connection between commutative semifield spreads and symplectic semifield spreads, using the Knuth cubical arrays. Here we consider these ideas from the matrix spreadset viewpoint. Although we not doing anything new, we do generalize these connections somewhat. So, in this section, we are interested in the iterative process of transposition and duality of a semifield spread with the initial assumption that the original spread is self-dual, which is always true if the semifield is commutative. Recall again that there is possible confusion between the concept of transposition, which refers to the dual spread in the projective space and the procedure of duality, which refers to the dual projective plane. To be clear, we first define a symplectic spread.

DEFINITION 21. *A 'symplectic spread' is a t-spread in a vector space of dimension $2t$, equipped with a symplectic form that leaves invariant each component of the spread (the subspaces are totally isotropic subspaces with respect to a symplectic polarity).*

REMARK 11. *With a suitable representation, a symplectic spread is obtained if and only if the associated matrices of a matrix spread set are symmetric. Hence, a symplectic spread is self-transpose but, of course, a self-transpose spread need not be symplectic.*

Here we use the column version of the representation of a semifield spread of order p^n and write the spread in the form:

(2.8) $x = 0, \; y = x[C_1w^t, C_2w^t, .., C_nw^t]$, for all w n-vectors where $C_1, C_2, .., C_n$ are non-singular $n \times n$ matrices over $GF(p)$, for all n-vectors w and x. Now we apply Theorem 9, that shows the spread corresponding to the dual pre-semifield is

(2.9) $\quad x = 0, \ y = x[C_1^t w_1^t, C_2^t w_1^t, .., C_n^t w_1^t]$, for all w_1 n-vectors.

In particular if,

(2.10) $\qquad x \circ w = x[C_1 w^t, C_2 w^t, .., C_n w^t],$

then $*$ defines the dual pre-semifield if

$$x * w = w[C_1 x^t, C_2 x^t, .., C_n x^t].$$

Now assume that the spread obtained from the dual pre-semifield is the same as the original spread then

(2.11) $\qquad x[C_1 w^t, C_2 w^t, .., C_n w^t] = x[C_1^t w_1^t, C_2^t w_1^t, .., C_n^t w_1^t],$

where the mapping $w \to w_1$ is a bijection. This means that $\{C_1, C_2, .., C_n\} = \{C_1^t, .., C_n^t\}$, (note here the subtle difference between self-dual and self-transpose), which we symbolize by

$$\{C_i; i = 1, 2, .., n\}^t = \{C_i; i = 1, 2, .., n\}.$$

If we now transpose and dualize, we see the corresponding spread is

$$x = 0, y = x \left(\sum_{i=1}^n \alpha_i C_i^t \right), \text{ for all } \alpha_i \in GF(p),$$

which shows that

$$\left(\sum_{i=1}^n \alpha_i C_i^t; \alpha_i \in GF(p) \right)^t = \left(\sum_{i=1}^n \beta_i C_i^t; \beta_i \in GF(p) \right).$$

In other words, the original spread is self-dual if and only if the transposed-dual spread is self-transpose. Note that a semifield spread defines a commutative pre-semifield if and only if it is individually self-dual and a semifield spread is symplectic (the components are totally isotropic subspaces of a symplectic form) if and only if the matrices are symmetric.

THEOREM 11. *(Compare with Kantor* [156] *in the symplectic case)* *Let* $(S, +, \circ)$ *be a pre-semifield (or semifield) of order* p^n *defining a self-dual semifield spread. Then take the semifield spread*

(2.12) $x = 0, y = x[C_1 w^t, C_2 w^t, .., C_n w^t] = x \circ w;$ *for all* n-*vectors* w,

where C_i *is a non-singular* $n \times n$ *matrix.*

(1) Then $\{C_i; i = 1, 2, .., n\}^t = \{C_i; i = 1, 2, .., n\}$, *and* $C_i = C_i^t$ *if and only the pre-semifield is commutative.*

(2) Transpose the matrices of the self-dual pre-semifield spread to

$$
obtain \; y = x
\begin{bmatrix}
wC_1 \\
wC_2 \\
wC_3 \\
\cdot \\
\cdot \\
\cdot \\
wC_n
\end{bmatrix}.
$$

(3) Dualize to obtain the spread

$$
y = x \left(\sum_{i=1}^{n} \alpha_i C_i \right), \; for \; all \; \alpha_i \in GF(p).
$$

Then note that $(\sum_{i=1}^{n} \alpha_i C_i^t; \alpha_i \in GF(p))^t = (\sum_{i=1}^{n} \beta_i C_i^t; \beta_i \in GF(p))$. *So, the spread itself is self-transpose. The spread is symplectic if and only if the original spread is commutative.*

(4) Conversely, if

$$
\left(\sum_{i=1}^{n} \alpha_i C_i^t; \alpha_i \in GF(p) \right)^t = \left(\sum_{i=1}^{n} \beta_i C_i^t; \beta_i \in GF(p) \right)
$$

defines a self-transpose semifield spread then the dual semifield spread is

$$
y = x
\begin{bmatrix}
wC_1 \\
wC_2 \\
wC_3 \\
\cdot \\
\cdot \\
\cdot \\
wC_n
\end{bmatrix}
$$

and the transpose of this spread is

(2.13) $$y = x[C_1 w^t, C_2 w^t, .., C_n w^t].$$

Now define a multiplication by

$$
x \circ w = x[C_1 w^t, C_2 w^t, .., C_n w^t]
$$

for all x, w *n-vectors over* $GF(p)$. *Since*

$$
\{C_i; i = 1, 2, .., n\}^t = \{C_i; i = 1, 2, .., n\},
$$

it is immediate that we obtain a self-dual pre-semifield, which is commutative if and only if $C_i^t = C_i$.

PROOF. Then simply apply Theorem 8 or Theorem 9. $\qquad \square$

Finally, we note that in our results, we have chosen a specific representation for an initial semifield spread; either in row or column form. The column form is best suited to determine whether a semifield is self-dual and the row form is ideal if asking if a semifield spread is self-transpose. Theorem 10 shows how to deal with the two opposite situations.

For example, if we have a representation

$$
x = 0, \; y = x \begin{bmatrix} wA_1 \\ wA_2 \\ wA_3 \\ . \\ . \\ . \\ wA_n \end{bmatrix}, \text{ for all } n\text{-vectors } x \text{ over } GF(p),
$$

and wish to determine if the semifield is self-dual or commutative, we note that by Theorem 10 then C_j^t is the set of ordered j-th columns of the $A_1, A_2, .., A_n$. Hence, the semifield is commutative if and only if the $C_j^t = C_j$ and is self-dual if and only if $\{C_j; j = 1, 2, .., n\}^t = \{C_j; j = 1, 2, .., n\}$.

8. Additive Focal-Spreads

We now return to the study of 'additive focal-spreads.' Again, we choose the standard representation for a focal-spread and consider the focus as $x = 0$, the unique t-dimensional subspace of the partition, and $y = 0$ a k-space. We have seen in Theorem 7 if there is a group of the focal-spread of order $q^t - 1$ that fixes a k-component pointwise, fixes the focus, and acts transitively on the remaining components then the focal-spread is a k-cut of a nearfield plane. However, if the group fixing the focus pointwise is transitive on the partial Sperner k-space and fixes $x = 0$ pointwise, it is not known that the focal-spread is a k-cut of a semifield plane or even that the group is an elation group. If we assume that a group of order q^t fixes $x = 0$ pointwise (and fixes no other points), then the group will act sharply transitive on the set of q^t k-spaces of the partial Sperner k-space.

DEFINITION 22. *A focal-spread of type (t, k) over $GF(q)$ shall be said to be 'additive' if there is an elation group E of order q^t with axis the focus.*

The following remark shows from where the term 'additive' arises.

REMARK 12. *Choose a standard representation for the focal-spread. Then if there is an elation group E of order q^t with axis the focus, then there is a matrix spread set for the focal-spread as follows:*

$$x = 0, y = xC; \ C \in \mathcal{M},$$

where $\mathcal{M}$ is an elementary Abelian p-group of order q^t of $k \times t$ matrices. The group $E = \left\langle e_C : (x, y) \to (x, xC + y) = (x, y) \begin{bmatrix} I_k & C \\ 0 & I_t \end{bmatrix} ; C \in \mathcal{M} \right\rangle$.

Thus, an additive focal-spread has a matrix representation whose matrix set is an additive group of $k \times t$ matrices.

So, natural generalizations of semifield spreads are additive focal-spreads.

Suppose that we have an additive focal-spread of type (t, k) over $GF(q)$. Then the focal-spread is $x = 0, y = xM$, where M is in a set S of q^t $k \times t$ matrices of rank k or the zero matrix, such that the difference of any two distinct matrices in S is also of rank k and, where S is an elementary Abelian p-group and hence a vector space over $GF(p)$.

The following result now follows immediately from Remark 12.

THEOREM 12. *A focal-spread of type (t, k) that admits an elementary Abelian matrix spread set is an additive focal-spread (admits an elation group q^t that fixes the focus pointwise and acts transitive on the partial Sperner k-spread).*

Although it may seem that additive focal-spreads are difficult objects to study, actually they are natural geometries that arise from additive partial spreads by a method analogous to that of dualization of a semifield spread, which apart from the wonderful matrix transformations describing the six connecting semifields, is the principal reason for including these ideas in the material on focal-spreads.

As we have mentioned previously for semifields, it is also true here for additive focal-spreads, it is helpful to work over the prime field, for the connections to additive partial spreads arises with this consideration, regardless of the kernel. We invite the reader to reread the nature of the dual of a semifield spread using the matrix approach and supply the proof of the following theorem (the proof, of course, can be read in the article [**77**]). We do supply the proof of the converse, that additive partial spreads do produce an additive focal-spread using a generalization of the dual semifield, which we term the 'companion.'

THEOREM 13. *(1) Each additive focal-spread of type (t, k) over $GF(p)$, for p a prime, gives rise to a partial spread of k $t \times t$ matrices,*

$$y = xA_i, \ for \ i = 1, 2, .., k,$$

where $A_1 = I$ and

$$x \left(\sum_{i=1}^{kr} \alpha_i A_i \right) = 0,$$

implies $\alpha_i = 0$, for $i = 1, 2, .., k$, for any x non-zero.

(2) Any additive partial spread that can be extended to some additive t-spread produces a $(t + k)$-additive focal-spread over $GF(p)$, which is a k-cut; the extension of the additive partial spread is equivalent to the extension of the additive-focal spread.

PROOF. Left to the reader as an exercise.   □

DEFINITION 23. *The partial spread obtained from an additive focal-spread as in the previous theorem shall be called the 'companion partial spread.' If we actually have an additive spread, that is, a semifield spread, we call this related semifield spread the 'companion semifield spread' and the previous chapter discusses this situation.*

THEOREM 14. *Additive focal-spreads of dimension $t + k$ with focus of dimension t over $GF(p)$ are equivalent to additive partial t-spreads of degree $1 + p^k$; the companion of the partial spread is an additive focal-spread.*

PROOF. It remains to show that an additive partial spread produces an additive focal-spread. Consider any additive partial spread of order p^t, and degree p^k. By definition, an additive spread is an elementary Abelian p-group so is a $GF(p)$-vector space and thus has a basis $\{A_1 = I, A_2, .., A_k\}$ of $t \times t$ matrices such that the partial spread is $\sum_{i=1}^{k} \alpha_i A_i$, where $\alpha_i \in GF(p)$. Now consider a $t + k$-vector space over $GF(p)$ with vectors

$$(x_1, x_2, .., x_k, y_1, y_2, .., y_t).$$

Regard $y = 0 = (y_1, .., y_t)$ to be in the original additive spread. Now form

$$x = 0, y = x \left[\begin{array}{cccc} w^T & (wA_2)^T & (wA_3)^T & , .., & (wA_k)^T \end{array} \right],$$

We claim that this is a $t + k$-dimensional additive focal-spread with focus of dimension t. To see this we note that $\sum_{i=1}^{k} \alpha_i A_i$ is a non-singular linear transformation, for α_i not all zero and so for w non-zero t-vectors then $w \sum_{i=1}^{k} \alpha_i A_i = 0$, if and only if $\alpha_i = 0$, but $w \sum_{i=1}^{k} \alpha_i A_i = \sum_{i=1}^{k} \alpha_i w A_i$, since the α_i are in $GF(p)$. Therefore, the matrices

$$\left[\begin{array}{cccc} w^T & (wA_2)^T & (wA_3)^T & , .., & (wA_k)^T \end{array} \right]^T$$

are all of rank k and therefore we have an additive focal-spread of dimension $t+k$ with focus of dimension t. This completes the proof.   □

9. Designs, and Multiple-Spreads

In this section, we give some constructions of designs and double and triple spreads that may be constructed from focal-spreads. In Chapter 0.1, we have shown how a double-spread may be obtained from a t-spread of an st-dimensional vector space and have discussed more generally partitions with a number of different dimensions of subspaces. We recall the definition of a double-spread.

DEFINITION 24. *A 'double-spread' of a vector space is a partition using subspaces of two distinct dimensions. So a focal-spread is also a double-spread. Similarly, a 'triple-spread' is a partition of a vector space into subspaces of three distinct dimensions.*

9.1. Design Lemma. The following lemma begins a procedure that constructs designs from focal-spreads of type $(1 + k, k)$, using hyperplanes that intersect the focus in a k-dimensional subspace.

LEMMA 6. *Suppose that B is a focal-spread of dimension $2k + 1$ of type $(1 + k, k)$ over $GF(q)$. Then each hyperplane that intersects the focus in a k-dimensional subspace induces a partition of a vector space of dimension $2k$ over $GF(q)$ by $q + 1$ subspaces of dimension k and $q^{k+1} - q$ subspaces of dimension $k - 1$. Hence, each hyperplane then produces a double-spread.*

PROOF. For a focal-spread of type $(1+k, k)$, with focus L, consider any hyperplane H, a subspace of dimension $2k$, that intersects L in a subspace of dimension k. Since $2k + k - (2k + 1) = k - 1$, then H intersects each k-component in at least a $k - 1$-subspace. Let a denote the number of k-dimensional intersections not equal to L, so that $q^{k+1} - a$ is the number of $k - 1$-dimensional intersections.

$$a(q^k - 1) + (q^{k+1} - a)(q^{k-1} - 1) = q^{2k} - q^k,$$

which implies that

$$aq^{k-1}(q^k - 1) = q^k(q^k - 1),$$

so that $a = q$.   □

In a similar manner, triple-spreads may be constructed. The following may be proved in a manner similar to the previous lemma and will be left to the reader.

LEMMA 7. *Given any focal-spread of type (t, k) with focus L. Then any subspace H_{2k-1} of dimension $2k - 1$ that intersects L in a k-dimensional subspace is partitioned by a subspace of dimension k, q^2 subspaces of dimension $k - 1$, and $(q^{k+1} - q^2)$ subspaces of dimension $k - 2$. Thus, in this way, a triple-spread is obtained.*

Using the design lemma, we construct resolvable $2 - (q^{k+1}, q, 1)$-designs.

9.2. The Design Theorem.

THEOREM 15. *Let $\mathcal{F}$ be any focal-spread of type $(1+k, k)$ with focus L. Denote by $\boldsymbol{D} = \boldsymbol{D}(\mathcal{F}) = (\boldsymbol{P}, \boldsymbol{B})$ the incidence structure whose point set $\boldsymbol{P} = \mathcal{F} - \{L\}$ and whose blocks $\boldsymbol{B}$ are the hyperplanes that do not contain L. Then D is an affine (or resolvable) $2 - (q^{k+1}, q, 1)$-design. Any parallel class is formed from the hyperplanes that intersect L in a common k-space.*

PROOF. By Lemma 6, every block contains q points. Since

$$\begin{aligned} |\boldsymbol{B}| &= (q^{2k+1} - 1)/(q-1) - (q^k - 1)/(q-1) \\ &= q^k(q^{k+1} - 1)/(q-1), \end{aligned}$$

there are exactly q^k hyperplanes that intersect L in a common k-space U. Since any point P lies in the hyperplane $P + U$, it follows that this set of hyperplanes covers the points. Furthermore, two hyperplanes containing U cannot share a point. In addition, the points P and Q lie in a unique hyperplane $P + Q$ of $\boldsymbol{B}$. This completes the proof.   □

10. Focal-Spreads from Designs

In the previous section, we have seen that given a focal-spread of type $(k+1, k)$, there is a corresponding $2 - (q^{k+1}, q, 1)$-design. Here we try to find a converse construction to recapture the focal-spread from a design with the indicated parameters.

Note that in a $2 - (q^{k+1}, q, 1)$-design, there are $q^k(q^{k+1} - 1)/(q-1)$ blocks, of which there are exactly $(q^{k+1} - 1)/(q-1)$ containing a given point. If the design can be embedded into a vector space of dimension $2k + 1$ over $GF(q)$, where the blocks are hyperplanes, there are exactly $(q^k - 1)/(q - 1)$ hyperplanes not in the block set. A similar proof to the following theorem is also found in Jha and Johnson [77].

10.1. Focal-Spread Reconstruction Theorem.

THEOREM 16. *Let $\boldsymbol{D} = (\boldsymbol{P}, \boldsymbol{B})$ be a $2 - (q^{k+1}, q, 1)$-design with the following properties: (we shall refer to the points of the design as 'd-points')*

(a) The d-points are subspaces of a $(2k + 1)$-dimensional $GF(q)$-space V.

(b) The blocks are hyperplanes of V.

Then $\boldsymbol{D} = \boldsymbol{D}(\mathcal{F})$, where $\mathcal{F}$ is a focal-spread of type $(k + 1, k)$.

The proof will be given by a series of lemmas.

LEMMA 8. *Every d-point space P has dimension k and all hyper-planes containing P are in **B**.*

PROOF. If P is an s-space, then there are $(q^{2k+1-s} - 1)/(q - 1)$ hyperplanes containing P. Let $\mathcal{H}$ denote the set of $q^k(q^{k+1} - 1)/(q-1)$ hyperplanes of the design. Since there are exactly $(q^{2k+1} - 1)/(q - 1)$ hyperplanes, there are exactly $(q^k - 1)/(q-1)$ hyperplanes that do not belong to $\mathcal{H}$. There are exactly $(q^{k+1} - 1)/(q - 1)$ hyperplanes of $\mathcal{H}$ that contain P. Therefore, the maximum number of hyperplanes that contain P is $(q^{k+1} - 1)/(q - 1) + (q^{k+1} - 1)/(q - 1)$. Hence,

$$(q^{k+1} - 1)/(q - 1) + (q^k - 1)/(q - 1) \geq (q^{2k+1-s} - 1)/(q - 1),$$

or equivalently,

$$q^{k+1} + q^k \geq q^{2k+1-s} + 1 > q^{2k+1-s}.$$

Therefore, $q^{k+1} + q^k > q^{2k+1-s}$ so that $q > q^{k+1-s} - 1$. It follows that $q \geq q^{k+1-s}$, which implies that $1 \geq k+1 - s$ or rather that $s \geq k$. But $s \not> k$ since otherwise there could not be $(q^{k+1} - 1)/(q - 1)$ hyperplanes containing P. This completes the proof. □

LEMMA 9. ***P** is a partial spread of k-spaces on V.*

PROOF. If not assume that the subspace $\langle P, Q \rangle$ is of dimension $s \leq 2k - 1$ so lies in exactly $(q^{2k+1-s} - 1)/(q - 1)$ hyperplanes. So, there are at least $(q^2 - 1)/(q - 1)$ hyperplanes containing P and Q and these are the hyperplanes of ***B*** that contain P. But, there is exactly one block containing P and Q. Therefore, the dimension of $\langle P, Q \rangle$ is $2k$, hence the proof is complete. □

LEMMA 10. *Let $\mathcal{N}$ denote the set of hyperplanes that do not be-long to **B**. Then $L = \cap_{H \in \mathcal{N}} H$ is a $(k+1)$-dimensional subspace that intersects trivially with all points.*

PROOF. Set $\Gamma = (V - \{0\}) - \cup_{P \in \mathbf{P}}(P - \{0\})$. Then $|\Gamma| = q^{k+1} - 1$ by Lemma 8. Moreover, $H \in \mathcal{N}$ does not contain any d-point by the same lemma. Therefore, $\dim H \cap P = k - 1$ for $P \in \mathbf{P}$. This shows

$$|(H - \{0\}) - \cup_{P \in \mathbf{P}}((P \cap H) - \{0\}| = q^{k+1} - 1,$$

as $\cup_{P \in \mathbf{P}}((P \cap H) - \{0\}$ has cardinality $q^{k+1}(q^{k-1} - 1)$.

Therefore, it follows that $\Gamma \subseteq H$, which implies that $\Gamma \subseteq L$. Therefore, the $\dim L = \ell \geq k + 1$. As L is contained in exactly $(q^{2k+1-\ell} - 1)/(q - 1)$ hyperplanes, but there are exactly $(q^k - 1)/(q-1)$ hyperplanes of V that are not in $\mathcal{H}$, therefore, $\dim L = k + 1$, which completes the proof of the lemma. □

The sets of lemmas then establishes the proof of the theorem.

10.2. Parallelisms of Designs. Given a $2 - (q^{k+1}, q, 1)$-design, there are $q^k(q^{k+1} - 1)/(q - 1)$ lines (or blocks). The point set can be covered by a set of q^{k+1} lines. If the line set can be partitioned into disjoint subsets, this is often called a 'resolution' of the design. Here we adopt the term 'parallelism' for this concept.

COROLLARY 2. *Assume that a* $2 - (q^{k+1}, q, 1)$-*design can be embedded into a vector space of dimension* $2k + 1$ *such that the points are vector subspaces and the lines are hyperplanes. Then there is a parallelism of the lines into* $(q^{k+1} - 1)/(q - 1)$ *parallel classes of* q^{k+1} *lines each.*

The reader interested in how isomorphic focal-spreads reflect isomorphic designs is directed to Jha and Johnson [**77**].

CHAPTER 3

Generalizing André Spreads

There is an important class of translation planes called the 'André planes,' that may be constructed from the Desarguesian plane by a set of mutually disjoint derivations, that is, by 'multiple derivation.' The representation of the spreads have the general form

$$\left\{ x = 0, y = 0, y = x^{q^i} m; m^{(q^n-1)/(q-1)} = \delta; m \in GF(q^n)^* \right\},$$

(i is fixed for each δ).

If a spread for a translation plane can be represented in the form

$$\left\{ x = 0, y = 0, y = x^{q^{\lambda(m)}} m; m \in GF(q^n)^* \right\},$$

where λ is a function of $GF(q^r)^*$ to $\{1, 2, .., n\}$, then the translation plane is called a 'generalized André plane.'

Although with the use of the term it might seem that the set of generalized André planes has been completely determined, actually the opposite is true—almost nothing is known and the complete set of generalized André planes is (probably) almost unknown.

Now our terminology becomes problematic as we wish to extend the ideas and construction of the André spreads from Desarguesian planes, where our generalization intends to mean to generalize from spreads defining translation planes to arbitrary t-spreads defining translation Sperner spaces. Now this might be considered acceptable, but now what does one call a generalization of a generalized André spread?

In any case we shall try to make it clear what we are trying to do as follows: As mentioned, we wish to generalize the construction of the finite André planes π with kernel containing $GF(q)$ from Desarguesian affine planes Σ_{q^n} of order q^n, where $q = p^r$, for p a prime, in such a way so that certain important collineation groups that are present in André planes are also present in these generalized structures.

In order to follow our generalizations from André planes, we give a primer on the construction on André planes. Some of this chapter is modified from the author's work [98].

1. A Primer on André Planes

We begin the constructions of the most basic type of André plane those that can be constructed from Desarguesian planes by multiple derivation of a certain set of regulus nets. These planes are the so-called 'subregular André planes.' It is not difficult to show that if a spread $\mathcal{S}$ in $PG(3,q)$, for $q > 2$, has the property that there is a regulus generated by any three distinct spread components and if this regulus is also in $\mathcal{S}$, the spread is forced to be Desarguesian. We also use the term 'regular spread' for a Desarguesian spread. We shall eventually discuss 'regular parallelisms' , which are parallelisms all of whose spreads are regular (Desarguesian).

In the following, we shall be considering planes of order q^2.

Assume that a spread for the Desarguesian plane Σ_{q^n} is given by

$$x = 0, y = xm; m \in GF(q^n)),$$

then an 'André partial spread' A_δ is defined by

$$A_\delta : \left\{ y = xm; m^{(q^n-1)/(q-1)} = \delta \right\},$$

$\delta \in GF(q)$. It is easy to show that A_δ is a replaceable partial spread with $n-1$ replacement nets regulus $A_\delta^{q^i}$ with partial spread

$$A_\delta^{q^i} : \left\{ y = x^{q^i} m; m^{(q^n-1)/(q-1)} = \delta \right\}.$$

When $n = 2$, the André partial spreads are reguli and the replacement partial spread A_δ^q is the opposite regulus to A_δ.

DEFINITION 25. *Any translation plane obtained from a Desarguesian plane of order q^2 by 'deriving' or 'multiply deriving' a set of André partial spreads is called an 'André plane of dimension two' or a 'sub-regular André plane.' (Note that the spread, in this case, is in $PG(3,q)$.)*

The reader might note that there is much more room for construction when n is large due to the large number $(n-1)$ of possible net replacements for each André net. The more general use of the term André plane is as follows:

DEFINITION 26. *The 'André planes of order q^{sn} and kernel containing $GF(q)$' are defined as follows. We let*

$$A_\delta = \left\{ y = xm; m^{(q^{sn}-1)/(q-1)} = \delta \right\}, \delta \in GF(q)$$

be called an 'André partial spread' of degree $(q^{sn}-1)/(q-1)$ and order q^{sn}. This partial spread is replaceable by any partial spread

$$A_\delta^{q^\lambda} = \left\{ y = x^{q^\lambda} m; m^{(q^{sn}-1)/(q-1)} = \delta \right\}, \delta \in GF(q), 0 \le \lambda \le sn - 1,$$

called an 'André replacement.' Hence, there are exactly $sn - 1$ non-trivial replacements and, of course, if $\lambda = 0$, the partial spread has not been replaced. There are exactly $q - 1$ André nets each admitting sn replacements. An 'André plane' is defined as any translation plane obtained with spread consisting of $q - 1$ André replacement partial spreads together with $x = 0, y = 0$. Therefore, there are $(sn)^{q-1} - 1$ distinct André spreads obtained from a given Desarguesian affine plane.

We now mention the most important collineation groups that general André planes admit.

DEFINITION 27. *The 'kernel homology group' of the Desarguesian plane Σ_{q^n}, is the group of central collineations with axis the line at infinity and center the zero vector of the associated vector space. This group of order $q^n - 1$ then will act on each André plane, but the replacement nets are now in an orbit of length $(q^n - 1)/(q - 1)$ under the kernel homology group.*

We now generalize the concept of an André plane as arising from a Desarguesian affine plane to analogous structures constructible from what are called 'Desarguesian t-spreads.' We shall be using the term 'extended André spreads' to describe this generalization. What we obtain are new Sperner spaces in a similar way that André planes are produced from Desarguesian affine planes. Although these Sperner spaces are natural generalizations, they do not seem to have been noticed previously, or at least have not generated significant applications.

When we consider the generalizations of a generalized André plane in the context of what we are doing here, we shall use the somewhat awkward term 'generalized extended André $r - (sn, q)$-spreads.' We shall also use the terms for the natural net replacements as 'extended André replacements' and 'generalized extended André replacements.'

The reader is directed to Biliotti, Jha and Johnson [21] for additional background on André and generalized André planes.

2. r-(sn, q)-Spreads

We shall now give the definitions of the type of general vector space spread that we shall be constructing.

Consider a field $GF(q^{rsn})$, where $q = p^z$, for p a prime. Then $GF(q^{rsn})$ is an r-dimensional vector space over $GF(q^{sn})$. More generally, let V be the r-dimensional vector space over $GF(q^{sn})$.

The most difficult part of the definition is that of a 'j-(0-set),' so we offer a few words on this concept. Suppose that we have a normal Desarguesian spread over a field K. We may consider this spread as a spread of 1-dimensional K-subspaces, with an associated vector space of dimension 2 over $GF(q^n)$, so that vectors are (x_1, x_2), for $x_i \in GF(q^n)$, for $i = 1, 2$. Now we normally designate $x_1 = 0$ as $x = 0$ and $x_2 = 0$ as $y = 0$, so that when we consider the spread over a subfield $GF(q^{n/k})$ of $GF(q^n)$, we represent the spread as $x = 0, y = 0, y = xM$, where M is in a certain set of $n/k \times n/k$ matrices. Now suppose we have the set of all vectors (x_1, x_2, x_3), where the x_i are vectors over a field $GF(q^n)$. We now have the sets $x_1 = x_2 = 0$, $x_1 = x_3 = 0$, and $x_2 = x_3 = 0$. We would call such subsets 2-(0-sets). From a matrix point of view, these would take the place of $x = 0$ and $y = 0$ in the standard case. Also, in the standard case, apart from $x = 0$, or $y = 0$, the remaining vectors are (x_1, x_2), where $x_1 x_2 \neq 0$. The corresponding set in this more general setting would be (x_1, x_2, x_3), where $x_1 x_2 x_3 \neq 0$. This set would be the 0-(0-set). So we would naturally decompose the set of all vector (x_1, x_2, x_3) into disjoint sets of one 0-(0-subset), three 1-(0-subsets), and three 2-(0-subsets). We then generalize this and consider that x_i are r-vectors over a field $GF(q^{sn})$.

DEFINITION 28. *A '1-dimensional r-spread' or 'Desarguesian* $r -$ *(sn, q)-spread' is defined to be a partition of V by the set of all 1-dimensional $GF(q^{sn})$-subspaces, where q is a prime power. In this case, the vectors are represented in the form $(x_1, x_2, .., x_r)$, where $x_i \in GF(q^{sn})$.*

Furthermore, the 1-dimensional $GF(q^{sn})$-subspaces may be partitioned into the following sets called 'j-(0-sets).' A 'j-(0-set)' is the set of vectors with exactly j of the entries equal to 0. For a specific set of j-zeros among the r elements, the set of such non-zero vectors in the remaining $r - j$ non-zero entries is called a '(j-(0-subset)).'

Note that there are exactly $\binom{r}{r-j} (q^{sn} - 1)^{r-j}$ vectors (non-zero vectors) in each j-(0-set) and exactly $(q^{sn} - 1)^{r-j}$ vectors in each of the $\binom{r}{r-j}$ disjoint $j - (0\text{-subsets})$.

Hence, $j = 0, 1, .., r - 1$, and we denote the $j - (0\text{-sets})$ by Σ_j, and by specifying any particular order, we index the $\binom{r}{r-j}$ $j - (0\text{-subsets})$ by $\Sigma_{j,w}$, for $w = 1, 2, .., \binom{r}{r-j}$. We note that

$$\cup_{w=1}^{\binom{r}{r-j}} \Sigma_{j,w} = \Sigma_j,$$

a disjoint union. Furthermore, the $(q^{rsn} - 1)$ non-zero vectors are partitioned into the $j - (0\text{-sets})$ by

$$(q^{rsn} - 1) = \sum_{j=0}^{r-1} \binom{r}{r-j} (q^{sn} - 1)^{r-j},$$

and the number of 1-dimensional $GF(q^{sn})$-subspaces is

$$(q^{rsn} - 1)/(q^{sn} - 1) = \sum_{j=0}^{r-1} \binom{r}{j} (q^{sn} - 1)^{r-j-1}.$$

Also, note that this is then also the number of sn-dimensional $GF(q)$-subspaces in an $r - (sn, q)$-spread.

Some established notation is convenient here. The reader might note that what we are trying to do is to parrot the Desarguesian case as much as possible. So, in the (x, y)-case, for vectors other than $x = 0$, and $y = 0$, a vector in the spread is (x, xm), or $y = xm$, for some element m of the associated field. Also, it is important to note that we shall be working in the various j-(0-sets), Σ_j, there the j zeros are omitted in the notation. We are still in the Desarguesian 1-dimensional r-spread situation.

NOTATION 2. *Consider a vector* $(x_1, x_2, .., x_r)$ *over* $GF(q^{sn})$, *we use the notation* (x_1, y) *for this vector. Consider a j-(0-set) Σ_j and let x_{j_1} denote the first non-zero entry. Then all of the other entries are of the form $x_{j_1} m$, for $m \in GF(q^{sn})$. For example, the elements of an element of a 0-(0-set) may be presented in the form* $(x_1, x_1 m_1, .., x_1 m_{r-1})$, *for x_1 non-zero and m_i also non-zero in $GF(q^{sn})$. That is,*

$$y = (x_1 m_1, .., x_1 m_{r-1}).$$

More importantly if we vary x_1 over $GF(q^{sn})$, then

$$y = (x_1 m_1, .., x_1 m_{r-1}),$$

is a 1-dimensional $GF(q^{sn})$-subspace. However, we now consider this subspace as an sn-dimensional $GF(q)$-subspace.

In this notation, a Desarguesian 1-spread leads to an affine Sperner space by defining 'lines' to be translates of the 1-dimensional $GF(q^{sn})$-subspaces.

DEFINITION 29. *In general, for $r > 2$, a Desarguesian r-spread leads to a 'Desarguesian translation Sperner space' (simply the associated affine space) by the same definition on lines. Every 1-dimensional $GF(q^{sn})$-space may be considered an sn-space over $GF(q)$. When this occurs we have an example of what we shall call an 'r-(sn-spread)' (or also a '$r - (sn, q)$-spread').*

We are now ready to give a formal definition of what we shall call an 'r-(sn, q)-spread.

DEFINITION 30. *A partition of an rsn-dimensional vector space over $GF(q)$ by mutually disjoint sn-dimensional subspaces shall be called* *an 'r-(sn, q)-spread.' In the literature, this is often called an 'sn-spread' or projectively on the associated projective space as an 'sn − 1-spread.'*

If $r = 2$, any '2-(sn, q)-spread' is equivalent to a translation plane of order q^{sn}, with kernel containing $GF(q)$.

We are looking for 'nice' groups; we begin with those of Desarguesian r-spreads, which are the sn-kernel and a natural generalization.

DEFINITION 31. *Let Σ be an $(r-(sn\text{-}spread)$.*
We define the 'collineation group' of Σ to be the subgroup of $\Gamma L(rsn, q)$ that permutes the spread elements (henceforth called 'components').
For a Desarguesian r-spread Σ, the subgroup K_{sn}^* with elements

$$(x_1, x_2, .., x_r) \longmapsto (dx_1, dx_2, .., dx_r)$$

for all d nonzero in $GF(q^{sn})$ is called the 'sn-kernel' subgroup of Σ. The group fixes each Desarguesian component and acts transitively on its points. The group K_{sn}^* union the zero mapping is isomorphic to $GF(q^{qn})$. K_{sn}^* has a subgroup K_s^*, where d above is restricted to $GF(q^s)^*$, and K_s^* union the zero mapping is isomorphic to $GF(q^s)$. K_s^* is called the 's-kernel subgroup.'
More generally, also for a Desarguesian spread Π, we note that the group $G^{(sn)^r}$ of order $(q^{sn} - 1)^r$ with elements

$$(x_1, x_2, .., x_r) \longmapsto (d_1 x_1, d_2 x_2, .., d_r x_r); d_i \in GF(q^{sn})^*, i = 1, 2, .., r$$

also acts as a collineation group of Π. We call $G^{(sn)^r}$, the 'generalized kernel group.'

If we have a Desarguesian 2-spread with vectors (x, y) and spread $x = 0, y = 0, y = xm$, for m in a field $GF(q^2)$, choose any subspace of the form $y = x^q m_1$, where m_1 is a fixed element of $GF(q^n)$. Now find the set of elements $y = xm$ that non-trivially intersect $y = x^q m_1$ by solving the equation

$$x^{q-1} = mm_1^{-1}.$$

In this case, we see that $(mm_1^{-1})^{q+1} = 1$ or equivalently that $m^{q+1} = m_1^{q+1}$, so this defines an associated André partial spread, one of whose replacement subspaces is $y = x^q m_1$. Now apply the kernel homology group with elements $(x, y) \to (dx, dy)$, for d in $GF(q^2)$, which maps $y = x^q m_1$ onto the subspace $y = x^q d^{1-q} m_1$ and notice that this provides the set of all $q + 1$ replacement subspaces. This is the approach that we shall take in finding the appropriate extension of an André partial spread.

Let Σ be a Desarguesian r-spread with vectors $(x_1, x_2, .., x_r)$. Consider any set j-(0-set) Σ_j, and suppress the set of j zeros and write vectors in the form $(x_1^*, x_2^*, .., x_{r-j}^*)$, in the order of non-zero elements within $(x_1, .., x_r)$. Assume that $j \leq r - 1$. We consider such vectors of the following form

$$(x_1^*, x_1^{*q^{\lambda_1}} m_1, .., x_1^{*q^{\lambda_{r-j}}} m_{r-j}).$$

If x_1^* varies over $GF(q^{sn})$, and if we consider Σ as an rsn-vector space over $GF(q)$, we then have an sn-vector subspace over $GF(q)$ that we call

$$y = (x_1^{*q^{\lambda_1}} m_1, .., x_1^{*q^{\lambda_{r-j-1}}} m_{r-j-1}),$$

where λ_i are integers between 0 and $sn - 1$. We are interested in the set of Desarguesian sn-subspaces

$$y = (x_1^* w_1, .., x_1^* w_{r-j-1})$$

(using the same notation) that can intersect

$$y = (x_1^{*q^{\lambda_1}} m_1, .., x_1^{*q^{\lambda_{r-j-1}}} m_{r-j-1}).$$

We note that we have a non-zero intersection if and only if

$$x_1^{*q^{\lambda_i}-1} = w_i/m_i, \text{ for all } i = 1, 2, .., r - j - 1.$$

DEFINITION 32. *The above set of non-zero intersections of Σ shall be called an 'extended André set of type $(\lambda_1, \lambda_2, .., \lambda_{r-j-1})$.' The set of all subspaces*

$$y = (x_1^{*q^{\lambda_1}} n_1, .., x_1^{*q^{\lambda_{r-j-1}}} n_{r-j-1}),$$

such that

$$x_1^{*q^{\lambda_i}-1} = w_i/n_i, \text{ for all } i = 1, 2, .., r - j - 1,$$

has a solution is called an 'extended André replacement.'

Using these ideas, we may formulate the main ideas on replacement in this generalized setting and shortly, we give a slightly different definition of extended André replacement.

2.1. Main Theorem on Replacements.

THEOREM 17. *Let Σ be a Desarguesian r-spread of order q^{sn}. Let $\Sigma_{j,w}$ be any j-(0-subset) for $j = 0, 1, 2, .., r - 1$. Choose any sn-dimensional subspace*

$$y = (x_1^{*q^{\lambda_1}} n_1, .., x_1^{*q^{\lambda_{r-j-1}}} n_{r-j-1});$$

where $n_i \in GF(q^{sn})^*$, $i = 1, 2, .., r - j - 1$. Let $d = (\lambda_1, \lambda_2, .., \lambda_{r-j-1})$, where $0 \leq \lambda_i \leq sn - 1$.

(1) Then $A_{(n_1,..,n_{r-j-1})} =$

$$\left\{ \begin{array}{c} y = (x_1^* w_1, .., x_1^* w_{r-j-1}); \\ \text{there is an } x_1^* \text{ such that } x_1^{*q^{\lambda_i}-1} = w_i/n_i, \\ \text{for all } i = 1, 2, .., r-j-1 \end{array} \right\}$$

is a set of $(q^{sn} - 1)/(q^d - 1)$ sn-subspaces, which is covered by the set of $(q^{sn} - 1)/(q^d - 1)$ sets $A_{(n_1,..,n_{r-j-1})}^{(\lambda_1,...,\lambda_{r-j-1})} =$

$$\left\{ \begin{array}{c} y = (x_1^{*q^{\lambda_1}} n_1 d^{1-q^{\lambda_1}}, .., x_1^{*q^{\lambda_{r-j-1}}} n_{r-j-1} d^{1-q^{\lambda_{r-j-1}}}); \\ d \in GF(q^{sn})^* \end{array} \right\}.$$

(2) Let $C_{(q^{sn}-1)/(q-1)}$ denote the cyclic subgroup of $GF(q^{sn})^*$ of order $(q^{sn} - 1)/(q - 1)$. Then, for each

$$y = (x_1^* w_1, .., x_1^* w_{r-j-1}),$$

there exists an element τ in $C_{(q^{sn}-1)/(q-1)}$ such that

$$w_i = n_i \tau^{(q^{\lambda_i}-1)/(q-1)}.$$

(3) The

$$(q^{sn} - 1)/(q^{(\lambda_1,...,\lambda_{r-j-1},sn)} - 1)$$

components of

$$\left\{ \begin{array}{c} y = (x_1^{*q^{\lambda_1}} n_1 d^{1-q^{\lambda_1}}, .., x_1^{*q^{\lambda_{r-j-1}}} n_{r-j-1} d^{1-q^{\lambda_{r-j-1}}}); \\ d \in GF(q^{sn})^* \end{array} \right\}$$

are in

$$\left((q^{sn} - 1)/(q^{(\lambda_1,...,\lambda_{r-j-1})} - 1) \right) / \left((q^s - 1)/(q^{(\lambda_1,\lambda_2,...,\lambda_{r-j-1},s)} - 1) \right)$$

orbits of length

$$(q^s - 1)/(q^{(\lambda_1,\lambda_2,...,\lambda_{r-j-1},s)} - 1)$$

under the s-kernel homology group K_s.

PROOF. There exists an integer i_0 such that $\lambda_{i_0} = d\rho_{i_0}$, for $(\rho_{i_0}, sn/d) = 1$. For fixed n_{i_0} non-zero in $GF(q^{sn})^*$, consider the set of all elements w_{i_0} such that $w_{i_0}/n_{i_o} = \tau_0$, for some τ_0 such that $\tau_0^{(q^{sn}-1)/(q^{d\rho_{i_0}}-1)} = 1$, and clearly there is a set of $(q^{sn}-1)/(q^d-1)$ such elements in $GF(q^{sn})^*$. There exists an element $x_1^{*(q-1)} = \tau$ so that

$$x_1^{*q^{\lambda_i}-1} = w_i/n_i = \tau^{(q^{\lambda_i}-1)/(q-1)}.$$

This proves part (2).

Now to show that the indicated set

$$\left\{ \begin{array}{c} y = (x_1^{*q^{\lambda_1}} n_1 d^{1-q^{\lambda_1}}, .., x_1^{*q^{\lambda_{r-j-1}}} n_{r-j-1} d^{1-q^{\lambda_{r-j-1}}}); \\ d \in GF(q^{sn})^* \end{array} \right\}$$

covers the set

$$\left\{ \begin{array}{c} y = (x_1^* n_1 \tau^{(q^{\lambda_1}-1)/(q-1)}, .., x_1^* n_{r-j-1} \tau^{(q^{\lambda_{r-j-1}}-1)/(q-1)}); \\ \tau \in C_{(q^{sn}-1)/(q-1)} \end{array} \right\}.$$

First note that the following sets are equal:

$$\left\{ x_1^{*q^{\lambda_i}} n_i d^{1-q^{\lambda_i}}; x_i^*, d \in GF(q^{sn})^* \right\}$$

$$= \left\{ x_1^* n_i \tau^{(q^{\lambda_i}-1)/(q-1)}; x_i \in GF(q^{sn}), \tau \in C_{(q^{sn}-1)/(q-1)} \right\}.$$

Assume for a fixed x_1^* that

$$x_1^{*(q-1)} d^{1-q} = \tau.$$

Then

$$(x_1^* d^{-1})^{(q^{\lambda_i}-1)} = \tau^{(q^{\lambda_i}-1)/(q-1)},$$

which is true if and only if

$$x_1^{*q^{\lambda_i}} n_i d^{1-q^{\lambda_i}} = x_1^* n_i \tau^{(q^{\lambda_i}-1)/(q-1)}.$$

Therefore, as

$$\left\{ x_1^{*q} \tau; \tau \in C_{(q^{sn}-1)/(q-1)} \right\}$$

covers

$$\left\{ x_1^* d^{1-q}; d \in GF(q^{sn})^* \right\},$$

it follows that

$$\left\{ \begin{array}{c} y = (x_1^* n_1 \tau^{(q^{\lambda_1}-1)/(q-1)}, .., x_1^* n_{r-j-1} \tau^{(q^{\lambda_{r-j-1}}-1)/(q-1)}); \\ \tau \in C_{(q^{sn}-1)/(q-1)} \end{array} \right\}.$$

is covered by

$$\left\{ \begin{array}{c} y = (x_1^{*q^{\lambda_1}} n_1 d^{1-q^{\lambda_1}}, .., x_1^{*q^{\lambda_{r-j-1}}} n_{r-j-1} d^{1-q^{\lambda_{r-j-1}}}); \\ d \in GF(q^{sn})^* \end{array} \right\}.$$

Now to determine the total number of components in the André net

$$(\lambda_1, \lambda_2, .., \lambda_{r-j-1}).$$

Thus, the question is, when is

$$\tau^{(q^{\lambda_i}-1)/(q-1)} = 1$$

for all $i = 1, 2, .., r-j-1$, where τ is an arbitrary element of $C_{(q^{sn}-1)/(q-1)}$? It is then clear that τ must have order

$$(q^{(\lambda_1,..,\lambda_{r-j-1})} - 1)/(q - 1).$$

In other words, there are exactly

$$(q^{sn} - 1)/(q^{(\lambda_1,..,\lambda_{r-j-1})} - 1)$$

components of the extended André set $A_{(\lambda_1,..,\lambda_{r-j-1})}$.

Now consider an orbit in the replacement set under the s-kernel homology group K_s. So,

$$y = (x_1^{*q^{\lambda_1}} n_1, .., x_1^{*q^{\lambda_{r-j-1}}} n_{r-j-1})$$

maps to

$$y = (x_1^{*q^{\lambda_1}} n_1^{1-q^{\lambda_1}} d^{1-q^{\lambda_1}}, .., x_1^{*q^{\lambda_{r-j-1}}} n_{r-j-1} d^{1-q^{\lambda_{r-j-1}}})$$

for $d \in GF(q^s)^*$. This orbit clearly has the cardinality indicated. $\square$

We are now in a position to define our generalizations. If we take $y = x^{q^\lambda} m_0$ for a fixed element of $GF(q^{sn})^*$, let N_{λ,m_0} define the set of components of the associated Desarguesian affine plane that non-trivially intersect this subspace. Then the set of images under the kernel homology group of order $(q^{sn} - 1)$ is

$$\left\{ y = x^{q^\lambda} m_0 d^{1-q^\lambda} \right\},$$

and we see that we obtain a net of degree

$$(q^{sn} - 1)/(q^{(\lambda, sn)} - 1).$$

The partial sn-spread

$$A = \left\{ \begin{array}{l} y = (x_1^* n_1 \tau^{(q^{\lambda_1}-1)/(q-1)}, .., x_1^* n_{r-j-1} \tau^{(q^{\lambda_{r-j-1}}-1)/(q-1)}); \\ \tau \text{ has order dividing } (q^{sn} - 1)/(q - 1). \end{array} \right\}$$

is a set of

$$(q^{sn} - 1)/(q^{(\lambda_1,\lambda_2,...\lambda_{r-j-1},sn)} - 1)$$

sn-subspaces, which is covered by the set of

$$(q^{sn} - 1)/(q^{(\lambda_1,\lambda_2,...\lambda_{r-j-1},sn)} - 1)$$

sn-subspaces

$$A_{(n_1,..,n_{r-j-1})}^{(\lambda_1,..,\lambda_{r-j-1})}$$

$$= \left\{ \begin{array}{c} y = (x_1^{*q^{\lambda_1}} n_1 d^{1-q^{\lambda_1}}, ..., x_1^{*q^{\lambda_{r-j-1}}} n_{r-j-1} d^{1-q^{\lambda_{r-j-1}}}); \\ d \in GF(q^{sn})^* \end{array} \right\}.$$

DEFINITION 33. *We shall call $A_{(n_1,..,n_{r-j-1})}$ an 'extended André partial spread' of degree $(q^{sn} - 1)/(q^{(\lambda_1,\lambda_2,...,\lambda_{r-j-1},sn)} - 1)$' and order q^{sn}. So, we note that $A_{(n_1,..,n_{r-j-1})}^{(\lambda_1,...,\lambda_{r-j-1})}$ is a replacement partial spread of the same degree and order called an 'extended André replacement.'*

2.2. Extended André Replacements. Suppose we have a given extended André replacement, how do we find others? For example, if we are looking for extended André partial spreads of the same degree as $A^{(\lambda_1,..,\lambda_{r-j-1})}_{(n_1,..,n_{r-j-1})}$, we take other exponent sets $\{\rho_1, \rho_2, .., \rho_{r-j-1}\}$ so that

$$\left(\rho_1, \rho_2, .., \rho_{r-j-1}, sn\right) = (\lambda_1, \lambda_2, .., \lambda_{r-j-1}, sn),$$

there are exactly

$$((q^{(\lambda_1,\lambda_2,..,\lambda_{r-j-1},sn)} - 1)(q^{sn} - 1)^{r-j-2}$$

possible extended André partial spreads of degree

$$(q^{sn} - 1)/(q^{(\lambda_1,\lambda_2,..,\lambda_{r-j-1},sn)} - 1).$$

Therefore, if we let $\rho_i = \lambda_i t_i$, for $0 \leq t_i \leq sn/(\lambda_i, sn) - 1$, then let

$$(\lambda_1, \lambda_2, .., \lambda_{r-j-1}, sn) = s^*.$$

It follows that

$$
\begin{aligned}
&\left(\rho_1, \rho_2, .., \rho_{r-j-1}, sn\right) \\
=\ & (\lambda_1 t_1, \lambda_2 t_2, .., \lambda_{r-j-1} t_{r-j-1}, sn) \\
=\ & (\lambda_1 t_1/s^*, \lambda_2 t_2/s^*, .., \lambda_{r-j-1} t_{r-j-1}/s^*, sn/s^*).
\end{aligned}
$$

Now assume that all $\lambda_i = 1$, for $i = 1, 2, .., r-j-1$. Then the extended André partial spread $A_{(n_1,..,n_{r-s-1})}$ has degree $(q^{sn} - 1)/(q - 1)$, and we have a replacement partial spread with components

$$
\begin{aligned}
& A^{(1,..,1)}_{(n_1,..,n_{r-j-1})} \\
=\ & \left\{ y = (x_1^q n_1 d^{1-q}, x_2^q n_2 d^{1-q}, .., x_{r-j-1}^q n_{r-j-1} d^{1-q}) \right\}.
\end{aligned}
$$

As noted,

$$
\begin{aligned}
& A_{(n_1\tau^{(q^{\lambda_1}-1)/(q-1)},..,n_{r-j-1}\tau^{(q^{\lambda_{r-j-1}}-1)/(q-1))},} \\
=\ & \left\{ \begin{array}{c} y = (x_1^* n_1 \tau^{(q^{\lambda_1}-1)/(q-1)}, .., x_1^* n_{r-j-1} \tau^{(q^{\lambda_{r-j-1}}-1)/(q-1))}); \\ \tau \text{ has order dividing } (q^{sn} - 1)/(q - 1). \end{array} \right\}
\end{aligned}
$$

is a set of

$$(q^{sn} - 1)/(q^{(\lambda_1,\lambda_2,..,\lambda_{r-j-1},sn)} - 1)$$

sn-subspaces, which is covered by the set of

$$(q^{sn} - 1)/(q^{(\lambda_1,\lambda_2,..,\lambda_{r-j-1},sn)} - 1)$$

sn-subspaces

$$
\begin{aligned}
& A^{(\lambda_1,..,\lambda_{r-j-1})}_{(n_1,..,n_{r-j-1})} \\
=\ & \left\{ \begin{array}{c} y = (x_1^{*q^{\lambda_1}} n_1 d^{1-q^{\lambda_1}}, .., x_1^{*q^{\lambda_{r-j-1}}} n_{r-j-1} d^{1-q^{\lambda_{r-j-1}}}); \\ d \in GF(q^{sn})^* \end{array} \right\}.
\end{aligned}
$$

Now consider

$$\left\{ (n_1 d^{1-q}, n_2 d^{1-q}, .., n_{r-j-1} d^{1-q}); d \in GF(q^{sn})^* \right\}.$$

If $\tau^{(q^{\lambda_k}-1)/(q-1)} = \tau^{(q^{\lambda_i}-1)/(q-1)}$, for all k, i, so, for example, $\lambda_i = \lambda_z$, for all $i = 1, 2, .., r - j - 1$, then we would obtain that there are $q^{sn} - 1)/(q^{(\lambda_1 sn)} - 1)$ components in this extended André partial spread. We now vary over the

$$(q^{sn} - 1)/(q - 1)/(q^{sn} - 1)/(q^{(\lambda_1 sn)} - 1) = (q^{(\lambda_1 sn)} - 1)/(q - 1)$$

cosets of the cyclic group of order

$$(q^{sn} - 1)/(q^{(\lambda_1 \cdot sn)} - 1)$$

with respect to the group of order

$$(q^{sn} - 1)/(q - 1).$$

Therefore, we would have

$$(n_i d^{1-q})^{(q^{sn}-1)/(q-1)} = n_i^{(q^{sn}-1)/(q-1)} = \delta_i.$$

Hence, we might then call this net

$$A_{\delta_1, \delta_2, .., \delta_{r-j-1}},$$

which is then covered by

$$A(S)_{\delta_1, .., \delta_{r-j-1}}^{(\lambda_1, .., \lambda_{r-j-1})}$$

$$= \left\{ \begin{array}{l} y = (x_1^{q^{\lambda_1}} n_1^*, x_1^{q^{\lambda_1}} n_2^*, .., x_1^{q^{\lambda_1}} n_{r-j-1}^*); \\ n_i^{*(q^{sn}-1)/(q-1)} = \delta_i, i = 1, 2, .., r - j - 1. \end{array} \right\}$$

So there are $sn - 1$ non-trivial extended André replacements for this particular extended André partial spread.

REMARK 13. *In this setting, there are*

$$(q^n - 1)^{r-j-2}(q - 1),$$

mutually disjoint extended André partial spreads and each admit $sn - 1$ non-trivial extended André replacements.

Finally, we consider the most general setting for André-like structures. To consider the analogous situation, for André planes, we note that we can make such replacements for all j-(0-subset), for each j such that $r - j \geq 2$.

DEFINITION 34. *Hence, we define an 'extended André r-(sn)-spread' to be any of the sn replacements for each of the $\binom{r}{r-j}$ j-(0-subsets), for each of the $j \geq r - 2$ j-(0-sets). This then constructs a set of*

$$\sum_{j=0}^{r-2} \binom{r}{r-j} (sn)^{(q^n-1)^{r-j-2}(q-1)} - 1$$

distinct non-trivial $r - (sn) -$ spreads, not equal to the original Desarguesian $r - (sn)-$ spread.

3. Multiple Extended Replacements

To begin the ideas of generalizing a generalized André plane, we note that from André planes, we may easily construct generalized André planes by varying the degrees of the replacement nets. For example, an André plane is obtained by making replacements all of the same degree, say $(q^{sn} - 1)/(q - 1)$. However, if we partition a given André partial spread into $(q^{s^*} - 1)/(q - 1)$ André partial spreads of degree $(q^{sn}-1)/(q^{s^*}-1)$, where s^* divides sn, in this setting a typical subspace of a replacement has the form $y = x^{h^i}m$, where $h = q^{s^*}$. Normally, one would take $0 \leq i \leq sn/s^* - 1$. Any of these André partial spreads may be further subdivided and André replacements of various different degrees may be considered. Any translation plane constructed by making André replacements of various degrees is not (always) an André plane but is a 'generalized André plane,' since the components of the constructed translation plane have the general form $y = x^{q^{\lambda(m)}}m$, where m is in $GF(q^{sn})$ and λ is a function from $GF(q^{sn})^*$ to the set of integers $0, 1, .., sn - 1$. We formally define this sort of generalized André plane.

DEFINITION 35. *The generalized André plane constructed using the method above of varying degrees is constructed by 'multiple André replacement.'*

We now consider an analogous construction procedure using a different approach. In the 'vary degrees approach,' we must make sure that we actually have a partition. In the more general setting of extended structures, this is more complicated.

PROPOSITION 4.

$$A_{(n_1,..,n_{r-j-1})}$$
$$= \left\{ \begin{array}{l} y = (x_1^* n_1 \tau^{(q^{\lambda_1}-1)/(q-1)}, x_1^* n_2 \tau^{(q^{\lambda_2}-1)/(q-1)}, \\ \quad ..., x_1^* n_{r-j-1} \tau^{(q^{\lambda_{r-j-1}}-1)/(q-1)}); \\ \tau \text{ has order dividing } (q^{sn} - 1)/(q - 1) \end{array} \right\},$$

and

$$A_{(n_1^*,..,n_{r-j-1}^*)}$$

$$= \left\{ \begin{array}{l} y = (x_1^* n_1^* \tau^{(q^{\lambda_1^*}-1)/(q-1)}, x_1^* n_2^* \tau^{(q^{\lambda_2^*}-1)/(q-1)}, \\ \quad ..., x_1^* n_{r-j-1}^* \tau^{(q^{\lambda_{r-j-1}^*}-1)/(q-1)}); \\ \tau \text{ has order dividing } (q^{sn}-1)/(q-1) \end{array} \right\},$$

share a component if and only if there exist elements τ_1 and τ_1^* of order dividing $(q^{sn}-1)/(q-1)$ such that

$$n_i \tau_1^{(q^{\lambda_i}-1)/(q-1)} = n_i^* \tau_1^{*(q^{\lambda_i^*}-1)/(q-1)}; i = 1, 2, .., r-j-1.$$

PROOF. Suppose in the first listed set, we have τ_1 in place of τ and in the second listed set, we have τ_1^* in place of τ, such that the corresponding components

$$\begin{aligned} y &= \left(x_1^* n_1 \tau_1^{(q^{\lambda_1}-1)/(q-1)}, x_1^* n_2 \tau_1^{(q^{\lambda_2}-1)/(q-1)}, \right. \\ &\quad \left. ..., x_1^* n_{r-j-1} \tau_1^{(q^{\lambda_{r-j-1}}-1)/(q-1)}\right) \\ &= y = \left(x_1^* n_1^* \tau_1^{*(q^{\lambda_1^*}-1)/(q-1)}, x_1^* n_2^* \tau_1^{*(q^{\lambda_2^*}-1)/(q-1)}, \right. \\ &\quad \left. ..., x_1^* n_{r-j-1}^* \tau_1^{*(q^{\lambda_{r-j-1}^*}-1)/(q-1)}\right) \end{aligned}$$

are equal. Then we must have

$$n_i \tau_1^{(q^{\lambda_i}-1)/(q-1)} = n_i^* \tau_1^{*(q^{\lambda_i^*}-1)/(q-1)}; i = 1, 2, .., r-j-1.$$

Since the extended André partial spread is an orbit, we may assume that $\tau_1 = 1$. So, the only way that this could occur is if $(n_1, .., n_{r-j-1})$ and $(n_1^*, n_2^*, .., n_{r-j-1}^*)$ are related by the set of equations above. This completes the proof of the lemma. □

There are algorithms developed in the author's article [98] that show how to construct an extremely large variety of 'multiple extended André replacements' and the interested reader is directed to this work for the details.

DEFINITION 36. *We call any spread obtained by multiple extended André replacement a 'generalized extended André spread.'*

We turn now to the groups that we wish to act on the extended structures that are analogous to groups acting on André planes.

4. Large Groups on t-Spreads

In this section, we consider again the generalized extended André t-spreads and focus on the collineation groups that can be obtained. This material will prepare the way for a study of 'subgeometry partitions' of projective spaces based on these t-spreads.

We note that the generalized kernel group $G_{sn,r}$ of order $(q^{sn} - 1)^r$ acts on the Desarguesian $r-(sn, q)$-spread and the kernel group of order $(q^{sn} - 1)$ also acts on each extended André replacement partial spread, as this is the way that the replacements are determined. Indeed, the generalized kernel group of order $(q^{sn}-1)^r$ is transitive on 1-dimensional $GF(q^{sn})$-subspaces.

Furthermore, the group $G_{sn,r}$ fixes each j-(0-subset) and acts transitively on each set of (non-zero) vectors. Furthermore, if we take a given extended André replacement set

$$A^{(\lambda_1,..,\lambda_{r-j-1})}_{(n_1,..,n_{r-j-1})}$$

$$= \left\{ \begin{array}{c} y = (x_1^{*q^{\lambda_1}} n_1 d^{1-q^{\lambda_1}}, .., x_1^{*q^{\lambda_{r-j-1}}} n_{r-j-1} d^{1-q^{\lambda_{r-j-1}}}); \\ d \in GF(q^{sn})^* \end{array} \right\},$$

the kernel group K_{sn} is transitive on the subspaces and the group R_{sn} of order $q^{sn} - 1$

$$\left\langle \begin{array}{c} (x_1, x_2, ...x_{r-j-1}) \longmapsto (x_1^* m_0, x_1^* m_0^{q^{\lambda_1}}, .., x_1^* m_0^{q^{\lambda_{r-j-1}}}); \\ m_0 \in GF(q^{sn}) \end{array} \right\rangle,$$

fixes each sn-dimensional subspace of $A^{(\lambda_1,..,\lambda_{r-j-1})}_{(n_1,..,n_{r-j-1})}$. Note that the remaining entries that are 0 are omitted. In order that this group acts on the $r - (sn, q)$-spread, choose exactly one j and exactly one $j - $ (0-subset). Then the resulting generalized extended André sn-spread will admit a group of order $(q^{sn} - 1)^r$ of which there is a group of order $(q^{sn} - 1)^2$ that acts transitively on the components of the replaced partial spread and there is a subgroup of order $(q^{sn} - 1)^j$ that fixes each vector of the $j - $ (0-subset (set)) (just take the j 0-entries to have arbitrary coefficients in $GF(q^{sn})^*$ and take the other coefficients to be 1). Note that the remaining sn-dimensional $GF(q)$ subspaces are actually 1-dimensional $GF(q^{sn})$-subspaces and since $G_{sn,r}$ just maps 1-dimensional $GF(q^{sn})$-subspaces to 1-dimensional $GF(q^{sn})$-subspaces, therefore there is an Abelian group of order $(q^{sn} - 1)^{j+2}$, acting on such an sn-spread.

4.1. Extended André Large Group Theorem.

THEOREM 18. *(i) Choose any subspace that generates*

$$A^{(\lambda_1,..,\lambda_{r-j-1})}_{(n_1,..,n_{r-j-1})}$$

$$= \left\{ \begin{array}{c} y = (x_1^{*q^{\lambda_1}} n_1 d^{1-q^{\lambda_1}}, .., x_1^{*q^{\lambda_{r-j-1}}} n_{r-j-1} d^{1-q^{\lambda_{r-j-1}}}); \\ d \in GF(q^{sn})^* \end{array} \right\},$$

that in turn generates

$$A_{(n_1,..,n_{r-j-1})}$$

$$= \left\{ \begin{array}{c} y = (x_1^* n_1 \tau^{(q^{\lambda_1}-1)/(q-1)}, x_1^* n_2 \tau^{(q^{\lambda_2}-1)/(q-1)}, \\ ..., x_1^* n_{r-j-1} \tau^{(q^{\lambda_{r-j-1}}-1)/(q-1)}); \\ \tau \text{ has order dividing } (q^{sn}-1)/(q-1) \end{array} \right\}.$$

(ii) Now in the $j-(0\text{-set})$, choose the same ordered set of coefficients and then take different ordered sets of coefficients $(n_1, n_2, .., n_{r-j-1})$ so as to completely partition the $j-(0\text{-subset})$. There are exactly

$$(q^{sn}-1)^{r-j-2}((q^{(\lambda_1,\lambda_2,..,\lambda_{r-j-1},sn)}-1)$$

possible extended André sets.

(iii) Now for each extended André set choose either $(\lambda_1,..,\lambda_{r-j-1})$ or $(0,0,0,..,0)$. Then construct the $r-(sn,q)$-spread obtained by replacing the various

$$A_{(n_1,..,n_{r-j-1})} \text{ by } A^{(\lambda_1,..,\lambda_{r-j-1})}_{(n_1,..,n_{r-j-1})},$$

or by $A_{(n_1,..,n_{r-j-1})}$, respectively, where the remaining sn-subspaces are the remaining uncovered 1-dimensional $GF(q^{sn})$-subspaces.

(iv) Then any such extended André spread admits an Abelian group of order $(q^{sn}-1)^{j+2}$, which is the direct product of $j+2$ cyclic groups of order $(q^{sn}-1)$, and $r-j \geq 2$.

(v) Let $N_{(\lambda_1,..,\lambda_{r-j-1})}$ denote the number of different ordered sets of exponents

$$(\lambda_1^*, \lambda_2^*, .., \lambda_{r-j-1}^*), \text{ such that } \gcd(\lambda_1^*, \lambda_2^*, .., \lambda_{r-j-1}^*, sn)$$

$$= \gcd(\lambda_1, \lambda_2, .., \lambda_{r-j-1}, sn).$$

There are then

$$\left(\binom{r}{r-j}\right) 2^{(q^{sn}-1)^{r-j-2}((q^{(\lambda_1,\lambda_2,...,\lambda_{r-j-1},sn)}-1)} - 1)N_{(\lambda_1,..,\lambda_{r-j-1})}$$

proper sn-spreads that admit Abelian groups of order $(q^{sn}-1)^{j+2}$.

As an application, it can be shown that the Ebert-Mellinger spreads [51] can be obtained from the previous theorem. Again, the details can be found in [98].

REMARK 14. *In our constructions of generalized extended André spreads, we have found replacements (the extended André replacements) of extended André partial spreads of various sizes using the kernel homology group of order $q^{sn} - 1$. Hence, all of our new spreads necessarily admit the kernel group of order $q^{sn} - 1$. It is an open question whether it might be possible to find replacements of the extended André partial spreads that do not admit this kernel group.*

REMARK 15. *We finally also note that from any of our $r - (sn, q)$-spreads from vector spaces of dimension rsn, we may use the construction of Section 0.3 to construct double-spreads of type $(sn, sn - 1)$ of $q^{r(sn-1)}$ subspaces of dimension $sn - 1$ and $(q^{r(sn-1)} - 1)/(q^{sn} - 1)$ subspaces of dimension sn.*

CHAPTER 4

The Going Up Construction for Focal-Spreads

We have introduced focal-spreads in the previous chapters and have noted that the known examples arise from k-cuts of translation planes. In this chapter, we provide some very general constructions of focal-spreads that do not appear to be k-cuts, although this is still an open question.

This process allows us to specify subspaces of order q^k, in a focal-spread of type $(k(s-1), k)$, where the focus has dimension $k(s-1) = t$ and we have an associated partial Sperner k-spread of q^t k-dimensional subspaces. Furthermore, if a focal-spread is defined more generally as a partition of a vector space of dimension $t + k$ over $GF(q)$ by one subspace of dimension t' and the remaining subspaces of dimension k, we say the focal-spread is of type (t, t', k), when $t' < t$. The going up process allows constructions of this more general type of focal-spreads.

If we begin with a vector space V_{tk} of dimension tk-over $GF(q)$, we may construct a k-spread in V_{tk} by a choice of any sequence of $\sum_{j=0}^{t-2} \binom{t}{j} (t - j - 1) = N_t$ translation planes of order q^k and associated spreads S_i, for $i = 1, 2, .., N_t$.

The reader is directed to the article by Jha and Johnson [**80**] for additional details.

The construction that we give basically relies on another construction of Sperner spreads from suitably many k-spreads. In this construction, we again use the concept of j-(0-sets).

0.2. Sperner Spread Construction Theorem. The idea of the construction is to piece together ordinary k-spreads 'planar k-spreads' corresponding to translation planes of dimension $2k$ over $GF(q)$ in order to construct a k-spread of a tk-dimensional vector space. The trick is that considering a ordinary spread in the form $x = 0, y = 0, y = xM$, we only use the $y = xM$'s and ignore the $x = 0$ and $y = 0$'s, and we do this by identifying all of the $x = 0$ and $y = 0$'s in all of the corresponding ordinary k-spreads. In the general case, this means that we ignore the sets $(x_1, 0, 0, .., 0)$, etc., and focus on the remaining j-(0-sets). The following gives the construction.

THEOREM 19. *(Jha and Johnson* [80]*) Let V be a tk-dimensional $GF(q)$-subspace and represent vectors in the form $(x_1, x_2, .., x_t)$, where the x_i are k-vectors, for $i = 1, 2, .., t$. Let $\mathcal{S}_t$ be a sequence of $\sum_{j=0}^{t-2} \binom{t}{j} (t - j - 1) = N_t$ planar k-spreads over $GF(q)$. We may represent each of the planar k-spreads as follows:*

We identify two common components $x = 0, y = 0$, and where x and y are k-vectors. Then if $\mathcal{M}_i$ is a planar k-spread, there is a set of $q^k - 1$ nonsingular matrices $M_{i,z}$, for $z = 1, 2, .., q^k - 1$, $i = 1, 2, .., N_t$, whose differences are also non-singular.

Hence, we represent the planar k-spread by $y = x M_{iz}$, for $z = 1, 2, .., q^k - 1$. Consider the $j-(0$-subsets$)$, for $j = 0, 1, 2, .., t - 2$, and assume an ordering for the N_t planar k-spreads. For each such $j-(0$-subset$)$, for $j > 2$, we eliminate the zero elements and write vectors in the form $(x_1, x_2, .., x_{t-j})$, where x_w, for $w = 1, 2, .., t - j$, are all non-zero k-vectors. We partition this set by

$$(x_1, x_1 M_{1,z_1}, x_1 M_{2,z_2}, x_1 M_{3,z_3}, .., x_1 M_{t-j,z_{t-j}}),$$

where M_{i,z_i} varies over $\mathcal{M}_i$, and where the indices z_i are independent of each other. By adjoining the zero vector, we then may consider

$$y = (x_1 M_{1,z_1}, x_1 M_{2,z_2}, x_1 M_{3,z_3}, .., x_1 M_{t-j,z_{t-j}}),$$

as a k-subspace for fixed M_{i,z_i}, for each $i = 1, 2, .., t-j$. Then, together with the $j-(0$-subsets$)$ adjoined by the zero vector, we obtain a spread of k-spaces of a tk-dimensional vector space over $GF(q)$ as a union of $j-(0$-subsets$)$ (with the zero vector adjoined to each).

PROOF. We begin with a vector space V_{tk} of dimension tk over $GF(q)$. Represent a vector in the form $(x_1, x_2, ..., x_t)$, where each x_i, for $i = 1, .., t$ in, in turn, a k-vector over $GF(q)$. Consider first $t = 4$.

With vectors written in the form (x_1, x_2, x_3, x_4), choose the $3k$-dimensional subspace given by equation $x_4 = 0$. If we were to construct a k-spread in a $4k$-space, we would use 17 translation planes of order q^k, as above. If we were to construct a k-spread in a $3k$-space, we would require 5 translation planes of order q^k. If we ignore the k-spread construction of x_4, then the remaining vectors can be covered using $17 - 5 = 12$ translation planes of order q^k. We obtain a focal-spread of type $(3k, k)$ as follows: Let $x_4 = 0$, denote the focus, a subspace of dimension $3k$. There are $(q^{4k} - 1)/(q^k - 1)$ k-subspaces in what would be a k-spread in the $4k$-dimensional vector space and there would be $(q^{3k} - 1)/(q^k - 1)$ k-subspaces in the k-spread for the $3k$-dimensional vector space $x_4 = 0$. Hence, there are

$$(q^{4k} - 1)/(q^k - 1) - (q^{3k} - 1)/(q^k - 1) = q^{3k}$$

k-subspaces in a partial spread distinct from $x_4 = 0$. Now simply generalize this idea. A planar k-spread is to be a k-spread of a $2k$-dimensional vector space over $GF(q)$, thus equivalent to a translation plane of order q^k with kernel containing $GF(q)$. We are working with vector spaces V of dimension tk over $GF(q)$, with vectors represented in the form $(x_1, x_2, .., x_t)$, where the x_i are k-vectors, for $i = 1, 2, .., t$. We partition the set of vectors into the $j-$(0-subsets), which are the sets of vectors with a set of j k-vectors equal to 0 in fixed locations, where the remaining entries are non-zero. For example, $t - 1-$(0-subsets) have the general form $(x_1, 0, 0, .., 0)$, where x_1 is a non-zero k-vector, and the $j = 1$-zero sets are sets of vectors with exactly one entry equal to 0, such as $\{(0, x_2, .., x_t); \ x_i$ are non-zero k-vectors, $i = 2, .., t\}$. The remaining details are left to the reader to complete. □

The reader is directed to the open problem section for problems related to this construction process.

REMARK 16. *From any of our k-spreads of tk-dimensional vector spaces, we may use the construction method of Section 0.3 to construct double-spreads of type $(k, k - 1)$ with $q^{t(k-1)}$ subspaces of dimension $k - 1$ and $(q^{t(k-1)} - 1)/(q^k - 1)$ subspaces of dimension k.*

1. The 'Going Up' Construction

Recall that we previously thought of defining a focal-spread more generally as a vector space partition as a double-spread with exactly one subspace of a particular dimension and the remaining subspaces of the other dimension. We now are in a position to find such generalizations to (t, k)-focal-spreads. When we find such generalizations, it is now a very interesting question as how these might be related to k-cuts or perhaps they are not so related. We consider the so-called (t, t', k)-focal-spreads, which are partitions of the vectors into one subspace of dimension t', the 'focus,' and the remaining subspaces of dimension k, where the dimension is $t + k$, for $t' \leq t$.

We call such partitions focal-spreads of type (t, t', k). In this setting, $q^{t+k} - q^{t'} = q^{t'}(q^{t-t'+k} - 1)$ must be divisible by $q^k - 1$ so that $t - t'$ is divisible by k.

We now use the ideas of the k-spread construction Theorem 19 to provide focal-spreads of type $(k(t - w), k)$. Choose the $k(t - w)$-dimensional subspace F given by the equation $x_1 = x_2 = ... = x_w = 0$.

To construct a k-spread requires

$$N_{t-w} = \sum_{j=0}^{t-(w+2)} \binom{t-1}{j}(t-j-(w+1))$$

spreads. This means that the remaining $q^{k(t-1)}(q^{kw}-1)/(q^k-1)$ k-components are obtained using a set of $N_t - N_{t-w}$ spreads as in the previous theorem. Ignoring the set of N_{t-w} spreads covering F, we obtain a focal-spread of type $(kt(t-1), k(t-w), k)$.

We obtain then the following result ((see Jha and Johnson [80] for a result similar to part (1), where only $t-1$ spreads were used in a construction). For part (2), the reader is directed to Jha and Johnson [80] for the analogous result for the constructed k-spreads in kt-dimensional vector spaces.

THEOREM 20. *Given a vector space of dimension kt over a field $GF(q)$. Choose any $k(t-w)$-dimensional subspace F. Now choose a set of $N_t - N_{t-w}$ planar k-spreads (spreads corresponding to translation planes of order q^k).*

(1) Then, using the method of the Sperner Spread Construction Theorem, we have constructed a focal-spread of type $(k(t-1), k(t-w), k)$.

(2) If, by identifying the components $x = 0, y = 0$, each k-spread admits a collineation group with elements of the general form $(x, y) \to (xA, yA)$, where A is a $k \times k$ matrix, then the focal-spread also admits this group.

The following corollary shows that either there are a large variety of focal-spreads that cannot be obtained as k-cuts, or there are translation planes of order $q^{k(t-1)}$ with widely variable translation subplanes of order q^k.

COROLLARY 3. *Assume that all focal-spreads of type (s, k) may be obtained as k-cuts from planar s-spreads (corresponding to translation planes of order q^s). Let $s = k(t-1)$. Now specify any given set of*

$$\sum_{j=0}^{t-2} \binom{t}{j}(t-j-1) - \sum_{j=0}^{t-3} \binom{t-1}{j}(t-j-2)$$

translation planes of order q^k. Then there exists a translation plane of order $q^{k(t-1)}$ containing the given set as a set of affine subplanes of order q^k.

PROOF. For an example, consider a focal-spread of type $(2k, k)$ obtained from the above process. Choose vectors (x_1, x_2, x_3), where

x_i are k-vectors over $GF(q)$. Let $x_1 = 0$, denote the focus, a $2k$-dimensional subspace. Then, the partial Sperner k-spread is given by

$$\{(x_1, x_1M_1, 0)\},\ \{(x_1, 0, x_1M_2)\},\ \{(x_1, x_1M_3, x_1M_4)\},\ \{(x_1, 0, 0)\},$$

for $M_i \in \mathcal{M}_i$, k-spreads, $i = 1, 2, 3, 4$. We claim that $\{(x_1, x_1M_1, 0)\}$ extends to a matrix k-spread and provides a subplane of order q^k in the associated translation plane of order q^{2k}. If we consider x_1 as $(z_1, z_2, .., z_k)$, for $z_i \in GF(q)$ then, in the extension, we take $x = (x_1, w_1, .., w_k)$, for $w_i \in GF(q)$. Considering $(x_1, x_1M_1, 0)$ as $y = x_1[M_1, 0]$, the latter matrix is a $k \times 2k$ matrix. We extend to $y = x \begin{bmatrix} M_1 & 0 \\ M_1^- & S_1 \end{bmatrix}$, a $2k \times 2k$ matrix of rank $2k$. Since M_1 has rank k, it follows that S_1 has rank k. Then, consider

$$y = (x_1, 0, 0, .., 0) \begin{bmatrix} M_1 & 0 \\ M_1^- & S_1 \end{bmatrix},\ \text{for all } M_1 \in \mathcal{M}_1,\ \text{and } x_1 = 0,$$

then provides a subplane of order q^k. More generally, in the focal-spread of type $(k(t - w), k)$, the M_1 would still be $k \times k$ but the S_1 would be $k(t - w) \times k(t - w)$. The more general proof is completely analogous and shall be left to the reader. $\square$

2. Generalization to Partitions with Many Spread Types

Previously, we gave a method to construct a k-spread of a vector space of dimension tk, using spreads from translation planes. In this section, we show that we may generalize the construction to produce partitions of a vector space of dimension tk by subspaces of various dimensions n, for any divisor of k.

In our construction method, we covered $j - (0\text{-subsets})$ using k-spreads. However, for any particular $j - (0\text{-subset})$, we may instead choose to cover this set using n-spreads, where n properly divides k. For this part, we think of the vector space of dimension tmn, where $mn = k$, and we wish to use n-spreads arising from translation planes instead of k-spreads. To give an illustration, assume that $k = 6$ and we have a vector space of dimension $t6$. Our previous construction uses N_t, six spreads arising from translation planes. Take a particular $j - (0\text{-subset})$ (x_1, x_2, x_3) (zeros are suppressed), where x_i is non-zero (for $t > 2$). There are $(q^k-1)^3$ vectors, and we would require three k-spreads to cover this set. Let $n = 3$ and $m = 2$, write $x_i = (x_{1i}, x_{2i})$, where the x_{1i} and x_{2i} are 3-vectors. Now further decompose this set into $j - (0\text{-subsets})$. For example, it is possible to have $(x_{11}, 0, x_{12}, x_{22}, 0, x_{22})$. Note that we could never have $(x_{11}, 0, 0, 0, 0, 0)$, since the x_i are non-zero. There is one $0 - (0\text{-subsets})$, exactly six $1 - (0\text{-subsets})$, twelve

$2 - $ (0-subsets) (choose two of x_1, x_2, x_3 then one of two 3-vectors), and eight $3 = 0$-sets. Hence, there are in total $1 + 6 + 12 + 8 = 27$ $j - (0$-subsets). We have one set with $(q^n - 1)^6$ vectors, 6 sets with $(q^n - 1)^5$ vectors, 12 sets with $(q^n - 1)^4$ vectors and 8 sets with $(q^n - 1)^3$ vectors.

Note that $(q^6 - 1)^3 = (q^3 - 1)^6 + 6(q^3 - 1)^5 + 12(q^3 - 1)^4 + 8(q^3 - 1)^3$. To see this, just note that $(q^6 - 1)^3 = (((q^3 - 1) + 1)^2 - 1)^3 = ((q^3 - 1)^2 + 2(q^3 - 1))^3$. Thus, we have a $j = 0$-set partition of n-vectors of the $j - (0$-subset) (x_1, x_2, x_3) of k-vectors. Now we may apply our construction using n-spreads from translation planes. That is, we would require $5 + 6 \cdot 4 + 12 \cdot 3 + 8 \cdot 2$ n-spreads.

In this way, beginning from a tk-dimensional vector space over $GF(q)$, then we may obtain a partition of the vector space by subspaces of dimensions n_i for any divisor n_i of k from n_i-spreads arising from translation planes. Of course, given any k-spread, choose any particular k-subspace and find a n_i-spread on that k-subspace. This simple idea will also produce partitions of vector spaces with subspaces of many dimensions. However, the construction method that we give here does not involve such refinements of the individual k-spaces.

But, suppose that we would wish to find a finite vector space partition by subspaces of dimensions n_i, for $i = 1, 2, ..., z$. Then take the product $\prod_{i=1}^{z} n_i = k$. Then take any vector space V_{tk} of dimension tk over $GF(q)$. Our construction then gives a partition of V_{tk} by subspaces not only of dimension n_i but also of dimension m for any divisor of k.

Just as in the extended André situation, there are algorithms that may be employed for the construction of Sperner spaces and of focal-spreads and the reader is directed to Jha and Johnson [80] for these algorithms and additional details.

2.1. sk-Generalization. In this section, we generalize the construction of the previous chapter by what might be considered a direct sum of focal-spreads, although the construction is much more general.

Choose a vector space V_{k+N} with vectors $(x_1, y_1, y_2, .., y_s)$, where x_1 is a k-vector and, where y_i is a t_i-vector over $GF(q)$. Hence, we have a $k + N$-dimensional vector space, where $N = \sum_{i=1}^{s} t_i$. Partition the $k + N$-dimensional vector space into $x_1 = 0$, an N-dimensional subspace F_N and partition the remaining vectors in $j - (0$-subsets), for $j = 0, 1, 2, 3, .., s - 1$. That is, we choose a set of j vectors y_i to be zero. Use the notation $(y_1, .., y_s)$ for the set of vectors $(x_1, y_1, .., y_s)$, for x_1 non-zero.

We consider individual focal-spreads of type (t_i, k), so we have the representation

$$x_1 = 0, y_i = 0, \ y = x_1 M_i \text{ for } M_i \in \mathcal{M}_i,$$

where y_i is a t_i-vector and M_i's are $k \times t_i$ matrices of rank k, whose differences are of rank k. Then we wish to partition the vectors of $V_{k+N} - F_N$ as follows: Start with the $0 - (0$-subset), where all vectors of $(y_1, .., y_s)$ are non-zero. Choose any sequence of (t_i, k)-focal spreads F_i and form the set

$$y = x_1[M_1, M_2, .., M_s], \text{ for } M_i \in \mathcal{M}_i^0,$$

where the indicated matrix is $k \times N$ and x_1 nonzero and for a set $\mathcal{M}_i^0$ of associated $k \times t_i$ matrices. Choose a $1 = 0$-set. There are $\begin{pmatrix} s \\ 1 \end{pmatrix}$, ways to choose such a set. For the $1 - (0$-subset) with 0 in the z-entry, choose any sequence of (t_i, k)-focal-spreads, for $i \neq z$, and form

$$y = x_1[M_1, M_2, .., M_s], \text{ for } M_i \in \mathcal{M}_i^{1,z},$$

for $\{\mathcal{M}_i^{1,z}\}$ a set of focal-spreads of type (t_i, k), for $i \neq z$.

Take $t_1 + t_2 = N$. Partition (x_1, y_1, y_2) for $x_1 \neq 0$ by:
$(x_1, x_1[M_1, M_2])$, $(x_1, x_1[M_1', 0])$, $(x_1, x_1[0, M_2'])$, and $(x_1, x_1[0, 0])$, which requires 4 focal-spreads two of type (t_1, k) and two of type (t_2, k). We are using $(q^{t_1} - 1)(q^{t_2} - 1) + (q^{t_1} - 1) + (q^{t_2} - 1) + 1$ k-components for this partition, the number of which is $q^{t_1+t_2} = q^N$. This set of k-components together with $x_1 = 0$, the focus of dimension N is a focal-spread of type (N, k).

Take $t_1 + t_2 + t_3 = N$ and partition (x_1, y_1, y_2, y_3) for $x_1 \neq 0$ into $j - (0$-subsets).

Anyway, we would need a focal-spread of type (t_i, k) for $i = 1, 2, 3$ to construct $y = x[M_1, M_2, M_3]$. This gives $(q^{t_1} - 1)(q^{t_2} - 1)(q^{t_3} - 1)$ k-components for this $j - (0$-subset), one each of the three types; $3 \begin{pmatrix} 3 \\ 0 \end{pmatrix}$ total.

Then the $1 - (0$-subsets) would give
$(q^{t_1} - 1)(q^{t_2} - 1) + (q^{t_1} - 1)(q^{t_3} - 1) + (q^{t_2} - 1)(q^{t_3} - 1)$ k-components. Also, we would need then 2 each of the three types; $(3-1) \begin{pmatrix} 3 \\ 1 \end{pmatrix}$ total.

Then the $2 - (0$-subsets) would give
$(q^{t_1} - 1) + (q^{t_2} - 1) + (q^{t_3} - 1)$ k-components, 1 each of the three types; $(3 - 2) \begin{pmatrix} 3 \\ 2 \end{pmatrix}$ total, then for the $3 - (0$-subset), we would not

need a focal-spread, since we obtain exactly one k-spread. Then the number of k-components obtained is:

$$(q^{t_1} - 1)(q^{t_2} - 1)(q^{t_3} - 1) +$$
$$(q^{t_1} - 1)(q^{t_2} - 1) + (q^{t_1} - 1)(q^{t_3} - 1) + (q^{t_2} - 1)(q^{t_3} - 1) +$$
$$(q^{t_1} - 1) + (q^{t_2} - 1) + (q^{t_3} - 1) + 1$$

$$= q^{t_1 + t_2 + t_3} \ k\text{-components}; \ \sum_{i=1}^{3} (3 - i) \binom{3}{i} = 3 + 6 + 3$$

$$= 12 \text{ total focal spreads,}$$

$$\left(\sum_{i=1}^{3} (3 - i) \binom{3}{i} \right) /3 \text{ of each type } (t_i, k).$$

In general, we would need

$$\sum_{i=1}^{s} (s - i) \binom{s}{i}$$

focal-spreads,

$$\left(\sum_{i=1}^{s} (s - i) \binom{s}{i} \right) /s$$

of each type.

We note the important special case: If all focal-spreads are taken to be planar k-spaces, we would need

$$\sum_{i=1}^{s} (s - i) \binom{s}{i}$$

planar k-spreads to construct the focal spread.

Take $(x_1, y_1) = 0$. This then is a $(t_2 + t_3 + \dots + t_s)$-dimensional vector space. We wish to cover the remaining vectors $(x_1, y_1, y_2, \dots, y_s)$, where either x_1 or y_1 is non-zero. The question is, how do we cover $(0, y_1, y_2, \dots, y_s)$, for y_1 non-zero? This could done if y_1 is a k-vector, as $(0, x_2, y_2, \dots, y_s)$, for $x_2 = y_1$ as a k-vector and x_2 non-zero. We would thus require

$$\sum_{i=1}^{s-1} (s - 1 - i) \binom{s-1}{i}$$

focal-spreads. In this way, we obtain a vector space of dimension $N + k$, with a focus of dimension $N - k$, and therefore, we obtain a focal-spread of type $(N, N - k, k)$.

So if y_1 is a k-vector, we can take $(x_1, y_1) = 0$, to obtain a $(t_2 + t_3 + \ldots + t_s)$-dimensional focus and cover the remaining objects in the same way as before.

3. Going Up–Direct Sums

Let S be a set of s focal-spreads $\mathcal{F}_i$ of type (t_i, k), for $i = 1, 2, .., s$ and let the foci be denoted by F_i, of dimension t_i, for $i = 1, 2, .., s$. Then the matrix representation of $\mathcal{F}_i$ be given by

$$x = 0, \ y = 0, \ y = xM_i, \text{ where } M_i \in \mathcal{M}_i$$

and where, in context, $x = 0$ is a t_i-dimensional subspace and $\mathcal{M}_i$ is a set of $k \times t_i$ matrices of rank k, whose differences are of rank k. Considering vectors as $(x_1, .., x_k, y_1, .., y_{t_i})$, where the x_i and y_i in $GF(q)$ then letting $x = (x_1, .., x_k)$, and $y = (y_1, .., y_{t_i})$, then $x = 0$ is a t_i-dimensional $GF(q)$-space and $y = 0$ is a k-dimensional $GF(q)$-space.

Now form the vector space $(y = 0) \oplus F_1 \oplus F_2 \oplus \ldots \oplus F_s$ of dimension $t_1 + t_2 + \ldots + t_s + k$. Let the vectors be denoted by $(x_1, x_2, .., x_k, y_1, .., y_N)$, where $t_1 + t_2 + \ldots + t_s = N$, where x_i are in $GF(q)$, and the y_i are t_i-vectors over $GF(q)$. Now let $(x_1, .., x_k) = x$ and let $y_i = (y_{1i}, y_{2i}, .., y_{t_i i})$, for $y_{ji} \in GF(q)$, and let $y = (y_{11}, .., y_{t_s s})$, which is an N-vector over $GF(q)$, for $N = \sum_{i=1}^{s} t_i$. Then $x = 0$ is the direct sum of $F_1 \oplus F_2 \oplus \ldots \oplus F_s$. Consider

$$x = 0, \ y = x[M_1, M_2, .., M_s]; M_i \in \mathcal{M}_i \cup \{0\},$$

where, in context, the adjoined 0-matrix is the $k \times t_i$-zero matrix. The indicated matrix is a $k \times N$-matrix. Then, since the difference of two of the indicated matrices is either of rank k or $y = x[0, 0, 0, .., 0] = 0$, we obtain a focal-spread of type $(t_1 + t_1 + \ldots, t_s, k)$. We call this construction the 'direct sum construction.' Note that we may relax the condition $t_i > k$, for certain of integers t_i and the construction still applies.

Assume that all of the focal-spreads are additive in the sense that the $\mathcal{M}_i \cup \{0\}$ are additive groups. Then, clearly, the direct sum focal spread is also additive.

THEOREM 21. *(1) Given any sequence of s focal-spreads of type (t_i, k) or spreads for translation planes of order q^k for $i = 1, 2, .., s$, then there are focal-spreads of type $(t_1 + t_2 + \ldots + t_s, k)$ obtained by the direct sum construction.*

(2) If the s focal-spreads of type (t_i, k) are all additive, then there are $s!$ additive focal-spreads of type $(t_1 + t_2 + \ldots + t_s, k)$ constructed.

(3) If the integers t_i are distinct, then there are $s!$ focal-spreads of type $(t_1 + t_2 + \ldots + t_s, k)$ that may be constructed in this manner.

(4) In the direct sum, if a planar k-spread is used in place of a focal-spread, and the direct sum is a k-cut, then there is an affine subplane of order q^k in the extended translation plane.

(5) If we have a direct sum of s semifield planar k-spreads, then an additive k-spread of an sk-dimensional vector space is obtained. In this case, the direct sum may be given by the going up procedure.

PROOF. For (4), see the proof of Corollary 3. We give a few remarks and leave the rest of the proof to the reader. Now start the k-spread construction, using the going up process to get focal-spreads of type $(k(t-1), k(t-w), k)$. Now vary the w's. So, we can get focal-speads of type $(k(t-1), k(t-w_i), k)$. In this setting, all of the vector spaces have the same dimension $k(t-1)$ but the focal-spreads have possibly different dimensions. In the $(k(t-1), k(t-w), k)$-focal-spreads, to illustrate the main idea, consider (x_1, x_2, x_3, x_4) all k-vectors. Then let $x_1 = x_2 = 0$, be a $2k$-dimensional vector space. We need to cover the remaining vectors (x_1, x_2, x_3, x_4), where either x_1 or x_2 is non-zero. If both are non-zero, we have $(x_1, x_1 M_1, x_1 M_2, x_1 M_3)$, or $(x_1, x_1 M_4, 0, x_1 M_5)$, or $(x_1, x_1 M_6, x_1 M_7, 0)$; that is, $y = x_1[M_1, M_2, M_3]$, $y = x[M_4, 0, M_5]$, $y = x_1[M_6, M_7, 0]$, where each matrix is a $k \times 3k$ matrix. If exactly one is non-zero, we have $(0, x_2, x_3, x_4) = (0, y = x_2[M_8, M_9])$ or $(x_1, 0, x_3, x_4)$ as $y = x_1[0, M_{10}, M_{11}]$. This sort of argument clearly generalizes and the reader should have no trouble providing the more general argument. □

Part 2

Subgeometry Partitions

In this part, we construct subgeometry partitions in a variety of ways: Using group actions on focal-spreads, using semifield planes, or using the groups that can act on generalized extended André spreads. All of these constructions use in an interesting way collineation groups acting in a particular manner. We use the ideas of 'fusion' of the nuclei of semifields as a manner of obtaining such groups. These ideas are generalized when we consider the so-called double-Baer groups. In all of these cases, we isolate on what we call 'retraction groups' for the construction. In addition, there are the important constructions of flag-transitive designs due to Kantor, and as is suggested, there are certain collineation groups that often can act as retraction groups.

As this is an important area of current interest, we give the theory of subgeometry partitions in two somewhat large chapters. In the first chapter, the focus is mostly on what can be said using focal-spreads, and in the second chapter, we consider subgeometry partitions that can be constructed using extended André spreads.

We begin by providing a general overview of the subject. We then discuss the difference between subgeometry and quasi-subgeometry partitions.

CHAPTER 5

Subgeometry and Quasi-subgeometry Partitions

A 'subgeometry partition of a finite projective space' is simply a partition of the projective space by subgeometries isomorphic to finite projective spaces of various dimensions over various fields.

The first examples of subgeometry partitions were partitions of $PG(2, q^2)$ by subgeometries isomorphic to $PG(2, q)$'s, the so-called Baer subgeometry partitions. These partitions originated in the text [73] by Hirschfeld and Thas. There is a so-called lifting process, which we term 'geometric lifting' that produces from a Baer subgeometry partition a planar-spread that corresponds to a translation plane of order q^3 with projective 2-spread in $PG(5, q)$. The associated translation planes of order q^3 and kernel containing $GF(q)$ will often admit a collineation group that is flag-transitive. Regarding the group action on the line at infinity, a flag-transitive group is simply a group that is transitive on the line at infinity. In this case, and in the cases of most interest, this collineation group is cyclic of order $q^3 + 1$. The subgroup of order $q + 1$ together with the kernel homology group of order $q - 1$ together with the zero mapping generate a field isomorphic to $GF(q)$, which contains a cyclic subgroup of order $q^2 - 1$ that acts on the translation plane with orbits of length $q + 1$. The existence of a group of order $q^2 - 1$ containing the kernel group of order $q - 1$, such that by adjoining the zero map one obtains a field R, over which the point set as a vector space is important to the theory. This retraction group is the key to the understanding of how to construct subgeometry partitions (see Johnson [113]). Basically, the orbits of length $q + 1$ 'retract' to subgeometries isomorphic to $PG(2, q)$ that cover the associated projective space (the 6-dimensional vector space over $GF(q)$ becomes 3-dimensional over R and produces the subgeometries isomorphic to $PG(2, q^2)$). The geometric lifting process generalizes to partitions of $PG(3, q^2)$ by $PG(1, q^2)$'s and $PG(3, q)$'s, the so-called mixed partitions and, in this case, produces translation planes of order q^4 with kernel containing $GF(q)$ and again there is an associated group of order $q^2 - 1$ containing the kernel homology group of order $q - 1$, and there is an associated field R, whose multiplicative group now has component orbits of length 1 or $q+1$. The orbits of length $q+1$ retract to subgeometries isomorphic to $PG(3, q)$, and the orbits of length 1 retract to subgeometries isomorphic to $PG(1, q^2)$. The existence of the retraction group with field isomorphic to $GF(q^2)$ was shown in Johnson [113].

Furthermore, in Jha and Johnson [84], it is shown that it is possible to use 'algebraic lifting' to produce from a translation plane of order q^2 with spread in $PG(3, q)$, a corresponding spread of order q^4 in $PG(3, q^2)$, wherein lies a retraction group. Considered over the subkernel isomorphic to $GF(q)$, it turns out that there is a retraction group of order $q^2 - 1$ with associated field isomorphic to $GF(q^2)$ that constructs an associated subgeometry partition of $PG(3, q^2)$ by $PG(3, q)$'s and $PG(1, q^2)$'s. We shall use a variation of this process to construct new subgeometry partitions from focal-spreads.

There are a variety of subgeometry partitions arising from semifields since with appropriate right and middle nuclei, it is possible to generate a retraction group. For examples of these, see Jha and Johnson [82]. Further, there are subgeometries arising from semifield planes admitting the so-called double-Baer groups, and there are new examples in Jha and Johnson [83] arising from new semifields due to Dempwolff [40].

There is a natural generalization of what might be called a retraction group in a translation plane with field isomorphic to $GF(q^w)$, for arbitrary w considered in Johnson (see [107], [106]). There is a more general construction partitioning $PG(z, q^w)$ into the so-called quasi-subgeometry partitions isomorphic to $PG(l, q^e)$, for e dividing w. There is a rather subtle difference between subgeometry partitions and quasi-subgeometry partitions. If we have a subfield $GF(q^e)$ of $GF(q^w)$, containing a kernel subfield $GF(q)$, where there is an component orbit Γ of length $(q^e - 1)/(q - 1)$, then the 'points' of the corresponding projective space $PG(z, q^w)$ correspond to the point orbits under the group of order $q^e - 1$. Then the lines of the putative subgeometry are the 2-dimensional subspaces over $GF(q^e)$ and the points on such lines are obtained by intersection with the point orbits of Γ. If this intersection gives $q + 1$ point orbits, a subgeometry isomorphic to $PG(l, q^e)$ is constructed. On the other hand, we may define the lines to be simply the 2-dimensional subspaces over $GF(q)$ and a quasi-subgeometry isomorphic to $PG(l, q^e)$ is constructed.

Hence, the theory may be carried out for general groups with the understanding that a more general partition may be constructed and when the intersection property mentioned above holds, a subgeometry partition may be constructed. Although we have been discussing such partitions as connected to translation planes, a geometric lifting from the associated partition of the projective space reveals that the main ingredient is not necessarily a translation plane but could be a t-spread defining it within an associated kt-dimensional vector space, where a translation plane is constructed if and only if $k = 2$, or even more general partitions of vector spaces. So Sperner spaces based on Sperner-spreads are prime candidates to construct subgeometry partitions. Ebert and Mellinger [51] have found a new class of subgeometry partitions that construct new Sperner-spreads by geometric lifting, where there are two types of subgeometries in the partitions. These Sperner-spreads have a somewhat natural construction, which we have generalized in this text in the manner of the author [98]. We shall see that since we have nice groups acting on the types of Sperner spreads constructed, there will be some very wonderful subgeometry partitions that can be constructed. For example, there are subgeometry partitions of $PG(k, q^w)$ by subgeometries isomorphic to $PG(l, q^e)$, with essentially no restriction on e other than e divides w.

Coming back to the beginning idea, that of Baer subgeometry partitions of $PG(2, q^2)$ by $PG(1, q)$'s, there is a construction of the author [100], of subgeometry partitions called of $PG(n - 1, q^m)$ by $PG(n - 1, q)$'s based on the flag-transitive designs of Kantor, which we shall give in this text. There is a further discussion of such subgeometry partitions in Johnson and Cordero [137].

This is essentially the set of known examples of subgeometry partitions. Now going back to the general lifting procedure from quasi-subgeometry partitions, we noted that it may be seen that such a projective partition lifts back to a general partition of a vector space. Hence, it is certainly possible that focal-spreads may be constructed from quasi-subgeometry or subgeometry partitions. But, also subgeometry partitions could conceivably be constructed from focal-spreads provided a retraction group is present.

The material give here is adapted, to some extent, from the work of the author in [99], and the reader is directed to this work for additional details. We first give the formal definitions of subgeometry and quasi-subgeometry partitions.

DEFINITION 37. *Let Ω be a projective space and let Π be a subset of points of Σ such that if a line ℓ of Ω intersects Π in at least two points A and B then we define a 'line' of Π to be the set of points $\ell \cap \Pi$. If the points of Π and the 'lines' induced from lines of Σ form a projective* *space, we say that* Π *is a 'subgeometry' of* Σ. *A 'subgeometry partition' of a projective space* Ω *is a partition of* Ω *into mutually disjoint subgeometries.*

As we have previously mentioned since the action of a retraction group of order $q^2 - 1$ is the active ingredient in the retraction process, it is a natural question to ask if one would use other groups of orders $q^k - 1$ with similar properties, whether subgeometry partitions of projective spaces with different subgeometries are possible to construct from translation planes. Since such groups are easily constructed, it makes some sense to formulate some sort of general theory with hopes (assuming this is what one is interested in) of finding methods of the construction of subgeometry partitions.

DEFINITION 38. *A 'quasi-subgeometry partition' of a projective space is a subset of points that can be made into a projective space such that the lines of the projective set are subsets of lines of the projective space.*

It was shown in Johnson [106] that any translation plane of order q^n with spread in $PG(2n - 1, q)$ that admits a fixed-point-free group of order $q^k - 1$, for k a divisor of n, containing the kernel homology group does actually produce quasi-subgeometry partitions of an associated projective space with a wide variety of possible quasi-subgeometry types involved in the partition. The following result from Johnson [106] details the possibilities:

COROLLARY 4. *(Johnson [106]) Assume a partial spread* $\mathcal{Z}$ *of order* q^{ds} *in a vector space of dimension* $2ds$ *over* K *isomorphic to* $GF(q)$ *admits a fixed-point-free field group* F_w^* *of order* $(q^w - 1)$ *containing* K^*, *for* $w = d$ *or* $2d$ *and* s *is odd if* $w = 2d$. *Then any component orbit* Γ *of length* $(q^w - 1)/(q^e - 1)$ *a '*q^e*-fan', for* e *a divisor of* d, *produces a quasi-subgeometry isomorphic to a* $PG(ds/e - 1, q^e)$ *in the corresponding projective space* $PG(2ds/w - 1, q^w)$, *considered as the lattice of* F_w*-subspaces of* V.

These results are also valid in 'generalized spreads' in the infinite case, over skewfields, and in the t-spreads or more generally in vector space partitions.

This background section is also given so that the reader might better understand when a quasi-subgeometry becomes a 'subgeometry' and when the partitions then become subgeometry partitions.

Most of the construction techniques have been previously given in Johnson [106]. As this material is quite general and suitable both for finite and infinite affine translation planes and more generally for generalized spreads, we repeat the main parts of these results here.

DEFINITION 39. *A 'q^e-fan' in an rds-dimensional vector space over K isomorphic to GF(q) is a set of $(q^w - 1)/(q^e - 1)$ mutually disjoint K-subspaces of dimension ds that are in an orbit under a field group F_w^* of order $(q^d - 1)$, where F_w contains K and such that F_w^* is fixed-point-free, and where $w = d$ or 2d and in the latter case s is odd.*

We use the term that $GF(q^w)$ 'acts' on the vector space, if the multiplicative group acts in a manner as in the previous definition.

THEOREM 22. *(Johnson [106]) (1) Let S be a t-spread of V of size (ds, q), and assume that $w = d$ or t^*d, where t^*d does not divide ds, and t^* divides t.*

Assume that $GF(q^w)$ 'acts' on S then there is a quasi-subgeometry partition of $PG(tds/w - 1, q^w)$ consisting of $PG(ds/e_i - 1, q^{e_i})$'s, where e_i divides w for $i \in \Lambda$.

(2) Conversely, any quasi-subgeometry partition of $PG(tds/w - 1, q^w)$ by $PG(ds/e_i - 1, q^{e_i})$'s for $i \in \Lambda$, and e_i a divisor of w, produces a t-spread of size (ds, q) that is a union of q^{e_i}-fans. The t-spread corresponds to a translation Sperner space admitting $GF(q^w)^*$ as a collineation group.

REMARK 17. *All of the constructions above involve 'fans.' A 'fan' produces a quasi-subgeometry and involves a process that is called 'folding the fan.' A quasi-subgeometry also produces a 'fan,' wherein the process is called 'unfolding the fan.' In other terminology, 'folding the fan' has been called 'retraction' in Johnson [113], and 'unfolding the fan' is called 'lifting' in Hirschfeld and Thas [73] and Ebert and Mellinger [51].*

0.1. When Is a Quasi-Subgeometry a Subgeometry? The following discussion, taken from the author's previous work and part of this discussion shows how an extended André spread may be used to construct a full-blown subgeometry partition.

Suppose one would consider $PG(1, q^8)$ and ask if it would be possible to construct a quasi-subgeometry of $PG(1, q^8)$? So, certainly it is possible that we could have a $PG(1, q^2)$, for example, or a $PG(0, q^8)$, but since there is exactly one line, it would be impossible to have a $PG(t - 1, q^j)$, for t not 2 or 1.

So, any non-trivial subgeometry partition must necessarily require that the subgeometries are $PG(0, q^8)$'s and $PG(1, q^k)$, for $k < 8$.

However, it is possible to construct quasi-subgeometries, as follows: Consider a 2-dimensional $GF(q^8)$ vector space V_2, and form the 8-dimensional subspace over $GF((q)$, $\left\{(x, x^{q^2}); x \in GF(8)\right\}$. Call this subspace $y = x^{q^2}$ and consider the orbit of this subspace under the group

$$G_2 = \left\langle (x, y) \longmapsto (xd, yd); d \in GF(q^8)^* \right\rangle .$$

We note that $y = x^{q^2}$ will map to $y = x^{q^2} d^{1-q^2}$, so there will be an orbit Γ of length $(q^8 - 1)/(q^2 - 1)$. We note that the point orbits that lie on Γ define 1-dimensional $GF(q^8)$-subspaces. Now define the set of 'points' to be the set of point-orbits under the group G that lie within Γ, and define a 'line' to be the set of 'points' corresponding to a 2-dimensional vector subspace over $GF(q^2)$, generated by two distinct orbits P and Q. That is, a 'line' is the set of points that lie in $\{\alpha P + \beta Q; \alpha, \beta \in GF(q^2)\}$. Then it is not difficult to show that this set of points and lines is isomorphic to $PG(3, q^2)$. If we note that

$$\left\{ y = x^{q^2} d^{1-q^2} \right\} = \left\{ \begin{array}{c} y = x\tau^{(q^2-1)/(q-1)}; \\ \tau \text{ has order dividing } (q^8 - 1)/(q-1) \end{array} \right\},$$

then we may replace this latter set by the former set. What we are actually doing is constructing an André plane from a Desarguesian plane by net replacement. If we consider

$$\left\{ y = x^q n d^{1-q} \right\}$$
$$= \left\{ y = x\tau n; \tau \text{ has order dividing } (q^8 - 1)/(q - 1) \right\},$$

then for appropriate n, the sets are disjoint on subspaces.

Hence, in this way, it is possible to obtain a quasi-subgeometry partition of $PG(1, q^8)$ by quasi-subgeometries isomorphic to $PG(3, q^2)$, $PG(1, q^4)$, and $PG(0, q^8)$'s.

Now consider $PG(2, q^8)$ and ask again if there are subgeometry partitions of $PG(2, q^8)$, by subgeometries isomorphic to projective spaces other than $PG(2, q^k)$, $PG(1, q^t)$, or $PG(0, q^z)$? Now since there are $1 + q^8 + q^{16}$ lines, it is at least possible to construct such partitions. If the following illustration needs more clarification, the reader is directed back to the chapter on extended André spreads.

So, consider a 3-dimensional $GF(8)$-vector space V_8 and consider the following subspace over $GF(q^2)$.

$$\left\{ (x, x^{q^2}, x^{q^6}); x \in GF(8) \right\} .$$

Call this subspace $y = (x^{q^2}, x^{q^6})$. Consider the group

$$G_3 = \left\langle (x, y, z) \longmapsto (xd, yd, zd); d \in GF(q^8)^* \right\rangle .$$

Then the image set Γ of this subspace under G_3 is

$$\left\{ y = (x^{q^2} d^{1-q^2}, x^{q^6} d^{1-q^6}); d \in GF(q^8)^* \right\}$$

and has orbit length

$$(q^8 - 1)/(q^{(2,6,8)} - 1) = (q^8 - 1)/(q^2 - 1).$$

We note that this set of 8-dimensional subspaces over $GF(q)$ is similarly covered by

$$\left\{ \begin{array}{c} y = (x\tau^{(q^2-1)/(q-1)}, x\tau^{(q^6-1)/(q-1)}); \\ \tau \text{ has order dividing } (q^8 - 1)/(q - 1) \end{array} \right\}.$$

We claim that, in this setting, the orbit forms a 'subgeometry' of $PG(2, q^8)$ that is isomorphic to $PG(3, q^2)$. The 'points' of the point-line incidence structure S_Γ are the point orbits under $GF(q^8)$ that lie within Γ. The 'lines' of S_Γ are the intersections in Γ with lines of $PG(2, q^8)$, that is, the intersections with Γ of 2-dimensional vector spaces over $GF(q^8)$, generated by two distinct 1-dimensional $GF(q^8)$, subspaces that lie within Γ. We note that there are $(q^8 - 1)/(q^2 - 1)$ 'points,' so if we were to obtain a projective space, it would be isomorphic to $PG(3, q^2)$.

So, take two distinct orbits P and Q under G_3 that lie within Γ. These become 1-dimensional $GF(q^8)$-spaces. Take a generator from each and form the 2-dimensional $GF(q^8)$-space. Since in P and Q there are always generators in $y = (x^{q^2}, x^{q^6})$, suppose

$$\left\langle (x_1, x_1^{q^2}, x_1^{q^6}) \right\rangle = P \text{ and } \left\langle (x_2, x_2^{q^2}, x_2^{q^6}) \right\rangle = Q.$$

Let $\alpha, \beta \in GF(q^8)$ and generate the two-dimensional vector space

$$\left\langle (x_1, x_1^{q^2}, x_1^{q^6}), (x_2, x_2^{q^2}, x_2^{q^6}) \right\rangle$$

$$= \left\{ \alpha(x_1, x_1^{q^2}, x_1^{q^6}) + \beta(x_2, x_2^{q^2}, x_2^{q^6}); \alpha, \beta \in GF(q^8) \right\}.$$

Now form

$$\left\{ \alpha(x_1, x_1^{q^2}, x_1^{q^6}) + \beta(x_2, x_2^{q^2}, x_2^{q^6}); \alpha, \beta \in GF(q^8) \right\} \cap \Gamma,$$

and let R be an orbit of Γ different from P and Q such that

$$R = \left\langle \alpha(x_1, x_1^{q^2}, x_1^{q^6}) + \beta(x_2, x_2^{q^2}, x_2^{q^6}) \right\rangle$$

for some fixed α, β, now both non-zero. Again, R contains a vector of $y = (x^{q^2}, q^{q^6})$ that generates R over $GF(q^8)$. Let

$$R = \left\langle (x_3, x_3^{q^2}, x_3^{q^6}) \right\rangle.$$

Then it follows that there is a non-zero element of $GF(q^8)$, so that

$$\rho(\alpha(x_1, x_1^{q^2}, x_1^{q^6}) + \beta(x_2, x_2^{q^2}, x_2^{q^6})) = (x_3, x_3^{q^2}, x_3^{q^6}).$$

Hence, ρ, α, β are all non-zero, and since scalar addition by non-zero elements of $GF(q^8)$ is given by the group G_3, we may incorporate ρ into α and β, and assume without loss of generality that

$$\begin{aligned}
&(\alpha(x_1, x_1^{q^2}, x_1^{q^6}) + \beta(x_2, x_2^{q^2}, x_2^{q^6})) \\
&= (\alpha x_1 + \beta x_2, \alpha x_1^{q^2} + \beta x_2^{q^2}, \alpha x_1^{q^6} + \beta x_2^{q}) \\
&= (x_3, x_3^{q^2}, x_3^{q^6}).
\end{aligned}$$

Hence,

$$\alpha x_1 + \beta x_2 = x_3, \quad \alpha x_1^{q^2} + \beta x_2^{q^2} = x_3^{q^2} \text{ and } \alpha x_1^{q^6} + \beta x_2^{q} = x_3^{q^6}.$$

This leads to the following two equations:

$$(*) : (\alpha - \alpha^{q^2})x_1^{q^2} + (\beta - \beta^{q^2})x_2^{q^2} = 0$$

and

$$(**) : (\alpha - \alpha^{q^6})x_1^{q^6} + (\beta - \beta^{q^2})x_2^{q^6} = 0.$$

Since x_1, x_2, x_3 are all non-zero, then $\alpha - \alpha^{q^2} = 0$ if and only if $\beta - \beta^{q^2} = 0$, which is valid if and only if α and β are in $GF(q^2)$. Similarly, $\alpha - \alpha^{q^6} = 0$ if and only if $\alpha^{q^6} = \alpha$, so that α has order dividing $(q^6 - 1, q^8 - 1) = (q^2 - 1)$, so again, α and β are in $GF(q^2)$. Hence, we may assume that $\alpha - \alpha^{q^2}$ and $\alpha - \alpha^{q^6}$ are both non-zero. So, we obtain

$$(*)' : \left(\frac{x_1}{x_2}\right)^{q^2} = \left(\frac{\beta^{q^2} - \beta}{\alpha - \alpha^{q^2}}\right), \text{ and}$$

$$(**)' : \left(\frac{x_1}{x_2}\right)^{q^6} = \left(\frac{\beta^{q^6} - \beta}{\alpha - \alpha^{q^6}}\right).$$

Raise $(*)'$ to the q^6-th power to obtain:

$$(*)' : \left(\frac{x_1}{x_2}\right)^{q^8} = \left(\frac{\beta^{q^8} - \beta^{q^6}}{\alpha^{q^6} - \alpha^{q^8}}\right) = \left(\frac{\beta - \beta^{q^6}}{\alpha^{q^6} - \alpha}\right) = \left(\frac{\beta^{q^6} - \beta}{\alpha - \alpha^{q^6}}\right).$$

Therefore, we arrive at the equation:

$$\left(\frac{x_1}{x_2}\right)^{q^6} = \left(\frac{x_1}{x_2}\right)^{q^8}.$$

This is equivalent to

$$\left(\frac{x_1}{x_2}\right)^{q^8 - q^6} = \left(\frac{x_1}{x_2}\right)^{q^6(q^2 - 1)} = 1.$$

Therefore, $\left(\frac{x_1}{x_2}\right)^{q^6}$ is in $GF(q^2)^*$, which implies that $\left(\frac{x_1}{x_2}\right)$ is in $GF(q^2)$. Let $x_1 = x_2\delta$, for $\delta \in GF(q^2)^*$ then $(x_1, x_1^{q^2}, x_1^{q^6}) = (\delta x_2, \delta x_2^{q^2}, \delta x_2^{q^6})$, since $\delta^{q^2} = \delta$. However, this means that $P = Q$, originally.

Hence, it can only be that $\alpha, \beta \in GF(q^2)^*$. Therefore, there are $(q^2)^2 + 1$ 'points' of intersection with a line of $PG(2, q^8)$ and S_Γ. It follows directly that S_Γ is isomorphic to $PG(3, q^2)$.

In a similar manner, we may obtain subgeometries isomorphic to $PG(1, q^4)$ and $PG(0, q^8)$. Therefore, we obtain a subgeometry partition of $PG(2, q^8)$, by subgeometries isomorphic to $PG(3, q^2)$, $PG(1, q^4)$, and $PG(0, q^8)$.

If it is objectionable to include 'points' within a partition then simply take $PG(3n - 1, q^8)$, for $n > 1$, and it is similarly possible to construct subgeometry partitions by subgeometries isomorphic to $PG(4n - 1, q^2)$, $PG(2n - 1, q^4)$, and $PG(n - 1, q^8)$.

In the section on subgeometries arising from extended André partitions and their generalizations, we may repeat these ideas in general and using the procedure in r-dimensional vector spaces over $GF(q^{sn})$, it is possible to construct various subgeometries isomorphic to $PG(sn/s^* - 1, q^{s^*})$, where s^* is any divisor of s. Hence, it is possible to obtain subgeometry partitions using a wide variety of different isomorphism types of projective spaces as subgeometry partitions.

Finally, since the focal-spreads that we have constructed admit groups of the type that are conceivably retraction groups, we now show how easy it is to find quite interesting subgeometry partitions from focal-spreads, their associated double-spreads and triple-spreads as well as from their associated designs.

CHAPTER 6

Subgeometries from Focal-Spreads

As an important application of the theory of focal-spreads, we shall be constructing new subgeometry portions. As mentioned in the introductory remarks, the most important and relevant construction device of subgeometry partitions from vector spaces uses an appropriate fixed-point-free group of order $q^d - 1$, containing a scalar group of order $q - 1$, which is a field when adjoining the zero mapping. That is, from such a group acting on a partition of a vector space, a 'retraction' procedure produces what is called a quasi-subgeometry partition by Johnson [**106**] and a subgeometry partition when $d = 2$. The quasi-subgeometries are subgeometries if and only if each 'line' is the full intersection with the set of 'points' of a line of the projective superspace. This is the beauty of using groups first constructing quasi-subgeometries providing the bulk of the effort with the only question remaining being whether we actually obtain subgeometries in the partition.

1. k-Cuts of Subgeometry Partitions

Initially, the subgeometries that were found 'lifted' to spreads producing translation planes. Conversely, as has been mentioned, translation planes having certain fixed-point-free groups admit a certain 'retraction' procedure by which subgeometry partitions of projective spaces may be constructed. Of course, lifted subgeometry partitions may construct more generally arbitrary spreads or, even more generally, various interesting partitions of vector spaces. However, the use of general partitions have not so far been used for such constructions.

Given a vector space V of dimension rt, admitting a t-spread, let W be a subspace V, then there is a partition of W induced from the t-spread. This simple idea is akin to a k-cut, producing a focal-spread. We shall simply call such a partition a 'generalized k-cut', even if the partitions induced has many different subspace dimensions. The key idea for us here is to ensure that if there is a retraction group acting on the t-spread that the group also acts on the generalized k-cut of the subspace W.

DEFINITION 40. *Let $\mathcal{P}$ be a subgeometry partition of a projective space Π and let Π' denote a projective subspace of Π. If there is a subgeometry partition $\mathcal{P}'$ of Π' induced from $\mathcal{P}$, we shall say that $\mathcal{P}'$ is a 'generalized k-cut' of $\mathcal{P}$.*

So, suppose that we start with a translation plane given by a t-spread that admits nice groups. We formulate the associated k-cuts and ask what happens in the projective space with regards to subgeometry partitions. As mentioned, it seems to be more convenient to start the other way—from the projective space first, which admits a subgeometry partition. So, assume that we have a subgeometry partition of $PG(z - 1, q^w)$ by subgeometries isomorphic to $PG(l - 1, q^e)$ for various values of l and e and assume that the subgeometry partition arises from a t-spread over $GF(q)$ as a k-cut. To avoid trivial situations, we would normally not allow 'points' to be considered subgeometries. Then $z = 2t/w$ and $l = t/e$. From any subgeometry partition of $PG(n, q^w)$, there is the nature lifting procedure 'unfolding the fans' that produces partitions of vector spaces in this setting. We note that the field over which the vector space is originally defined for the t-spread might actually be changed.

In any case, we see that there is a collineation group G of order $q^w - 1$, with component orbits of length $(q^w - 1)/(q^e - 1)$, for various values of e, which induces the spread components (these are the 'fans'). If there is a value for e equal to w, then there is a fixed component (a 'trivial fan').

Assume that there are at least two fixed components, say, $x = 0$ and $y = 0$. On one of these fixed components, say $y = 0$, choose any set of orbits Γ under the group. Each of these orbits is a 1-dimensional $GF(q^w)$-subspace. Let V_k denote the subspace of dimension k generated by the set Γ. Now form the $t+k$-dimensional subspace $V_{t+k} = \langle x = 0, V_k \rangle$. This subspace is left invariant by the group G and hence is invariant under the field $G \cup \{0\}$. We may then form the focal-spread. We claim that the group G produces another subgeometry partition. Now here is the beautiful part! Because the group inherits as a collineation group of the focal-spread, we clearly obtain a so-called quasi-subgeometry partition. So, the only question is whether we actually obtain subgeometries in the partition.

To see that we do, take any orbit O of length $(q^w - 1)/(q^e - 1)$ of the group G (there will be at least one such orbit for each $PG(l - 1, q^e)$). Take two $GF(q^w)$-linearly independent point orbits of G lying on the k-component orbit O and form the 2-dimensional $GF(q^w)$-space (a line in the projective superspace). Let O^+ denote the t-component orbit in the t-spread. Since G leaves invariant the focal-spread, we see that the 2-dimensional $GF(q^w)$-space is a set of vectors within V_{t+k}. The intersection of the 2-dimensional $GF(q^w)$-space with O^+ is then the intersection within O. Hence, we obtain a 2-dimensional $GF(q^e)$-space as a space of intersection. So we still obtain a subgeometry. However, in this setting, instead of a $PG(t/e - 1, q^e)$, we obtain a $PG(k/e - 1, q^e)$— truly an amazing construction. So, in a sense, we have a 'k-cut of a subgeometry partition,' but we now have some new subgeometry partitions..

THEOREM 23. *Assume that there is a subgeometry partition of $PG(2t/w - 1, q^w)$ arising from a translation plane of order q^t, and kernel containing $GF(q)$, for $t = sw$, by subgeometries isomorphic to $PG(t/e - 1, q^e)$, for $e \in D_w$, a set of divisors of w such that there are at least two $PG(t/w - 1, q^w)$'s.*

Then for $t = sw$, and $k = w, 2w, 3w, .., (s - 1)w$, there is a focal-spead of type (t, k) of a vector space with focus of dimension t that constructs a subgeometry partition of $PG((t + k)/w - 1, q^w)$ by one $PG(t/w - 1, q^w)$, at least one $PG(k/w - 1, q^w)$ and remaining $PG(k/e - 1, q^e)$'s.

2. Additive k-Cuts

In this section, we consider subgeometry partitions that can arise from semifield t-spreads, which might be called 'additive subgeometry partitions,' as well as the k-cut subgeometry partitions that can arise from k-cuts of the semifield spreads.

We begin with the subgeometry partitions that can arise directly. So, suppose that we have a semifield t-spread of order q^t, with kernel containing $GF(q)$ admitting right and middle nuclei isomorphic to $GF(q^k)$. Using Jha and Johnson [87], we fuse the kernel and nuclei isomorphic to $GF(q)$, as well as fuse the middle and right nuclei. Then $x = 0$ and $y = 0$ are pointwise fixed, respectively, by affine homology groups corresponding to the middle and right nuclei, respectively. The semifield is a vector space over the middle nucleus isomorphic to $GF(q^k)$. Consider the set of vectors generated by $x = 0$ and any $GF(q^k)$ orbit on $y = 0$ of the middle nucleus. Then $x = 0$ and the orbit on $y = 0$ generate a vector space of dimension $t + k$. This vector space is fixed by both associated homology groups, as the right nucleus fixes $x = 0$ and fixes $y = 0$ pointwise and the middle nucleus fixes $x = 0$ pointwise and fixes the indicated orbit. Now form the focal-spread of dimension $t + k$ admitting the product of the homology groups. In the product, there are field groups that are fixed point-free of the form

$$\left\langle \begin{bmatrix} A & 0 \\ 0 & A^\sigma \end{bmatrix} ; A \in GF(q^k)^* \right\rangle$$

and σ an automorphism of $GF(q^k)$, where σ fixes $GF(q)$-pointwise.

THEOREM 24. *Assume that corresponding to a semifield spread of order q^t, there is a semifield with right and middle nuclei isomorphic to $GF(q^k)$ and left nucleus containing a field isomorphic to $GF(q)$.*

(1) Using the retraction group

$$\left\langle \begin{bmatrix} A & 0 \\ 0 & A^\sigma \end{bmatrix} ; A \in GF(q^k)^* \right\rangle,$$

either a quasi-subgeometry partition or a subgeometry partition is obtained.

(2) Assume that we have a subgeometry (or quasi-subgeometry) partition of $PG(2t/k - 1, q^k)$ by subgeometries (or quasi-subgeometries) isomorphic to $PG(t/k - 1, q^k)$ or $PG(t - 1, q)$ (assuming that the component orbits have lengths 1 or $(q^k - 1)/(q - 1)$). Then the k-cut subgeometry (or quasi-subgeometry) partition associated with the additive focal-spread of type (t, k) has subgeometries (or quasi-subgeometries) isomorphic to $PG(t/k - 1, q^k)$, $PG(0, q^k)$, and $PG(k - 1, q)$.

3. Right/Left Focal-Spreads

From semifields that admit appropriately fused nuclei there are associated subgeometry partitions. For the right and middle nuclei, the associated collineation groups are affine homology groups that operate on the right or left, respectively. Since we are considering focal-spreads analogous to spreads describing translation planes, we see that it is possible to generalize the left and right mappings to produce associated subgeometry partitions. We first note that groups of order $q^2 - 1$ with the required properties always produce subgeometry partitions.

THEOREM 25. *Suppose that a finite vector space partition over $GF(q)$ admits a group of order $q^2 - 1$, such that unioning the zero vector, the associated vector space becomes a $K \simeq GF(q^2)$-vector space. So, we have a vector space over the 'field' K. If a subgroup of order at least q fixes a k-space over $GF(q)$, and if g in $K - GF(q)$ fixes a $GF(q)$-subspace T, then T is fixed by the $GF(q) \langle g \rangle$-module and hence is fixed by the full group. Then the orbit lengths of the generalized components are either 1 or $q+1$. In this case, we obtain a retraction group and an associated subgeometry partition. If a given orbit of length $q+1$*

arises from a z-component, then there is a subgeometry isomorphic to
$PG(z - 1, q)$ *in the associated projective space.*

PROOF. By the work of the author on quasi-subgeomety partition
[**106**], it is quite clear that we obtain a quasi-subgeometry partition, so
we need only show that we obtain a subgeometry partition. However,
the point-orbits of length $q^2 - 1$ in a z-component orbit of length $q + 1$
form a subplane covered net. In this setting, the subplanes of order
q become points and the orbits of the z-component of order q^z form
subspaces isomorphic to $PG(z-1, q)$ of the associated projective space.

□

We now consider the right and left homology groups of focal-spreads.
Suppose we choose a focal-spread of dimension $t + k$ over $GF(q)$, with
focus of dimension t, which admits also right and left 'homology groups'
of order $q - 1$. Let q be a square h^2. Then consider the vector space of
dimension $2(t + k)$ over $GF(h)$. Form the group with elements:

$$\begin{bmatrix} \alpha I_k & 0 \\ 0 & \beta I_t \end{bmatrix}; \alpha, \beta \in GF(h^2)^*.$$

Choose a subgroup, where $\beta = \alpha^h$. Note that when $\alpha \in GF(h)$, we
would have the scalar group, which implies that the orbit length is
either 1 or at most $h + 1$ and hence exactly $h + 1$ in this last case by a
simple argument.

Therefore, from any orbit of length $h + 1$ of a k-dimensional subspace, we obtain a subgeometry isomorphic to $PG(k - 1, h)$. We then
will obtain a retraction group of order $h^2 - 1$, which constructs a subgeometry partition of $PG(t + k - 1, h^2)$ by subgeometries with one
$PG(t - 1, h^2)$, at least one $PG(k - 1, h^2)$, and various $PG(k - 1, h)$'s.

Suppose that there are no $PG(k - 1, h)$'s. Then the full group of
order $(h^2 - 1)^2$ fixes each k-component. Consider a general component
$y = xM$, where x is a k-vector, M is a $k \times t$-matrix, and y is a t-vector.
Then $y = xM$ is mapped to $y = x\alpha^{-1}I_k M \beta I_t = xM\alpha^{-1}\beta I_t$, for all
M. If we allow originally that one M has 1's in the (i, i), entries and
0's in the others, then $\alpha = \beta$, a contradiction. Therefore, we have the
following result, as the converse follows immediately from the results
on quasi-subgeometries.

THEOREM 26. *Any right and left focal-spread of dimension $t + k$
over $GF(q)$, with $q = h^2$ and admitting left and right homology groups
isomorphic to $GF(q)^*$ produces a subgeometry partition of $PG(t + k -
1, h^2)$ by subgeometries isomorphic to one $PG(t - 1, h^2)$, at least one
$PG(k - 1, h^2)$'s and at least one $PG(k - 1, h)$, and no other types.*

Conversely, any projective space $PG(t + k - 1, h^2)$ that admits a subgeometry partition by one $PG(t - 1, h^2)$, at least one $PG(k - 1, h^2)$ and at least one $PG(k - 1, h)$, and no other types 'lifts' to a focal-spread of dimension $2(t + k)$ over $GF(h)$, with focus of dimension $2t$, with group G of order $h^2 - 1$ that fixes the focus and one other $2k$-component and has at least one orbit of length $h + 1$.

To note some examples that have groups required in the above theorem, choose any translation plane of order q^t and kernel containing $GF(q)$ that has symmetric homology groups of order $q - 1$ with axes $x = 0$, and/or $y = 0$. It follows essentially immediately that any k-cut focal-spread of type (t, k) will admit these groups. If $q = h^2$, then we may construct subgeometry partitions as in the previous theorem. The typical types of planes that have such groups are semifield planes, and nearfield planes. The following section gives a different slant on the use of right and left groups by constructing hyperplanes that have this property. That is, it might be possible to use hyperplanes in the manner that we have seen previously in the study of focal-spreads to construct associated subgeometry partitions.

4. Hyperplane Constructions

Suppose we have a focal-spread admitting a left and right homology group, each isomorphic to $GF(q)^*$, choose any k-dimensional $GF(q)$-space on $x = 0$ and the k-dimensional $GF(q)$-space on $y = 0$ and generate the hyperplane H.

It follows immediately that the hyperplane admits such groups as collineation groups. Suppose we choose a coordinate structure so that H contains the $y = x$ k-space and if we agree to do this originally, we find a hyperplane that has one orbit of k-subspaces of length $h + 1$. What this means is that we have found a hyperplane that admits both homology groups and there is a set of $q - 1$ components that are permuted non-trivially into orbits of lengths $h + 1$ and 1.

Here is the resulting subgeometry structure in the focal-spread: The orbits of length 1 will produce subgeometries isomorphic to $PG(k - 1, h^2)$, the orbits of $h + 1$ k-components will produce subgeometries isomorphic to $PG(k - 1, h^2)$, and there are a total of $q + 1$ of these subgeometries.

Now we utilize the induced generalized cut-like subgeometry from the hyperplane: From the intersections of dimension $k-1$ of H with the partial Sperner space, we obtain $PG(k-2, h^2)$'s and $PG(k-2, h)$'s. So, we obtain a subgeometry partition of $PG(2k - 1, h^2)$ by subgeometries isomorphic to $PG(k-1, h^2)$, $PG(k-1, h)$, $PG(k-2, h^2)$, and $PG(k-2, h)$'s. Hence, we obtain:

THEOREM 27. *Any focal-spread of type $(k+1, k)$ over $GF(q)$, with $q = h^2$ and admitting left and right homology groups isomorphic to $GF(q)$ produces a generalized k-cut subgeometry partition of $PG(2k-1, h^2)$ by subgeometries isomorphic to: $PG(k-1, h^2)$, $PG(k-1, h)$, $PG(k-2, h^2)$, and $PG(k-2, h)$, by the use of any hyperplane sharing a k-space with the focus.*

4.1. Using Desarguesian Planes. While it might seem that the subgeometries arising from k-cuts and hyperplane constructions originating from a Desarguesian plane would be fairly uninteresting, actually, the subgeometry partitions are quite diverse.

So, assume that we have a focal-spread of dimension $2k + 1$ with focus of dimension $k + 1$ that arises from a Desarguesian $(k + 1)$-spread. The homology group with elements $(x, y) \to (xm, ym)$, where $m \in GF(q)^*$ acts on the focal-spread. If we choose the focus to be represented by $y = 0$, then the homology group with elements $(x, y) \to (xm, y)$ and $m \in GF(q)^*$, is a collineation of the associated k-cut. The group fixes $x = 0$ pointwise, and so induces a collineation group on the focal-spread generated by any k-subspace of $x = 0$ and $y = 0$. Furthermore, the group $(x, y) \to (x, y\alpha)$, for all $\alpha \in GF(q)^*$, is clearly a collineation group of the focal-spread. In this setting, take the group G of order $(q-1)^2$ generated by the two types of affine homology group collineations for $m \in GF(q)$.

In the associated Desarguesian plane, a given component not equal to $x = 0$ or $y = 0$ is fixed by a subgroup of order $q^{k+1} - 1$. The group of order $(q-1)^2$ is isomorphic to $Z_{q-1} \times Z_{q-1}$, which means that the intersection of G with a cyclic subgroup of order $q^{k+1} - 1$ is exactly of order $q - 1$. Hence, the given group has orbits of length $q - 1$. In the focal-spread, G will act as a collineation group fixing precisely $x = 0$ and $y = 0$ and having $(q^{k+1} - 1)/(q - 1)$ orbits of length $q - 1$. Fix a k-space on $y = 0$, which is then fixed by G, and take the hyperplane H generated by this k-space and $x = 0$, so G acts on this hyperplane. Furthermore, G has the remaining orbits of length $q - 1$, which means that one of these orbits is the remaining set of $q - 1$ k-components of intersection of H with the focal-spread k-components.

Now we apply the results of the previous section and use hyperplane constructions as in Theorem 27. Now consider the field group T with elements $(x, y) \to (x\alpha, y\alpha^h); \alpha \in GF(q)^*$, for $q = h^2$, which is a subgroup of G. The non-trivial orbit lengths of this group are then $(q - 1)/(h - 1) = h + 1$. What this means is that with retraction, we may construct a subgeometry partition of $PG(2k - 1, h^2)$ by two $PG(k - 1, h^2)$'s, $h + 1$, $PG(k - 1, h)$'s, and $(h^{2(k+1)} - h^2)/(h + 1)$ $PG(k - 2, h)$'s. Note that the same argument is valid for nearfield planes of order q^{k+1} with kernels containing $GF(q)$.

THEOREM 28. *Any k-cut focal-spread of type $(k + 1, k)$ from a Desarguesian $(k + 1)$-spread or from a nearfield plane of order q^{k+1} with kernel containing $GF(q)$ with $q = h^2$ produces a subgeometry partition of $PG(2k - 1, h^2)$ by subgeometries isomorphic to (two) $PG(k - 1, h^2)$'s, $h + 1$ $PG(k - 1, h)$'s, and $((h^{2(k+1)} - h^2)/(h + 1))$ $PG(k - 2, h)$'s, using any hyperplane that shares a k-space with the focus.*

Coming from the subgeometry point of view, it is a difficult problem to determine when numbers and types of subgeometries of a partition classify the type of spread (or vector space partition) or translation plane that could arise.

There is actually much more than can be said relating focal-spreads and subgeometries, but most of the theory requires material on Baer groups and algebraic lifting. So, we shall return to this theory in Part 14 after the requisite background is established.

We turn now to subgeometry partitions that can arise from extended André spreads.

CHAPTER 7

Extended André Subgeometries

In Chapter 3, we developed the notion of an extended André spread, and we noted there were normally nice groups attached with such partitions. Since subgeometry partitions arise directly due to certain fixed-point-free groups, it certainly is possible to construct subgeometry partitions from extended André spreads. There is really an embarrassing wealth of wonderful subgeometry partitions that can be generated and we sketch only a few of these in this text.

Let Σ be a Desarguesian r-spread of order q^{sn}. We may regard Σ as an rsn-vector space over $GF(q)$. Moreover, since the kernel homology group K_s determines a field $K_s \cup \{0\} = K_s^+$, we may also regard Σ as an rn-vector space over K_s^+: Letting

$$\tau_d \quad : \quad (x_1, x_2, .., x_r) \longmapsto (x_1 d, x_2 d, .., x_r d);$$
$$d \quad \in \quad GF(q^s), \text{ with } \tau_0, \text{ the zero mapping,}$$

then $K_s^+ = \langle \tau_d; d \in GF(q^s) \rangle$. Define $\tau_d v = d \odot v = (x_1 d, x_2 d, .., x_r d)$, for $v = (x_1, x_2, .., x_r)$, then clearly Σ is an rn-dimensional K_s^+-space. Let the lattice of subspaces be denoted by $PG(rn - 1, q^s)$. Then Σ is an rn-dimensional subspace over K_s^+ and the lattice of vector subspaces forms a projective space $PG(rn - 1, q^s)$. The Main Theorem on Extended André Subgeometries is as follows:

THEOREM 29. *Let the lattice of vector subspaces of Σ over K_s^+ be denoted by $PG(rn - 1, q^s)$. Take any*

$$A_{(n_1,..,n_{r-j-1})}^{(\lambda_1,..,\lambda_{r-j-1})}$$
$$= \left\{ y = (x_1^{*q^{\lambda_1}} n_1 d^{1-q^{\lambda_1}}, .., x_1^{*q^{\lambda_{r-j-1}}} n_{r-j-1} d^{1-q^{\lambda_{r-j-1}}}); d \in GF(q^{sn})^* \right\}.$$

Again, there are

$$\left((q^{sn} - 1)/(q^{(\lambda_1,..,\lambda_{r-j-1})} - 1) \right) / \left((q^s - 1)/(q^{(\lambda_1,\lambda_2,..,\lambda_{r-j-1},s)} - 1) \right)$$

component orbits of length

$$(q^s - 1)/(q^{(\lambda_1,\lambda_2,..,\lambda_{r-j-1},s)} - 1)$$

under K_s.

91

Let $s^* = (\lambda_1, \lambda_2, .., \lambda_{r-j-s})$. Assume that within $(\lambda_1, \lambda_2, ..., \lambda_{r-j-1})$, there are elements λ_l and λ_k such that $\lambda_l = s^*\rho$ and $\lambda_k = s^*\rho(s-1)$. If $s = 2$, λ_l and λ_k are equal, and it is then possible that $r - j - 1 = 1$. Each such orbit becomes a projective subgeometry of $PG(rn - 1, q^s)$, isomorphic to

$$PG(sn/(\lambda_1, \lambda_2, .., \lambda_{r-j-1}, s) - 1, q^{(\lambda_1, \lambda_2, .., \lambda_{r-j-1}, s)}),$$

for $r - j - 1 \geq 2$, provided $s > 2$ and $r - j - 1 \geq 1$ for $s = 2$.

PROOF. From what we have developed previously in this text, we do obtain a quasi-subgeometry from the group, so it remains to prove that we actually obtain a subgeometry. The proof is a general version of the discussion given in the primer on quasi-subgeometries, so the reader who has read Subsection 0.1 may skip the following proof.

Here is the general argument. Consider

$$L : y = (x^{q^{\lambda_1}} m_1, x^{q^{\lambda_2}} m_2 ...)$$

in the K_s-orbit. Choose two vectors v_1 and v_2 that lie in the union of the subspaces in the orbit and form $\langle v_1, v_2 \rangle_{K_s^+}$, the 2-dimensional K_s^+-subspace, assuming that $\{v_1, v_2\}$ is linearly independent over K_s^+. Since each 1-dimensional K_s^+-subspace non-trivially intersects L, we may assume that v_1 and v_2 lie on L. Hence, let

$$v_i = (x_i, x_i^{q^{\lambda_1}} m_1, x_i^{q^{\lambda_2}} m_2 ...), \text{ for } i = 1, 2.$$

To show that we obtain a subgeometry, we need to show that the intersection of $\langle v_1, v_2 \rangle_{K_s^+}$, with 1-dimensional K_s^+-subspaces that lie in the orbit of L may be given via scalars in $GF(q^{(\lambda_1, .., \lambda_{r-j-1}, s)}) = GF(q^{s^*})$. Any intersection of $\langle v_1, v_2 \rangle_{K_s^+}$ contains a vector on L and in that particular 1-dimensional K_s^+-subspace. Hence, assume that

$$\alpha v_1 + \beta v_2 = v_3,$$

where v_3 is also on L. We need to show that α and β are in $GF(q^{s^*})$.

So, let

$$v_i = (x_i, x_i^{q^{\lambda_1}} m_1, x_i^{q^{\lambda_2}} m_2 ...), \text{ for } i = 1, 2, 3.$$

Then

$$\alpha(x_1, x_1^{q^{\lambda_1}} m_1, x_1^{q^{\lambda_2}} m_2 ...) + \beta(x_2, x_2^{q^{\lambda_1}} m_1, x_2^{q^{\lambda_2}} m_2 ...)$$

$$= (x_3, x_3^{q^{\lambda_1}} m_1, x_3^{q^{\lambda_2}} m_2 ...)$$

if and only if

$$(\alpha x_1 + \beta x_2)^{q^{\lambda_i}} m_i = \alpha x_1^{q^{\lambda_i}} m_i + \beta x_1^{q^{\lambda_i}} m_i,$$
$$\text{for all } i = 1, 2, .., r - j - 1.$$

This leads to the following equivalent set of equations:

$$x_1^{q^{\lambda_i}}(\alpha - \alpha^{q^{\lambda_i}}) + x_2^{q^{\lambda_i}}(\beta - \beta^{q^{\lambda_i}})$$
$$= 0, \text{ for all } i = 1, 2, .., r - j - 1.$$

We now restrict to the two equations (or one if $s = 2$) for $i = l$ and k:

$$(*) : x_1^{q^{s*}\rho}(\alpha - \alpha^{q^{s*}\rho}) + x_2^{q^{s*}\rho}(\beta - \beta^{q^{s*}}) = 0,$$

$$(**) : x_1^{q^{s*}\rho(s-1)}(\alpha - \alpha^{q^{s*}\rho(s-1)}) + x_2^{q^{s*}\rho(S-1)}(\beta - \beta^{s*\rho(s-1)}) = 0.$$

Note that $\alpha, \beta \in GF(q^{s*})$ if and only if $\alpha - \alpha^{q^{s*}\rho} = \beta - \beta^{q^{s*}\rho} = 0$. Moreover, the previous two equations show that $\alpha - \alpha^{q^{s*}\rho} = 0$ if and only if $\beta - \beta^{q^{s*}\rho} = 0$. Since this is what we would like to prove, assume otherwise, that $\alpha - \alpha^{q^{s*}} \neq 0$. Then,

$$(*)' : \left(\frac{x_1}{x_2}\right)^{q^{s*}\rho} = \frac{\beta^{q^{s*}\rho} - \beta}{\alpha - \alpha^{q^{s*}\rho}}$$

and

$$(**)' : \left(\frac{x_1}{x_2}\right)^{q^{s*}(s-1)} = \frac{\beta^{q^{s*}(s-1)} - \beta}{\alpha - \alpha^{q^{s*}(s-1)}}.$$

From $(*)'$, we obtain

$$\left(\left(\frac{x_1}{x_2}\right)^{q^{s*}\rho}\right)^{q^{s*}\rho(s-1)} = \left(\frac{\beta^{q^{s*}\rho} - \beta}{\alpha - \alpha^{q^{s*}\rho}}\right)^{q^{s*}\rho(s-1)} = \left(\frac{\beta^{q^{s*}\rho s} - \beta^{q^{s*}\rho(s-1)}}{\alpha^{q^{s*}\rho(s-1)} - \alpha^{q^{s*}\rho s}}\right)$$

$$= \left(\frac{\beta - \beta^{q^{s*}\rho(s-1)}}{\alpha^{q^{s*}\rho(s-1)} - \alpha}\right) = \left(\frac{\beta^{q^{s*}\rho(s-1)} - \beta}{\alpha - \alpha^{q^{s*}\rho(s-1)}}\right).$$

That is,

$$(***)' : \left(\frac{x_1}{x_2}\right)^{q^{s*}\rho s} = \left(\frac{\beta^{q^{s*}\rho(s-1)} - \beta}{\alpha - \alpha^{q^{s*}\rho(s-1)}}\right).$$

Using, $(**)'$, we have

$$\left(\frac{x_1}{x_2}\right)^{q^{s*}\rho s} = \left(\frac{x_1}{x_2}\right)^{q^{s*}\rho(s-1)},$$

or equivalently,

$$\left(\frac{x_1}{x_2}\right)^{q^{s*}\rho s s* \rho - q^{s*}\rho(s-1)} = 1 = \left(\frac{x_1}{x_2}\right)^{q^{s*}\rho(s-1)(q^{s*}\rho-1)}.$$

But, this says that

$$\left(\frac{x_1}{x_2}\right)^{q^{s^*}\rho(s-1)} \in GF(q^{s^*}),$$

which, in turn, implies that

$$\left(\frac{x_1}{x_2}\right) \in GF(q^{s^*}).$$

But we know that

$$(*)' : \left(\frac{x_1}{x_2}\right) = \left(\frac{x_1}{x_2}\right)^{q^{s^*}\rho} = \frac{\beta^{q^{s^*}\rho} - \beta}{\alpha - \alpha^{q^{s^*}\rho}},$$

and hence

$$\left(\frac{x_1}{x_2}\right)^{q^{\lambda_i}} = \frac{\beta^{q^{s^*}\rho} - \beta}{\alpha - \alpha^{q^{s^*}\rho}}, \text{ for all } i,$$

which is equivalent to

$$(*)^+ \; : \; x_1^{q^{\lambda_i}}(\alpha - \alpha^{q^{s^*}\rho}) + x_2^{q^{\lambda_i}}(\beta - \beta^{q^{s^*}\rho})$$
$$= \; 0, \text{ for all } i = 0, 1, .., r - j - 1.$$

Now consider

$$(\alpha - \alpha^{q^{s^*}\rho})(x_1, x_1^{q^{\lambda_1}} m_1, x_1^{q^{\lambda_2}} m_2...) +$$
$$(\beta - \beta^{q^{s^*}\rho})(x_2, x_2^{q^{\lambda_1}} m_1, x_2^{q^{\lambda_2}} m_2...)$$

and note that we obtain 0, in each coordinate: Hence,

$$(\alpha - \alpha^{q^{s^*}\rho})(x_1, x_1^{q^{\lambda_1}} m_1, x_1^{q^{\lambda_2}} m_2...)$$
$$+(\beta - \beta^{q^{s^*}\rho})(x_2, x_2^{q^{\lambda_1}} m_1, x_2^{q^{\lambda_2}} m_2...)$$
$$= \; (0, 0, 0, ...0).$$

Since the two vectors are linearly independent over $GF(q^s)$, it follows that

$$(\alpha - \alpha^{q^{s^*}\rho}) = 0 = (\beta - \beta^{q^{s^*}\rho}),$$

a contradiction, and hence we have the proof that we obtain a subgeometry in this situation. □

Using the previous theorem, we give a method to find subgeometry partitions. We are assuming in the following theorem that we are within an extended André spread or constructing one using the extended André replacements.

THEOREM 30. *Let D_s denote the set of divisors of s (including 1 and s). When a replacement set of $(q^{sn} - 1)/(q^{s^*} - 1)$ sn-spaces is obtained for $s^* \in D_s$, let k_{s^*} denote the number of different and mutually disjoint replacement sets of $(q^{sn} - 1)/(q^{s^*} - 1)$ sn-spaces (k_{s^*} could be 0). Then we merely require that*

$$\sum_{s^* \in D_s} \left(\frac{q^{sn} - 1}{q^{s^*} - 1} \right) k_{s^*} = (q^{sn} - 1)^{r-j-1}.$$

Now we can actually do this for each of the $\binom{r}{r-j}$ $j - (0$-subsets). Let $k_{s^*,j,w}$ be the number of different and mutually disjoint replacement sets of $(q^{sn} - 1)/(q^{s^*} - 1)$ sn-spaces in $\Sigma_{j,w}$. Then, considering the $j - (0$-sets), for $r - j \geq 3$, we require

$$\sum_{j=0}^{r-3} \sum_{w=1}^{\binom{r}{r-j}} \sum_{s^* \in D_s} (\frac{q^{sn} - 1}{q^{d^*} - 1}) k_{d^*}$$

$$= (q^{rsn} - 1)/(q^{sn} - 1) - \binom{r}{r-2}(q^{sn} - 1) - \binom{r}{r-1}.$$

Let $s^* = \gcd(\lambda_1, \lambda_2, .., \lambda_{r-j-1}, s)$. Assume that within $(\lambda_1, \lambda_2, ..., \lambda_{r-j-1})$, there are elements λ_l and λ_k such that $\lambda_l = s^* \rho$ and $\lambda_k = s^* \rho(s - 1)$. If $s = 2$, λ_l and λ_k are equal, and it is then possible that $r - j - 1 = 1$.

Then each orbit under the kernel group of order $q^s - 1$ becomes a projective subgeometry of $PG(rn - 1, q^s)$, isomorphic to

$$PG(sn/(\lambda_1, \lambda_2, .., \lambda_{r-j-1}, s) - 1, q^{(\lambda_1, \lambda_2, .., \lambda_{r-j-1}, s)}),$$

for $r - j - 1 \geq 2$, provided $s > 2$ and $r - j - 1 \geq 1$ for $s = 2$.

Therefore, the associated $PG(rn - 1, q^s)$ is partitioned by subgeometries as follows: There are

$$\sum_{j=0}^{r-1} \sum_{w=1}^{\binom{r}{r-j}} k_{s^*,j,w}$$

subgeometries isomorphic to $PG(sn/s^* - 1, q^{s^*})$ (the $j = (r - 1) - (0$-subsets) always lead to $\binom{r}{r-1} = r$ $PG(n - 1, q^s)$'s).

The reader is also directed to Johnson [98] for an algorithmic approach to the selection of subgeometries within the partition.

0.2. The Ebert-Mellinger r-(rn,q)-Spreads.

Ebert and Mellinger [51] construct a new $r - (rn, q)$-spread admitting an Abelian group of order $(q^{rn} - 1)^2$ that may be constructed with the methods of the previous theorem. The construction in Ebert and Mellinger begins with the construction of new subgeometry partitions in $PG(rn - 1, q^r)$ by subgeometries isomorphic to $PG(rn - 1, q)$ and $PG(n - 1, q^r)$. The following connection shows that these subgeometries may be obtained from the above theorem.

THEOREM 31. *(Johnson [98]) The $r - (rn, q)$-spreads of Ebert and Mellinger are extended André spreads. When $r = 2$, the spreads correspond to André planes of order q^{2n}.*

So, there is a group of order $q^r - 1$ that arises from the Ebert-Mellinger subgeometry partitions. Unfolding the fan(s) shows that the $PG(rn - 1, q)$'s unfold to an orbit of rn-dimensional vector subspaces over $GF(q)$ of length $(q^r - 1)/(q - 1)$, and the $PG(n - 1, q^r)$'s unfold to rn-dimensional vector subspaces over G. Now begin with a Desarguesian $r - (rn, q)$-spread and construct the generalized André type covers,

$$A_{(n_1,..,n_{r-j-1})}^{(\lambda_1,..,\lambda_{r-j-1})}$$

$$= \left\{ y = (x_1^{*q^{\lambda_1}} n_1 d^{1-q^{\lambda_1}}, .., x_1^{*q^{\lambda_{r-j-1}}} n_{r-j-1} d^{1-q^{\lambda_{r-j-1}}}); d \in GF(q^{sn})^* \right\},$$

where we only use the $j = 0$-(0-sets), and where we take $\lambda_1 = 1, \lambda_2 = 2, .., \lambda_{r-1} = r - 1$.

Therefore, the replacement sets 'all' have the general form:

$$A_{(n_1,..,n_{r-1})}^{(1,2,..,r-1)}$$

$$= \left\{ y = (x^q n_1 d^{1-q}, .., x^{q^{r-1}} n_{r-11} d^{1-q^{r-j-1}}); d \in GF(q^{rn})^* \right\}.$$

Furthermore, the other $j - (0\text{-sets})$ are not replaced. In particular, we recall our main theorem on subgeometries.

In this setting, we have $s = r$ and $j = 0$, $\lambda_i = i$, so

$$(q^s - 1)/(q^{(\lambda_1,\lambda_2,..,\lambda_{r-j-1},s)} - 1) = (q^r - 1)/(q - 1).$$

So, we obtain a subgeometry partition of $PG(rn - 1, q^r)$ by $PG(rn - 1, q)$'s and $PG(n - 1, q^r)'s$. These are the subgeometry partitions constructed by Ebert and Mellinger. This also means that since the group arises from a Desarguesian $r - (rn, q)$-spread, all remaining components of the spread are fixed by the group.

The reader is also directed back to Theorem 242 for more of the details. But, note that by making different replacements of the indicated (generalized) extended André spreads and also by making replacements in other $j - (0\text{-sets})$, a great variety of subgeometry partitions may be obtained, with varying and many different types of subgeometries.

CHAPTER 8

Kantor's Flag-Transitive Designs

In this chapter, constructions of Kantor [156] are given of transitive t-spreads that admit a retraction group that construct interesting subgeometry partitions. Actually, these t-spreads also produce flag-transitive designs. Although there are variations on the types given, there are four main classes of flag-transitive designs. In this chapter, we shall discuss the two flag-transitive designs that admit cyclic transitive linear subgroups. In the next chapters, the partitions that admit cyclic transitive t-spreads are then used to construct new subgeometry partitions. The reader is directed to Kantor [156] for a description of the other two main classes. These classes admit a transitive linear group on the associated t-spread that contains a cyclic subgroup of index 2.

1. Kantor's Class I

The following construction of Kantor [156] of flag-transitive designs shall be called 'Kantor's Class I.' The part of the flag-transitive group that we shall be interested in is the linear part in $GL(V, q)$, which defines the associated t-spreads of the designs, whereas the full group is in $AG(V, q)$.

THEOREM 32. *Let* $m, n > 1$, *such that* $(m, n) = 1$. *Let* m *divide* $q - 1$. *Let* $b \in GF(q^{mn})^*$ *such that* $b^{q^n} = b\omega$, *where* ω *has order* m. *Let* σ *be any non-identity automorphism of* $GF(q^n)$ *fixing* $GF(q)$ *pointwise and let* $h(x) = x - bx^\sigma$ *for* x *in* $GF(q^{mn})$. *Let* $\langle s \rangle$ *denote the cyclic subgroup of* $GF(q^{mn})$ *of order* $(q^{mn} - 1)(q - 1)/(q^n - 1)$. *Then*

$$\mathcal{S}_{b,\sigma} = \left\{ s^i h(GF(q^n)); i = 0, 1, .., (q^{mn} - 1)/(q^n - 1) \right\}$$

defines a flag-transitive design, which also gives an $m - (n, q)$-*spread that admits a cyclic transitive group on the* $(q^{mn} - 1)/(q^n - 1)$ *components (n-spaces over* $GF(q)$).

The flag-transitive group of the design is given by

$$\langle z \to fz + w; w \in GF(q^{mn}) \text{ and } f \in \langle s \rangle \rangle.$$

PROOF. We have an mn-dimensional vector space $V = GF(q^{mn})$ over $GF(q)$, and we wish to show that $\mathcal{S}_{b,\sigma}$ is a set of n-dimensional subspaces, whose non-zero vectors partition V. Since the group $\langle s \rangle$ is clearly transitive, we need only show that we do obtain a set of n-subspaces and that $h(GF(q^n)) \cap s^i h(GF(q^n)) = 0$. By transitivity, we claim that $h(x) = x - bx^\sigma$ is a subspace, since the set is additive, we need only show that $h(x) = 0$ implies that $x = 0$. But, $h(x) = 0$ implies, for $x \neq 0$, that $b \in GF(q^n)$, so that $b^{q^n} = b\omega$ implies $\omega = 1$. Therefore, $h(GF(q^n))$ is a subspace of dimension n over $GF(q)$. Now assume that $h(x) = s^i h(y) \in h(GF(q^n)) \cap s^i h(GF(q^n))$, for $x, y \in GF(q^n)^*$ and $0 < i < (q^{mn} - 1)/(q^n - 1)$.

The idea of the proof showing that this leads to a contradiction is to apply the norm mapping N from $GF(q^{mn})$ to $GF(q^n)$. So, we note that $x^m - b^m = \prod_{i=0}^{m-1} (x - b^{q^{ni}})$, which is valid since both polynomials have the roots $\omega^i b$, for all $i = 1, 2, .., m$, noting that $b^{q^{ni}} = \omega^i b$. So,

$$N(x - bx^\sigma) = \prod_{i=0}^{m-1} (x - bx^\sigma)^{q^{ni}} = \prod_{i=0}^{m-1} (x - b^{q^{ni}} x^\sigma) = x^m - b^m x^{\sigma m},$$

which is, in turn, equal to

$$\prod_{i=0}^{m-1} (s^i (y - by^\sigma))^{q^{ni}} = N(s^i) \prod_{i=0}^{m-1} (y - b^{q^{ni}} y^\sigma) = N(s^i)(y^m - b^m y^{\sigma m}).$$

We note that $N(s^i) = s^{i(1 + q^n + q^{2n} + ... + q^{n(m-1)})} = s^{i(q^{mn} - 1)}$, so that $N(s^i) \in GF(q)$. We then obtain

$$x^m - N(s^i)y^m = b^m (x^m - N(s^i)y^m)^\sigma.$$

Assume that the left hand side is not zero and note that m divides $\sigma - 1$, so that

$$1 = b(x^m - N(s^i)y^m)^{(\sigma-1)/m} \in \langle \omega \rangle \subseteq GF(q),$$

which forces b to be in $GF(q^n)$, a contradiction. Hence,

$$\left(\frac{x}{y} \right)^m = N(s^i) \in GF(q),$$

which implies that

$$\left(\frac{x}{y} \right)^{m(q-1)} = 1.$$

We note that

$$(q^n - 1, m(q - 1)) = (q - 1)(1 + q + q^2 + ... + q^{n-1}, m),$$

which is

$$(q-1)(q-1+q^2-1+...+q^{n-1}-1+n,m)$$
$$= (q-1)(n,m) = (q-1).$$

This means that $\frac{x}{y} = \alpha$ is in $GF(q)$, so that

$$h(x) = s^i h(y) = \alpha(y - by^\sigma) = s^i(y - by^\sigma),$$

so that $s^i \in GF(q)$. But, the order of s^i divides $(q^{mn}-1)(q-1)/(q^n-1)$, a contradiction, since $0 < i < (q^{mn}-1)/(q^n-1)$. Therefore, we have an $m-(n,q)$-spread as maintained. □

REMARK 18. *When $m = 2$, an $m-(n,q)$-spread is defined on a $2n$-dimensional vector space over $GF(q)$ and thus defines a flag-transitive translation plane.*

2. Kantor's Class II

Also in Kantor [**156**], there is another class of flag-transitive designs admitting a cyclic transitive linear group.

The description is as follows:

THEOREM 33. *Let p be a prime and let $q > 1$ and $m > 1$ be p-powers. Let n be an integer > 1 such that $(p,n) = 1$. We consider the field $GF(q^{mn})$, and note, in particular, $(m,n) = 1$. Assume that $((q^n-1)/(q-1), m-1) = 1$. Let N_0 denote the norm map from $GF(q^m)$ to $GF(q)$, let T denote the trace map from $GF(q^n)$ to $GF(q)$ and T_0 the trace map from $GF(q^m)$ to $GF(q)$. Choose $r \in GF(q^m) - GF(q)$ so that the polynomial*

$$N_0(x+r) = x^m + T_0(N_0(r)/r)x + N_0(r)$$

holds in $GF(q)[x]$. Let $s \in GF(q^{mn})^*$ have order $(q^{mn}-1)(q-1)/(q^n-1)$. Then

$$S_r = \{s^i(KerT + rGF(q)); 0 \le i \le (q^{mn}-1)/(q^n-1)\}$$

produces a design with flag-transitive group that also gives an $m-(n,q)$-spread admitting a cyclic transitive group on the components. The flag-transitive group of the design is given by

$$\langle z \to gz + w; w \in GF(q^{mn}), g \in \langle s \rangle \rangle.$$

PROOF. We first note that $KerT + rGF(q)$ has $q^{n-1}q = q^n$ elements, since $KerT \cap rGF(q) = \langle 0 \rangle$, as $r \in GF(q^m)$, and $(m,n) = 1$.

Furthermore, clearly $KerT + rGF(q)$ is a $GF(q)$-subspace. So, to show that we obtain an $m - (n, q)$-spread, it suffices to show that

$$(KerT + rGF(q)) \cap s^i(KerT + rGF(q)) = \langle 0 \rangle,$$

if s^i is not a scalar. Note that if $\langle s^j \rangle$ is in $GF(q)$, then we see that such a group acts as a $GF(q)$-scalar group of order $q - 1$, that fixes each component. Hence, assume that $u, v \in KerT$ and $rk, rw \in rGF(q)$ so that

$$u + rk = s^i(v + rw),$$

and s^i is not a scalar. Let N denote the norm function from $GF(q^{mn})$ to $GF(q^n)$, noting that

$$N_0(x + r) = x^m + T_0(N_0(r)/r)x + N_0(r),$$

to obtain

$$
\begin{aligned}
N(u + rk) &= u^m + T_0(N_0(r)/r)uk^{m-1} + N_0(r)k^m \\
&= N(s^i(v + rw)) = N(s^i)N(v + rw) \\
&= N(s^i)(v^m + T_0(N_0(r)/r)vw^{m-1} + N_0(r)w^m).
\end{aligned}
$$

Hence, we have

$$(*) : (u^m - N(s^i)v^m) + T_0(N_0(r)/r)(uk^{m-1} - N(s^i)vw^{m-1}) +$$

$$+ N_0(r)(k^m - N(s^i)w^m) = 0.$$

Note that $N(s^i) \in GF(q)$, and $(u^m - N(s^i))v^m$ and $uk^{m-1} - N(s^i)vw^{m-1}$ are in $KerT$, as $(KerT)^m = KerT$ and $N_0(r) \in GF(q)$. Therefore, the trace T maps the previous expression to $nN_0(r)(k^m - N(s^i)w^m) = 0$. Hence, $(\frac{k}{w})^m \in GF(q)$, so that $k = yw$, where $y^m = N(s^i)$. The expression $(*)$ then becomes

$$(u - yv)^m + T_0(N_0(r)/r)(u - yv)(yw)^{m-1}) = 0.$$

If $u - yv$ is not zero, then

$$(u - yv)^{m-1} = -T_0(N_0(r)/r)(yw)^{m-1}),$$

which implies that $(u - yv)^{(m-1)(q-1)} = 1$. But,

$$
\begin{aligned}
&(q^n - 1, (m-1)(q-1)) \\
&= (q-1)((q^n - 1)/(q-1), m-1) = (q-1).
\end{aligned}
$$

Hence, $u - yv$ is in $GF(q) \cap KerT = \langle 0 \rangle$. This completes the proof. $\square$

REMARK 19. *When $m = 2$, we obtain a $2 - (n, q)$, spread (for q even), which then constructs a flag-transitive translation plane.*

3. m^{th}-Root Subgeometry Partitions

In this section, we shall restrict ourselves to the two of the flag-transitive designs constructed by Kantor, that admit cyclic transitive t-spreads.

As mentioned previously, a Baer subgeometry partition of a finite projective space $PG(n-1, q^2)$ is a partition by subgeometries isomorphic to $PG(n-1, q)$, the so-called 'Baer subgeometries' of $PG(n-1, q^2)$. The known Baer subgeometry partitions arise from flag-transitive translation planes that admit a cyclic group acting transitively on the components of the spread. That is, there is a cyclic collineation group of order $q^n + 1$ acting transitively on the components. When n is odd, there is a group of order $q^2 - 1$ containing the kernel homology group of order $q - 1$ such that are $(q^n + 1)/(q + 1)$ orbits under this group. Therefore, it is possible to 'retract' the translation plane to construct a Baer subgeometry partition of $PG(n-1, q^2)$ by Baer subgeometries, which admits a collineation group transitive on the Baer subgeometries.

In this section, generalizations of Baer subgeometry partitions are constructed. In particular, constructions are given of subgeometry partitions of $PG(n-1, q^m)$ by subgeometries all isomorphic to $PG(n-1, q)$, where $(m, n) = 1$ and m is not always 2. We call such subgeometries m^{th}-root subgeometries and the associated partition an 'm^{th}-root subgeometry partition.'

As we have noted in the previous chapter, constructions of flag-transitive translation planes have been generalized by Kantor [**156**] and, in particular, there are two general classes admitting cyclic groups, that we have called Kantor's classes I and II. From a vector space of dimension mn over $GF(q)$, there are constructions of partitions of the vector space by n-dimensional $GF(q)$-spaces; n-spreads (or $m - (n, q)$-spreads) that admit a cyclic group acting on the 'components' (the n-dimensional vector spaces of the partition). For essentially arbitrary m, we show in the following that we may obtain subgeometry partitions from these spreads in an analogous manner as when $m = 2$, thereby constructing subgeometry partitions of $PG(n-1, q^m)$ by subgeometries isomorphic to $PG(n-1, q)$. The following material is adapted from the author's article [**100**].

4. Subgeometries from Kantor's Class I

We first consider Kantor's class I and the reader is directed back to the previous sections for the description.

REMARK 20. *Note that since* $(q^m - 1, q^n - 1) = (q-1)$, *then* $((q^m - 1)/(q-1), (q^n-1)/(q-1)) = 1$, *and* $((q^m - 1)/(q - 1)) \cdot ((q^n-1)/(q-1))$ *divides* $(q^{mn} - 1)/(q-1)$. *So,* $(q^m - 1)/(q-1)$ *divides* $(q^{mn} - 1)/(q^n - 1)$.

LEMMA 11. *Consider the group* $\langle s^{(q^{mn}-1)/(q^n-1)/((q^m-1)/(q-1))} \rangle$. *Since the original group is cyclic, we have exactly* $(q^{mn} - 1)/(q^n - 1)/((q^m - 1)/(q - 1))$ *orbits of length* $(q^m - 1)/(q - 1)$.
If we let $GF(q^m)^*$ *act on the spread, then this group contains the scalar group of order* $q - 1$. *Furthermore, each spread orbit under* $GF(q^m)^*$ *is a fan of length* $(q^m - 1)/(q - 1)$.

PROOF. Since the order of s is $(q^{mn}-1)(q-1)/(q^n-1)$, we note that the stabilizer of a spread component has order $q-1$, which implies the stabilizer of any component is $GF(q)^*$. Note that for k in $GF(q^m)^*$ then $ks^i h(GF(q^n)) = s^i h(GF(q))$, if and only if $kh(GF(q^n)) = h(GF(q^n))$ and $k \in GF(q)^*$. Furthermore, the group $GF(q^m)^*$ acting on the spread has order dividing $(q^m - 1)/(q - 1)$, and since this group order divides $(q^{mn} - 1)/(q^n - 1)$, it follows that we may regard the space as a vector space over $GF(q^m)$, each orbit under $GF(q^m)^*$ has length $(q^m - 1)/(q - 1)$ and each orbit of this length is a 'fan.' □

LEMMA 12. *We may regard* $GF(q^m)^*$ *as a collineation group of the associated spread (design), each of whose orbits has length* $(q^m-1)/(q-1)$.
Considering the point-orbits of this group as 1-dimensional $GF(q^m)$-*subspaces, we form the corresponding* $PG(n - 1, q^m)$. *Let* Γ *be any component orbit of length* $(q^m - 1)/(q - 1)$.
Then each such orbit becomes a subgeometry isomorphic to $PG(n-1, q)$.

PROOF. Let Γ be a fan (orbit of length $(q^m - 1)/(q - 1)$ under $GF(q^m)^*$). Choose any two point-orbits P and Q under $GF(q^m)^*$ that lie in Γ and form the 2-dimensional $GF(q^m)$-space $\langle P, Q \rangle$. Since Γ is a fan, given any component L of Γ, we may take elements of P and Q to lie on L. Now choose a point orbit, one of whose 'points R' is in $\langle P, Q \rangle$, where the orbit $\langle R \rangle$ contains a point on L that generates the same 1-dimensional $GF(q^m)$-space.
We need only to show that the 2-dimensional $GF(q^m)$-subspace generated by two point-orbits by representatives P and Q in the same component L of the orbit Γ of length $(q^m - 1)/(q - 1)$ intersects Γ in a 2-dimensional $GF(q)$-subspace. That is, we have a quasi-subgeometry at this point.
First, we note that the order of b is $(q^n - 1)m$.

Consider an orbit Γ. Since we are dealing with a group that is transitive on the components, we may, without loss of generality, take the orbit containing $h(GF(q^n)) = L$, and choose two 1-dimensional $GF(q^m)$-subspaces $\langle P \rangle$ and $\langle Q \rangle$ on L, say, $\langle P \rangle = \langle x_1 - bx_1^\sigma \rangle$ and $\langle Q \rangle = \langle x_2 - bx_2^\sigma \rangle$. Therefore, as we may also choose a representative generator from $h(GF(q^n))$, assume that we take the 2-dimensional $GF(q^m)$-space

$$\langle x_1 - bx_1^\sigma, x_2 - bx_2^\sigma \rangle.$$

Consider any 1-dimensional $GF(q^m)$-subspace $\langle R \rangle$ not equal $\langle P \rangle$ or $\langle Q \rangle$ of this 2-space. Since $\langle R \rangle$ lies in Γ, we may assume that it is generated by a point $\langle R \rangle = \langle x_3 - bx_3^\sigma \rangle$ on L. Therefore, we obtain an equation

$$\alpha(x_1 - bx_1^\sigma) + \beta(x_2 - bx_2^\sigma) = x_3 - bx_3^\sigma, \text{ for } \alpha, \beta \in GF(q^m)^*.$$

Since $(m, n) = 1$, if ρ is a generating automorphism of the Galois group of $GF(q^n)$ over $GF(q)$ then $\langle \rho \rangle = \langle \rho^m \rangle$. Then we have

$$(\alpha x_1 + \beta x_2 - x_3) = b(\alpha x_1^\sigma + \beta x_2^\sigma - x_3^\sigma) = b(\alpha x_1 + \beta x_2 - x_3)^{q^{mi}},$$

for some integer i.

This says that either $(\alpha x_1 + \beta x_2 - x_3) = 0$, or b has order dividing $(q^{nm} - 1)/(q^{m(i,n)} - 1)$, which divides $(q^{mn} - 1)/(q^m - 1)$. But we know that b has order $(q^n - 1)m$. Note that since m divides $q - 1$ then m divides $q^{km} - 1$, for any integer k, which implies that $(m, (q^{nm} - 1)/(q - 1)) = (m, n) = 1$. Therefore, $(q^n - 1)m$ cannot divide $(q^{nm} - 1)/(q^m - 1)$. Hence, we must have $(\alpha x_1 + \beta x_2 - x_3) = 0$. This implies that $\alpha x_1 + \beta x_2$ is in $GF(q^n)$, for α, β in $GF(q^m)^*$ and x_1, x_2, x_3 in $GF(q^n)$. Thus, $\alpha + \beta x_2/x_1 \in GF(q^n)$. Let $z = x_2/x_1$. Therefore, we have $\alpha + \beta z \in GF(q^n)$. So $\alpha + \beta z$ and $\alpha + \beta z^{q^m}$ are in $GF(q^n)$, which implies that $\beta(z^{q^m} - z)$ is in $GF(q^n)$.

If $z^{q^m} = z$, then z is in $GF(q^n) \cap GF(q^m) = GF(q)$, but then $x_2/x_1 = z$ is in $GF(q)$. Hence, $x_1 - bx_1^{q^m} = z(x_2 - bx_2^{q^m})$, so that $x_1 - bx_1^\sigma$ and $x_2 - bx_2^\sigma$ are not linearly independent over $GF(q^m)$. Therefore, this shows that β is in $GF(q^n)$, which, in turn, implies that α is in $GF(q^n)$, so that $\alpha, \beta \in GF(q^m) \cap GF(q^n) = GF(q)$. This completes the proof of the lemma.   □

Since the image of a subgeometry is also a subgeometry, we therefore have proved the following theorem.

THEOREM 34. *Let m and $n > 1$ be positive integers such that $(m, n) = 1$ and q a prime power such that m divides $q - 1$. Then a transitive subgeometry partition of*

$$PG(n - 1, q^m) \text{ by } PG(n - 1, q)\text{'s}$$

is obtained from any design of Kantor's class I.

5. Subgeometries from Kantor's Class II

The reader is again directed back to the previous sections for the description of Kantor's class II. We begin by a discussion of the orbit lengths of components under the cyclic group in question.

We note that

$$(q^m - 1)/(q - 1) \text{ divides } (q^{mn} - 1)/(q^n - 1),$$

which implies that, under the group $GF(q^m)^*$, we have a set of

$$((q^{mn} - 1)/(q^n - 1))/((q^m - 1)/(q - 1))$$

orbits of length $(q^m - 1)/(q - 1)$.

Now form the associated $PG(n - 1, q^m)$.

By transitivity, we consider the $GF(q^m)^*$-orbit containing the n-dimensional $GF(q)$-subspace $\{(KerT + rGF(q)\}$. The 'points' are the $GF(q^m)^*$-orbits that lie in Γ.

We now claim that each orbit Γ of length $(q^m - 1)/(q - 1)$ defines a subgeometry isomorphic to $PG(n - 1, q)$.

Again, we always obtain a quasi-subgeometry, so it remains to show the condition on the 2-dimensional subspaces. Specifically, we need to show that the 2-dimensional $GF(q^m)$-subspace $\langle P, Q \rangle$ generated by two of these points P and Q, $GF(q^m)^*$-orbits in Γ then intersects Γ in exactly $q + 1$ 'points.'

We follow the ideas presented in the previous sections. We may assume that Γ contains $L = \{(KerT + rGF(q)\}$, that $\langle P \rangle = \langle x_1 + r\alpha_0 \rangle$, $\langle Q \rangle = \langle x_2 + r\beta_0 \rangle$, for $x_1, x_2 \in KerT$, and $\alpha_0, \beta_0 \in GF(q)$.

Take $\langle x_1 + r\alpha_0, x_2 + r\beta_0 \rangle$. Suppose for $\alpha, \beta \in GF(q^m)$, $\alpha\beta \neq 0$, we have an intersection of $\langle R \rangle$ distinct from $\langle P \rangle$ or $\langle Q \rangle$ on L of Γ, so we may assume that $\langle R \rangle = \langle x_3 + r\gamma_0 \rangle$ and

$$\alpha(x_1 + r\alpha_0) + \beta(x_2 + r\beta_0) = x_3 + r\gamma_0, \text{ for } \alpha, \beta \in GF(q^m)^*.$$

Then

$$(*) : (\alpha x_1 + \beta x_2 - x_3) = r(\gamma_0 - (\alpha\alpha_0 + \beta\beta_0)).$$

Since $(\alpha x_1 + \beta x_2 - x_3)$ and $(\alpha^{q^n} x_1 + \beta^{q^n} x_2 - x_3)$ are both in $GF(q^m)$, we have

$$(\alpha x_1 + \beta x_2 - x_3) - (\alpha^{q^n} x_1 + \beta^{q^n} x_2 - x_3) \in GF(q^m).$$

Therefore, we have $(\alpha - \alpha^{q^n})x_1 + (\beta - \beta^{q^n})x_2 \in GF(q^m)$.

We claim that neither x_1 or x_2 is in $GF(q^m)$. To see this note that, if, say, x_1 is in $GF(q^n) \cap GF(q^m) = GF(q)$, this would imply that $T(x_1) = nx_1 = 0$, but since $(p, n) = 1$, this implies that $x_1 = 0$. But, if $x_1 = 0$, then since $x_1 + r\alpha_0$ and $x_2 + r\beta_0$ are linearly independent over $GF(q^m)$, then x_2 cannot be zero, implying that since $\beta - \beta^{q^n}$ is in $GF(q^m)$, either $\beta - \beta^{q^n} = 0$ or x_2 is in $GF(q^m)$. However, if x_2 is in $GF(q^m)$, this also would imply that $x_2 = 0$. Therefore, $\beta = \beta^{q^n}$.

Hence,

$$\beta \in GF(q^n) \cap GF(q^m) = GF(q), \text{ if } x_1 = 0.$$

We claim that $\beta x_2 - x_3 = 0$. Suppose not! Then the right hand side of $(*)$ is also not 0.

Then $\beta x_2 - x_3$ is in $GF(q^n)$, and r is not in $GF(q^n)$, so $\beta x_2 - x_3$ is in $GF(q^m)$, implying that $\beta x_2 - x_3 = \varepsilon_0 \in GF(q)$. But then since the trace of the left hand side is 0 and the trace of the right hand side is $n\varepsilon_0$, this implies that $\varepsilon_0 = 0$.

Therefore, $\beta x_2 = x_3$, which implies that $\beta \in GF(q^m) \cap GF(q^n) = GF(q)$. Therefore, $\langle x_3 + r\gamma_0 \rangle = \left\langle x_2 + r\frac{\gamma_0}{\beta} \right\rangle$.

It is assumed that the 1-dimensional $GF(q^m)$-space $\langle x_3 + r\gamma_0 \rangle$ is not either $\langle x_1 + r\alpha_0 \rangle$ or $\langle x_2 + r\beta_0 \rangle$. Therefore,

$$\langle x_1 + r\alpha_0, x_2 + r\beta_0 \rangle = \langle x_3 + r\gamma_0, x_2 + r\beta_0 \rangle = \langle \beta x_2 + r\gamma_0, x_2 + r\beta_0 \rangle,$$

for $\beta \in GF(q)$, $\beta \neq 0$. But, if we take for $\alpha_1, \beta_1 \in GF(q)$, $\alpha_1 \neq \beta_1$, the 2-subspace

$$\langle x_2 + r\alpha_1, x_2 + r\beta_1 \rangle_{GF(q^m)} \cap \Gamma$$

and, if for $\alpha^*, \beta^* \in GF(q^m)$,

$$\alpha^*(x_2 + r\alpha_1) + \beta^*(x_2 + r\beta_1) = x_4 + r\gamma_1,$$

we would obtain

$$(**): (\alpha^* + \beta^*)x_2 - x_4 = r(\gamma_1 - (\alpha^*\alpha_1 + \beta^*\beta_1)).$$

Since the right hand side is in $GF(q^m)$, then so is

$$((\alpha^* + \beta^*)x_2 - x_4) - (\alpha^* + \beta^*)^{q^n} x_2 - x_4.$$

Therefore, we have that $(\alpha^* + \beta^*)^{q^n} = (\alpha^* + \beta^*)$, which, in turn, implies that $(\alpha^* + \beta^*) \in GF(q)$. As the left hand side has trace 0 and only the element 0 of $GF(q)$ has trace 0, it follows that $x_4 = (\alpha^* + \beta^*)x_2$, and $r(\gamma_1 - (\alpha^*\alpha_1 + \beta^*\beta_1)) = 0$, so that $(\gamma_1 - (\alpha^*\alpha_1 + \beta^*\beta_1)) = 0$, where $\alpha_1, \beta_1, \gamma_1 \in GF(q)$.

Recall that either α_1 or β_1 is non-zero since the subspace

$$\langle x_2 + r\alpha_1, x_2 + r\beta_1 \rangle$$

is 2-dimensional and, of course, $\alpha_1 \neq \beta_1$. Therefore, assume α_1 is not zero so that

$$(\alpha^* + \beta^*(\beta_1/\alpha_1)) \in GF(q).$$

But since $(\alpha^* + \beta^*) \in GF(q)$, this forces $\beta^*(\beta_1/\alpha_1 - 1) \in GF(q)$. Since $\beta_1/\alpha_1 - 1$ is non-zero, we have $\beta^* \in GF(q)$, which implies that $\alpha^* \in GF(q)$.

Hence, in particular, we have proved the following two lemmas:

LEMMA 13. *If x_1 or x_2 is zero (but not both), then*

$$\langle x_1 + r\alpha_0, x_2 + r\beta_0 \rangle_{GF(q^m)} \cap \Gamma$$

consists of exactly $q + 1$ $GF(q^m)$-subspaces.

LEMMA 14. *If x_2 is non-zero and $\alpha_0 \neq \beta_0$, then*

$$\langle x_2 + r\alpha_0, x_2 + r\beta_0 \rangle_{GF(q^m)} \cap \Gamma$$

consists of exactly $q + 1$ $GF(q^m)$-subspaces. Using the previous analysis and these lemmas, we now prove that subgeometries are obtained from orbits.

LEMMA 15. *Each orbit of components Γ of length $(q^m - 1)/(q - 1)$ defines a subgeometry isomorphic to $PG(n - 1, q)$.*

PROOF. So, consider

$$\langle x_1 + r\alpha_0, x_2 + r\beta_0 \rangle_{GF(q^m)} \cap \Gamma.$$

By Lemmas 13 and 14, we may assume that $x_1 x_2 \neq 0$ and $x_1 \neq \rho x_2$, for $\rho \in GF(q)$. Now come back to equation

$$(*) : (\alpha x_1 + \beta x_2 - x_3) = r(\gamma_0 - (\alpha \alpha_0 + \beta \beta_0)),$$

and then

$$(\alpha - \alpha^{q^n}) x_1 + (\beta - \beta^{q^n}) x_2 \in GF(q^m).$$

Ideas similar to the previous general analysis apply to finish the proof, recalling that we already have a quasi-subgeometry. These details are left to the reader to complete.   □

Hence, we have proved the following theorem.

THEOREM 35. *Let $q = p^r$, for p a prime. Let m be a power of p and $n > 1$ a positive integer such that $(n, p) = 1$. Then a transitive subgeometry partition of*

$$PG(n - 1, q^m) \text{ by } PG(n - 1, q)\text{'s}$$

is obtained from any design of Kantor's class II.

5.1. Additional Constructions. Consider one of the Kantor designs of class I or II. There is an associated $m - (n, q)$-spread admitting a cyclic group acting transitively on the spread, where $(m, n) = 1$, producing an mth root subgeometry partition of $PG(n - 1, q^m)$ by $PG(n - 1, q)$'s.

Hence, we have a group of order $(q^m - 1)$ with orbits of lengths $(q^m - 1)/(q - 1)$. In such a situation, we have $GF(q^m)^*$ acting as a collineation group of the spread S. Let ℓ be a divisor of m. Consider $GF(q^\ell)^*$ acting on S and let Γ^* be a $GF(q^m)^*$-orbit. Then, we have within Γ^* exactly $(q^m - 1)/(q^\ell - 1)$ orbits under $GF(q^\ell)^*$ of length $(q^\ell - 1)/(q - 1)$. Since $(\ell, n) = 1$, our arguments establishing that a 2-dimensional $GF(q^m)$-space generated by two $GF(q^m)^*$-orbits also applies with ℓ replacing m and shows that the intersection with Γ^* has exactly $q + 1$ $GF(q^\ell)$-spaces. In this way, a given orbit under $GF(q^m)^*$ produces subgeometries of $PG(mn/\ell - 1, q^\ell)$ by $PG(n - 1, q)$'s. Note that now $(mn/\ell, \ell)$ possibly may not be 1. For $\ell \neq m$, the associated subgeometry partitions obtained are not isomorphic to any one obtained directly from one of Kantor's classes (see remarks below).

Moreover, more generally, the following theorem may be proved. The reader is encouraged to start a proof of this theory, which also appears in the author's work [100].

THEOREM 36. *From any m^{th}-root subgeometry partition of $PG(n - 1, q^m)$ by $PG(n - 1, q)$'s then, for any divisor ℓ of m, it is possible to construct a subgeometry partition of $PG(mn/\ell - 1, q^\ell)$ by $PG(n - 1, q)$'s.*

REMARK 21. *(1) Note that the subgeometry partitions constructed are all partitioned by $PG(n - 1, q)$'s, whereas some of the partitions are of $PG(n - 1, q^m)$ and others of $PG(mn/\ell - 1, q^\ell)$. Although the subspaces of the two partitions are projectively the same, the lifting process from such subgeometry partitions constructs different sets depending on the group $GF(q^f)^*$, where $f = m$ or ℓ, as the group produces a $GF(q^f)$-fan of $(q^f - 1)/(q - 1)$ components.*

(2) For different divisors ℓ of m, none of these subgeometry partitions are considered to be isomorphic to each other. So, for integers with a large number of divisors, there are a correspondingly large number of mutually non-isomorphic subgeometry partitions.

Consider that we have an $n(m/\ell) = n^*$-dimensional vector space over $GF(q^\ell)$. Suppose that $(n^*, \ell) = 1$, then there are subgeometry partitions constructed from Kantor's class I or II, if, respectively, m divides $q - 1$, or when $q = p^r$, then m is a p-power and in either cases $(m, n) = 1$. These geometries provide partitions of $PG(n^* - 1, q^\ell)$ by

$PG(n^* - 1, q)'s$. However, the construction above gives partitions of $PG(n^* - 1, q^\ell)$ by $PG(n - 1, q)'s$. Therefore, we note the following.

REMARK 22. *The subgeometry partitions for ℓ properly dividing m cannot be considered isomorphic to any partitions obtained directly from Kantor's classes I or II, even if $(n(m/\ell), \ell) = 1$.*

Actually, the proof of the previous theorem shows that the following more general theorem is also valid.

THEOREM 37. *Consider any subgeometry partition of $PG(rn-1, q^s)$ by a set of subgeometries isomorphic to $PG(sn/t - 1, q^t)$, for $t \in D_s^-$, where D_s^- is a subset of the set of divisors of s. Let ℓ be any divisor of s. Then there is a corresponding subgeometry partition of $PG(rn/\ell-1, q^\ell)$ by subgeometries isomorphic to $PG(sn/(t, \ell) - 1, q^{(t,\ell)})$'s for $t \in D_s^-$.*

PROOF. Note that $PG(sn/t - 1, q^t)$ corresponds to an orbit Γ of length $(q^s - 1)/(q^t - 1)$ under $GF(q^s)^*$. Now take $GF(q^\ell)^*$. The stabilizer of a component L in Γ has order $(q^t - 1)$ in $GF(q^s)^*$. Therefore, the stabilizer of L in $GF(q^\ell)^*$ has order $(q^{(t,\ell)} - 1)$. Hence, the orbit lengths relative to $GF(q^\ell)^*$ are $(q^m - 1)/(q^{(t,\ell)} - 1)$. □

5.2. Field Subgeometry Partitions. For some reason, these subgeometry partitions have been overlooked, but from any field whose of order q^{mn}, it is possible to obtain very interesting subgeometry partitions.

Consider $GF(q^{mn})^*$. Define points as elements of $GF(q^{mn})^*$ and components as the images of $GF(q^n)$ under the group $GF(q^{mn})^*$. Then there are $(q^{mn} - 1)/(q^n - 1)$ components, so we have a transitive spread of n-dimensional $GF(q)$-subspaces of the associated m-dimensional vector space over $GF(q^n)$. Consider the group $GF(q^m)^*$. This group acts on the spread and the subgroup that fixes the component $GF(q^n)$ is $GF(q^{(m,n)})^*$. Since the original group is cyclic and transitive, we see that $GF(q^{(m,n)})^*$ fixes each component and is then considered the 'kernel' of the group. Take an orbit Γ of length $(q^n - 1)/(q^{(m,n)} - 1)$ under $GF(q^m)^*$. Form the projective space $PG(n - 1, q^m)$. Then Γ becomes a subgeometry isomorphic to $PG(n/(m, n) - 1, q^{(m,n)})$, by the quasi-subgeometry type of arguments.

THEOREM 38. *From any field $GF(q^{mn})$, there exists a subgeometry partition of $PG(n-1, q^m)$ by subgeometries isomorphic to $PG(n/(m, n)-1, q^{(m,n)})$.*

The reader is also directed to Johnson and Cordero [**137**] for additional details.

REMARK 23. *We note from Section 0.3, for every n-spread of an mn-dimensional vector space, we obtain a $(n, n-1)$ double spread of $q^{m(n-1)}$ subspaces of dimension $n-1$ and $\frac{q^{m(n-1)}-1}{q^n-1}$ subspaces of dimension n.*

CHAPTER 9

Maximal Additive Partial Spreads

In Chapter 2 on the basic theory of focal-spreads, we showed that additive focal-spreads are equivalent to additive partial t-spreads in vector spaces of dimension $2t$, wherein lives, of course, translation planes of order q^t. It is still an open question whether every focal-spread of type (t, k) arises as a k-cut of a t-spread of a translation plane. In the additive case, if the 'companion' partial t-spread is not maximal over the prime field, then there is a semifield plane of order q^t that extends the partial spread and the dual semifield plane then extends the additive focal-spread. There are maximal additive partial spreads in $PG(3, q)$ (see, e.g., Jha and Johnson [**90**]), and although there is no semifield with spread in $PG(3, q)$ that extends the additive partial spreads, it is not known whether there are spreads in larger dimensional projective spaces within which an additive partial spread in $PG(3, q)$ may be extended to a semifield spread (the reader is directed to the open problem section for this partial spread).

In this chapter, we construct a number of very interesting additive partial t-spreads. What makes these partial spreads interesting involves the 'subplane dimension question': Given any translation plane of order p^s and any affine subplane of order p^z must z divide s? It is easy to show that any affine subplane of a (affine) translation plane is also a translation plane of the same characteristic. The reader probably recalls the existence of Fano subplanes (subplanes of order 2) in translation planes of odd characteristic, but these subplanes are subplanes of the 'projective translation plane.'

There are a few situations where the subplane dimension question can be answered in the affirmative. For example, if π is a Desarguesian plane coordinatized by a finite field isomorphic to $GF(p^t)$, then any affine subplane π_0 is also Desarguesian and may be coordinatized by a subfield isomorphic to $GF(p^k)$, so k does divide t in this case. More generally, if a finite translation plane π of order p^t admits a collineation of order p that fixes an affine subplane π_0 of order p^k pointwise, then Foulser [**52**] has shown that k must, in fact, divide t. However, this is the extent of the knowledge regarding the subplane dimension question.

So, how does this idea impact what we are trying to say with regards to additive partial t-spreads and their companion additive focal-spreads of type (t, k)? We begin by showing how to construct a tremendous variety of additive partial t-spreads that admit extremely unusual affine subplanes, ones for which the subplane dimension question is answered in a resounding 'no'! If any of these additive partial t-spreads can be extended they may also be extended to semifield t-spreads admitting the same affine subplanes. What this means is that either there are undiscovered semifield planes about which we basically know nothing or there are additive maximal partial spreads about which nothing is known. Of course, the problem is, so far we can't be certain which alternative might be valid.

It is apparently not known that additivity of partial spreads can be arranged to be inherited. That is, given an additive partial spread that is not maximal, there is, of course, a partial spread containing the original partial spread, but we can also arrange it so that super partial spread is also additive.

DEFINITION 41. *We define an additive partial spread S to be 'additively maximal' if and only if there is not an additive partial spread properly containing S. Note that we may always consider the subspace $x = 0$ adjoined to any additive partial spread.*

We regard all partial spreads to be finite and considered over the prime field $GF(p)$. Now we note

THEOREM 39. *An additively maximal additive partial spread is a maximal partial spread. Any additive partial spread that is not maximal may be extended to an additive partial spread.*

PROOF. Let S be any additively maximal additive partial spread and assume that it is not maximal. We then obtain a subspace $y = xM$, where M is non-singular, that is, not in

$$x = 0, y = x \sum_{i=1}^{k} \alpha_i A_i,$$

where

$$S = \left\{ \sum_{i=1}^{k} \alpha_i A_i; \text{ for all } \alpha_i \in GF(p) \right\}.$$

Therefore, we have that

$$M - \sum_{i=1}^{k} \alpha_i A_i$$

is non-singular for $\alpha_i \in GF(p)$. Thus,

$$\beta M - \sum_{i=1}^{k} \alpha_i A_i$$

is non-singular for all β, $\alpha_i \in GF(p)$, where at least one of β or α_i, $i = 1, 2, .., k$ is non-zero. Hence, this means that letting $M = A_{k+1}$, then we have

$$S \cup \{M\} = \left\{ \sum_{i=1}^{k+1} \alpha_i A_i; \text{ for all } \alpha_i \in GF(p) \right\},$$

is an additive partial spread of degree p^{k+1}. This proves the theorem.
$\square$

Now the wonderfully simple corollary:

COROLLARY 5. *Any additive partial spread (with $x = 0$ adjoined), may be extended either to a proper maximal partial spread that is additively maximal or extended to a semifield spread.*

So, as mentioned, we may consider this idea algorithmically to construct maximal additive partial spreads (or semifield spreads). We use the usual setup for spreads, choosing a basis in the appropriate way, and all that is really required is any partial spread of at least three t-subspaces in a $2t$-dimensional vector space over the prime field $GF(p)$. Specifically, if L, M, and N are three mutually disjoint t-dimensional subspaces, choose a basis for the vector space so that the three subspaces are $x = 0, y = 0, y = x$, writing vectors as $(x_1, .., x_t, y_1, .., y_t)$, $x = (x_1, .., x_t)$ and $y = (y_1, .., y_t)$. Use $y = x$ to generate an additive partial spread $y = x\ iI_t$, for all $i \in GF(p)$, where I_t is the $t \times t$ identity matrix. (The reader is left the verification that a partial spread is obtained as an exercise.) Either this partial spread is maximal or not. If not, choose any t-subspace that is disjoint from the existing additive partial spread and note that the subspace must be of the form $y = xA$, where A is non-singular. Take $x = 0, y = x(iA + jI_t)$, for all $i, j \in GF(p)$. A simple check shows that this is a partial spread of degree $p^2 + 1$. Clearly, this process will terminate either at an additive maximal partial spread or a semifield t-spread. If we do not obtain a semifield the process will terminate at an additive maximal partial spread of degree at most $p^{t-1} + 1$.

COROLLARY 6. *Suppose that S is an additive partial t-spread over $GF(p)$ that cannot be extended to a semifield spread. Then there is a maximal partial t-spread of degree $\leq p^{t-1} + 1$.*

1. Direct Sums of Semifields

In this section, we concentrate on additive maximal partial spreads, with the goal of constructing exotic affine subplanes, which, in turn, gives some impetus to the study of additive focal-spreads. However, there is really nothing about additivity that is essential to the analysis and could easily be applied to arbitrary partial spreads. The reader is also directed to the article by Jha and Johnson [**79**].

The following lemma will set the stage for our constructions.

LEMMA 16. *Let π be a finite translation plane of order p^t. Then there is a set of $t \times t$ non-singular matrices $S_{Mat}^{t \times t}$ of cardinality $p^t - 1$, whose distinct differences are also non-singular and such that given any non-zero t-vector w, there is a unique matrix $M_w^{t \times t}$ such that the first row of $M_w^{t \times t}$ is w.*

PROOF. Note that since there are exactly $p^t - 1$ non-singular matrices, whose distinct differences are also non-singular, it follows that the set $S_{Mat}^{t \times t}$ is necessarily sharply transitive on the set of non-zero vectors. The lemma is now clear, modulo a basis change. □

DEFINITION 42. *Any matrix spread for a translation plane π of order p^t chosen as in Lemma 16 shall be called a 'standard matrix spread set.' We shall use the term 'matrix t-spread set' when it is necessary to specify the dimension of the matrices. Our components for the associated translation plane are*

$$x = 0, y = 0, y = x M_w^{t \times t}; \; M_w^{t \times t} \in S_{Mat}^{t \times t}.$$

This set shall be called a 'matrix spread' for π.

THEOREM 40. *Choose any standard matrix c-spread set $S_{Mat}^{c \times c}$ of $c \times c$ matrices and any standard matrix d-spread set $S_{Mat}^{d \times d}$ of $d \times d$ matrices for $c > d$. Select any $c - d$ entries to be 0 in a c-vector, then there is a subspread set S_{Mat}^{c-d} of $S_{Mat}^{c \times c}$ of cardinality $p^d - 1$, whose matrices have their first rows with this same set of $c - d$ entries all zero. Let the d-vector w represent rows in both S_{Mat}^{c-d} and $S_{Mat}^{d \times d}$.*

Form the bijective correspondence between the subspread S_{Mat}^{c-d} of $S_{Mat}^{c \times c}$ and $S_{Mat}^{d \times d}$, by mapping $M_w^{c \times c}$ onto $M_w^{d \times d}$ in the notation of Lemma 16. Form the set

$$\mathcal{P} = \left\{ x = 0, y = 0, \; y = x \begin{bmatrix} M_w^{c \times c} & 0 \\ 0 & M_w^{d \times d} \end{bmatrix} ; \; w \text{ a d-vector} \right\},$$

for all $M_w^{c \times c} \in S_{Mat}^{c-d}$ and $M_w^{d \times d} \in S_{Mat}^{d \times d}$.

Then $\mathcal{P}$ is a partial spread of order p^{d+c} and degree $1+p^d$ that contains a translation subplane of order p^d isomorphic to the translation plane given by the d-spread set $S_{Mat}^{d \times d}$.

PROOF. It follows easily that $M_w^{c \times c}$ is the zero matrix if and only if $M_w^{d \times d}$ is the zero matrix. Therefore, we have a set of non-singular matrices $\begin{bmatrix} M_w^{c \times c} & 0 \\ 0 & M_w^{d \times d} \end{bmatrix}$. Now take the difference of two of these matrices

$$\begin{bmatrix} M_w^{c \times c} & 0 \\ 0 & M_w^d \end{bmatrix} - \begin{bmatrix} M_{w*}^{c \times c} & 0 \\ 0 & M_{w*}^d \end{bmatrix}$$

$$= \begin{bmatrix} M_w^{c \times c} - M_{w*}^{c \times c} & 0 \\ 0 & M_w^d - M_{w*}^d \end{bmatrix}.$$

Since $M_w^{c \times c} - M_{w*}^{c \times c}$ and $M_w^d - M_{w*}^d$ are both non-singular for $w \neq w^*$ and w and w^* non-zero vectors (adjoin the zero-entries, when appropriate), we have that

$$x = 0, y = x \begin{bmatrix} M_w^{c \times c} & 0 \\ 0 & M_w^{d \times d} \end{bmatrix}; \ w \text{ a } d\text{-vector}$$

is a partial spread of degree $1 + p^d$. The associated vector space is $2(d+c)$-dimensional over $GF(p)$, and let the $2(d+c)$-vectors be denoted by

$$(x_1, x_2, .., x_{d+c}, y_1, y_2, .., y_{d+c}).$$

Now let $\pi_0 =$

$$\left\{ \begin{array}{c} (0, 0, .., 0, x_{c+1}, x_{c+2}, .., x_{c+d}, 0, 0, .., 0, y_{c+1}, y_{c+2}, .., y_{c+d}); \\ x_i, y_i \in GF(p), \ i = c+1, ...c+d \end{array} \right\}.$$

Note that π_0 is a vector space of dimension $2d$ over $GF(p)$ and intersects $x = 0$ and $y = 0$ in d-dimensional subspaces. Furthermore, the intersection with

$$y = x \begin{bmatrix} M_w^{c \times c} & 0 \\ 0 & M_w^d \end{bmatrix}$$

is

$$(0, 0, .., 0, y_{c+1}, y_{c+2}, .., y_{c+d})$$

$$= (0, 0, .., 0, x_{c+1}, x_{c+2}, .., x_{c+d}) \begin{bmatrix} M_w^{c \times c} & 0 \\ 0 & M_w^d \end{bmatrix}$$

$$= (0, 0, .., 0, (x_{c+1}, x_{c+2}, .., x_{c+d})M_w^d),$$

which is also a d-dimensional $GF(p)$-subspace for each non-zero d-vector w. Hence, we have a spread of $1 + p^d$ d-dimensional subspaces of π_0, so that π_0 becomes an affine subplane of order p^d, which is isomorphic to the original d-spread. This completes the proof. $\qquad\square$

The following corollaries then speak to the subplane dimension question and additive partial spreads.

COROLLARY 7. *If the subplane dimension question is answered affirmatively then* $\mathcal{P}$ *cannot be extended to a matrix* $(c + d)$-*spread.*

COROLLARY 8. *If* $S_{Mat}^{c \times c}$ *and* $S_{Mat}^{d \times d}$ *are semifield spreads (additive), then the subspread* S_{Mat}^{c-d} *is additive and the partial spread* $\mathcal{P}$ *is an additive partial spread. Furthermore, there is a semifield subplane of order* p^d *isomorphic to the semifield plane with matrix spread set given by* $S_{Mat}^{d \times d}$.

COROLLARY 9. *If the subplane dimension question is answered affirmatively for order* p^{d+c} *semifield planes, the previous additive partial spread can be embedded into a maximal additive partial spread of degree* $1 + p^{d+e} \leq 1 + p^{d+c-1}$, *which is a maximal partial spread.*

COROLLARY 10. *One of the following situations must occur:*
   (i) there is a maximal partial spread of order p^{d+c} *and deficiency at least* $p^{d+c} - p^{d+c-1}$, *or*
   (ii) every such partial spread may be extended to a semifield plane of order p^{d+c} *that contains a subplane of order* p^d, *where d does not divide* $d + c$.
   (iii) In this setting, the affirmation of the subplane dimension question also says that given any semifield plane π_0 *of order* p^d *then there is a semifield plane of order* p^{d+c}, *for* $c > d$, *such that d does not divide c, that contains a subplane of order* p^d *isomorphic to the semifield plane given by the original d-spread.*

The proofs to these corollaries are reasonably straightforward, and we invite the reader to complete at least one and then realize that all four will be more or less completed.

We may consider the subplane dimension question in another way: Are there known classes of semifield planes of order p^z that admit affine subplanes of order p^s, where s does not divide z? In the following, we shall concentrate on the binary Knuth semifield planes and the generalization due to Kantor both of even order 2^n that admit affine subplanes of order 2^2. There are many open questions here, some of which the reader can find (and hopefully solve) in the open problems material of Chapter 40.

2. Subfields of Order 4 in Knuth Semifields

Let $x, y \in GF(2^n)$, for n odd, then the following defines the multiplication for a commutative pre-semifield due to Knuth, called the 'commutative binary Knuth pre-semifield of order 2':

$$x \circ y = xyc + (xT(yc) + yT(x))^2,$$

where T is the trace function from $GF(2^n)$ to $GF(2)$. It will turn out that some of semifield planes coordinatized by such semifields admit subplanes of order 4. The operative word is 'some' since the arguments that there are subplanes of order 4 seem to always require a partial basis argument. Although this is certainly not required, we determine a unit element for the pre-semifield thus giving a semifield within which a subfield $GF(4)$ is located. This will say that there is an affine subplane of order 4. Since it also seems that our calculations do not work in commutative pre-semifields, we consider instead a corresponding isotopic pre-semifield.

The following result that sets up our examples given in the next section is as follows. In the statement of the theorem, the pre-semifield multiplication when $b = c = 1$ produces the commutative binary Knuth pre-semifield.

THEOREM 41. *Consider the pre-semifield multiplication*

$$x \circ y = xbyc + (xbT(yc) + ycT(xb))^2,$$

where b and c are constants in $GF(2^n)$, for n odd, and T is the trace function from $GF(2^n)$ to $GF(2)$. Choose any nonzero element e and form the semifield

$$(x \circ e) * (e \circ y) = x \circ y.$$

If

$$T(ec) = T(b) = T(eb) = 0,$$

$$T(c) = 1, \quad \frac{e^2}{e+1} = \frac{b}{c} + 1,$$

then there exists a subfield isomorphic to $GF(4)$ in $(S, +, *)$.

The corresponding semifield plane is the commutative binary Knuth semifield plane of order 2^n and would then admit a subsemifield plane of order 2^2.

PROOF. In the semifield with multiplication $(x \circ e) * (e \circ y) = x \circ y$, then $e \circ e$ becomes the unit element. Take $x = 1$ and consider when

$(1 \circ e) = (e \circ y)$. Hence, we need to solve:

$$1 \circ e = bec + (bT(ec) + ycT(b))^2$$
$$= e \circ y = ebyc + (ebT(yc) + ycT(eb))^2.$$

Assume the following conditions: $T(ec) = T(b) = T(eb) = 0$, $T(yc) = 1$. Then we have $y = 1 + e\frac{b}{c}$. This forces $T(yc) = T(c + eb) = T(c) = 1$. Now we ask for values of e, b, c such that

$$1 \circ y \; = \; 1 \circ (1 + e\frac{b}{c}) = (1 \circ e) * (1 \circ e)$$
$$= \; 1 \circ e + e \circ e = (1 + e) \circ e.$$

This would say that considering juxtaposition to be $*$-multiplication and realizing that $e \circ e$ becomes the 1 in the associated semifield, letting $d = 1 \circ e$, we would have then $d^2 = d + 1$. Then $\{0, 1, d, d^2\}$ becomes a subfield isomorphic to $GF(4)$ in the semifield $(S, +, *)$. Therefore, we have

$$1 \circ (1 + e\frac{b}{c}) = b(1 + e\frac{b}{c})c + (bT((1 + e\frac{b}{c})c) + (1 + e\frac{b}{c})cT(b))^2$$
$$= (1 + e) \circ e$$
$$= (1 + e)bec + ((1 + e)bT(ec) + ecT((1 + e)b)^2.$$

This produces the equivalent equation:

$$b(1 + e\frac{b}{c})c + b^2 = (1 + e)bec.$$

We then obtain:

$$e^2(bc) = e(bc + b^2) + (bc + b^2) = (e + 1)(bc + b^2).$$

Choose $e \neq 1$ to obtain the following equation $\frac{e^2}{e+1} = \frac{b}{c} + 1$. Thus, we require the following equations to ensure that there is a subfield of order 4:

$$T(ec) = T(b) = T(eb) = 0,$$
$$T(c) = 1, \quad \frac{e^2}{e+1} = \frac{b}{c} + 1.$$

This completes the proof of the theorem.   □

Now the results become more specialized. In particular, we show that for n divisible by 5 or 7, then there are subplanes of order 4 in the commutative binary Knuth semifield planes of order 2^n, n odd. Here are the specific ways of choosing e, b, c in these cases.

COROLLARY 11. *(1) Let $n = 5k$, for k odd, and in $GF(2^5)$, let $x^5 + x^2 + 1 = 0$ be the irreducible polynomial. If $e = 1 + x + x^3, b = x^2, c = x^3$, then the semifield $(S, +, *)$ of order 2^{5k} admits a subfield isomorphic to $GF(4)$. Hence, every commutative binary Knuth semifield plane of order 2^{5k}, for k odd, admits a Desarguesian subplane of order 4.*

(2) Let $n = 7k$, for k odd, and in $GF(2^7)$, let $x^7 + x^4 + x^3 + x^2 + 1$ be the irreducible polynomial. If $e = 1 + x^7, b = x^7$, and $c = x^3$, then the semifield $(S, +, *)$ of order 2^{7k} admits a subfield isomorphic to $GF(4)$. Hence, every commutative binary Knuth semifield plane of order 2^{7k}, for k odd, admits a Desarguesian subplane of order 4.

PROOF. Once the reader gets the idea of how the proof proceeds, the calculations are straightforward. So, we give just the sketch of the proof.

We claim that if z is in $GF(2^5)$ then $T(z) = kT_5(z)$, where T_5 is the trace function of $GF(2^5)$ over $GF(2)$. Recall that we are working in $GF(2^{5k})$, so k is 1 modulo 2. $T(z) = \sum_{i=0}^{5k-1} z^{2^i}$. But, if $z \in GF(2^5)$ then $\sum_{i=0}^{5-1} z^{2^i} = T_5(z)$ is in $GF(2)$. And

$$\sum_{i=0}^{5k-1} z^{2^i} = T_5(z) + (z^{2^5} + z^{2^6} + z^{2^7} + z^{2^8} + z^{2^9}) +$$

$$\ldots + (z^{2^{5(k-1)}} + z^{2^{5k-4}} + z^{2^{5k-3}} + z^{2^{5k-2}} + z^{2^{5k-1}})$$

$$= kT_5(z) = T(z), \text{ for } k \equiv 1 \bmod 2.$$

If, for elements e, b, c in $GF(2^5)^*$, we have

$$T_5(ec) = T_5(b) = T_5(eb) = 0,$$

$$T_5(c) = 1, \quad \frac{e^2}{e+1} = \frac{b}{c} + 1,$$

then

$$T(ec) = T(b) = T(eb) = 0,$$

$$T(c) = 1.$$

So, if we have a subfield isomorphic to $GF(4)$ of the sub-semifield of order 2^5, we then have a subfield isomorphic to $GF(4)$ of the semifield of order 2^{5k}, for k odd. We first note that if $e = 1 + x + x^3, b = x^2, c = x^3$, then $\frac{e^2}{e+1} = \frac{b}{c} + 1$. This follows by an easy calculation. We now verify

$$T_5(ec) = T_5(b) = T_5(eb) = 0,$$

$$T_5(c) = 1.$$

First,

$$T_5(ec) = T_5((1 + x + x^3)x^3) = T_5(x^3 + x^4 + x^6)$$
$$= T_5(x^3 + x^4 + x(x^2 + 1))$$
$$= T_5(x^3 + x^4 + x^3 + x) = T_5(x^4) + T_5(x)$$

and since $T_5(x^{2^a}) = T_5(x)$, we see that $T_5(ec) = 0$. Then, we see that

$$T_5(x^3) = x^3 + x^6 + (x^6)^2 + (x^6)^4 + (x^6)^8$$
$$= x^3 + (x + x^3) + (x + x^2 + x^3) + (x + x^2 + x^3 + x^4)$$
$$+ (x^2 + x^4 + x + x^3 + 1 + x^2 + x^3) = 1.$$

Also,

$$T_5(eb) = T_5((1 + x + x^3)x^2) = T_5(x^2 + x^3 + x^5))$$
$$= T_5(x^2 + x^3 + x^2 + 1)$$
$$= 0, \text{ since } T_5(x^3) = T_5(1) = 1.$$

$$T_5(b) = T_5(x^2) = x^2 + x^4 + x^8 + x^{16} + x^{32}.$$

We are working in $GF(2^5)$, so $x^{32} = x$. Then we obtain:

$$x^2 + x^4 + (1 + x^2 + x^3) + (1 + x + x^3 + x^4) + x = 0.$$

This completes the proof for 2^5.

The proof for 2^7 is very similar once the correct choices of e, b, c are found. The proof is left for the reader to complete. $\square$

Considering the transposed and dualized spreads, these also contain subplanes of order 4, as can easily be established (the reader should ask 'why?'). Hence, we also obtain:

THEOREM 42. (1) *The transposed commutative binary Knuth semi-field planes of order 2^n, for $n = 5k$, or $7k$, for k odd, admit Desargue-sian subplanes of orders 2^2.*

(2) *The transposed then dualized commutative binary Knuth semi-field planes of order 2^n for $n = 5k$, or $7k$, for k odd, are symplectic and admit Desarguesian subplanes of orders 2^2.*

3. The Commutative Kantor Semifields

There are generalizations of the binary Knuth commutative semi-fields due to Kantor [156]. These have the following construction: Let F be a finite field of characteristic 2, fix a subfield F_n isomorphic to $GF(q)$, and let $F = F_0 \supset F_1 \supset F_2 \supset ... \supset F_n$ such that $[F : F_n] = k$ is odd. Choose a set of elements $\zeta_i \in F^*$, for $i = 1, 2, ..., n$. Let T_i denote the trace map from F to F_i. Then

$$x \circ y = xy + \left(x \sum_{i=1}^{n} T_i(\zeta_i y) + y \sum_{i=1}^{n} T_i(\zeta_i x) \right)^2$$

defines a commutative pre-semifield of order $q^k = q^{n_1 n_2 ... n_n}$, where $[F_i : F_{i+1}] = n_i$, so all n_i are odd.

THEOREM 43. *Let* $F \simeq GF(2^{5k})$, *for* $5k = n_1 n_2 ... n_n$, k *odd with sequence* $(\zeta_1, \zeta_2, ... \zeta_{n-1}, 1)$ *such that* $F_n = GF(2)$ *and* $F_{n-1} \simeq GF(2^5)$. *Let* e, b, c *be elements of* $GF(2^5)^*$ *such that for* $x^5 + x^2 + 1$ *is an irreducible polynomial for* $GF(2^5)$. *Let*

$$e = 1 + x + x^3,$$
$$b = x^2$$
$$c = x^3$$

be elements of $GF(2^5)$. *Choose the sequence* $(\zeta_1, \zeta_2, ... \zeta_{n-1}, 1)$ *so that*

$$\sum_{i=1}^{n-1} T_i(\zeta_i) = 0.$$

Then there is a binary Knuth commutative semifield of order 2^5 *contained as a sub-semifield in the corresponding Kantor commutative semifield.*

Hence, there is an isotope of the Kantor commutative semifield that contains a field isomorphic to $GF(4)$.

These Kantor commutative semifield planes of order 2^{5k}, *for* k *odd, admit subplanes of order 4.*

PROOF. Note that for $1 \leq i \leq n - 1$, $T_i(\zeta_i r) = rT_i(\zeta_i)$, for $r \in F_{n-1} \simeq GF(2^5)$. If

$$\sum_{i=1}^{n-1} T_i(\zeta_i) = 0,$$

then

$$\sum_{i=1}^{n} T_i(\zeta_i r) = T_n(\zeta_n r) = T_n(r).$$

This says that there is a binary Knuth commutative semifield that is a sub-semifield of the Kantor commutative semifield. □

COROLLARY 12. *The transposed and transposed-dual (symplectic) semifields of the Kantor commutative semifields corresponding to the* *semifields of Theorem 43 of order 2^{5k}, for k odd, have isotopes that contain a field isomorphic to $GF(4)$. The corresponding semifield planes of order 2^{5k} admit subplanes of order 4.*

Here are a few examples of Kantor commutative semifield planes of the required orders admitting the subplane in question.

EXAMPLE 1. (1) *Assume that we have a Kantor commutative semifield of order $2^{5^2 \cdot 7}$. Take $n = 3$, $F_3 \simeq GF(2)$, $F_2 \simeq GF(2^5)$, $F_1 \simeq GF(2^{5^2})$, and $F_0 = F \simeq GF(2^{5 \cdot 7})$. Assume that ζ_i's are all in F_2. Then*

$$\sum_{i=1}^{3} T_i(\zeta_i ed) = T_n(ed), \text{ for } e \text{ and } d \text{ in } F_2. \text{ Similarly, } \sum_{i=1}^{n} T_i(\zeta_i c) = T_n(c).$$

(2) More generally, if n is odd, then there are an even number of proper subfields containing $F_{n-1} \simeq GF(2^5)$. So take $F_{n-1} \simeq GF(2^5)$, and assume that all elements ζ_i are in F_{n-1}^* and $\zeta_n = 1$. In this setting,

$$\sum_{i=1}^{n-1} T_i(\zeta_i) = 0,$$

since $T_i(\zeta_i) = n_i \zeta_i = \zeta_i$, where n_i is odd.

Hence, in either of these two situations, we obtain an isotope that contains a field isomorphic to $GF(4)$.

THEOREM 44. *Let $F \simeq GF(2^{7k})$, for $7k = n_1 n_2 ... n_n$, odd, with sequence $(\zeta_1, \zeta_2, ... \zeta_{n-1}, 1)$ such that $F_n = GF(2)$ and $F_{n-1} \simeq GF(2^7)$. Let e, b, c be elements of $GF(2^7)^*$ such that $x^7 + x^4 + x^3 + x^2 + 1$ is an irreducible polynomial for $GF(2^7)$. Let*

$$e = 1 + x^7$$

$$b = x^7$$

$$c = x^3$$

be elements of $GF(2^7)$. Choose the sequence $(\zeta_1, \zeta_2, ... \zeta_{n-1}, 1)$ so that

$$\sum_{i=1}^{n-1} T_i(\zeta_i) = 0.$$

Then there is a binary Knuth commutative semifield of order 2^7 contained as a sub-semifield in the corresponding Kantor commutative semifield.

Hence, there is an isotope of the Kantor commutative semifield that contains a field isomorphic to $GF(4)$.

The associated Kantor semifield plane of order 2^{7k} admits subplanes of order 4.

COROLLARY 13. *The transposed and transposed-dual (symplectic) semifields of the Kantor commutative semifields corresponding to the semifields of Theorem 44 of orders 2^{7k}, for k odd, have isotopes that contain a field isomorphic to $GF(4)$. The semifield planes of order 2^{7k} admit subplanes of order 4.*

Clearly, we have merely scratched the surface of what subplanes lie within any given translation plane as our results seem to imply.

Part 3

Subplane Covered Nets and Baer Groups

This part concerns how the theory of subplane covered nets may be applied for the construction of parallelisms. A general study of Baer groups is given and applied to the theory of flocks of quadratic cones and flocks of hyperbolic quadrics.

CHAPTER 10

Partial Desarguesian t-Parallelisms

In this chapter, we consider partial Desarguesian parallelisms by viewing such structures through a corresponding direct sum, which become nets that are often subplane covered nets. Before we begin our study, we provide a short review of the necessary background.

1. A Primer on Subplane Covered Nets

In this chapter, we give a primer of sorts on 'subplane covered nets.' Of course, the reader may read the full account of such point-line geometries in the author's text [**114**]. We shall give here the statements of the complete theorems, which ultimately deal with right and left vector spaces over skewfields. The general case is quite complicated since for derivable nets, the components of the net are right subspaces, whereas the Baer subplanes of the net are left subspaces. Although this text is primarily about finite incidence geometry, the ideas are really quite general and the reader interested in considering the material over infinite fields or even skew fields will find many topics of general interest. For the most part and whenever convenient, we shall keep our theory as general as possible.

DEFINITION 43. *A 'net' $\mathcal{N} = (\mathcal{P}, \mathcal{L}, \mathcal{C}, \mathcal{I})$, is an incidence structure with a set $\mathcal{P}$ of 'points,' a set $\mathcal{L}$ of 'lines,' a set $\mathcal{C}$ of 'parallel classes,' and a set $\mathcal{I}$, which is called the 'incidence set' such that the following properties hold:*

(i) Every point is incident with exactly one line from each parallel class, each parallel class is a cover of the points, and each line of $\mathcal{L}$ is incident with exactly one of the classes of $\mathcal{C}$ (parallelism is an equivalence relation and the equivalence classes are called 'parallel classes').

(ii) Furthermore, lines from two different parallel classes have exactly one common incident point.

(iii) Two distinct points are incident with exactly one line of $\mathcal{L}$ or are not incident.

DEFINITION 44. *A 'subplane covered net' $\mathcal{N} = (\mathcal{P}, \mathcal{L}, \mathcal{B}, \mathcal{C}, \mathcal{I})$ is a net $(\mathcal{P}, \mathcal{L}, \mathcal{C}, \mathcal{I})$ together with a set $\mathcal{B}$ of affine subplanes, such that*

*given any two distinct points a and b, there is an subplane $\pi_{a,b}$ of $\mathcal{B}$
containing a and b, whose parallel class set is exactly $\mathcal{C}$.*

A 'Baer subplane' π_0 *of a net is an affine subplane with the property
that every point a of the net is incident with a line of the subplane π_0
and every line ℓ of the net is incident with a point of π_0.*

A 'derivable net' *is a subplane covered net whose subplanes in $\mathcal{B}$ are
all Baer subplanes.*

There are two main theorems for subplane covered nets, the first
of which shows how to embed such nets in projective spaces and the
second of which shows that subplane covered nets are rational Desarguesian nets. But, first we define the derivable subnets of a subplane
covered net.

Let L and N be any two affine points of a subplane covered net
that are not collinear. Let x be any line incident with N. For the
intersection $L\beta \cap x$, if x does not lie in the parallel class $\beta \in \mathcal{C}$, where
$L\beta$ is the line of β incident with L, further, determine the corresponding
subplane $\pi_{L,L\beta \cap x}$ as in Definition 44. This subplane contains all of the
points $L\delta \cap x$ so any set of intersection points together with L will
completely determine the subplane. We then use the notation $\pi_{L,x}$
for $\pi_{L,L\beta \cap x}$. In addition, we shall use the notation $\pi_{A,z}$ to denote a
subplane defined by a line through a point A and a line z incident with
a point B, where A and B are non-collinear. The point B incident
with z shall be understood by context.

DEFINITION 45. *Let L and N be any two non-collinear points of
the subplane covered net $\mathcal{R} = (\mathcal{P}, \mathcal{L}, \mathcal{B}, \mathcal{C}, \mathcal{I})$. The point-line geometry
structure $\mathcal{S}_L^N$ is defined as follows:*

*The 'points' are the points of $\cup_N \pi_{L,x}$, where x varies over the set
of lines incident with N. The 'lines' are the lines of a subplane $\pi_{L,x}$.
Note that the points of $\mathcal{S}_L^N$ are defined as intersections of non-parallel
lines of the subplanes $\pi_{L,x}$ for various lines x.*

DEFINITION 46. *A 'subnet' of a subplane covered net is a triple
of subsets of the sets of points, lines and parallel classes defined as
follows: The 'lines' of the subnet will be the lines of the subplanes $\pi_{L,x}$
for x incident with N, where L and N are non-collinear points. The
'points' of the subnet shall be the intersections of lines of the subplanes
indicated. The subset of lines of each parallel class $\alpha \in \mathcal{C}$ is the union
of the sets of lines belonging to the subplanes $\pi_{L,x}$, which lie in α.*

*It will turn out that $\mathcal{S}_L^N = \mathcal{S}_L^Q$ for all points Q of $\mathcal{S}_L^N$ that are non-
collinear with L. We now use the notation $\mathcal{S}_L$ for $\mathcal{S}_L^N$.*

THEOREM 45. *(see 15.19 of Johnson* [**114**]*) The structures $\mathcal{S}_L$ are derivable subnets; the structures $\mathcal{S}_L$ are subnets with parallel class $\mathcal{C}$ and the subplanes contained within the structures are Baer subplanes of $\mathcal{S}_L$.*

1.1. Classification of Subplane Covered Nets.

THEOREM 46. *(see 15.5.1 of Johnson* [**114**]*)*

Let $\mathcal{R} = (\mathcal{P}, \mathcal{L}, \mathcal{B}, \mathcal{C}, \mathcal{I})$ *be a subplane covered net. Define the point-line geometry $\Sigma_{\mathcal{R}}$ as follows:*

Call the lines of a given parallel class 'class lines' and call the lines of a given parallel class of a derivable subnet 'class subplanes.' Note that there are equivalence classes of both the set of class lines and on the set of class subplanes. Call equivalence classes of the class lines 'infinite points' and call the equivalence classes of the class subplanes 'infinite lines.'

The infinite points and infinite lines form a projective space $\mathcal{N}$.

The 'points' of $\Sigma_{\mathcal{R}}$ are of two types:

(i) the 'lines' $\mathcal{L}$ of the net $\mathcal{R}$, and

(ii) the infinite points as defined above.

The lines of $\Sigma_{\mathcal{R}}$ are of three types:

(i) the set of lines incident with an affine point (identified with the set $\mathcal{C}$),

(ii) the class lines extended by the infinite point containing the class line, and

(iii) the lines of the projective subspace $\mathcal{N}$.

The 'planes' of $\Sigma_{\mathcal{R}}$ are of three types:

(i) subplanes of $\mathcal{B}$ extended by the infinite point on the equivalence class lines of each particular subplane, where the points and lines of the subplane are now considered as above (another interpretation is this is the dual of the subplane extended),

(ii) the affine planes whose 'points' are the lines of a new parallel class and 'lines' the class lines of a derivable subnet of the new parallel class extended by the infinite points and infinite line, and

(iii) the projective planes of the projective space $\mathcal{N} = \mathcal{N}^{\alpha}$, for all $\alpha \in \mathcal{C}$.

The hyperplanes of $\Sigma_{\mathcal{R}}$ that contain $\mathcal{N}$ are the parallel classes $\mathcal{C}$, extended by the infinite points and infinite lines of $\mathcal{N}$. Then $\mathcal{N}$ becomes a co-dimension two projective subspace of $\Sigma_{\mathcal{R}}$.

THEOREM 47. *(see 14.14 Johnson* [**114**]*) Let W be a left vector space over a skewfield K. Let $V = W \oplus W$. Let*

$$(x = 0) \equiv \{(0, y); y \in W\},$$

and

$$(y = \delta x) \equiv \{x, \delta x\}; x \in W, \text{ for a fixed } \delta \in K\}.$$

Define a point-line geometry $\mathcal{R} = (\mathcal{P}, \mathcal{L}, \mathcal{B}, \mathcal{C}, \mathcal{I})$ as follows: The set of 'points' $\mathcal{P}$ is V, the set of lines $\mathcal{L}$ is the set of translates of $x = 0$ and $y = \delta x$, for all $\delta \in K$ and these lines incident with $(0, 0)$ are representatives for the set of parallel classes $\mathcal{C}$.

The set of subplanes $\mathcal{B}$ is the set of translates of the following set of subplanes $\pi_w = \{(\alpha w, \beta w); \alpha, \beta \in K\}$ for fixed $w \in W - \{(0, 0)\}$ and incidence $\mathcal{I}$ is the naturally induced incidence set.

Then $\mathcal{R}$ is a subplane covered net, which is called a 'pseudo-regulus net.' When the associated skewfield K defining the pseudo-regulus net is a field, then we call the subplane covered net a K-regulus net.

The following is the main theorem on Subplane Covered Nets.

THEOREM 48. *(15.40 and 15.41 of Johnson [114])*
(i) A subplane covered net is a pseudo-regulus net.
(ii) A finite subplane covered net is a regulus net.

2. Projective Spreads and Affine Nets

A t-parallelism is a partition of the set of t-subspaces of a vector space of dimension zt over a skewfield K by a set of mutually disjoint t-spreads. When $t = 2$, there are a number of such parallelisms and in Johnson and Montinaro [144], there are new classes of transitive 2-parallelisms given in $PG(2r - 1, 2)$ that generalizes the Baker parallelisms, which we also construct in this text. However, in that article, it is also shown that transitive t-parallelisms can exist only if $t = 2$. The reader is directed to Chapter 40, Theorem 307. Previously, Johnson and Montinaro [143] have shown that the only doubly transitive t-parallelisms are the two regular parallelisms in $PG(3, 2)$, and here again $t = 2$. So, the question is, 'are there any t-parallelisms, for $t > 2$'? Perhaps the more interesting question is, 'do Desarguesian t-parallelisms exist, for $t > 2$'?

One of the reasons that Desarguesian 2-parallelisms in $PG(3, q)$ are particularly important is that they construct rare translation planes of order q^4 with spreads in $PG(7, q)$, that admit $GL(2, q)$ as a collineation group, and whose spread is covered by a set of $1+q+q^2$ derivable partial spreads that share a regulus spread of degree $1 + q$. The theory that supports this construction is due to M. Walker [181] and independently to G. Lunardon [166]. Here, it is shown that, at least when t is a prime, partial Desarguesian t-parallelisms correspond to partial spreads that consist of a set of Desarguesian partial spreads of degree q^t+1 that share a regulus partial spread of degree $1 + q$. The constraint of the partial spread will provide a bound on the cardinality of the Desarguesian partial spread, which will imply, in turn, that Desarguesian t-spreads for t an odd prime cannot exist.

We will also connect partial Desarguesian t-parallelisms, at least for t a prime, with certain translation nets whose partial spreads are covered by Pappian partial spreads. Because of the inevitable confusion between 'left' and 'right' spreads, we work through this material first when K is a field.

DEFINITION 47. A 't-spread' S of a vector space V_{zt} of dimension zt over a field K is a partition of the non-zero vectors by 2-dimensional subspaces of W_k. A 'Pappian 2-spread,' sometimes called a 'regular t-spread' is a t-spread with the property that there is a t-dimensional field extension K^+ of K such that the t-dimensional subspaces over K of S are 1-dimensional subspaces over K^+.

In the finite case, or when K is a skewfield admitting a t-dimensional skewfield extension K^+, we shall use the term 'Desarguesian t-spread.'

DEFINITION 48. A 'K-regulus' of V_{2k} is a subset $\mathcal{R}$ of k-dimensional subspaces over K with the property that there is a set of 2-dimensional K-subspaces that partition the non-zero vectors of the set. Using the language of partial spreads, a K-regulus is a partial k-spread that is partitioned by a partial 2-spread. We note that it is an assumption that all vectors of the partial 2-spread are covered by vectors of $\mathcal{R}$. If there is a field J so that V_{2k} is a 4-dimensional J-vector space then a J-regulus is called simply a 'regulus.' There are no K-reguli if K is a skewfield that is not a field.

Equivalently, a 'projective regulus' is a set of lines of $PG(3, J)$ that is covered by a corresponding set of lines, which also form a regulus. This second regulus is called the 'opposite regulus' or 'opposite projective regulus' of the first regulus.

Of course, we shall be using the concept of a 'subplane covered net,' and loosely speaking, such a point-line geometry becomes a translation net over a vector space such that there is a set of Desarguesian affine subplanes incident with the zero vector, all of which have the complete set of components as their defining spread set. In this section, we shall be considering only situations where the Desarguesian subplanes are Pappian, for example, this always occurs in the finite case. In the next section, we consider all of this more generally over arbitrary skewfields K.

To better grasp the idea of the following, we consider a translation plane of dimension $2tk$ over a field K.

DEFINITION 49. *Let V_{2z} be a vector space over a field L. Represent vectors as (x, y), where x and y are z-vectors over a field L. A 'rational Pappian partial z-spread' $\mathcal{R}^+$ is a partial z-spread that may coordinatized by a field L in the following manner: There exists a field $\mathcal{M}$ of $z \times z$ matrices over L, so that*

$$\mathcal{R}^+ : x = 0, y = xM, M \in \mathcal{M}.$$

We shall be assuming also that there is a field $\{\alpha I_z ; \alpha \in L\} = \widehat{L}$, which is a subfield of $\mathcal{M}$. Note that every L-regulus $\mathcal{R}$ may be coordinatized so that

$$\mathcal{R} : x = 0, y = xM, M \in \widehat{L}.$$

We shall say that the rational Pappian partial spread is 'coordinatizable by $\mathcal{M}$.' These partial spreads also called 'reguli' over $\mathcal{M}$.

CONVENTION 1. *We shall always assume that the field of matrices of a rational Pappian partial spread is finite dimensional over $\widehat{L}$.*

Therefore, the rational Pappian partial z-spreads that we shall be considering will all contain the L-regulus partial spread as a subspread. The theory in Johnson [114] clearly shows that rational Pappian partial spreads define W-regulus nets for some field W and hence are subplane covered nets.

CONVENTION 2. *In this section, all of the translation nets that we consider shall be 'vector space translation nets', in the sense that there is a translation group on the net corresponding to the translation group of an associated vector space. Furthermore, given a partial spread, there is a corresponding translation net whose points are the vectors and whose lines are the vector space translates of the components of the partial spread. Conversely, translation nets in the context considered here also produce partial spreads and hence such geometric structures shall be considered equivalent in this text.*

The pertinent lemma here is as follows.

LEMMA 17. *The translation nets produced by any rational Pappian partial z-spread $\mathcal{R}^+$ of a $2z$-dimensional vector space over a field L are subplane covered nets. There exists a set of Pappian affine translation subplanes, whose associated spreads have kernel J, a t-dimensional field extension of L. The subplanes are in the associated rational Pappian translation net.*

In the finite case, for L isomorphic to $GF(q)$, $q = p^r$, p a prime, then the L-regulus has $1 + q$ components, and the rational Pappian (Desarguesian) partial spread has $1 + q^t$ components. We also use the term 'rational Desarguesian translation net.'

The connection with partial Pappian t-parallelisms and with translation nets covered by rational Pappian partial k-spreads is as follows.

THEOREM 49. *Let π be a translation net with projective $k - 1$-spread in $PG(2k - 1, K)$, where K is a field. Assume that the vector k-spread $\mathcal{S}$ is the union of a set of rational Pappian partial k-spreads coordinatizable by a field isomorphic to a field $\mathcal{M}$ that mutually share precisely a K-regulus partial spread $\mathcal{R}$ and is t-dimensional over K. Choose any component of A of $\mathcal{R}$.*

(1) Each rational Pappian partial k-spread produces a translation net, which is a subplane covered net whose subplanes intersect A in a t-spread of A considered as a k-dimensional K-vector space.

(2) The set of t-spreads as defined in part (1) form a partial Pappian t-parallelism.

(3) If K is isomorphic to $GF(q)$, each rational Desarguesian k-spread has degree $1+q^t$ and whose components share a K-regulus partial spread of $1 + q$ components. The translation net then has order q^k and degree $1 + q + m(q^t - q)$. Choose any component A of the K-regulus partial spread. If there are m rational Desarguesian k-spreads all with $1 + q^t$ components, there is an associated partial Desarguesian t-parallelism of m t-spreads.

PROOF. Let $\mathcal{R}_t$ denote a partial rational Pappian partial spread coordinatizable by a field that is t-dimensional over K. Let A denote a component of the regulus partial spread $\mathcal{R}$. A is a k-dimensional K-subspace. Since $\mathcal{R}_t$ is covered by Pappian subplanes coordinatizable by a field K^+, which is a t-dimensional extension of K, the intersections of these subplanes with A are t-dimensional subspaces that are 1-dimensional K^+-subspaces. Since the Pappian subplanes are disjoint on points (apart from the zero vector), it follows that the set of Pappian subplanes of $\mathcal{R}_t$ induces a Pappian t-spread on A. □

So, the hope then would be to consider partial Pappian t-parallelisms and then to be able to connect translation nets whose partial spreads are covered by rational Pappian partial spreads. In general, we show the following theorem:

THEOREM 50. *The set of Pappian t-spreads over a field K in a k-dimensional vector space V_k over a field K is equivalent to the set of rational Pappian partial spreads in a vector space of dimension $2k$ over K, each of which is coordinatizable by a field extension of degree t over K.*

Specifically, the construction is given as follows.

Let $\mathcal{S}$ be a Pappian *t*-spread of a $k = zt$-dimensional vector space V_{zt} over a field K. Let K^+ denote the associated *t*-dimensional field extension of K such that the elements of the *t*-spread are 1-dimensional K^+-subspaces.

Embed V_{zt} in a vector space V_{2zt} of dimension $2zt$ over K. Choose a basis for V_{2zt}, where vectors are (x, y), where x and y are zt-vectors over K and so that V_{zt} is represented as $x = 0$.

Choose three mutually disjoint zt-dimensional subspaces as follows: $x = 0, y = 0, y = x$. We note that the 'unique' K-regulus partial spread $\mathcal{R}$ containing $x = 0, y = 0, y = x$ is represented as follows:

$$x = 0, y = x\delta I_{zt}; \delta \in K.$$

For $\mathcal{S}$ a Pappian *t*-spread of $x = 0$, use the mapping $\sigma: \begin{bmatrix} 0 & I \\ I & 0 \end{bmatrix}$, of V_{2zt}, which fixes $y = x$ pointwise, and which maps $\mathcal{S}$ to a Pappian *t*-spread $\mathcal{S}^\sigma$ of $y = 0$. Then $\mathcal{S} \oplus \mathcal{S}^\sigma = \{L \oplus L^\sigma; L \in \mathcal{S}\}$ is the set of Pappian subplanes of a rational Pappian net $\mathcal{N}_t$ sharing the translation K-regulus net $\mathcal{N}$ defined by $\mathcal{R}$.

PROOF. L is a 1-dimensional K^+-subspace and $L \oplus L^\sigma$ is a 2-dimensional vector space over K^+, under the natural definition of scalar multiplication extended to the direct sum. We see that $0 \oplus L^\sigma$ is a subspace that intersects $x = 0$ in a 1-dimensional K^+-subspace. Similarly, $L \oplus L^\sigma$ intersects $x = 0, y = 0, y = x$ in 1-dimensional K^+-subspaces. Considering V_{zt} as a vector space over K^+, all other 1-dimensional K^+-subspaces have the form $\langle (x_0, x_0\delta) \rangle$, where δ is a fixed non-zero element of K^+, for x_0 a non-zero vector, where the notation indicates that the subspace is generated over K^+ by $(x_0, x_0\delta)$. What this means is that this 2-dimensional K^+-subspace lies on the K^+-regulus generated by $x = 0, y = 0, y = x$.

It is immediate that the subspaces $L \oplus L^\sigma$ and $M \oplus M^\sigma$ for $M \neq L$ in $\mathcal{S}$ are disjoint.

Note that it follows that the K-regulus partial spread $\mathcal{R}$ is contained in the rational Desarguesian partial spread $\mathcal{R}_t$ defining the net $\mathcal{N}_t$. $\square$

Now if there are two *t*-spreads that are disjoint on their *t*-subspaces, $\mathcal{S}_1$ and $\mathcal{S}_2$, although rational Pappian partial spreads $\mathcal{R}_{1,t}$ and $\mathcal{R}_{2,t}$ are constructed, it is not clear that these two partial spreads are disjoint on their zt-components that are not in the common K-regulus.

In Jha and Johnson [**93**], there is a general study of the connection between Desarguesian *t*-spreads in $PG(zt - 1, q)$ and rational Desarguesian partial spreads of degree $1 + q^t$ and order q^{zt} in $PG(2zt - 1, q)$, and our treatment above is similar in the finite case.

There is also a general correspondence between partial Desarguesian 2-parallelisms and translation planes covered by rational Desarguesian nets, also established by the authors in [**93**] that generalizes unpublished work of Prohaska and Walker, and is inspired by the work of Walker [**181**], and Lunardon [**166**].

We now are able to generalize this work for t a prime and K an arbitrary field. First we state a lemma, which we leave to the reader for verification. Coordinate structures for translation planes are called 'quasifields' and the reader unfamiliar with such algebraic systems is directed to the Foundations book [**21**] for necessary background. The following material is adapted from the author's work [**97**].

LEMMA 18. *Let K and L be skewfields. Assume there is a quasifield Q such that $K \subseteq Q \subseteq L$. Then Q is a skewfield.*

THEOREM 51. *Let V be a vector space of dimension $2tz$ over a field K and let $\mathcal{R}$ be a regulus of V (of $PG(2tz - 1, K)$). Let Γ be a set of rational Pappian partial spreads, each coordinatizable by a field extension of K of degree t and containing $\mathcal{R}$.*

If t is a prime, then

$$\cup(\Gamma - \mathcal{R}) \cup \mathcal{R}$$

is a partial spread if and only if for any choice of component A of $\mathcal{R}$, considered as a tz-dimensional K-space, Γ induces a Pappian partial t-parallelism of A.

PROOF. By Theorems 50 and 49, it remains to prove that two t-spreads of a Pappian partial t-parallelism construct partial spreads whose union is also a partial spread.

Let A be a zt-dimensional vector subspace of a $2zt$-dimensional vector space V over a field K. Let $\mathcal{S}_1$ be a Pappian t-spread of A and by Theorem 50 form the associated rational partial zt-spread $\mathcal{R}_1$ in V that contains the K-regulus partial spread $\mathcal{R}$. Now take a second Pappian t-spread of A $\mathcal{S}_2$ and form the rational partial zt-spread $\mathcal{R}_2$ containing $\mathcal{R}$. The proof of our result is finished if it could be shown that the two rational partial spreads do not share any points outside of the regulus $\mathcal{R}$. So, assume that Q is a common point of $\mathcal{R}_1$ and $\mathcal{R}_2$ that does not lie in $\mathcal{R}$ (on a component of $\mathcal{R}$). Since rational Pappian nets are subplane covered nets, covered by Pappian subplanes, therefore, there is a subplane π_1 of $\mathcal{R}_1$ a subplane π_2 of $\mathcal{R}_2$ that share the point Q. Since π_1 is a $2t$-dimensional K-subspace generated any two t-intersections of components of the regulus $\mathcal{R}$, take $x = 0, y = 0, y = x$ as three components of $\mathcal{R}$, say taking A as $x = 0$. Then there are points on $x = 0, y = 0, y = x$ in π_1 as follows: $P_{x=0} + P_{y=0} = Q = Z_{x=0} + Z_{y=x} =$

$W_{y=0} + W_{y=x}$, where the subscripts indicate in what component the points are located.

Take the subspace $\Lambda = \langle P_{x=0}, P_{y=0}, Z_{x=0}, Z_{y=x}, W_{y=0}, W_{y=x}\rangle$, which is at least 2-dimensional over K. If the subspace is 2-dimensional over K, then since $\mathcal{R}$ is a regulus and since Λ contains points of $x = 0, y = 0, y = x$, then Λ must intersect all components of $\mathcal{R}$, which means that Q cannot be in Λ, as the subspace would define a subplane of the regulus net. Therefore, Λ has dimension at least 3 over K. Let $U = \pi_1 \cap \pi_2$ as a subspace of dimension at least 3. We claim that the dimension is at least 4. Actually, no two of the subspaces $\langle P_{x=0}, P_{y=0}\rangle$, $\langle Z_{x=0}, Z_{y=x}\rangle$, $\langle W_{y=0}, W_{y=x}\rangle$ can be equal since otherwise a 2-dimensional subspace would non-trivially intersect three components of a regulus, and the same contradiction applies. By the construction given to establish Theorem 50, then π_1 and π_2 admit the group element $\begin{bmatrix} 0 & I \\ I & 0 \end{bmatrix}$, which means that $\pi_1 \cap \pi_2$ also admits this group. Therefore, this implies that the intersection on $x = 0$ is at least 2-dimensional. Hence, Λ is 4-dimensional, as it is generated by two mutually disjoint 2-dimensional K-subspaces.

Let $P^*_{x=0} + P^*_{y=0} = Q$, where the points are in π_2. Then $P_{x=0} + P_{y=0} = P^*_{x=0} + P^*_{y=0} = Q$, clearly implies that $P_{x=0} = P^*_{x=0}$, $P_{y=0} = P^*_{y=0}$ so that Λ is common to $\pi_1 \cap \pi_2$.

Define a point-line geometry as follows: The 'points' are the points of U, the 'lines' are the lines PQ, of both π_1 and π_2, where P, Q are distinct points of U. We claim then U becomes an affine plane and then an affine subplane of both π_1 and π_2. The reader is invited to prove this fact. $U = \pi_1 \cap \pi_2$ is an affine translation subplane and as a subspace is of dimension at least 4. But π_1 is a Pappian affine plane coordinatized by a field extension K of degree t and U is an affine subplane of π_1 that strictly contains the Pappian subplane coordinatized by K and clearly U is a translation subplane and therefore coordinatizable by a quasifield. Let ρ_K denote the common Pappian subplane of π_1 and π_2 coordinatized by K and let K_i^+ denote the t-dimensional extension field of K that coordinatizes π_i, for $i = 1, 2$. Now ρ_K, U, and π_1 may be simultaneously coordinatized by K, a quasifield $\mathcal{U}$ and a field K_1^+, respectively, which makes $\mathcal{U}$ a field by Lemma 18. Since $[K_1^+ : K] = t = [K_1^+ : \mathcal{U}][\mathcal{U} : K]$, $\mathcal{U}$ properly contains K and t is a prime so that $\mathcal{U} = K_1^+$ and similarly $\mathcal{U} = K_2^+$, so that $\pi_1 = \pi_2$, a contradiction as this would imply that the original t-spreads $\mathcal{S}_1$ and $\mathcal{S}_2$ have a common t-space in their spreads.

This completes the proof of the theorem. $\square$

It remains to consider the collineation group of the partial spread Γ of Theorem 51.

THEOREM 52. *Let Γ denote a partial spread of $PG(2zt - 1, K)$, which is covered by rational Pappian t-spreads, each of which may be coordinatized by a t-dimensional field extension of K, and which contain a fixed K-regulus partial spread $\mathcal{R}$.*

Then Γ admits a collineation group isomorphic to $\Gamma L(2, K)$, where the normal subgroup $GL(2, K)$ is generated by central collineations and scalar collineations, with axes components of $\mathcal{R}$, and centers on the infinite points of $\mathcal{R}$. In particular, $SL(2, K)$ is generated by elations. $\Gamma L(2, K)$ leaves invariant each Pappian subplane of each rational Pappian spread.

PROOF. The proof is straightforward and left to the reader to complete.   □

2.1. A Bound on Partial Desarguesian t-Parallelisms. So, a partial Desarguesian t-parallelism in a vector space of dimension zt over $GF(q)$ of cardinality m induces a partial spread of degree

$$1 + q + m(q^t - q).$$

The maximum partial spread has total degree $q^{zt} - q + 1 + q = 1 + q^{zt}$. Therefore,

$$m \leq (q^{zt-1} - 1)/(q^{t-1} - 1).$$

We then have the following corollary:

COROLLARY 14. *If t is a prime, the maximum cardinality of a Desarguesian partial t-parallelism in a vector space of dimension zt over $GF(q)$ is*

$$\left[(q^{zt-1} - 1)/(q^{t-1} - 1)\right]$$

and if this bound is taken on then $t - 1$ must divide $zt - 1$.

The following two theorems discuss possible Desarguesian t-parallelisms.

THEOREM 53. *If $t = 2$, then $(q^{2r-1} - 1)/(q - 1)$ is the number of spreads of a 2-parallelism.*

If t is an odd prime, then to achieve a parallelism, we require

$$\frac{(q^{zt} - 1)(q^{zt-1} - 1)...(q^{zt-t-1} - 1)}{(q^t - 1)(q^{t-1} - 1)...(q - 1)}$$

t-spaces and hence

$$\frac{(q^{zt-1} - 1)...(q^{zt-t-1} - 1)}{(q^{t-1} - 1)...(q - 1)}$$

t-spreads.

Therefore, a Desarguesian t-parallelism exists for t a prime only if $t = 2$.

THEOREM 54. *A Desarguesian 2-parallelism in $PG(3,q)$ then constructs a translation plane of order q^4 whose spread is covered by a set of $(q^3 - 1)/(q-1) = 1 + q + q^2$ rational Desarguesian partial spreads of degree $1 + q^2$ that share a regulus partial spread of degree $1 + q$.*

2.2. Generalization to Arbitrary Skewfields. Now the reader might ask if all of the preceding is valid over arbitrary skewfields K. In a word, 'yes' and see the author's article [**97**]. On the other hand, it is a good exercise for the reader to work out this material for oneself. Here is where you are going:

THEOREM 55. *The set of Desarguesian left t-spreads over a skewfield K in a k-dimensional vector space V_k over a skewfield K is equivalent to the set of rational Desarguesian right partial spreads in a vector space of left dimension $2k$ over K, each of which is coordinatizable by a left skewfield extension of degree t over K.*

Specifically, the construction is given as follows.

Let S be a Desarguesian left t-spread of a $k = zt$-dimensional vector space V_{zt} over a field K. Let K^+ denote the associated left t-dimensional field extension of K such that the elements of the t-spread are left 1-dimensional K^+-subspaces.

Embed V_{zt} in a vector space V_{2zt} of dimension $2zt$ over K. Choose a basis for V_{2zt}, where vectors are (x, y), where x and y are zt-vectors over K, and so that V_{zt} is represented as $x = 0$.

Choose three mutually disjoint zt-dimensional subspaces as follows: $x = 0, y = 0, y = x$, and consider the standard K-pseudo-regulus $\mathcal{R}$

$$x = 0, y = \delta x; \delta \in K.$$

For S a Desarguesian left t-spread of $x = 0$, use the mapping σ: $\begin{bmatrix} 0 & I \\ I & 0 \end{bmatrix}$,

of V_{2zt}, which fixes $y = x$ pointwise, and which maps S to a Desarguesian left t-spread S^σ of $y = 0$. Then $S \oplus S^\sigma = \{L \oplus L^\sigma; L \in S\}$ is the set of left Desarguesian subplanes of the right K^+-pseudo-regulus net $\mathcal{N}_t$ defined by

$$\mathcal{R}_t : x = 0, y = \delta x; \delta \in K^+.$$

sharing the right K-pseudo-regulus $\mathcal{N}$ defined by $\mathcal{R}$.

Also, we note the following generalization to skewfields.

THEOREM 56. *Let π be a translation net with right projective $k - 1$-spread in $PG(2k - 1, K)$, where K is a skewfield. Assume that the* vector k-spread $\mathcal{S}$ is the union of a set of rational Desarguesian right partial k-spreads each coordinatizable by a skewfield extension of K of left dimension t. Such right partial spreads will mutually share precisely a right K-pseudo-regulus partial spread $\mathcal{R}$. Choose any component of A of $\mathcal{R}$.

(1) Each rational Desarguesian partial k-spread produces a translation net, which is a subplane covered net whose subplanes intersect A in a left t-spread of A considered as a left k-dimensional K-vector space.

(2) The set of left t-spreads as defined in part (1) form a left partial Desarguesian t-parallelism.

THEOREM 57. Let V be a vector space of left dimension $2tz$ over a skewfield K, and let $\mathcal{R}$ be the standard right pseudo-regulus of V (of $PG(2tz-1,K)$). Let Γ be a set of rational Desarguesian right partial spreads, each coordinatizable by a left skewfield extension of K of degree t and containing $\mathcal{R}$.

If t is a prime, then

$$\cup(\Gamma - \mathcal{R}) \cup \mathcal{R}$$

is a partial spread if and only if for any choice of component A of $\mathcal{R}$, considered as a left tz-dimensional K-space, Γ induces a Desarguesian left partial t-parallelism of A.

REMARKS 1. We have connected translation nets covered by right pseudo-regulus nets sharing a common right K-pseudo-regulus net and each corresponding to a left t-dimensional skewfield extension of K with Desarguesian partial left t-parallelisms. However, in the infinite case it is not clear that there are not Desarguesian t-parallelisms. Certainly, a translation plane of this type will produce a maximal partial Desarguesian t-parallelism.

CHAPTER 11

Direct Products of Affine Planes

In Chapter 10, we gave a construction of partial t-spreads starting from Desarguesian partial left t-parallelisms, which would construct first the putative left subplanes of a right pseudo-regulus net. In this chapter, we consider, conversely, a construction that gives the right pseudo-regulus net directly. The main problem with reversing these ideas to construct the right pseudo-regulus net directly involves taking a t-spread of a vector space of dimension zt over a skewfield K and forming the 'direct product' to hopefully obtain a net. However, if z is not 2, then there will be a translation Sperner space constructed from a t-spread by taking translates of the t-subspaces of the spread, which is not a net. These ideas will be revisited and considered in a more general manner when we discuss general spreads and the affine structures that they generate. In any case, for this chapter, we will consider t-spreads in vector spaces of dimension $2t$ over skewfields K. This will allow that any t-spread will define a translation plane. The direct product of two nets is also a net, and so the direct product of two affine translation planes will also be a net. This will allow the consideration of the arbitrary product of affine translation planes thus constructing various nets.

Many of these ideas have been considered in the Subplanes text [**114**], when $t = 2$ and the material there considers the direct product of two isomorphic translation planes (the 2-fold product). Here we consider a more general theory of n-fold products of n isomorphic translation planes. Some of the early material in this text is also in the Subplanes text. There are a variety of ways that one might define a 'direct sum' or 'direct product' of point-line geometries. In our chapters on Sperner spaces and focal-spreads and their construction, we give other definitions of direct sums. So, the reader might be aware that there could more than one meaning for these terms.

We first shall consider the direct product of two affine planes. This may be generalized to more general products of nets and other structures.

DEFINITION 50. *Let π_1 and π_2 be affine planes equal to $(\mathcal{P}_i, \mathcal{L}_i, \mathcal{C}_i, \mathcal{I}_i)$ for $i = 1, 2$, respectively, where $\mathcal{P}_i$ denotes the set of points, $\mathcal{L}_i$ denotes the set of lines, $\mathcal{C}_i$ denotes the set of parallel classes, and $\mathcal{I}_i$ denotes the incidence relation of π_i for $i = 1, 2$.*

Assume that there exists a $1 - 1$ correspondence σ from the set of parallel classes of π_1 onto the set of parallel classes of π_2.

We denote the following incidence structure $(\mathcal{P}_1 \times \mathcal{P}_2, \mathcal{L}_1 \times_\sigma \mathcal{L}_2, \mathcal{C}_\sigma, \mathcal{I}_\sigma)$ by $\pi_1 \times_\sigma \pi_2$, and called the 'direct product of π_1 and π_2' defined as follows:

The set of 'points' of the incidence structure is the ordinary direct product of the two point sets $\mathcal{P}_1 \times \mathcal{P}_2$.

The 'line set' of the incidence structure $\mathcal{L}_1 \times_\sigma \mathcal{L}_2$ is defined as follows: Let ℓ_1 be any line of π_1 and let ℓ_1 be in the parallel class α. Choose any line ℓ_2 of π_2 of the parallel class $\alpha\sigma$ and form the cross product $\ell_1 \times \ell_2$ of the set of points of the two lines. The 'lines' of $\mathcal{L}_1 \times_\sigma \mathcal{L}_2$ are the sets of the type $\ell_1 \times \ell_2$.

The set of 'parallel classes' $\mathcal{C}_\sigma$ is defined as follows: Two lines $\ell_1 \times \ell_2$ and $\ell_1' \times \ell_2'$ are parallel if and only if ℓ_1 is parallel to ℓ_1'. Note that this implies that ℓ_2 is parallel to ℓ_2' since both lines are in the parallel class $\alpha\sigma$.

Parallelism is an equivalence relation and the set of classes is $\mathcal{C}_\sigma$.

Incidence $\mathcal{I}_\sigma$ is defined by the incidence induced in the above defined sets of points, lines, and parallel classes.

DEFINITION 51. *A net is a 'vector space net' if and only if there is a vector space over a skewfield K such that the points of the net are the vectors and the vector translation group makes the net into an Abelian translation net.*

THEOREM 58. *(1) If π_1 and π_2 are affine planes and σ is a $1 - 1$ correspondence from the parallel classes of π_1 onto the parallel classes of π_2, then*

$\pi_1 \times_\sigma \pi_2$ is a net whose parallel class set is in $1 - 1$ correspondence with the parallel class set of either affine plane. Furthermore, this net admits Baer subplanes isomorphic to π_1 and π_2 on any given point.

(2) If π_1 and π_2 both have order n, then the order of $\pi_1 \times_\sigma \pi_2$ is n^2 and the degree (number of parallel classes) is $n + 1$.

(3) If π_1 and π_2 are both translation planes with translation groups T_i for $i = 1, 2$ respectively, then $\pi_1 \times_\sigma \pi_2$ is an Abelian translation net with translation group $T_1 \times T_2$.

PROOF. It follows easily that since parallelism is an equivalence relation on the line set of π_i for $i = 1, 2$, then parallelism is an equivalence relation in the line set of $\pi_1 \times_\sigma \pi_2$. Hence, if α is a parallel class of π_1, we may denote by $\alpha \times \alpha\sigma$ or $(\alpha, \alpha\sigma)$ the corresponding parallel class of $\pi_1 \times_\sigma \pi_2$. It is then clear that this mapping is a $1-1$ correspondence from the parallel classes of π_1 onto the parallel classes of $\pi_1 \times_\sigma \pi_2$. Given any point (P_1, P_2) of $\pi_1 \times_\sigma \pi_2$, where $P_i \in \mathcal{P}_i$ for $i = 1, 2$, choose any parallel class $(\alpha, \alpha\sigma)$ of $\mathcal{C}_\sigma$. Then exists a unique line ℓ_{α, P_1}, $\ell_{\alpha\sigma, P_2}$ of $\mathcal{L}_i$ for $i = 1, 2$ respectively, that contains P_1, P_2, respectively. Thus, there is a unique line $\ell_{\alpha, P_1} \times \ell_{\alpha\sigma, P_2}$ of $\mathcal{L}_1 \times_\sigma \mathcal{L}_2$ containing (P_1, P_2). This proves that the incidence structure is a net and furthermore proves (1) and (2) since it follows that the subplanes themselves become Baer subplanes of the net. To prove (3), let (P_1, P_2) denote points of the direct product for $P_i \in \mathcal{P}_i$, $i = 1, 2$. Then, it is easy to verify that the mapping $\tau : (\tau_1, \tau_2)$ for $\tau_i \in T_i$, $i = 1, 2$ defined by $(P_1, P_2)\tau = (P_1\tau_1, P_2\tau_2)$ is a collineation that fixes each parallel class of the direct product. Furthermore, since T_i acts regularly on the points of $\mathcal{P}_i$ for $i = 1, 2$, it follows that $T_1 \times T_2$ acts regularly on the points of the direct product. Hence, the direct product net is a translation net and since T_i are both Abelian, the direct product net is an Abelian translation net. This completes the proof of (3). □

REMARK 24. *In part (3) of the previous theorem, if the two translation planes are defined over the same prime field, then, the direct product net is a vector space net, which is an elementary Abelian translation net. On the other hand, one cannot guarantee this with the existence of two Baer subplanes within a net. However, when there are three Baer subplanes incident with a given point of an Abelian translation net, one always obtains an elementary Abelian translation net as seen below.*

DEFINITION 52. *A direct product net, where σ is induced from an isomorphism of π_1 onto π_2 is called a 'regular direct product' by π_1. Identifying π_1 and π_2, we also refer to $\pi_1 \times \pi_1$ as a '2-fold product.'*

COROLLARY 15. *A regular direct product net contains points incident with at least three mutually isomorphic Baer subplanes.*

PROOF. Note that as we have the regular direct product $\pi \times \pi$, we also obtain the Baer subplane defined by $\{(P, P); P \text{ a point of } \pi\}$ as well as $\pi \times 0$ and $0 \times \pi$. □

0.3. n-Fold Products. For n finite, we now define the n-fold product inductively and denoted by $\Pi_n\pi$. Also we consider π a translation plane given by left t-spread over a skewfield K. Hence, the point set is the direct product $\Pi_n\mathcal{P}$, where $\mathcal{P}$ is the point set of π. Assume also that π is given by left t-spread in a vector space V_{2t} of left dimension over a skewfield K. Now that $\pi \times \pi$ is a net we consider the direct product with π, the 3-fold product. In this case, the points of the product are points of $(\pi \times \pi) \times \pi$. If π is finite of order n, then we similarly obtain a net of degree $1 + n$ and order n^3. In this setting, we still obtain subplanes of the net isomorphic to π. The following generalization is left for the reader to verify. It is possible to consider arbitrary products of nets again producing nets. These ideas are continued in Johnson and Ostrom [145]. We indicate how this might be completed by the direct product of n affine planes, particularly when the affine planes are all Desarguesian. In the Subplanes text [114], the main emphasis was on nets containing Baer subplanes, so there the primary objects demanded 2-fold products. Here we wish to include a more general setting, so we generalize this work to n-fold products.

As mentioned, some of the previous theory was taken from the author's text [114]. In that text, the question of whether Abelian nets that contain sufficiently many point-Baer subplanes turn out to be direct product nets. In a word, they do! We state the so-called Reconstruction Theorem here, as it has general use. The reader is directed to Theorem 17.7 and Corollary 17.8 of [114] for the proof.

0.4. Reconstruction of Direct Product Nets.

THEOREM 59. *Let M be an Abelian translation net.*

(1) If M contains two point-Baer subplanes incident with a point whose infinite points are the infinite points of M, then M is a direct product net.

(2) If M contains three such point-Baer subplanes incident with a point, then M is a regular direct product net and each pair of the planes are isomorphic.

(3) Furthermore, M is then a vector space net over a field L, and the point-Baer subplanes may be considered L-subspaces.

(4) If one of the subplanes π_0 has kernel K_0 and M is isomorphic to $\pi_0 \times \pi_0$, then M is a K_0-vector space nets. At least three of the point-Baer subplanes of the net, which share an affine point and all of their parallel classes are K_0-subspaces, but not all point-Baer subplanes are necessarily K_0-subspaces.

1. Desarguesian Products

We consider the nature of the lines of n-fold products (regular n-fold direct products) of isomorphic Desarguesian affine planes π_i, $i = 1, 2, .., n$. We identify π_i and π_j for all i, j and consider the corresponding regular direct product or n-fold product $\Pi_n \pi$.

Let π_i have points (x_i, y_i), where x_i, y_i are in a skewfield K_{π_1} coordinatizing π_i. Hence, the points are $((x_1, y_1), (x_2, y_2), .., (x_n, y_n))$ which we rewrite as $(x_1, y_1, x_2, y_2, .., x_n, y_n)$, for all $x_i, y_i \in K_{\pi_1}$ and $i = 1, 2, .., n$. We consider the lines of the n-fold cross product incident with $(0, 0, 0, 0, .., 0)$ ($2n$ zeros).

Now assume that the associated vector space is a left K_{π_1}-vector space. Recall for Desarguesian planes, the 'slopes' are then written on the right to ensure that the components are left K_{π_1}-subspaces. Choose a line $y_1 = x_1 \alpha$ for $\alpha \in K_{\pi_1}$. The parallel class of π_j associated with α is then α and the unique line incident with $(0, 0)$ of π_j is $y_j = x_j \alpha$.

Hence, the line of the cross product is

$$\{(x_1, x_1\alpha, x_2, x_2\alpha, .., x_n, y_n\alpha); x_i \in K_{\pi_1} \text{ for } i = 1, 2, .., n\}.$$

Similarly, the line of the cross product incident with $(0, 0, 0, 0, .., 0)$ in the parallel class $((\infty), (\infty)\sigma_1, .., (\infty)_{n-1})$ (where the σ_i mappings are isomorphisms connecting the slope set of the n isomorphic Desarguesian affine planes) is

$$\{(0, y_1, 0, y_2, .., 0, y_n); y_i \in K_{\pi_1} \text{ for } i = 1, 2, .., n\}.$$

Since the plane π_i admits the translations $(x_i, y_i) \longmapsto (x_i + a, y_i + b)$ for all $a, b \in K_{\pi_1}$, $i = 1, 2, .., n$, it follows that the mappings

$$(x_1, y_1, x_2, y_2, .., x_n, y_n)$$
$$\longmapsto (x_1 + a_1, y_1 + b_1, x_2 + a_2, y_2 + b_2, .., x_n + a_n, y_n + b_n)$$

are collineations, for each a_i, b_i of K_{π_1}, which fix each parallel class of the net.

Hence, the net admits a group of translations and it follows to specify the lines, we need only specify the lines incident with the zero vector. Hence, the components of the translation net are the left K-subspaces.

$$\{(x_1, x_1\alpha, x_2, x_2\alpha, .., x_n, x_n\alpha);$$
$$x_i \in K_{\pi_1} \text{ for } i = 1, 2, .., n\}, \text{ for } \alpha \in K_{\pi_1}$$

and

$$\{(0, y_1, 0, y_2, .., 0, y_n);$$
$$y_i \in K_{\pi_1} \text{ for } i = 1, 2, .., n\}.$$

Now consider the mapping

$$(x_1, y_1, .., x_n, y_n) \to (x_1, .., x_n, y_1, .., y_n),$$

which will change the form of the components of the translation net but normalizes the translation group. Hence, with the notation change, we obtain the components in the following form:

$$(x, x\alpha), \ y = x\alpha, \text{ and } x = 0,$$

where

$$x = (x_1, x_2, .., x_n) \ and \ x\alpha = (x_1\alpha, .., x_n\alpha).$$

We are now considering the n-fold direct product of the original vector space over K_{π_1}. Then we obtain the left pseudo-regulus net:

$$x = 0, y = x\alpha; \alpha \in K_{\pi_1}.$$

It follows that since we have a subplane covered net the net is covered by subplanes isomorphic to π. Also, note that while the components are left K_{π_1}-subspaces, the subplanes of the direct product net (the original π's) are natural 'right' K_{π_1}-subspaces.

Now suppose that π is given by a left t-spread over a skewfield K.

We formalize this, noting that we have proved all parts except part (5) of the following:

THEOREM 60. *Let π be a Desarguesian affine plane defined by a spread of left t-dimensional vector subspaces of a left $2t$-dimensional K-vector space V, where K is a skewfield.*

(1) Then, corresponding to π is a dimension t extension skewfield K_π of K considered as a left K-vector space.

(2) An n-fold product $\Pi_n\pi$ produces a right pseudo-regulus net co-ordinatizable by K_π, whose components (line incident with the zero vector) are left n-dimensional K_π-subspaces and nt-dimensional left K-subspaces of a left $2nt$-dimensional left vector space V_{2nt} over K. This net contains the K-pseudo-regulus net.

(3) There is a natural definition of the set of vectors of $\Pi_n\pi$ as a right $2n$-dimensional K_π-subspace and as a right $2nt$-dimensional K-subspace. Furthermore, the subplanes incident with the zero vector are left 2-dimensional K_π-subspaces and left $2t$-dimensional K-subspaces.

(4) Choose any component A of the K-regulus net. Then the right pseudo-regulus net induces a left Desarguesian t-spread of A. Any right pseudo-regulus net contains a derivable subnet coordinatized by a quadratic skewfield extension K_2 of K. This subnet induces the original left Desarguesian t-spread of V, for V a subspace of A.

(5) Conversely, any left Desarguesian t-spread in an nt-dimensional vector space V_{nt} over K may be considered induced from an n-fold product of a Desarguesian affine plane given by a left Desarguesian t-spread of a $2t$-dimensional vector space V_{2t}.

PROOF. (5): For example, if we start with a vector space V_{2t} of dimension $2t$, and a left Desarguesian t-spread, we would induce back a vector space V_{nt} of dimension nt, and a left Desarguesian t-spread of V_{nt} that contains the original left Desarguesian t-spread of V_{2t}. The question of part (5), is the converse also true? That is, take a left Desarguesian t-spread of V_{nt} over K, can this spread be considered as induced from an n-fold product $\Pi_n \pi$, where π is a left Desarguesian affine plane given by a t-spread in a $2t$-dimensional K-vector space? Consider again the structure of $\Pi_n \pi$ as a right pseudo-regulus net over K_π. This is simply a subplane covered net covered by subplanes isomorphic to π. We know from Theorem 55 that from a left Desarguesian t-spread of V_{nt} there is a corresponding rational Desarguesian left partial spread coordinatized by K_π. Since this partial spread may be considered a n-fold product of any subplane of the corresponding net, we see that these concepts are equivalent. □

REMARKS 2. *The previous result might seem to say something about partial parallelisms in the following sense: Suppose there are two Desarguesian 2-spreads of V_4 over K, S_1 and S_2 that do not share a component. Construct the corresponding n-fold products $\Pi_n \pi_1$, and $\Pi_n \pi_2$, where π_1 and π_2 are Desarguesian affine planes given by the spreads S_1 and S_2. The question is are $\Pi_n \pi_1$ and $\Pi_n \pi_2$ disjoint on components outside of the K-pseudo-regulus partial spread? If not then the induced spreads share a 2-dimensional K-space. However, this is at least possible, since all we know is that the induced spreads in V_4 within the vector space V_{2n} are disjoint, but the induced spreads in V_{2n} may not be disjoint.*

2. Left Spreads and Left Partial Spreads

Now we shall want to consider some 'left' partial spreads in $PG(4t-1, K)_\mathcal{L}$ defined by left 2-spreads in $PG(2t-1, K)_\mathcal{L}$, whether the spread is Desarguesian or not.

Let V, U, W be any three mutually disjoint 4-dimensional left subspaces of a 8-dimensional left K-vector space V_8.

Decompose V_8 relative to V and W so that writing vectors of V_8 in the form (x, y), we assume that V, W, and U have the equations $x = 0$, $y = 0$, $y = x$.

LEMMA 19. *Consider any 2-dimensional left K-subspace X of V. Let X^σ denote the image under the mapping $(x, y) \longmapsto (y, x)$, where x, y are 4-vectors over K.*

Then $X \oplus X^\sigma$ is a 4-dimensional left K-subspace of V_8, which intersects $x = 0$, $y = 0$, $y = x$ in 2-dimensional left K-subspaces. Conversely, any 4-dimensional left K-subspace that intersects $x = 0$, $y = 0$, $y = x$ in 2-dimensional left K-subspaces is of this form.

PROOF. Let P be any point of a 4-dimensional left K-subspace with the indicated properties and which is not incident with $x = 0, y = \gamma x$ for any $\gamma \in K$. By uniqueness of representation, there exist unique nonzero vectors on $x = 0$ and $y = 0$ such that

$$P = (0,0,0,0,y_1,y_2,y_3,y_4) + (x_1,x_2,x_3,x_4,0,0,0,0).$$

In particular, for any vector $(0,0,0,0,y_1,y_2,y_3,y_4)$, there is a vector

$$(x_1,x_2,x_3,x_4,0,0,0,0),$$

which is not a linear combination of

$$(y_1,y_2,y_3,y_4,0,0,0,0),$$

as the intersection on $y = 0$ is 2-dimensional. Similarly, there exist unique nonzero vectors on $x = 0, y = x$ such that

$$P = (0,0,0,0,z_1,z_2,z_3,z_4) + (w_1,w_2,w_3,w_4,w_1,w_2,w_3,w_4),$$

and there exist unique nonzero vectors on $y = x, y = 0$ such that

$$P = (s_1,s_2,s_3,s_4,s_1,s_2,s_3,s_4) + (r_1,r_2,r_3,r_4,0,0,0,0).$$

It follows that (w_1,w_2,w_3,w_4) is

$$= (x_1,x_2,x_3,x_4) = (s_1+r_1, s_2+r_2, s_3+r_3, s_4+r_4)$$

and (y_1,y_2,y_3,y_4) is

$$= (z_1+w_1, z_2+w_2, z_3,+w_3, z_4+w_4) = (s_1,s_2,s_3,s_4).$$

Thus, the left 4-dimensional space generated by these vectors is generated by

$$(0,0,0,0,y_1,y_2,y_3,y_4), (0,0,0,0,x_1,x_2,x_3,x_4),$$
$$(y_1,y_2,y_3,y_4,0,0,0,0), (x_1,x_2,x_3,x_4,0,0,0,0)$$

and is of the form $X \oplus X^\sigma$ for X the left 2-dimensional K-space

$$\langle (0,0,y_1,y_2,y_3,y_4), (0,0,x_1,x_2,x_3,x_4) \rangle.$$

Note that we know that $(y_1,y_2,y_3,y_4) \neq \gamma(x_1,x_2,x_3,x_4)$, for $\gamma \in K$ as otherwise P would be on $y = \gamma x$ contrary to assumption. Hence, the subspace must be generated by the four linearly independent vectors. Since this space is of the form $X \oplus X^\sigma$, we have completed the proof. $\square$

LEMMA 20. *Define a 'right' 8-dimensional K-space relative to $V \oplus W$ so that V, U, W are both right and left K-subspaces. Let $\mathcal{N}_K$ denote the right pseudo-regulus net with components $x = 0, y = \alpha x$, for all $\alpha \in K$. Note that the set of components contains right 4-dimensional K-subspaces and contains V, U, W. Let $\mathcal{P}(\mathcal{N}_K)$ denote the set of vectors that lie on the components of $\mathcal{N}_K$. If P is any vector not in $\mathcal{P}(\mathcal{N}_K)$*

then there exists a unique 2-dimensional left K-subspace of V, X, such that P is incident with $X \oplus X^\sigma$; there is a unique 4-dimensional left K-subspace containing P that intersects V, U, W in 2-dimensional left K-subspaces.

PROOF. The previous lemma shows that there is a unique such 4-dimensional left subspace containing P that intersects V, U, W in 2-dimensional left K-subspaces, and it has the form $X \oplus X^\sigma$. □

2.1. 2-Fold Products.

THEOREM 61. *Given a left spread S of V of 2-dimensional left K-vector subspaces, $\{X \oplus X^\sigma; X \in S\} = \mathcal{P}_S$ is a partial spread of 4-dimensional left K-subspaces, which contains the pseudo-regulus partial spread defining $\mathcal{N}_K$ as a set of Baer subplanes.*

Let π_S denote the affine translation plane determined by the spread S. Considering σ as an isomorphism, we may form the regular direct product $\pi_S \times_\sigma \pi_S$. This partial spread $\mathcal{P}_S$ is defined by the 2-fold product $\pi_S \times \pi_S$, where π_S is the translation plane determined by the spread S.

PROOF. Let ρ denote the affine translation plane determined by the spread S. We note that ρ has kernel containing K and may be considered a left 4-dimensional subspace. Form $\rho \times \rho$ to obtain a net. But note that this is not necessarily a derivable net, as ρ is not necessarily a Desarguesian affine plane. Now clearly, $\rho \times 0, 0 \times \rho$ and $\{(P, \alpha P); \alpha \in K\}$ are Baer subplanes of the net, which are isomorphic to ρ. We can make the space into a right 8-dimensional K-subspace so that these Baer subplanes are in the net and are 4-dimensional right K-subspaces. The lines of the net incident with the zero vector are the lines of the regular direct product net and are 4-dimensional left K-subspaces of the form $X \oplus X^\sigma$ for all $X \in K$ by the previous uniqueness result. This proves the theorem. □

THEOREM 62. $\cup\{X \oplus X^\sigma - \mathcal{P}(\mathcal{N}_K); \forall$ *2-dimensional left subspaces X of $V\}$ is a disjoint cover of the vectors of V_8 that are not in $\mathcal{P}(\mathcal{N}_K)$.*

PROOF. Note that if two left spaces $X \oplus X^\sigma$ and $Y \oplus Y^\sigma$ have a point P in common outside of $\mathcal{N}_K$ then the previous argument shows that $X = Y$. Hence, we have a disjoint cover as maintained. □ □

COROLLARY 16. *The net $\rho \times \rho$ containing the pseudo-regulus net $\mathcal{N}_K$ (as a set of Baer subplanes) admits a collineation group G isomorphic to $GL(2, K)$ defined as follows:*

$$\left\langle (x, y) \longmapsto \left(\begin{bmatrix} \alpha & \beta \\ \delta & \gamma \end{bmatrix} (x, y); \forall \alpha, \beta, \delta, \gamma \in K; \alpha\beta - \delta\gamma \neq 0 \right\rangle$$

(the elements of K act on the left by scalar multiplication).

Furthermore, G leaves invariant each 4-dimensional K-subspace $X \oplus X^\sigma$ for X a 2-dimensional K–subspace of V and acts on the points of $X \oplus X^\sigma - \mathcal{P}(\mathcal{N}_K)$.

PROOF. Represent $\mathcal{N}_K$ in the form

$$x = 0, \ y = x\alpha I_4 \text{ for all } \alpha \in K.$$

Suppose $(x, y) = (x, 0) + (0, y)$ is in $X \oplus X^\sigma$ so that both x and y are in X. Then $\alpha x + \beta y, \delta x + \gamma y$ are both in X. Hence

$$(\alpha x + \beta y, \delta x + \gamma y) = \begin{bmatrix} \alpha & \beta \\ \delta & \gamma \end{bmatrix} (x, y)$$

is in $X \oplus X^\sigma$ for all $\alpha, \beta, \delta, \gamma \in K$. $\square$

3. Desarguesian Left Parallelisms–Right Spreads

THEOREM 63. *Let $\mathcal{P}$ be a left Desarguesian parallelism of $PG(3, K)_\mathcal{L}$, where K is a skewfield. Then there is a corresponding right spread $\pi_\mathcal{P}$ in $PG(7, K)_\mathcal{R}$ consisting of derivable right partial spreads containing the right pseudo-regulus defined by K.*

Conversely, a right spread in $PG(7, K)_\mathcal{R}$, which is a union of derivable right partial spreads containing the right K-pseudo-regulus produces a Desarguesian left parallelism of $PG(3, K)_\mathcal{L}$.

PROOF. Given a left parallelism $\mathcal{P}$ of $PG(3, K)_\mathcal{L}$, there is a corresponding translation net $\pi_\mathcal{P}$. We assert that $\pi_\mathcal{P}$ is a translation plane; that the derivable right partial spreads cover the vectors. If not, there exists a point P outside of the points of $\mathcal{N}_K$ so there is a unique 4-dimensional left vector space $X \oplus X^\sigma$ containing P and intersections $x = 0, y = 0, y = x$ in 2-dimensional left K-subspaces by Lemma 19. Since $\mathcal{P}$ is a left parallelism, there is a Desarguesian spread S containing X and let π denote the Desarguesian affine plane determined by S. Then in $\pi \times \pi$, there is a unique Baer subplane of the net containing P. But, then the derived net of $\pi \times \pi$ is within the structure $\pi_\mathcal{P}$. Now assume that Σ is a spread in $PG(7, K)_\mathcal{R}$ consisting of derivable right partial spreads containing the K-pseudo-regulus. Let V be a component that is also a left 4-dimensional subspace. Then the set of Baer subplanes of each derivable net defines a left 2-spread on V. Since no two derivable nets can share a left 2-space on V, it follows that there is a corresponding Desarguesian left partial parallelism $\mathcal{P}_\Sigma$. Assume that there is a left 2-space X over K, which is not covered by $\mathcal{P}_\Sigma$. Then $X \oplus X^\sigma$ is a 4-dimensional left K-subspace. If P is a point of $X \oplus X^\sigma$, which is not a point of a component of the K-pseudo-regulus, then there exists a unique derivable net containing P. The Baer subplane containing of the derivable net is a 4-dimensional left K-subspace containing P and by uniqueness must be $X \oplus X^\sigma$, which is a contradiction. Hence, $\mathcal{P}_\Sigma$ is a Desarguesian left parallelism. $\qquad\square$

3.1. Deficiency. The Johnson parallelisms of Chapter 28, are parallelisms whose spreads are all Hall save one, so the fact that all but one of the spreads are of a given isomorphism type probably does not say much about the remaining spread. Here we consider the problem from another angle. Suppose we have a partial parallelism in $PG(3, q)$ with say $1 + q + q^2 - t$ spreads. Is it possible to extend to a 'larger' partial parallelism? Suppose $t = 1$, for example. In this case, it is always possible to extend to a unique parallelism. Now try this problem: If $t = 1$ and all of the spreads are Desarguesian, then, is the adjoined spread of the extended parallelism also Desarguesian? Everyone who ever encountered this problem thought so, but no one has been able to give a proof. The reader is directed to Chapter 40 to find this problem listed as open. Please prove it!

DEFINITION 53. *A partial left parallelism in $PG(3, K)_\mathcal{L}$ has deficiency $t < \infty$ if and only if, for each point P, there are exactly t 2-dimensional left K–subspaces containing P that are not in a left spread of the partial left parallelism or, equivalently, there are exactly t lines not incident with P.*

Thus, in the finite case, a partial left parallelism is of deficiency t if and only if the number of spreads is $1 + q + q^2 - t$.

The proof of the following result is left as an exercise for the reader.

THEOREM 64. *Let $\mathcal{P}_{-1}$ denote a partial left parallelism of deficiency one in $PG(3, K)_\mathcal{L}$ for K a skewfield.*

Then, there is a unique extension of $\mathcal{P}_{-1}$ to a left parallelism of $PG(3, K)_\mathcal{L}$.

We now consider Desarguesian partial parallelisms, which have extensions.

THEOREM 65. *Let $\mathcal{S}_{-t}$ be a Desarguesian partial parallelism of deficiency $t < \infty$ in $PG(3, K)$, for K a skewfield.*

Let $\pi_{\mathcal{S}_{-t}}$ denote the corresponding translation net defined by a partial spread in $PG(7, K)_\mathcal{R}$, consisting of derivable right partial spreads sharing the K–pseudo-regulus $\mathcal{N}_K$.

(1) If $\mathcal{S}_{-t}$ can be extended to a partial left parallelism of deficiency $t - 1$ by the adjunction of a left spread S, let π_S denote the translation plane determined by the spread.

Then a partial spread Σ in $PG(V_8, Z(K))$ may be defined as follows:

$$\Sigma = (\pi_{S_{-t}} - \mathcal{N}_K) \cup (\pi_S \times \pi_S),$$

where the notation is intended to indicate the right 4-dimensional K–subspace components of the net $\pi_{S_{-t}}$ that are not in $\mathcal{N}_K$, union the 4-dimensional left K–subspace components of $\pi_S \times \pi_S$.

(2) Σ admits $GL(2, K)$ as a collineation group, where the subgroup $SL(2, K)$ is generated by collineations that fix Baer subplanes pointwise.

PROOF. We have seen previously that $\pi_S \times \pi_S$ is a partial spread containing $\mathcal{N}_K$ as defining Baer subplanes. If a component ℓ of $\pi_{S_{-t}}$ not in $\mathcal{N}_K$ and $\pi_S \times \pi_S$ share a common point P, then there is a unique Baer subplane π_ℓ of one of the derivable nets incident with ℓ that contains P. Since the Baer subplanes incident with the zero vector, assume that two 4-dimensional left K–subspaces intersect three common components of $\mathcal{N}_K$ in 2-dimensional left K–subspaces, then it follows by uniqueness of representation that the two 4-dimensional subspaces are equal, which is a contradiction that S is a spread that extends the parallelism. This proves (1). To prove (2), note that we now are using a partial spread $X \oplus X^\sigma$ for $X \in S$ as components of the new translation net. The components are each invariant under $GL(2, K)$. The union of these partial spreads $X \oplus X^\sigma$ for $X \in S$ cover the components of the regulus net $\mathcal{N}_K$ and the elation groups fix the components of the regulus net pointwise. In the new net, the components of the regulus net now appear as subplanes of the union of the set of partial spreads $X \oplus X^\sigma$ for $X \in S$ and hence are Baer subplanes of the new translation net. This completes the proof of (2) and the proof of the theorem. $\square$

3.2. Deficiency One Partial Parallelisms.

THEOREM 66. *Let $\mathcal{S}_{-1}$ denote a Desarguesian left partial parallelism of deficiency one in $PG(3, K)_{\mathcal{L}}$, where K is a skewfield. Let S denote the spread extending $\mathcal{P}_{-1}$, and let π_S denote the affine plane given by the spread S.*

Then there is a corresponding translation plane with spread that may be defined as follows $\Sigma = (\pi_{S_{-1}} - \mathcal{N}_K) \cup (\pi_S \times \pi_S)$, where the notation is intended to indicate the right 4-dimensional K–subspace components of the net $\pi_{S_{-1}}$ that are not in $\mathcal{N}_K$ together with the 4-dimensional left K–subspace components of $\pi_S \times \pi_S$. The spread then is in $PG(V_8, Z(K))$.

(2) Conversely, a spread with right and left 4-spaces defined as above produces a Desarguesian left partial parallelism of deficiency one.

(3) The unique extension of a Desarguesian left partial parallelism of deficiency one to a left parallelism is Desarguesian if and only if the net $\pi_S \times \pi_S$ defined by the extending left spread is derivable.

PROOF. Let $\mathcal{S}_0 = \mathcal{S}_{-1} \cup \{S\}$ denote the unique parallelism extending $\mathcal{S}_{-1}$. If Σ is not a spread, then there is a point P that is not covered by the components. Hence, there is a unique left 4-dimensional space $X \oplus X^\sigma$ containing P that intersects $x = 0$, $y = 0$, $y = x$ in 2-dimensional left K–subspaces. But X is in a unique spread S' of $\mathcal{S}_0$ so is either in $\mathcal{S}_{-1}$ or is in S. The converse is now essentially immediate. If the extension parallelism is Desarguesian, then $\pi_S \times \pi_S = \mathcal{D}$ is a derivable net and derivation produces the type of translation plane associated initially with a Desarguesian parallelism. Now assume that the net $\mathcal{D}$ is derivable. Then there is a skewfield F so that the Baer subplanes of the net are all 2-dimensional right F-subspaces. Since all of the Baer subplanes are isomorphic Desarguesian subplanes coordinatizable by F, it follows that F must be an extension of K. Furthermore, the components are 2-dimensional left F-subspaces. But the Baer subplanes of $\mathcal{N}_K$ are 4-dimensional right K–subspaces, so it must be that F is a quadratic extension of K and that all Baer subplanes of $\mathcal{D}$ are 4-dimensional right K–subspaces. Note for 'derivation' in the infinite case to produce an affine plane (a translation plane) the subplanes that are Baer in $\mathcal{D}$ must be Baer within the containing plane. That is, the subplane is automatically 'point-Baer' as the points of the net are the points of the translation plane. The question is then whether the subplane is also a 'line-Baer' subplane of the translation plane. In this case, the Baer subplanes are right 4-dimensional K–subspaces and the components not in the net $\mathcal{D}$ are right 4-dimensional K–subspaces. We shall see in the next chapter that, if any point-Baer subplane and any disjoint component direct sum to the ambient vector space then the subplane is also line-Baer. Hence, derivation produces a translation plane whose spread is covered by derivable right partial spreads containing the K-pseudo-regulus partial spread which, in turn, produces a Desarguesian left parallelism extending $\mathcal{P}_{-1}$. $\qquad\square$

In the finite case, we may obtain a characterization by a collineation group:

THEOREM 67. *(1) Given a Desarguesian partial parallelism $\mathcal{P}_{-1}$ of deficiency one in $PG(3,q)$, there is a corresponding translation plane Σ with spread in $PG(7,q)$.*

(2) The translation plane admits a collineation group isomorphic to $GL(2,q)$ which contains a subgroup isomorphic to $SL(2,q)$ that is generated by Baer collineations.

Furthermore,

(2)(a) there are exactly $(1+q)$ Baer axes which belong to a common net $\mathcal{N}^*$ of degree $(1+q^2)$.

(2)(b) $SL(2,q)$ has exactly $q(q+1)$ component orbits of length $q(q-1)$. These orbits union the $(1+q)$ Baer axes define derivable nets.

The $q(q+1)$ derivable nets induce a Desarguesian parallel parallelism of deficiency one on any Baer axis considered as $PG(3,q)$.

(2)(c) The Desarguesian partial parallelism may be extended to a Desarguesian parallelism if and only if the net $\mathcal{N}^*$ is derivable.

We shall give the converse in the next chapter (see Theorem 69).

CHAPTER 12

Jha-Johnson $SL(2, q) \times$ C-Theorem

In Chapter 10, a strong connection was shown between partial Desarguesian t-parallelisms over $GF(q)$ and translation nets of order q^{zt} covered by rational Desarguesian nets of degree $1 + q^t$ that share a $GF(q)$-regulus net. The known Desarguesian 2-parallelisms are in $PG(3, q)$ and admit a cyclic group C of order $1 + q + q^2$ acting transitively on the set of $1 + q + q^2$ 2-spreads of the parallelism, and where $q \equiv 2 \bmod 3$, so that $(\frac{(q^3-1)}{(q-1)}, q - 1) = (q - 1, 3) = 1$. Also, there are transitive 2-parallelisms in $PG(2r - 1, 2)$, which we will present in a subsequent chapter. So all transitive 2-parallelisms are in $PG(2r-1, q)$, and in these cases $(\frac{(q^{2r-1}-1)}{q-1}, q - 1) = (2r - 1, q - 1) = 1$. The following was proved by Jha and Johnson in [93], [94], without the assumption $(2r - 1, q - 1) = 1$ (and without the assumption in the version of the theorem listed below) so there is potentially a gap in the proof given there. We shall give a somewhat brief version of the proof, using results of Lüneburg, Hering, and Ostrom.

LEMMA 21. *In translation planes of order q^{2r},*
(1) if $2r = 4$ and $d = (3, q-1)$, then $(q+1)/(q+1, d/t - 1) \nmid p^k - 1$, for t dividing d, $t \neq d$, and there is an element g in $SL(2, q)$ such that $|g| \mid (q + 1)/(q + 1, d/t - 1)$ but $\nmid p^k - 1$, for t dividing d, $t \neq d$.

PROOF. It remains to consider $t = 1$ and $(q + 1)/(q + 1, 2)$, which is $(q + 1)$ if q is even and $(q + 1)/2$ if q is odd. If q is odd and $(q + 1)/2$ divides $p^k - 1$, which implies that $q + 1 = 4$, but then $d = 1$. If $q^2 - 1$ has a p-primitive divisor, we are finished. If not, assume that $q = 8$, and take an element of order 9. If $q = p$ and $p + 1 = 2^a$, choose an element of order 8. $\qquad\square$

THEOREM 68. *Let π be a translation plane of order q^{2r}, $r > 1$, $p^s = q$, where $(2r-1, q-1) = d$ and assume that there is a collineation g_t of $SL(2, q)$ of order dividing $(q + 1, d/t - 1) \nmid p^k - 1$, for t any proper divisor of d. Assume that π admits a collineation group in the translation complement isomorphic to $SL(2, q) \times C$, where C is a cyclic group of order $(q^{2r-1} - 1)/(q - 1)$. Then, the following are all true:*
(1) The p-elements of $SL(2, q)$ are elations.

(2) The kernel is isomorphic to $GF(q)$, and the partial spread of elation axes is a regulus $\mathcal{R}$ in $PG(2r-1,q)$.

(3) The spread is a union of rational Desarguesian partial spreads of order q^{2r} and degree $1 + q^2$ that contain the regulus $\mathcal{R}$.

(4) On any elation axis A, the intersection with each rational Desarguesian partial spread induces a 2-spread of A, and the set of the 2-spreads form a Desarguesian 2-parallelism.

The proof shall be given as a series of lemmas.

We begin by assuming that V is a vector space of dimension $4sr$ over $GF(p)$ and assume that u is a prime divisor of $p^{s(2r-1)} - 1$ that does not divide $p^z - 1$, for $z < s(2r-1)$, for $r - 1 \geq 2$. Noting that since $q^{2r-1} \neq 2^6$, there is always a p-primitive divisor, since $s(2r - 1) > 2$, as $r > 1$. Let U be a non-trivial u-subgroup of order u^t and consider $FixU$. Let τ be an element of U of order u and note that

$$(p^{4rs} - 1, p^{s(2r-1)} - 1)$$
$$= p^{(4sr, s(2r-1))} - 1 = p^{s(4r, 2r-1)} - 1 = p^{s(r, 2r-1)=s} - 1.$$

Therefore, τ must have fixed points on V, so let $Fix\tau$ denote the $GF(p)$-subspace pointwise fixed by τ. Let L be a component of π fixed by τ and let $Fix_L\tau$ denote the subspace on L pointwise fixed by τ. Let z denote the dimension of $Fix_L\tau$. Let M be a τ-invariant Maschke complement of $Fix_L\tau$ on M. So, τ induces a group on M, which contains $p^{2rs-z} - 1$ nonzero vectors. Assume that $z > 1$, so u divides

$$(p^{2rs-z} - 1, p^{s(2r-1)} - 1) = p^{(2rs-z, s(2r-1))} - 1.$$

This implies that $(2rs - z, s(2r-1)) = s(2r-1)$, and therefore $s(2r-1)$ divides $2rs - z$. Then s divides $z = sb$, implying that $2r - 1$ divides $2r - b$, a contradiction unless $b = 1$, so that $z = s$. Thus, τ fixes exactly $p^s = q$ points on any invariant component. Suppose that τ fixes exactly one component, then the order of τ divides the remaining q^{2r} components, a contradiction. Hence, τ fixes at least two components, leaving $q^{2r} - 1$, but since $(q^{2r} - 1, q^{2r-1} - 1) = 1$, τ must fix at least three components and fixes exactly q points on any component. Hence, τ fixes a subplane π_0 of order q pointwise. Now let τ be in the center of U, so that U leaves invariant the subplane π_0. Let g be any element of U of order u^e. Then, clearly, g fixes π_0 pointwise. If g fixes a point P not in π_0, then an element of order u fixes P, a contradiction to the previous argument.

Hence, we have proved the first part of the following lemma.

LEMMA 22. *(1) Then $FixU = q = p^s$ and U is semi-regular on $V/FixU$.*

(2) $V = FixU \oplus C_U$, *where* C_U *is the unique Maschke complement of* $FixU$.

PROOF. It remains to prove part (2). Suppose that C_1 and C_2 are Maschke complements of $FixU$, so each have dimension $4rs - s = s(4r - 1)$. Then u must divide $C_1 \cap C_2$, of order $p^w - 1$, so clearly $C_1 \cap C_2 = \langle 0 \rangle$ or $C_1 = C_2$. In the first case, then $C_1 + C_2$ has dimension $2s(4r - 1) > 4rs$, for $(4r - 1) > 2r$, a contradiction, as $2r > 1$. This completes the proof. $\square$

LEMMA 23. *The p-elements of $SL(2, q)$ are elations, and there is a partial spread of degree $1 + q$ consisting of elation axes. The corresponding elation net $\mathcal{E}$ is a rational Desarguesian net of order q^{2r} and degree $1 + q$.*

PROOF. Let σ be an element of $SL(2, q)$ of order u. Since σ commutes with U, then σ leaves invariant $FixU = \pi_0$ and C_U and therefore, fixes non-zero points of each. So σ will either fix a component pointwise and so would be an elation, fix a subplane $Fix\sigma$ pointwise, or fix a proper subspace X_L pointwise on a unique component L. Therefore, $SL(2, q)$ acts on π_0, an affine translation plane of order q. The structure of translation planes of order q admitting $SL(2, q)$ faithfully is known by Lüneburg [**168**] the plane π_0 is Desarguesian and the p-elements are elations. However, it is possible that $SL(2, q)$ fixes π_0 pointwise. If $SL(2, q)$ fixes π_0 pointwise, then $Fix\sigma$ contains π_0 and is a proper affine subplane of order p^k. Therefore, U must fix $Fix\sigma$ pointwise or k is a multiple of $s(2r - 1)$. In a translation plane of order p^{2rs}, the order of any proper subplane is $\leq p^{rs}$, so since $s(2r - 1) > rs$, for $r > 1$, this cannot occur. Therefore, U will fix $Fix\sigma$ pointwise so that $Fix\sigma = \pi_0 = FixSL(2, q)$, since $SL(2, q)$ is generated by its p-elements. Moreover, if any element g of $SL(2, q)$ fixes a point not in π_0, then τ will leave invariant $Fixg$, a subplane of order p^k, and it follows almost immediately that k is a multiple of $s(2r - 1)$, but the largest possible subplane has order p^{sr} in a translation plane of order p^{2sr} and $2r - 1 > r$, for $r > 1$. Therefore, $SL(2, q)$ is semi-regular on the $q^{2r} - q = q(q^{2r-1} - 1)$ points not in π_0 on any component of π_0, implying that $q(q^2 - 1)$ must divide $q(q^{2r-1} - 1)$, a contradiction. Therefore, the p-elements are elations acting on π_0. If the p-elements are planar, then U must fix $Fix\sigma - \pi_0$, which would imply that U has fixed points exterior to π_0. Similarly, if σ fixes pointwise a proper subspace X_L of a component L, then, clearly, U fixes X_L pointwise or X_L has dimension $s(2r - 1)$ over $GF(p)$ and is disjoint from $\pi_0 \cap L$, again a contradiction. So $X_L = \pi_0 \cap L$. But, X_L must have non-zero fixed points on C_U. Therefore, this means that the p-elements are elations of π. Hence, $SL(2, q)$ is generated by affine elations and there is a set $\mathcal{E}$ of $1 + q$ components, which are left invariant by $SL(2, q)$, each of which is fixed pointwise by exactly one of the $1 + q$ Sylow p-subgroups, and this also implies that π_0 is an affine subplane of the net generated by $\mathcal{E}$, which we call the elation net. The structure of elation nets are known by Theorem 290. This result shows that elation nets are rational Desarguesian nets and in this case of order q^{2r} and degree $1 + q$. $\quad\square$

LEMMA 24. *(1) The group*

$$SL(2, q) = \left\langle \begin{array}{l} \left[\begin{array}{cc} A & B \\ C & D \end{array}\right] \\ ; AB - CD = 1, \ A, B, D, C \in K \simeq GF(q) \end{array} \right\rangle$$

and the components of the elation partial spread $\mathcal{E}$ have the form

$$x = 0, y = xA; A \in K.$$

(2) There is a collineation g in $SL(2, q)$ such that $|g| \mid q + 1$, $|g| \nmid p^z - 1$, for $z < 2s$, for $q = p^s$. If g fixes a component $y = xT$ of $\pi - \mathcal{E}$. Then $K[T]$ is a field isomorphic to $GF(q^2)$.

There is a rational Desarguesian partial spread of degree $1 + q^2$ containing $y = xT$ and the $y = xT$-orbit under $SL(2, q)$ is of length $q(q - 1)$. Hence, if there is a collineation of the order of g above that fixes a component $y = xT$, there is a group of order $q + 1$ that fixes $y = xT$.

PROOF. For (1), the reader is directed to the Foundations text [21].

For (2), since the order of $SL(2, q)$ is $q(q^2 - 1)$, it is known that either there is a collineation g of the type required or $q = 8$ or $q = p$ and $p + 1 = 2^a$. If $q = 8$, let g be an element of order 9. If $q = p$, choose h of order 8 and let $g = h^2$.

Now g is in $GL(2, K)$ and has a minimum polynomial $f(x)$ of degree ≤ 2 over K. Let $L(g)$ denote the matrix ring generated over $GF(p)$ by g as a matrix $\left[\begin{array}{cc} A & B \\ C & D \end{array}\right]$. We consider the minimal polynomial $f(x)$ of g over K. Since the order of g cannot divide $q(q - 1)$, it follows that $f(x)$ is irreducible over K so that the module generated by g over K is a field F isomorphic to $GF(q^2)$, which contains $L(g)$. But g cannot be in a proper subfield so that $L(g) = F$. Now noticing that $y = xT$ is fixed by g means that $L(g)$ also fixes g. We note that $L(g)$ contains

$$\widetilde{K} = \left\{ \left[\begin{array}{cc} A & 0 \\ 0 & A \end{array}\right] ; A \in K \right\}.$$

This means that $\widetilde{K}$ also fixes $y = xT$, which is equivalent to the fact that T is centralized by K. If $g = \begin{bmatrix} A & B \\ C & D \end{bmatrix}$, then it follows, since T is centralized by K, that $T^2 C = T(D - A) + B$. If $C = 0$, then $D = A^{-1}$ and g is in a subgroup of order $q(q - 1)$. Therefore, $C \neq 0$, and, therefore, T satisfies a polynomial of degree 2 over K. Since $y = xT$ is not in $\mathcal{E}$, it follows that the polynomial is irreducible so that $K[T]$ is a field isomorphic to $GF(q^2)$. Therefore, there is a rational Desarguesian partial spread containing the elation net admitting $SL(2, q)$. It follows that there is one orbit in this net outside of the elation net.   □

LEMMA 25. *There is an orbit under $SL(2, q)$ of $q^2 - q$ components of π that together with the the elation partial spread forms a rational Desarguesian partial spread of degree $1 + q^2$. There is a C-orbit of $((q^{2r-1} - 1)/(q - 1)) \frac{t}{d}$, rational Desarguesian partial spreads of degree $1 + q^2$, each containing the partial spread of elation axes, where $d = (2r - 1, q - 1)$ and t divides d.*

PROOF. By Lemma 24, to show that there is a Desarguesian partial spread of degree $q^2 + 1$ containing the elation axes $\mathcal{E}$, we need to only show that there is an element g in $SL(2, q)$ of order dividing $q^2 - 1$ but not dividing $p^z - 1$, for $z < 2s$, and fixing a component not in $\mathcal{E}$. We note that $(q^2 - 1, q^{2r} - q) = (q^2 - 1, q^{2r-1} - 1) = (q - 1)$. Hence, there is a group of order $q + 1$ fixing a component $y = xT$. Let $\mathcal{R}_2$ denote the orbit of length $q^2 - q$ under $SL(2, q)$. If h in C moves a component $\mathcal{R}_2$ to a component not in $\mathcal{R}_2$, then since $SL(2, q)$ is normal in $\langle SL(2, q), h \rangle$, it follows that there is a second orbit of length $q^2 - q$ under $SL(2, q)$ and hence a second rational Desarguesian net of order q^{2r} and degree $1 + q^2$ containing the elation net $\mathcal{E}$. Now let g of prime order in C leave $\mathcal{R}_2$ invariant. Since g commutes with all elations, it follows immediately that g fixes all infinite points of the elation net. Since $SL(2, q)$ is transitive on the components not in the elation net, we may assume that g fixes a component of $\mathcal{R}_2 - \mathcal{E}$, and then again using the fact that g commutes with $SL(2, q)$ implies that g fixes all components of $\mathcal{R}_2$. Since g leaves π_0 invariant (as C is cyclic) and there is a unique Desarguesian subplane π_1 of order q^2 of $\mathcal{R}_2$ containing π_0, then π_1 is also invariant under g. Therefore, either g fixes π_1 pointwise or induces a kernel homology on π_1. In the latter case, the order of g has prime order w dividing $(q^2 - 1, (q^{2r-1} - 1)/(q - 1))$. Noting that $(q^2 - 1, q^{2r-1} - 1) = (q - 1)$, then w divides $q - 1$. Now assume that g fixes π_1 pointwise. Therefore, $Fixg$ is an affine subplane of order p^k, which must be fixed by U, which again implies that k is a multiple of $s(2r - 1)$, whereas $k \leq rs$.

Now let h be any element of C that leaves $\mathcal{R}_2$ invariant. Then it follows that h will leave invariant the Desarguesian subplane π_1 and induce a faithful group on π_1, and it is now clear that h will act as a kernel homology and have order dividing $q - 1$. Hence, there are at least $\frac{(q^{2r-1}-1)}{(q-1)}$, rational Desarguesian partial spreads of degree $1 + q^2$ that share the set of $1 + q$ elation axes, where $d = (q - 1, 2r - 1) = 1$. Assume that C has an orbit Γ of $\frac{(q^{2r-1}-1)}{(q-1)} t$ rational Desarguesian partial spread of degree $1 + q^2$. Since the orbit length divides C, it follows that t divides d, which completes the proof of the lemma. $\qquad\square$

LEMMA 26. *There are exactly* $(q^{2r-1} - 1)/(q - 1)$ *rational Desarguesian spreads of degree* $1 + q^2$ *containing the elation axes.*

PROOF. We now note that there are

$$q\frac{\frac{(q^{2r-1}-1)}{(q-1)}}{d}(d - t)(q - 1)$$

infinite points outside of the set of infinite points of the orbit Γ. So, if $t = d$, we have a complete cover of the components by rational Desarguesian partial spreads containing the elation axes $\mathcal{E}$. Therefore, assume that $t < d$. Consider $SL(2, q)$ acting on this set. Note that $q+1$ is relatively prime to $\frac{(q^{2r-1}-1)}{(q-1)} \over d$ and $(q + 1, q - 1) = 1$ or 2, respectively, as q is even or odd. Therefore, we have a group fixing a component not in Γ of order divisible by $N = (q + 1)/(d/t - 1, q + 1)$. According to our assumptions, $N = (q + 1)/(q + 1, d/t - 1)$ is 0 or $\nmid p^k - 1$, $k < 2s$, for t a proper divisor of d. Thus, we have a collineation subgroup H of $SL(2, q)$ of order divisible by N, but $|H|$ does not divide $p^k - 1$, for $k < 2s$, that fixes a component $y = xT$. Furthermore, H does not contain a p-element, since all p-elements are elations. We note that we have a collineation $\left\langle \rho = \begin{bmatrix} -I & 0 \\ 0 & -I \end{bmatrix} \right\rangle$ that fixes all components. By our assumptions, there is always an element g_t, whose order divides N but does not divide $p^k - 1$, for $k < 2s$. Hence, we have another rational Desarguesian partial spread by Lemma 24. Therefore, we have another orbit under C of rational Desarguesian partial spreads of length $\frac{(q^{2r-1}-1)}{(q-1)} \over d} t_1$, and, subtracting these two orbit lengths from the number of components, we see that we may apply the argument again. Therefore, by induction, we have a set of $(q^{2r-1} - 1)/(q - 1)$ rational Desarguesian spreads. $\qquad\square$

LEMMA 27. *The kernel is isomorphic to* $GF(q)$.

PROOF. Each rational Desarguesian partial spread may be coordinatized by a field extension of K, and therefore, each component is a K-space.   □

LEMMA 28. *The set of* $(q^{2r-1}-1)/(q-1)$ *rational Desarguesian partial spreads induces a Desarguesian 2-parallelism on any elation axis.*

PROOF. Each rational Desarguesian net induces a Desarguesian 2-spread on any elation axis. There are $(q^{2r-1}-1)/(q-1)$ rational Desarguesian nets inducing the same number of Desarguesian 2-spreads. If any of these 2-spreads share a 2-space, then there is a unique subplane of order q^2 containing this subspace, so we obtain a parallelism.   □

COROLLARY 17. *Let* π *be a translation plane of order* q^4 *admitting a collineation group in the translation complement isomorphic to* $SL(2,q) \times C$, *where* C *is a cyclic group of order* $(q^3-1)/(q-1)$. *Then, the following are all true:*

(1) The p-elements of $SL(2,q)$ *are elations.*

(2) The kernel is isomorphic to $GF(q)$, *and the partial spread of elation axes is a regulus* $\mathcal{R}$ *in* $PG(3,q)$.

(3) The spread is a union of rational Desarguesian partial spreads of order q^{2r} *and degree* $1+q^2$ *that contain the regulus* $\mathcal{R}$.

(4) On any elation axis A, *the intersection with each rational Desarguesian partial spread induces a 2-spread of* A *and the set of the 2-spreads form a Desarguesian parallelism.*

PROOF. Apply Theorem 68 and Lemma 21.   □

There is another interesting theorem similar to the Jha-Johnson $SL(2,q)\times C$-theorem that probably belongs in this chapter. From Chapter 3.1, there was mention of Desarguesian partial parallelisms with deficiency. Suppose we have a finite Desarguesian deficiency one partial parallelism in $PG(3,q)$, then there is a translation plane of order q^4 with kernel $GF(q)$ admitting a collineation group isomorphic to $SL(2,q)$, generated by Baer collineations such that each orbit of length $q^2 - q$ together with a set of $q+1$ Baer subplanes of the component-fixed net are derivable nets. If the net is derivable, then we obtain a translation plane of order q^4 admitting $SL(2,q)$, generated by elation groups. There is a set of $q+1$ Sylow p-subgroups of $SL(2,q)$, each of which fix a Baer subplane pointwise and this set of $q+1$ Baer subplanes is in an orbit under $SL(2,q)$. Consider any one of these subplanes π_0, so there is a group of order $q(q-1)$ that fixes π_0.

We claim that $SL(2,q)$ is normal in the collineation group of the derived plane π^*. If not then there is a set of elation axes larger than

$q + 1$. If any of these elation axes are not in the derived net N^*, then there are at least $q + 1 + q^2 - q$ elation axes. We know that in this case, the group generated by the elations is isomorphic to $SL(2, p^z)$, where $q = p^r$, and there are $1 + p^z$ elation axes. Hence, p^z is at least q^2. Therefore, we have a translation plane of order q^4 admitting a collineation group isomorphic to $SL(2, p^z)$, where $p^z \geq q^2$. It is not difficult to conclude that $p^z = q^2$ or q^4, and in both cases, the plane is Desarguesian (see, e.g., Chapter 41), a contradiction by the kernel assumption. A similar argument applies if there are $> q+1$ elation axes, inducing $SL(2, q^2)$. Hence, $SL(2, q)$ is normal in the full collineation group of π^*. Therefore, the kernel leaves invariant each of the $q+1$ Baer subplanes that are pointwise fixed by p-collineation (that is, fix one, fix all applies here). Hence, induced on any one of the Baer subplanes π_0 as a 4-dimensional $GF(q)$-subspace is a parallelism. We consider the derived net N^* again and consider the remaining $q^2 - q$ components that are not elation axes. We leave to the reader to determine that it is true that these components are in an orbit under $SL(2, q)$. This means that there is a quadratic extension field $GF(q^2)$, so that N^* is a rational Desarguesian net. This now implies immediately that the new spread induced on π_0 is Desarguesian. This proves the following theorem.

THEOREM 69. *Let π be a translation plane of order q^4 and kernel $GF(q)$ admitting a collineation group isomorphic to $SL(2, q)$, generated by Baer collineations, all of whose non-trivial orbits are of length $q^2 - q$. Let N denote the net of degree $q^2 + 1$ that is component-wise fixed by $SL(2, q)$.*

(1) There are at least $q + 1$ Baer subplanes of N that are incident with the zero vector and on any such Baer subplane π_0, considered as a 3-dimensional projective space over $GF(q)$, there is an induced Desarguesian partial parallelism of deficiency one.

(2) The Desarguesian partial parallelism of deficiency one may be extended to a Desarguesian partial parallelism if and only if the net N is derivable.

CHAPTER 13

Baer Groups of Nets

First, we offer a few words about why in a book on general spreads and parallelisms there would be included general theory of Baer groups. Of course, the previous chapter shows that partial Desarguesian parallelisms of deficiency one in $PG(3, q)$ are directly connected to translation planes admitting $SL(2, q)$ generated by Baer collineations. Also, it is well known that given a flock of a quadratic cone in $PG(2, K)$, for K a field, there is an associated translation plane with spread in $PG(3, K)$, such that the spread is a union of reguli that mutually share exactly one component ℓ. The translation plane admits an elation group E with axis ℓ, which fixes each regulus partial spread and acts transitively on the components not equal to ℓ. Since such a translation plane is then derivable by any of the associated regulus nets, we see that any derived translation plane admits a 'Baer group' that fixes the infinite points of the derived net, fixes a Baer subplane, and fixes each original regulus partial spread, whose component orbits give the components of each regulus partial spread other than ℓ. Conversely, if there is a translation plane with spread in $PG(3, Z)$, for Z a skewfield, such that there is a Baer group whose orbits union the fixed point subplane (the 'Baer axis') define reguli, then there is a partial flock of a quadratic cone with one conic missing; a partial flock of a quadratic cone of deficiency one. This theory also works for flocks of hyperbolic quadrics, where now there is a translation plane with spread in $PG(3, Z)$, for Z a skewfield, admitting a Baer group such that the net containing the Baer subplane in question (the 'Baer axis') admits a second Baer subplane left invariant by the Baer group (the 'Baer coaxis'), and whose component orbits union the Baer axis and Baer coaxis are reguli, and again, in this case, we obtain a partial flock of a hyperbolic quadric of deficiency one.

It is an extremely important problem to determine whether the net defined by the Baer group in either the conical flock type situation or the hyperbolic flock type situation is a regulus net, for then upon derivation of the regulus net, we obtain a flock. Hence, the question is whether partial flocks of quadratic cones or of hyperbolic quadrics of deficiency one may be extended to flocks. In the infinite case, the existence of Baer groups of this type, implies K to be a field. In the finite case, there are deficiency one partial flocks of hyperbolic quadrics, but the extension theory of Payne and Thas [**173**] shows that every partial flock of a quadratic cone of deficiency one in $PG(3, q)$ may be extended uniquely to a flock. Moreover, the recently emerging general theory of flocks over arbitrary cones provides new translation planes admitting Baer groups, where now the spread is not in $PG(3, q)$. So, the problem is to determine the structure of the components that do not intersect the corresponding Baer subplane.

Much of the following material is taken more or less from our previous texts, and we include this here for the benefit of the reader. Also, since whenever a Baer group exists, there is always the question of whether the corresponding Baer net defined by the components of the Baer subplane is derivable. We have previously discussed regular direct product nets and noted that they are, in fact, derivable nets. Therefore, a natural starting place for the study of Baer groups would be to study Baer subplanes in nets. However, now the subplane, which is Baer in the net might not be Baer in the affine plane containing the Baer net and even if the Baer net is derivable, in the general case there is no assurance that the affine plane is derivable. Indeed, in the infinite case, if the kernel is a skewfield that is not a field, there are problems with left and right vector spaces.

We wish to include enough material on Baer groups and flocks of quadric sets so as to understand the nature of the problem both in the infinite and finite cases and also to give the theorem of Johnson, Payne-Thas mentioned above. This theorem can be generalized to more general settings such as flocks of α-cones, the theory of which we sketch in this text. Furthermore, we maintain that the interchange between finite and infinite point-line geometries provides considerable insights into both areas and whenever possible, we develop theory ostensible about finite geometry somehow related to finite fields in a completely general manner in the infinite case and somehow related to skewfields.

So we begin with the definition of a Baer subplane in the general case. Of course, in a finite projective plane of order n, a Baer subplane is simply a subplane of order $\sqrt{n}$, thus requiring n to be a square. Baer subplanes can also be characterized in the finite case either as subplanes such that each point of the superplane is incident with a line of the subplane or that each line of the subplane is incident with a point of the subplane. The wonderful thing is, in the infinite case, one of these properties does not imply the other. So, one property gives what we call a 'point-Baer' subplane and the other gives a 'line-Baer' subplane. If we have both properties, we obtain a 'Baer subplane.'

DEFINITION 54. *A Baer subplane of a net (including the case that the net is an affine plane) has two properties: The point-Baer property that every point of the net is incident with a line of the Baer subplane and the line-Baer property that every line of the net is incident with a point of the Baer subplane.*

A point-Baer subplane is a subplane of the net that has the point-Baer property and a line-Baer subplane is a subplane of the net that has the line-Baer property.

We include the following section to somewhat ease into the complications of what comes later—the analysis of Baer groups within translation planes. As mentioned, we decided to leave this material as general as possible, following the ideas in the Subplane Covered Nets text, as there is some emerging theory of general flocks, for which a more general theory of Baer groups on nets may be applicable.

1. Left and Right in Direct Product Nets

We first consider Desarguesian spreads over skewfields and worry about commutativity or the lack thereof in K. We then consider arbitrary translation planes π with kernel K and consider the kernel mappings acting on the left, whereas elements of the group $GL(\pi, K)$ necessarily act on the right when π is a left K–subspace. We complicate the issue by considering a point-Baer translation subplane π_o with its own kernel K_o and ask when it is possible that there is a direct product net $\pi_o \times \pi_o$ sitting in the translation plane π. We have seen that there is a natural group $GL(2, K_o)$, which is defined via scalar mappings of π_o so this group must act on the 'left.' This group is generated by certain collineations that fix Baer subplanes pointwise and, in particular, certain collineations of the translation planes could fix π_o pointwise and then simultaneously be in $GL(\pi, K)$ and $GL(2, K_o)$ posing a potential problem as one group acts on the right while the other group acts on the left. Such situations seem to imply some sort of commutativity of K in the case when π_o is a K–subspace of π.

Actually, the use of the notation is a bit problematic as the elements of the group are not necessarily K-linear mappings in the traditional sense.

Consider a Desarguesian affine plane (x, y), as a 2-dimensional left vector space over a skewfield $(K, +, \cdot)$. Since we may also consider the affine plane as a 2-dimensional right space over K, we take components to have the form $y = x\alpha$ for α in K and $x = 0$, and note that $y = x\alpha$

and $x = 0$ are 1-dimensional left K–subspaces. We may consider the mappings called the kernel mappings $T_\beta : (x, y) \to (\beta x, \beta y)$. It follows easily that $\{T_\beta ; \beta \text{ in } K\}$ forms a field isomorphic to $K \equiv (K, +, \cdot)$ and fixes each component of the Desarguesian plane.

Define the determinant of $\begin{bmatrix} a & b \\ c & d \end{bmatrix}$ as $ac^{-1}d - b$ if $c \neq 0$ and ad otherwise. Now consider the mappings $(x, y) \to (xa + yb, xc + yd)$ such that the corresponding determinant of $\begin{bmatrix} a & b \\ c & d \end{bmatrix}$ is nonzero. Then it follows easily that each mapping is a $\{T_\beta\}$-linear mapping. Hence, we may justify the designation $GL(2, K)$.

Previously, we have defined the kernel of a translation plane as the set of endomorphisms, which leave each component of the plane invariant.

Hence, $\{T_\beta \mid \beta \in K\} = K^{end}$ is the kernel of the Desarguesian plane π. Furthermore, the full collineation group of π, which fixes the zero vector (the translation complement), is $\Gamma L(2, K)$ or is $\Gamma L(2, K^{end})$. Since the use of K or $K^{end} \simeq K$ is merely in the distinction between K and the associated kernel mappings, we also refer to K as the kernel of the plane. So, considering the translation complement of π as $\Gamma L(2, K)$, then $K^{end*} = K^{end} - \{T_o\}$ is a group of semilinear mappings, and each collineation T_β is in $\Gamma L(2, K)$ but is in $GL(2, K)$ if and only if β is in $Z(K_o)$. That is to say that the elements of $GL(2, K)$ are the elements of $\Gamma L(2, K)$, which commute with K^{end} and T_β commutes with K^{end} if and only if β is in $Z(K)$. The notation can be particularly tricky if one considers βx as a linear mapping over the prime field P of K. For example, βx is normally written $x\beta$ when considering β as a P-linear endomorphism. Then considering an element u in K as a P-linear endomorphism it follows then that notationally $\beta u = u\beta$ when considering the elements as linear endomorphisms, whereas it is not necessarily the case that $\beta u = u\beta$ when the operation juxtaposition is considered as skewfield multiplication.

Note that elements of $GL(2, K)$ act on the elements (x, y) on the right, whereas T_β acts on the elements (x, y) on the left.

We now generalize to an arbitrary affine translation plane with kernel K as follows: Let X be a left K–subspace and form $V = X \oplus X$. We denote points by (x, y) for x, y in X.

When we have an affine translation plane Σ with kernel K, we similarly consider the lines through $(0, 0)$ (components) in the form $x = 0, y = xM$, where M is a K-linear transformation. In the finite dimensional case, we may take M as a matrix with entries in K say as

$[a_{ij}]$ and define $xM = (x_1, x_2, ... x_n)M$ as $(\Sigma x_i a_{i1}, .., \Sigma x_i a_{in})$. It follows that M becomes a left K-linear mapping with scalar multiplication defined by $\beta x = (\beta x_1, \beta x_2, .., \beta x_n)$ and furthermore, $\{(x, xM)\}$ is a left K-linear subspace. In this case K^{end} is a skewfield isomorphic to K and as a collineation group of Σ, K^{end*} is a semilinear K-group. Similar to the Desarguesian case, one may consider the left scalar multiplication as a linear endomorphism over the prime field P of K. When we do this, we shall use the notation K_L. Hence, the $M's$ now commute with the elements of K_L.

To be clear, we now have three different uses of the term 'kernel' of a translation plane. We always consider the translation plane as $X \oplus X$, where X is a left vector space over a skewfield K, the kernel mappings are denoted by K^{end} and the component kernel mappings thought of as prime field endomorphisms are denoted by K_L. All three skewfields are isomorphic and each is called the kernel of the translation plane where context usually determines which skewfield we are actually employing. Of course, there is a fourth usage of 'kernel' if we are to include the kernel of an associated quasifield.

We now consider a coordinate set for a regular direct product net.

We may identify any point-Baer subplane as π_o within the direct product so that the points of the net have the general form (p_1, p_2) for p_1 and p_2 in π_o and the lines have the form $L_1 \times L_2$ for L_1 and L_2 parallel lines of π_o. It follows that the net M is $\pi_o \times \pi_o$ with the identity mapping defined on the set of parallel classes.

Considering the translation plane π_o with kernel K_o, we specify two lines incident with the zero vector as $x_o = 0$ and $y_o = 0$. Recall, that we have a group $GL(2, K_o)$ acting on the direct product net with elements from K_o acting on the 'left.'

We further decompose π_o in terms of these two subspaces and write the elements of π_o as (x_o, y_o), where x_o, y_o are in a common K_o-subspace W_o. We may take $y_o = x_o$ as the equation of a line of π_o incident with the zero vector so that the remaining lines are of the general form $y_o = x_o M$, where M is a K_o-linear transformation of W_o for M in a set Π_o.

The points of the net now have the general form (x_o, y_o, x_1, y_1), where x_o, y_o, x_1, y_1 are in W_o. The lines of the net are as follows: $(y_o = x_o M + c_o) \times (y_o = x_o M + c_1)$ for all M in Π_o containing I and O and $(x_o = c_o) \times (x_o = c_1)$.

Note change bases by the mapping

$$\chi : (x_o, y_o, x_1, y_1) \longmapsto (x_o, x_1, y_o, y_1).$$

Finally, we write $(x_o, x_1) = x$ and $(y_o, y_1) = y$ when (x_o, y_o, x_1, y_1) is a original point of the net or (x_o, x_1, y_o, y_1) is a point after the basis change.

Note that, before the basis change χ, the lines of the net are sets of points

$$\{(x_o, x_o M + c_o, x_1, x_1 M + c_1) \text{ for all } x_o, x_1 \text{ in } W_o\},$$
$$\text{for fixed } c_o \text{ and } c_1 \text{ in } W_o$$

and

$$\{(c_o, y_1, c_1, y_2) \text{ for all } y_1, y_2 \text{ in } W_o\},$$
$$\text{for fixed } c_o \text{ and } c_1 \text{ in } W_o.$$

Hence, after the basis change, the lines of the net have the basic form $x = (c_o, c_1)$ and $y = x \begin{bmatrix} M & 0 \\ 0 & M \end{bmatrix} + (c_o, c_1)$.

Before the basis change χ, the point-Baer subplanes incident with the zero vector, which are in a $GL(2, K_o)$ orbit of π_o, have the following form:

$$\rho_\infty = \{(0, 0, x_1, y_1) \text{ for all } x_1, y_1 \text{ in } W_o\} \text{ and } \rho_\alpha = \{(x_o, x_1, \alpha x_o, \alpha x_1)$$
for all x_o, x_1 in $W_o\}$ for each α in K_o.

DEFINITION 55. *We shall call these subplanes ρ_∞, or ρ_α the 'base subplanes.'*

REMARK 25. *We now observe that the group $GL(2, K_o)$ acting on the left is represented by mappings of the form*

$$(x_o, x_1, y_1, y_2)$$
$$\longmapsto (ax_1 + bx_2, cx_1 + dx_2, ay_1 + by_2, cy_1 + dy_2).$$

We now connect the ideas of coordinates of a translation plane, which contains a regular direct product net and coordinates of the assumed point-Baer subplanes.

Now assume that we have a translation plane Σ with kernel K and there are at least three point-Baer subplanes as above with kernel K_o, which are left invariant under the mappings K^{end^*} or equivalently are K–subspaces. Then there is a regular direct product net N isomorphic to $\pi_o \times \pi_o$ embedded in Σ. The translation complement of Σ is a subgroup of $\Gamma L(\Sigma, K)$ with the linear elements acting on the right. Furthermore, there is a group of the direct product net N, which is isomorphic to $GL(2, K_o)$ and naturally embedded in $GL(4, K)$ with the elements acting on the left. It is easy to see that if a collineation g of Σ fixes a K–subspace π_o pointwise then g is in $GL(\Sigma, K)$ and hence commutes with the mappings T_β. Now any kernel homology group K^{end^*} induces a faithful kernel group on any invariant point-Baer subplane so K may be considered a subskewfield of K_o.

1.1. When K is K_o. We specialize to the case that $K = K_o$. Then Σ is a $K = K_o$-vector space, and the above decomposition preserves the structure of the translation plane. In order that g be in $GL(\Sigma, K)$ and $GL(2, K_o)$, it must be in

$$\left\langle Diag \begin{bmatrix} 1 & \lambda \\ 0 & 1 \end{bmatrix} ; \lambda \in K_o \right\rangle \cdot$$

$$\cdot \left\langle Diag \begin{bmatrix} \delta & 0 \\ 0 & 1 \end{bmatrix} ; \delta \in K_o - \{0\} \right\rangle$$

acting on the left. On the other hand, each such mapping is in $GL(\Sigma, K)$ acting on the right.

For example, assuming that $\delta = 1$, we would have

$$g : (x_1, x_2, y_1, y_2) \longmapsto (x_1, \lambda x_1 + x_2, y_1, \lambda y_1 + y_2)$$

is in $GL(\Sigma, K)$. In order that the mapping commutes with each T_β it must be that

$$(\beta x_1, \beta(\lambda x_1) + \beta x_2, \beta y_1, \beta(\lambda y_1) + \beta y_2)$$
$$= (\beta x_1, \lambda(\beta x_1) + \beta x_2, \beta y_1, \lambda(\beta y_1) + \beta y_2)$$

so that λ must be in $Z(K)$. Similarly, if $Diag \begin{bmatrix} \delta & 0 \\ 0 & 1 \end{bmatrix} = g$, then δ is in $Z(K)$. Alternatively, the above mapping is K-linear for λ in K if and only if the mapping $x_1 \longmapsto \lambda x_1$ is K-linear if and only if λ is in $Z(K)$.

We note that although we are assuming that $K = K_o$, there is a distinction between K_o^{end} and K^{end} as they are the associated kernel mappings on π_o and on the translation plane Σ, respectively, and $K^{end} \mid \pi_o = K_o^{end}$.

So, we have noted that when there are at least three point-Baer subplanes sharing an affine point and all of their infinite points, which are left invariant by the kernel K of the super-translation plane, then any point-Baer perspectivity induces a degree of commutativity in K in the case when K is (i.e. isomorphic to) the kernel of the subplane.

Of course, it is not necessarily the case that any given point-Baer subplane which is pointwise fixed by a collineation of a translation plane with kernel K is automatically necessarily a K–subspace. Thus, part of our analysis in the next section details situations when such point-Baer subplanes are, in fact, kernel subspaces. Moreover, we shall expect that certain large point-Baer collineation groups in translation planes with kernel K could induce a certain degree of commutativity in K.

2. Point-Baer Subplanes in Vector Space Nets

In the following three subsections, we are interested in what is implied of a translation plane due to the existence of a Baer subplane. Since the points of a translation plane form a vector space, we relax this condition and assume only that there is a vector space net admitting a point-Baer subplane. The theory decomposes naturally into three separate cases: When there are at least three point-Baer subplanes incident with the zero vector and sharing all parallel classes, when there are exactly two such point-Baer subplanes and when there is but one.

Apart from a general theory, our goal is to apply these results to Baer or point-Baer subplanes within translation planes and particularly those with spreads in $PG(3, K)$, for K a skewfield.

2.1. Three Point-Baer Subplanes. In this subsection, we show that if there are at least three point-Baer subplanes incident with the zero vector and sharing all parallel classes with a vector space net then the 'base' subplanes form the entire set of point-Baer subplanes incident with the zero vector and in the net (sharing all parallel classes with the net).

REMARK 26. *We adopt the notation of the previous preliminary section. In particular, we adopt the notation $x = (x_o, x_1)$ and $y = (y_o, y_1)$ if and only if (x_o, y_o, x_1, y_1) is a point originating from the direct product notation.*

We shall use the notation (∞) to denote the parallel class containing the line $x = 0$ and (0) to denote the parallel class containing the line $y = 0$. Note that, in our notation, the points on $x = 0$ are the points $(0, y_o, 0, y_1)$, and the points on $y = 0$ are the points $(x_o, 0, x_1, 0)$. Similarly, points on $y = x$ have the representation (x_o, x_o, x_1, x_1). We shall use both the original direct product point notation and the notation after the basis change χ more or less simultaneously. After our main structure theorem, we shall use the representation after the basis change exclusively.

LEMMA 29. *Let ρ be any point-Baer subplane incident with the zero vector and sharing all parallel classes with the net. Then $(0, x_o, 0, x_1)$ is in $\rho \cap (x = 0)$ if and only if $(x_o, 0, x_1, 0)$ is in $\rho \cap (y = 0)$.*

PROOF. Let the infinite points of $x = 0, y = x \begin{bmatrix} M & 0 \\ 0 & M \end{bmatrix}$ be denoted by (∞) and (M), respectively. Let $(x_o, 0, x_1, 0)$ be a point of

$\rho \cap (y = 0)$. Form the line $(\infty)(x_o, 0, x_1, 0) \equiv (x = (x_o, x_1))$ and intersect the line $y = x$ to obtain (x_o, x_o, x_1, x_1). Since all such lines are lines of ρ, the intersection is a point of ρ. Now form the line of ρ, $(0)(x_o, x_o, x_1, x_1)$ and intersect $x = 0$ to obtain $(0, x_o, 0, x_1)$ in $\rho \cap (x = 0)$. $\qquad \square$

LEMMA 30. *Now assume that ρ is any point-Baer subplane.*
For $(0, x_o, 0, x_1)$ in $\rho \cap (x = 0)$ define a mapping λ on W_o that maps x_o to $x_1 (\lambda x_o = x_1)$.
Then λ is a bijective, additive transformation of W_o.
Furthermore, $\rho = \{(x_o, y_o, \lambda x_o, \lambda y_o) \text{ for all } x_o, y_o \text{ in } W_o\}$.
Moreover, $\lambda \in K_o$ and ρ is a base-subplane.

PROOF. We noted previously that no two distinct point-Baer subplanes incident with a common affine point and sharing all of their parallel classes can share two distinct affine points. Hence, $x_o = 0$ if and only if $x_1 = 0$, when $(0, x_o, 0, x_1)$ is a point of ρ and ρ is not the base subplane ρ_∞ or ρ_o. It follows that the subplane ρ is a translation affine subplane and hence a subspace of the underlying vector space taken over at least the prime field. Hence, it follows that λ is $1-1$ since the intersections with any of the base subplanes contain exactly the zero vector, and it is also now clear that λ is additive as ρ is additive. It remains only to show that λ is an onto mapping. From the above remarks, any two distinct point-Baer subplanes sharing a common affine point and their infinite points sum to the vector space and their intersections with a line incident with the common point sum to the line. Hence, given any element x_1^* of W_o, we consider the vector $(0, 0, 0, x_1^*)$. There exist vectors $(0, x_o^*, 0, 0)$ in $\rho_o \cap (x = 0)$ and $(0, x_o, 0, x_1)$ in $\rho \cap (x = 0)$ such that $(0, 0, 0, x_1^*) = (0, x_o^*, 0, 0) + (0, x_o, 0, x_1)$. It follows that $x_1 = x_1^*$, so there exists a vector $(0, x_o, 0, x_1^*)$ in ρ. Therefore, the mapping λ : $x_o \to x_1$ is onto. If $(0, x_o, 0, \lambda x_o, 0)$ is in $\rho \cap (x = 0)$, then $(x_o, 0, \lambda x_o, 0)$ is in $\rho \cap (y = 0)$ by the previous lemma so that $(x_o, y_o, \lambda x_o, \lambda y_o)$ is in ρ for all x_o, y_o in W_o as ρ is the direct sum of any two components. Let $(x_o^*, y_o^*, x_1^*, y_1^*)$ be any point of ρ then it follows that ρ also contains $(0, 0, \lambda x_o^* - x_1^*, \lambda y_o^* - y_1^*)$, and since $\rho \cap \rho_\infty = (0, 0, 0, 0)$ this forces $x_1^* = \lambda x_o^*$ and $y_1^* = \lambda y_o^*$. It remains to show that λ is in K_o.
For $y = x \begin{bmatrix} M & 0 \\ 0 & M \end{bmatrix}$ a line of the net and $(x_o, 0, x_1, 0)$ in ρ then $(x_o, x_o M, x_1, x_1 M)$ is also in ρ. We have seen this previously in the preliminary section. We form $(x_o, 0, x_1, 0)(\infty) \equiv (x = (x_o, x_1))$ and intersect $y = x \begin{bmatrix} M & 0 \\ 0 & M \end{bmatrix}$ to obtain the point $(x_o, x_o M, x_1, x_1 M)$. Since all of the points and lines are points and lines of Σ, it follows that the

intersection point is also in ρ. Again, we know that if $(x_o, 0, x_1, 0)$ is in ρ, then so is $(x_o, x_o M, x_1, x_1 M)$, which, in turn, implies that $(0, x_o M, 0, x_1 M)$ is in ρ. However, also we have that $x_1 = \lambda x_o$ and we know that $(0, x_o M, 0, \lambda(x_o M))$ is in ρ. Subtracting, since ρ is additive, we have that $(0, 0, 0, (\lambda x_o)M - \lambda(x_o M))$ is in ρ for all x_o. Since $\rho \cap \rho_\infty = (0, 0, 0, 0)$, it follows that $(\lambda x_o)M = \lambda(x_o M)$. Let L_o be any skewfield such that $\{M \text{ for } M \text{ in } \Pi_o\}$ is a set of L_o-linear transformations. Then it follows that L_o must be contained in the kernel K_o of $\pi_o = \rho_\infty$. Hence, λ is in $L_o \subseteq K_o$. $\square$

Therefore, we have proved the following result:

THEOREM 70. *Let M be any Abelian net, which contains three point-Baer subplanes that share the same affine point and share all of their parallel classes.*

Then there is a skewfield K_o such that M is a K_o-vector space net and there is a K_o-space W_o such that the points of M may be identified with $W_o \oplus W_o \oplus W_o \oplus W_o$. The set of all point-Baer subplanes of M that share the zero vector is isomorphic to the set whose subplanes are

$$\{(0, 0, y_o, y_1) \text{ for all } (y_o, y_1) \text{ in } W_o \oplus W_o\}$$
$$\cup_{\alpha \epsilon K_o}\{(x_o, y_o, \alpha x_o, \alpha y_o) \text{ for all } (x_o, y_o) \text{ in } W_o \oplus W_o\}.$$

Furthermore, there is a collineation group Γ of the net isomorphic to $GL(2, K_o)$ that fixes $(0, 0, 0, 0)$ and all parallel classes and acts triply transitively on the set of all point-Baer subplanes incident with $(0, 0, 0, 0)$. Moreover, if B denotes the set of all point-Baer subplanes of M and $\Gamma_{[\pi_o]}$ is the pointwise stabilizer of a subplane π_o of B then $\Gamma = \langle \Gamma_{[\pi_o]} \mid \pi_o \epsilon B \rangle$.

COROLLARY 18. *Let M be any Abelian net which contains three point-Baer subplanes that share the same affine point P and all of their parallel classes.*

If one of the point-Baer subplanes has kernel K_o, then the set of all point-Baer subplanes of M incident with P is isomorphic to $PGL(1, K_o)$.

PROOF. We consider the above representation after the basis change χ. The group

(13.1) $$\left\langle Diag \begin{bmatrix} 1 & \lambda \\ 0 & 1 \end{bmatrix} ; \lambda \in K_o \right\rangle \cdot$$

(13.2) $$\cdot \left\langle Diag \begin{bmatrix} \delta & 0 \\ 0 & 1 \end{bmatrix} ; \delta \in K_o - \{0\} \right\rangle$$

fixes $\pi_o = \rho_\infty$ pointwise and acts doubly transitively on the remaining point-Baer subplanes. $\square$

Below, we completely determine the collineation group of a net of type in the statement of the above theorem. We first note the following:

REMARK 27. *Let R be any Abelian net, which contains three point-Baer subplanes that share the same affine point A and all of their parallel classes. Let π_o be any point-Baer subplane incident with A.*

Then π_o is an affine translation plane with kernel K_o.

Let G_{π_o} denote the full translation complement of π_o.

Then there is a collineation group of R isomorphic to G_{π_o} which leaves π_o invariant.

PROOF. We have noted that R is a regular direct-product net. The result is then not difficult and is left for the reader to verify. □

REMARK 28. *In the net R defined as a regular direct product of a translation plane as a left vector space, the components of the net R become 'left' subspaces, whereas the Baer subplanes incident with the zero vector become natural 'right' subspaces. Of course, the original translation plane may be consider both a left and a right space. Furthermore, any three given Baer subplanes may be chosen so that the three subplanes are both left and right spaces.*

THEOREM 71. *Let R be any Abelian net, which contains three point-Baer subplanes that share the same affine point P and all of their parallel classes. Let π_o be any point-Baer subplane incident with P.*

Then π_o is an affine translation plane with kernel K_o. Let G_{π_o} denote the full translation complement of π_o obtained as a collineation group of R that leaves π_o invariant.

Then the full collineation group of R that fixes P is isomorphic to the product of G_{π_o} by $GL(2, K_o)$. The two groups intersect in the group kernel of π_o naturally extended to a collineation group of R.

PROOF. The group $GL(2, K_o)$ acts 3-transitively on the point-Baer subplanes of the net R and fixes R componentwise. Hence, we may assume that a collineation fixes the zero vector and permutes the point-Baer subplanes $\pi_\infty = \{(0,p)$ such that $p \in \pi_o\}, \pi_\lambda = \{(p, \lambda p)$ such that $p \in \pi_o$ and λ in the kernel of $\pi_o\}$ (when $\lambda = 0$, the subplane π_o is identified with $\pi_o \times 0$). So, if g is a collineation of R that fixes the zero vector, then we may assume that g leaves π_∞, π_o, and π_1 invariant. Hence, g is in G_{π_o} as it acts faithfully on π_o. We now consider intersections. Since $GL(2, K_o)$ fixes R componentwise, assume g fixes R componentwise. Then g induces the kernel mappings on π_o and on π_1 and is fixed-point-free as it also leaves π_∞ invariant. Thus, the faithful stabilizer of π_o in $GL(2, K_o)$ that fixes π_∞, π_o, and π_1 is

$$\left\langle \begin{bmatrix} \beta & 0 \\ 0 & \beta \end{bmatrix} \text{ such that } \beta \in K_o \right\rangle$$

in this representation. It then follows that the collineation group of R is the product as maintained.   □

COROLLARY 19. *Let N be a vector space net, which contains three point-Baer subplanes that share the same parallel classes and have a common affine point.*

If one of the subplanes is Desarguesian, then the net N is derivable.

PROOF. Under the assumptions, it follows that if K_o is the kernel of one of the subplanes π_o then π_o is a left 2-dimensional K_o-subspace. In the direct product net, generally Baer subplanes are right spaces where the components are left spaces and note the direct product net becomes a left 4-dimensional K_o-subspace. However, by three-transitivity, any three point-Baer subplanes may be considered as left K_o-subspaces in the sense that the components of the net are left K_o^{end}-subspaces (under what we have been calling $\{T_\beta$ for all β in $K_o\}$). Take any component T of the net. Since the intersections of T with any two of the point-Baer subplanes sum (direct sum or product) to the component and there are two of the point-Baer subplanes that are left 1-dimensional K_o-subspaces, it follows as before that each component may be considered a 2-dimensional left K_o-subspace. Since $GL(2, K_o)$ acts transitively on any left 2-dimensional K_o-subspace that it leaves invariant, it follows that the net N is covered by point-Baer subplanes. To see this in coordinate notation, after the basis change χ, we have points of the net (x_0, x_1, y_0, y_1), where (x_0, y_0) denote the points of π_o. The group $GL(2, K_o)$ acts on the left mapping (x_0, y_0, x_1, y_1) onto $(\alpha(x_0, y_0) + \delta(x_1, y_1), \beta(x_0, y_0) + \gamma(x_1, y_1))$. Now $x = 0$ is represented after the basis change by the points $(0, 0, y_0, y_1)$, which are mapped by $GL(2, K_o)$ onto the points $(0, 0, \alpha y_0 + \delta y_1, \beta y_0 + \gamma y_1)$. Since π_o is 2-dimensional, we may take $x_i, y_i \in K_o$. In particular, $(0, 0, 0, 1)$ maps to $(0, 0, \delta, \gamma)$ for all δ, γ. Hence, the net is derivable.   □

2.2. Nets with Two Point-Baer Subplanes. In this section, we consider the situation where there are exactly two point-Baer subplanes that share the same parallel classes and an affine point.

If we consider that the net involved is a vector space net, then it follows from the Reconstruction Theorem 59 that the net is a direct product net by the two point-Baer subplanes. However, the net is not necessarily a regular direct product net. Still, all of the above lemmas involving decomposition of the space via the point-Baer subplanes remain valid. We may write the space relative to a vector space W_o as $W_o \oplus W_o \oplus W_o \oplus W_o$. We choose coordinates so that $(0, x_2, 0, y_2)$ defines a point-Baer subplane π_o and $(x_1, 0, y_1, 0)$ defines a point-Baer subplane π_1. Since the net is a direct product net, we may set up a choice of coordinates for the point-Baer subplanes so that the $x = 0's, y = 0's$, and $y = x's$ correspond under the correspondence between parallel classes. Note that we have a subtle distinction between having a net that contains two point-Baer subplanes sharing a point and all parallel classes and realizing the net as a direct product net $\pi_o \times_\sigma \pi_1$ where σ is a bijection from the parallel classes of π_o onto the parallel classes of π_1. In the latter case, we choose coordinates for the spreads as noted directly above. If the components for π_o are, say, $x_1 = 0, y_1 = x_1 M_o$ and the components for π_1 are $x_2 = 0, y_2 = x_2 M_1$ then we may agree that σ maps $x_1 = 0$ onto $x_2 = 0$, $y_1 = 0$ onto $y_2 = 0$ and $y_1 = x_1$ onto $y_2 = x_2$.

In this way, we may then rewrite the equations for π_o and π_1 so that allowing that (x_1, x_2, y_1, y_2) now represent points of the net then π_o has equation $x_1 = 0 = y_1$ and π_1 has equation $x_2 = 0 = y_2$.

If one considers how components of the direct product net are formulated with the switch of coordinates mentioned above, then it follows that components of the net have the general form $(x_1, x_2) = (0, 0), (y_1, y_2) = (0, 0)$ and $(y_1, y_2) = (x_1, x_2) \begin{bmatrix} M_1 & 0 \\ 0 & M_o \end{bmatrix}$.

THEOREM 72. *If a vector space net contains two point-Baer subplanes sharing the same parallel classes and an affine point then the net is a direct product net by point-Baer subplanes. Let the two point-Baer subplanes be denoted by π_o and π_1 and let the kernels of π_o and π_1 be denoted by K_o and K_1, respectively, and assume that both planes are left vector spaces over their respective kernels.*

If the net does not contain three point-Baer subplanes sharing an affine point and all parallel classes, then the collineation group of the net that fixes π_o pointwise induces the full kernel homology group K_1 on π_1 and may be represented in the form

$$\left\langle \begin{bmatrix} \lambda & 0 & 0 & 0 \\ 0 & I & 0 & 0 \\ 0 & 0 & \lambda & 0 \\ 0 & 0 & 0 & I \end{bmatrix} \; for \; all \; \lambda \; in \; K_1^* \right\rangle.$$

PROOF. Clearly, if there are not three point-Baer subplanes, the group fixing π_o pointwise must also leave π_1 invariant so that the group has the general form and certainly induces kernel homologies on the subplane π_1. From the general set up of the direct product net, it is clear that any kernel homology of π_1 can be used to define a collineation of the net which fixes π_o pointwise. We consider the mappings acting on the left. □

2.3. Decomposable by a Point-Baer Subplane. We assume that there is but one point-Baer subplane within the vector space net.

In this situation, we must make an assumption which is automatically valid in the finite-dimensional case.

We assume that it is possible to decompose the space as (x_1, x_2, y_1, y_2) for x_i, y_i in W_o for $i = 1, 2$ such that a point-Baer subplane π_o is given by $x_1 = y_1 = 0$ while the components of the net have the form

$$x = 0, y = 0, y = x \begin{bmatrix} A & B \\ 0 & M_o \end{bmatrix},$$

where M_o defines a component of π_o and the elements A, B, M_o are considered as prime field linear mappings of W_o. Furthermore, suppose γ is a prime field linear endomorphism that commutes with all such $M_o's$. Then the mapping $(x_o, y_o) \longmapsto (x_o\gamma, y_o\gamma)$ will fix each component $x_o = 0, y_o = 0, y_o = x_o M_o$ of π_o. Hence, defining left scalar multiplication by $\gamma x_o = x_o\gamma$ (recall that the right hand notation considers the image of x_o as a prime field endomorphism), then the mapping is T_γ as previously discussed so that γ is considered within the kernel skewfield K_o of π_o.

In this situation, we say that the space is 'decomposable by a point-Baer subplane.'

THEOREM 73. *Let N be a vector space net that is decomposable by a point-Baer subplane π_o. Recall that we denote by $(K_o)_L$ the kernel of π_o taken as the centralizer of the set $\{M_o\}$ such that $y_o = x_o M_o$ defines a component of π_o and π_o is a left K_o-vector space.*

Assume that N contains exactly one point-Baer subplane incident with a given point and sharing its parallel classes.

If G is the full point-Baer central collineation group of the net that fixes π_o pointwise and the central planes are the translates of π_o, then G may be represented in the form

$$\left\langle \begin{bmatrix} I & \beta & 0 & 0 \\ 0 & I & 0 & 0 \\ 0 & 0 & I & \beta \\ 0 & 0 & 0 & I \end{bmatrix} \text{ for all } \beta \text{ in } K_o \right\rangle \text{ acting on the left.}$$

PROOF. Under the given initial set up, any collineation fixing π_o pointwise may be represented in the form

$$\begin{bmatrix} \alpha & \delta & 0 & 0 \\ 0 & I & 0 & 0 \\ 0 & 0 & \alpha & \delta \\ 0 & 0 & 0 & I \end{bmatrix},$$

where $\alpha, \delta, 0, I$ are considered as linear endomorphisms over the prime field written on the right. Note that we will be showing that δ is in $(K_o)_L$. By assumption, each element of the group leaves invariant each translate of π_o. This implies that for any elements c, d of W_o then $\{(c, x_2, d, y_2)\}$ is left invariant. Hence, it follows that $\alpha = I$ in all cases. Now represent the components of π_o in the form $x = 0 = (x_1, x_2)$ and $y = x \begin{bmatrix} A & B \\ 0 & C \end{bmatrix}$, where A, B, C are considered as linear endomorphisms over the prime field. Since, the indicated collineation fixes each component, it follows that

$$\begin{bmatrix} I & -\delta \\ 0 & I \end{bmatrix} \begin{bmatrix} A & B \\ 0 & C \end{bmatrix} \begin{bmatrix} I & \delta \\ 0 & I \end{bmatrix}$$
$$= \begin{bmatrix} A & B - \delta C + A\delta \\ 0 & C \end{bmatrix},$$

so it follows that $\delta C = A\delta$. Now, suppose that δ is singular. Then the collineation fixes points outside the point-Baer subplane. Let b be such a fixed point not in π_o. However, it was noted previously that π_o is also a Baer subplane so $\langle b, \pi_o \rangle$ is the entire vector space as it is a subplane. That is, the collineation acts trivially. Thus, δ is nonsingular as a prime field P-linear endomorphism. For some fixed collineation, change bases over the prime field P by

$$\begin{bmatrix} T & 0 & 0 & 0 \\ 0 & T & 0 & 0 \\ 0 & 0 & T & 0 \\ 0 & 0 & 0 & T \end{bmatrix},$$

where T is a P-linear automorphism so that the group

$$\left\langle \begin{bmatrix} I & \beta & 0 & 0 \\ 0 & I & 0 & 0 \\ 0 & 0 & I & \beta \\ 0 & 0 & 0 & I \end{bmatrix} \right\rangle$$

becomes

$$\left\langle \begin{bmatrix} I & T^{-1}\beta T & 0 & 0 \\ 0 & I & 0 & 0 \\ 0 & 0 & I & T^{-1}\beta T \\ 0 & 0 & 0 & I \end{bmatrix} \right\rangle.$$

Choose a particular δ so that $T^{-1}\delta T = I$. The components of π_o now have the form

$$x = 0, y = x \begin{bmatrix} T^{-1} & 0 \\ 0 & T^{-1} \end{bmatrix} \begin{bmatrix} A & B \\ 0 & C \end{bmatrix} \begin{bmatrix} T & 0 \\ 0 & T \end{bmatrix},$$

which is

$$y = x \begin{bmatrix} T^{-1}AT & T^{-1}BT \\ 0 & T^{-1}CT \end{bmatrix} = \begin{bmatrix} A^* & B^* \\ 0 & C^* \end{bmatrix}.$$

Since we now have an element of the form

$$\begin{bmatrix} I & I & 0 & 0 \\ 0 & I & 0 & 0 \\ 0 & 0 & I & I \\ 0 & 0 & 0 & I \end{bmatrix},$$

it follows that $A^* = C^*$. Following the previous argument, we have that

if $\begin{bmatrix} I & \gamma & 0 & 0 \\ 0 & I & 0 & 0 \\ 0 & 0 & I & \gamma \\ 0 & 0 & 0 & I \end{bmatrix}$ is an element of the group after the basis change then

γ is nonsingular and $\gamma A^* = \gamma C^* = C^* \gamma$. We are now thinking of the elements γ as linear endomorphisms over the prime field. The points of the subplane π_o have the form (x_o, y_o), and components have the general form $x_o = 0, y = x_o C^*$, where C^* is a P-linear automorphism of W_o. Therefore, γ is in the kernel of π_o considered as an element of $(K_o)_L$. Defining left scalar multiplication as in the preliminary section, we may consider the above matrix group acting on the left. This proves the theorem. $\qquad\square$

3. Translation Planes with Baer Groups

In this section, we consider when an Abelian net with three point-Baer subplanes can be embedded in a translation plane with kernel K in such a way so that K may be considered to be contained in the kernel K_o of a given point-Baer subplane.

DEFINITION 56. *Let D be a vector space net. Let π_o be a translation point-Baer subplane with kernel K_o.*

A collineation group of D fixing π_o pointwise and which may be identified with

$$\left\langle \begin{bmatrix} I & \gamma \\ 0 & I \end{bmatrix} \text{ for all } \gamma \text{ in } K_o \right\rangle$$

is called a 'full K_o-point-Baer elation group.'

A collineation group of D fixing π_o pointwise, which may be identified with

$$\left\langle \begin{bmatrix} \gamma & 0 \\ 0 & I \end{bmatrix} \text{ for all } \gamma \text{ in } K_o \setminus \{0\} \right\rangle$$

is called a 'full K_o-point-Baer homology group.'

The following somewhat omnibus theorem tries to detail various situations in which a point-Baer subplane of a translation plane with kernel K is left invariant under the kernel mappings K^{end}. The proofs to all parts of the theorem will be given collectively.

THEOREM 74. *Let π be a translation plane with kernel K. Assume that σ is a point-Baer collineation and $Fix\sigma = \pi_o$. Let K_o denote the kernel of π_o. Let G denote a collineation group of π in the translation complement that fixes π_o pointwise.*

Then π_o is a K-subspace in any of the following situations.

(i) K is a field and $|G| > 2$.

(ii) The characteristic is not 2 and σ is a point-Baer elation.

(iii) G is a full K_o-point-Baer elation group and $|K_o| > 2$.

(iv) L_o is a subskewfield of K_o, $|L_o| > 3$, and G is a full L_o-point-Baer homology group.

(iv)' More precisely,

If the collineation group of π that fixes π_o pointwise contains a point-Baer homology group isomorphic to the multiplicative group of a subskewfield of K_o and has order > 2, then π_o is a K-subspace, and there exists another point-Baer subplane of R_{π_o} that shares an affine point with π_o and is invariant under the point-Baer group and which is also a K-subspace.

PROOF. Either π_o is a K-subspace and we are finished, or not every kernel homology T_β leaves π_o invariant. First, assume that there are exactly two point-Baer subplanes π_o, π_1 incident with the zero vector that share the infinite points of π_o and g in the kernel of π interchanges π_o and π_1. Let H denote the subgroup of the multiplicative group fixing π_o. Then $H \cup gH = K^* = K \setminus \{0\}$. We assert that $H \cup \{0\}$ is a skewfield. To see this, let h, k be in H. Then $h + k$ is either in H or gH. However, if $\alpha \in \pi_o$, then $\alpha(h + k) = \alpha h + \alpha k$, and since both αh and αk are in π_o and π_o is a subspace over the prime field, it follows that $h + k$ is in $H \cup \{0\}$. It is immediate that -1 in K must leave every subspace over the prime field invariant so that it follows that h in H implies that $-h$ is in H. Hence, $H \cup \{0\}$ is an additive group, H is a multiplicative group so H must be a subskewfield as the remaining skewfield properties are inherited from those of K. This implies that $gH \cup \{0\}$ is additive. Now consider $gh + h = (g + 1)h$ for h in H and note that this element is either in H or gH. If $(g+1)h$ is in H, then $g+1$ is in H, but -1 is in H and H is additive so that g is in H. If $(g + 1)h$ is in gH then, since gH is additive, this forces h to be in gH. Thus, we obtain a contradiction in either situation. Hence, we may assume that there are at least three point-Baer subplanes incident with the zero vector that share the infinite points of π_o. By the previous material, we have a complete determination of R_{π_o}. Recall that we obtain the representation of the points of the net in the form (x_1, y_1, x_2, y_2), where π_o is given by $x_1 = y_1 = 0$. Change bases by χ as before so as to represent the points in the form (x_1, x_2, y_1, y_2) for all x_i, y_i in W_o for $i = 1, 2$ and W_o a left K_o-subspace now considered as a P-subspace, where P is the prime field. Now we may allow that the equations

$$x = 0, \ y = 0, \ y = xM \text{ for } M = \begin{bmatrix} m_1 & m_2 \\ m_3 & m_4 \end{bmatrix} \text{ define the components of}$$

the plane, where the entries m_i are P-linear transformations of W_o. We may also assume that $y = x$ is one of the lines of the plane. The point-Baer subplanes of R_{π_o} incident with the zero vector are now π_o and π_λ

$$= \{(x_1, \lambda x_1, y_1, \lambda_1 y_1) \text{ for all } x_1, y_1 \text{ in } W_o \text{ for each } \lambda \text{ of } K_o\}.$$

Furthermore, the components of R_{π_o} are $x \begin{bmatrix} w & 0 \\ 0 & w \end{bmatrix}$, for a set of P-linear automorphisms $\{w\}$, such that the components of π_o are $x_1 = 0$ and $y_1 = x_1 w$. Now what has occurred is that we have changed bases with respect to the prime field P to obtain this representation. Let τ denote the basis change. Hence, there is an isomorphic translation plane $\pi\tau$ and subplane $\pi_o\tau$. The original kernel mappings T_β are represented as P-linear mappings, so let $\beta x = xB$ over the prime field, where B commutes with M when $y = xM$ is a component. Since we are free to choose $x = 0, \ y = 0, \ y = x$ without destroying the original kernel K, we now have a kernel of the form

$$\left\langle S_\beta = \begin{bmatrix} S^{-1}BS & 0 \\ 0 & S^{-1}BS \end{bmatrix} \text{ for all } \beta \text{ in } K \right\rangle.$$

We further denote the new kernel by $S^{-1}KS$. Hence, there exists a field of P-linear automorphisms, $\left\langle \begin{bmatrix} B\hat{} & 0 \\ 0 & B\hat{} \end{bmatrix} \right\rangle$ isomorphic to K that fixes all components of $R_{\pi_o}\tau$. We will show that this kernel for $\pi\tau$ leaves $\pi_o\tau$ invariant. Hence, this implies that the original subplane π_o is a K-space. We have determined the full collineation group $G_{\pi_o} \cdot GL(2, K_o)$ of the net. Note that the subgroup of G_{π_o} that fixes each component is actually in $GL(2, K_o)$. Suppose that σ is an element of the kernel of the translation plane. Since σ is in $G_{\pi_o} \cdot GL(2, K_o)$, it follows that we may assume that the kernel is in $GL(2, K_o)$ in this representation. Notice that the group $GL(2, K_o)$ of the net R_{π_o} in this representation is

$$\left\langle Diag \begin{bmatrix} \alpha & \beta \\ \delta & \gamma \end{bmatrix} \text{ for all } \alpha, \beta, \delta, \gamma \text{ in } K_o \right\rangle,$$

such that the mappings are nonsingular

and where the entry elements ρ of K_o are written as left K_o-scalar mappings and then, in turn, as P-linear mappings acting on W_o. In other words, the action of the element $Diag \begin{bmatrix} \alpha & \beta \\ \delta & \gamma \end{bmatrix}$ on a point (x_1, x_2, y_1, y_2) is

$$(\alpha x_1 + \delta x_2, \beta x_1 + \gamma x_2, \alpha y_1 + \delta y_2, \beta y_1 + \gamma y_2).$$

Furthermore, such an element leaves π_o ($x_1 = y_1 = 0$) invariant if and only if $\delta = 0$. Hence, each element $B\hat{} = \begin{bmatrix} \alpha & \beta \\ \delta & \gamma \end{bmatrix}$ for some $\alpha, \beta, \delta, \gamma$ of K_o. We note that the set is an additive set of matrices with entry elements thought of as P-linear automorphisms. Each collineation of the plane that fixes the zero vector is an element of $\Gamma L(\pi, K^S)$ (a semilinear automorphism over the kernel). Each collineation that fixes π_o pointwise in this representation is in the group

$$\left\langle Diag \begin{bmatrix} 1 & \lambda \\ 0 & 1 \end{bmatrix} \text{ such that } \lambda \in K_o \right\rangle.$$

$$\left\langle Diag \begin{bmatrix} \beta & 0 \\ 0 & 1 \end{bmatrix} \text{ such that } \beta \in K_o \setminus \{0\} \right\rangle.$$

Furthermore, each such collineation normalizes K^S, which is a sub-skewfield of

$$\left\langle Diag \begin{bmatrix} \alpha & \beta \\ \delta & \gamma \end{bmatrix} ; \forall \alpha, \beta, \delta, \delta \in K_0 \right\rangle,$$

such that the mappings are nonsingular.

We point out that all of the operations are considered on the left. First, assume that a point-Baer collineation σ of the plane is $Diag \begin{bmatrix} 1 & \lambda \\ 0 & 1 \end{bmatrix}$.

Suppose that $k = Diag \begin{bmatrix} \alpha & \beta \\ \delta & \gamma \end{bmatrix}$ is in the kernel of the translation plane. Then consider the upper 2×2 submatrix of $\sigma k \sigma^{-1} - k$ (note that $\sigma k \sigma^{-1} - k$ must be in the kernel of the translation plane):

$$\begin{bmatrix} 1 & \lambda \\ 0 & 1 \end{bmatrix} \begin{bmatrix} \alpha & \beta \\ \delta & \gamma \end{bmatrix} \begin{bmatrix} 1 & -\lambda \\ 0 & 1 \end{bmatrix} - \begin{bmatrix} \alpha & \beta \\ \delta & \gamma \end{bmatrix}$$

$$= \begin{bmatrix} \lambda\delta & -(\alpha + \lambda\delta)\lambda + \lambda\gamma \\ 0 & -\delta\lambda \end{bmatrix}.$$

Apply the conjugate subtraction construction again to obtain the element

$$\begin{bmatrix} 0 & -2\lambda\delta\lambda \\ 0 & 0 \end{bmatrix}$$

so that for $\lambda \neq 0$, it must be that $\delta = 0$ or the characteristic is two. Again note that if the characteristic is not two then since the subplane π_o has the representation $\{(0, x_2, 0, y_2)\}$, it follows that any kernel element represented on the net in the form $Diag \begin{bmatrix} \alpha & \beta \\ 0 & \gamma \end{bmatrix}$ must leave π_o invariant. Hence, when the characteristic is not two and there is a point-Baer elation with fixed point set π_o, it follows that the kernel must leave π_o invariant and therefore π_o is a K–subspace. This proves part (ii). Now assume that K is a field. Note that we do not necessarily then have that K_o is a field. First, assume that there is a point-Baer elation. We may assume that the characteristic is two by the above argument. Also, by the previous argument, we consider the following element:

$$\begin{bmatrix} \lambda\delta & -(\alpha + \lambda\delta)\lambda + \lambda\gamma \\ 0 & -\delta\lambda \end{bmatrix} \begin{bmatrix} \alpha & \beta \\ \delta & \gamma \end{bmatrix}$$

$$= \begin{bmatrix} \alpha & \beta \\ \delta & \gamma \end{bmatrix} \begin{bmatrix} \lambda\delta & -(\alpha + \lambda\delta)\lambda + \lambda\gamma \\ 0 & -\delta\lambda \end{bmatrix}.$$

Then

$$\begin{bmatrix} \lambda\delta & -(\alpha + \lambda\delta)\lambda + \lambda\gamma \\ 0 & -\delta\lambda \end{bmatrix} \begin{bmatrix} \alpha & \beta \\ \delta & \gamma \end{bmatrix}$$

$$= \begin{bmatrix} \lambda\delta\alpha + ((\alpha + \lambda\delta)\lambda + \lambda\gamma)\delta & \begin{array}{c} \lambda\delta\beta \\ +((\alpha + \lambda\delta)\lambda + \lambda\gamma)\gamma \end{array} \\ \delta\lambda\delta & \delta\lambda\gamma \end{bmatrix}$$

and

$$
\begin{bmatrix} \alpha & \beta \\ \delta & \gamma \end{bmatrix} \begin{bmatrix} \lambda\delta & -(\alpha + \lambda\delta)\lambda + \lambda\gamma \\ 0 & -\delta\lambda \end{bmatrix}
$$
$$
= \begin{bmatrix} \alpha\lambda\delta & \alpha((\alpha + \lambda\delta)\lambda + \lambda\gamma) + \beta\delta\lambda \\ \delta\lambda\delta & \delta((\alpha + \lambda\delta)\lambda + \lambda\gamma) + \gamma\delta\lambda \end{bmatrix}.
$$

Equating the $(2,2)$-entries, we obtain $\delta((\alpha + \lambda\delta)\lambda + \lambda\gamma) + \gamma\delta\lambda = \delta\lambda\gamma$ so that $\delta((\alpha + \lambda\delta)\lambda) = \gamma\delta\lambda$, which implies that $\delta(\alpha + \lambda\delta) = \gamma\delta$. This is to say that whenever there is a point-Baer elation of the form $Diag \begin{bmatrix} 1 & \lambda \\ 0 & 1 \end{bmatrix}$ then $\delta(\alpha + \lambda\delta) = \gamma\delta$. So, if there are two point-Baer elations, then there exists another element ρ of $K_o \setminus \{\lambda\}$ such that $\delta(\alpha + \rho\delta) = \gamma\delta = \delta(\alpha + \lambda\delta)$, which implies that $\delta\rho\delta = \delta\lambda\delta$. Hence, either $\delta = 0$ or $\rho = \lambda$. So, if the point-Baer elation group has order > 2 and K is a field, we have a contradiction. Thus, assume that K is a field and there is a point-Baer homology. We may assume that there is a point-Baer homology, which fixes π_o pointwise of the following form: $Diag \begin{bmatrix} \lambda & 0 \\ 0 & 1 \end{bmatrix}$ for $\lambda \neq 0$. Form

$$
\begin{bmatrix} \lambda & 0 \\ 0 & 1 \end{bmatrix} \begin{bmatrix} \alpha & \beta \\ \delta & \gamma \end{bmatrix} \begin{bmatrix} \lambda^{-1} & 0 \\ 0 & 1 \end{bmatrix} - \begin{bmatrix} \alpha & \beta \\ \delta & \gamma \end{bmatrix}
$$
$$
= \begin{bmatrix} \lambda\alpha\lambda^{-1} - \alpha & (\lambda - 1)\beta \\ \delta(\lambda^{-1} - 1) & 0 \end{bmatrix}.
$$

Since K is a field then K^S is a field. (Note again that we are not assuming that K_o is a field.) Then

$$
\begin{bmatrix} \lambda\alpha\lambda^{-1} - \alpha & (\lambda - 1)\beta \\ \delta(\lambda^{-1} - 1) & 0 \end{bmatrix} \begin{bmatrix} \alpha & \beta \\ \delta & \gamma \end{bmatrix}
$$
$$
= \begin{bmatrix} \alpha & \beta \\ \delta & \gamma \end{bmatrix} \begin{bmatrix} \lambda\alpha\lambda^{-1} - \alpha & (\lambda - 1)\beta \\ \delta(\lambda^{-1} - 1) & 0 \end{bmatrix}.
$$

Hence, we have

$$
\begin{bmatrix} \left(\lambda\alpha\lambda^{-1} - \alpha\right)\alpha + (\lambda - 1)\beta\delta & \left(\lambda\alpha\lambda^{-1} - \alpha\right)\beta + (\lambda - 1)\beta\gamma \\ \delta\left(\lambda^{-1} - 1\right)\alpha & \delta\left(\lambda^{-1} - 1\right)\beta \end{bmatrix}
$$
$$
= \begin{bmatrix} \alpha\left(\lambda\alpha\lambda^{-1} - \alpha\right) + \beta\delta\left(\lambda^{-1} - 1\right) & \alpha(\lambda - 1)\beta \\ \delta\left(\lambda\alpha\lambda^{-1} - \alpha\right) + \gamma\delta\left(\lambda^{-1} - 1\right) & \delta(\lambda - 1)\beta \end{bmatrix}.
$$

Equating the $(2,2)$-entries, we have $\delta(\lambda - 1)\beta = \delta(\lambda^{-1} - 1)\beta$. If $\delta\beta \neq 0$, then $\lambda = \lambda^{-1}$ so that the group has order 2. If $\delta\beta = 0$, assume

$\beta = 0$. Then the element

$$Diag \begin{bmatrix} \lambda\alpha\lambda^{-1} - \alpha & 0 \\ \delta\left(\lambda^{-1} - 1\right) & 0 \end{bmatrix}$$

is in K^S, which implies $\delta = 0$ as $\lambda^{-1} \neq 1$ and $\lambda\alpha\lambda^{-1} = \alpha$. Hence, $\beta = 0 = \delta = 0$, which implies that π_o is left invariant by the kernel mappings of the translation plane. So, if there is a point-Baer homology, K is a field, and if the fixed-point subspace is not a K–subspace, then the only point-Baer homology is

$$Diag \begin{bmatrix} -1 & 0 \\ 0 & 1 \end{bmatrix}.$$

Thus, there must be point-Baer elations if the group has order larger than 2. However, by previous arguments, the characteristic must be 2 and the condition $\lambda^{-1} = \lambda$ provides a contradiction unless $\delta = 0$. Hence, in the case that K is a field, and the group fixing π_o pointwise has order > 2, then π_o is a K–subspace. This proves part (i). We now assume that there is a full point-Baer elation group

$$\left\langle Diag \begin{bmatrix} 1 & \lambda \\ 0 & 1 \end{bmatrix} \text{ for } \lambda \text{ in } K_o \right\rangle.$$

The previous arguments show that the following elements are obtained

$$\begin{bmatrix} \lambda\delta & -(\alpha + \lambda\delta)\lambda + \lambda\gamma \\ 0 & -\delta\lambda \end{bmatrix}$$

for all λ in K_o so that

$$Diag \begin{bmatrix} \lambda\delta & -(\alpha + \lambda\delta)\lambda + \lambda\gamma \\ 0 & -\delta\lambda \end{bmatrix}$$

is in the kernel. Moreover, we may assume that the characteristic is 2. We have

$$\begin{bmatrix} \lambda\delta & -(\alpha + \lambda\delta)\lambda + \lambda\gamma \\ 0 & -\delta\lambda \end{bmatrix}$$
$$+ \begin{bmatrix} \rho\delta & -(\alpha + \rho\delta)\rho + \rho\gamma \\ 0 & -\delta\rho \end{bmatrix}$$
$$= \begin{bmatrix} (\lambda + \rho)\delta & -(\alpha + (\lambda + \rho)\delta)(\lambda + \rho) + (\lambda + \rho)\gamma \\ 0 & -\delta(\lambda + \rho) \end{bmatrix},$$

which implies (using the $(1, 2)$ elements) that

$$\rho\delta\lambda = -\lambda\delta\rho \text{ for all } \rho, \lambda \text{ in } K_o.$$

Hence, K_o is a field or $\delta = 0$. Assume the former. Then

$$\begin{bmatrix} \lambda\delta & -(\alpha + \lambda\delta)\lambda + \lambda\gamma \\ 0 & -\delta\lambda \end{bmatrix} \begin{bmatrix} \rho\delta & -(\alpha + \rho\delta)\rho + \rho\gamma \\ 0 & -\delta\rho \end{bmatrix}$$

$$= \begin{bmatrix} \lambda\rho\delta^2 & \lambda\delta(-(\alpha + \rho\delta)\rho + \rho\gamma) \\ & +(-(\alpha + \lambda\delta)\lambda + \lambda\gamma)(-\delta\rho) \\ 0 & \lambda\rho\delta^2 \end{bmatrix}.$$

On the other hand, we already would have an element constructed from a suitable point-Baer elation of the form

$$\begin{bmatrix} (\lambda\rho\delta)\delta & -(\alpha + (\lambda\rho\delta)\delta)(\lambda\rho\delta) + (\lambda\rho\delta)\gamma \\ 0 & -\delta(\lambda\rho\delta) \end{bmatrix}.$$

This implies that

$$\lambda\delta(-(\alpha + \rho\delta)\rho + \rho\gamma)) + (-(\alpha + \lambda\delta)\lambda + \lambda\gamma)(-\delta\rho)$$
$$= -(\alpha + (\lambda\rho\delta)\delta)(\lambda\rho\delta) + (\lambda\rho\delta)\gamma.$$

Since K_o is a field, we are reduced to

$$\lambda\delta^2\rho^2 + \lambda^2\delta^2\rho = \alpha\lambda\rho\delta + \lambda^2\rho^2\delta^3 + \lambda\rho\delta\gamma \text{ for all } \lambda, \rho \text{ in } K_o.$$

Keep λ fixed and nonzero then ρ satisfies a quadratic in K_o for all ρ in K_o. By our assumptions, this implies that $\lambda\delta^2 = \lambda^2\delta^3$ for all λ in K_o so that $\delta = 0$. Hence, π_o is a K-subspace in this situation. This proves part (iii). Now assume that K is not necessarily a field and assume the conditions of part (iv) so that every element fixing π_o pointwise has the form $Diag\begin{bmatrix} \lambda & 0 \\ 0 & 1 \end{bmatrix}$ for all $\lambda \neq 0$ in a subskewfield L_o of K_o. So, there are at least two point-Baer subplanes that share the same infinite points and exactly one affine point. Either both are K-subspaces or we have at least three point-Baer subplanes that share an affine point and all parallel classes. Hence, we may assume the latter case. Conjugate the kernel element $Diag\begin{bmatrix} \alpha & \beta \\ \delta & \gamma \end{bmatrix}$ and subtract to obtain that

$$Diag\begin{bmatrix} \lambda\alpha\lambda^{-1} - \alpha & (\lambda - 1)\beta \\ \delta(\lambda^{-1} - 1) & 0 \end{bmatrix}$$

is in the kernel for all elements $\lambda \neq 0$ in L_o and fixed β, δ, γ. Assume that $\beta\delta \neq 0$. Fix λ as $\lambda_o \neq 1$ and take the inverse of this element and note it is $DiagD$, where D is

$$\begin{bmatrix} 0 & F \\ G & H \end{bmatrix},$$

where

$$F = \left(\delta\left(\lambda^{-1} - 1\right)\right)^{-1},$$

$$G = \left((\lambda - 1)\beta\right)^{-1},$$

$$H = \left((\lambda - 1)\beta\right)^{-1}\left(-\left(\lambda a\lambda^{-1} - a\right)\right)\left(\delta\left(\lambda^{-1} - 1\right)\right)^{-1}$$

$$= Diag\begin{bmatrix} 0 & f \\ g & e \end{bmatrix}.$$

Fix the element, letting $\lambda = \lambda_o$. Apply the conjugation-subtraction method again using a general $Diag\begin{bmatrix} \rho & 0 \\ 0 & 1 \end{bmatrix}$ as above to obtain an element in the kernel of the form

$$Diag\begin{bmatrix} 0 & (\rho - 1)f = b \\ g(\rho^{-1} - 1) = d & 0 \end{bmatrix},$$

where $g = \left((\lambda_o - 1)\beta\right)^{-1}$ and $f = \left(\delta\left(\lambda_o^{-1} - 1\right)\right)^{-1}$. Assume $bd \neq 0$. Since

$$Diag\begin{bmatrix} \lambda & 0 \\ 0 & 1 \end{bmatrix}\begin{bmatrix} 0 & b \\ d & 0 \end{bmatrix}\begin{bmatrix} \lambda^{-1} & 0 \\ 0 & 1 \end{bmatrix}$$

$$= Diag\begin{bmatrix} 0 & \lambda b \\ d\lambda^{-1} & 0 \end{bmatrix}$$

is in the kernel, we have that

$$Diag\left(\begin{bmatrix} 0 & \lambda b \\ d\lambda^{-1} & 0 \end{bmatrix}\begin{bmatrix} 0 & d^{-1} \\ b^{-1} & 0 \end{bmatrix}\right)$$

$$= Diag\begin{bmatrix} \lambda & 0 \\ 0 & d\lambda^{-1}d^{-1} \end{bmatrix}$$

is in the kernel for all elements λ in L_o^*. Since the kernel is fixed-point-free, it follows that $DiagE$, where E is

$$\left(\begin{bmatrix} \lambda & 0 \\ 0 & d\lambda^{-1}d^{-1} \end{bmatrix}\begin{bmatrix} \rho & 0 \\ 0 & d\rho^{-1}d^{-1} \end{bmatrix}\right.$$
$$\left. -\begin{bmatrix} \lambda\rho & 0 \\ 0 & d(\lambda\rho)^{-1}d^{-1} \end{bmatrix}\right)$$

$$= Diag\begin{bmatrix} 0 & 0 \\ 0 & d\lambda^{-1}\rho^{-1}d^{-1} - d\rho^{-1}\lambda^{-1}d^{-1} \end{bmatrix}.$$

Hence, we have $\lambda^{-1}\rho^{-1} = \rho^{-1}\lambda^{-1}$ so that L_o is a field provided $bd \neq 0$. Thus, there is a function $\sigma : L_o \rightarrow K_o$ such that $\lambda^\sigma = d\lambda^{-1}d^{-1}$ for $\lambda \neq 0$ and where we may assume that $0^\sigma = 0$. Since σ is 1–1 and onto and preserves addition and multiplication, it follows that σ is an isomorphism of L_o into K_o. For fixed d, and $\tau \in K_o$, the mapping

$\beta : K_o \to K_o$ such that $\tau^\beta = d\tau d^{-1}$ is an inner automorphism of K_o. Hence, $\lambda^{-\beta} = \lambda^\sigma$ so that $\lambda^{-1} = \lambda^{\sigma\beta^{-1}}$. Note that $L_o^\beta = L_o^\sigma$. Therefore, $\sigma\beta^{-1} = \theta$ is an automorphism of L_o such that $\lambda^\theta = \lambda^{-1}$ for all $\lambda \neq 0$ of L_o, and if $bd \neq 0$, L_o is a field. In this latter case, it follows easily that $L_o \cong GF(2)$, $GF(3)$, or $GF(4)$. However, the group is assumed to have order > 2, so we have a contradiction unless we have $GF(4)$. If L_o is isomorphic to $GF(4)$, let the elements be denoted by $0, 1, \theta$, and $\theta^2 = \theta + 1$. Since we have the elements

$$Diag \begin{bmatrix} \lambda & 0 \\ 0 & d\lambda^{-1}d^{-1} \end{bmatrix},$$

we apply the argument again for a possibly different d, say, d^*. It then follows that $d = d^*$ since the kernel is additive. Recall that $d = g(\rho^{-1} - 1) = ((\lambda_o - 1)\beta)^{-1}(\rho^{-1} - 1)$. Fix ρ for this particular d as ρ_o. We could have obtained an element d^* by fixing say λ_1 initially and then isolate on ρ_1. Hence, it follows that

$$((\lambda_o - 1)\beta)^{-1}(\rho_o^{-1} - 1) \;=\; ((\lambda_1 - 1)\beta)^{-1}(\rho_1^{-1} - 1)$$

and, hence, for β nonzero, we have

$$(\lambda_o - 1)^{-1}(\rho_o^{-1} - 1) \;=\; (\lambda_1 - 1)^{-1}(\rho_1^{-1} - 1)$$

for all $\lambda_o, \lambda_1, \rho_o, \rho_1$ in $GF(4)$ not equal to 0 or 1. Since $\tau^{-1} = \tau^2$, we obtain

$$((\lambda_o + 1)(\rho_o + 1))^2 = ((\lambda_1 + 1)(\rho_1 + 1))^2.$$

Let $\lambda_o = \rho_o = \theta$ and $\lambda_1 = \rho_1 = \theta + 1$ to obtain $\theta^4 = (\theta + 1)^4$, a contradiction. Hence, the situation $GF(4)$ does not occur. So, $bd = 0$, and note that $b = 0$ if and only if $d = 0$. Thus, $((\lambda_o - 1)\beta)^{-1}(\rho^{-1} - 1) = 0 = (\rho - 1)(\delta(\lambda_o^{-1} - 1))^{-1}$ for all such non-zero elements of L_o. This implies $\delta = 0 = \beta$. Thus, the kernel K must leave π_o invariant. This completes the proof to part (iv). So, this completes the proofs to all parts of the theorem. $\qquad \square$

4. Spreads in $PG(3, K)$

In this section, we shall show that spreads in $PG(3, K)$, for K a skewfield, that admit full K_o-Baer elation or homology groups can only exist if the skewfield K is a field. More precisely,

THEOREM 75. *Let π be a translation plane with spread in $PG(3, K)$ for K a skewfield.*
Assume that π_o is a point-Baer subplane with kernel K_o.
If π admits a full K_o-point-Baer elation group of order > 2 or a full K_o-point-Baer homology group of order > 2 fixing π_o pointwise, then

(1) K_o *is isomorphic to* K *and*
(2) K *is a field.*

PROOF. By the previous section, π_o is a K–subspace. Hence, it follows that π_o is a 2-dimensional K–subspace. The components of π_o are 2-dimensional K–subspaces so that there is an induced Desarguesian spread, which forces π_o to be a Desarguesian affine subplane. Hence, the kernel of π_o is isomorphic to K. Choose coordinates so that π is $\{(x_1, x_2, y_1, y_2)$ for x_i, y_i in K and $i = 1, 2\}$, where scalar multiplication is on the left

$$\gamma(x_1, x_2, y_1, y_2) = (\gamma x_1, \gamma x_2, \gamma y_1, \gamma y_2)$$

and π_o is $\{(0, x_2, 0, y_2)$ for x_2, y_2 in $K\}$ and where components of π have the general form

$$y = x \begin{bmatrix} f(u,t) & g(u,t) \\ t & u \end{bmatrix},$$

$$x = 0 \; \forall t, u \in K \text{ and } f, g \text{ functions} : K \times K \text{ into } K.$$

It is easy to see that the components of π_o are

$$x = 0, \text{ and } y = x \begin{bmatrix} f(u,0) & g(u,0) \\ 0 & u \end{bmatrix} \forall u \in K.$$

Let $f(u, 0) = f(u)$ and $g(u, 0) = g(u)$. Assume that π admits a full K_o-point-Baer elation group B with axis π_o and center C the set of translates of π_o. Then B fixes π_o pointwise and since π_o is a K–subspace, it follows that B is a linear group representable by a 4×4 matrix with elements in K and acting on the right. Furthermore, B leaves invariant every subplane of C. The group that fixes π_o pointwise is K-linear and has the matrix form

$$\left\langle \begin{bmatrix} a & b & 0 & 0 \\ 0 & 1 & 0 & 0 \\ 0 & 0 & a & b \\ 0 & 0 & 0 & 1 \end{bmatrix} \text{ for all } a, b \text{ in } K \right\rangle.$$

But each point-Baer elation fixes each translate of π_o so that it follows that $a = 1$ in all cases. We emphasize that this matrix group is acting on the right, whereas previously the action was defined on the left when we were considering the group acting on the net and elements defined via the kernel of the subplane K_o. It is this fact that we shall see will force the commutativity of K. The point-Baer elation group fixes each component $y = x \begin{bmatrix} f(u) & g(u) \\ 0 & u \end{bmatrix}$ for u in K. Hence, we must have

$$\begin{bmatrix} 1 & -b \\ 0 & 1 \end{bmatrix} \begin{bmatrix} f(u) & g(u) \\ 0 & u \end{bmatrix} \begin{bmatrix} 1 & b \\ 0 & 1 \end{bmatrix}$$

$$= \begin{bmatrix} f(u) & (g(u) - bu) + f(u)b \\ 0 & u \end{bmatrix}$$

$$= \begin{bmatrix} f(u) & g(u) \\ 0 & u \end{bmatrix}.$$

Thus, $bu = f(u)b$ for all b, u in K. Therefore, $f(u) = u$ for all u in K and $bu = ub$ for all b, u in K. Hence, K is commutative. Now assume that there is a full K_o-point-Baer homology group B fixing π_o pointwise. By the previous section, π_o is a K–subspace. By results of the previous chapter, we see that there are two point-Baer subplanes π_o and $\pi(\sigma - 1)$ sharing the same parallel class both of which are K– subspaces for σ in $B - \{1\}$. Note that $\pi(\sigma - 1)$ is clearly σ-invariant and distinct from π_o. Hence, we may choose bases as above but now $\pi(\sigma - 1)$ is represented by $\{(x_1, 0, y_1, 0)$ for all x_1, y_1 in $K\}$. It follows that the collineation group B is K-linear and has the matrix form

$$\left\langle \begin{bmatrix} a & b & 0 & 0 \\ 0 & 1 & 0 & 0 \\ 0 & 0 & a & b \\ 0 & 0 & 0 & 1 \end{bmatrix} \text{ for all } a \neq 0, b \text{ in } K \right\rangle$$

acting on the right. It is immediate that, in this case, $b = 0$ for all elements. The components of π_o now have the form $y = x \begin{bmatrix} f(u) & 0 \\ 0 & u \end{bmatrix}$, $x = 0$ for all u in K. Then,

$$\begin{bmatrix} a^{-1} & 0 \\ 0 & 1 \end{bmatrix} \begin{bmatrix} f(u) & 0 \\ 0 & u \end{bmatrix} \begin{bmatrix} a & 0 \\ 0 & 1 \end{bmatrix}$$

$$= \begin{bmatrix} a^{-1}f(u)a & 0 \\ 0 & u \end{bmatrix} = \begin{bmatrix} f(u) & 0 \\ 0 & u \end{bmatrix}.$$

So, we have

$$a^{-1}f(u)a = f(u)$$

for all $a \neq 0, u$ in K. Since $\{f(u)$ for all u in $K\} = K$, it follows that $av = va$ for all a, v in K so that K is commutative. This proves the theorem. □

We now prove that a point-Baer homology group becomes a Baer homology group under certain circumstances.

THEOREM 76. *Let π be a translation plane with spread in $PG(3, K)$, for K a field.*

Then, any point-Baer homology group of order > 2 is a Baer homology group.

PROOF. Let B denote the point-Baer homology group and let $\pi_o = FixB$. Then, by convention, $\pi(\sigma - 1) = \pi_1 = \pi(\tau - 1)$ for all σ, τ in B. We know from the previous results that π_o is a K-space. Moreover, the previous proof shows that the unique 0-coaxis of B (unique point-Baer subplane incident with the zero vector) is also left invariant by K. The action of B on $\pi(\sigma - 1)$ induces a kernel homology subgroup on the subplane. On the other hand, $\pi(\sigma - 1)$ is a Pappian plane with kernel K. Hence, there exists a subgroup of the kernel homology group that induces the same action as does B on $\pi(\sigma - 1)$. Hence, $\pi(\sigma - 1)$ is fixed pointwise by a nontrivial collineation group, which forces the subplane to be Baer. This completes the proof of the theorem. $\qquad\square$

COROLLARY 20. *Let π be a translation plane with spread in $PG(3, K)$, for K a skewfield.*

Let π_o be a point-Baer subplane with kernel K_o that is fixed pointwise by a full K_o-point-Baer group B of order > 2 (either a point-Baer elation or a point-Baer homology group).

Then K is a field and B is a Baer elation or a Baer homology group.

PROOF. We may assume that B is a full K_o-point-Baer homology group. Since K is now a field by the above remarks and π_o is a K-subspace, it follows from the above theorem that B is a Baer elation or Baer homology group. $\qquad\square$

THEOREM 77. *Let π be a translation plane with spread in $PG(3, K)$, for K a skewfield. Let π_o be a point-Baer subplane with kernel K_o.*

If B is a full K_o-point-Baer elation group of order > 2, then K is a field and one of the following three situations occur:

(i) The net defined by the components of π_o is a regulus net relative to $PG(3, K)$.

(ii) The net defined by the components of π_o contains exactly one point-Baer subplane incident with the zero vector, the net is not derivable, and the full collineation group which fixes π_o pointwise is

$$\left\langle \begin{bmatrix} 1 & a & 0 & 0 \\ 0 & 1 & 0 & 0 \\ 0 & 0 & 1 & a \\ 0 & 0 & 0 & 1 \end{bmatrix} \ for\ all\ a\ in\ K \right\rangle.$$

(iii) The net defined by the components of π_o is a derivable net and there is exactly one point-Baer subplane incident with the zero vector, which is a K–subspace.

In this case, the components of the net have the following form:

$$x = 0, y = x \begin{bmatrix} u & Au - uA \\ 0 & u \end{bmatrix} \text{ for all } u \text{ in } K \text{ and where } A \text{ is a}$$

linear transformation considered over the prime field P of K such that $Au - uA$ is in K for all u in K.

PROOF. We see that K is a field isomorphic to K_o. Furthermore, π_o is a K-space so is 2-dimensional and hence we have the decomposable situation previously described. Either there exists another point-Baer subplane incident with the zero vector and sharing the same parallel classes or we may apply the results of the previous section. Since B acts regularly on the nonfixed 1-dimensional K–subspaces that are on components of π_o, it follows that BK^* acts transitively on the set of points not in π_o on components of π_o. If there exists another point-Baer subplane, there is a covering by point-Baer subplanes and the net is derivable. Hence, it remains to consider the situation when the net is derivable but not a regulus net. The derivable net must have the basic components $x = 0, y = x \begin{bmatrix} u & g(u) \\ 0 & u \end{bmatrix}$, for all u in K, where g is a function on K. And we note from a previous theorem that K is a field. If the net is a derivable net then there exists a Baer group that fixes π_o and any particular vector on, say, $x = 0$. We assume that, considered over the prime field P of K, a basis change leaves π_o invariant as well as $x = 0, y = 0, y = x$ and is thus represented as

$$\begin{bmatrix} T_1 & T_2 & 0 & 0 \\ 0 & T_4 & 0 & 0 \\ 0 & 0 & T_1 & T_2 \\ 0 & 0 & 0 & T_4 \end{bmatrix},$$

where T_i represent nonsingular linear mappings over the prime field P. Furthermore, we assume that the standard representation of the derivable net is now obtained. That is, we have

$$\text{Let } M_1 = \left(\begin{bmatrix} T_1 & T_2 \\ 0 & T_4 \end{bmatrix}^{-1} = \begin{bmatrix} T_1^{-1} & -T_1^{-1}T_2T_4^{-1} \\ 0 & T_4^{-1} \end{bmatrix} \right). \text{ Then}$$

$$M_1 \begin{bmatrix} u & g(u) \\ 0 & u \end{bmatrix} \begin{bmatrix} T_1 & T_2 \\ 0 & T_4 \end{bmatrix}$$

$$= \begin{bmatrix} M & 0 \\ 0 & M \end{bmatrix}.$$

Hence, it follows after a bit of calculation, which the reader should complete, that

$$g(u) = T_2 T_4^{-1} u - u T_2 T_4^{-1}$$

for all u in K. Letting $A = T_2 T_4^{-1}$ shows that case (iii) is the only remaining possibility. Conversely, any set of matrices of the form in case (iii) determine a derivable net by the basis change

$$\begin{bmatrix} I & A & 0 & 0 \\ 0 & I & 0 & 0 \\ 0 & 0 & I & A \\ 0 & 0 & 0 & I \end{bmatrix}.$$

This completes the proof of the theorem.   □

COROLLARY 21. *Under the assumptions of the above theorem, if K is finite, then $Au - uA = 0$ for all u in K and case (iii) does not occur.*

PROOF. $g(u) = Au - uA$ implies that $vg(u) + ug(v) = g(uv)$. Hence, $2ug(u) = Au^2 - u^2 A$. In particular, for characteristic two perfect fields, it follows that $g(w) = 0$ for all w in K. An easy induction argument shows that g acts like a formal derivative so that $g(u^r) = ru^{r-1}g(u)$ for all integers r. Hence, if K is $GF(q)$, then $g(u) = g(u^q) = qu^{q-1}g(u) = 0$ for all u in K.   □

EXAMPLE 2. *Let L be a field of characteristic two, which is not perfect. Let γ denote a nonsquare in L. Then $\left(\begin{bmatrix} u & \gamma t \\ t & u \end{bmatrix} ; u, t \in L \right)$ defines a quadratic extension field K of L.*

Furthermore, for any 2×2 matrix A over L,

$$x = 0, y = x \begin{bmatrix} w & Aw - wA \\ 0 & w \end{bmatrix} \quad \forall w \in K$$

defines a derivable net, which is a regulus net in $PG(3, K)$ if and only if A is in K.

PROOF. It suffices to show that for any 2×2 matrix A over L then $Aw - wA$ is in K. Let $A = \begin{bmatrix} a_1 & a_2 \\ a_3 & a_4 \end{bmatrix}$. Then, for $w = \begin{bmatrix} u & \gamma t \\ t & u \end{bmatrix}$ and u, t in L, we obtain

$$Aw - wA = \begin{bmatrix} (a_2 + \gamma a_3)t & \gamma(a_4 + a_1)t \\ (a_4 + a_1)t & (a_2 + \gamma a_3)t \end{bmatrix},$$

which is always in K. This expression is identically zero if and only if $a_2 = \gamma a_3$ and $a_4 = a_1$, which is valid if and only if A is in K.   □

COROLLARY 22. *Let K be any field that admits a nontrivial derivation δ. Then there is a derivable partial spread in $PG(3, K)$, which is not a regulus and whose corresponding derivable net contains exactly one K–subspace point-Baer subplane.*

THEOREM 78. *Let π be a translation plane with spread in $PG(3, K)$, for K a skewfield.*

If π admits a point-Baer subplane π_o with kernel K_o and a collineation group of order > 2 in the translation complement, which is a full K_o-point-Baer homology group, then one of the following occurs:

(i) There are exactly two point-Baer subplanes incident with the zero vector, the net is not derivable, and the full collineation group, which fixes π_o pointwise, is

$$\left\langle \begin{bmatrix} \lambda & 0 & 0 & 0 \\ 0 & I & 0 & 0 \\ 0 & 0 & \lambda & 0 \\ 0 & 0 & 0 & I \end{bmatrix} \right\rangle.$$

for all λ in K^*

(ii) There are exactly two point-Baer subplanes incident with the zero vector, which are also K–subspaces, but the net defined by π_o is a derivable net. The components may be represented in the form $x = 0, y = x \begin{bmatrix} u^\sigma & 0 \\ 0 & u \end{bmatrix}$ for all u in K and $\sigma \neq 1$ in $\mathrm{Aut} K$.

(iii) There are at least three point-Baer subplanes incident with the zero vector, which are K–subspaces, and the net defined by π_o is a regulus net (corresponds to a regulus in $PG(3, K)$).

PROOF. By previous results, there are at least two point-Baer subplanes that are K–subspaces, K is a field and K_1 is isomorphic to K, where K_1 denotes the kernel of $\pi(\sigma - 1) = \pi_1$ and where σ is a point-Baer collineation. The group B mentioned in part (i) is clearly a collineation group of the plane. Moreover, the group BK^* acts regularly on the points on component of π_o that are not in $\pi_o \cup \pi_1$. Hence, if there exists a third point-Baer subplane, the net is derivable. If there exists a third point-Baer subplane that is a K–subspace, the net is a regulus net. Finally, when there are exactly two K–subspace point-Baer subplanes, one can diagonalize the space and when the net is derivable, the associated matrices of the spread components form a field. Hence, (ii) follows immediately. $\qquad \square$

CHAPTER 14

Ubiquity of Subgeometry Partitions

In this chapter, we use the theory that we have developed regarding Baer groups and their representation. We also connect algebraic lifting and geometric lifting, which are as disparate concepts as one could imagine: one process, that of lifting a spread in $PG(3, q)$ defining a translation plane of order q^2 to a corresponding spread in $PG(3, q^2)$ that provides a translation plane of order q^4, and the second process, that of lifting a subgeometry partition to a translation plane. We also are able to use the algebraic lifting process in combination with our Sperner spread construction theorem presented in Chapter 4 for the construction of still more, interesting subgeometry partitions. There is a fusion process in the theory of semifields that allows a coordinate change that fuses certain subfields of the nuclei. Here we consider a similar fusion process for Baer groups, noting that a Baer group will arise from a group associated with a nucleus upon derivation (usually of a semifield plane). Using these ideas, there will be a ubiquity of subgeometry partitions that arise from the concept of double-Baer groups.

In the previous chapters, we have tied together a number of various spreads that admit retraction groups, thereby constructing subgeometry partitions. The natural question is how general are the spreads that give rise to subgeometry partitions. Or, we may ask the opposite question: given a t-spread, is there a subgeometry partition that gives rise to it?

As we know, the main issue is when there is a suitable group, a 'retraction group' that can create a quasi-subgeometry partition with the hope that the quasi-subgeometries are actually subgeometries. Recall that the intrinsic character is that the translation plane has order q^t with subkernel K isomorphic to $GF(q)$ and admits a collineation group (on the nonzero vectors), which contains the scalar group K^*, written as GK^*, such that GK^* together with the zero mapping is a field isomorphic to $GF(q^z)$.

So, since subgeometry partitions often first arise directly from translation planes or more generally from t-spreads in kt-dimensional vector spaces, it is essentially an open question to find geometric conditions that can squeeze a retraction group from the spread or more generally from an arbitrary partition. We have seen, for example, that often semifield planes that admit certain right and middle nuclei admit retraction groups by looking at the fusion of their nuclei. If we can derive a net preserved by the affine homology groups associated with right and middle nuclei, we obtain Baer groups that basically been 'fused.' So, we start in this chapter with the idea of fusing Baer groups with the hope that a retraction group will pop up. Of course, the question of what sorts of translation planes admit such Baer groups becomes of major interest. In this regard, there is a very unlikely connection with what are called 'algebraically lifted planes' with the algebraic construction procedure for spreads that is called 'algebraic lifting' by which a spread in $PG(3, q)$ may be lifted to a spread in $PG(3, q^2)$. More precisely, this construction is a construction on the associated quasifields for the spread and different quasifields may produce different algebraically lifted spreads. This material is explicated in Biliotti, Jha and Johnson, and the reader is referred to this text for additional details and information (see p. 437 [21]).

The objective of this chapter is to connect two very different construction techniques and to show how ubiquitous are certain basic subgeometry partitions. The work of the chapter follows articles by Jha and Johnson [84], [83], with various changes appropriate for this text and uses material from the chapters on Baer groups acting on translation planes.

1. Jha-Johnson Lifting Theorem

Jha and Johnson [84] have been able to connect subgeometry partitions of $PG(3, q^2)$ with two types of subgeometries isomorphic to $PG(3, q)$ and $PG(1, q^2)$ with every spread in $PG(3, q)$. In other words, these 'mixed subgeometry partitions' are basically equivalent to the set of all spreads in $PG(3, q)$. The Jha-Johnson lifting theorem establishes that fact as follows:

THEOREM 79. *(Jha and Johnson [84]) Let S be any spread in $PG(3, q)$. Then there is a mixed subgeometry partition of $PG(3, q^2)$, which geometrically lifts to a spread in $PG(3, q^2)$ that algebraically contracts to S.*

COROLLARY 23. *(Jha and Johnson [84]) The set of mixed subgeometry partitions of a 3-dimensional projective space $PG(3, k^2)$ constructs all spreads of $PG(3, k)$.*

THEOREM 80. *From any quasifield, with spread in $PG(3, q)$, there is a process that lifts and derives to a spread permitting retraction, which, in turn, produces a mixed partition of $(q - 1)$ $PG(3, q)$'s and $q^4 - q$ lines of $PG(3, q^2)$.*

We begin with some preliminaries.

The following theorem recalls the main ideas of algebraic lifting. The reader is directed to the foundation's text by Biliotti, Jha, and Johnson [21] for details on algebraic lifting. Actually, this concept originated and is due to Hiramine-Matsumoto-Oyama [69].

1.1. Hiramine-Matsumoto-Oyama Algebraic Lifting. The retraction part of the following theorem is due to the author and established in this text.

THEOREM 81. *Let π be translation plane with spread S in $PG(3, q)$. Let F denote the associated field of order q and let K be a quadratic extension field with basis $\{1, \theta\}$ such that $\theta^2 = \theta\alpha + \beta$ for $\alpha, \beta \in F$. Choose any quasifield and write the spread as follows:*

$$x = 0, y = x \begin{bmatrix} g(t, u) & h(t, u) - \alpha g(t, u) = f(t, u) \\ t & u \end{bmatrix}$$

$$\forall t, u \in F$$

where g, f are unique functions on $F \times F$ and h is defined as noted in the matrix, using the term α.

Define $F(\theta t + u) = -g(t, u)\theta + h(t, u)$. Then

$$x = 0, y = x \begin{bmatrix} \theta t + u & F(\theta s + v) \\ \theta s + v & (\theta t + u)^q \end{bmatrix} \forall t, u, s, v \in F$$

is a spread S^L in $PG(3, q^2)$ called the spread 'algebraically lifted' from S. We note that there is a derivable net

$$x = 0, y = x \begin{bmatrix} w & 0 \\ 0 & w^q \end{bmatrix} \forall w \in K \simeq GF(q^2)$$

with the property that the derived net (replaceable net) contains exactly two Baer subplanes, which are $GF(q^2)$-subspaces and the remaining $q^2 - 1$ Baer subplanes form $(q - 1)$ orbits of length $q + 1$ under the kernel homology group.

Hence, we obtain a mixed partition of $(q - 1)$ $PG(3, q)$'s and $q^4 - q$ lines of $PG(3, q^2)$. In this case, one retraction provides a mixed subgeometry partition of $q^2 + 1$ lines and $q^2(q - 1)$ $PG(3, q)$'s of a $PG(3, q^2)$.

Considered vectorially, retraction may be considered more generally over infinite vector spaces, even infinite vector spaces of infinite dimension over their kernels. Hence, in essence, geometric lifting may be more generally considered, even though in this text, we restrict ourselves to the finite situation.

We have used ideas of retraction throughout this text in a very general manner. Now we intend to be more concrete and actually define a retraction group of order $q^2 - 1$.

First, the pertinent theorem:

1.2. Johnson's Retraction Theorem.

THEOREM 82. *(Johnson* [113]*) Let π be a translation plane of order $q^{2^a r}$ and kernel containing $GF(q)$. Let G be a collineation group of order $q^2 - 1$ containing the kernel homology group of order $q - 1$ and assume that $G \cup \{0\}$ (the zero mapping) is a field K.*

(1) Then the component orbit lengths of G are either 1 or $q + 1$. Forming the projective space $PG(2^a r - 1, q^2)$, the orbits of length 1 become projective subgeometries isomorphic to $PG(2^{a-1}r - 1, q^2)$ and the orbits of length $q + 1$ become projective subgeometries isomorphic to $PG(2^{a-1}r - 1, q)$.

(2) The set of subgeometries partition the points of the projective space providing a 'mixed subgeometry partition.'

We have discussed previously the 'unfolding of a fan,' that is, the geometric lifting process that constructs translation planes of order $q^{2^a r}$ and kernel containing $GF(q)$ from any mixed subgeometry partition. Of course, there is a result in Johnson [113] for translation planes of order q^t, for t odd but admitting collineation groups of order $q^2 - 1$ having the properties of the group in the previous theorem. In this setting, all component orbits will have length $q + 1$, and the associated subgeometry partition is said to be a Baer subgeometry partition. Previously, we have used the term 'retraction' and 'folding the fan' to mean the same process.

Specifically, for this setting, we define the process of retraction as follows:

DEFINITION 57. *Assume that a finite translation plane with kernel containing $GF(q)$ admits a collineation group of order $q^2 - 1$ that contains the kernel homology group of order $q - 1$ such that together with the zero mapping, a field isomorphic to $GF(q^2)$ is obtained. Then the group is said to be a 'retraction group.'*

DEFINITION 58. *A retraction group is said to be 'trivial' in the projective space $PG(2n - 1, q^2)$ associated with the spread, if the subgeometries are all isomorphic to $PG(n - 1, q^2)$.*

REMARK 29. *It is immediate that a retraction group is trivial if and only if it is a kernel homology group of the translation plane.*

DEFINITION 59. *The associated process of constructing the subgeometry partition from a non-trivial retraction group is said to be 're-traction' (or 'spread retraction'). If there are two distinct retraction groups G_1 and G_2 that centralize each other, we shall say that we have 'double-retraction.' More generally, if there are k distinct retraction that centralize all other associated groups, we say that we have 'k-retraction.' We note that we are allowing 'double-retraction' to include the possibility that one of the associated fields is a kernel field of the translation plane.*

Of course, in the finite case, we recall that a 'Baer group' in a translation plane of order h^2 is a group in the translation complement that fixes a Baer subgroup pointwise. Indeed, such Baer groups have orders that divide $h(h - 1)$. Here we let $h = q^{2^a r}$ and consider the possibility of having two Baer groups of order $q + 1$. The reason for consideration of such Baer groups may be seen from algebraic lifting, where a spread in $PG(3, q)$ algebraically lifts to a spread in $PG(3, q^2)$ admitting a Baer group of order $q + 1$. In this setting, the order of the translation plane has order q^4. We now are able to give the proof of the Jha-Johnson lifting theorem.

PROOF. We note that in any algebraically lifted spread, the following B_1 is a Baer group of order $q + 1$.

$$B_1 = \langle \mathrm{diag}(e^{-1}, 1, 1, e^{-1}); \ e \text{ has order } q + 1 \rangle.$$

Now B_1 maps a basic spread component

$$y = x \begin{bmatrix} u & F(z) \\ z & u^q \end{bmatrix}$$

onto

$$
\begin{aligned}
y &= x \begin{bmatrix} e & 0 \\ 0 & 1 \end{bmatrix} \begin{bmatrix} u & F(z) \\ z & u^q \end{bmatrix} \begin{bmatrix} 1 & 0 \\ 0 & e^{-1} \end{bmatrix} \\
&= y = x \begin{bmatrix} ue & F(z) \\ z & (ue)^q \end{bmatrix},
\end{aligned}
$$

and so this also is a verification that B_1 is a collineation group of the algebraically lifted translation plane. Furthermore, B_1 fixes the vector subspace

$$FixB_1 = \langle (0, x_2, y_1, 0); x_2, y_1 \in GF(q^2) \rangle$$

pointwise, which emphasizes that B_1 is a Baer group of order $q+1$. The algebraically lifted translation plane has order q^4 and kernel $GF(q^2)$ (the spread is in $PG(3, q^2)$). Since we have the kernel $GF(q^2)$ in the lifted plane, it follows that we have a second Baer group

$$B_2 = \langle \text{diag}(1, e, e, 1); \ e \text{ has order } q+1 \rangle$$

fixing the vector subspace

$$\langle (x_1, 0, 0, y_2); \ x_1, \ y_2 \in GF(q^2) \rangle$$

pointwise. Furthermore, B_1 and B_2 commute and note that B_1 is contained in the group $B_2 K^{2*}$, where K^{2*} denotes the kernel homology group of order $q^2 - 1$. We now locate our retraction group within $\langle B_1, B_2 \rangle K^{2*}$. We now claim that $\langle B_1, B_2 \rangle K^{2*}$ is

$$D = \left\langle \begin{array}{c} \text{diag}(a, b, b, a); \ a, \ b \in GF(q^2)^* \\ \text{such that } a^{q+1} = b^{q+1} \end{array} \right\rangle$$

of order $(q + 1)(q^2 - 1)$. Note that the order of $\langle B_1, B_2 \rangle K^{2*}$ is $(q + 1)^2(q^2 - 1)/I$, where I is the intersection with $B_1 B_2$ and K^{2*}. This group leaves invariant the subplane $FixB_1$, B_1 fixes it pointwise, B_2 induces a kernel group on it as does K^{2*}. Hence, the group induced on $FixB_1$ is isomorphic to K^{2*}. But since B_1 of order $q + 1$ fixes $FixB_1$ pointwise, it follows that the group has order $(q + 1)(q^2 - 1)$. When $a = b$, D contains K^{2*} and when $a = 1$, D contains B_1, similarly, D contains B_2. Since D has a K^{2*} as a normal subgroup of index $q + 1$, we see that $B_1 B_2 K^{2*} = D$. Let $q = p^r$, for p a prime and form the following set of fields L_σ of order q^2, for $\sigma = p^i$ for i dividing $2r$.

$$L_\sigma = \langle \text{diag}(a, a^\sigma, a^\sigma, a); \ a \in GF(q^2)^* \rangle \cup \{0\}.$$

To obtain L_σ^* in our group, then $a^{\sigma(q+1)} = a^{q+1}$, which implies that if $\sigma = p^i$, then q divides p^i, so let $\sigma = q$ or 1. Now each of the fields L_σ satisfies the hypothesis of the retraction theorem so that we obtain a double-retraction. However, when $\sigma = 1$, in this particular setting, the subgeometry partition is essentially trivial, as K_1 is the kernel homology group of the associated translation plane so that we do not actually obtain a mixed partition as all subgeometries are isomorphic to $PG(1, q^2)$'s. This then completes the proof of the Jha-Johnson Lifting Theorem.   □

2. Double-Baer Groups

In the proof of the Jha-Johnson lifting theorem, from one Baer group B_1, we used the kernel homology group to locate another group B_2, and from these two groups, we found a retraction group. We wish now to generalize these ideas, and here we make no assumptions on the spread, that is, no assumptions on the kernel of the translation plane π, merely that we have a translation plane of order $q^{2^a r}$, with unspecified kernel. Basically, if one thinks about what happens in an algebraically lifted plane, this will be the situation we are trying to replicate.

There are two principal situations where the use of Baer groups provides double-retraction. First, two Baer groups of order $q + 1$ in translation planes that admit a subkernel isomorphic to $GF(q^2)$ and two Baer groups of order $(2, q - 1)(q + 1)$ in translation planes that admit a subkernel isomorphic to $GF(q)$. Of course, when q is even, we may use either of these conditions for our results.

DEFINITION 60. *Let B_1 and coB_1 be distinct commuting Baer groups of the same order and in the same net of degree $q^{2^{a-1}r} + 1$ in a translation plane π of order $q^{2^a r}$ and kernel containing K for $(r, 2) = 1$. If either K is isomorphic to $GF(q^2)$ and B_1 is divisible by $q + 1$ or if K is isomorphic to $GF(q)$ and B_1 is divisible by $(2, q - 1)(q + 1)$, the pair (B_1, coB_1) shall be called a 'double-Baer group.' The 'order' of the double-Baer group is the order of either Baer group in the set.*

DEFINITION 61. *Any Baer group whose order is that of a double-Baer group is said to be 'critical.'*

We continue the notation established in the previous definition for all sections of this chapter.

Now assume that we have a double-Baer group of order divisible by $q + 1$ and a translation plane of order $q^{2^a r}$, where there is a subkernel of K_{q^2} isomorphic to $GF(q^2)$.

LEMMA 31. *We may choose coordinates so that B_1 and coB_1 have the following representation:*

$$B_1 = \langle\, diag(I, A, I, A); A \in K_1^+ \text{ of order dividing } q + 1 \,\rangle$$

and

$$coB_1 = \langle\, diag(C, I, C, I); C \in K_2^+ \text{ of order dividing } q + 1 \,\rangle,$$

where both K_1^+ and K_2^+ are fields isomorphic to $GF(q^2)$, and K_i^+ is the K_q-module generated over K_q by A or C, respectively, for $i = 1, 2$.

PROOF. We shall leave this proof to the reader, merely mentioning that Baer groups are linear over K_q. $\qquad\square$

Therefore, we have the following representation for $B_1 co B_1 K_{q^2}^*$:

$$B_1 co B_1 K_{q^2}^* = \langle \text{diag}(C\alpha, A\alpha, C\alpha, A\alpha) \rangle,$$
$$A \in K_1^+, C \in K_2^+ \text{ of orders dividing } (q+1)$$
$$\text{and } \alpha \in K_{q^2} - \{0\}.$$

We note our previous argument in the dimension 2 algebraically lifting situation applies directly here to show that the order of this group is $(q+1)(q^2-1)$.

$$B_1 co B_1 K_{q^2}^* = \langle \text{diag}(C\alpha, A\alpha, C\alpha, A\alpha) \rangle$$
$$= \left\langle \begin{array}{c} \text{diag}(D, E, D, E); D^{q+1} = E^{q+1}, \\ \text{for } D \in K_1^+, E \in K_2^* \end{array} \right\rangle.$$

Since K_1^+ and K_2^+ are isomorphic fields of matrices containing K_q, they are conjugate by an K_q-matrix H. So, let $K_2^{+H} = K_1^+$ and consider the following fields:

$$L_\sigma = \langle \text{diag}(D, D^{\sigma H}, D, D^{\sigma H}); D \in K_1^+, \sigma \text{ automorphism of } K_1^+ \rangle.$$

By our previous argument L_σ^* belongs to our group if and only if $\sigma = 1$ or q. We now have the conditions for double-retraction as given in Theorem 82. Of course, in this setting, A and C are in the kernel group K_{q^2} but in a related setting to follow, this will not be the case.

Hence, we have the following theorem.

THEOREM 83. *Let π be a translation plane of order $q^{2^a r}$ for $(r, 2) = 1$, admitting a double-Baer group of order divisible by $q+1$ with kernel containing K_{q^2} isomorphic to $GF(q^2)$.*

From the double-Baer group, there are exactly two retraction fields L_σ, for $\sigma = 1$ or q, for the translation plane so we have double-retraction.

We now consider a somewhat related situation, that of a translation plane of order $q^{2^a r}$ with kernel containing K_q isomorphic to $GF(q)$ that admits a double-Baer group of order divisible by $(2, q-1)(q+1)$. Now just as before we see that we can still create two fields L_σ, for $\sigma = 1$ or q. In particular, we have

$$B_1 co B_1 K_q^* = \left\langle \begin{array}{c} \text{diag}(C\alpha, A\alpha, C\alpha, A\alpha); \\ C \text{ and } A \text{ have orders divisible by} \\ (2, q-1)(q+1)z \text{ and } \alpha \in K_q^* \end{array} \right\rangle.$$

The order of this group is $((2, q-1)(q+1))^2(q-1)/I$, where I is the intersection of $B_1 co B_1$ and K_q^*. In order that an element of $g \in B_1 co B_1$

be also in K_q^*, then $C = A$ has order dividing $(q-1)$. It is not difficult to establish the following: We have the subgroup of $B_1 co B_1 K_q^*$

$$\langle \mathrm{diag}(D, E, D, E); D^{q+1} = E^{q+1}, \text{ for } D \in K_1^+, E \in K_2^* \rangle$$

of order $(q+1)(q^2-1)$, and we obtain the fields L_σ, for $\sigma = 1$ or q, exactly as in the previous situation.

However, in this setting, the field L_1 need not be a kernel homology group. Therefore, we obtain the following theorem:

THEOREM 84. *Let π be a translation plane of order $q^{2^a r}$ for $(r, 2) = 1$, admitting a double-Baer group of order divisible by $(2, q-1)(q+1)$ with kernel containing K_q isomorphic to $GF(q)$.*

(1) From the double-Baer group, we may construct two retraction fields L_σ, for $\sigma = 1$ or q, for the translation plane so we have double-retraction.

(2) If the plane has kernel $GF(q^2)$, but L_1 is not a kernel homology group, we obtain triple-retraction.

3. Fusion of Baer Groups

We have mentioned previously that some of the ideas of double-Baer groups comes from what happens when deriving semifield planes, where the derived net is invariant under the affine homology groups associated with the fused right and middle nuclei. (The reader is directed to the article by Jha and Johnson [87] for more on fusion of nuclei). Here is the important point in theorem form.

THEOREM 85. *Let π be a semifield plane of order $q^{2^a r}$ and kernel containing K isomorphic to $GF(q)$, for $(r, 2) = 1$. Assume that the right nucleus and middle nucleus of the associated semifield contain fields isomorphic to $GF(q^2)$. If there is a derivable net that contains a middle nucleus net invariant under the associated affine homology group, then the derived plane admits a double-Baer group.*

DEFINITION 62. *Let π be a translation plane of order $q^{2^a r}$ admitting a double-Baer group. Then both associated Baer groups are represented by fields K_1^+ and K_2^+ isomorphic to $GF(q^2)$. If these fields can be identified, we shall say that the Baer groups are 'fused.'*

Let B_1 and B_2 denote the two Baer groups of the double group. If there is a field L^+ isomorphic to $GF(q^2)$ whose multiplicative group is a collineation group such that $B_2 \subseteq B_1 L^{+*}$, we shall say that L^+ 'intertwines' the double group, or that L^+ is an 'intertwining field' for the double group.

We note that when the fields corresponding to the Baer groups in a double-Baer group are fused, we still might not have the situation that one of the fields constructed above L_σ, for $\sigma = 1$ or q is a kernel homology group.

We want now to somehow find a retraction group from double-Baer groups, just as we did in the Jha-Johnson lifting theorem. We note that fusion is obtained when a double-Baer group actually commutes with a retraction field, but this is not a necessary condition.

THEOREM 86. *Let π of order $q^{2^a}\,r$ and kernel containing K isomorphic to $GF(q)$ admit spread-retraction relative to the group G_1K^*. Assume that a double Baer group B_1B_2 commutes with G_1K^*. Let N denote the net containing the axes of the Baer groups B_i, for $i = 1, 2$.*

(1) Then $G_1K^*B_1B_2$ fixes at least two components of the net N containing the axes of B_i, $i = 1, 2$.

(2) Assume that G_1K^* fixes all components of the net N.

    (a) Then the kernel of the Baer subplane $FixB_i$ contains

$$G_1K^* \cup \{0\} = L^+.$$

    (b) Furthermore, the double Baer group may be fused and L^+ is an intertwining field for the double group.

PROOF. Here is a sketch of the proof. The full explanation is given in Jha and Johnson [83]. The double Baer group fixes a set of $q^{2^{a-1}r} + 1$ components, which are necessarily permuted by G_1K^*, a group of order $q^2 - 1$. The component orbit lengths of G_1K^* are 1 and $q + 1$. Since $q + 1$ divides $q^{2^{a-1}r} - 1$, it follows that there must be a common set of at least two fixed components, say, L and M. Decompose both L and M relative to K_1^+ and an easy argument shows that B_1 and B_2 must diagonalize relative to K_1^+ in the same manner on both L and M. Furthermore, the group element acting on a 1-dimensional K_1^+-subspace on L or M must act on that 1-space as a K_1^+-scalar mapping. Assume that on L, B_1 has the general form:

$$(x_1, .., x_{q^{2^{a-1}}r}) \longmapsto (a_1x_1, a_2x_2, a_3x_3, a_4x_4, ...)$$

for $a_i \in K_1^+$ of orders dividing $q + 1$. It then is not difficult to show that a generator h_1 for B_1 maps on L

$$(x_1, .., x_{q^{2^{a-1}}r}) \longmapsto (cx_1, x_2, c^{q^{\lambda(3)}}x_3, x_4, c^{q^{\lambda(5)}}x_5, ..),$$

where c has order $q + 1$. The analogous decomposition may be assumed on M.

Then a simple counting argument (left to the reader to complete) will show that either we have all $c's$, all $c^{q'}s$, or half $c's$ and half $c^{q'}s$ relative to B_1.

However, we have assumed that the group G_1K^* fixes all components on the net N defined by π_1 and π_2. Since G_1K^* leaves π_1 and π_2 invariant and fixes all components on N, it must induce a kernel homology group of order $q^2 - 1$ on both subplanes.

In this case, then choosing a third component of N as $y = x$, the representation already given for G_1K^* is basically valid over the translation plane, since now $x = 0, y = 0, y = x$ are fixed. However, the representation of B_1 must then have all $c's$ or all $c^{q'}s$. In other words, we may represent the group B_1 as previously

$$B_1 = \langle \operatorname{diag}(I, A, I, A); \ A \in K_1^+ \text{ of order dividing } q + 1 \rangle.$$

Now since the kernel acting on π_2 is $\operatorname{diag}(A, A)$, it follows that L^+ acting on π_1 is K_1^+. It now follows that $A = a^{q-1}I_{2^{a-2}r}$ (recalling that $a \in K_1^+$ isomorphic to $GF(q^2)$). Furthermore, we now have that $a^{1-q}I_{2^a r}B_1$ is B_2. Hence, the double group may be fused and L^+ is an intertwining group. $\qquad \square$

The following two corollaries and theorem now follow fairly directly.

COROLLARY 24. *Let π be a translation plane of order $q^{2^a\ r}$ and kernel containing K isomorphic to $GF(q)$. If π admits spread-retraction with a group G_1K^* and admits a Baer group B_1 of order $q + 1$ such that B_1 and G_1K^* commute and G_1K^* fixes all components of the net N containing $Fix B_1$ then there is a double group of order $q + 1$ and two intertwining fields.*

Hence, the plane admits double-retraction.

COROLLARY 25. *Let π be a translation plane of order $q^{2^a\ r}$ and kernel containing F isomorphic to $GF(q^2)$.*

(1) If π admits a Baer group of order $q + 1$, then a double group exists with intertwining field F.

(2) Furthermore, then the three fields, kernel, and two associated Baer-fields may be fused.

(3) Double-retraction exists.

THEOREM 87. *Let π be a translation plane of order $q^{2^a r}$, $(r, 2) = 1$ with subkernel F isomorphic to $GF(q^2)$ containing K isomorphic to $GF(q)$.*

(1) If π admits t distinct commuting Baer subgroups of order $q + 1$, then π admits $(t + 1)$-retraction and has kernel a subfield of $GF(q^{2r})$, where r is odd.

(2) Moreover, for $2^{a-1}r = n$ and if $t = 2^n - 1$, then π admits symmetric homology groups of orders $q + 1$.

4. Double-Homology Groups

Now the idea is to show some sort of converse, something like double retraction is equivalent to the existence of a double-Baer group, and this is almost true. We need to add in the concept of a 'double-homology group.'

THEOREM 88. *Let π be a translation plane of order $q^{2^a r}$ and kernel containing K isomorphic to $GF(q)$.*

(1) When q is even, a double-Baer group of order $q + 1$ implies double-retraction.

(2) For arbitrary order, a double-Baer group of order $q + 1$ with intertwining field implies double-retraction.

(3) A double-Baer group of order $(2, q - 1)(q + 1)$ implies double-retraction.

(4) Double-retraction implies a double-Baer group of order $q + 1$ or two commuting homology groups of order $q + 1$; a 'double-homology group' of order $q + 1$.

(5) A double-homology group of order $q + 1$ in a semifield plane implies double-retraction.

PROOF. It remains to prove (4) and (5). Assume that we have double-retraction. Write one of the fixed-point-free groups as $G_1 K^*$ and decompose the space so that this group looks like a scalar group of order $q^2 - 1$ generated by $z \longmapsto az$. Then, we know that the second group $G_2 K^*$ must look either like $x_i \longmapsto ax_i$ or $x_j \longmapsto a^q x_j$ and there are exactly half $a's$ and half $a^q s$. It follows that by multiplication, we may obtain a commuting pair of homology groups of order $q + 1$ or a double-Baer group of order $q + 1$, which proves (4).

Now assume that we have a semifield plane and we have a double-homology group of order $q+1$. Hence, we must have a double-homology group of order $q^2 - 1$. Moreover, we can 'fuse' the nuclei. The possible products produce two fixed-point-free fields of order $q^2 - 1$. We need to show that these fields contain the K-kernel homology group of order $q - 1$. Let a subgroup of the product of the two homology groups be given by:

$$\mathcal{F}^{q^\lambda} : \left\langle (x, y) \longmapsto (xA, yA^{q^\lambda}); A \in F^{+*} \right\rangle,$$

where F^{+*} is a field of order q^2. However, since the homologies are K-linear groups, it follows that F^{+*} commutes with the kernel homology group K^* so that by Schur's lemma, $\langle F^{+*}, K^* \rangle$ is contained in a field.

It follows by uniqueness of cyclic groups that $K \subseteq F^+$. Hence, both $\mathcal{F}^{q^\lambda}$ for $\lambda = 0$ or 1 produce retraction. Therefore, double-retraction exists. This proves (5). □

We give some formal definitions.

DEFINITION 63. *Let π be a translation plane of order q^{2n} and kernel containing K isomorphic to $GF(q)$. A 'double-homology group' is a pair of commuting homology groups of orders divisible by the order of a critical Baer order. A 'doubly-generalized central group' is either a double-homology group or a double-Baer group.*

COROLLARY 26. *Let π be a semifield plane of even order q^{2n} and kernel containing K isomorphic to $GF(q)$.*

Then double-retraction is equivalent to a double-generalized central group of order $(q + 1)$, when the kernel contains $GF(q^2)$.

We now assume that we have a translation plane of order $q^{2^a r}$, which has kernel isomorphic to $GF(q^2)$ and also assume that we have right and middle subnuclei isomorphic to this subkernel F. Assume that we can fuse these nuclei (this is possible, for example, when the plane is a semifield by Jha and Johnson [87]). Let Q denote a quasi-field coordinatizing the plane π. Let K be a subkernel subfield of F and isomorphic to $GF(q)$. If K commutes with Q, then there is an associated K-regulus in the spread. Furthermore, we assume that F does not commute over the associated quasifield. Then, we take the subkernel F acting as $z \longmapsto az$ for all vectors z, writing the vector space over F. Moreover, if the axis and coaxis are taken as $x = 0$ and $y = 0$, we have the action of either homology group as $z \longmapsto a^t z$, where $t = 0, 1$ or q, assuming that K commutes with the quasifield.

We claim that from the two homology groups, we obtain two distinct retraction-groups not equal to the kernel group. Each of these groups provides, in turn, a double-Baer group of order $q + 1$. Hence, we obtain

THEOREM 89. *Let π be a translation plane of order $q^{2^a r}$, $(r, 2) = 1$, with subkernel F isomorphic to $GF(q^2)$. Assume that we have right and middle subnuclei, which are both fields, and, which are field-isomorphic to this subkernel F. Let K denote the subfield isomorphic to $GF(q)$.*

If the associated quasifield commutes over K but does not commute over F, then we obtain commuting double-Baer groups providing double-retraction, as well as double-homology groups that provide double-retraction.

Now applying this to semifield planes, we obtain:

THEOREM 90. *Let π be a semifield plane of order $q^{2^a r}$, for $(r, 2) = 1$, admit subnuclei, left, right and middle all isomorphic to $GF(q^2)$. Assume also that there is a double-Baer group of order $q+1$ that is not in the group generated by the three associated homology groups.*

Then π admits triple-retraction.

PROOF. Since we have a semifield plane, we know that we can fuse the nuclei. By the above theorem, a double-Baer group must be commensurate with a kernel group of order $q^2 - 1$. $\square$

5. Dempwolff's Double-Baer Groups

Here we mention some new classes of translation planes that admit double-Baer groups due to Dempwolff [**39**]. Consider the following matrices:

$$\begin{bmatrix} u & 0 & at + bt^q & (ct + dt^q)^q \\ 0 & u^q & ct + dt^q & (at + bt^q)^q \\ t & 0 & u^q & 0 \\ 0 & t^q & 0 & u \end{bmatrix} ; u, t \in GF(q^2),$$

a, b, c, d constants.

Dempwolff shows that it is always possible to find situations so that $acd \neq 0$, and when this happens, one obtains a completely new class of semifields.

Furthermore, the semifield plane is derivable, and a right and middle homology group corresponding to the right and middle nuclei of order $q - 1$ will become into Baer groups in the derived plane. Therefore, the derived plane of order q^4 admits two Baer groups of order $q - 1$. Now we need to modify the orders to fuse the Baer groups into a double-Baer group. This happens where $q = p^{2^a r}$ and $(2, r) = 1$; then we have a plane of order $p^{2^{a+2} r}$ with two Baer groups of order $p^{2^a r} - 1$. Assume that $q = h^2$, so we have two Baer groups of order $h + 1$ in a translation plane of order h^8 with kernel $GF(h^2)$. Therefore, we obtain double-Baer groups that are fused, and hence, we have double-retraction.

THEOREM 91. *The class of Dempwolff semifield planes π of order q^4, for $acd \neq 0$, listed above are derivable.*

(1) When q is a square, let π^* denote the translation plane obtained by derivation of the net

$$x = 0, y = x \; diag(u, u^q, u^q, u); \; u \in GF(q^2).$$

There is a collineation group of order $q = h^2$ that contains a kernel subgroup of order $h - 1$ and is fixed-point-free, and we obtain a double-Baer group and double-retraction.

(2) Hence, there is a retraction to a mixed subgeometry partition of $PG(7, h^2)$ by subgeometries isomorphic to $PG(7, h)$ or $PG(3, h^2)$.

Dempwolff considers generalized Knuth semifields defined as follows: Let $q = p^k$ for $m = 2k$ and define multiplication as follows:

$$(u, v) * (x, y) = (ux + gv^{p^a} y^{p^b}, y + vx^{p^k}), \text{ for } u, v, x, y \in GF(q^2 = p^m),$$

where $a, b \in \{0, 1, .., m - 1\}$ and such that $p^m \neq p^{m(p^b + 1 + p^{|a-b|} + 1 + p^{k+b} - 1)}$ and g in $GF(p^m)$ and of order not $p^{m(p^b + 1 + p^{|a-b|} + 1 + p^{k+b} - 1)}$. We note that the kernel is $GF(p^{(a,m)})$, the middle nucleus is $GF(p^{(k+a-b,m)})$, and the right nucleus is $GF(p^{(k-b,m)})$. In all cases, there is a Baer group of order $q + 1 = p^k + 1$. Moreover, the semifield plane is derivable with a net of the form

$$x = 0, y = x \text{ diag}(u, u^q, u^q, u); u \in GF(q^2).$$

First, assume that a is even so that we have a subkernel $GF(p^{2(a/2,k)})$. Now consider $GF(p^{2(a/2,k)}) \cap GF(q^2) = GF(p^{2(a/2,k,2k)})$. Assume that k is odd, so since $(a/2, k, 2k)$ divides k, we see that we have a sub-Baer group of order $p^{(a/2,k,2k)} + 1 = p^{(a/2,k)} + 1$. This means we have a double-Baer group of order $p^{(a/2,k)} + 1$, and hence by Corollary 25, we have double-retraction. Therefore, we have the following theorem:

THEOREM 92. *Consider the semifield*

$$(u, v) * (x, y)$$
$$= (ux + gv^{p^a} y^{p^b}, y + vx^{p^k}), \text{ for } u, v, x, y \in GF(q^2 = p^m).$$

Assume that a is even and k is odd ($a \leq 2k - 1$). Then there is a double-Baer group of order $p^{(a/2,k)} + 1$, which may be fused so that double-retraction occurs.

When we derive the net, we obtain Baer groups of orders $p^{(k+a-b,m)} - 1$ and $p^{(k-b,m)} - 1$, and we may fuse the obvious intersection Baer groups of orders $p^{((k-a-b,2k),(k-b,2k))} - 1$. In this case, we have double-Baer groups of this order. The kernel of the derived plane is

$$GF(q) \cap GF(p^{(a,2k)}) = GF(p^{(a,2k,k)}) = GF(p^{(a,k)}).$$

Furthermore, we may fuse all subnuclei isomorphic to

$$GF(p^{(a,2k)}) \cap GF(p^{(k+a-b,2k)}) \cap GF(p^{(k-b,2k)})$$
$$= GF(p^{((a,2k),(k+a-b,2k),(k-b,2k))}).$$

And if the subfield is a square $GF(h^2)$, then we obtain double-retraction.

THEOREM 93. *The derivation of the semifield plane listed in the previous theorem admits two fused Baer groups of order*

$$p^{((k-a-b,2k),(k-b,2k))} - 1.$$

If

$$((a,2k),(k+a-b,2k),(k-b,2k))$$

is even, then we obtain double-Baer groups and double-retraction.

6. Subgeometry Partitions from Going Up

So, in the previous sections, we have come down, so to speak, from a t-spread in a $2t$-dimensional vector space to a focal-spread of type (t,k), and with the appropriate retraction group inherited by the k-cut, we may construct interesting subgeometry partitions from existing ones. We now start from below and 'go up' with the same goal. We wish to construct subgeometries from focal-spreads arising from the going up process. So, we wish to consider all of this from the context of various new subgeometry partitions constructed using the k-spreads in tk-dimensional vector spaces, basically using a variation of the Sperner spread construction process.

We begin again with a subgeometry partition and ask what such a going-up process might imply for subgeometries. Recall that we construct a k-spread from the Sperner spread construction process from suitably many k-spreads in $2k$-dimensional subspaces (that is, from translation planes of order q^k), and we then use this construction to construct our focal-spreads as well as our generalized focal-spreads (where the dimension is larger than $t+k$, in the type (t,k) situation).

So assume that we have a subgeometry partition of $PG(z-1, q^w)$ by subgeometries isomorphic to $PG(l-1, q^e)$, for various values of l and e and assume that the subgeometry partition arises from a k-spread over $GF(q)$. If we begin with a vector space of dimension ks over $GF(q)$, then for w dividing ks, we would have a subgeometry partition of $PG(ks/w - 1, q^w)$ by subgeometries isomorphic to $PG(l-1, q^e)$, again for many various values of l and e. In the context of focal-spreads of type $(k(s-1), k)$ arising from the going up process, the question is whether the subgeometry partitions of the k-spread may be modified to construct subgeometry partitions of the associated focal-spread. All of the subgeometry partitions arise from groups acting on the individual translation planes of order q^k that are used in the construction of the k-spread in the ks-dimensional vector space over $GF(q)$, and which fix two components $x = 0, y = 0$ that we identify so that a group G of type given in Theorem 20 acts on the focal-spread. This is the key issue—to make this group a retraction group that produces at least a quasi-subgeometry partition. Therefore, if the vectors are $(x_1, x_2, .., x_s)$, where x_i are k-vectors, for $i = 1, 2, .., s$, and the focus is the $k(s-1)$ dimensional vector space with equation $x_s = 0$, the group G of order $q^w - 1$, for w dividing k, will leave $x_s = 0$ invariant, and we may consider one of the subgeometries in a subgeometry partition to be $PG(k(s-1)/w - 1, q^w)$. Now noting that the same construction will work in focal-spreads of type $(k(t-1), k(t-m), k)$, we obtain the following theorem:

THEOREM 94. *Using the Sperner k-spread construction Theorem, if a subgeometry partition of $PG(ks/w - 1, q^w)$, for w dividing k, by subgeometries of type $PG(l-1, q^e)$ for various values of l and e, is constructed, then using the going up process, we also obtain a subgeometry partition with the same sort of collection of subgeometries but with an extra type of subgeometry $PG(k(s-m)/w - 1, q^w)$, where m is any integer so that $s - m > 1$.*

Of course, we may consider all of these ideas in a much more general setting, and the reader is directed to [99] for the construction of a great many subgeometry partitions of projective spaces. In particular, if the retraction group is of order $q^w - 1$, then for each divisor e of w, assume that it is possible to find subgeometry partitions of $PG(ks/w-1, q^w)$ by subgeometries isomorphic to $PG(l-1, q^e)$. Then the bonus is that there are related subgeometry partitions with an extra type of subgeometry $PG(k(s-m)/w - 1, q^w)$.

7. Algebraic Lifting of Focal-Spreads

We may now use the ideas presented in the Jha-Johnson lifting theorem coupled with k-cuts of subgeometry partitions to obtain some new subgeometry partitions.

Here are the main points: given a spread in $PG(3, q)$, algebraically lift to a spread in $PG(3, q^2)$ and then derive the plane of order q^4. Locate and fuse the Baer groups. There is now an associated retraction group, which has $q^2 + 1$ invariant components and $(q^4 - q^2)/(q+1) = q^2(q^2 - 1)/(q+1) = q^2(q-1)$ orbits of length $q + 1$.

By Theorem 23, noting that we have a 4-spread, we choose $x = 0$ and a 1-space on $y = 0$ over the retraction group/field isomorphic to $GF(q^2)$. Then we construct a focal-spread of dimension $4+2$ with focus of dimension 4 that admits a retraction field isomorphic to $GF(q^2)$. This constructs a subgeometry partition of $PG(6/2-1, q^2) = PG(2, q^2)$ with one subgeometry isomorphic to $PG(4/2 - 1, q^2) = PG(1, q^2)$, q^2

subgeometries isomorphic to $PG(2/2-1, q^2) = PG(0, q^2)$ and $q^2(q-1)$ subgeometries isomorphic to $PG(2-1, q) = PG(1, q)$. Here is the outcome of similar analysis of k-cuts spreads, thereby creating k-cuts of subgeometry partitions.

THEOREM 95. *From any spread in* $PG(3, q)$, *and* $q = p^r$ *and* e *a proper odd prime divisor, let* $2^j r = 2^j se$ *and* $k = 2ie$, *for* $i = 2, 3, .., 2^{j-1}s - 1$, *where* $j \geq 2$. *Then it is possible to construct subgeometry partitions of*

$$PG((2^j r/e + k)/2 - 1, p^{2e})$$

by

$$one\ PG(2^j r/e - 1, p^{2e}),$$
$$p^{2r}\ PG(k/2, p^{2e})\text{'s, and}$$
$$p^{2r}(p^r - 1)/(p - 1)\ PG(k - 1, p^e)\text{'s.}$$

PROOF. The process of algebraic lifting from a spread in $PG(3, q)$ to a spread in $PG(3, q^{2^2})$ can be repeated to construct a spread in $PG(3, q^{2^2 2})$, and so on. Hence, j is any integer at least 2.   □

Now there is a very general concept of general algebraically lifting, which we shall call 'lifting and twisting,' by which we may algebraically lift $(2, k)$-spreads and then twist the groups so that we obtain a group acting in the manner of Theorem 20. This then will imply that there is a retraction group action on the constructed k-spread in a tk-dimensional vector space, which, in turn, will act on the associated focal-spread obtained using the going up procedure, which will then act on the focal-spread obtained by the going up procedure. The main result in this regard is as follows. The reader is directed to the article by Jha and Johnson [80], but if one would think about the natural action of the group inherited from a lifting procedure and the requirement that the group act on the constructed spread the concept of the 'twist' would become apparent.

THEOREM 96. *(Jha and Johnson* [80]*) Choose any set of* $\sum_{j=0}^{z-2} \binom{z}{j} (z - j - 1) = N_z$ *spreads in* $PG(3, q)$. *Algebraically lift and twist each of these spreads. Apply the 2-spread construction to construct a 4-spread of an $8z$ dimensional $GF(q)$-space, with*

$$(q^{4z} - 1)/(q^4 - 1)$$

2-spaces over $GF(q^2)$.

(1) This 2-spread over $GF(q^2)$ admits a retraction group G of order $q^2 - 1$, such that the union of the zero vector gives a field F over which the ambient space is an F-space.

(2) The group G fixes exactly $(q^{2z} - 1)/(q^2 - 1)$ components. If K^* denotes the kernel homology group of order $q^2 - 1$, then in GK^*, there is a group of order $q + 1$ that fixes pointwise a subspace of dimension $4z$ over $GF(q)$. We call this a 'Baer subspace.'

(3) Since the group G contains the scalar homology group of order $q - 1$, we may 'retract' the spread to produce a subgeometry partition of $PG(4z - 1, q^2)$ by subgeometries isomorphic to $PG(3, q)$ and $PG(1, q^2)$. In particular, there are exactly

$$(q^{2z} - 1)/(q^2 - 1)$$

$PG(1, q^2)$*'s and*

$$\left((q^{4z} - 1)/(q^4 - 1) - (q^{4t} - 1)/(q^2 - 1)\right)/(q + 1)$$

$PG(3, q)$*'s.*

(4) We may derive any of the 2-spreads by the subspace given by

$$x = 0, y = 0, y = x \begin{bmatrix} 0 & u^q \\ u & 0 \end{bmatrix}; u \in GF(q^2)^*.$$

We note that the group G leaves this spread invariant, fixes two components, and has $(q - 1)$ orbits of length $q + 1$. The group fixes two Baer subplanes of the net and has $q - 1$ orbits of length $q + 1$. To re-represent this spread, we obtain the derived spread in the form:

$$x = 0, y = 0, y = x^q \begin{bmatrix} 0 & u \\ u & 0 \end{bmatrix}; u \in GF(q^2)^*.$$

The spread

$$x = 0, y = x \begin{bmatrix} u & F(t) \\ t & u^q \end{bmatrix}, \text{ for all } u, t \in GF(q^2),$$

when derived and twisted is

$$x = 0, y = 0,$$

$$y = x \begin{bmatrix} F(t) - t^{-1}u^{q+1} & -ut^{-1} \\ t^{-1}u^q & t^{-1} \end{bmatrix},$$

for all $u, t \neq 0 \in GF(q^2)$,

$$y = x^q \begin{bmatrix} 0 & u \\ u & 0 \end{bmatrix}; u \in GF(q^2)^*,$$

where x^q means to apply the automorphism to the elements of the 2-vector.

The group G acts on the derived spread and has the same form. The group maps

$$y = x \begin{bmatrix} F(t) - t^{-1}u^{q+1} & -ut^{-1} \\ t^{-1}u^q & t^{-1} \end{bmatrix}$$

$$to\ y = x \begin{bmatrix} F(t) - t^{-1}u^{q+1} & -(a^{q-1}u)t^{-1} \\ t^{-1}(a^{q-1}u)^q & t^{-1} \end{bmatrix}.$$

Hence, any of these general spreads also admit the retraction group $\bar{G}$ and produce additional examples of subgeometry partitions of $PG(4t - 1, q^2)$ by $PG(3, q)$'s and $PG(1, q^2)$'s. G acts as a 'Baer group' of the 'multiply-derived' spreads.

Note that if we derive any of the spread, we obtain a set of 4-dimensional subspaces over $GF(q)$. Hence, the derivation of any spread produces a 4-spread in the 8z-dimensional space over $GF(q)$.

(5) If we choose the N_z spreads in $PG(3, q)$ to be mutually non-isomorphic, then the lifted and twisted spreads are mutually non-isomorphic. Therefore, the collineation group of the constructed 2-spreads or 4-spreads must leave invariant each of the 2-spreads (4-spreads).

(6) To construct the spreads, we use the notion of a j-(0-set). We may choose any ordering we like and the constructed 2-spreads will not normally be isomorphic. Hence, we obtain $N_z!$ 2-spreads of dimension 8z over $GF(q)$. Then, for each of these spreads, we may choose any set to derive. Hence, there are

$$2^{N_z!}$$

possible ways to do this for each of the $N_z!$ spreads. Therefore, we obtain

$$N_z!2^{N_z}$$

possible mutually non-isomorphic spreads, all of which give rise to subgeometry partitions of $PG(4z - 1, q^2)$ by subgeometries isomorphic to $PG(3, q)$'s and $PG(1, q^2)$'s.

The reader interested in continuing reading about these ideas is directed to the article by Jha and Johnson [**80**].

Part 4

Flocks and Related Geometries

In this part, we consider flocks and the translation planes connected with them, as well as generalizations of these ideas. As it is not much more difficult to consider spreads covered by pseudo-reguli as opposed to spreads covered by reguli, we give the more general theory in this first chapter. The ideas presented here will also pave the way for the study of flocks of α-cones.

CHAPTER 15

Spreads Covered by Pseudo-Reguli

In this chapter, we consider spreads in $PG(3, K)$ that are covered by pseudo-reguli that share a given line, which we call 'conical spreads' in $PG(3, K)$, and we consider spreads which are covered by pseudo-reguli that share two given lines, which we call 'ruled spreads' and formulate the corresponding theory. In the following chapter, we consider conical and hyperbolic flocks that correspond to spreads that are unions of reguli, sharing either one or two common components. However, there are translation planes whose spreads are unions of derivable nets that are not given by reguli over the field or projective space in question. This is true both for the finite and infinite cases. Thus, we include here a study of a more general situation than encountered in spreads corresponding to flocks.

Although it seems natural enough to consider this study in the context of derivable nets, this material originated not with derivable nets but with the consideration of flocks of quadric sets. In the next chapter, we sketch part of the theory interconnecting coverings of quadrics by planes to the analysis of spreads covered by reguli. However, we choose to work from the general to the specific in this instance. This and the following chapters on flocks of hyperbolic and quadratic cones are also in the author's Subplanes text with a few changes in form. We include these chapters here for convenience and completeness.

1. Pseudo-Reguli

There are some technical problems forming unions of pseudo-reguli that can occur because of the possible non-commutativity of multiplication of K so we consider what are called 'normal sets' of pseudo-reguli.

We have discussed a geometric version of a 'regulus' in $PG(3, q)$, and a pseudo-regulus is only defined algebraically. Furthermore, we have defined a K-regulus as a pseudo-regulus over a field and have given a geometric definition.

We now consider a possible geometric definition of a pseudo-regulus. This is not completely satisfactory, but it gives some insight as to the difficulties in working with non-commutative geometries.

It has been shown that given a derivable net N algebraically represented as a pseudo-regulus net with reference to a skewfield K then the
derived net N^* may be algebraically represented with reference to the
skewfield K^{opp} where multiplication $\circ$ in K^{opp} is defined by $a \circ b = ba$
where juxtaposition denotes multiplication in K.

We recall that if the vector space V of points is a left vector space
over the skewfield K then the lines of the derivable net incident with
the zero vector are not necessarily always 2-dimensional left K-vector
subspaces although they are natural 2-dimensional right K-vector subspaces. However, the Baer subplanes incident with the zero vector are
left 2-dimensional left K-vector spaces.

For the derived net, the situation is reversed. The lines of the derived net incident with the zero vector are not always 2-dimensional left
K^{opp}-vector subspaces but they are natural 2-dimensional right K^{opp}-
vector subspaces as they are always 2-dimensional left K–subspaces.
Similarly, the Baer subplanes incident with the zero vector of the derived net are 2-dimensional left K^{opp}-vector spaces as they are the
lines of the original net incident with the zero vector, which are 2-
dimensional left K-vector subspaces.

Since we would like to represent the lines of our derivable net within
the lattice of left subspaces of a 4-dimensional left vector space, we
dualize everything and note the following:

REMARK 30. *Let N denote a derivable net with lines incident with
the zero vector represented in the form $y = x\delta$ for all δ in a skewfield
J, where the associated vector space V is a 4-dimensional left J-space
and the lines indicated are 2-dimensional left J-subspaces.*

*So, the lines of N incident with the zero vector become lines in the
projective space Σ isomorphic to $PG(3, J)_{\mathcal{L}}$ defined as the lattice of left
vector J-subspaces.*

*In terms of a given basis, V may also be defined as a 4-dimensional
right or left J^{opp}-vector space V^* and the lines of N^* incident with
the zero vector become lines in the projective space Σ^* isomorphic to
$PG(3, J^{opp})_{\mathcal{L}}$ defined as the lattice of left vector J^{opp}-subspaces.*

*Choose a left K-basis $B = \{e_i$ for i in $\lambda\}$. For a vector $\Sigma x_i e_i$, x_i
in K for $i = 1, 2, 3, 4$, a left space over K^{opp} may be defined as follows:
$u \circ \Sigma x_i e_i = \Sigma x_i u e_i = \Sigma(u \circ x_i)e_i$. So, there are many ways to form a
projective space $PG(3, K^{opp})$, if K is a non-commutative skewfield.*

REMARK 31. *In the following, we shall be considering always 'left'
vector spaces over either K or K^{opp}, so we shall use simply $PG(3, K)$
to denote $PG(3, K)_{\mathcal{L}}$.*

Note that a right 1-dimensional K–subspace may not be associated with a 'point' of $PG(3, K)$, but it would be a 'point' of $PG(3, K^{opp})$.

If the associated vector space is considered over $Z(K)$, we call 1-dimensional $Z(K)$-subspaces, '$Z(K)$-projective points.'

DEFINITION 64. *Let S be any set of mutually skew lines of $PG(3, K)$. A 'vector-transversal' L to S is a line of some $PG(3, K^{opp})$ of $Z(K)$-projective points with the property that L as a left $Z(K)$-subspace has a nontrivial vector intersection with each line of S as a left $Z(K)$-subspace such that the direct sum of any two such intersections is L. A 'point-transversal' to S is a line of $PG(3, K)$ that is also a vector-transversal.*

A 'projective pseudo-regulus' $R = R_{\{L,M,N\}}^{\{U,V,W\}}$ in $PG(3, K)$ is a set of lines (as left 2-dimensional K-vector spaces) containing $\{L, M, N\}$ with a set of points $\{U, V, W\}$ of L such that any line T that intersects L in either U, V, or W and also intersects M and N intersects each line of R and T is contained in the set of these intersections. So, any such line T becomes a point-transversal.

REMARK 32. *We have previously defined the pseudo-regulus net. Here we consider the corresponding net defined by a projective pseudo-regulus also called a pseudo-regulus net. We shall show that there is no distinction between the two nets and thus the terminology is justified. Furthermore, when it does not present problems, we shall use the same notation for the pseudo-regulus and the corresponding net and allow context to dictate which is under consideration.*

THEOREM 97. *Choose any three mutually skew lines L, M, N of $PG(3, K)$ and let U, V, W be any three distinct points on L.*

(1) Then there exists a unique projective pseudo-regulus $R_{\{L,M,N\}}^{\{U,V,W\}}$ in $PG(3, K)$, which contains L, M, N and which has point-transversals intersecting L in U, V, and W.

Furthermore, there is a unique basis such that L, M, N may be represented in the form $y = x$, $x = 0$, $y = 0$, respectively where $L = \langle U \rangle \oplus \langle V \rangle$ and $W = \langle U + V \rangle$.

The pseudo-regulus then has the form

$$x = 0, y = x \begin{bmatrix} u & 0 \\ 0 & u \end{bmatrix}$$

for all u in K.

(2) Choose any set $\{L^*, M^*, N^*\}$ of three mutually skew lines of the pseudo-regulus $R_{\{L,M,N\}}^{\{U,V,W\}}$. Then there exist points U^*, V^*, W^*, on L^*

such that

$$R_{\{L^*,M^*,N^*\}}^{\{U^*,V^*,W^*\}} = R_{\{L,M,N\}}^{\{U,V,W\}}.$$

(3) *Any pseudo-regulus net is a derivable net (is a standard pseudo-regulus net).*

PROOF. Consider the associated 4-dimensional left K-vector space V. Choose a basis for L as $\langle U, V \rangle$. Then $W = \alpha U + \beta V$ for α, β in K. Choose $U^* = \alpha U$ and $V^* = \beta V$. Hence, projectively, we may assume without loss of generality that $W = U + V$. Represent $V_4 = M \oplus N$, then there exist unique elements m_u, m_v of M, and n_u, n_v of N such that $U = m_u + n_u$ and $V = n_v + n_u$. Then

$$\begin{aligned} W &= U + V = (m_u + n_u) + (m_v + n_v) \\ &= ((m_u + m_v) = m_w) + ((n_u + n_v) = n_w). \end{aligned}$$

It is immediate that $\{m_u, m_v\}$ is a basis for M and $\{n_u, n_v\}$ is a basis for N. Now form

$$T_U = \langle m_u, n_u \rangle, T_V = \langle m_v, n_v \rangle, T_W = \langle m_w, n_w \rangle.$$

Clearly, $T_U \cap L = \langle U \rangle$, $T_V \cap L = \langle V \rangle$ and $T_W \cap L = \langle W \rangle$. Choose a basis $\{m_u, m_v, n_u, n_v\}$ for V_4. In terms of this basis,

$$V_4 = \{(x_1, x_2, y_1, y_2) \mid x_i, y_i \in K, \, i = 1, 2\}.$$

Let $x = (x_1, x_2)$ and $y = (y_1, y_2)$. Then L, M, N are $y = x$, $y = 0$, $x = 0$, respectively. Furthermore,

$$\begin{aligned} T_U &= \{(x_1, 0, x_2, 0) \mid x_i \in K, \, i = 1, 2\}, \\ T_V &= \{(0, y_1, 0, y_2) \mid y_i \in K, \, i = 1, 2\}, \\ T_W &= \{(x_1, x_1, y_1, y_1) \mid x_1, y_1 \in K\}. \end{aligned}$$

It follows that any component of the pseudo-regulus is of the form $y = x \begin{bmatrix} a & b \\ c & d \end{bmatrix}$ for a, b, c, d in K and the intersection with T_U, T_V, and T_W shows that $a = d = u$ and $b = c = 0$. Let the pseudo-regulus R be represented in the form $x = 0, y = 0, y = x \begin{bmatrix} u & 0 \\ 0 & u \end{bmatrix}$ for $u \in \lambda \subseteq K$. Now in order that T_U is contained in $\{T_U \cap Z \mid z \in R\}$, we have

$$\{(x_1, 0, x_1 u, 0) \mid x_1 \in K\} = T_U \; \forall u \in \lambda.$$

It clearly follows that this forces $\lambda = K$ so that the pseudo-regulus has the required form. It follows immediately that any pseudo-regulus is a derivable partial spread. The derivable net R defined by

$$x = 0, y = x \begin{bmatrix} u & 0 \\ 0 & u \end{bmatrix} \; \forall u \in K$$

has Baer subplanes

$$\rho_{a,b} = \{(a\alpha, b\alpha, a\beta, b\beta) \ \forall \alpha, \beta \in K\}$$

and $(a, b) \neq (0, 0)$, and by such a choice of basis for the vector space, we see that there are at least three 2-dimensional K^{opp}-subspaces that are vector-transversals, which are also point-transversals, namely, $\rho_{0,1}$, $\rho_{1,0}$, and $\rho_{1,1}$. Note that if $Z(K)$ is isomorphic to $GF(2)$, there are exactly three lines of $PG(3, K)$, which are point-transversals to this net. Now assume that there is another pseudo-regulus satisfying these conditions. Assume that exist three point-transversals $\gamma_0, \gamma_1, \gamma_2$ (2-dimensional left K–subspaces) such that $\gamma_o \cap L = U, \gamma_1 \cap L = V$ and $\gamma_2 \cap L = W$. The question is whether γ_0 actually turns out to be T_U above. Since it intersects L in the same subspace, we may assume that γ_0 contains $(1, 0, 1, 0)$. Let (x_1, x_2, y_1, y_2) and $(1, 0, 1, 0)$ generate γ_0. Since γ_0 intersects $x = 0$ and $y = 0$ in a 1-dimensional K–subspaces, it follows that the only way to manage this is for $x_2 = 0$ and for $y_2 = 0$. The two indicated vectors are linearly independent so it is clear that $\gamma_0 = \langle(1, 0, 0, 0), (0, 0, 1, 0)\rangle = T_U$. Similarly, $\gamma_1 = T_V$ and $\gamma_2 = T_U$. In other words, any two sets of point-transversals of three elements to $\{L, M, N\}$ that intersect L in the same set of points are identical. Since these point-transversals determine a unique representation, we have a uniquely defined pseudo-regulus. We have noted that there are at least three lines of $PG(3, K)$ that are point-transversals to the pseudo-regulus $R_{\{L,M,N\}}^{\{U,V,W\}}$. By the structure theory for derivable nets previously determined, there exists a collineation group of the pseudo-regulus net, which is triply transitive of the components of the pseudo-regulus and fixes each Baer subplane incident with the zero vector and hence every vector-transversal. Thus, there exists a collineation σ of the net that carries $\{L, M, N\}$ onto $\{L^*, M^*, N^*\}$ as ordered sets. Choose U^*, V^*, W^* as $U\sigma, V\sigma, W\sigma$, respectively. Then the above construction is merely a basis change, so that the two pseudo-regulus nets are identical. $\square$

REMARK 33. *Any pseudo-regulus in $PG(3, K)$ has a set of transversal lines in $1 - 1$ correspondence with a set of cardinality $Z(K) + 1$.*

We note that since the vector-transversals define Baer subplanes of the net incident with the zero vector, it is not necessarily true that every Baer subplane incident with the zero vector intersects each component in a 1-dimensional left K–subspace (point of $PG(3, K)$).

In fact, the point-transversals are determined by any one intersection.

COROLLARY 27. *Let R be any pseudo-regulus in $PG(3, K)$. If a vector-transversal intersects some line of R in a point (a 1-dimensional left K-space), then the vector-transversal is a point-transversal (line).*

PROOF. We may represent R in the standard form

$$x = 0, y = x \begin{bmatrix} u & 0 \\ 0 & u \end{bmatrix} \forall u \in K.$$

The vector-transversals are exactly the Baer subplanes $\rho_{a,b}$. Suppose $\rho_{a,b}$ intersects $x = 0$ in a 1-dimensional left K-space. Then, it follows that the intersection is $\{(0, 0, a\alpha, b\alpha) \ \forall \ \alpha \text{ in } K\}$. However, this is a 1-dimensional left K-space if and only if a and b are in $Z(K)$, which implies that the intersection with $y = x \begin{bmatrix} u & 0 \\ 0 & u \end{bmatrix}$ is

$$\{(a\alpha, b\alpha, a\alpha u, b\alpha u) \forall \alpha \in K\}$$

and is a 1-dimensional left K-space. Similarly, if any such intersection

$$\{(a\alpha, b\alpha, a\alpha u, b\alpha u) \forall \alpha \in K\}$$

is a 1-dimensional left K-space, then a and b are in $Z(K)$ so that all intersections are 1-dimensional left K-spaces. □

This also proves the following well-known corollary, which we repeat in order to emphasize the ensuing remarks.

COROLLARY 28. *If K is a field, then there is a unique regulus containing any three mutually skew lines L, M, and N.*

PROOF. Choose any three points on L and construct the corresponding regulus net. Any vector-transversal is a point-transversal and corresponds to a Baer subplane of the net. As any three points on L correspond to unique Baer subplanes of this regulus net, it follows that the choice of three points on L is arbitrary. □

REMARK 34. *To illustrate that the previous corollary is not necessarily valid for pseudo-reguli, suppose K is a skewfield such that $K \neq Z(K) \simeq GF(2)$. Then, the above result shows that for any three distinct points U, V, W on a line L of $PG(3, K)$, there is a unique pseudo-regulus $R^{\{U,V,W\}}_{\{L,M,N\}}$ containing $\{L, M, N\}$. Furthermore, there are exactly three Baer-transversals, which are point-transversals to this pseudo-regulus. So, take any three distinct points U^*, V^*, W^* on L such that $\{U, V, W\} \neq \{U^*, V^*, W^*\}$. Then, it is not possible that*

$$R^{\{U,V,W\}}_{\{L,M,N\}} = R^{\{U^*,V^*,W^*\}}_{\{L,M,N\}}.$$

Hence, there exist skewfields K such that there are infinitely many pseudo-reguli that share any three mutually skew lines in $PG(3, K)$.

2. Conical and Ruled Spreads over Skewfields

Now suppose that D is a pseudo-regulus net represented in standard form. Suppose that a subplane $\rho_{a,b}$ intersects $x = 0$ in $\{(0, 0, a\beta, b\beta)$ $\forall \beta \in K\}$. Assume that a and b are not both in $Z(K)$. Choose the vector $(0, 0, a, b)$ and let $\langle (0, 0, a, b) \rangle = U$ denote the left 1-dimensional K–subspace generated by $(0, 0, a, b)$. Assume that $\{x = 0, M, N\}$ is a set of skew lines in $PG(3, K)$ that is not in D. Choose any other points V, W of $x = 0$ and form $R_{\{x=0,M,N\}}^{\{U,V,W\}}\} = R$. Then R and D are pseudo-reguli, which share at least one line but do not have the property that the vector-transversals to the two pseudo-reguli partition $x = 0$ in the same set of sublines. We shall be interested in situations where there is such a partition, and, to this end, we formulate the following definition.

DEFINITION 65. *Let D_1 and D_2 be any two pseudo-regulus nets whose union defines a partial spread in $PG(3, K)$ that share either one or more components. Assume that on one of the common components L there exists two points of $PG(3, K)$ (1-dimensional left K–subspaces), each of which are in point-transversals to D_i for $i = 1, 2$, respectively.*

Any two pseudo-reguli sharing one or more two lines, whose nets satisfy the above property shall be said to be 'normalizing'. Furthermore, any set of pseudo-reguli sharing one or more lines each pair of which satisfies the above property with respect to the same two points shall be said to be 'normal set.' Any such point shall be said to be a 'normal point.'

REMARK 35. *If K is a field, any two reguli satisfying the above conditions are normalizing pseudo-reguli.*

THEOREM 98. (1) *Two normalizing pseudo-reguli in $PG(3, K)$ share one or two lines.*

(2) *If two normalizing pseudo-reguli share exactly one line then there is an elation group with axis the common line that acts regularly on the remaining lines of each pseudo-regulus. Furthermore, the vector-transversals to each pseudo-regulus net induce the same partition on this common line.*

(3) *If two normalizing pseudo-reguli share at least two lines then there is a homology group with axis and coaxis the two common lines that acts regularly on the remaining lines of each pseudo-regulus. Furthermore, on one of the common lines, the vector-transversals to the various pseudo-reguli induce the same partition on this line.*

PROOF. Assume the hypothesis of (2). By appropriate choice of coordinates, any given pseudo-regulus may be brought into the form

$$x = 0, y = x \begin{bmatrix} u & 0 \\ 0 & u \end{bmatrix} \forall u \in K.$$

Let D_1 have this standard form. We assume that $x = 0$ is the common line. The Baer subplanes incident with the zero vector are

$$\pi_{a,b} = \{(a\alpha, b\alpha, a\beta, b\beta) \forall \alpha, \beta \in K\}.$$

Note the Baer subplanes that are transversal lines are $\pi_{a,b}$ where both a and b are in $Z(K)$. Furthermore, we may assume that the two 1-dimensional left K–subspaces on $x = 0$ that belong to transversal lines of each of the two pseudo-reguli have the general form $\langle(0,0,1,0)\rangle$ and $\langle(0,0,0,1)\rangle$. Note that the Baer subplanes of D_1 containing the indicated 1-dimensional left K–subspaces are $\pi_{1,0}$ and $\pi_{0,1}$ respectively. Assume that $y = xT_i$ for i in λ and $x = 0$ are the lines of the second pseudo-regulus net D_2. Select two distinct values c, d of λ and change bases by the mapping

$$(x, y) \rightarrow (x, -xT_c + y).$$

This mapping fixes $x = 0$ pointwise and carries $y = xT_c$ onto $y = 0$. Now change bases again by the mapping

$$(x, y) \rightarrow (x(T_d - T_c), y).$$

It follows, after the basis change, that the components of the second pseudo-regulus net are

$$x = 0, y = x(T_d - T_c)^{-1}(T_i - T_c).$$

In particular, $x = 0, y = x, y = 0$ are components of the second net D_2. Since $x = 0$ is fixed pointwise by both of the above basis changes, it follows that there are two transversal lines L, M to D_2 intersecting $x = 0$ in $\langle(0,0,0,1)\rangle$ and $\langle(0,0,1,0)\rangle$, respectively such that $L \cap (y = 0) = \langle(m,n,0,0)\rangle$ and $M \cap (y = 0) = \langle(s,t,0,0)\rangle$. Since L and M both intersect $y = x$ in a 1-dimensional left K–subspace, it follows that $m = 0 = t$. Hence, L and M are $\pi_{0,1}$ and $\pi_{1,0}$, respectively. It then easily follows since D_2 is a derivable net with partial spread in $PG(3, K)$, that the components of the net have the general form

$$x = 0, \quad y = x \begin{bmatrix} u^\sigma & 0 \\ 0 & u \end{bmatrix} \forall u \in K$$

and σ an automorphism of K. However, there is at least a third transversal line to the derivable net. Since the Baer subplanes of the net in question now have the form $\rho_{a,b} = \{(a\alpha^\sigma, b\alpha, a\beta^\sigma b, \beta)\}$, then there exist a, b such that $ab \neq 0$ and a and b are in $Z(K)$ so it must be that $\sigma = 1$. Hence, the set of images

$$x = 0, \ y = 0, \ y = x(T_d - T_c)^{-1}(T_i - T_c)$$

is equal to

$$x = 0, \ y = 0, \ y = x \begin{bmatrix} u & 0 \\ 0 & u \end{bmatrix} \forall u \in K.$$

Thus,

$$T_i = T_c + (T_d - T_c)uI \forall u \in K.$$

To prove part (2), it suffices to show that $T_d - T_c = v_o I$ for some v_o in K. Since $x = 0$ is fixed pointwise, it would then also follow that the vector-transversals to both pseudo-regulus nets share the same points on $x = 0$. Hence, we have a partial spread

$$x \ = \ 0, y = x \begin{bmatrix} u & 0 \\ 0 & u \end{bmatrix} \forall u \in K,$$

$$y \ = \ x(T_c + (T_d - T_c)vI)\forall v \in K.$$

Let $T_d = \begin{bmatrix} a_1 & b_1 \\ c_1 & d_1 \end{bmatrix}$ and $T_c = \begin{bmatrix} a_2 & b_2 \\ c_2 & d_2 \end{bmatrix}$. It then follows that all matrix differences are nonsingular or zero so that we must have

$$\begin{bmatrix} a_2 + (a_1 - a_2)v - u & b_2 + (b_1 - b_2)v \\ c_2 + (c_1 - c_2)v & d_2 + (d_1 - d_2)v - u \end{bmatrix},$$

which is nonsingular $\forall \ u, v \in K$. Assume that $(b_1 - b_2) \neq 0$. Then choose v so that

$$b_2 + (b_1 - b_2)v = 0.$$

Then there exists a $u \in K$ such that

$$a_2 + (a_1 - a_2)v - u = 0.$$

This is a contradiction so $b_1 = b_2$ and similarly $c_1 = c_2$. Note that if $c_2 = 0$ then choosing $u = a_2$ shows the matrix difference to be singular. Hence, $c_2 d_2 \neq 0$. Also, since $c_1 = c_2$ and $T_d - T_c$ is nonsingular, it follows that $a_1 - a_2 \neq 0$. Then, for each u in K, we may choose v in K so that

$$a_2 + (a_1 - a_2)v - u = 1.$$

This leaves us with the matrix

$$\begin{bmatrix} 1 & b_2 \\ c_2 & 1 - a_2 + d_2 + ((d_1 - d_2) - (a_1 - a_2))v \end{bmatrix}.$$

Then, in order that the matrix be nonsingular, we must have

$$b_2 - c_2^{-1}(1 - a_2 + d_2 + ((d_1 - d_2) - (a_1 - a_2))v)$$
$$\neq \ 0 \ \forall v \in K.$$

Clearly, this forces $(d_1 - d_2) - (a_1 - a_2) = 0$. Hence,

$$T_d - T_c = \begin{bmatrix} a_1 - a_2 & 0 \\ 0 & a_1 - a_2 \end{bmatrix}.$$

Thus, there is an associated elation group E of the form

$$\left\langle \begin{bmatrix} 1 & 0 & u & 0 \\ 0 & 1 & 0 & u \\ 0 & 0 & 1 & 0 \\ 0 & 0 & 0 & 1 \end{bmatrix} \forall\, u \in K \right\rangle.$$

Assume the conditions of (3). Let D_1 be represented in the form

$$x = 0, y = x \begin{bmatrix} u & 0 \\ 0 & u \end{bmatrix} \forall u \in K$$

and choose the two common components to be $x = 0, y = 0$. Let D_2 have components $y = xT_i$ and $x = 0, y = 0$. Further, assume the two 1-dimensional left K subspaces lie on $x = 0$ and are $< (0, 0, 1, 0) >$ and $< (0, 0, 0, 1) >$. Choose a new basis by $(x, y) \to (xT_d, y)$. Then there are Baer subplanes in the image of D_2 that are 2-dimensional left $K-$ subspaces that share $x = 0$, $y = 0$, $y = x$ and such that the subplanes intersect $x = 0$ in $\langle (0, 0, 1, 0) \rangle$ and $\langle (0, 0, 0, 1) \rangle$. It follows, similarly as in the previous argument, that the two Baer subplanes have the form

$$\langle (1, 0, 0, 0), (0, 0, 1, 0) \rangle \quad and \quad \langle (0, 1, 0, 0), (0, 0, 0, 1) \rangle.$$

Hence, as before, D_2 is now represented as

$$x = 0, y = 0, y = x \begin{bmatrix} u & 0 \\ 0 & u \end{bmatrix} \forall u \in K - \{0\}.$$

Thus, $T_i = T_d uI$. Hence, there is an associated homology group H of the form

$$\left\langle \begin{bmatrix} 1 & 0 & 0 & 0 \\ 0 & 1 & 0 & 0 \\ 0 & 0 & u & 0 \\ 0 & 0 & 0 & u \end{bmatrix} \forall\, K - \{0\} \right\rangle.$$

Since $y = 0$ is pointwise fixed by the basis change above, this proves (3). Furthermore, if D_2 shares at least three components with D_1, then this will force $T_d = v_oI$ so that the two pseudo-regulus nets are identical. This proves (1). $\qquad\square$

Thus, we have the following result:

THEOREM 99. (1) *Let π be a translation plane with spread in $PG(3, K)$, for K a skewfield, which is a union of a normal set of pseudo-reguli that* share exactly one line L. Then there is an elation group E with axis L of π that acts regularly on lines $\neq L$ of each pseudo-regulus.

(2) Let π be a translation plane with spread in $PG(3, K)$, which is a union of a normal set of pseudo-reguli that share exactly two lines L and M. Then there is a homology group H of π with axis and coaxis L and M that acts regularly on lines $\neq L$ or M of each pseudo-regulus.

DEFINITION 66. Let π be a translation plane with spread in $PG(3, K)$, for K a skewfield.

If the spread for π is a normal set of pseudo-reguli sharing exactly one line L, we shall call π a 'conical translation plane' and the corresponding spread is said to be a 'conical spread.'

If the spread for π is a normal set of pseudo-reguli sharing exactly two lines M and N, we shall call π a 'ruled translation plane' or a 'hyperbolic translation plane' and the corresponding spread is said to be a 'hyperbolic' or 'ruled' spread.

THEOREM 100. (1) If π is a conical translation plane, then the spread for π may be represented in the form

$$x = 0, \ y = x \begin{bmatrix} u + g(t) & f(t) \\ t & u \end{bmatrix} \forall t, u \in K,$$

and for g, f functions on K.

(2) f and g are functions on K such that $x^2 t + x g(t) - f(t) = \phi_x(t)$ is bijective $\forall x$ in K if and only if the functions define a spread of the form in (1).

(3) If π is a ruled translation plane, then the spread for π may be represented in the form

$$x = 0, \ y = 0, \ y = x \begin{bmatrix} v & 0 \\ 0 & v \end{bmatrix},$$

$$y = x \begin{bmatrix} g(t)u & f(t)u \\ tu & u \end{bmatrix} \forall t, v, u, ut \neq 0 \in K,$$

where g, f are functions on K.

(4) If π is a conical translation plane with line L and a ruled translation plane with lines L and M such that two normal points (1-dimensional left K–subspaces) lie on L, then the spread for π may be represented in the form

$$x = 0, y = x \begin{bmatrix} u + gt & ft \\ t & u \end{bmatrix} \forall t, u \in K,$$

where g and f are constants in K.

PROOF. If π is a conical translation plane, choose coordinates so that the common component is $x = 0$, and one of the pseudo-reguli has the form

$$x = 0, y = x \begin{bmatrix} u & 0 \\ 0 & u \end{bmatrix} \forall u \in K.$$

We shall refer to this as the standard form. It follows from the above results that each remaining pseudo-regulus has the form $x = 0, y = x(S + uI)$ for a set $\{S\}$ of 2×2 K-matrices. Recall that the set of components must be of the general form

$$x = 0, \; y = x \begin{bmatrix} G(t, u) & F(t, u) \\ t & u \end{bmatrix} \forall t, u \in K$$

and for functions G and F from $K \times K$ to K. Since the group E exists as a collineation group, the result (1) now directly follows. If π is a ruled translation plane, choose coordinates so that the two common components are $x = 0$, $y = 0$ and that the two 1-dimensional left K–subspaces referred to in the statement lie on $x = 0$. Choose one pseudo-regulus to have the standard form. Since the plane now admits the group H listed previously, the result (3) now follows immediately. If π is a ruled translation plane of the type listed in statement (4), use the form of (1) and apply the group H to obtain the conclusion that

$$g(t)v = g(tv) \text{ and } f(t)v = f(tv) \; \forall t, v \neq 0 \in K.$$

Hence, letting $g(1) = g$ and $f(1) = f$, (4) is now clear. Assume the conditions of (2). We have that

$$x = 0, y = x \begin{bmatrix} u + g(t) & f(t) \\ t & u \end{bmatrix} \forall t, u \in K$$

and for $g(t), f(t) \in K$ for all $t \in K$, defines a spread if and only if, for each nonzero vector (a, b, c, d) such that not both a and b zero, there exists a unique pair (u, t) such that

$$\begin{aligned} a(u + g(t)) + bt &= c \text{ and} \\ af(t) + bu &= d. \end{aligned}$$

If a is zero, then there is a unique such pair, namely, $(b^{-1}d, b^{-1}c)$. If $b = 0$, then since $f(t)$ is bijective, the unique pair is

$$(f^{-1}(a^{-1}d), a^{-1}c - g(f^{-1}(a^{-1}d))).$$

If $ab \neq 0$ then

$$\begin{aligned} b^{-1}af(t) - g(t) - a^{-1}bt &= b^{-1}d - a^{-1}c \text{ if and only if} \\ z^2t + zg(t) - f(t) &= a^{-1}d - a^{-1}ba^{-1}c \text{ for } z = a^{-1}b. \end{aligned}$$

Hence, if ϕ_z is bijective then, for any given d,c, there is a unique t in K that satisfies the above equation and defining

$$u = -a^{-1}b - a^{-1}bt - g(t) = b^{-1}d + b^{-1}af(t);$$

there is a unique pair (u, t), which satisfies the first system of equations.

$\square$

CHAPTER 16

Flocks

We have considered affine planes that are covered by subplane covered nets and, in the last chapter, we studied translation planes covered by pseudo-reguli in two ways. In this chapter, we restrict the skew-field K to be a field and formulate some of the related theory of flocks of quadric sets.

1. Conical Flocks

DEFINITION 67. *Represent $PG(n-1, K)$, where K is a field, by 1-dimensional vector subspaces of a n-dimensional vector space V_n over K where n is finite.*

We may represent points by nonzero vectors $(x_0, x_1, .., x_{n-1})$ if we agree to identify vectors in the same 1-dimensional vector subspace.

A 'symmetric bilinear form' is a mapping f from $V_n \times V_n$ into K such that

$f(\alpha v + \beta w, u) = \alpha f(v, u) + \beta f(w, u)$ and $f(v, u) = f(u, v)$ for all $\alpha, \beta \in K$ and for all $u, v, w \in V_n$.

A 'quadratic form' Q is a mapping of $V_n \times V_n$ into K with the following properties:

(1) $Q(\alpha v) = \alpha^2 Q(v)$ and

(2) $Q(v + w) = Q(v) + Q(w) + f(v, w)$ where f is a symmetric bilinear form for all $\alpha \in K$ and for all $v, w \in V_n$.

A 'quadric' is the set of points $v \in PG(n-1, K)$ such that $Q(v) = 0$ for some quadratic form. A 'nondegenerate quadric' is a quadric such that $Q(v) = 0$ and $f(v, u) = 0$ for all $u \in V_n$ implies that $v = 0$.

A 'conic' is a quadric in $PG(2, K)$.

If K is finite and isomorphic to $GF(q)$, the number of points of a nondegenerate conic is $q + 1$.

DEFINITION 68. *Let Σ be a 3-dimensional projective geometry isomorphic to $PG(3, K)$, where K is a field. Choose a plane π_o (so isomorphic to $PG(2, K)$) and a point v_0 of $\Sigma - \{\pi_o\}$ and let C be a nondegenerate conic of π_o.*

   We define the 'quadratic cone' defined by C and v_0 as the set of points that lie on lines v_0c for all $c \in C$. We call v_0 the 'vertex' of the cone.

   A 'flock' of a quadratic cone is a set of mutually disjoint conics in planes of Σ whose union is $C - \{v_o\}$.

   For convenience, we represent the points of Σ by (x_0, x_1, x_2, x_3) for all $x_i \in K$, not all $x_i = 0$ for $i = 1, 2, 3, 4$ and the points of π_o by $(x_0, x_1, x_2, 0)$. Then, we may choose v_0 as $(0, 0, 0, 1)$.

It turns out that there are canonical forms for conics in $PG(2, K)$ and quadrics in $PG(3, K)$, but we shall not be overly concerned with such forms. What we are trying to do is to show that conical translation planes when K is a field and flocks of quadratic cones are in 1–1 correspondence.

LEMMA 32. *Given $PG(2, K)$, and points (x_0, x_1, x_2), then*

$$x_o x_1 = x_2^2$$

is the equation of the non-degenerate conic with defining quadratic form $Q : Q(x_0, x_1, x_2) = x_0 x_1 - x_2^2$.

PROOF. The associated bilinear form is

$$f : f((x_0, x_1, x_2), (x_0^*, x_1^*, x_2^*)) = x_0 x_1^* + x_0 x_1^* - 2x_2 x_2^*.$$

If $Q(x_0, x_1, x_2) = 0$, then letting (x_0^*, x_1^*, x_2^*) equal $(1, 0, 0)$, $(0, 1, 0)$ and $(0, 0, 1)$ forces $x_0 = x_1 = x_2 = 0$ so that the conic is non-degenerate. □

THEOREM 101. *Represent the points of $PG(3, K)$ by homogeneous coordinates (x_o, x_1, x_2, x_3). A flock of a quadratic cone in $PG(3, K)$ with conic $x_0 x_1 = x_2^2$ in the plane whose points are given by the equation $x_2 = 0$ and with vertex $v_0 = (0, 0, 0, 1)$ has planes represented in the following form:*

$$tx_0 + G(t)x_1 + F(t)x_2 + x_3 = 0$$

for all $t \in K$ and for functions G, F on K.

PROOF. Every plane which is not incident with $(0, 0, 0, 1)$ has the basic form $ax_0 + bx_1 + cx_2 + x_3 = 0$ for $a, b, c \in K$. If there is a flock and two planes have the same $a's$ then their intersection is of the general form $(b - b^*)x_1 + (c - c^*)x_2 = 0$. In addition, their intersection on the quadratic cone whose equation is given by $x_0 x_1 = x_2^2$ contains the point $(0, (c - c^*)/(b - b^*), 1, 0)$ unless $b = b^*$. Similarly, $c = c^*$ and the two planes are identical. Hence, the $a's$ must vary over K. It is equally straightforward to see that the $b's$ and $c's$ are functions of $a's$. □

We now show the equivalence between flocks of quadratic cones and conical translation planes.

THEOREM 102. *For every translation plane with spread in $PG(3, K)$ with spread*

$$\left\{ x = 0, \ y = x \left[\begin{array}{cc} u + g(t) & f(t) \\ t & u \end{array} \right] ; u, t \in K \right\},$$

where g, f are functions from $K \times K \to K$, there is an associated flock of a quadratic cone defined as follows: Let (x_0, x_1, x_2, x_3) denote homogeneous coordinates for $PG(3, K)$, and consider the conic C with equation $x_0 x_1 = x_2^2$ in the plane $x_3 = 0$. Let $v_o = (0, 0, 0, 1)$ and form the cone $v_o P$ where $P \in C$.

Then the following planes define, by intersection, a flock of the cone:

$$\pi_t : t x_0 - f(t) x_1 + g(t) x_2 + x_3 = 0.$$

Conversely, a flock of a cone in $PG(3, K)$ with equation $x_0 x_1 = x_2^2$ and vertex $(0, 0, 0, 1)$ has plane equations of the form given above and defines a translation plane with spread in $PG(3, K)$ with components defined as above.

Let

$$\mathcal{R}_t \ : \ \left\{ x = 0, \ y = x \left[\begin{array}{cc} u + g(t) & f(t) \\ t & u \end{array} \right] ; u \in K \right\},$$

t fixed in K.

Then the spread is $\cup_{t \in K} \mathcal{R}_t$ and $\mathcal{R}_t \cup \mathcal{R}_s = (x = 0)$, for $t \neq s$. The spread admits an elation group E of order q, where

$$E = \left\langle \left[\begin{array}{cccc} 1 & 0 & u & 0 \\ 0 & 1 & 0 & u \\ 0 & 0 & 1 & 0 \\ 0 & 0 & 0 & 1 \end{array} \right] ; u \in K \right\rangle.$$

E is said to be a 'regulus-inducing group elation group.'

PROOF. We shall first show the converse. Suppose that $\{\pi_t | t \in K\}$ with the given form provides a disjoint cover of the cone of points $\neq v_o$ (vertex). This means that for any z_0, z_1, z_2 such that $z_0 z_1 = z_2^2$ and $\delta \in K$ then there is exactly one plane π_t that contains the point (z_o, z_1, z_2, δ) of the line $< (z_0, z_1, z_2, 0), (0, 0, 0, 1) >$. That is, we have the following: For any point (z_0, z_1, z_2, δ) such that $z_0 z_1 = z_2^2$, where $\delta \in K$, there is a unique $t \in K$ such that $t z_0 - f(t) z_1 + g(t) z_2 + \delta = 0$. If $z_2 = 0$, then we have the points $(0, 1, 0, \delta)$ and $(1, 0, 0, \delta)$ so that $-f(t) + \delta = 0$, which implies that $f(t)$ is bijective. If z_2 is not zero consider the point $(z_2^2, 1, z_1, \delta)$, by coordinate change we may assume that $g(0) = f(0) = 0$. Now assert that $u^2 + ug(t) - tf(t) \neq 0$, unless $u = t = 0$. Consider the point $(z_2^2, 1, z_2, 0)$.

There is a unique value t, such that $tz_2^2 - f(t) + g(t)z_2 = 0$.

In general, let $u = tz_2$ to obtain the equation above by multiplying the equation by t to obtain $u^2 - tf(t) + g(t)u = 0$. Now in order that

$$x = 0, y = x \begin{bmatrix} u + g(t) & f(t) \\ t & u \end{bmatrix} \; \forall \, u, t \subset K$$

is a spread in $PG(3, K)$, we need only to show that, for any nonzero vector (v_1, v_2, w_1, w_2), there is a unique pair (u, t) such that the vector is on the indicated component (note that the sets above are 2-dimensional K–subspaces). That is, we must show that when v_1, v_2, w_1, w_2 are fixed and not all zero then the following system has a unique solution for u and t:

$$\begin{aligned} w_1 &= v_1(u + g(t)) + v_2 t, \\ w_2 &= v_1 f(t) + v_2 u. \end{aligned}$$

Note that if $v_1 = v_2 = 0$ then since $u^2 + ug(t) - tf(t) \neq 0$ for $(u, t) \neq (0, 0)$, it follows that $x = 0$ contains the indicated vector and is the only component that does. Assume that $v_1 = 0$ but $v_2 \neq 0$. Then $t = w_1/v_2$ and $u = w_2/v_2$. If $v_2 = 0$, but $v_1 \neq 0$ then $w_2/v_1 = f(t)$ and since f is 1–1 and onto, this uniquely specifies t. Then $u = w_1/v_1 - v_1 g(t)$. Hence, assume that $v_1 v_2 \neq 0$. Then

$$w_1 v_2 - w_2 v_1 = v_1 v_2 u + v_1 v_2 g(t) + v_2^2 t - (v_1^2 f(t) + v_1 v_2 u),$$

which is equal to

$$v_1 v_2 g(t) + v_2^2 t - v_1^2 f(t).$$

Now call $(w_1 v_2 - w_2 v_1)/v_2^2 = -\delta$ and $v_1/v_2 = w$ to obtain the equation

$$t - f(t)w^2 + g(t)w + \delta = 0.$$

By the above, for given w and δ in K, there is a unique $t \in K$ that satisfies this equation. Note that

$$(w_1 - v_1 g(t) - v_2 t)/v_1 = (w_2 - v_1 f(t))/v_2$$

then uniquely defines u. Hence, we obtain a spread from a flock. To show that a spread of the form listed above produces a flock, it is possible to essentially reread to the proof to show that a spread produces a cover of the associated cone and such details are left to the interested reader to verify. Furthermore, the statement regarding the group E is also easily proved and again is left to the reader to complete. $\square$

2. Hyperbolic Flocks

DEFINITION 69. *Consider Σ isomorphic to $PG(3, K)$ and represent points homogeneously in the form (x_0, x_1, x_2, x_3) for $x_i \in K$, $i = 0, 1, 2, 3$.*

Then $x_0 x_3 = x_1 x_2$ represents a nondegenerate quadric $\mathcal{H}$ in $PG(3, K)$, which we call a 'hyperbolic quadric' $Q : Q(x_o, x_1, x_2, x_3) = x_o x_3 - x_1 x_2$.

A 'flock of a hyperbolic quadric' is a set of mutually disjoint nondegenerate conics in planes of Σ whose union is $\mathcal{H}$.

We are interested in the 'ruling' lines of the hyperbolic quadric.

These are the lines that lie within $\mathcal{H}$. There are two designated classes, and it is not difficult to show that the classes are reguli, which are mutually opposite.

Thus, a hyperbolic or 'ruled' quadric is the set of points of a regulus, and the lines of the quadric are the lines of a regulus and those of its opposite regulus.

In the finite case when K is isomorphic to $GF(q)$, $\mathcal{H}$ has $(q + 1)^2$ points, and there are $q + 1$ conics in a flock as each regulus has $q + 1$ lines. Also, a finite hyperbolic quadric has $2(q + 1)$ lines contained in it.

We now show that there are connections with hyperbolic translation planes and hyperbolic flocks in a manner similar to that of conical translation planes and conical flocks.

THEOREM 103. *(1) Let F be a flock of the hyperbolic quadric $x_1 x_4 = x_2 x_3$ in $PG(3, K)$, whose points are represented by homogeneous coordinates (x_1, x_2, x_3, x_4), where K is a field.*

Then the set of planes that contain the conics in F may be represented as follows:

$$\rho \; : \; x_2 = x_3,$$
$$\pi_t \; : \; x_1 - t x_2 + f(t) x_3 - g(t) x_4 = 0$$

for all t in K, where f and g are functions of K such that f is bijective.

(2) Corresponding to the flock F is a translation plane π_F with spread in $PG(3, K)$ written over the corresponding 4-dimensional vector space V_4 over K as follows: Let $V_4 = (x, y)$, where x and y are 2-vectors over K.

Then the spread may be represented as follows:

$$y \ = \ x \begin{bmatrix} f(t)u & g(t)u \\ u & tu \end{bmatrix},$$

$$y \ = \ x \begin{bmatrix} v & 0 \\ 0 & v \end{bmatrix}, x = 0, \ for \ all \ t, v, u \neq 0 \ \in K.$$

Define

$$R_t \ = \ \{y = x \begin{bmatrix} f(t)u & g(t)u \\ u & tu \end{bmatrix}, \ x = 0 \mid u \in K\},$$

$$R_\infty \ = \ \{y = x \begin{bmatrix} v & 0 \\ 0 & v \end{bmatrix} \mid v \in K\}.$$

Then $\{R_t, R_\infty\}$ is a set of reguli that share two lines (components $x = 0, y = 0$).

The translation plane admits the collineation group

$$\left\langle \begin{bmatrix} v & 0 & 0 & 0 \\ 0 & v & 0 & 0 \\ 0 & 0 & u & 0 \\ 0 & 0 & 0 & u \end{bmatrix} \mid v, u \ \in K - \{0\} \right\rangle$$

that contains two affine homology groups whose component orbits union the axis and coaxis define the reguli (regulus nets).

(3) A translation plane with spread in $PG(3, K)$, which is the union of reguli that share two components may be represented in the form (2).

Equivalently, a translation plane with spread in $PG(3, K)$, which admits a homology group one of whose component orbits union the axis and coaxis is a regulus may be represented in the form (2). In either case, such a translation plane produces a flock of a hyperbolic quadric in $PG(3, K)$.

PROOF. Suppose that a translation plane with spread in $PG(3, K)$ admits an affine homology group, one of whose component orbits union the axis and coaxis is a regulus R in $PG(3, K)$. Choose a representation so that the axis is $y = 0$, the coaxis $x = 0$ and $y = x$ is a component (line) of the regulus R. Then R is represented by the partial spread $x = 0, y = x \begin{bmatrix} v & 0 \\ 0 & v \end{bmatrix}$ for all $v \in K$. Moreover, the homology group takes the matrix form:

$$\left\langle \begin{bmatrix} 1 & 0 & 0 & 0 \\ 0 & 1 & 0 & 0 \\ 0 & 0 & u & 0 \\ 0 & 0 & 0 & u \end{bmatrix} \mid u \ \in \ K - \{0\} \right\rangle.$$

There are functions f and g on K and components of the following matrix form:

$$y = x \begin{bmatrix} f(t) & g(t) \\ 1 & t \end{bmatrix}; \ t \in K.$$

Note that, in particular, this says that the function f is $1 - 1$ as otherwise differences of certain corresponding matrices are singular and nonzero, contrary to the assumption that the components form a unique cover of the vector space. The homology group maps these components into $y = x \begin{bmatrix} f(t)u & g(t)u \\ u & tu \end{bmatrix}$ for all nonzero u in K. Hence, the regulus R and these components for all $v, t, u \neq 0$ in K define the spread in $PG(3, K)$. Take any value a in K and consider the vector $(1, -a, 0, 1)$. Since this vector is not on $x = 0$ or $y = xvI$ and we are assuming a 'cover,' there is a unique pair (u, t) with u nonzero such that $(1, -a, 0, 1)$ is incident with the component

$$y = x \begin{bmatrix} f(t)u & g(t)u \\ u & tu \end{bmatrix}.$$

Hence, we have $f(t)u - au = 0$ and $g(t)u - atu = 1$. In particular, since u is nonzero, we must have $f(t) = a$. Hence, f is 'onto.' In order to see that the planes listed in the theorem intersected with the hyperbolic quadric in $PG(3, K)$ form a unique cover of the hyperbolic quadric and hence define a hyperbolic flock, we must show that for all points (a, b, c, d) for $b \neq c$ and $ad = bc$, there is a unique t in K such that the point is on the plane π_t. Since we have a cover of the 4-dimensional vector space, we know that for a vector (e, h, m, n), where not both e and h are zero and $\langle(m, n)\rangle$ is not in $\langle(e, h)\rangle$, there is a unique ordered pair (t, u) such that (e, h, m, n) is on the component

$$y = x \begin{bmatrix} f(t)u & g(t)u \\ u & tu \end{bmatrix}.$$

To distinguish between points of $PG(3, K)$ that relate to the flock and vectors of V_4 that relate to the translation plane, we shall use the terms 'points' and 'vectors,' respectively. That is, for all e, h, m, n such that not both e and h are zero and the vector (m, n) is not in the 1-space generated by (e, h), there is a unique ordered pair (t, u) such that

$(2.2): \ ef(t)u + hu = m$ and $eg(t)u + htu = n$.

The point (a, b, c, d) is on π_t if and only if $a - bt + f(t)c - g(t)d = 0$. First assume that $bc \neq 0$. Then, without loss of generality, we may take $b = 1$ so that $ad = c$ (recall that the point is considered homogeneously)

Then the above equation becomes

$$(2.3) \quad : \quad ad - dt + f(t)cd - g(t)d^2$$
$$= \quad 0 = c - dt + f(t)cd - g(t)d^2.$$

From (2.2), we may multiply the second equation by an element r in K and subtract to obtain

$$(2.4) : (-m + nr + hu) - t(rhu) + f(t)(eu) - g(t)(reu) = 0.$$

Hence, clearly we may solve the equation

$$(2.5) : c - dt + f(t)cd - g(t)d^2 = 0 \text{ for some } t = t_o \in K.$$

First, assume that $f(t_o)d + 1 = 0 = z$. Then $g(t_o)d + t = 0 = w$ and the vector $(1, d^{-1}, 0, 0)$ is on the component

$$y = x \begin{bmatrix} f(t_o)d & g(t_o)d \\ d & t_o d \end{bmatrix} \text{ and } y = 0,$$

which is a contradiction. Hence, $zw \neq 0$. So, $w = zcd^{-1}$, and the vector $(1, d^{-1}, z, zcd^{-1})$ is on the component

$$y = x \begin{bmatrix} f(t_o)d & g(t_o)d \\ d & t_o d \end{bmatrix}.$$

Now assume that there exists another element s_o such that

$$c - ds_o + f(s_o)cd - g(s_o)d^2 = 0.$$

Then $f(s_o)d + 1 = z^* \neq 0$, and there exists an element v in K such that $z^*v = z$. So, the vector

$$(1, d^{-1}, z, zcd^{-1}) = (1, d^{-1}, z^*v, z^*vcd^{-1})$$

is also on

$$y = x \begin{bmatrix} f(s_o)dv & g(s_o)dv \\ dv & s_o dv \end{bmatrix}.$$

By uniqueness of the vector space cover, it follows that $(t_o, d) = (s_o, dv)$. Hence, there is a unique plane π_t containing the point (a, b, c, d) such that $b \neq c$ and $ad = bc$, where bc is nonzero. Now assume that $bc = 0$. If $b = 0$ and $d = 0$, then, without loss of generality, we may take $c = 1$ so we are considering the point $(a, 0, 1, 0)$. We need to determine an element t in K such that $a + f(t) = 0$. Since f is bijective as noted above, there exists a unique value t which solves this equation and hence a unique plane π_t containing the point $(a, 0, 1, 0)$. If $b = 0$ and $a = 0$ and $c = 1$, it is required to uniquely cover the point $(0, 0, 1, d)$ by a plane so we require a unique solution to the equation $f(t) - g(t)d = 0$.

By (2.3), certainly there is a solution t_1. Moreover, $f(t_1) = z_1 \neq 0$ as otherwise the spread would contain

$$y = x \begin{bmatrix} 0 & 0 \\ 1 & t_1 \end{bmatrix}.$$

Hence, the vector

$$(1, 0, z_1, z_1/d) \ is \ on \ y = x \begin{bmatrix} f(t_1) & g(t_1) \\ 1 & t_1 \end{bmatrix}.$$

If there exists another solution s_1, then $f(s_1) = z_1^*$ and there exists an element w of K such that $z_1^* w = z_1$. Therefore, the previous vector also belongs to

$$y = x \begin{bmatrix} f(s_1)w & g(s_1)w \\ w & s_1 w \end{bmatrix},$$

which, by uniqueness of the vector space cover, implies that $(t_1, 1) = (s_1, w)$. If $c = 0$, then $a = 0$ or $d = 0$ and $b = 1$ without loss of generality. We are trying to show that there is a unique solution to $a - t + g(t)d = 0$. If $d = 0$, this is trivial. Thus, assume that $a = 0$. By (2.3), there is a solution t_2 to $-t + g(t)d = 0$. Let $f(t_2)d - 1 = z_2$, then the vector $(1, -d^{-1}, z_2, 0)$ is on the component

$$y = x \begin{bmatrix} f(t_2)d & g(t_2)d \\ d & t_2 d \end{bmatrix},$$

so clearly $z_2 \neq 0$. If there is another solution s_2, then let $f(s_2)d - 1 = z_2^*$ so that there exists an element w such that $z_2^* w = z_2$. Hence, the previous point is also on the component

$$y = x \begin{bmatrix} f(s_2)dw & g(s_2)dw \\ dw & s_2 dw \end{bmatrix},$$

and by uniqueness, we must have $(t_2, d) = (s_2, dw)$. Thus, a translation plane with spread in $PG(3, K)$, which admits an affine homology group of the type listed above produces a flock of a hyperbolic quadric. To complete the proof of part (3), we must show that if a translation plane has its spread in $PG(3, K)$ and the spread is a union of reguli sharing two components, then there is a homology group of the type mentioned above. However, this follows from the relevant theorem in the previous chapter as the reguli are normalizing. To prove (1), we may choose a basis so that a given plane of the flock has equation $x_2 = x_3$. From here, it is fairly direct that we may represent the flock in the form given. The function $f(t)$ is $1 - 1$ to avoid intersections and must be onto in order to ensure a cover. The proof of (2) follows along the lines of the proof (3) and is left to the reader. $\square$

3. Partial Flocks of Deficiency One

In this section, we consider structures, which are essentially flocks 'missing' one conic—partial flocks of deficiency one. We have considered partial parallelisms of deficiency one previously. The general idea for the study of deficiency one comes about when deriving an affine plane covered by derivable nets. The derived plane, by the derivation of one of these nets, now is not covered by derivable nets but a good share of its parallel classes still are connected to derivable nets although it is apparent that the remaining lines of a given derivable net are now Baer subplanes of the derived plane. If one forgets that the derived net is, in fact, derivable, it is natural to consider such planes abstractly and ask what geometric point-line geometries are associated and what is implied of such structures when the corresponding net does turn out to be derivable.

We first consider the import of the theory of Baer groups for spreads in $PG(3, K)$.

4. Point-Baer Groups and Partial Flocks

THEOREM 104. *Let π be a translation plane with spread in $PG(3, K)$, for K a skewfield. Note that the following are considered the kernel mappings:*

$$(x_1, x_2, y_1, y_2) \to (kx_1, kx_2, ky_1, ky_2), \text{ for } k \in K - \{0\},$$

whereas linear collineations may be represented by 4×4 matrices acting on the right and components have the general form $x = 0, y = xM$, where M is a 2×2 matrix acting on the right.

(1) Let π admit a nontrivial point-Baer elation group B, which fixes the point-Baer subplane π_o pointwise. Let the kernel of π_o be K_o.

If B is a full K_o-point-Baer elation group of order > 2, then π_o is a K-space, K is a field isomorphic to K_o, and B may be represented in the form

$$\left\langle \begin{bmatrix} 1 & \beta & 0 & 0 \\ 0 & 1 & 0 & 0 \\ 0 & 0 & 1 & \beta \\ 0 & 0 & 0 & 1 \end{bmatrix} \; \forall \beta \in K \right\rangle.$$

(2) If π admits a nontrivial point-Baer homology group C, then let $FixC = \pi_o$ have kernel K_o. If C acts as the full K_o-point-Baer homology group of the net containing $FixC$ and has order > 2, then K

is a field isomorphic to K_o and C may be represented in the form

$$\left\langle \begin{bmatrix} \lambda & 0 & 0 & 0 \\ 0 & I & 0 & 0 \\ 0 & 0 & \lambda & 0 \\ 0 & 0 & 0 & 1 \end{bmatrix} \quad \forall \lambda \in K^* \right\rangle.$$

Note, in this case, there is another point-Baer subplane sharing its infinite points with $FixC$ and invariant under C. This second point-Baer subplane is called $coFixC$.

(3) If L is any 2-dimensional left K–subspace that is disjoint from $FixB$ then the orbit of L under B union $FixB$ is a regulus net in $PG(3, K)$.

(4) If L is any 2-dimensional left K–subspace that is disjoint from $FixC$ or $coFixC$, then the orbit of L under C union $FixC$ and $coFixC$ is a regulus net in $PG(3, K)$.

PROOF. (1) and (2) follow directly and are left to the reader to complete. Now assume the conditions of (3). Change coordinates so that $FixB$ is represented by $x = 0$. Clearly, B now has the following form

$$\left\langle \begin{bmatrix} 1 & 0 & u & 0 \\ 0 & 1 & 0 & u \\ 0 & 0 & 1 & 0 \\ 0 & 0 & 0 & 1 \end{bmatrix} \quad \forall u \in K \right\rangle.$$

The set of B-images of $y = 0$ is $y = x \begin{bmatrix} u & 0 \\ 0 & u \end{bmatrix} \forall u \in K$. We may assume that a 2-dimensional subspace disjoint from $x = 0$ has the general form $y = xN$, where N is a 2×2 matrix (possibly singular). By a coordinate change of the form $\begin{bmatrix} I & -N \\ 0 & I \end{bmatrix}$, we may take the 2-dimensional subspace as $y = 0$ without changing the new form of the group B. This then proves (3). Assume the conditions of (4). Change bases over the prime field so that the fixed subplanes have the form $x = 0$ and $y = 0$. The group now has the form

$$\left\langle \begin{bmatrix} u & 0 & 0 & 0 \\ 0 & u & 0 & 0 \\ 0 & 0 & 1 & 0 \\ 0 & 0 & 0 & 1 \end{bmatrix} \quad \forall u \in K^* \right\rangle.$$

Any 2-dimensional K–subspace disjoint from $x = 0$ and $y = 0$ has the form $y = xN$, where N is a nonsingular 2×2 matrix. Change bases by $\begin{bmatrix} I & 0 \\ 0 & N^{-1} \end{bmatrix}$. This basis change leaves the new form of the

group invariant and the images of the 2-dimensional subspace are $y =$
$x \begin{bmatrix} u & 0 \\ 0 & u \end{bmatrix} \forall\, u \in K^*$. Hence, we have the proof to (4). $\square$

5. Deficiency One Partial Conical Flocks

DEFINITION 70. *Let C_o be a non-degenerate conic in a plane of $PG(3, K)$, for K a field. Let v_o be a point exterior to C_o and form the quadratic cone.*

A 'partial flock of the quadratic cone' is a set of mutually disjoint conics that lie in the $C_o - \{v_o\}$.

A 'partial flock of deficiency one' is a partial flock such that for each line of the cone, the union of the conics cover all but exactly one of the non-vertex points.

REMARK 36. *We have seen previously how to connect a flock of a quadratic cone with a conical spread in $PG(3, K)$. Each plane of the flock corresponds to a conic of the cone, which, in turn, corresponds to a regulus net of the conical spread. Two conics of the flock correspond to two regulus nets, but the points of the conics do not correspond to the partial spread components that share a line but to the set of Baer subplanes incident with the zero vector of the regulus nets. Under this sort of correspondence of points on the cone with Baer subplanes of the regulus nets, the cone points on a line correspond to the set of Baer subplanes that share a given 1-dimensional K–subspace on the common line.*

When we have a partial flock, we have the same sort of connections with cone points and Baer subplanes. Note, if the planes of a partial flock cover all but one of the cone points on a line, then there would be a set of regulus nets containing a set of Baer subplanes that intersect a given 1-dimensional K-space on the common line and, as a set of 2-dimensional K–subspaces would be one short of containing all such 2-dimensional K–subspaces with a given mutual 1-space intersection.

THEOREM 105. (1) *The set of partial flocks of a quadratic cone of deficiency one in $PG(3, K)$, for K a field, is equivalent to the set of translation planes with spreads in $PG(3, K)$ that admit a point-Baer elation group B, which acts transitively on non-fixed 1-dimensional K–subspaces of components of the fixed point subplane.*

(2) *A partial flock of a quadratic cone of deficiency one in $PG(3, K)$, for K a field, may be extended to a flock if and only if, in the corresponding translation plane that admits a point-Baer elation group B as in (1), the net defined by the point-Baer affine plane $\mathrm{Fix}B$ defines a regulus in $PG(3, K)$.*

PROOF. Assume that such a translation plane π exists. We point out that when we use the term Baer subplane of a regulus net we are not asserting that the subplane indicated is a Baer subplane of any other net containing the regulus net as we have seen that this is not always the case. Since K is a field, the point-Baer subplane is always a 2-dimensional K–subspace. Hence, the group has the following representation:

$$\left\langle \begin{bmatrix} 1 & \beta & 0 & 0 \\ 0 & 1 & 0 & 0 \\ 0 & 0 & 1 & \beta \\ 0 & 0 & 0 & 1 \end{bmatrix} \forall \beta \in K \right\rangle.$$

The components of the net N containing π_o have the basic form

$$x = 0, y = x \begin{bmatrix} u & b(u) \\ 0 & u \end{bmatrix} \forall u \in K$$

and b a function on K. Since the kernel may now be given unambiguously by the mappings

$$\begin{bmatrix} \beta & 0 & 0 & 0 \\ 0 & \beta & 0 & 0 \\ 0 & 0 & \beta & 0 \\ 0 & 0 & 0 & \beta \end{bmatrix} \forall \beta \in K,$$

then the components of the spread, which are not in the net N have the following general form:

$$y = x \begin{bmatrix} G(t, u) & F(t, u) \\ t & u \end{bmatrix}$$
$$\forall t \neq 0, u \in K,$$

where G and F are functions of $K \times K$. Change coordinates by $(x_1, x_2, y_1, y_2) \to (x_1, y_1, x_2, y_2)$ so that the group has the following form:

$$\left\langle \begin{bmatrix} 1 & 0 & \beta & 0 \\ 0 & 1 & 0 & \beta \\ 0 & 0 & 1 & 0 \\ 0 & 0 & 0 & 1 \end{bmatrix} \forall \beta \in K \right\rangle.$$

The components that are not in the net have the form:

$$y = x \begin{bmatrix} -G(t, u)t^{-1} & F(t, u) - G(t, u)t^{-1}u \\ t^{-1} & t^{-1}u \end{bmatrix}$$
$$\forall t \neq 0, u \in K.$$

The reader should try to prove this last statement.

Hence, the components not in N have the general form

$$y \ = \ x \left[\begin{array}{cc} G^*(t,u) & F^*(t,u) \\ t & u \end{array} \right]$$

$$\forall t \ \neq \ 0, u \in K$$

for functions G^* and F^* on $K \times K$. Since the plane admits the group listed as above, it follows that $G^*(t,u) = g(t) + u$ and $G^*(t,u) = f(t)$ for functions g, f on K. It is now clear that there is a set of regulus nets R_t with partial spread $x = 0$ (which is the new equation for π_o) and

$$y \ = \ x \left[\begin{array}{cc} g(t) + u & f(t) \\ t & u \end{array} \right]$$

$$\forall u \ \in \ K \text{ and } t \text{ fixed in } K - \{0\}.$$

We are not claiming that the matrices involved are nonsingular. However, we may select one of the matrices and change bases without changing $x = 0$ so that this matrix is zero. The resulting partial spread has the form R_t with partial spread $x = 0$, which is the new equation for π_o and

$$y \ = \ x \left[\begin{array}{cc} g_1(t) + u & f_1(t) \\ t & u \end{array} \right]$$

$$\forall u \ \in \ K \text{ and } t \text{ fixed in } K - \{t_o\}, \text{ for } t_o \neq 0$$

where the matrices and differences are either zero or nonsingular. Now given any 1-dimensional K–subspace X, if X is in the original net N, then the B-orbit of X generates a 2-dimensional K–subspace, which is a component of the net, and which intersects π_o in a 1-dimensional subspace. If X is not in the original net N then the B-orbit of X generates a 2-dimensional K–subspace that again intersects π_o in a 1-dimensional K–subspace. Take the set of 2-dimensional K–subspaces, each of which is invariant under B, and each of which intersects π_o in the same 1-dimensional K–subspace P_o. Then, there is exactly one such subspace that defines a component of the net N. It then follows directly that any other such 2-dimensional subspace is a Baer subplane of one of the regulus nets R_t. We consider the following partial flock of a quadratic cone in $PG(3, K)$ defined by equation $x_o x_1 = x_2^2$ in the plane $x_3 = 0$ given by homogeneous coordinates (x_o, x_1, x_2, x_3) with vertex $(0, 0, 0, 1)$. The planes containing the conics of intersection are

$$\rho_t : x_o t - x_1 f(t) + x_2 g(t) + x_3 = 0.$$

It follows easily that $\{\rho_t \ \forall \ t \neq 0 \}$ defines a partial flock of the quadratic cone. Recall that the points on the lines of the cone correspond to

Baer subplanes incident with the zero vector of the regulus nets and each of these is a 2-dimensional K–subspace that is invariant under B. Furthermore, for each 1-dimensional K–subspace on π_o, there is exactly one component containing this 1-dimensional K–subspace. Thus, it follows that there is exactly one point on each line of the cone that is not covered by the partial flock. This shows that the deficiency is one. Now assume that there is a partial flock F of a quadratic cone of deficiency one in $PG(3, K)$. Again, note that there is a corresponding partial spread P_F in $PG(3, K)$ that is the union of a set of reguli that share a common component and admit a group B. The points on a line of the cone, which are covered by the partial flock, correspond to Baer subplanes of the regulus nets of the translation net corresponding to the partial spread. Hence, for each line of the cone, there is exactly one point that then corresponds to a 2-dimensional K–subspace that is B-invariant in the corresponding vector space (that is, the ambient space of the translation plane), and which is not a Baer subplane of one of the regulus nets. The points on the line of the cone correspond to the set of all B-invariant 2-dimensional K-spaces that intersect the common component L in a fixed 1-dimensional K–subspace. Hence, for each 1-dimensional K–subspace Z of L, there is a unique 2-dimensional K–subspace π_Z containing Z that is B-invariant, and which does not lie as a Baer subplane in one of the regulus nets of the partial spread. So, there is a set $R = \{\pi_Z$; Z is a 1-dimensional K–subspace of $L\}$ of B-orbits, which then must consist of mutually disjoint 2-dimensional K–subspaces. It also clearly follows that these subspaces are disjoint from the partial spread, excluding the common component L. Clearly, $R \cup (P_F - L)$ covers the set of all 1-dimensional K–subspaces and is a set of 2-dimensional K–subspaces. Hence, we have a translation plane with spread in $PG(3, K)$, and which admits B as a collineation group that fixes L pointwise. Since L is a 2-dimensional K–subspace, L becomes an affine plane of the new translation plane Σ. We note that L defines a line-Baer subplane of Σ that admits a collineation group B fixing it pointwise. We have also noted that this implies that the subplane is Baer and hence point-Baer. This completes the proof of part (1) of the theorem. If the partial flock can be extended to a flock, then there is a translation plane π^+ corresponding to the flock that admits B as an elation group. By (1), there is a translation plane π corresponding to the partial flock that admits B as a point-Baer group and that shares all components with π^+ that do not lie in the net N defined by the components of $\text{Fix}B$. Let N^+ denote the subnet of π^+ that replaces N. We notice that the components of N are generated by B-orbits.

Moreover, it follows that N^+ is a regulus net so that B-orbits of 1-dimensional K–subspaces within N^+ generate the Baer subplanes of N^+ that are the components of N. Hence, N is the opposite regulus net of N^+. Now assume that the plane π is derivable with net N. Then the net contains at least three point-Baer subplanes so it follows from the preceding section that each of the point-Baer subplanes is a K–subspace. This means the net N is a regulus net. We have seen that the derivation of this net determines a translation plane with spread in $PG(3, K)$ that admits an elation group B such that any component orbit union the axis of the group forms a regulus net. It follows that the derived translation plane corresponds to a flock of a quadratic cone that extends the original partial flock. This completes the proof of the theorem.   □

6. Deficiency One Partial Hyperbolic Flocks

DEFINITION 71. *Let H be a hyperbolic or ruled quadric of the form $x_1 x_4 = x_2 x_3$ in $PG(3, K)$, which is represented by homogeneous coordinates (x_1, x_2, x_3, x_4). A partial flock of H is a set of mutually disjoint conics that lie in H.*

A 'partial flock of deficiency one' is a partial flock such that on any line of either ruling, there is exactly one point, which is not covered by the conics of the partial flock.

REMARK 37. *We have mentioned that given a hyperbolic quadric in $PG(3, K)$, there are two 'ruling' classes of lines, which are reguli. When we considered flocks of hyperbolic quadrics, we found that there are associated hyperbolic spreads in $PG(3, K)$. The conics are associated with regulus nets, and the points on the conics are associated with the Baer subplanes incident with the zero vector of the regulus nets. Hence, we have connected the points of a hyperbolic quadric with the Baer subplanes incident with the zero vector of a set of regulus nets sharing two common components. Each line of a given ruling class (one of the reguli of the hyperbolic quadric) corresponds to the set of Baer subplanes that are incident with a given 1-dimensional K–subspace on one of the common lines of the regulus nets. A line of the other ruling class (the other of the reguli of the hyperbolic quadric) corresponds to the set of Baer subplanes that are incident with a given 1-dimensional K–subspace on the other of the common lines of the regulus nets. Now these same connections hold for partial flocks. In particular, if the planes of the partial flock cover all but one of the points on a line of one of the ruling classes, there is a corresponding set of Baer subplanes, which, as 2-dimensional K–subspaces, share a common 1-space*

such that as a subset of the set of all 2-dimensional K–subspaces shar-
ing that common 1-space, there is exactly one missing 2-dimensional
K-space.

THEOREM 106. (1) *The set of partial flocks of ruled quadrics in*
$PG(3, K)$, for K a field, of deficiency one is equivalent to the set of
translation planes with spreads in $PG(3, K)$ which admit a point-Baer
homology group B that is transitive on the non-fixed 1-dimensional K–
subspaces on any component of $Fix B$.

(2) *A partial flock of deficiency one of a ruled quadric may be ex-*
tended to a flock if and only if the net defined by $Fix B$ of the corre-
sponding translation plane is a regulus net.

PROOF. A partial flock of a ruled quadric in $PG(3, K)$ gives rise to
a partial spread in $PG(3, K)$ that is the union of a set of reguli that
share two lines. Furthermore, a flock corresponds to a translation plane
whose spread has the same property. In addition, the translation plane
admits an affine homology group B, each of whose component orbits
union the axis and coaxis is a regulus net. This is also true of any
such corresponding partial spread. In general, the Baer subplanes of
the regulus nets correspond to points on the ruling lines. Again, we are
not necessarily assuming that the subplanes are Baer in any net other
than the regulus net in question. Furthermore, the set of points on
a given ruling line correspond to the set of Baer subplanes that share
a given 1-dimensional K-space on a given common component of the
partial spread. The corresponding Baer subplanes are 2-dimensional
K–subspaces generated by point orbits under B. Let the two sets of
ruling lines be denoted by R_+ and R_-. Let J_+ be a line of R_+. The
points on J_+ correspond to the 2-dimensional K–subspaces that are
generated by B-orbits of 1-dimensional K–subspaces, which are not
on the common components L and M, and which intersect on one
of the common components, say, L, in a particular 1-dimensional K–
subspace. First, let π be a translation plane with spread in $PG(3, K)$
that admits a point-Baer homology group so there are at least two
point-Baer subplanes incident with the zero vector that share the same
infinite points. By the theory of the previous chapters, since K is a
field, each point-Baer subplane is a K–subspace and the group B may
be represented in the form

$$\left\langle \begin{bmatrix} \lambda & 0 & 0 & 0 \\ 0 & I & 0 & 0 \\ 0 & 0 & \lambda & 0 \\ 0 & 0 & 0 & I \end{bmatrix} \ \forall \lambda \in K^* \right\rangle.$$

It follows that each component orbit disjoint from the two subplanes π_o and π_1 under B union π_o and π_1 defines a K-regulus in $PG(3, K)$. Let N denote the net containing the two point-Baer subplanes. It then follows that corresponding to the partial spread $\{\pi - N\} \cup \{L, M\}$ is a partial flock since $\{\pi - N\} \cup \{L, M\}$ admits the appropriate 'homology group' with axis π_o and coaxis π_1 if $FixB = \pi_o$ and $\{L, M\} = \{\pi_o, \pi_1\}$. The 2-dimensional K–subspaces that are generated by point orbits under B intersect π_o and π_1 in 1-dimensional K–subspaces. Each such point-orbit of a point that is not on a component of N lies in a Baer subplane of a K-regulus net that corresponds to a conic. For a given 1-dimensional K–subspace of π_o, there is a unique B-orbit 2-dimensional K–subspace that is not in one of the regulus nets. This unique B-orbit is the component of N containing the subspace in question. This component also non-trivially intersects π_1. Hence, it follows that, for each line of either ruling of the ruled quadric, there is a unique point of the line that is not covered by conics corresponding to the regulus nets of $\{\pi - N\} \cup \{L, M\}$. Thus, the partial spread has deficiency one. Now assume that a partial spread has deficiency one. Then, there is a partial spread of the form $\mathcal{P} \cup \{L, M\}$ admitting a homology group B that may be represented in the form

$$\left\langle \begin{bmatrix} \lambda & 0 & 0 & 0 \\ 0 & \lambda & 0 & 0 \\ 0 & 0 & I & 0 \\ 0 & 0 & 0 & I \end{bmatrix} \; \forall \lambda \in K^* \right\rangle$$

and such that each component orbit of $\mathcal{P}$ union $\{L, M\}$ defines a regulus in $PG(3, K)$. Similarly as above, the points of the ruling lines correspond to the 2-dimensional K–subspaces that are B-invariant and not equal to L or M. The deficiency one assumption implies that for each 1-dimensional K–subspace of L, there is a unique 2-dimensional K–subspace invariant under B that is not in a regulus net of the partial spread. The union of the set S of these 2-dimensional subspaces cover both L and M and are mutually disjoint. Form $\mathcal{P} \cup S$. Clearly, this partial spread completely covers all 1-dimensional K–subspaces and consists of 2-dimensional K–subspaces so that a translation plane with spread in $PG(3, K)$ is obtained. Since the components are defined via B-orbits, B is a collineation group of the constructed translation plane. However, now L and M are 2-dimensional K–subspaces that then define line-Baer affine planes. Since

$$\left\langle \begin{bmatrix} \lambda & 0 & 0 & 0 \\ 0 & \lambda & 0 & 0 \\ 0 & 0 & I & 0 \\ 0 & 0 & 0 & I \end{bmatrix} \forall \lambda \in K^* \right\rangle$$

is a collineation group of the translation plane that fixes L pointwise and

$$\left\langle \begin{bmatrix} \lambda & 0 & 0 & 0 \\ 0 & \lambda & 0 & 0 \\ 0 & 0 & \lambda & 0 \\ 0 & 0 & 0 & \lambda \end{bmatrix} \forall \lambda \in K^* \right\rangle$$

defines the kernel homology group, it follows that

$$\left\langle \begin{bmatrix} I & 0 & 0 & 0 \\ 0 & I & 0 & 0 \\ 0 & 0 & \lambda & 0 \\ 0 & 0 & 0 & \lambda \end{bmatrix} \forall \lambda \in K^* \right\rangle$$

is also a collineation group of the translation plane that fixes M pointwise. We have noted in the chapter on point-Baer collineations that any point-Baer or line-Baer subplane that is pointwise fixed by a collineation is a Baer subplane. This completes the proof of part (1). Now assume that there is a partial flock of deficiency one that may be extended to a flock. Let π denote the translation plane admitting the point-Baer group B and let Σ denote the translation plane corresponding to the flock. Let G denote the regulus-inducing homology group of Σ. We note that the subplanes of Σ, which are the Baer subplanes of the regulus nets sharing two components L and M, correspond to the 2-dimensional K-spaces that are G-invariant. Hence, all of the components of the net N of π are now various of these subplanes of Σ. Hence, $G = B$. In other words, N is a regulus net and π and Σ are derivates of each other by replacement of N. Conversely, if N is a regulus net in π, then it must be that derivation by N produces a translation plane Σ and B becomes a homology group in Σ of the correct form to produce a flock of a hyperbolic quadric. $\square$

DEFINITION 72. *The translation planes corresponding to partial flocks of deficiency one of either quadratic cones or hyperbolic quadrics shall be called 'deficiency one translation planes.'*

Just as with Desarguesian partial parallelisms of deficiency one and Desarguesian parallelisms, little is known about deficiency one partial flocks in the infinite case. However, there cannot be proper finite partial conical flocks of deficiency one in $PG(3,q)$ by results of Payne and Thas [**173**], which we shall now present.

7. The Theorem of Johnson, Payne-Thas

The Payne-Thas theorem, that a partial flock of a quadratic cone of deficiency one in $PG(3,q)$ has a unique extension to a flock is given in two parts, q even and q odd, whose proofs were, respectively, ingenious but elementary, for q is even, and quite an intricate and involved method valid only when q is odd (see, e.g., [**173**]).

The following proof has an interesting history. I learned of this proof from a lecture by Stan Payne for the International Conference on Finite Geometry, San Antonio, March 2009, Normfest. Apparently, Jef Thas showed Stan Payne the proof in 2004, and attributed the proof to Peter Sziklai. I am grateful to have the opportunity to share this beautiful result and thank Stan Payne for allowing me to include the proof in this text.

First, the theorem of Payne-Thas:

7.1. Payne-Thas Extension Theorem. Let $\mathcal{C}$ be a quadratic cone in $PG(3,q)$ with vertex $v_0 = (0,0,0,1)$, with points (x_0, x_1, x_2, x_3) on the cone if and only if $x_0 x_2 = x_1^2$. A plane not containing the vertex v_0 has equation $ax_0 + bx_1 + cx_2 + x_3 = 0$, which is designated by $[a_i, b_i, c_i, 1] = \pi_i$, for $i = 2, .., q$. So, let $\mathcal{P} = \{\pi_i; i = 2, .., q\}$ be a partial flock of deficiency one, whose conics are defined by intersection $\pi_i \cap \mathcal{P}$. Let $F_i(t) = a_i + b_i t + c_i t^2$, the point of π_i on the line $\ell_t = \langle (1, t, t^2, 0), (0, 0, 0, 1) \rangle$. Since we have a partial flock, for each t, the mapping $\phi_t : i \to a_i + b_i t + c_i t^2$ is an injective mapping on $GF(q)$. Hence, for each t, there is exactly one element of $GF(q)$ that is not in the image of ϕ_t. But $(1, t, t^2, -F_i(t))$ constitutes $q - 1$ points on ℓ_t and for $q > 2$; the sum of the elements of $GF(q)$ is 0. Hence, the sum $-\sum_{i=2}^{q} F_i(t)$ is the missing element of $GF(q)$. Therefore, the point $(1, t, t^2, \sum_{i=2}^{q} F_i(t))$ is the $q+1$'st point on the line ℓ_t; the missing point. Similarly, the point of π_i on $\langle (0,0,1,0), (0,0,0,1) \rangle = \ell_\infty$ is $(0,0,1,-c_i)$, and the missing point on ℓ_∞ is $(0,0,1, \sum_{i=2}^{q} c_i)$. We note that all of these $q+1$ points are on the plane $\pi_1 = [-\sum a_i, -\sum b_i, -\sum c_i, 1]$, and therefore by intersection with the cone, we have another conic $\pi_1 \cap \mathcal{P}$. Therefore, we have the following theorem:

THEOREM 107. *(Payne-Thas [**173**]) A partial flock of a quadratic cone of deficiency one in $PG(3,q)$ has a unique extension to a flock of a quadratic cone.*

Now consider Theorem 105, which was proved by the author in [**124**] for K isomorphic to $GF(q)$. In this setting, we have $q - 1$ conics of a flock of a quadratic cone, which may be extended to a flock by the Theorem of Payne-Thas.

Therefore, we have:

7.2. Johnson, Payne-Thas Theorem.

THEOREM 108. *Let π be a translation plane of order q^2 with spread in $PG(3, q)$. If π admits a Baer group of order q, then the partial spread of degree $q + 1$ whose components of the Baer axis is a regulus partial spread. Derivation of this spread constructs a translation plane of order q^2 with spread in $PG(3, q)$, admitting an elation group of order q whose orbits together with the elation axis are reguli; a conical translation plane.*

Therefore, translation planes of order q^2 with spread in $PG(3, q)$ admitting Baer groups of order q are equivalent to flocks of quadratic cones in $PG(3, q)$.

CHAPTER 17

Regulus-Inducing Homology Groups

In the previous chapters, we have developed the ideas of flocks of quadratic cones using translation planes whose spreads in $PG(3, q)$ are unions of reguli that share a component. Such a translation plane admits what we call a 'regulus-inducing' elation group. It is an amazing fact that it turns out that there is another type of translation plane with a regulus-inducing homology group that produces a flock of a quadratic cone, in perhaps an oblique manner. In this chapter and in the following, we shall be connecting various translation planes with flocks of quadratic cones and more generally with what shall be called flocks of α-cones. The present chapter generally follows the author's work in [**103**] but modified for this text.

The main thrust of this chapter is that there are two disparate classes of translation planes of order q^2 with spreads in $PG(3, q)$, which are equivalent to the set of flocks of quadratic cones in $PG(3, q)$. We have seen that if π is a translation plane of order q^2 that admits an elation group E of order q such that one component orbit together with the axis is a regulus in $PG(3, q)$ then all component orbits union the axis are reguli, so there is a set of q regulus nets sharing one parallel class. There are a total of $q(q+1)$ Baer subplanes that are incident with the zero vector, and these subplanes may be mapped to the set of non-vertex points of a quadratic cone in $PG(3, q)$. So, translation planes with such 'regulus-inducing' elation groups are equivalent to flocks of quadratic cones.

Now consider a completely different sort of translation plane of order q^2, one that admits a cyclic affine homology group of order $q + 1$. It turns out that the component orbits of such a group are reguli in $PG(3, q)$, so there is a set of $q - 1$ reguli together with two components that constitute the spread for this translation plane. It is a very mysterious fact that these two classes of translation planes are equivalent, one giving rise to the other, and each of these classes is then equivalent to the class of flocks of quadratic cones.

We shall approach this material from the point of view of the translation plane and cyclic homology groups. However, we shall consider the theory more generally over arbitrary projective spaces $PG(3, K)$, where K is an arbitrary field. The question the reader might ask is what made anyone think that such classes of translation planes might be equivalent. The answer basically is that the idea for such came not from the theory of translation planes but from what is called a 'hyperbolic fibration.'

DEFINITION 73. *A 'hyperbolic fibration' is a set Q of $q-1$ hyperbolic quadrics and two carrying lines L and M such that the union $L \cup M \cup Q$ is a cover of the points of $PG(3, q)$. (More generally, one could consider a hyperbolic fibration of $PG(3, K)$, for K an arbitrary field, as a disjoint covering of the points by a set of hyperbolic quadrics union two carrying lines.) The term 'regular hyperbolic fibration' is used to describe a hyperbolic fibration such that for each of its $q - 1$ quadrics, the induced polarity interchanges L and M. When this occurs, and (x_1, x_2, y_1, y_2) represent points homogeneously, the hyperbolic quadrics have the form*

$$V(x_1^2 a_i + x_1 x_2 b_i + x_2^2 c_i + y_1^2 e_i + y_1 y_2 f_i + y_2^2 g_i)$$

for $i = 1, 2, .., q - 1$ (the variety defined by the fibration). When $(e_i, f_i, g_i) = (e, f, g)$ for all $i = 1, 2, \ldots, q - 1$, the regular hyperbolic quadric is said to have 'constant back half.'

The main theorem of Baker, Ebert, and Penttila [3] is equivalent to the following.

0.3. Baker, Ebert, Penttila Hyperbolic Fibration Theorem.

THEOREM 109. (Baker, Ebert, Penttila [3])
(1) *Let $\mathcal{H} : V(x_1^2 a_i + x_1 x_2 b_i + x_2^2 c_i + y_1^2 e + y_1 y_2 f + y_2^2 g)$ for $i = 1, 2, \ldots, q - 1$ be a regular hyperbolic fibration with constant back half.*
Consider $PG(3, q)$ as (x_1, x_2, x_3, x_4), and let C denote the quadratic cone with equation $x_1 x_2 = x_3^2$.
Define

$$\pi_0 : x_4 = 0, \ \pi_i : x_1 a_i + x_2 c_i + x_3 b_i + x_4 = 0 \quad \text{for } 1, 2, \ldots, q - 1.$$

Then

$$\{\pi_j, j = 0, 1, 2, \ldots, q - 1\}$$

is a flock of the quadratic cone C.
(2) *Conversely, if $\mathcal{F}$ is a flock of a quadratic cone, choose a representation $\{\pi_j, j = 0, 1, 2, .., q - 1\}$ as above. Choose any convenient constant back half (e, f, g), and define $\mathcal{H}$ as $V(x_1^2 a_i + x_1 x_2 b_i + x_2^2 c_i +$*

$y_1^2 e + y_1 y_2 f + y_2^2 g)$ *for* $i = 1, 2, \ldots, q-1$. *Then* $\mathcal{H}$ *is a regular hyperbolic fibration with constant back half.*

Once the ideas of Baker, Ebert and Penttila are properly interpreted in terms of spreads of translation planes, it is possible to isolate the particular types of translation planes that produce hyperbolic fibrations, thereby producing flocks of quadratic cones. But, since flocks of quadratic cones are, in turn, equivalent to translation planes admitting regulus-inducing elation groups, we obtain a direct connection between these two general types of translation planes.

Now for each of the $q-1$ reguli, choose one of the two reguli of totally isotropic lines. Such a choice will produce a spread and a translation plane. Hence, there are potentially 2^{q-1} possible translation planes obtained in this way.

In this chapter, as mentioned, we intend to connect flocks of quadratic cones with the translation planes obtained from regular hyperbolic quadrics with constant back halves in a more direct manner. In fact, the conical flock associated with a hyperbolic fibration will arise directly by consideration of the associated spreads. Again, all of this may be carried out over arbitrary fields K, except that we obtain instead a partial flock of a quadratic cone from a hyperbolic fibration. Then the question is whether there is an extension to a flock.

There even is some sort of connection with parallelisms with these translation planes admitting cyclic homology groups of order $q + 1$. How this might occur is as follows: It will turn out that a parallelism in $PG(3, q)$ that admits a collineation group of order $q^2 + q$ that fixes one spread and acts transitively on the remaining spreads of the parallelism will force the non-fixed spreads to correspond to flocks of quadratic cones. Specifically, given any conical flock translation plane, the spread is a union of q reguli that share a component. One may derive any of these reguli to obtain a 'derived conical flock plane.' Any such non-fixed spread of our parallelism will be a derived conical flock spread, as there will be an induced Baer group of order q acting on each translation plane of each non-fixed spread of the parallelism. So, we have a set of $q^2 + q$ isomorphic derived conical flocks planes. The group will map the derivable net in question for one plane to the derivable net in question for a second plane. This will imply that the conical flock planes are also isomorphic and the same collineation group will permute transitively a set of $q^2 + q$ conical flock planes. Each conical flock plane will produce a set of affine homology planes. Each of these planes will admit a cyclic affine homology group of order $q + 1$, which will produce a set of $q - 1$ mutually disjoint reguli. Now take any subset of this set of reguli and multiply derive the set. Then the affine homology group still acts on the multiply derived plane. What this means is there are 2^{q-1} possible ways of constructing planes admitting cyclic affine homology groups of order $q + 1$ from a given flock of a quadratic cone. However, each pair of such translation planes share at least two components. On the other hand, it could be asked if there is a choice of one spread from each such class so that the union of these spreads form a partial parallelism of deficiency one. We leave this as an open question and the reader is directed to the appendix on open problems for a related statement on these ideas.

Another interesting connection with spreads in $PG(3, K)$ and flocks will be consider in a forth coming chapter and relates flocks of α-cones to certain translation planes.

DEFINITION 74. *Let $\mathcal{D}$ be any finite derivable partial spread in $PG(3, K)$, where K is a field isomorphic to $GF(q)$. There is a basis change so that*

$\mathcal{D}$ has the form $x = 0, y = x \begin{bmatrix} u & 0 \\ 0 & u^\alpha \end{bmatrix}$; $u \in K$, for some $\alpha \in GalK$.

Any derivable net whose partial spread may be transformed in the above form shall be called an 'α-derivable net.'

DEFINITION 75. *An 'α-hyperbolic fibration' is a set $\mathcal{Q}$ of $q - 1$ α-derivable partial spreads and two carrying lines L and M such that the union $L \cup M \cup \mathcal{Q}$ is a cover of the points of $PG(3, K)$.*

So, the question would be somehow to connect α-hyperbolic fibrations to generalizations of flocks of quadratic cones. This idea will be briefly revisited in the chapter on α-flocks and α-flokki planes.

When we work generally in $PG(3, K)$, we actually will end up re-proving the Baker, Ebert Penttila Theorem, but only in a sense. What we actually do is to show that for arbitrary fields K, translation planes with spreads in $PG(3, K)$ that admit regulus-inducing affine homology groups analogous to cyclic groups of order $q + 1$, in the case where K is isomorphic to $GF(q)$, are exactly equivalent to regular hyperbolic fibrations with constant back half. But, in the general case, such hyperbolic fibrations do not necessarily always produce flocks of quadratic cones. If K is a field that admits a quadratic extension field K^+, let σ denote the involution in $Gal_K K^+$ and consider $K^{+(\sigma+1)}$. In the finite case, this is merely K again, but in genera, it could be a proper subset of K and this is critical, for what occurs is the hyperbolic fibration corresponds to a partial flock that is indexed by $K^{+(\sigma+1)}$.

So we intend to prove the following result.

0.4. Johnson's Homology Theorem.

THEOREM 110. *Let π be a translation plane with spread in $PG(3, K)$, for K a field. Assume that π admits an affine homology group H so that some orbit of components is a regulus in $PG(3, K)$.*

(1) Then π produces a regular hyperbolic fibration with constant back half.

(2) Conversely, each translation plane obtained from a regular hyperbolic fibration with constant back half admits an affine homology group H, one orbit of which is a regulus in $PG(3, K)$.

H is isomorphic to a subgroup of the collineation group of a Pappian spread Σ, coordinatized by a quadratic extension field K^+,

$$H \simeq \langle h^{\sigma+1}; h \in K^+ - \{0\}\rangle,$$

where σ is the unique involution in $Gal_K K^+$.

(3) Let $\mathcal{H}$ be a regular hyperbolic fibration with constant back half of $PG(3, K)$, for K a field. The subgroup of $\Gamma L(4, K)$ that fixes each hyperbolic quadric of the regular hyperbolic fibration $\mathcal{H}$ and acts trivially on the front half is isomorphic to $\langle \rho, \langle h^{\sigma+1}; h \in K^+ - \{0\}\rangle\rangle$, where ρ is defined as follows: If $e^2 = ef + g$, f, g in K and $\langle e, 1\rangle_K = K^+$, then ρ is $\begin{bmatrix} I & 0 \\ 0 & P \end{bmatrix}$, where $P = \begin{bmatrix} 1 & 0 \\ g & -1 \end{bmatrix}$.

In particular, $\langle h^{\sigma+1}; h \in K^+ - \{0\}\rangle$ fixes each regulus and opposite regulus of each hyperbolic quadric of $\mathcal{H}$ and ρ inverts each regulus and opposite regulus of each hyperbolic quadric.

Now once this general theorem is proved, we may prove the finite version as follows:

THEOREM 111. *Let π be a translation plane with spread in $PG(3, q)$. Assume that π admits a cyclic affine homology group of order $q + 1$.*

Then π produces a regular hyperbolic fibration with constant back half.

PROOF. By Jha and Johnson [86], any orbit of components Γ is a derivable net sitting in a Desarguesian spread Σ also containing $x = 0, y = 0$, disjoint from Γ. This is a unique Desarguesian spread so the group H acts as a group of the Desarguesian spread. Let K^* denote the kernel homology group of π. Then, K^* fixes each component of the derivable net and $x = 0, y = 0$. It follows easily that K^* is a collineation group of Σ that fixes at least $q + 3$ components. Clearly, this means that K^* fixes all components of Σ, as Σ is Desarguesian and K^* then sits in $\Gamma L(2, q^2)$. So K^* is a kernel homology group of Σ. But this means that Σ is coordinatized by a field extension of the kernel K

isomorphic to $GF(q)$. Hence, Γ is a regulus in $PG(3, K)$ and our more general result applies. $\square$

Now since a regular hyperbolic fibration with constant back half produces translation planes with spreads in $PG(3, q)$ admitting cyclic affine homology groups of order $q+1$, we obtain the following corollary.

COROLLARY 29. *Finite regular hyperbolic fibrations with constant back half are equivalent to translation planes with spreads in $PG(3, q)$ that admit cyclic homology groups of order $q + 1$.*

DEFINITION 76. *Any translation plane with spread in $PG(3, K)$, where K is a field that arises from a hyperbolic fibration is said to be a 'hyperbolic fibration translation plane.' If the hyperbolic fibration is regular, we use the term 'regular hyperbolic fibration plane.'*

Once all of the theory connecting translation planes admitting regulus-inducing affine homology groups is connected up with regular hyperbolic fibrations with constant back half, we then can see how this ties up with partial flocks of quadratic cones. We begin with some fundamental lemmas, which ultimately will be used to prove Theorem 110. The first lemma merely states what we have already discussed.

LEMMA 33. (1) *Let $\mathcal{H}$ be a hyperbolic fibration of $PG(3, K)$, for K a field (a covering of the points by a set λ of mutually disjoint hyperbolic quadrics union two disjoint lines having an empty intersection with any of the quadrics). For each quadric in λ, choose one of the two reguli (a regulus or its opposite). The union of these reguli and the "carrying lines" form a spread in $PG(3, K)$.*
(2) *Conversely, any spread in $PG(3, K)$ that is a union of hyperbolic quadrics union two disjoint carrying lines (i.e., two lines not contained in any of the reguli) produces a hyperbolic fibration.*

What is most effective when dealing with translation planes is to try to find an associated Pappian plane, from which information involving groups and partial spreads can be obtained. In our regulus-inducing homology translation planes, the idea is simply to take a regulus orbit and the axis and coaxis and find a unique Pappian plane containing this partial spread.

LEMMA 34. *Let π be a translation plane with spread in $PG(3, K)$, for K a field, that admits a homology group H, such that some orbit of components is a regulus in $PG(3, K)$. Let Γ be any H-orbit of components.*
Then there is a unique Pappian spread Σ containing Γ and the axis and coaxis of H.

PROOF. Note that any regulus net and any component disjoint from the elements of that regulus may be embedded into a unique Pappian spread. Such a Pappian spread Σ may be coordinatized by a field extension of K, say K^+. Then there is a representation of that regulus within Σ as follows:

$$y = xm; m^{\sigma+1} = 1, \text{ for } m \in K^+,$$

where σ is the involution in $Gal_K K^+$. Let the homology group H have coaxis $x = 0$ and axis $y = 0$ in the associated 4-dimensional vector space. Thus, we may represent the group H by

$$\left\langle \begin{bmatrix} I & 0 \\ 0 & T \end{bmatrix}; T^{\sigma+1} = I, T \in K^+ \right\rangle.$$

Let $y = xM$ be any component, where M is a non-singular matrix. We claim that there is unique associated Pappian spread containing

$$\{x = 0, y = 0, y = xMT; T^{\sigma+1} = I\}.$$

To see this, change bases by $\begin{bmatrix} M^{-1} & 0 \\ 0 & I \end{bmatrix}$ to obtain Σ as

$$x = 0, y = xm; m \in K^+,$$

then change bases back to obtain the field MK^+M^{-1} and the associated Pappian spread $\begin{bmatrix} M & 0 \\ 0 & I \end{bmatrix} \Sigma$ containing the indicated set. Note that this is a regulus with $x = 0, y = 0$ adjoined in that Pappian spread. Since this Pappian spread may be coordinatized by an extension field of the kernel K, it follows that any such regulus (image under H) is a regulus in $PG(3, K)$. $\square$

The following lemma is now clear, but note that what we are actually doing is somehow connecting one orbit of the homology group with an André net of the Pappian plane that we just found. The second part of the lemma is quite important and fundamental. The reader should make an effort to prove it and is directed to the author's article [103] for additional details (as well as the proof).

LEMMA 35. *(1) Under the assumptions of the previous lemma, we may represent the coaxis, axis, and Γ as follows:*

$$x = 0, y = 0, y = xm; m^{\sigma+1} = 1; m \in K^+,$$

where m is in the field K^+, a 2-dimensional quadratic extension of K, and σ is the unique involution in $Gal_K K^+$.

A basis may be chosen so that Σ may be coordinatized by K^+ as

$$\begin{bmatrix} u & t \\ ft & u + gt \end{bmatrix} \text{ for all } u, t \text{ in } K, \text{ for suitable constants } f \text{ and } g.$$

(2) Under the previous assumptions, if $\{1, e\}$ is a basis for K^+ over K, then $e^2 = eg + f$, and $e^\sigma = -e + g$, $e^{\sigma+1} = -f$. Furthermore, $(et + u)^{\sigma+1} = 1$ if and only if in matrix form $et + u = \begin{bmatrix} u & t \\ ft & u + gt \end{bmatrix}$, such that $u(u + gt) - ft^2 = 1$.

The opposite regulus

$$y = x^\sigma m; m^{\sigma+1} = 1$$

may be written in the form

$$y = x \begin{bmatrix} 1 & 0 \\ g & -1 \end{bmatrix} \begin{bmatrix} u & t \\ ft & u + gt \end{bmatrix}; u(u + gt) - ft^2 = 1.$$

The following lemma now just reformulates the idea of the André net and what this implies regarding the representation of the homology group.

LEMMA 36. If π is a translation plane of order q^2 with spread in $PG(3, K)$ admitting a homology group H such that one component is a regulus in $PG(3, K)$, then choosing the axis of H as $y = 0$ and the coaxis as $x = 0$, we have the following form for the elements of H

$$\begin{bmatrix} I & 0 \\ 0 & T \end{bmatrix}; T^{\sigma+1} = I.$$

Furthermore, we may realize the matrices T in the form $\begin{bmatrix} u & t \\ tf & u + gt \end{bmatrix}$ such that $u(u + gt) - t^2 f = 1$.

When considering a hyperbolic quadric as a variety, we then also obtain the following nice representation of the regulus corresponding to the orbit of $y = x$ under H.

LEMMA 37. The associated hyperbolic quadric corresponding to $(y = x)H$: has the following form:

$$V[1, g, -f, -1, -g, f].$$

Using this homology group, we obtain:

PROPOSITION 5. If π is a translation plane with spread in $PG(3, K)$, and K admits a quadric extension field K^+, let σ denote the involution in $Gal_K K^+$.

Then we may choose a representation for the associated spread as

$$x = 0, y = 0, y = xM_i \begin{bmatrix} u & t \\ ft & u + gt \end{bmatrix}; u(u + gt) - ft^2 = 1$$

and M_i a set of 2×2 matrices over K, where $i \in \rho$, some index set. Let

$$R_i = \{y = xM_iT; T^{\sigma+1} = 1\} \quad \text{for } i \in \rho, \text{ where } M_1 = I,$$

is a regulus in $PG(3, K)$ and

$$R_i^* = \left\{ y = xM_i \begin{bmatrix} 1 & 0 \\ g & -1 \end{bmatrix} T; T^{\sigma+1} = I \right\}$$

represents the opposite regulus to R_i.

  Therefore, there is a corresponding set of ρ reguli, where $x = 0$ and $y = 0$ are interchanged by the polarity of the associated hyperbolic quadric.

PROOF. Since $GL(2, K^+)$, where K^+ is the quadratic extension field of K coordinatizing Σ, is triply transitive on the components of Σ, this implies that any orbit Γ of π may be chosen to have the form $y = xm; m^{\sigma+1} = 1$. It is not difficult to show that any Baer subplane π_0, which is a K–subspace and also disjoint from $x = 0, y = 0$, of the associated Pappian plane Σ has the following form:

$$y = x^\sigma m + xn; \ m \neq 0, n \in K^+.$$

This will mean that

$$y = x^\sigma m \ m^{\sigma+1} = 1 \in K^+$$

is the set of replacement Baer subplanes of the image of $y = x$ under the homology group of Σ with axis $x = 0, y = 0$. So, this means that we may choose the coaxis, axis and Γ as

$$x = 0, y = 0, y = xm; m^{\sigma+1} = 1,$$

without loss of generality.
  The quadratic form for Γ is

$$V\left(x \begin{bmatrix} 1 & g \\ 0 & -f \end{bmatrix} x^t - y \begin{bmatrix} 1 & g \\ 0 & -f \end{bmatrix} y^t \right),$$

where $x = (x_1, x_2), y = (y_1, y_2)$, and (x_1, x_2, y_1, y_2) is the representation in $\{\{1, e\}, \{1, e\}\}$ on $x = 0, y = 0$.

Recall that now Γ is $y = x \begin{bmatrix} u & t \\ ft & u + gt \end{bmatrix}$; $u(u + gt) - ft^2 = 1$.
Now directly check that

$$x \begin{bmatrix} 1 & g \\ 0 & -f \end{bmatrix} x^t$$

$$- x \begin{bmatrix} u & t \\ ft & u + gt \end{bmatrix} \begin{bmatrix} 1 & g \\ 0 & -f \end{bmatrix} \begin{bmatrix} u & t \\ ft & u + gt \end{bmatrix}^t x^t$$

$$= 0.$$

Also, note that

$$x \begin{bmatrix} 1 & g \\ 0 & -f \end{bmatrix} x^t - x \begin{bmatrix} 1 & g \\ 0 & -1 \end{bmatrix} \begin{bmatrix} 1 & g \\ 0 & -f \end{bmatrix} \begin{bmatrix} 1 & g \\ 0 & -1 \end{bmatrix}^t x^t$$

$$= x \begin{bmatrix} 1 & g \\ 0 & -f \end{bmatrix} x^t - x \begin{bmatrix} 1 & g \\ 0 & -f \end{bmatrix}^t x^t$$

and since

$$x \begin{bmatrix} 1 & g \\ 0 & -f \end{bmatrix}^t x^t = x \begin{bmatrix} 1 & g \\ 0 & -f \end{bmatrix} x^t,$$

we have

$$y = x^\sigma \begin{bmatrix} u & t \\ ft & u + gt \end{bmatrix} = x \begin{bmatrix} 1 & 0 \\ g & -1 \end{bmatrix} \begin{bmatrix} u & t \\ ft & u + gt \end{bmatrix},$$

where $u(u + gt) - ft^2 = 1$, is the set of opposite lines and these are also isotropic under the indicated quadratic form. Finally, we know that the opposite regulus to $R_1 = \{y = xT; T^{\sigma+1} = 1\}$ is

$$\left\{ y = x^\sigma T = x \begin{bmatrix} 1 & 0 \\ g & -1 \end{bmatrix} T; T^{\sigma+1} = 1 \right\}.$$

Consider $R_i = \{y = xM_iT; T^{\sigma+1} = 1\}$ and change bases by τ_{M_i} : $(x, y) \longmapsto (x, yM_i^{-1})$ to change R_i into R_1. Then R_1^* maps to R_i^* under $\tau_{M_i}^{-1}$. This implies that

$$R_i^* = \left\{ y = xM_i \begin{bmatrix} 1 & 0 \\ g & -1 \end{bmatrix} T; T^{\sigma+1} = 1 \right\}.$$

□

We now determine the associated quadratic forms. We use the notation of the previous proposition.

LEMMA 38. *The quadratic form for R_i is*

$$V \left(xM_i \begin{bmatrix} 1 & g \\ 0 & -f \end{bmatrix} M_i^t x^t - y \begin{bmatrix} 1 & g \\ 0 & -f \end{bmatrix} y^t \right).$$

PROOF.

$$x M_i \begin{bmatrix} 1 & g \\ 0 & -f \end{bmatrix} M_i^t x^t - x M_i T \begin{bmatrix} 1 & g \\ 0 & -f \end{bmatrix} T^t M_i^t x^t$$

$$= x M_i \begin{bmatrix} 1 & g \\ 0 & -f \end{bmatrix} M_i^t x^t - x M_i \begin{bmatrix} 1 & g \\ 0 & -f \end{bmatrix} M_i^t x^t.$$

This shows that the components of R_i are isotropic subspaces. Now note that the opposite regulus R_i^* to R_i has components

$$y = x M_i \begin{bmatrix} 1 & 0 \\ g & -1 \end{bmatrix} T; T^{\sigma+1} = 1.$$

Then

$$x M_i \begin{bmatrix} 1 & g \\ 0 & -f \end{bmatrix} M_i^t x^t$$

$$- x M_i \begin{bmatrix} 1 & 0 \\ g & -1 \end{bmatrix} T \begin{bmatrix} 1 & g \\ 0 & -f \end{bmatrix} T^t \begin{bmatrix} 1 & 0 \\ g & -1 \end{bmatrix}^t M_i^t x^t. \text{ So}$$

$$x M_i \begin{bmatrix} 1 & g \\ 0 & -f \end{bmatrix} M_i^t x^t$$

$$- x M_i \begin{bmatrix} 1 & 0 \\ g & -1 \end{bmatrix} \begin{bmatrix} 1 & g \\ 0 & -f \end{bmatrix} \begin{bmatrix} 1 & 0 \\ g & -1 \end{bmatrix}^t M_i^t x^t$$

$$= x M_i \begin{bmatrix} 1 & g \\ 0 & -f \end{bmatrix} M_i^t x^t - x M_i \begin{bmatrix} 1 & g \\ 0 & -f \end{bmatrix}^t M_i^t x^t.$$

Note that

$$x M_i \begin{bmatrix} 1 & g \\ 0 & -f \end{bmatrix}^t M_i^t x^t \text{ is self-transpose and}$$

$$\text{thus equal to } x M_i \begin{bmatrix} 1 & g \\ 0 & -f \end{bmatrix}^t M_i^t x^t.$$

This proves the lemma.   □

LEMMA 39. *The regulus*

$$R_i = \{y = x M_i T; \ T^{\sigma+1} = 1\} \quad \textit{for } i \in \rho, \textit{ where } M_1 = I$$

and its opposite regulus

$$R_i^* = \left\{ y = x M_i \begin{bmatrix} 1 & 0 \\ g & -1 \end{bmatrix} T; T^{\sigma+1} = 1 \right\}$$

are interchanged by the mapping

$$\rho = \begin{bmatrix} I & 0 \\ 0 & P \end{bmatrix}, \ where$$

$$P = \begin{bmatrix} 1 & 0 \\ g & -1 \end{bmatrix}; \ (x, y) \longmapsto (x, y^\sigma).$$

Now referring back to the statement of Theorem 110, it should be clear that all parts are proven with the exception of part (3).

So now choose $x = 0, y = 0$ as the two lines (components) interchanged by the associated polarity of each hyperbolic quadric of $\mathcal{H}$, a regular hyperbolic quadric with constant back half. Fix a quadric and choose either regulus of this quadric. We have seen above that there is a unique Pappian spread containing the regulus and $x = 0$, $y = 0$. Choose the quadratic extension field of K with basis $\{e, 1\}$, where $e^2 = ef + g$. Then the quadric has the following form:

$$V \left(x \begin{bmatrix} 1 & g \\ 0 & -f \end{bmatrix} x^t - y \begin{bmatrix} 1 & g \\ 0 & -f \end{bmatrix} y^t \right).$$

As any quadric of $\mathcal{H}$ has constant back half, we note that

$$\begin{bmatrix} u & t \\ ft & u + gt \end{bmatrix} \begin{bmatrix} 1 & g \\ 0 & -f \end{bmatrix} \begin{bmatrix} u & t \\ ft & u + gt \end{bmatrix}^t = \begin{bmatrix} 1 & g \\ 0 & -f \end{bmatrix},$$

where $u(u + gt) - ft^2 = 1$. Furthermore,

$$y \begin{bmatrix} 1 & 0 \\ g & -1 \end{bmatrix} \begin{bmatrix} 1 & g \\ 0 & -f \end{bmatrix} \begin{bmatrix} 1 & 0 \\ g & -1 \end{bmatrix}^t y^t$$

$$= y \begin{bmatrix} 1 & g \\ 0 & -f \end{bmatrix}^t y^t = y \begin{bmatrix} 1 & g \\ 0 & -f \end{bmatrix} y^t.$$

Since the quadric has constant back half, we see that the mappings by elements of K^+ of determinant 1 of the first type occur as homology groups of each translation plane obtained from the hyperbolic fibration, each of whose orbits define reguli. If $y = xN_i$ for $i \in \tau$ is a regulus of some hyperbolic fibration in some spread, then since we now have a homology group acting on the spread, it follows that

$$\{y = xN_i \ for \ i \in \tau\}$$
$$= \left\{ y = xM_i \begin{bmatrix} u & t \\ ft & u + gt \end{bmatrix}; u(u + gt) - ft^2 = 1 \right\},$$

for $i \in \rho$, where ρ is an appropriate index set. We have noted previously that

$$\left\{ \begin{array}{c} y = xM_i \begin{bmatrix} 1 & 0 \\ g & -1 \end{bmatrix} \begin{bmatrix} u & t \\ ft & u+gt \end{bmatrix} \\ ; u(u+gt) - ft^2 = 1 \end{array} \right\}$$

is the associated opposite regulus and that the mapping ρ will interchange the regulus and the opposite regulus. Note that ρ is just the matrix version of x^σ. Now let k be an element of $\Gamma L(4, K)$ that fixes each hyperbolic quadric of $\mathcal{H}$ and acts trivially on the front half of each quadric. Thus, we may represent k in the form

$$k : (x, y) \rightarrow (x, y^\tau Q),$$

where τ is an automorphism of the field K^+ coordinatizing Σ and Q is a non-zero element of K^+. We will see that if $Q^{\sigma+1} = 1$ or $\tau = \sigma$, the unique involution in $Gal_K K^+$, then k certainly satisfies the conditions. However, we see that we must have

$$y^\tau Q \begin{bmatrix} 1 & g \\ 0 & -f \end{bmatrix} Q^t y^{\tau t} = y \begin{bmatrix} 1 & g \\ 0 & -f \end{bmatrix} y^t \quad \forall y \in K^+.$$

Now we note that k will fix each hyperbolic quadric of $\mathcal{H}$. We also know that the mapping $(x, y) \rightarrow (x, y^\sigma)$ interchanges the regulus with its opposite regulus in each such quadric. Suppose that k fixes all reguli of the hyperbolic quadrics. Then k will be a collineation of each corresponding spread. This means that since $y = 0$ is fixed pointwise, the element k is an affine homology of any of the spreads. But this means that k is in $GL(4, K)$, when acting on the spread. This, in turn, implies that τ is 1 or σ. But, if $\tau = \sigma$, k does not act on a spread as a collineation. But then

$$Q \begin{bmatrix} 1 & g \\ 0 & -f \end{bmatrix} Q^t = \begin{bmatrix} a & b \\ c & d \end{bmatrix}.$$

If $\tau = 1$, an easy calculation shows that $a = 1, d = -f$ and $b + c = g$. Working out the form of

$$Q \begin{bmatrix} 1 & g \\ 0 & -f \end{bmatrix} Q^t$$

shows that the element a is the determinant of Q. Hence, if $\tau = 1$, we obtain the det 1 group; i.e., $Q^{\sigma+1} = 1$. So, we may assume that k does not fix all reguli of each quadric. If k fixes one regulus of a quadric, then g is a collineation group of an associated Pappian spread defined by the regulus and $x = 0$ and $y = 0$. The above remarks show that τ is still one in this setting. Thus, we may assume that k interchanges each regulus and opposite regulus of each hyperbolic quadric. If we follow the mapping $(x, y) \rightarrow (x, y^\sigma)$ by k, it follows that $Q^{\sigma+1} = 1$. But then it also now is immediate that $\tau = \sigma$ or 1.

This now completes the proof of the theorem, which establishes the connection between translation planes admitting regulus-inducing homology groups and hyperbolic fibrations with constant back half. In this next chapter, we see the connections between partial flocks of quadratic cones and hyperbolic fibrations.

CHAPTER 18

Hyperbolic Fibrations and Partial Flocks

Let π be a translation plane with spread in $PG(3, K)$ that produces a hyperbolic fibration with carriers $x = 0, y = 0$. We are assuming that K is a field that admits a quadratic extension K^+. What we do in this chapter is show that the way that the reguli are represented as hyperbolic quadrics induces certain functions $\mathcal{F}$ and $\mathcal{G}$ on $K^{+(\sigma+1)}$ that will represent the partial flock of a quadratic cone in $PG(3, K)$. Recalling that $K^{+(\sigma+1)}$ corresponds to the set of determinants of the field of matrices representing K^+, we see that there is an intrinsic problem of representation. In the finite case, this never becomes an issue since $K^{+(\sigma+1)}$ is simply K isomorphic to $GF(q)$.

We begin with a general representation of the spread of π as

$$x = 0, y = 0, y = x \begin{bmatrix} u & t \\ F(u,t) & G(u,t) \end{bmatrix}; u, t \in K,$$

for functions F and G on $K \times K$ to K.

Let

$$\delta_{u,t} = \det \begin{bmatrix} u & t \\ F(u,t) & G(u,t) \end{bmatrix}.$$

We have K^+ as

$$\left\{ \begin{bmatrix} u & t \\ tf & u + gt \end{bmatrix}; u, t \in K \right\}.$$

We may assume that when $u(u + gt) - ft^2 = 1$ then $F(u,t) = ft$ and $G(u,t) = u + gt$. We now compute

$$V\left(xM_i \begin{bmatrix} 1 & g \\ 0 & -f \end{bmatrix} M_i^t x^t - y \begin{bmatrix} 1 & g \\ 0 & -f \end{bmatrix} y^t \right).$$

The operand

$$xM_i \begin{bmatrix} 1 & g \\ 0 & -f \end{bmatrix} M_i^t x^t - y \begin{bmatrix} 1 & g \\ 0 & -f \end{bmatrix} y^t$$

for $M_i = \begin{bmatrix} u & t \\ F(u,t) & G(u,t) \end{bmatrix}$ is easily calculated as follows: Let

$$A_{11} = \delta_{u,t} = u^2 + (ug - tf)t,$$

$$A_{12} = F(u,t) + (ug - tf)G(u,t),$$

$$A_{21} = F(u,t)u + (gF(u,t) - fG(u,t))t,$$

$$A_{22} = F(u,t)^2 + (gF(u,t) - fG(u,t))G(u,t).$$

Therefore, we are considering:

$$x \begin{bmatrix} A_{11} & A_{12} \\ A_{21} & A_{22} \end{bmatrix} x^t - y \begin{bmatrix} 1 & g \\ 0 & -f \end{bmatrix} y^t.$$

To find functions connecting a partial flock, we claim that the $(2,2)$-entry in the front half of the quadric is a function $\mathcal{F}(\delta_{u,t})$ of $\delta_{u,t}$. The proof is merely by reductio ad absurdum and is left to the reader. Similarly, if the sum of the $(1,2)$- and the $(2,1)$-entries of the front half is not a function of $\delta_{u,t}$ then there would be two distinct sums for a given $\delta_{u,t}$. But this again would say that $(x_1, 0, (x_1, 0)M_i)$ would be in two quadrics. Hence, the sum of the $(1,2)$- and $(2,1)$-elements is a function of $\delta_{u,t}$, say, $\mathcal{G}(\delta_{u,t})$. Consider a corresponding translation plane with components $y = xM$ and $y = xN$. Notice that the $(2,2)$-element of the previous matrix for the front half is

$$\det \begin{bmatrix} F(u,t) & G(u,t) \\ fG(u,t) & F(u,t) + gG(u,t) \end{bmatrix}.$$

THEOREM 112. *A regular hyperbolic fibration with constant back half in $PG(3,K)$, K a field, with carrier lines $x = 0, y = 0$, may be represented as follows:*

$$V \left(x \begin{bmatrix} \delta & \mathcal{G}(\delta) \\ 0 & -\mathcal{F}(\delta) \end{bmatrix} x^t - y \begin{bmatrix} 1 & g \\ 0 & -f \end{bmatrix} y^t \right)$$

for all δ in $\det K^+ =$

$$\left\{ \begin{bmatrix} u & t \\ ft & u+gt \end{bmatrix}^{\sigma+1} ; \atop u,t \in K, (u,t) \neq (0,0) \right\},$$

where

$$\delta \in \left\{ \begin{bmatrix} \delta & \mathcal{G}(\delta) \\ 0 & -\mathcal{F}(\delta) \end{bmatrix} ; \atop \begin{bmatrix} u & t \\ ft & u+gt \end{bmatrix}^{\sigma+1} ; \atop u,t \in K, (u,t) \neq (0,0) \right\}$$

$$\cup \left\{ \begin{bmatrix} 0 & 0 \\ 0 & 0 \end{bmatrix} \right\}$$

corresponds to a partial flock of a quadratic cone in $PG(3, K)$, and where $\mathcal{F}$ and $\mathcal{G}$ are functions on $\det K^+$.

PROOF. Let λ be a subset of K such that for each t in $K - \{0\}$, the function $\phi_s(t) = s^2 t + s\mathcal{G}(t) - \mathcal{F}(t)$ is injective for each element s in K. Then there is a corresponding partial flock of a quadratic cone in $PG(3, K)$. The partial flock is a flock if and only if $\phi_s(t)$ is bijective for all s in K. For $\lambda = \{\det K^+\}$, assume that $\phi_s(t)$ is not injective for some element s, so assume that $\phi_s(t) = \phi_s(t^*)$, for $t \neq t^*$. Then, consider $(s, 1, y_1, y_2)$, where $y = (y_1, y_2)$, and this point is on the hyperbolic quadric corresponding to t. It is easy to check that this point would also be on the hyperbolic quadric corresponding to t^*, a contradiction.

This, combined with our previous comments, proves everything in part (1), with the exception of the surjectivity of the functions ϕ_s on $\det K^+$. We note that $y \begin{bmatrix} 1 & g \\ 0 & -f \end{bmatrix} y^t$ maps $y = (y_1, y_2)$ onto $\det \begin{bmatrix} y_1 & y_2 \\ f y_2 & y_1 + g y_2 \end{bmatrix}$. Hence, the functions listed must be surjective on $\det K^+$ in order that the hyperbolic fibration cover $PG(3, K)$. $\square$

Our preliminary remarks prove the following theorem:

0.5. The Correspondence Theorem. The following theorem basically sums up the connections between hyperbolic fibrations and partial flocks of quadratic cones.

THEOREM 113. *The correspondence between any spread π in $PG(3, K)$ corresponding to the hyperbolic fibration and the partial flock of a quadratic cone in $PG(3, K)$ is as follows:*
If π is

$$x = 0, y = 0, y = x \begin{bmatrix} u & t \\ F(u, t) & G(u, t) \end{bmatrix},$$

then the partial flock is given by $\begin{bmatrix} \delta_{u,t} & \mathcal{G}(\delta_{u,t}) \\ 0 & -\mathcal{F}(\delta_{u,t}) \end{bmatrix}$ *with*

$$\delta_{u,t} = \det \begin{bmatrix} u & t \\ ft & u + gt \end{bmatrix},$$

$$\mathcal{G}(\delta_{u,t})$$
$$= g(uG(u, t) + tF(u, t)) + 2(uF(u, t) - tfG(u, t)),$$
$$- \mathcal{F}(\delta_{u,t})$$
$$= \delta_{F(u,t),G(u,t)},$$

where

$$\delta_{F(u,t),G(u,t)}$$

$$= \det \begin{bmatrix} F(u,t) & G(u,t) \\ fG(u,t) & F(u,t) + gG(u,t) \end{bmatrix} \in \det K^+.$$

It is noted that the homology group, one orbit of which is a regulus in $PG(3,K)$, is given by

$$H = \left\{ T = \begin{bmatrix} u & t \\ ft & u + gt \end{bmatrix} ; \delta_{u,t} = 1 \right\}.$$

If we have a hyperbolic fibration in $PG(3,K)$, there are corresponding functions given in the previous theorem such that the corresponding functions

$$\phi_s(t) = s^2 t + sG(t) - F(t)$$

are injective for all s in K and for all $t \in \det K^+$.

Indeed, the functions restricted to $\det K^+$ are surjective on $\det K^+$.

Conversely, any partial flock of a quadratic cone in $PG(3,K)$, with defining set λ (i.e., so t ranges over λ and planes of the partial flock are defined via functions in t) equal to $\det K^+$, whose associated functions on $\det K^+$, as above, are surjective on $\det K^+$ (K^+ some quadratic extension of K), produces a regular hyperbolic fibration in $PG(3,K)$ with constant back half.

PROOF. We have a partial flock of a quadratic cone, indexed by $\det K^+$, for some quadratic extension of K, where the functions ϕ_s are bijective on $\det K^+$. Use K^+ to define the elements f and g as above. For any given function $y = xM_i$, disjoint from $x = 0, y = 0$, we know that

$$xM_i \begin{bmatrix} 1 & g \\ 0 & -f \end{bmatrix} M_i^t x^t = x \begin{bmatrix} \delta_{u,t} & G(\delta_{u,t}) \\ 0 & -F(\delta_{u,t}) \end{bmatrix} x^t,$$

where $\mathcal{F}$ and $\mathcal{G}$ and the functions defined in the previous theorem, for some elements u, t. Construct a partial spread from $y = xM_i$ by applying the mappings $(x, y) \rightarrow (x, yQ)$, where $Q^{\sigma+1} = 1$, Q in K^+. It will then follow that this is a regulus, and its opposite regulus defines a hyperbolic quadric, whose polarity interchanges $x = 0$ and $y = 0$.

We claim that, in this way, we obtain a spread and hence a hyperbolic fibration. Consider any point not on $x = 0, y = 0$. We may always start with the Pappian spread Σ coordinatized by K^+ and choose $x = 0, y = 0, y = x$ in Σ to begin the process.

Each element $t \in \det K^+$, defines a hyperbolic quadric of the form

$$V_t = V(x_1^2 t + x_1 x_2 \mathcal{G}(t) - x_2^2 \mathcal{F}(t) - (y_1^2 + y_1 y_2 g - f y_2^2)),$$

where $\mathcal{F}(1) = f$ and $\mathcal{G}(1) = g$ and one regulus R_1 together with $x = 0, y = 0$ define a unique Pappian spread Σ admitting K^+, as above, as a coordinatizing field. Suppose that a point (x_o, y_o) lies in V_t and V_{t^*}, for x_o and y_o both non-zero vectors. Then clearly $t = t^*$, as we have a partial flock of a quadratic cone. It remains to show that we have a cover of $PG(3, K)$. Choosing any regulus for each quadric certainly produces a partial spread. Furthermore, the partial spread admits the 'homology' group $(x, y) \rightarrow (x, yQ)$, such that $Q^{\sigma+1} = 1$. If there is a 2-dimensional subspace disjoint from this partial spread, we may write the subspace in the form $y = xM$, where M is a non-singular 2×2 matrix over K. We may then use this subspace to construct an extension to the partial flock, also admitting the homology group but still indexed by $\det K^+$, a contradiction. Hence, we have a maximal partial spread. Choose any point (x_1, x_2, y_1, y_2) such that $x_2 \neq 0$. Note that $y_1^2 + y_1 y_1 g - f y_2^2 = c$ is in $\det K^+$, as is x_2^2. Since the mappings ϕ_s are onto functions, there exists a unique t in $\det K^+$ such that $(x_1/x_2)^2 t + (x_1/x_2)\mathcal{G}(t) - \mathcal{F}(t) = c/x_2^2$. This means that if $x_2 \neq 0$ then (x_1, x_2, y_1, y_2) is covered by the $\det K^+$-set of hyperbolic quadrics. Now assume that $x_2 = 0$. Then $x_1^2 t = c$ has a unique solution in $\det K^+$. Hence, this completes the proof of the theorem. $\qquad\square$

CHAPTER 19

j-Planes and Monomial Flocks

The idea of a j-plane arose from Kantor's work on the slicing of ovoids [**157**], when the author used this analysis to find representations of the translation planes that arose from the slices. Using the basic representation of the spreads, Johnson, Pomareda and Wilke [**155**], construct j-planes and develop the theory in the finite case, in particular for $j = 1, 2$, where $j = 1$-planes may be obtained by slicing. In this text, we will generalize the ideas and consider j-planes over any field K. Here is the formal definition.

DEFINITION 77. *Let K be a field and K^+ a quadratic field extension of K represented as follows:*

$$K^+ = \left\{ \begin{bmatrix} u & t \\ ft & u + gt \end{bmatrix} ; u, t \in K \right\}.$$

Consider the following group:

$$G_{K^+, j} = \left\{ \begin{bmatrix} 1 & 0 & 0 & 0 \\ 0 & \delta_{u,t}^{-j} & 0 & 0 \\ 0 & 0 & u & t \\ 0 & 0 & ft & u + gt \end{bmatrix} ; \atop u, t \in K, \ (u, t) \neq (0, 0) \right\},$$

where j is a fixed integer and $\delta_{u,t} = \det \begin{bmatrix} u & t \\ ft & u + gt \end{bmatrix}$. A 'j-plane' is any translation plane π containing $x = 0, y = 0$ and $y = x$ that admits $G_{K,j}^+$ as a collineation group acting transitively on the components of $\pi - \{x = 0, y = 0\}$.

Consider the subgroup when $\delta_{u,t} = 1$. We see that this group is an affine homology group and any image together with $x = 0, y = 0$ is a regulus in $PG(3, K)$. This means that there is an associated regular hyperbolic fibration with constant back-half. Furthermore, this also implies that there is an associated partial flock of a quadratic cone in $PG(3, K)$, and in the finite case, there is an associated flock of a quadratic cone in $PG(3, q)$, for K isomorphic to $GF(q)$, since in this setting the affine homology group is cyclic.

So, the question is to determine the class of conical flocks that correspond to j-planes. Actually, all of the known j-planes construct flocks that were also previously known and all of these are monomial in the sense that the functions $f(t)$ and $g(t)$ representing the flock are monomial functions.

In this chapter, we continue the ideas presented in the chapter on hyperbolic fibrations in that translation planes with spreads in $PG(3, K)$ admitting regulus-inducing elation groups are somehow connected to translation planes with spreads in $PG(3, K)$ admitting regulus-inducing homology groups. The material given in this chapter somewhat follows the work of the author [101].

For j-planes the group in question is isomorphic to the multiplicative group of a field L. In general, when this occurs, we use the term 'H-group', and the plane an 'H-plane,' where H is a multiplicative endomorphism of $L - \{0\}$.

DEFINITION 78. *Let K be a field that admits a quadratic extension field K^+. Consider the following group*

$$\left\{ \begin{bmatrix} 1 & 0 & 0 & 0 \\ 0 & H(\delta_{u,t})^{-1} & 0 & 0 \\ 0 & 0 & u & t \\ 0 & 0 & ft & u+gt \end{bmatrix} ; \atop \begin{array}{c} x^2 + xg - f \text{ is } K\text{-irreducible}, u,t \in K, \text{ not both } 0, \\ H \text{ an endomorphism on} \\ \left\{ \delta_{u,t} = \det \begin{bmatrix} u & t \\ ft & u+gt \end{bmatrix} ; u,t \in K \right\} \end{array} \right\}.$$

If

$$\left\{ x = 0, y = 0, y = x \begin{bmatrix} 1 & 0 \\ 0 & H(\delta_{u,t}) \end{bmatrix} \begin{bmatrix} u & t \\ ft & u+gt \end{bmatrix} ; u,t \in K \right\}$$

is a spread in $PG(3, K)$, we call this an H-spread and the corresponding translation plane, an H-plane. Clearly, there is an associated affine homology group obtained from post-multiplication of $\begin{bmatrix} u & t \\ ft & u+gt \end{bmatrix}$ of determinant 1.

Hence, the spread components other than $x = 0, y = 0$ are

$$y = x \begin{bmatrix} u & t \\ H(\delta_{u,t})ft & H(\delta_{u,t})(u+gt) \end{bmatrix} T,$$

where T is the group G' of field matrices of determinant 1. There is, therefore, an associated partial $\det K^+$-partial flock.

For j-planes and H-planes, here are the connections to partial flocks (or flocks when the field is finite).

THEOREM 114. *(1) A j-plane with spread set*

$$\left\{ y = x \begin{array}{c} x = 0, y = 0, \\ \left[\begin{array}{cc} 1 & 0 \\ 0 & \delta^j_{u,t} \end{array} \right] \left[\begin{array}{cc} u & t \\ ft & u + gt \end{array} \right] ; u, t \in K, \\ (u, t) \neq (0, 0), K \ a \ field \end{array} \right\},$$

produces a monomial $\det K^+$*-partial flock of a quadratic cone with monomial functions*

$$f(\delta_{u,t}) = f\delta^{2j+1}_{u,t}, \ g(\delta_{u,t}) = g\delta^{j+1}_{u,t}.$$

(2) An H-plane with spread set

$$\left\{ y = x \begin{array}{c} x = 0, y = 0, \\ \left[\begin{array}{cc} 1 & 0 \\ 0 & H(\delta^j_{u,t}) \end{array} \right] \left[\begin{array}{cc} u & t \\ ft & u + gt \end{array} \right] ; u, t \in K, \\ (u, t) \neq (0, 0), K \ a \ field \end{array} \right\},$$

produces a $\det K^+$*-partial flock of a quadratic cone with functions:*

$$\begin{aligned} g(\delta_{u,t}) &= gH(\delta_{u,t})\delta_{u,t} \\ f(\delta_{u,t}) &= fH(\delta_{u,t})^2\delta_{u,t}. \end{aligned}$$

PROOF. Let π be a j-plane, so that there is a spread set of the following form:

$$\left\{ y = x \begin{array}{c} x = 0, y = 0, \\ \left[\begin{array}{cc} 1 & 0 \\ 0 & \delta^j_{u,t} \end{array} \right] \left[\begin{array}{cc} u & t \\ ft & u + gt \end{array} \right] ; u, t \in K, \\ (u, t) \neq (0, 0), \end{array} \right\},$$

where $\delta_{u,t} = \det \left[\begin{array}{cc} u & t \\ ft & u + gt \end{array} \right]$.

By Theorem 0.5, we have

$$\begin{bmatrix} \delta_{u,t} & g(\delta_{u,t}) \\ 0 & -f(\delta_{u,t}) \end{bmatrix}; \delta_{u,t}$$

$$= \det \begin{bmatrix} u & t \\ ft & u+gt \end{bmatrix},$$

$$g(\delta_{u,t}) = g(uG(u,t) + tF(u,t)) + 2(uF(u,t) - tfG(u,t)),$$

$$-f(\delta_{u,t}) = \delta_{F(u,t),G(u,t)},$$

where

$$\delta_{M_i} = \det M_i,$$

$$\text{and } \delta_{F(u,t),G(u,t)} = \det \begin{bmatrix} F(u,t) & G(u,t) \\ fG(u,t) & F(u,t)+gG(u,t) \end{bmatrix}.$$

In this case,

$$F(u,t) = \delta_{u,t}^j ft, \ G(u,t) = \delta_{u,t}^j (u+gt).$$

Hence,

$$\begin{aligned}
g(\delta_{u,t}) &= g(uG(u,t) + tF(u,t)) + 2(uF(u,t) - tfG(u,t)) \\
&= g(u\delta_{u,t}^j(u+gt) + t\delta_{u,t}^j ft) + 2(u\delta_{u,t}^j ft - tf(\delta_{u,t}^j(u+gt))) \\
&= g\delta_{u,t}^{j+1}.
\end{aligned}$$

Also,

$$\begin{aligned}
-f(\delta_{u,t}) &= \delta_{F(u,t),G(u,t)} \\
&= F(u,t)(F(u,t) + gG(u,t)) - fG(u,t)^2 \\
&= \delta_{u,t}^j ft(\delta_{u,t}^j ft + g(\delta_{u,t}^j(u+gt))) - f(\delta_{u,t}^j(u+gt))^2 \\
&= \delta_{u,t}^{2j} f(ft^2 + g(u+gt)t - f(u^2 + 2ugt + g^2t^2) \\
&= \delta_{u,t}^{2j} f(-f(u^2 + ugt + -ft^2) \\
&= -f\delta_{u,t}^{2j+1}.
\end{aligned}$$

Therefore,

$$f(\delta_{u,t}) = f\delta_{u,t}^{2j+1}.$$

This proves the result.

For H-planes, the reader may supply the similar proof with $H(\delta_{u,t})$ in place of $\delta_{u,t}^j$, to obtain

$$\begin{aligned}
g(\delta_{u,t}) &= gH(\delta_{u,t})\delta_{u,t} \\
-f(\delta_{u,t}) &= -fH(\delta_{u,t})^2\delta_{u,t}.
\end{aligned}$$

$\square$

This theorem leads to the following classification result.

0.6. Classification of Finite Even Order j-Planes. Assume that we have a j-plane of order q^2, where $q = 2^r$. The monomial flocks are completely determined in Penttila-Storme [**175**], where is it shown that $j = 0, 1, 2$ and correspond to the linear, Betten and Payne flocks, respectively. This proof is beyond the scope of this text and the reader is directed to the article for the intricate proof. Hence, the associated j-planes are also completely determined. We note that originally the j-planes for $j = 1$ are constructed in Johnson [**121**] and are due to Kantor as a particular slice of a unitary ovoid (see, in particular, section 3 and (3.6)). Furthermore, for $j = 2$, the planes are constructed in Johnson-Pomareda-Wilke [**155**].

Hence, we have the following theorem.

0.7. Johnson, Penttila-Storme Theorem.

THEOREM 115. *Let π be a j-plane of even order q^2. Then $j = 0, 1$ or 2 and the plane is one of the following types of planes:*

(1) Desarguesian ($j = 0$ and corresponding to the linear flock),

(2) Kantor-slice of a unitary ovoid ($j = 1$ and corresponding to the Betten flock), or

(3) The Johnson-Pomareda-Wilke $j = 2$-plane (corresponding to the Payne flock).

Hyperbolic fibrations may be considered over any field K that admits a quadratic field extension K^+, so, for example, any subfield of the field of real numbers would work. In the following chapter, we connect flocks of quadratic cones and hyperbolic fibrations over the field of real numbers.

1. Hyperbolic Fibrations over the Reals

Let $H(u) = u^r$, for $u > 0$, and equal to 0 when $u = 0$. Define $H(-u) = -u^r$, for $u > 0$. Consider putative associated functions

$$g(\delta_{u,t}) = gH(\delta_{u,t})\delta_{u,t} = g\delta_{u,t}^{r+1},$$
$$f(\delta_{u,t}) = fH(\delta_{u,t})^2\delta_{u,t} = f\delta_{u,t}.$$

Now we consider the function $K \to \det K^+$, when K is a field of real numbers. The elements in this set are $u^2 + ugt - ft^2$, where $g^2 + 4f < 0$. Hence, $-f > 0$ so that we know that $u = t = 0$, 0 is in the $\det K^+$ and if $t = 0$, we obtain the positive real numbers and for $t \neq 0$, we see that

$$y = x^2 + xg - f > 0, \text{ for all } x.$$

Hence, $\det K^+$ is the set of positive real numbers when $(u, t) \neq (0, 0)$. So, the question is whether the functions ϕ_s restricted to the positive reals are bijective on the positive real numbers, for any $s \in K$.

But, we let us begin by finding flocks of quadratic cones over the field of real numbers as these are amazingly easy to construct.

Let $\mathcal{R}$ denote the field of real numbers. Let $-f(t)$ denote a continuous, strictly increasing function on the reals such that $f(0) = 0$, such that $\lim_{t \to \pm\infty}(-f(t)) = \pm\infty$. To obtain a flock of a quadratic cone, it suffices to show that the functions

$$\phi_u : t \to u^2 t - f(t)$$

are bijective for all u. For example, if the function is differentiable, then $u^2 t + f(t)$ has derivative $u^2 - f'(t) > 0$, which implies that the functions are injective. More generally, if $s < t$, then $-f(s) < -f(t)$ so that $u^2(s - t) \neq -f(t) + f(s)$, so that the functions are injective. The limit conditions ensure bijectivity. So, there is an associated flock of a quadratic cone in $PG(3, R)$ and a conical flock plane with spread

$$x = 0, y = x \begin{bmatrix} u & f(t) \\ t & u \end{bmatrix} ; u, t \in \mathcal{R}.$$

Let $-f(1) = -f$. Then

$$\left\{ \begin{bmatrix} u & t \\ ft & u \end{bmatrix} ; u, t \in R \right\},$$

is a field, and note that the determinant is $u^2 - t^2 f$, which is non-negative, since $-f(1) > 0$. Furthermore, $t > 0$ if and only if $-f(t) > 0$, so it follows the conditions of the theorem of the previous section are valid. Thus, there is an induced hyperbolic fibration over the reals $\mathcal{R}$.

We first observe that if

$$C_1 = \{(v, s); v^2 + -fs^2 = 1\},$$

is an ellipse with center $(0, 0)$ in the real affine plane Π and $P = (u, t)$ is any point of Π then the line joining (u, t) and $(0, 0)$ non-trivially intersects C_1 in a point (v, s). This means that there is a real number k such that $k(v, s) = (u, t)$. This implies that any spread

$$x = 0, y = 0, y = x \begin{bmatrix} u & t \\ F(u, t) & G(u, t) \end{bmatrix} ; t, u \in \mathcal{R}$$

in $PG(3, \mathcal{R})$ that admits an affine homology group that maps $y = xM$ onto $y = xMT$, where

$$T \in G' = \left\{ \begin{bmatrix} u & t \\ ft & u + gt \end{bmatrix} ; \delta_{u,t} = 1 \right\},$$

corresponds to a j-plane. In this case, the spread is $x = 0, y = 0$, and a union of reguli D_k, for $k \in \mathcal{R}$:

$$D_k; y = x \begin{bmatrix} k & 0 \\ F(k) & G(k) \end{bmatrix} T; T \in G',$$

where we choose the notation so that $F(k, 0) = F(k)$ and $G(k, 0) = G(k)$. Using Theorem 0.5, a straightforward calculation then follows that $F(u) = 0$ for all u and $G(u)^2(-f) = -f(u^2)$.

If we would derive a given regulus net, what happens is that $G(u, t)$ becomes $-G(u, -t)$ and $F(u, t)$ remains the same, since the involution σ in the Galois group over the field of real numbers is given by the matrix $\begin{bmatrix} \begin{bmatrix} 1 & 0 \\ 0 & -1 \end{bmatrix} \end{bmatrix}$. What this means is that as one spread corresponding to the hyperbolic fibration has:

$$G(u) = \sqrt{f(u^2)/f}.$$

Furthermore,

$$\begin{bmatrix} u & 0 \\ 0 & G(u) \end{bmatrix} \text{ and } \begin{bmatrix} u & 0 \\ 0 & G(u) \end{bmatrix} \begin{bmatrix} -1 & 0 \\ 0 & -1 \end{bmatrix} = \begin{bmatrix} -u & 0 \\ 0 & -G(u) \end{bmatrix},$$

are in the same regulus (hence, we require $G(-u) = -G(u)$. That is, we may assume that u is positive. Therefore, we may assume that we also have that

$$\begin{bmatrix} -u & 0 \\ 0 & -\sqrt{f((-u)^2)/f} \end{bmatrix}$$

defines a component within the spread. Therefore, we obtain the following theorem:

THEOREM 116. *Let $-f(x)$ be any continuous strictly increasing function on the reals $\mathcal{R}$, such that $f(0) = 0$, and $\lim_{x \to \pm\infty} -f(x) = \pm\infty$.*

Then there is a spread of $PG(3, \mathcal{R})$

$$(*): x = 0, y = 0, y = x \begin{bmatrix} u & 0 \\ 0 & \sqrt{f(u^2)/f} \end{bmatrix} T \in G', \ u \geq 0,$$

where $f = f(1)$ and

$$G' = \left\{ T = \begin{bmatrix} u & t \\ ft & u \end{bmatrix}; \delta_{u,t} = 1 \right\}.$$

Furthermore, the spread produces a hyperbolic fibration.

We now consider arbitrary spreads of the homology group type $(*)$. Now let $h(u) = \sqrt{f(u^2)/f}$ for $u > 0$ and assume that $h(-u) = -h(u)$. Then

$$\begin{bmatrix} k & 0 \\ 0 & h(k) \end{bmatrix} \begin{bmatrix} v & s \\ fs & v \end{bmatrix} = \begin{bmatrix} kv & ks \\ h(k)fs & h(k)v \end{bmatrix},$$

where the determinant of $\begin{bmatrix} u & t \\ ft & u \end{bmatrix}$ is 1.

In general, the determinant of $\begin{bmatrix} u & t \\ ft & u \end{bmatrix}$ is positive, hence, for $k > 0$, there is a given matrix with determinant $k = \delta_{u,t}$.

We now consider the H-planes. So, assume that H is a multiplicative homomorphism on $\mathcal{R} - \{0\}$. Then

$$G = \left\{ \begin{bmatrix} 1 & 0 & 0 & 0 \\ 0 & H(\delta_{u,t})^{-1} & 0 & 0 \\ 0 & 0 & u & t \\ 0 & 0 & ft & u \end{bmatrix} ; u, t \in \mathcal{R} - \{0\} \right\}$$

is a group isomorphic to $\mathcal{R}^+ - \{0\}$. Moreover, the matrix

$$\begin{bmatrix} 1 & 0 \\ 0 & H(\delta_{u,t}) \end{bmatrix} \begin{bmatrix} u & t \\ ft & u \end{bmatrix} = \begin{bmatrix} u & t \\ H(\delta_{u,t})ft & H(\delta_{u,t})u \end{bmatrix},$$

has determinant $\delta_{u,t} H(\delta_{u,t}) = kH(k^2)$. Now let $kv = u$ and $ks = t$, and $h(k) = kH(k^2)$. This means that

$$f(k^2) = fk^2 H(h^2)^2,$$

or rather that

$$f(v) = fvH(v)^2, \; v > 0.$$

In other words, if H is a multiplicative endomorphism defined on the positive real numbers, then there is an associated collineation group G isomorphic to $\mathcal{R} - \{0\}$, which fixes two components $x = 0, y = 0$ and acts transitively on the remaining components of the spread. Conversely, let h be an multiplicative endomorphism on the positive real numbers such that $h(u) > 0$, where $u > 0$.

Now to connect to the notation:

$$x = 0, y = 0, y = x \begin{bmatrix} u & 0 \\ 0 & \sqrt{f(u^2)/f} \end{bmatrix} T, \; u \geq 0,$$

we would require that

$$f(v) = fvH(v)^2, \; v > 0.$$

is continuous on the positive reals, is non-decreasing and has range the set of all positive reals. If H is a function differentiable on the positive reals, then $H(v) = v^j$, for some real number. In this case, we consider

$$f(v) = fv^{2j+1} \text{ for } v > 0.$$

To be non-decreasing, we would require that $f(2j+1) > 0$ and to have that range, the set of all positive reals requires that $f > 0$ and so

$$2j + 1 > 0.$$

As this function may be extended to a continuous function with the property required, we obtained an associated hyperbolic fibration plane.

THEOREM 117. *Let h be an endomorphism on the positive real numbers such that $f \cdot h(\sqrt{v})^2$, for $v > 0$ is continuous, strictly increasing and surjective on the positive real numbers. Then, there is an associated flock and partial flock and therefore a corresponding hyperbolic fibration.*

We have seen that the finite j-planes of even order are completely classified, but there might be no hope of obtaining anything close to a classification for odd order j-planes. Furthermore, we have seen that there are many H-planes and j-planes over the field of real numbers, and it might be that there again is little hope for a classification. However, in the next chapter, we actually are able to classify the real j-planes.

2. Classification of the Real j-Planes

In this section, we completely classify the real j-planes as follows: (We continue with the notation established in the previous section).

THEOREM 118. *A translation plane π is a real j-plane if and only if j is a real number and $j > -1/2$. In all cases, there is a partial monomial flock over the non-negative real numbers. The partial monomial flock may be extended to a monomial flock over the field of real numbers if and only if $(-1)^j$ is defined when g is not zero and if and only if $(-1)^{2j+1}$ is defined when $g = 0$.*

PROOF. We know that a real j-plane produces a $\det K^+$ partial monomial flock with the following functions:

$$f(t) = ft^{2j+1}, g(t) = gt^{j+1},$$

where $t \in \det K^+$. Furthermore, we note that

$$\begin{bmatrix} u & t \\ ft & u + gt \end{bmatrix} ; u, t \in K,$$

forces

$$u^2 + ugt - ft^2 = 0$$

if and only if $u = t = 0$. When $t = 0$, and K is the field of real numbers, $\det K^+$ contains the non-negative reals. And since $g^2 + 4f < 0$ (the discriminant must be negative), it follows that $\det K^+$ is the set of all non-negative reals. Thus, consider the functions:

$$\phi_s : \phi_s(t) = s^2 t + g(t)s - f(t).$$

For each $s \in K$, ϕ_s must be injective restricted to $t > 0$, and $f(0) = 0$, $g(0) = 0$ (which we require if the original functions are defined only for $t > 0$) and surjective on $\det K^+$.

Therefore, we consider

$$\phi_s(t) = s^2 t + gt^{j+1}s - ft^{2j+1}.$$

We note that

$$\phi_s(t) = t((s + gt^j/2)^2 - (g^2 + 2f)t^{2j}/4) > 0, \; for \; t > 0.$$

The given function is differentiable on $t > 0$, regardless of j, hence, the derivative is

$$s^2 + (j+1)gt^j s - (2j+1)ft^{2j}.$$

We claim that for each s, the derivative is ≥ 0. Furthermore, when $s = 0$, $-ft^{2j+1}$ is clearly injective for $t > 0$. For s non-zero, the derivative at 0 is $s^2 > 0$. Since the derivative function is continuous for $t > 0$, assume that for some positive value of t, the derivative is negative. Then there is a root of

$$s^2 + (j+1)gt^j s - (2j+1)ft^{2j}.$$

But note considering the functions as polynomials in s, we assert that the discriminant is negative. To see this simply note that

$$((j+1)g)^2 + 4(2j+1)f$$
$$= (j+1)^2(g+4f) - 4j^2 < 0$$

since $(g + 4f) < 0$. Hence, the derivative is ≥ 0, which implies that ϕ_s is injective. In order that the function be surjective on $\det K^+$, we see that we must have

$$\lim_{t \to \infty} \phi_s(t) = \infty \text{ and } \lim_{t \to 0^+} \phi_s(t) = 0.$$

We consider the cases: $j > 0$ and $j < 0$ ($j = 0$ produces a Pappian affine plane). First, assume that $j > 0$. Then the two required limits are clearly valid, since

$$\phi_s(t) = t((s + gt^j/2)^2 - (g^2 + 2f)t^{2j}/4) > 0, \; for \; t > 0.$$

Now assume that $j < 0$ then $\lim_{t \to \infty} \phi_s(t) = \infty$. Consider

$$\phi_s(t) = s^2 t + t^j(gt - ft^{j+1}).$$

Then we require

$$\lim_{t \to 0^+} t^j(gt - ft^{j+1}) = 0.$$

Note that when $s = 0$, we require that $\lim_{t \to 0^+}(-f(t) = -ft^{2j+1}) = 0$. Therefore, $2j + 1 > 0$. Thus, we must also have $\lim_{t \to 0^+} gt^{j+1} = 0$, but if $2j + 1 > 0$, then $j > -1/2$ so that $j + 1 > 1/2$. This completes the proof that a j-plane is obtained if and only if $j > -1/2$. The question remains if we may extend the partial monomial flock to a monomial flock. If so, then the same functions must be used and must be defined on the negative real numbers. Hence, the question is whether $(-1)^{j+1}$ and $(-1)^{2j+1}$ are defined, for g non-zero and whether $(-1)^{2j+1}$ is defined for $g = 0$. In the former case, the necessary and sufficient condition is whether $(-1)^j$ is defined. If g is non-zero and $(-1)^j$ is defined then $(-t)^j = (-1)^j t^j$, for $t > 0$, implies that $(-t)^j$ is defined. It now follows analogously as in the previous argument that the functions ϕ_s are defined for all real elements t and that these functions are bijective on K, implying that there is an associated monomial flock of a quadratic cone. $\qquad\qquad\square$

Now assume that we have a real H-plane, so that H is an endomorphism of the multiplicative group of non-negative real numbers, which we are assuming is Lebesque integrable. In this setting, we know that $H(t) = t^j$, for j a real number. Hence, we have the following corollary.

COROLLARY 30. *The H-planes over the field of real numbers are completely determined as j planes for $j > -1/2$, provided H is Lebesque integrable.*

2.1. Extension of Partial $\det K^+$-Partial Flocks. We have noted that a partial monomial flock may not always be extended to a monomial flock at least when K is the field of real numbers. However, we will see that there is always an extension to some flock and the wonderful thing is that there are non-countably many ways to do this!

THEOREM 119. *Assume that K is the field of real numbers. Every partial $\det K^+$-partial flock may be extended to a flock in non-countably infinitely many ways.*

PROOF. A $\det K^+$-partial exists if and only if the functions $\phi_s(t) = s^2 t + g(t)s - f(t)$ are injective and surjective onto the non-negative real numbers, where $t \geq 0$. Take any function $f_1(t)$, which is defined on

$(-\infty, 0]$ such that $f_1(0) = 0$ and f_1 is continuous and non-decreasing and $\lim_{t \to -\infty} f_1(t) = -\infty$. Then define

$$
\begin{aligned}
g_2(t) &= \quad 0 \text{ if } t \leq 0 \text{ and } g_1(t) = g(t) \text{ for } t > 0, \\
f_2(t) &= \quad f_1(t) \text{ for } t \leq 0 \text{ and } f_2(t) = f(t), \text{ for } t > 0.
\end{aligned}
$$

Then, clearly,

$$
\psi_s(t) = s^2 t + g_2(t)s \quad f_2(t)
$$

is bijective on the set of real numbers.   □

So, the staggering result (at least to this writer) of this is summed up using net replacement.

THEOREM 120. *Any conical flock plane defined on the real numbers, let $\mathcal{P}$ and $\mathcal{N}$ denote the components with positive and negative slopes. Then there are non-countably infinitely many replacements of $\mathcal{N}$ producing conical flock planes.*

Part 5

Derivable Geometries

In this part, we consider first a generalization of flocks of quadratic cones, called α-flocks. We also consider parallelisms of α-cones and parallelisms of generalizations of hyperbolic flocks, as well as general consideration of extensions of derivable nets from the standpoint of their embedding into 3-dimensional projective space.

CHAPTER 20

Flocks of α-Cones

If one is considering a point-line geometry, then a 'spread' is defined as a set of lines that covers each point exactly once and a 'parallelism' is then a set of spreads that covers each line exactly once. Of course, a refinement of these definitions then might be required to accommodate various geometries.

We have been discussing flocks of quadratic cones, flocks of hyperbolic quadrics, and flocks of elliptic quadrics. Considered as point-line geometries, we could also consider flocks of their circle plane generalizations: Laguerre planes, Minkowski planes, and Inversive planes, respectively. Also, we have previously discussed focal-spreads, where certain of these give rise to 2-designs that admit 'resolutions,' where our terminology of 'spreads' and 'parallelisms' could be utilized.

For example, if we identify the terms 'flock' and 'spread,' then a parallelism of a quadratic cone is a partition of the circles (conics) not incident with vertex v_0 by flocks ('spreads'). A parallelism of a hyperbolic quadric is a partition of the conics by flocks. For elliptic quadrics, the situation is more problematic since it is possible to partition certain infinite elliptic quadrics by conics with two points, one point, or no points omitted. Perhaps, we should call associated flocks, '2-point elliptic flocks,' '1-point elliptic flocks,' and '0-point elliptic flocks'; i-point elliptic flocks, where $i = 2, 1, 0$, is the number of points not covered by the set of conics. Let N and S be two points of an elliptic quadric $\mathcal{Q}$ in $PG(3, K)$, where K is a field.

DEFINITION 79. *A partition of the conics not containing N or S by i-point elliptic flocks shall be called a '$N - S$ i-point elliptic parallelism,' for $i = 2, 1, 0$.*

DEFINITION 80. *A partition of the conics by i-point elliptic flocks shall be a called a 'i-point elliptic parallelism.'*

REMARK 38. *In our constructions of infinite regular parallelisms in $PG(3, K)$, we show that there are 2-point elliptic parallelisms for essentially any infinite field K.*

REMARK 39. *In the finite case, as there are $q^2 + 1$ points on an elliptic quadric in $PG(3, q)$, and $q + 1$ points on a conic, it is combinatorially only possible to have 2-point elliptic parallelisms.*

The Johnson, Payne-Thas Theorem 108 connects translation planes with spreads in $PG(3, q)$ admitting Baer groups of order q with flocks of quadratic cones, but more precisely, with partial flocks of deficiency one. Similarly, translation planes with spreads in $PG(3, q)$ admitting Baer groups of order $q - 1$ are equivalent to partial flocks of hyperbolic quadrics of deficiency one. In this chapter, we generalize some of these connections.

In this chapter, we adopt a slightly different representation method for α-flocks and their associated translation planes, hereafter called 'flokki planes,' the name being coined by Tim Penttila, celebrating his heritage (Finnish).

We then are interested in extending the ideas of deficiency one flocks of quadratic cones in $PG(3, q)$ and the associated Baer group theory of translation planes with spreads in $PG(3, q)$ to their generalizations flocks of α-cones and the corresponding 'flokki' planes.

As we have pointed out previously, the Payne-Thas theorem in the odd order case involves the idea of derivation of a conical flock, whereas the proof in the even order case used ideas from extensions of k-arcs. However, a proof attributed to Peter Sziklai of the Payne-Thas proof is independent of order. Here, it is realized that this proof may be adapted to prove the same theorem for α-flocks. The Baer group theory that applies is then an extension of the work of Johnson [124].

We also consider q-cones in $PG(3, q^2)$ and algebraically lifting of spreads in $PG(3, q)$. Such lifted spreads automatically give rise to q-flocks of q-cones, and hence are a-flokki (also see Kantor and Penttila [158]). A 'bilinear' flock of an α-cone is a flock whose associated planes of intersections are incident with two lines. A result of Thas [180] shows that flocks of quadratic cones whose planes share a point must be linear and hence bilinear flocks of quadratic cones do not exist, at least in the finite case. However, Biliotti and Johnson [23], show that bilinear flocks can exist in $PG(3, K)$, where K is a infinite field. In fact, the situation is much more complex for infinite flocks; for example, there are n-linear flocks for any positive integer n (the planes share exactly n lines). Recently, Cherowitzo [34] found an extremely interesting bilinear q-flock using ideas from blocking sets. It might be suspected that the Cherowitzo q-flokki might be algebraically lifted from a translation plane, and indeed, this is the case. Since all of the results are direct generalizations of the analogous situations for conical flocks, we give only sketches of the proofs or omit the proofs entirely and direct the reader to the articles by Cherowitzo and Johnson [**33**], [**30**],[**31**], [**32**].

0.2. Elation Groups and Flokki Planes. Let π denote a translation plane of order q^2 with spread in $PG(3, q)$ that admits an elation group E such that some orbit Γ union the axis is a derivable partial spread. We know from Johnson [**132**] that the derivable net may be represented in the form

$$\left\{ x = 0, y = x \begin{bmatrix} u & 0 \\ 0 & u^\alpha \end{bmatrix} ; u \in GF(q) \right\},$$

where α is an automorphism of $GF(q)$, when we choose the axis of E to be $x = 0$, and Γ to contain $y = 0$ and $y = x$, where $x = (x_1, x_2)$, $y = (y_1, y_2)$, for $x_i, y_i \in GF(q)$, $i = 1, 2$ and vectors in the 4-dimensional vector space over $GF(q)$ are (x_1, x_2, y_1, y_2). Since Γ is an orbit, this means that E has the form

$$\left\langle \begin{bmatrix} 1 & 0 & u & 0 \\ 0 & 1 & 0 & u^\alpha \\ 0 & 0 & 1 & 0 \\ 0 & 0 & 0 & 1 \end{bmatrix} ; u \in GF(q) \right\rangle.$$

Let $y = x \begin{bmatrix} g(t) & f(t) \\ t & 0 \end{bmatrix}$ be a typical component for $t \in GF(q)$, with the $(2, 2)$-entry zero, where g and f are functions on $GF(q)$, where $g(0) = f(0) = 0$. This is always possible by a basis change allowing that $y = 0$ and $y = x$ represent components of the translation plane. Hence, the group action of E on $y = x \begin{bmatrix} g(t) & f(t) \\ t & 0 \end{bmatrix}$ produces components

$$y = x \begin{bmatrix} u + g(t) & f(t) \\ t & u^\alpha \end{bmatrix} ; t, u \in GF(q).$$

Now assume that π is a translation plane with spread in $PG(3, K)$, where K is an infinite field, and π admits an elation group such that the axis and some orbit Γ is a derivable partial spread. We have seen in the material on Baer groups and derivable nets that a derivable net has the form

$$\left\{ x = 0, y = x \begin{bmatrix} u & A(u) \\ 0 & u^\alpha \end{bmatrix} ; u \in GF(q) \right\},$$

where α is an automorphism of K and such that $\left\{ \begin{bmatrix} u & A(u) \\ 0 & u^\alpha \end{bmatrix} ; u \in K \right\}$ is a field. If there are at least two Baer subplanes incident with the zero vector that are K–subspaces, then $A \equiv 0$.

DEFINITION 81. *A translation plane π with spread in $PG(3,K)$, where K is a field is said to be an 'α-flokki plane' if and only if there are functions g and f on K so that*

$$(*) : y = x \begin{bmatrix} u + g(t) & f(t) \\ t & u^\alpha \end{bmatrix} ; t, u \in K$$

is the spread for π, and α is an automorphism of K.

If there is a representation $(*)$ of a partial spread in $PG(3,K)$, then this partial spread is a maximal partial spread. In this case, the maximal partial spread is said to be an 'injective but not bijective' partial spread.

It is left to the reader to ascertain why the partial spread is actually maximal.

Thus, we have immediately

THEOREM 121. *A translation plane π with spread in $PG(3,K)$, for K a field, is an α-flokki plane if and only if there is an elation group E of one of whose orbits is a derivable partial spread containing at least two Baer subplanes that are K–subspaces.*

1. Maximal Partial Flokki and α-Flocks

In this section, we connect α-flokki translation planes and α-flokki maximal partial spreads (which are injective but not bijective) with α-flocks. The ideas presented here originate from Cherowitzo [34], and Kantor-Penttila [158], in the finite case.

DEFINITION 82. *Let K be any field and let α be an automorphism of K. Considering homogeneous coordinates (x_0, x_1, x_2, x_3) of $PG(3,K)$, we define the 'α-cone' as $x_0^\alpha x_1 = x_2^{\alpha+1}$, with vertex $v_0 = (0,0,0,1)$. An 'α-flock' is a covering of non-vertex points of the α-cone by plane sections. The intersections are called 'α-conics.'*

Now we show that there are maximal partial spreads in $PG(3,K)$, associated with an α-flock, which are 'α-flokki partial spreads' or simply 'flokki' partial spreads, if the context is clear. When K is finite, these partial spreads are the α-flokki spreads. The proof is omitted again, but it might be noted that when $\alpha = 1$, that is, in the conical flock situation, it can be shown that dual spreads are always maximal using the Thas-Walker construction utilizing the Klein quadric (see Chapter 42).

THEOREM 122. *(1)(a) The set of injective α-flokki maximal partial spreads*

$$x = 0, y = x \begin{bmatrix} u + g(t) & f(t) \\ t & u^\alpha \end{bmatrix} ; u, t \in K,$$

for g and f functions from K to K, is equivalent to injectivity of each of the corresponding sets of functions

$$\phi_u : t \to t - u^{\alpha+1} f(t)^\alpha + u g(t)^\alpha,$$

for all $u \in K$.
    (1)(b)

$$x = 0, y = x \begin{bmatrix} u + g(t) & f(t) \\ t & u^\alpha \end{bmatrix} ; u, t \in K,$$

is injective if and only if

$$x = 0, y = x \begin{bmatrix} u + g(t) & t \\ f(t) & u^\alpha \end{bmatrix} ; u, t \in K,$$

is injective.
    (2)(a) The injectivity of each function ϕ_u is equivalent to a partial α-flock with defining equations for the planes as follows:

$$\rho_t : x_0 t - x_1 f(t)^\alpha + x_2 g(t)^\alpha + x_3 = 0 \text{ for all } t \in K,$$

when representing the α-cone as $x_0^\alpha x_1 = x_2^{\alpha+1}$, with vertex $v_0 = (0, 0, 0, 1)$.
    (2)(b) The two sets of functions

$$\mathcal{F} = \{\phi_u : \phi_u : t \to t - u^{\alpha+1} f(t)^\alpha + u g(t)^\alpha, \text{ for all } u \in K\}$$

$$\mathcal{F}^\perp = \{\phi_u^\perp : \phi_u^\perp : t \to f(t) - u^{\alpha+1} t^\alpha + u g(t)^\alpha, \text{ for all } u \in K\}$$

both consist of injective functions if and only if one set consists of injective functions.
    (3) An α-flock is obtained if and only if ϕ_u is bijective for all $u \in K$.
    (4) When K is finite, the set of α-flokki translation planes is equivalent to the set of flocks of the α-cone.
    (5) If $\alpha^2 = 1$ and $g(t) = 0$, and ϕ_u is bijective, then for K finite or infinite, this subset of α-flokki planes is equivalent to the corresponding set of flocks of the α-cone.

The following corollary is essentially immediate.

COROLLARY 31. *Injective partial α-flocks are equivalent to injective maximal partial α-flokki.*

By the previous theorem, an α-flock will produce a maximal partial α-flokki but to ensure that α-flocks and α-flokki are equivalent we need the concept of a dual spread, which we have previously given when discussing semifield spreads but shall be repeated somewhat here for convenience of the reader. Given a spread $\mathcal{S}$ in $PG(3, K)$, for K a field, applying a polarity $\perp$ to $PG(3, K)$ transforms $\mathcal{S}$ to a set of lines $\mathcal{S}^\perp$, with the property that each plane of $PG(3, K)$ contains exactly one line of $\mathcal{S}^\perp$, which is the definition of a 'dual spread.' In the finite case, dual spreads are also spreads, which may be seen by an easy counting argument. However, when K is infinite, there are spreads that are not dual spreads and dual spreads that are not spreads.

DEFINITION 83. *Let K be a field and α an automorphism of K. Choose functions f and g on K and consider the set of functions*

$$\mathcal{F} = \{\phi_u : \phi_u : t \rightarrow t - u^{\alpha+1}f(t)^\alpha + ug(t)^\alpha, \text{for all } u \in K\}.$$

Then we define the 'dual of $\mathcal{F}$,' $\mathcal{F}^\perp$ as follows:

$$\mathcal{F}^\perp = \{\phi_u^\perp : \phi_u^\perp : t \rightarrow f(t) - u^{\alpha+1}t^\alpha + ug(t)^\alpha, \text{for all } u \in K\}.$$

Assume that both $\mathcal{F}$ and $\mathcal{F}^\perp$ consist of bijective functions then there are corresponding α-flocks by Theorem 122. In this case, we shall also say that $\mathcal{F}$ and $\mathcal{F}^\perp$ are flocks, and furthermore use the terminology that the 'dual α-flock is an α-flock' in this context.

REMARK 40.

$$\{\phi_u : \phi_u : t \rightarrow t - u^{\alpha+1}f(t)^\alpha + ug(t)^\alpha, \text{for all } u \in K\}$$

is a set of bijective functions if and only if

$$\{\Gamma_u : \Gamma_u : t \rightarrow tu^{\alpha+1} - f(t)^\alpha + u^\alpha g(t)^\alpha, \text{for all } u \in K\}$$

is a set of bijective functions. Similarly,

$$\{\phi_u^\perp : \phi_u^\perp : t \rightarrow f(t) - u^{\alpha+1}t^\alpha + ug(t)^\alpha, \text{for all } u \in K\}$$

is a set of bijective functions if and only if

$$\{\phi_u^\perp : \Gamma_u^\perp : t \rightarrow f(t)u^{\alpha+1} - t^\alpha + u^\alpha g(t)^\alpha, \text{for all } u \in K\}$$

is a set of bijective functions.

PROOF. Simply note that $u^{\alpha+1}\Gamma_{u^{-1}} = \phi_u$.   □

The idea of applying a polarity to a spread in $PG(3, K)$ will produce a dual spread, which in the α-flokki case, may not be a spread. It turns out that the dual spread may be coordinatized by the transpose of matrices defining the spread. In other words, if

$$\pi = \left\{ x = 0, y = x \begin{bmatrix} u + g(t) & f(t) \\ t & u^\alpha \end{bmatrix} ; u, t \in K \right\}$$

is an α-flokki spread then the dual spread $\pi^\perp$ is (isomorphic to)

$$\pi^\perp = \left\{ x = 0, \begin{bmatrix} u + g(t) & t \\ f(t) & u^\alpha \end{bmatrix} ; u, t \in K \right\}.$$

We call $\pi^\perp$ the 'transposed spread' of π to avoid confusion with the dual of a projective plane. The connections are as follows:

THEOREM 123. *Let K be a field and let $\mathcal{F}$ be an injective partial α-flock, denoting the maximal partial α-flokki by $\pi_\mathcal{F}$.*

Then $\mathcal{F}$ is bijective and the dual flock is bijective if and only if $\pi_\mathcal{F}$ is a spread and the transposed spread $\pi_\mathcal{F}^\perp$ is a spread.

Again, we omit the argument and ask that the reader attempt the proof.

REMARK 41. *Every α-flokki plane is isomorphic to an α^{-1}-flokki plane. In particular,*

$$\begin{bmatrix} u + g(t) & f(t) \\ t & u^\alpha \end{bmatrix} \text{ and } \begin{bmatrix} u + g(t)^\alpha & t \\ f(t) & u^{\alpha^{-1}} \end{bmatrix}$$

give isomorphic flokki planes.

The transpose of this α^{-1}-flokki plane is determined by

$$\begin{bmatrix} u + g(t)^\alpha & f(t) \\ t & u^{\alpha^{-1}} \end{bmatrix}.$$

2. Deficiency One and Baer Groups

Any finite α-flock consists of q planes of intersection. A partial α-flock with $q - 1$ planes of intersection is said to be of 'deficiency one.' When $\alpha = 1$, the Payne-Thas theorem shows that there is a unique intersection. We extend this theorem to α-flocks, simply noting that the proof of Peter Sziklai is valid also in the situation, and the reader is invited to re-read the proof of Payne-Thas theorem in this more general setting.

THEOREM 124. *A finite deficiency one α-flock may be extended to a unique α-flock.*

2.1. Baer Groups on α-Partial Flokki.

Now the Baer group theory for translation planes with spreads in $PG(3, q)$ shows that Baer groups of order q produce partial flocks of quadratic cones of deficiency one. Given any conical flock plane, there is an elation group E, whose orbits union the axis form reguli in $PG(3, q)$. Derivation of one of these regulus nets produces a translation plane with spread in $PG(3, q)$ admitting a Baer group of order q. Now consider any given α-flock and corresponding α-flokki plane. Again, there is an elation group E, whose orbits union the axis form derivable partial spreads. However, derivation of one of the derivable nets produces a translation plane admitting a Baer group of order q, but the spread for this plane π^* is no longer in $PG(3, q)$, for $\alpha \neq 1$. The components not in the derivable net are still subspaces in $PG(3, q)$, as are the Baer subplanes of the derivable net of π^*. Therefore, we would expect that Baer groups of order q in such translation planes might also produce deficiency one partial α-flocks.

THEOREM 125. *Let π be a translation plane of order q^2 that admits a Baer group B of order q. Assume that the components of π and the Baer axis of B are lines of $PG(3, q)$.*

(1) Then π corresponds to a partial α-flock of deficiency one.

(2) Therefore, the Baer partial spread defined by the Baer group is derivable and the Baer subplanes incident with the zero vector are also lines in $PG(3, q)$. The derived plane is the unique α-flokki translation plane associated with the extended α-flock.

The reader is directed to Cherowitzo and Johnson [**31**], noting that the argument parallels that of the original proof for Baer groups in spreads in $P(3, K)$.

3. K-Flokki and Algebraic Lifting

We have that a q-flock in $PG(3, q^2)$ has an associated flokki plane with spread

$$x = 0, y = x \begin{bmatrix} u + g(t) & f(t) \\ t & u^q \end{bmatrix} ; t, u \in GF(q^2).$$

More generally we define K-flocks.

DEFINITION 84. *Let F be a quadratic extension field of the field K, and let σ denote the unique involution in $Gal_K F$. We call any maximal partial spread of the form*

$$x = 0, y = x \begin{bmatrix} u + g(t) & f(t) \\ t & u^\sigma \end{bmatrix} ; t, u \in F,$$

a K-flokki partial spread and a K-flokki spread if the maximal partial spread is, indeed, a spread.

Now we ask when a α-flokki plane is an algebraic lifted plane of a plane with spread in $PG(3, q)$. We know (see, e.g., Johnson, Jha and Biliotti [**138**], Theorem 35.18 for the finite case) that a translation plane with spread in $PG(3, q^2)$ is an algebraically lifted plane if and only if the plane has an elation group E of order q^2, one of whose orbits together with the elation axis is a derivable partial spread and a Baer group B of order > 2, such that E and B do not centralize each other and B normalizes E.

It turns out that existence of such a Baer group implies that the order is divisible by $q + 1$. By appropriate coordinate change, we may assume that

$$\left\langle \begin{bmatrix} 1 & 0 & 0 & 0 \\ 0 & e & 0 & 0 \\ 0 & 0 & e & 0 \\ 0 & 0 & 0 & 1 \end{bmatrix} ; e^{q+1} = 1 \right\rangle$$

is B. Any elation group E may be coordinatized then in the following form:

$$E = \left\langle \begin{bmatrix} I_2 & M \\ 0_2 & I_2 \end{bmatrix} ; M \in \mathcal{A} \right\rangle,$$

where $\mathcal{A}$ is an additive group of 2×2 matrices, and the non-zero matrices are non-singular. For $M = \begin{bmatrix} a & b \\ c & d \end{bmatrix}$ then

$$\begin{bmatrix} 1 & 0 \\ 0 & e^{-1} \end{bmatrix} \begin{bmatrix} a & b \\ c & d \end{bmatrix} \begin{bmatrix} e & 0 \\ 0 & 1 \end{bmatrix} = \begin{bmatrix} ae & b \\ c & de^{-1} \end{bmatrix} = \begin{bmatrix} ae & b \\ c & de^q \end{bmatrix}.$$

Since the α-flokki planes admit an elation group of order q of the correct type, it follows that from the α-flokki plane is an algebraically lifted plane if and only if $G(s) = 0$, $\alpha = q$, where the from for spread must then be, by an appropriate change, $G(r) = 0$.

$$x = 0, y = x \begin{bmatrix} u + G(r) & F(r) \\ r & u^q \end{bmatrix} ; r, u \in GF(q^2).$$

Hence, we obtain (see Handbook, 35.18, p. 268):

THEOREM 126. *A translation plane with spread*

$$x = 0, y = x \begin{bmatrix} u + G(s) & F(s) \\ s & u^\alpha \end{bmatrix} ; s, u \in GF(q^2),$$

is algebraically lifted if and only if there is a coordinate change so that $G(s) = 0$, $\alpha = q$, *for all r in $GF(q)$ and for all e of order dividing $q + 1$.*

4. Net Replacement in the Hughes-Kleinfeld Planes

First, we consider the q-Flokki of Cherowitzo (see Cherowitzo [34]). We have a flokki plane with spread

$$x = 0, y = x \begin{bmatrix} u & \gamma(s^q)^{(q^2+1)/2} \\ s & u^q \end{bmatrix} ; s, u \in GF(q^2),$$

where q is odd and γ is a non-square such that γ^2 is a non-square in $GF(q)$. Therefore, this is a lifted plane. The corresponding flock of the q-cone turns out to be bilinear. Actually, the following may be proved using the Thas, Bader-Lunardon Theorem (see Chapter 16 for the statement of this theorem).

THEOREM 127. *The flokki planes of Cherowitzo of order q^4 may be lifted from regular nearfield planes of order q^2.*

Now realizing the regular nearfield planes are André planes, we may also realize other bilinear flocks in q-cones, in both odd and even order as follows: Consider an André plane of odd order. Recalling that we have a Desarguesian affine plane of order q^2 with spread

$$x = 0, y = xm; m \in GF(q^2).$$

The André partial spreads A_β are

$$\left\{ y = xm; m^{q+1} = \beta \right\},$$

which have replacement partial spreads (the derived partial spreads)

$$A_\beta^q = \left\{ y = x^q m; m^{q+1} = \beta \right\}.$$

Now choose any subset of $GF(q)^*$, λ and construct an André plane with spread as follows:

$$\begin{aligned} x &= 0, y = x^q m; m^{q+1} \in GF(q) - \lambda, \\ y &= xn; n^{q+1} \in \lambda. \end{aligned}$$

Let $\{1, \theta\}$ be a $GF(q)$-basis for $GF(q^2)$, where $\theta^2 = \theta\alpha + \beta$. Let the matrix spread set representing the Desarguesian affine plane be

$$x = 0, y = x \begin{bmatrix} u + \alpha t & \beta t \\ t & u \end{bmatrix}; u, t \in GF(q).$$

Represent $x = x_1\theta + x_2$, for $x_i \in GF(q)$, $i = 1, 2$ and note that $x^q = -x_1\theta + x_2 = (x_1, x_2) \begin{bmatrix} -1 & \alpha \\ 0 & 1 \end{bmatrix}$. Also, recall that if $m = \begin{bmatrix} u & \beta t \\ t & u \end{bmatrix}$, then $m^{q+1} = u^2 + \alpha u t - \beta t^2$. Note that $y = x^q m = x \begin{bmatrix} -u & -\beta t + \alpha t \\ t & u \end{bmatrix}$.

Hence, the André spread is

$$x = 0, y = x \begin{bmatrix} u + \alpha t & \beta t \\ t & u \end{bmatrix}; u, t \in GF(q),$$

when for $u^2 + \alpha u t - \beta t \in \lambda$ and

$$x = 0, y = x \begin{bmatrix} -u & -\beta t + \alpha u \\ t & u \end{bmatrix}; u, t \in GF(q),$$

for $u^2 - \beta t \notin \lambda$. Now assume that $q - 2 > |\lambda| > 1$.

Now algebraically lift to the plane of order q^4 with spread:

$$x \;=\; 0, y = x \left[\begin{array}{cc} w & (\theta(u + \alpha t) + \beta t)^q \\ \theta t + u & w^q \end{array} \right];$$

$$u, t \;\in\; GF(q), \; w \in GF(q^2),$$

for $u^2 + \alpha u t - \beta t \;\in\; \lambda$ and

$$y \;=\; x \left[\begin{array}{cc} w & (\theta(-u) - \beta t + \alpha u)^q \\ \theta t + u & w^q \end{array} \right]$$

for $u^2 + \alpha u t - \beta t \;\notin\; \lambda$.

Now note that $\theta(u + \alpha t) + \beta t = \theta(\theta t + u)$ and $\theta(-u) - \beta t + \alpha u = (-\theta + \alpha)(\theta t + u)$. The corresponding flock of the q-cone is given by the planes:

$$\pi_{(\theta t+u)} \;:\; x_0(\theta t + u) - \theta(\theta t + u)x_1 + x_3; u^2 + \alpha u t - \beta t \in \lambda$$

$$\rho_{(\theta t+u)} \;:\; x_0(\theta t + u) - (-\theta + \alpha)(\theta t + u)x_1 + x_3; u^2 + \alpha u t - \beta t \notin \lambda.$$

In general, the sets of planes $\pi_{(\theta t+u)}$ for all t, u in $GF(q)$ and $\rho_{(\theta t+u)}$ are linear flocks; it follows that the q-flokki plane is a bilinear flock plane.

THEOREM 128. *Any André plane order q^2 that is constructed by the multiple derivation of λ André nets, when $q - 2 > |\lambda| > 1$ algebraically lifts to a q-flokki plane of order q^4 whose corresponding flock in the q-cone is bilinear.*

REMARK 42. *For the Cherowitzo q-flokki, take $\theta = \gamma$, $\alpha = 0$, and $\beta = \gamma^2$, a non-square in $GF(q)$.*

The truly wild thing is that from the Hughes-Kleinfeld semifield planes there is a very interesting net replacement that actually constructs the entire set of André planes.

We note that

$$x \;=\; 0, y = x \left[\begin{array}{cc} w & \theta^q(\theta t + u)^q \\ \theta t + u & w^q \end{array} \right];$$

$$u, t \;\in\; GF(q), \; w \in GF(q^2)$$

and

$$x \;=\; 0, y = x \left[\begin{array}{cc} w & (-\theta + \alpha)^q(\theta t + u)^q \\ \theta t + u & w^q \end{array} \right];$$

$$u, t \;\in\; GF(q), \; w \in GF(q^2)$$

are both Hughes-Kleinfeld semifield spreads. Therefore, we see that we obtain a net replacement in one of these by the partial spread in the other (either corresponding to λ or $GF(q) - \lambda$). What this means is that there are net replacements in certain Hughes-Kleinfeld planes that are lifted planes that retract to André planes.

To see exactly what is going on, suppose we see what happens when exactly one André net A_δ is derived to A_δ^q. Then we obtain the following q-flokki spread, noting that $\theta^q = -\theta + \alpha$,

$$x \; - \; 0, y = x \begin{bmatrix} w & \theta^q(\theta t + u)^q \\ \theta t + u & w^q \end{bmatrix};$$

$$u, t \; \in \; GF(q), \; w \in GF(q^2), \; for \; u^2 + \alpha ut - \beta t \neq \delta \; and$$

$$y \; = \; x \begin{bmatrix} w & \theta(\theta t + u)^q \\ \theta t + u & w^q \end{bmatrix} \; for \; u^2 + \alpha ut - \beta t = \delta.$$

But in the original Hughes-Kleinfeld semifield plane there are exactly $q^2(q+1)$ components left, which then must be covered by one partial spread of the same degree $q^2(q+1)$ of the second Hughes-Kleinfeld semifield plane. This gives the following result. However, the reader is directed to the article by Cherowitzo and Johnson [33] for the unbelievably simple net replacement of the Hughes-Kleinfeld semifield planes that produces the entire set of André planes of dimension two.

THEOREM 129. *(1) For any $\delta \in GF(q)$, the partial spread of degree $q^2(q+1)$ given by*

$$\mathcal{H}_\delta : y = x \begin{bmatrix} w & \theta^q(\theta t + u)^q \\ \theta t + u & w^q \end{bmatrix} \; for \; (\theta t + u)^{q+1} = \delta$$

has a replacement partial spread of degree $q^2(q+1)$ given by

$$\mathcal{H}_\delta^* : y = x \begin{bmatrix} w & \theta(\theta t + u)^q \\ \theta t + u & w^q \end{bmatrix} \; for \; (\theta t + u)^{q+1} = \delta.$$

(2) Given the Hughes-Kleinfeld semifield plane with spread π

$$x \; = \; 0, y = 0, y = x \begin{bmatrix} w & \theta^q(\theta t + u)^q \\ \theta t + u & w^q \end{bmatrix};$$

$$u, t \; \in \; GF(q), w \in GF(q^2),$$

u, t, w not all zero, then

$$\pi = \{x = 0, y = 0\} \cup_{\delta \in GF(q)} \mathcal{H}_\delta,$$

and each $\mathcal{H}_\delta$ may be replaced by $\mathcal{H}_\delta^*$ or not replaced to create a set of $2^{(q-1)}$ q-flokki planes, of which exactly two are Hughes-Kleinfeld semifield planes. The remaining $2^{q-1} - 2$ q-flokki planes produce bilinear flocks of the cone $\mathcal{C}_q$.

(3) Each of these planes may be retracted to construct an André plane with spread in $PG(3, q)$ and each André plane may be constructed in this manner.

There is even an infinite version of this theorem and the reader is directed to the article by Cherowitzo and Johnson [**31**] for the statement and more details.

4.1. The Group Permuting the α-Derivable Nets. Consider a q-flokki plane with spread

$$x = 0, y = x \begin{bmatrix} u + b^q t^q & a^q t^q \\ t & u^q \end{bmatrix}, u, t \in GF(q).$$

Then the corresponding flock of the α-cone is linear; that is, the planes of the flock share a line.

When we have a finite α-flokki plane

$$x = 0, y = x \begin{bmatrix} u + g(t) & f(t) \\ t & u^\alpha \end{bmatrix} ; u, t \in GF(q),$$

the question is when is the elation group

$$E = \left\langle \begin{bmatrix} I & \begin{bmatrix} u & 0 \\ 0 & u^\alpha \end{bmatrix} \\ 0 & I \end{bmatrix} ; u \in GF(q) \right\rangle$$

normalized by the full group. This is always the case for $q > 5$, unless the conical flock plane is Desarguesian. Since Desarguesian planes correspond to linear flocks, we would guess that the Hughes-Kleinfeld planes would take the plane of Desarguesian in this setting. And, in fact, this is what occurs. Again, the reader is directed to the article [**31**] by Cherowitzo and the author. The reader is also directed to the Handbook [**138**] (91.22)) for discussion of the Hughes-Kleinfeld semifield planes.

The following are the main results established in the article by Cherowitzo and the author.

THEOREM 130. *Let π be an α-flokki plane of order q^2, where $q > 5$. Then one of the following occurs:*

(1) the full collineation group permutes the q α-derivable nets,

(2) $\alpha = 1$ and the plane is Desarguesian or

(3) $\alpha \neq 1$ and the α-flokki plane is a Hughes-Kleinfeld semifield plane that corresponds to a linear flock.

Using these ideas, isomorphisms between α and δ-flokki can be obtained.

THEOREM 131. *Let π_1 and π_2 be α-flokki and δ-flokki planes, respectively, of order q^2.*

If π_1 and π_2 are isomorphic, then $\delta = \alpha^{\pm 1}$.

Much more can be said of α-flokki and flocks of α-cones and again the interested reader is directed to the work of Cherowitzo and the author for many other ideas and directions for research.

CHAPTER 21

Parallelisms of Quadric Sets

In this chapter, we consider the possibility of covering the conics of a quadric set. This material is loosely based on the article by Cherowitzo and the author [**32**] and modified for this text.

Normally, we would consider a quadric set in $PG(3, K)$, for K a field, as either an elliptic quadric, a hyperbolic quadric, or a quadratic cone. In this chapter, we consider also α-cones as well and could further consider oval-cones or ovoids and ask in what sense one could consider a parallelism of such geometric objects.

Naturally, there are some problems when trying to consider a general definition of a parallelism. If $\mathcal{G}$ is an elliptic quadric, hyperbolic quadric, or quadratic cone, we would require the plane intersections to be conics. But, if $\mathcal{G}$ is an ovoid, which is not an elliptic quadric, the plane intersections would only be required to be ovals. If $\mathcal{G}$ is an oval-cone, the plane intersections would be also be ovals, but if $\mathcal{G}$ is an α-cone, we would merely require the plane intersections not to contain a generating line (this would also work as a requirement for an oval-cone).

There are some additional complications in the infinite case for ovoids, since a flock could conceivably miss either 0, 1, or 2 points of $\mathcal{G}$. If a parallelism is a set of flocks, should we distinguish between flocks that miss 0 points and those that miss 2 points within the set of flocks of a parallelism? Initially, at least, we shall choose to ignore this potential problem. Also, we might wish to consider 'maximal partial parallelisms' , and again we delay such discussions. We shall also be considering flocks of Minkowski planes in a later chapter.

DEFINITION 85. *The type of planes of $PG(3, K)$ discussed above shall be called the 'planes of intersection' .*

DEFINITION 86. *Let $\mathcal{G}$ be either an ovoid, hyperbolic quadric, or oval-cone or α-cone in $PG(3, K)$, where K is a field. In all of these structures $\mathcal{G}$, a 'flock' is a set of planes of intersection of $PG(3, K)$, which are mutually disjoint and cover the points of $\mathcal{G}$.*

Consider first the finite case and K isomorphic to $GF(q)$. The number of planes in $PG(3,q)$ is $1 + q + q^2 + q^3$. We know that an elliptic quadric or ovoid has $q^2 + 1$ points and a non-tangent plane intersection has $q + 1$ points, therefore a flock would consist of $q - 1$ planes. Now in order that a parallelism exist we would require that $q - 1$ must divide the number of planes of intersection, which is $q + q^3$, but $(q-1, q+q^3) = (q-1, 2)$, so that the only possibility is when $q = 3$, so we are considering flocks and parallelisms of an elliptic quadric in $PG(3,3)$. We actually know by the Thas-Walker Theorem (see Chapter 42) that have associated translation planes of order 9, which, in fact, must be Desarguesian for a flock to exist. In this setting, each of the flocks of the elliptic quadric will turn out to share a line of $PG(3,3)$, and such flocks are said to be 'linear' . We would thus require 15 flocks to obtain a parallelism. We leave it to the reader to figure out why this can't work.

Therefore, parallelisms of finite ovoids cannot occur. However, we show later that there are really a vast number of infinite parallelisms of elliptic quadrics in $PG(3,K)$, where K is an infinite field.

Now consider a finite hyperbolic quadric in $PG(3,q)$, consisting of $(q + 1)^2$ points. Since a plane of intersection will now be a conic of $q + 1$ points, we see that we would require $q + 1$ flocks that are mutually disjoint on planes of intersection to obtain a parallelism. This might be a good time to mention the beautiful theorem of Thas, Bader-Lunardon, which classifies all flocks of finite hyperbolic quadrics. By the Thas-Walker construction (see Chapter 42), there are associated translation planes, for which the classification theorem actually proves are always nearfield planes with spreads in $PG(3,q)$. We also know that these planes all admit regulus-inducing homology groups of order $q-1$. It turns out that there are three irregular nearfield planes of orders 11^2, 23^2, and 59^2 that admit regulus-inducing homology groups of order $q - 1$, we know from the material on flocks of hyperbolic quadrics is equivalent to having a flock of a hyperbolic quadric. The results on the irregular nearfields were independently discovered by Bader [5], Johnson [128], and for order 11^2 and 23^2 by Baker and Ebert [4].

1. The Thas, Bader-Lunardon Theorem

THEOREM 132. *A flock of a hyperbolic quadric in $PG(3,q)$ is either*
(1) linear, corresponding to the Desarguesian affine plane,
(2) a Thas flock, corresponding to the regular nearfield planes, or
(3) a Bader-Baker-Ebert-Johnson flock of order p^2, corresponding to the irregular nearfield planes of orders 11^2, 23^2, and 59^2.

As noted by Bonisoli [25], using the Thas, Bader-Lunardon Theorem, it is possible to see that every flock of a hyperbolic quadric lies in a transitive parallelism, as every sharply 1-transitive set constructing an associated nearfield translation plane is a group. So, all finite flocks of hyperbolic quadrics lie in a transitive parallelism. But, suppose we were trying to consider parallelisms in general or we were not aware of the Thas, Bader-Lunardon Theorem, is there a 'geometric' way to see that there are parallelisms of the hyperbolic quadric in $PG(3, K)$?

2. Parallelisms of Hyperbolic Quadrics

Now just as in the finite case, we try to find what might be called a 'transitive parallelism' of a hyperbolic quadric in $PG(3, K)$, regardless of the field K. Note again in the finite case that we will need $q(q - 1)$ flocks, so we shall be looking for a group analogous to a group of order $q(q - 1)$. However, we will consider this generally over any field, but this will tip off the reader on what our group might look like.

Let F be a flock of the hyperbolic quadric $x_1 x_4 = x_2 x_3$ in $PG(3, K)$, whose points are represented by homogeneous coordinates (x_1, x_2, x_3, x_4) where K is a field. Then the set of planes that contain the conics in F may be represented as follows:

$$\rho \quad : \quad x_2 = x_3,$$
$$\pi_t \quad : \quad x_1 - t x_2 + f(t) x_3 - g(t) x_4 = 0$$

for all t in K, where f and g are functions of K such that f is bijective. We first point out that there is a natural collineation group of the hyperbolic quadric

$$G = \left\langle \tau_{a,b} = \begin{bmatrix} 1 & b & 0 & 0 \\ 0 & a & 0 & 0 \\ 0 & 0 & 1 & b \\ 0 & 0 & 0 & a \end{bmatrix} ; a \neq 0, b \in K \right\rangle.$$

Note that in the finite case, this is a group of order $q(q - 1)$. To see that the group leaves invariant the hyperbolic quadric, we note that a point (x_1, x_2, x_3, x_4) maps to $(x_1, x_2 a + x_1 b, x_3, x_4 a + x_3 b)$ so if the point is on the hyperbolic quadric then $x_1 x_4 = x_2 x_3$, and the image point is on the hyperbolic quadric if and only if $x_1(x_4 a + x_3 b) = (x_2 a + x_1 b) x_3$, which is clearly valid. Now consider the hyperbolic quadric as a regulus net. This insight will mean that the group G will end up fixing a Baer subplane of this regulus net pointwise. We can now give the proof of the following remarkable result.

THEOREM 133. *Let F be any flock of a hyperbolic quadric in $PG(3, K)$, for K a field. Then there is a transitive parallelism containing F:*

$$\boldsymbol{P}_F = \{F\tau_{a,b}; \tau_{a,b} \in G\}.$$

PROOF. So, $\tau_{a,b}$ will map the flock F onto another flock $F\tau_{a,b}$ and $\boldsymbol{P}_F$ is a parallelism of the hyperbolic quadric if and only if F and $F\tau_{a,b}$ share no plane.

Now the Baer subplane Σ_0 fixed pointwise by G is a ruling line of the hyperbolic quadric and hence each plane of the flock intersects Σ_0 in exactly one point. Now consider any plane η of F and assume that η contains the point P of Σ_0, then the image of η also contains the point P.

This will mean that $\eta\tau_{a,b}$ cannot belong to F, unless $\eta\tau_{a,b} = \eta$. Indeed, $\rho\tau_{a,b}$ is the plane

$$\langle(1, b, 0, 0), (0, a, 1, a), (0, 0, 0, 1)\rangle,$$

which is clearly not π_t, for any t and is ρ is and only if $a = 1$ and $b = 0$, so that $\tau_{1,0}$ is the identity mapping.

A basis for π_t is

$$\{(-f(t), 0, 1, 0), (t, 1, 0, 0), (g(t), 0, 0, 1)\},$$

which maps under $\tau_{a,b}$ to

$$\{(-f(t), -f(t)b, 1, b), (t, tb + a, 0, 0), (g(t), g(t)b, 0, a)\}.$$

If $\pi_t\tau_{a,b}$ is ρ, then b cannot be 0, which implies that $g(t) = 0$. Now consider the associated spread

$$y = x\begin{bmatrix} f(t)u & g(t)u \\ u & tu \end{bmatrix}, \quad y = x\begin{bmatrix} v & 0 \\ 0 & v \end{bmatrix},$$

$$x = 0, \text{ for all } t, v, u \neq 0 \in K.$$

If $g(t) = 0$, then subtracting the matrices

$$\begin{bmatrix} f(t) & 0 \\ 1 & t \end{bmatrix} - \begin{bmatrix} t & 0 \\ 0 & t \end{bmatrix} = \begin{bmatrix} f(t) - t & 0 \\ 1 & 0 \end{bmatrix},$$

which cannot be the case. So, finally, assume that $\pi_t\tau_{a,b}$ is π_s and by the above note, we may assume that $s = t$. Therefore, we obtain the following requirements:

$$-f(t) - sf(t)b + f(s) - g(s)b = 0,$$
$$t - s(tb + a) = 0,$$
$$g(t) - sg(t)b - g(s)a = 0.$$

If $tb + a = 0$, then $t = 0$ but then $(t, tb + a, 0, 0)$ is the zero vector. Hence, $s = t/(tb + a) = t$, so that $tb + a = 1$. Therefore, $f(t)(tb) =$

$g(t)b$. If $b \neq 0$ then $f(t)t - g(t) = 0$, a contradiction to the fact that
$\begin{bmatrix} f(t) & g(t) \\ 1 & t \end{bmatrix}$ is non-singular. Hence, $\boldsymbol{P}_F$ is a partial parallelism.
But, given a ruling line ℓ that is left invariant by G, the group G acts doubly transitively on the non-fixed points of ℓ.

It now follows that each point of each ruling line of the hyperbolic quadric is covered by $\boldsymbol{P}_F$.

Choose two conics C_1 and C_2 and assume that C_1 is the conic of intersection of a plane of F and assume without loss of generality that C_1 and C_2 share point P_1 on Σ_0. Let Q_1 and R_1 be points on ruling lines ℓ_1 and ℓ_2 of C_1 and let Q_2 and R_2 be points on lines ℓ_1 and ℓ_2 of C_2, where P_1, Q_1, R_1 uniquely defines C_1 and P_1, Q_2, R_2 uniquely defines C_2. Let $\tau \in G$ map Q_1 to Q_2 (even if $Q_1 = Q_2$). If $R_1\tau = R_2$, then τ maps C_1 to C_2. We note that $R_1\tau$ and Q_2 cannot be in the same Baer subplane, since $P_1, R_1\tau$ and Q_2 lie on a conic. Similarly, C_2 is a conic, then R_2 is not incident with P_1Q_2. Therefore, there is a collineation subgroup of G that fixes Σ_0 and fixes the point Q_2 and acts transitively on the points on $\ell_1 - \{Q_2, \ell_1 \cap \Sigma_0\}$. This means that τ' may be considered a Baer collineation that fixes the 1-dimensional subspace generated by P_1, fixes Q_2 and maps $R_1\tau$ to R_2. Then $\tau\tau'$ will map C_1 to C_2. Hence, given any conic C_2, there is a conic C_1 of F (of the conics of intersection), and a group element g of G so that $C_1g = C_2$. Hence, we obtain a parallelism in $PG(3, K)$, for any field K. □

3. Bol Planes

In the infinite case, we wondered if there are non-nearfield planes corresponding to flocks of hyperbolic quadrics, and this led to considering the theorem on hyperbolic quadrics in a much more different light. So, we ask again how the theorem was proven in the finite case?

The amazing insight of Jef Thas is the key element here, where Thas showed that given a finite hyperbolic flock F in $PG(3, q)$, given any conic C of F, there is an involution that fixes the conic pointwise. Sound familiar? Probably not! However, if we try to understand what this means in the associated translation plane, we see that there is an affine involutory homology that interchanges the two common components of the regulus nets in question, and which fixes a component pointwise of the regulus net corresponding to C. In the language of translation planes, this means that the translation plane is a Bol plane —this was how Bader and Lunardon realized that the theorem on hyperbolic flocks would hinge on proving that Bol planes are always nearfield planes.

The Bol planes are a class of planes investigated by Burn [28] in the general case. In the finite case, Bol planes of order q^2 were reasonably well known during the time that Thas was working on this theory, and all planes with spreads not in $PG(3, q)$, with a couple of exceptions, were known or proved to be nearfield planes. We now actually know that the theory has now been completed and 'finite Bol planes are nearfield planes' (e.g., see Johnson [102] and Draayer, Johnson, Pomareda [49]). But, back to the situation at hand: we are interested in Bol planes with spreads in $PG(3, q)$, where Bader and Lunardon [6], were able to complete this theory in $PG(3, q)$.

All Bol planes (defined using involutory homologies) satisfy the Bol axiom: $a(b \cdot ac) = (a \cdot ba)c$, for all a, b, c in an associated quasifield coordinatizing the plane. Of course, any nearfield satisfies this axiom, as nearfields have associative multiplication, as they are, of course, multiplicative groups.

However, when K is an infinite field, there are a great variety of interesting hyperbolic flocks and, consequently, transitive hyperbolic parallelisms. The most basic question that can be raised in general, is: Are the translation planes corresponding to flocks of an infinite hyperbolic quadric always Bol planes, or nearfield planes? Since all but three of the finite nearfield planes that are Bol planes are also André planes, it might be possible to find André planes that admit regulus-inducing homology groups that are not nearfield but are Bol, and/or to find such planes that Bol and not nearfield. Indeed, there are ruled planes corresponding to hyperbolic flocks that are not Bol.

Let K be a field that contains non-squares, let γ be a non-square and let $F = K[\sqrt{\gamma}]$. We consider the Pappian spread given by $x = 0, y = xm; m \in F$ and define an André partial spread A_δ as follows:

$$A_\delta = \left\{ y = xm; m^{1+\sigma} = \delta \in K^- \right\},$$

where $K^- = F^{1+\sigma}$ and where σ is the involution in $Gal_K F$. A_δ is a regulus and has opposite regulus

$$A_\delta^D = \left\{ y = x^\sigma m; m^{1+\sigma} = \delta \in K^- \right\}.$$

In this setting, for spreads in $PG(3, K)$, an André translation plane is any plane that is obtain by derivation or multiple derivation of André partial spreads (is sub-regular). Hence, we have the following multiplication of an André plane:

$$x * m = x^{\sigma^{(m^{1+\sigma}g)}} m;$$

where g is any mapping from K^- into Z_2, where $K^- = F^{*(1+\sigma)}$ and $m \in F$, such that $1g = 0$ (so that $y = x$ is a component of each André net.

In order to describe the set of André planes that produce flocks of hyperbolic quadrics, we need to find those that are invariant under a regulus-inducing affine homology group $H : \langle(x, y) \rightarrow (x, yu); u \in K^*\rangle$. Since A_δ under H is mapped to $A_{\delta u^2}$, as $(m^{1+\sigma}u) = m^{1+\sigma}u^2$), then if $m^{1+\sigma}g = 1$, A_δ is replaced by A_δ^D, which implies that we must also replace $A_{\delta u^2}$, for all $u \in K^*$ in order to preserve the action of the group.

Assume that $\{1, e\}$ is a K-basis for F so that for $m = e\alpha + \beta$, for $\alpha, \beta \in K$, we see that $m^{1+\sigma} = \beta^2 + \gamma\alpha^2$. For example, when K is the field of real numbers, K^- is the set of all positive elements and hence, H acts transitively on the set of all André nets, so that we would not obtain a hyperbolic flock in this case.

Let S denote the set of squares of K^- and note that K^- is a multiplicative group. We consider the quotient group K^-/S. Noting any $k \in K^-$, $m^{1+\sigma} = \beta^2 + \gamma\alpha^2 = k$, then $k^2 = m^{2(1+\sigma)} \in S$. Therefore, K^-/S is an elementary Abelian 2-group, which we may then consider as a $GF(2)$-vector space. A nearfield André multiplication is given if and only if $(x * m) * n = (x * (m * n))$, if and only if

$$(x * m) * n = x^{\sigma^{(m^{1+\sigma}g)}} m * n = (x^{\sigma^{(m^{1+\sigma}g)}})^{\sigma^{(n^{1+\sigma}g)}} m^{\sigma^{(n^{1+\sigma}g)}} n$$

and

$$(x * (m * n)) = x * (m^{\sigma^{(n^{1+\sigma}g)}} n) = x^{\sigma^{((m^{\sigma^{(n^{1+\sigma}g)}}n))^{1+\sigma}g)}} m^{\sigma^{(n^{1+\sigma}g)}} n.$$

Now we require that

$$(x^{\sigma^{(m^{1+\sigma}g)}})^{\sigma^{(n^{1+\sigma}g)}} = x^{\sigma^{((m^{\sigma^{(n^{1+\sigma}g)}}n))^{1+\sigma}g)}}.$$

Note that $(m^{\sigma^{(n^{1+\sigma}g)}})^{1+\sigma} = m^{1+\sigma}$, since $\sigma = 1$. Hence, we have

$$\sigma^{(m^{1+\sigma}g)}\sigma^{(n^{1+\sigma}g)} = \sigma^{(mn)^{1+\sigma}g},$$

which implies that

$$(m^{1+\sigma}g) + (n^{1+\sigma}g) = (m^{1+\sigma}n^{\sigma+1})g.$$

This means that g is a homomorphism from K^-/S into $GF(2)$.

THEOREM 134. *Let $F = K[\sqrt{\gamma}]$, where γ is a non-square in a field K. Let $K^- = F^{*(1+\sigma)}$, where σ is the involution in $Gal_K F$ and let S denote the non-squares in K^-. Consider K^-/S as a $GF(2)$-vector space.*

(1) The set of André translation planes that correspond to flocks of a hyperbolic quadric in $PG(3, K)$ is in $1 - 1$ correspondence with the set of functions g from K^-/S into $GF(2)$, such that $1g = 0$.

(2) The set of André nearfield translation planes that correspond to flocks of a hyperbolic quadric in $PG(3, K)$ is the set of linear functionals from K^-/S into $GF(2)$.

COROLLARY 32. *The dimension of K^-/S over $GF(2)$ is $log_2(K^-/S)$. The number of non-nearfield hyperbolic flock planes constructed as in the previous theorem is $2^{|K^-/S|-1} - 2^{\log_2(K^-/S)}$.*

Hence, if the dimension of K^-/S is at least 2, there are non-nearfield André planes that produce hyperbolic flocks in $PG(3, K)$.

REMARK 43. *The reader may also verify that all André quasifields presented here are also Bol quasifields. It is also true that the original Bol planes of Burn are André planes (for additional details, the reader is directed to Johnson [117]).*

Now the question of whether there exist translation planes that produce hyperbolic quadrics but which are not Bol planes is answered by some examples of Riesinger [177].

3.1. The Hyperbolic Flocks of Riesinger. Consider the hyperbolic quadric $x_1 x_4 = x_2 x_3$ in $PG(3, K)$, whose points are represented by homogeneous coordinates (x_1, x_2, x_3, x_4), where K is the field of real numbers. Then the set of planes that contain the conics in F may be represented as follows:

$$\rho \; : \; x_2 = x_3,$$
$$\pi_t \; : \; x_1 - tx_2 + f(t)x_3 - g(t)x_4 = 0,$$

where

$$f(t) = t(t^2 + \alpha t + 1)/(t^2 + \alpha + 1) \text{ and } g(t) = -f(t)/t,$$

where α is a real number of absolute value less that 0.08 (see Johnson [117], section 6). To see that the associated translation plane is not a Bol plane, we note that this basically is a coordinate problem in that a Bol quasifield may not be the quasifield that it used in the definition of the translation plane. However, it is possible to show that in the translation plane with spread

$$y = x \begin{bmatrix} f(t)u & g(t)u \\ u & tu \end{bmatrix}, \; y = x \begin{bmatrix} v & 0 \\ 0 & v \end{bmatrix},$$
$$x = 0, \text{ for all } t, v, u \neq 0 \in K,$$

the components $x = 0$ and $y = 0$ are either fixed or interchanged by the full collineation group, which means that if a coordinate quasifield is a Bol quasifield then it is a Bol quasifield with respect to $x = 0$ and $y = 0$; every component is the axis of an affine involution interchanging $x = 0$ and $y = 0$. But it is not difficult to show that the latter cannot occur. Hence, the flocks of Riesinger do not correspond to a Bol translation plane.

We finally consider parallelisms of α-cones. The reader is directed to the open problem chapter for an open question on oval-cones.

4. Parallelisms of α-Cones

In this section, we turn to the consideration of parallelisms of conical flocks. We will generally follow the article by Cherowitzo and the author [32] with suitable changes for this text.

We have previously shown that given any flock of a hyperbolic quadric in $PG(3, K)$, for K any field, there is always a transitive parallelism that contains the given flock. Here we are also cognizant of the idea of a transitive parallelism, and there is a group, but this time the group will not leave the cone invariant.

Given any flock of an α-cone, by Theorem 122, there is a corresponding flokki translation plane. Therefore, it follows that a parallelism of an α-flock can be determined by a set of flokki spreads of the form

$$x = 0, y = x \left[\begin{array}{cc} u + g(t) & f(t) \\ t & u^{\alpha} \end{array} \right] ; u, t \in K,$$

if and only if the α-flock is

$$x_0 t - x_1 f(t)^{\alpha} + x_2 g(t)^{\alpha} + x_3 = 0 \text{ for all } t \in K,$$

when representing the α-cone as $x_0^{\alpha} x_1 = x_2^{\alpha+1}$, with vertex $(0, 0, 0, 1)$.

On the other hand, this will not be an essential point of our proof. What matters mostly is that although it certainly is possible to initially assume or choose the functions f and g so that $f(0)$ or $g(0)$ is 0, this is clearly not required for the existence of a flock. Normally, we would allow that $y = 0$ is a component, which ensures that the matrices associated with the components are non-singular. However, this is not required for the connections. What is intended is there is a group that acts transitively on the set of functions that describe either the flock or the α-flokki translation plane. So, when we say we obtain a transitive parallelism, the understanding will be that the group acts on the sets of pairs of associated functions. In the finite case, there are a total of $1 + q + q^2 + q^3$ planes in $PG(3, q)$ and the number of planes that share the vertex is $1 + q + q^2$, so we would need q^2 flocks of q planes to form a parallelism, so we will be looking for a general analogue of a group of order q^2.

Now, in particular, for points $(z_2^{1+\alpha}, 1, z_2, \delta)$, $(1, 0, 0, \tau)$, for $\delta, \tau \in K$, we see that

$$tz_2^{1+\alpha} - f(t)^\alpha + z_2 g(t)^\alpha$$

is required to be bijective for all z_2 in K and this is necessary and sufficient for the existence of an α-flock. We note that

$$\phi_u : t \to tu^{1+\alpha} - f(t)^\alpha + ug(t)^\alpha$$

is bijective if and only if

$$\phi_u : t \to tu^{1+\alpha} - (f(t) + a)^\alpha + u(g(t) + b)^\alpha$$

is bijective, for all $a, b \in K$.

The group G that we are looking for has the following elements:

$$\tau_{a,b} = \begin{bmatrix} I & \begin{bmatrix} a & b \\ 0 & 0 \end{bmatrix} \\ 0 & I \end{bmatrix}$$

(note the order of the group containing these elements has order q^2 in the finite case). We consider the action on the set of functions (f, g). The group G will map

$$x = 0, y = x \begin{bmatrix} u + g(t) & f(t) \\ t & u^\alpha \end{bmatrix} ; u, t \in K,$$

to

$$x = 0, y = x \begin{bmatrix} u + g(t) + a & f(t) + b \\ t & u^\alpha \end{bmatrix} ; u, t \in K,$$

so the functions $(f(t), g(t))$ defining the α-flock become $(f(t)^\alpha + a^\alpha, g(t)^\alpha + b^\alpha)$. We note translation planes map to translation planes under $\tau_{a,b}$, which means that α-flocks map to α-flocks under $\tau_{a,b}$. We represent the planes $\pi_{a,b,c}$, for $a, b, c \in K$, that do not contain the vertex in the form

$$x_0 c + x_1 b + x_2 a + x_3 = 0.$$

An α-flock is then represented in the form

$$x_0 t - x_1 f(t) + x_2 g(t) + x_3 = 0,$$

for all $t \in K$. We also have an associated flokki translation plane π with spread

$$x = 0, y = x \begin{bmatrix} u + g(t) & f(t) \\ t & u^\alpha \end{bmatrix} ; u, t \in K.$$

Now consider the group

$$G = \left\langle \tau_{a,b} = \begin{bmatrix} I & \begin{bmatrix} a & b \\ 0 & 0 \end{bmatrix} \\ 0 & I \end{bmatrix} ; a, b \in K \right\rangle.$$

We note that $\pi \tau_{a,b}$ is a flokki plane isomorphic to π. Therefore, there is a corresponding α-flock. Note that $\pi \tau_{a,b}$ has the following spread

$$x = 0, y = x \begin{bmatrix} u + g(t) + a & f(t) + b \\ t & u^{\alpha} \end{bmatrix} ; u, t \in K.$$

Clearly, none of the associated derivable nets in $\pi \tau_{a,b}$ can be equal to the derivable nets of π. Therefore, none of the planes

$$x_0 t - x_1 (f(t)^{\alpha} + b^{\alpha}) + x_2 (g(t)^{\alpha} + a^{\alpha}) + x_3 = 0,$$

are equal to any of the planes of the α-flock, and also the associated image is also an α-flock. Since all planes that do not contain the vertex have the form

$$x_0 c + x_1 b + x_2 a + x_3 = 0,$$

we see that we have partitioned the α-conics by the α-flocks associated with πG. Hence, each α-flock belongs to a transitive parallelism—a remarkable fact!

Our main theorem on parallelisms of flocks of quadratic cones or of α-cones is the following:

THEOREM 135. *Every α-flock in $PG(3, K)$, for K a field, is in a transitive parallelism.*

REMARK 44. *The reader interested in 'maximal partial parallelisms' of α-cones is directed to the article by Cherowitzo and the author [32].*

CHAPTER 22

Sharply k-Transitive Sets

The set of points, conics of intersection, and ruling lines of a hyperbolic quadric in $PG(3, K)$ as a geometry may be generalized to a 'Minkowski plane,' where flocks and parallelisms may be revisited in a more general context.

DEFINITION 87. *A Minkowski plane is a set of 'points' $\mathcal{P}$ and subsets of the power set of $\mathcal{P}$ called 'circles,' $+$ generators and $-$ generators with the following properties: (Two distinct points P and Q are $+$ parallel or $-$ parallel if and only if they are incident with a $+$ generator or a $-$ generator, respectively.)*

1. Given a point P, there is a unique $+$ generator and a unique $-$ generator incident with P.

2. Every generator ($+$ or $-$) intersects every circle in exactly one point, and every $+$ generator and every $-$ generator uniquely intersect in a point.

3. Every three distinct, mutually non-parallel points are incident with a unique circle.

4. Let c be a circle and let P and Q be points incident with c and not incident with c, respectively. Then there exists a unique circle t incident with both P and Q such that $t \cap c = \{P\}$.

5. There is a circle c containing at least three points and not all points are incident with c.

In the hyperbolic quadric, of course, the circles are the conics of plane intersection.

In the finite case, a Minkowski plane may be identified with a sharply 3-transitive set $\mathcal{C}$ of mappings on a set $\mathcal{P}$. Where the circles are the sets $\{(x, x\tau); \tau \in \mathcal{C}\}$. Of course, every finite affine plane is equivalent to a sharply 2-transitive set of mappings. In particular, consider a finite translation plane π of order q^r, with a spread set

$$x = 0, y = 0, y = xM; M \in \mathcal{M},$$

where $\mathcal{M}$ is a set of $q^r - 1$ nonsingular matrices over $GF(q)$, and where the difference of each distinct pair of elements is also nonsingular. Considering each matrix M as a mapping from the x-axis to the y-axis and furthermore identifying these sets of points, we have that $\mathcal{M}$ is a sharply 1-transitive set of mappings on a set $\mathcal{P}$. Then if $\mathcal{T}$ is the translation group, $\mathcal{MT}$ is a sharply 2-transitive set of mappings that becomes the set of lines. So, if we have a sharply 2-transitive set of mappings, on a set of $q^r = n$ elements, we may define a 'flock' as a sharply 1-transitive set, as this corresponds to a set of lines that cover the points other than a 'zero' point. Then a 'parallelism' is a set of mutually disjoint flocks that cover the line set. In the case of an affine plane, the set of flocks is the set of lines incident with a particular point P, and this parallelism has basically nothing to do with the natural set of parallel classes of lines.

More generally, we define flocks and parallelisms of sharply k-transitive sets as follows.

DEFINITION 88. *Let X be a set and let $\mathcal{S}_k$ denote a sharply k-transitive set of mappings of X, for $k \geq 2$. A 'flock' is a sharply 1-transitive subset of $\mathcal{S}_k$ and a 'parallelism' is a set of disjoint flocks whose union is $\mathcal{S}_k$.*

We define the 'point set' as $X \times X$, and the 'block set' as the sets $\{(x, x\tau); \tau \in \mathcal{S}_k\}$.

DEFINITION 89. *When $k = 3$, we may define a $+$-parallelism as follows: (x, y) and (x', y') are $+$ parallel if and only if $x = x'$, and similarly the points are $-$ parallel if and only if $y = y'$.*

For Minkowski planes obtained using sharply 3-transitive sets, finiteness is required to ensure axiom (4) of Definition 87 is satisfied, so this axiom must be added in the infinite case. Also, the reader should not confuse the use of the term 'parallelism' with $+$ or $-$ parallelism. The term 'resolution' is often used in place of our term parallelism.

1. Subsets of $P\Gamma L(n, K)$

Ultimately, the Minkowski planes that we wish to consider shall be based on sharply 3-transitive subsets G of $P\Gamma L(2, p^m)$, and we shall be considering a 'flock' as a sharply 1-transitive subset of G. So, before we consider parallelisms of finite Minkowski planes, we ask what other geometries can be associated with sharply transitive subsets of $P\Gamma L(2, q)$. There are $q(q^2 - 1)$ conics of intersection and $PGL(2, q)$ is sharply 3-transitive on the conics and a sharply 1-transitive set corresponds to a flock of a hyperbolic quadric, which in turn corresponds to a translation plane whose spread in $PG(3, q)$ is a union of $q + 1$ reguli that share two components, and which corresponds to a nearfield plane by the Theorem of Thas/Baer-Lunardon.

In [**163**], Knarr considered more generally if there are geometries associated with sharply 1-transitive subsets of $P\Gamma L(2, q)$, and, of course, one could consider this more generally for sharply 1-transitive subsets of $P\Gamma L(n, q)$, for $n \geq 2$. For this subject, the reader is also directed to the author's articles [**120**] and [**119**]. Our treatment in this text is more general, considering partially sharp subsets of $P\Gamma L(n, K)$, where K is a skewfield and n is a positive integer ≥ 2. When applying these results for infinite skewfields K and $n = 3$, the tangency axiom must be added.

We begin with the fundamental connection.

THEOREM 136. *Let K be any skewfield and let V be a left n-dimensional vector space over K. Assume that Λ is a partially sharp subset of $P\Gamma L(n, K)$ acting on the left 1-dimensional K–subspaces.*

(1) Then there is a partially sharp subset of $\Gamma L(n, K)$, which defines a translation net N_Λ over $K \oplus K \oplus ... \oplus K$, (2n summands). The net N_Λ is a union of translation nets N_i, $i \in \lambda$ that mutually share two components L and M.

(2) Each net N_i is a subplane covered translation net and corresponds to a pseudo-regulus in some projective space $PG(2n - 1, K_i)$, where $K_i \simeq K$.

(3) The net N_Λ admits a collineation group H_L that fixes the components L and M and fixes one of them pointwise (say L) and which acts sharply transitive on the set of components not equal to L and M of each net. If K is a field, there is also a related collineation group H_M that fixes the components L and M and fixes M pointwise.

PROOF. Let $g \in \Lambda$ and choose a preimage g^+ in $\Gamma L(n, K)$ and represent g^+ as follows:

$$(x_1, x_2, .., x_n) \to (x_1^{\sigma_T}, x_2^{\sigma_T}, .., x_n^{\sigma_T})T,$$

where T is in $GL(n, K)$, and σ_T is an automorphism of K, and $x_i \in K$. Now form

$$(x_1, x_2, .., x_n) \to uI_n(x_1^{\sigma_T}, x_2^{\sigma_T}, .., x_n^{\sigma_T})T, \ u \in K - \{0\}.$$

Note that the mapping

$$(x_1, x_2, .., x_n) \to uI_n(x_1, .., x_n), \ u \in K - \{0\}$$

is not necessarily linear but does fix all 1-dimensional left K–subspaces so the image is the identity in $P\Gamma L(n, K)$. We easily see that

$$\langle g^+, \tau_u; g \in \Lambda, u \in K - \{0\}\rangle,$$

where g^+ is any preimage of g in $\Gamma L(n, K)$, is a partially sharp set acting on the non-zero vectors of V.

Form the translation net on $V \oplus V$, with vectors

$$(x = x_1, .., x_n, y = y_1, y_2, .., y_n)$$

$$x = 0, y = uI_n x^{\sigma_T} T; u \in K,$$

where we are adjoining $x = 0$ and $y = 0$. Now change bases and represent this net, say, N_T as

$$x = 0, y = ux; u \in K.$$

The reader may easily verify that this is a subplane covered net and hence a pseudo-regulus net over the skewfield K. If we let L be denoted by $x = 0$, and M be denoted by $y = 0$, and

$$
\begin{aligned}
H_L \quad &= \quad \tau_u : (x_1, x_2, .., x_n y_1, y_2, .., y_n) \\
&\longrightarrow \quad (ux_1, ux_2, .., ux_n, y_1, .., y_n),
\end{aligned}
$$

for $u \in K - \{0\}$, then H_L will fix $x = 0$ pointwise.

Finally, assume that K is a field. Then we do not need to distinguish between left and right in K, which means that a component $y = (x_1^{\sigma_T}, .., x_n^{\sigma_T})T$ is a left and right K–subspace. This means that there are also collineations $\rho_u : (x, y) \rightarrow (x, yu)$, for $u \in K - \{0\}$. This proves that there are two homology groups when K is a field. $\qquad\square$

COROLLARY 33. *If Λ is a sharply 1-transitive subset of $P\Gamma L(n, K)$, acting on the left 1-dimensional subspaces of an n-dimensional vector space V over a skewfield K, then there is a sharply 1-transitive subset of $\Gamma L(n, K)$, acting on the non-zero vectors of V and a corresponding translation plane π_Λ, whose ambient vector space is $V \oplus V$, as a left space over K and whose spread is a union of subplane covered nets (pseudo-reguli) sharing two components.*

When the dimension $n = 2$, the spread is a union of derivable nets.

When the dimension $n = 2$ and Λ is in $PGL(2, K)$, then the spread is a union of reguli that share two components and hence corresponds to a flock of a hyperbolic quadric.

PROOF. Form the translation net π_Λ

$$x = 0, y = 0, y = uI_n x^{\sigma_T} T; u \in K - \{0\}, g^+ : x \rightarrow x^{\sigma_T} T; g \in \Lambda.$$

If we identify $x = 0$ and $y = 0$ as the set of 1-spaces being permuted by Λ, then given any left 1-space of $y = 0$, $\langle y_0 \rangle$ and any left 1-space of $x = 0$, $\langle x_0 \rangle$, there is a unique element h of Λ that maps $\langle x_0 \rangle$ to $\langle y_0 \rangle$. This means that given any vector (x_0, y_0), both x_0 and y_0 non-zero n-vectors, then $y_0 = u_0 I_n x_0^{\sigma_{T_0}} T_0$, for some unique non-zero u_0 and T_0 in the notation above. This implies immediately that the translation net is a translation plane. Now assume that $n = 2$, and choose any two pseudo-reguli

$$D_1 \;:\; x = 0, y = 0, y = uI_2 x^{\sigma T_1} T_1; u \in K - \{0\} \text{ and}$$
$$D_2 \;:\; x = 0, y = 0, y = uI_2 x^{\sigma T_2} T_2; u \in K - \{0\}.$$

If the reader rereads the definition of normalizing pseudo-reguli Definition 65, it becomes clear that we have a ruled translation plane in this case. $\qquad\qquad\qquad\qquad\qquad\qquad\qquad\qquad\qquad\qquad\qquad\qquad\qquad\quad$ □

2. Subsets in $P\Gamma L(n, q)$ of Deficiency 1

Now all of this theory can be generalized as follows: Instead of a sharply 1-transitive set of $q + 1$ elements of $P\Gamma L(2, q)$, take a partially sharp subset of q elements of $P\Gamma L(2, q)$. If the partially sharp set lies in $PGL(2, q)$, then we will end up with a partial flock of a hyperbolic quadric of deficiency one. The way this works is through the use of the group H_L, which now still acts on the associated translation net, but the net shall be interpreted so that the group acts as a Baer group. There are partial flocks of deficiency one that cannot be extended to flocks so that theorem of Thas/Bader-Lunardon does not apply in the sense that there are associated translation planes admitting a Baer group that are not derived from finite ruled translation planes. In this section, we develop the theory of deficiency 1 only in the finite case. We leave open the problem of considering this theory generally over skewfields and the reader is directed to the Open Problems Chapter in the Appendix for a hint on how to define deficiency 1 in the general case.

It turns out that for $n > 2$, partially sharp subsets of $(q^n - 1)/(q - 1) - 1$ elements of $P\Gamma L(n, q)$ can always be extended (Johnson[119]) so that Corollary 33 applies for the connection to translation planes. Hence, we consider here the case $n = 2$ and let Λ denote a partial sharp subset of q elements of $P\Gamma L(2, q)$ acting on $PG(1, q)$. From Theorem 136, there is a corresponding translation net of degree $q(q - 1) + 2$ consisting of derivable nets sharing two components, which we shall denote by $x = 0, y = 0$ and such that the translation net admits two cyclic groups of order $q-1$, $H_{x=0}$, and $H_{y=0}$, both of which fix $x = 0$ and $y = 0$, where $H_{x=0}$ fixes $x = 0$ pointwise and $H_{y=0}$ fixes $y = 0$ pointwise (since $GF(q)$ is commutative, the vector space is both a right and left $GF(q)$-space with the standard scalar multiplication). Now consider one of the derivable nets

$$N_T : x = 0, y = 0, y = uI_2 x^{\sigma T} T; \; u \in K - \{0\}.$$

This is a regulus net in

$$PG\left(3, \left\langle \begin{bmatrix} u & 0 & 0 & 0 \\ 0 & u & 0 & 0 \\ 0 & 0 & u^{\sigma_T} & 0 \\ 0 & 0 & 0 & u^{\sigma_T} \end{bmatrix} ; u \in K \right\rangle = K_{\sigma_T}\right),$$

and if $x^{\sigma_T}T - xM$, where M is a non-singular matrix, change bases by $(x, y) \rightarrow (x, yM^{-1})$ to represent N_T as a regulus over K. The Baer subplanes incident with the zero vector are thus 2-dimensional K–subspaces of the form

$$\pi_{a,b} = \{(a\alpha, b\alpha, a\beta, b\beta); \alpha, \beta \in K\},$$

where not both a and b are zero. Now change bases back to obtain the Baer subplanes of N_T in the form

$$\pi^*_{a,b} = \{(a\alpha, b\alpha, (a^{\sigma_T}\beta^{\sigma_T}, b^{\sigma_T}\beta^{\sigma_T})T); \alpha, \beta \in K\},$$

and notice that $\pi^*_{a,b}$ is also a 2-dimensional K–subspace. Hence, we have a translation net of degree $q(q-1)+2$ consisting of q derivable nets that are 2-dimensional K_{σ_T}-subspaces, where the Baer subplanes of the derivable nets are 2-dimensional K–subspaces, as well as 2-dimensional K_{σ_T}-subspaces, under different scalar product definitions.

Now there are $q + 1$ 1-dimensional K–subspaces on each of $x = 0$ and $y = 0$, and each pair generates a 2-dimensional K–subspace, of which there are $q(q+1)$ of these that define Baer subplanes within the various q derivable nets N_T. This leaves $q + 1$ of these 2-dimensional K–subspaces and it is easily verified (the reader should try this) that the union of these K–subspaces contains both $x = 0$ and $y = 0$ (covers the 1-dimensional K–subspaces on each of these two components). We claim that these new 2-dimensional subspaces are mutually disjoint from the translation net of degree $q(q-1)$

$$N_T : x = 0, y = 0, y = uI_2x^{\sigma_T}T; u \in K - \{0\}, \ g \text{ in } \Lambda,$$

using the previous notation. Let $\mathcal{N}$ denote this new set of $q + 1$ 2-dimensional K–subspaces.

To prove our assertion, suppose not. Let X_1 and Y_1 be the 1-dimensional K–subspaces on $x = 0, y = 0$, respectively. Then there is a 1-dimensional K–subspace W that lies in $\langle X_1, Y_1 \rangle$ and in one of the q derivable nets. Take the unique Baer subplane of this net that contains W. We may assume that this Baer subplane is $\langle X_2, Y_2 \rangle$, where X_2 and Y_2 are the 1-dimensional K–subspaces of intersection on $x = 0, y = 0$, respectively. Hence, we have that $W = \alpha X_1 + \beta Y_1 = \alpha^* X_2 + \beta^* Y_2$, for $\alpha, \beta, \alpha^*, \beta^* \in K$. Since $\alpha X_1 - \alpha^* X_2 = \beta^* Y_2 - \beta Y_1$ and the left-hand side is in $x = 0$, and the right-hand side is in $y = 0$, this implies that $X_1 = X_2$ and $Y_1 = Y_2$, since all are 1-dimensional K–subspaces. However, this is not the pairing of 1-dimensional K–subspaces in $\mathcal{N}$. Hence, $\mathcal{N}$ forms a partial spread of 2-dimensional K–subspaces that cover $x = 0, y = 0$. What this means is that we have constructed a translation plane of order q^2 that admits two Baer groups B_1 and B_2, which have the same component orbits and such that the non-trivial orbits of length $q - 1$ together with $Fix B_i$, $i = 1, 2$ form derivable nets. Hence, we have part of the following theorem.

THEOREM 137. *Let* Λ *be a partially sharp subset of* $P\Gamma L(2, q)$ *of* q *elements acting on* $PG(1, q)$.

(1) Then there is a corresponding translation plane of order q^2 *admitting two Baer groups of order* $q - 1$, *whose component orbits are identical and such that the non-trivial orbits union* $Fix B_i$, $i = 1, 2$ *define a set of* q *derivable nets.*

(2) The set Λ *may be extended to a sharply 1-transitive subset of* $P\Gamma L(2, q)$ *if and only if the net* N_B *of degree* $q + 1$ *that contains the Baer subplanes fixed by the Baer groups is a derivable net.*

PROOF. It remains to prove part (2). First, assume that the net is derivable. We have noted that the components of this net are 2-dimensional K–subspaces. Hence, if we coordinatize so that we have $x = 0, y = 0, y = x$ as components of the net N_B, then the partial spread for the net has the following form

$$x = 0, y = x \begin{bmatrix} u & 0 \\ 0 & u^\sigma \end{bmatrix} ; u \in K.$$

The reader might like to try to prove this result as an exercise (or look in Johnson [**132**]). Now derive this net to obtain the derived net in the following form:

$$x = 0, y = x^\sigma u; u \in K.$$

If we map back by a 2×2 matrix S with elements in K, then we have the following spread:

$$
\begin{aligned}
x &= 0, y = 0, y = x^{\sigma S} S u I_2, \\
y &= u I_n x^{\sigma T} T; u \in K - \{0\}, \ g^+ : x \to x^{\sigma T} T; g \in \Lambda.
\end{aligned}
$$

It is now clear that we have extended Λ to a sharply 1-transitive subset of $P\Gamma L(2, q)$.

Conversely, if Λ can be extended, there is a new derivable net and again we arrive at the spread in the form directly above. This completes the proof of the theorem.   □

The converse is also valid.

THEOREM 138. *Let π be a translation plane of order q^2 that admits two cyclic Baer groups B_1 and B_2 of order $q - 1$ that have identical component orbits. Then there is a field K isomorphic to $GF(q)$ and a partially sharp subset of q elements of $P\Gamma L(2, q)$ such that π may be reconstructed using the previous theorem.*

PROOF. We shall give a sketch of the proof with enough of the ideas so that the reader should be able to work through the proof. The reader is directed to Johnson [119] for additional details.

Since the Baer groups have the same component orbits, they lie in the same net $\mathcal{N}$ of degree $q + 1$. The reader is directed to the chapter on Baer groups, particularly when there are two Baer groups defining two Baer subplanes that lie in the same net. The pertinent theorem is Theorem 72. The analogous coordinate set up follows and the reader is encouraged to work through the details. We know under these conditions that both Baer subplanes are Desarguesian. Furthermore, if there are three Desarguesian Baer subplanes on the same net then the net is derivable, as the Baer group of order $q - 1$ will map at least one of the other two Baer subplanes onto a set of $q - 1$ Baer subplanes. Hence, we may assume that the two Baer subplanes are left invariant by both Baer groups. Let $q = p^r$, for p a prime. Therefore, we have the following representation for the net

$$\mathcal{N} = \{(x_1, x_2, y_1, y_2); x_i, y_i \ are \ r - vectors \ over \ GF(p),$$
$$for \ i = 1, 2\}.$$

Then

$$(i) FixB_1 = \{(0, x_2, 0, y_2); x_2, y_2 \ are \ r - vectors \ over \ GF(p)\}$$
$$= coFixB_2,$$
$$FixB_2 = \{(x_1, 0, y_1, 0); x_1, y_1 \ are \ r - vectors \ over \ GF(p)\}$$
$$= coFixB_1.$$

(ii) The components of $\mathcal{N}$ have the following form:

$$x = 0, y = x \begin{bmatrix} A_1 & 0 \\ 0 & A_4 \end{bmatrix},$$

where A_1 and A_4 are $2r \times 2r$ matrices over $GF(p)$. The sets $\{A_1\}$ and $\{A_4\}$ corresponding to the components of $\mathcal{N}$ are both irreducible and the respective centralizers K_1 and K_4 are fields of matrices isomorphic to $GF(q)$. It follows that K_1 is the kernel of $FixB_1$ and K_4 is the kernel of $FixB_2$. Let T be a matrix so that $T^{-1}K_1T = K_4$. Rechoose a basis so that $FixB_1 = (x = 0)$, $FixB_2$ is $(y = 0)$.

Initially, we had the following representation for the Baer groups:

$$B_1 = \left\langle \begin{bmatrix} I & 0 & 0 & 0 \\ 0 & C & 0 & 0 \\ 0 & 0 & I & 0 \\ 0 & 0 & 0 & C \end{bmatrix} ; C \in K_4^* \right\rangle,$$

$$B_2 = \left\langle \begin{bmatrix} D & 0 & 0 & 0 \\ 0 & I & 0 & 0 \\ 0 & 0 & D & 0 \\ 0 & 0 & 0 & I \end{bmatrix} ; D \in K_1^* \right\rangle.$$

After the basis change, we have B_1 fixing $x = 0$ pointwise, and B_2 fixing $y = 0$ pointwise, and the vectors have the form (x_1, x_2, y_1, y_2); $x_i \in K_4$ and $y_i \in K_1$. If σ is an isomorphism from K_1 to K_4, then we may define a scalar multiplication as follows $(x_1, x_2, y_1, y_2)D = (x_1 D^\sigma, x_2 D^\sigma, y_1 D, y_2 D)$, for $D \in K_1$. If we change bases again by $(x, y) \rightarrow (x, yT)$, we have essentially identified K_1 and K_4 as K.

We now have a translation net of degree $q(q - 1) + 2$ and of order q^2 of the following form:

$$x = 0, y = 0, y = xuIM,$$

where $\{M\}$ is a set of $(q - 1)$ non-singular matrices over $GF(p)$, whose distinct differences are also non-singular. We claim that this set determines a partially sharp subset of $P\Gamma L(2, K)$, where K is isomorphic to $GF(q)$. We also have two Baer groups that have identical orbits. What this means is that each matrix $y = xM$ is fixed by a group of order $q - 1$, of the form

$$\left\langle \begin{bmatrix} D & 0 & 0 & 0 \\ 0 & D & 0 & 0 \\ 0 & 0 & C & 0 \\ 0 & 0 & 0 & C \end{bmatrix} ; D, C \in K - \{0\} \right\rangle,$$

where C is a function of D. In fact, we have

$$\begin{bmatrix} D^{-1} & 0 \\ 0 & D^{-1} \end{bmatrix} M \begin{bmatrix} C & 0 \\ 0 & C \end{bmatrix} = M$$

so that

$$M \begin{bmatrix} C & 0 \\ 0 & C \end{bmatrix} M^{-1} = \begin{bmatrix} D^{-1} & 0 \\ 0 & D^{-1} \end{bmatrix}.$$

It follows easily that there is an associated automorphism σ_M of K associated with M so that

$$M \begin{bmatrix} C & 0 \\ 0 & C \end{bmatrix} M^{-1} = \begin{bmatrix} C^{\sigma_M} & 0 \\ 0 & C^{\sigma_M} \end{bmatrix}.$$

Now define $h(x) = xM$. We claim that h is a semi-linear mapping, i.e., in $\Gamma L(2, K)$. First, $h(xD) = x \begin{bmatrix} D & 0 \\ 0 & D \end{bmatrix} M = xM \begin{bmatrix} D^\sigma & 0 \\ 0 & D^\sigma \end{bmatrix} = xMD^\sigma$, and since the mapping is clearly additive and non-singular, we have that h is in $\Gamma L(2, K)$. If we realize M semilinearly over K, then $y = xM$ is $y = x^{\sigma_T} T$, where T is a 2×2 matrix with elements in K. We then obtain a set of $q - 1$ elements of $P\Gamma L(2, K)$, and it is immediately that this set is a partially sharp subset.   □

REMARK 45. *There are partially sharp subsets of q elements $PGL(2, q)$ that cannot be extended to sharply 1-transitive subsets of $PGL(3, q)$. These are the deficiency one hyperbolic flocks. Some of the examples actually can be extended to a sharply 1-transitive subset of $P\Gamma L(2, q)$. The reason for the extension is that in the corresponding translation plane of order q^2 admitting two Baer groups of order $q - 1$, the Baer net is actually derivable but the net is not a regulus net.*

With our work connecting translation planes with sharply 1-transitive subsets of $P\Gamma L(2, q)$, we now turn to the parallelisms of Bonisoli.

3. The Parallelisms of Bonisoli

Let $G(p^m, \sigma)$ for p an odd prime, and σ an automorphism of $GF(p^m)$, denote the subset of elements of $P\Gamma L(2, p^m)$ acting on $PG(1, q)$, which are defined by the following mappings:

$$x \to \frac{x^{\sigma^i} a + b}{x^{\sigma^i} c + d}; a, b, c, d \in GF(p^m);$$

where $i = 0$ or 1 exactly when $ad - bc$ is a square or non-square in $GF(p^m)$. $G(p^m, \sigma)$ is a sharply 3-transitive set of mappings and thus a Minkowski plane $\mathcal{M}(p^m, \sigma)$ is determined. Hence, there are $q(q^2 - 1)$ circles and a flock, is a 1-transitive set on $q + 1$ elements so a parallelism is a set of $q(q - 1)$ mutually disjoint flocks.

Consider now that preimage group in $\Gamma L(2, p^m)$ acting an a 2-dimensional $GF(q)$-subspace X and form $X \oplus X$, the 4-dimensional $GF(q)$-vector space. Every element of $G(p^m, \sigma)$ is an element of $GL(4, q)$. We see that for every flock of a Minkowski plane $G(p^m, \sigma)$, there is an associated translation plane of order q^2 that is covered by a set of $q + 1$ derivable nets sharing two components $x = 0, y = 0$, such that the translation plane admits two affine homologies of order $q - 1$ whose orbits define the derivable nets. If $m > 1$ and $p^m \equiv -1 \bmod 4$, then it turns out that $PSL(2, q)$ admits a sharply 1-transitive subgroup (Bonisoli [25]).

REMARK 46. *If* Λ *is a sharply 1-transitive group, then the associated translation plane is a nearfield plane.*

To obtain a sharply 1-transitive subset of $P\Gamma L(2, q)$, which is not a group, we require that σ is not of order 2. The idea is to take a subgroup E of order $(q + 1)/2$ of $PSL(2, q)$, then find an element g of $P\Gamma L(2, q)$ normalizing E so that $E \cup Eg$ is a sharply 1-transitive subset.

THEOREM 139. *Let E be a semi-regular subgroup of $PSL(2, q)$ of order $(q + 1)/2$, g an element of $P\Gamma L(2, q)$ in $G(p^m, \sigma) - PSL(2, q)$ that normalizes E such that $E \cup Eg$ is a sharply 1-transitive subset of $P\Gamma L(2, q)$, then there is a transitive parallelism of $G(p^m, \sigma)$ containing $E \cup Eg$.*

PROOF. The main idea of the proof is that the image set of $E \cup Eg$ under $PSL(2, q)$ has precisely $q(q - 1)$ images, which are mutually disjoint. Each set defines a flock so that there are $q(q - 1)$ mutually disjoint flocks, thereby giving a transitive parallelism. We see that the order of $PSL(2, q)$ is $q(q^2 - 1)/2$ and has E has index $q(q - 1)$ in $PSL(2, q)$. Since the image set is an orbit, assume that $k \in (E \cup Eg)h \cap (E \cup Eg)$. Let $k = eg^j = e'g^ih$, where e is in E and j and $i \in \{0, 1\}$ and $h \in PSL(2, q)$. Assume that $j = i = 1$, so that $h \in E$. Since g normalizes E, then $(E \cup Eg)h = (E \cup Eg)$, this is also true if $j = i = 0$. So, let $j = 1$ and $i = 0$. Then it follows that $g \in PSL(2, q)$, a contradiction and this completes the proof. □

The parallelisms found by Bonisoli occur when E is cyclic.

More generally, the translation planes corresponding to parallelisms of Minkowski planes using $E \cup Eg$ can be determined, and we shall consider the more general situation as follows, where the particular sharply 3-transitive set is not specified or may not exist.

THEOREM 140. *Let π be a translation plane of odd order q^2 arising from a sharply 1-transitive set of $P\Gamma L(2, q)$, of the form $E \cup Eg$, where E is a cyclic subgroup of $PSL(2, q)$ of order $(q + 1)/2$, and $g \in P\Gamma L(2, q) - PSL(2, q)$, which normalizes E. Let the preimage of E in $SL(2, q)$ be denoted by $\overline{E}$. Let g^+ be a preimage of g in $\Gamma L(2, q)$ of the form $(x_1, x_2) \to (x_1^\sigma, x_2^\sigma)T$, where $T \in GL(2, q)$, as a 2×2 matrix over $GF(q)$.*

(1) The spread for π is then

$$x = 0, y = 0, y = x^\sigma TeuI_2, y = xe'u'I_2;$$

$e, e' \in \overline{E}$ *and* $u, u' \in GF(q)^*$.

(2) π admits symmetric cyclic affine homology groups $H_{x=0}$ and $H_{y=0}$ of orders $(q^2 - 1)/2$ as follows:

$$H_{y=0} \;:\; \langle (x,y) \to (x, yeuI_2); e \in \overline{E}, u \in GF(q)^* \rangle ,$$
$$H_{x=0} \;:\; \langle (x,y) \to (xeuI_2, y); e \in \overline{E}, u \in GF(q)^* \rangle .$$

(3) π is a union of two Desarguesian nets of degree $(q^2 - 1)/2 + 2$ with partial spreads in $PG(3, K)$ and

$$PG \left(3, PG \left(3, \left\langle \begin{bmatrix} u & 0 & 0 & 0 \\ 0 & u & 0 & 0 \\ 0 & 0 & u^\sigma & 0 \\ 0 & 0 & 0 & u^\sigma \end{bmatrix} ; u \in K \right\rangle \right) \right) .$$

(4) π is an André plane of order $q^2 = h^{2w}$, with kernel containing $Fix\sigma$, where $Fix\sigma$ is isomorphic to $GF(h)$, and $q = h^w$. Indeed, the spread for π may be represented as follows:

$$x \;=\; 0, y = 0, y = x^\rho m; y = xn; \; m \text{ is a non-square,}$$
$$\text{and } n \text{ is a non-zero square in } GF(q^2),$$

where ρ is an automorphism of $GF(q^2)$, such that the restriction to $GF(q)$ is σ.

PROOF. Since E is cyclic, it follows that $\overline{E}GF(q)^*I_2$ is Abelian of order $(q^2 - 1)/2$. But since this group is in $GL(2, q)$, and g normalizes E, it follows immediately that we have two cyclic homology groups of order $(q^2 - 1)/2$. Consider first the partial spread

$$x = 0, y = 0, y = xe'u'I_2; e' \in \overline{E}, u \in GF(q)^*.$$

Since $H_{y=0}$ is cyclic, the generator is a q-primitive divisor of $(q^2 - 1)$, the centralizer becomes a field isomorphic to $GF(q^2)^*$ contains the field isomorphic to $GF(q)$ in question. Hence, there is a Desarguesian affine plane Σ coordinatized by $GF(q^2)$ that contains the partial spread. The same argument applies to

$$\Phi : x = 0, y = 0, y = x^\sigma TeuI_2; e \in \overline{E}, u \in GF(q)^*.$$

Now write the vector space with vectors $GF(q^2) \oplus GF(q^2)$, so that

$$(*) : \langle (x,y) \to (xd^2, yc^2), \text{ for all } c, d \in GF(q^2)^* \rangle$$

now represents the direct product of $H_{x=0}$ and $H_{y=0}$. We note that $y = (x_1^\sigma, x_2^\sigma)T$, is a $Fix\sigma$ subfield of $GF(q)$. We may represent the vector space by $GF(q^2) \oplus GF(q^2)$, and extending σ to an automorphism of $GF(q^2)$, we have the representation

$$y = x^\sigma M,$$

where x is in $GF(q^2)$, and σ is an automorphism of $GF(q^2)$, and M is a 2×2 matrix over $GF(q)$. Hence, we have that $Fix\sigma$ may be considered a kernel subfield of π. The kernel homology group of the Desarguesian plane with elements $(x, y) \rightarrow (kx, ky)$, for $k \in GF(q^2)^*$, maps $y = x^\sigma M$ to $y = x^\sigma k^{1-\sigma} M$, and since $k^{1-\sigma}$ is square, it follows that $y = x^\sigma k^{1-\sigma} M$ is a component of π, by $(*)$. Moreover, since $y = x^\sigma M$ intersects $y = x\omega f^2$, for some set of squares f^2. Let $Fix\sigma = GF(h)$, where $q = h^w$. Then $k^{1-\sigma}$ has order dividing $(q^2 - 1)/(h - 1) = (q + 1)((q - 1)/(h - 1))$. Therefore, $k^{1-\sigma}$ is in $\overline{E}GF(q)I_2$ and M normalizes this group. However, this implies that M is in $\Gamma L(1, q^2)$. Hence, we have that the component is of the form

$$y = x^p m;$$

$m \in GF(q^2)$, and since this is disjoint from

$$x = 0, y = 0, y = xe'u'I_2; e' \in \overline{E}, u \in GF(q)^*,$$

it follows that m is a non-square in $GF(q^2)$. This completes all parts of the theorem. $\square$

Hence, to find a parallelism of the Minkowski plane $G(p^m, \sigma)$, it is required to determine those André planes as in the previous theorem that produce a sharply 1-transitive set $E \cup Eg$. Actually, Bonisoli [25], provides examples of situations when this occurs and by the above analysis, it will turn out that Bonisoli will have completely determined all possibilities when E is cyclic of order $(q + 1)/2$. The reader is directed to Bonisoli's article for the explicit constructions.

CHAPTER 23

Transversals to Derivable Nets

In this chapter, we consider a transversal to a derivable net interpreted in context of the author's embedding theorem of derivable nets. We recall that the author's work on derivable nets shows that every derivable net is combinatorially equivalent to a 3-dimensional projective space over a skewfield K. More precisely, the points and lines of the net become the lines and points skew to a fixed line N. Now consider any affine plane π containing the derivable net and choose any line ℓ of π, which is not a line of the derivable net. This set of points then becomes a set of lines in the combinatorially equivalent structure. For example, if the order of π is finite q^2, then this provides q^2 lines and adjoining N, we have $q^2 + 1$ lines. Furthermore, Knarr [163] proved that, every line ℓ not belonging to the derivable net embeds to a set of lines in the projective space such this set union N becomes a spread $S(\ell)$ of $PG(3, K)$, which is also a dual spread.

Now generalize Knarr's insight but merely assume that we have a derivable net $\mathcal{D}$ that admits a transversal T. Basically, the same ideas will show that in the embedding, we also obtain a spread and hence a corresponding translation plane, but which is not necessarily a dual spread. Now since I was a student of T. G. Ostrom, I had studied what could be said of the geometries that one might obtain from a finite derivable net that admits a transversal. What happens is that there is a more-or-less direct construction of a dual translation plane. So, on the one hand, Knarr's ideas give a translation plane from the transversal to a derivable net and Ostrom's ideas give a dual translation plane. How are the two affine planes related? The word 'dual' translation plane might suggest that the two planes are duals of each other when considered as projective planes.

DEFINITION 90. *(1) A spread of a 3-dimensional projective space over a skewfield shall be said to be a 'transversal spread' if and only if it arises from a transversal to a derivable net by the embedding process mentioned above.*

(2) A transversal spread shall be said to be a 'planar transversal-spread' if and only if there is an affine plane π containing the derivable net such that the transversal to the derivable net is a line of π.

Now since transversal spreads seem to be quite special we would expect some nice classification. But it turns out that all spreads in a three-dimensional projective space are transversal spreads! And, yes, it does turn out that the two methods of extension of a derivable net and a transversal do produce affine planes whose projective planes are dual to each other!

DEFINITION 91. *Let $\mathcal{D}$ be a derivable net with transversal T. If there is a dual translation plane $\pi_{\mathcal{D}}^A$ extending $\mathcal{D} \cup \{T\}$, $\pi_{\mathcal{D}}^A$ is said to be an 'algebraic extension of $\mathcal{D}$.' If there is a translation plane $\pi_{\mathcal{D}}^G$ obtained by the embedding process using $\mathcal{D} \cup \{T\}$, $\pi_{\mathcal{D}}^G$ is said to be a 'geometric extension of $\mathcal{D}$.'*

1. Algebraic Extensions of Derivable Nets

As is well known, the concept of the derivation of a finite affine plane was conceived by Ostrom in the 1960s. During this period, one of the associated problems that Ostrom considered concerned the extension of the so-called derivable nets to either a supernet or to an affine plane. At that time, coordinate geometry was the primary model in which to consider extension questions. With a particular vector-space structure assumed for a derivable net, Ostrom ([**172**]) was able to show that any transversal to such a finite derivable net allowed its embedding into a dual translation plane. We now know from the embedding theory of the author's that the vector-space structure is also the model for a derivable net. So, we may consider all of this more generally over any skewfield K. This material generally follows the author's article [**109**] suitably modified for this text.

DEFINITION 92. *Let K be a skewfield and V a right 2-dimensional vector space over F. A 'vector-space derivable net' $\mathcal{D}$ is a set of 'points' $(x, y) \; \forall x, y \in V$ and a set of 'lines' given by the following equations:*

$$x = c, \; y = x\alpha + b \; \forall c, b \in V, \; \forall \alpha \in F.$$

DEFINITION 93. *A 'transversal' T to a net $\mathcal{N}$ is a set of net points with the property that each line of the net intersects T in a unique point and each point of T lies on a line of each parallel class of $\mathcal{N}$.*

A 'transversal function' f to a vector-space derivable net is a bijective function on V with the following properties:

(i) $\forall c, d, \; c \neq d$ of V, $f(c) - f(d)$ and $c - d$ are linearly independent,
(ii) $\forall \alpha \in F$ and $\forall b \in V$, there exists a $c \in V$ such that $f(c) = c\alpha + b$.

It is not difficult to show that transversals and transversal functions to vector-space derivable nets are equivalent, each giving rise to the other, and this is left to the reader to verify.

It should be noted that everything can be phrased over the 'left' side as well. That is, a 'right vector-space net' over a skewfield F is naturally a 'left vector-space net' over the associated skewfield F^{opp}, where multiplication • in F^{opp} is defined by $a • b = ba$ where juxtaposition denotes multiplication in F.

We generally consider a left vector space setting $x*m = m \circ x$ where the multiplication $m \circ x$ arises from the translation plane associated with the spread. If the coordinate quasifield for the translation plane is a 'left quasifield,' then the dual translation plane may be coordinatized by the 'right quasifield' $(K \oplus K, +, *)$. How this works geometrically is to consider a translation plane with spread $x = 0, y = xM$, $M \in \mathcal{M}$, as a projective plane and M is a 2×2 matrix over K. If the 2-vector m occurs in the second row of M, then $xM = x \circ m$. Now if we dualize the projective plane letting (∞) become the line at infinity ℓ_∞, then removal of this line produces the affine dual translation plane with coordinate right quasifield as above. In this instance, the coordinate structure for the dual translation plane becomes a right 2-dimensional vector space over K and $m = e * \alpha + \beta$, where $\{1, e\}$ is a right basis. Then $y = x * m + b = (x * e) * \alpha + x * \beta + b$ and with $f(x) = x * e$, we obtain the form demanded of the extension process. We have then given a sketch of the following result.

THEOREM 141. *Let $\mathcal{D}$ be a vector-space derivable net and let T be a transversal. Then there is a transversal function f on the associated vector space V such that $\mathcal{D}$ may be extended to a dual translation plane with lines given as follows:*

$$x = c, y = f(x)\alpha + x\beta + b \ \forall \alpha, \beta \in K \ and \ \forall b, c \in V.$$

Conversely, any dual translation plane whose associated translation plane has its spread in $PG(3, K)$ may be constructed from a transversal function as above.

Viewed in this manner, the similar definition of algebraic extension takes the following form:

DEFINITION 94. *Let $\mathcal{D}$ be a (right) vector-space derivable net with transversal function f then the dual translation plane with lines given by*

$$x = c, y = f(x)\alpha + x\beta + b \ \forall \alpha, \beta \in K \ and \ \forall b, c \in V$$

shall be called the 'algebraic extension' of $\mathcal{D}$ by f and the set of such shall be termed the set of 'algebraic extensions of $\mathcal{D}$.'

2. Geometric Extension of Derivable Nets

We begin by giving a generalization of Knarr's theorem. Since we are interested in the more general situation, we assume only that there is a transversal T to the derivable net $\mathcal{D}$, which defines a simple net extension $\mathcal{D}^{+T}$. In Johnson [**116**], it is pointed out that it is possible to embed any derivable net into an affine plane where the affine plane may not be derivable itself.

DEFINITION 95. *An algebraic extension is said to be a 'derivable algebraic extension' if and only if the associated dual translation plane is derivable.*

Hence, we distinguish between having a net extension and having a 'derivable-extension' by which we mean that each Baer subplane of the net remains Baer when considered within the extension net; each point is on a line of each subplane, taken projectively (the subplane structure is 'point-Baer') and each line is incident with a point of each subplane, taken projectively (the subplane structure is 'line-Baer'). In essence, we would merely require that T intersect each Baer subplane.

2.1. Knarr's Theorem on Geometric Extension.

THEOREM 142. *(see Knarr [**163**]) Let $\mathcal{D}$ be a derivable net and assume that T is a transversal to $\mathcal{D}$ defining a extension net $\mathcal{D}^{+T}$.*

(1) Then the points of T determine a spread $S(T)$ of lines in the projective space Σ associated with $\mathcal{D}$ that contains the special line N.

(2) If the net extension is a derivable-extension, then $S(T)$ is a dual spread.

(3) Conversely, if $S(T)$ is a dual spread, for each line of $T - \mathcal{D}$, then the net extension is a derivable extension.

PROOF. A point of $\Sigma - N$ is a line of the net $\mathcal{D}$ that must intersect T in a unique point as $\mathcal{D}^{+T}$ is a net. Hence, every point of $\Sigma - N$ is incident with a unique line of $S(T) - N$, and it thus follows that every point of Σ is incident with a unique line of $S(T)$. This proves (1).

Now assume that the net extension is a derivable extension. To show that $S(T)$ is a dual spread, we need to show that every plane contains a unique line of $S(T)$. Since the planes not containing N correspond to Baer subplanes of $\mathcal{D}$, the question becomes whether each Baer subplane shares a unique net-point of T. Since each line of the net that is not in $\mathcal{D}$ shares a point, taken projectively, with each Baer subplane, it is immediate that this point is affine; i.e., an actual point of T. Hence, each plane of Σ contains exactly one line of $S(T)$; $S(T)$ is a dual spread.

It also follows that if $S(T)$ is a dual spread then the line T must share a net-point with each Baer subplane. Furthermore, since $\mathcal{D}$ is a derivable net, every point is incident with a line of each Baer subplane. Hence, it is now immediate that $S(T)$ is a dual spread for each line of T exterior to the derivable net $\mathcal{D}$ if and only if the net extension is a derivable-extension.   □

Now we see that any derivable net is, in fact, a vector-space derivable net so the two approaches merge.

We now further refine the concept of a geometric extension.

DEFINITION 96. *Let $\mathcal{D}$ be any derivable net. Then $\mathcal{D}$ may be considered a 'left' vector-space net over a skewfield K. Let Σ denote the 3-dimensional projective space $PG(3, K)$ for K a skewfield, with special line N defined combinatorially by $\mathcal{D}$ and so that $\mathcal{D}$ may be embedded in Σ. Let T be any transversal to $\mathcal{D}$ and let $S(T)$ denote the spread of Σ defined by the net-points of T as lines of Σ together with the line N. Let $\pi_{S(T)}$ denote the associated translation plane and let $\pi^D_{S(T)}$ denote any affine dual translation plane whose projective extension dualizes to $\pi_{S(T)}$, taken projectively. Then $\pi^D_{S(T)}$ contains a derivable net isomorphic to $\mathcal{D}$ but considered as a 'right' vector-space net over the skewfield K^{opp}.*

We shall call $\pi^D_{S(T)}$ a 'geometric extension of $\mathcal{D}$' by $S(T)$.

Hence, given a derivable net $\mathcal{D}$ with transversal T, we may consider two possible situations. First of all, we know that $\mathcal{D}$ may be considered a right vector-space net over a skewfield F, and there is an associated transversal function, which we may use to extend $\mathcal{D}$ to a dual translation plane π^D_f (the algebraic extension). On the other hand, we may consider $\mathcal{D}$ as a left vector-space net over $F^{opp} = K$, embed the net combinatorially into a (left) 3-dimensional projective space Σ isomorphic to $PG(3, K)$, with distinguished line N and then realize that the transversal T, as a set of points of $\mathcal{D}$, is a set of lines whose union with N, is a spread of Σ, which defines a translation plane with an associated dual translation plane $\pi^D_{S(T)}$ (the geometric extension).

We now establish the connection with spreads in 3-dimensional projective space and transversals to derivable nets. However, the reader might note that there are possible left and right transversal functions depending on whether the vector-space derivable net is taken as a left or a right vector space and this will play a part in our discussions.

THEOREM 143. *The set of spreads in 3-dimensional projective spaces is equivalent to the set of transversals to the set of derivable nets; every spread is a transversal spread.*

PROOF. Let K be any skewfield and let Σ be isomorphic to $PG(3, K)$, and let S be a spread of Σ. Choose any line N of S and form the corresponding derivable net $\mathcal{D}$ defined combinatorially with lines the points of $\Sigma - N$. Then $S - N$ is a set of lines of Σ and hence points of $\mathcal{D}$ such that each point of $\Sigma - N$ 'line of $\mathcal{D}$' is incident with a unique line 'point of $\mathcal{D}$.' Hence, $S - N$, as a set of net points, is a transversal to the derivable net $\mathcal{D}$. $\square$

In the next sections, we show that the algebraic and geometric extensions processes are equivalent and examine the nature of the transversal extensions.

3. Planar Transversal Extensions

THEOREM 144. *(see Knarr [163] (2.7)) Let $\mathcal{P}$ be any spread in $PG(3, K)$, for K a skewfield. Let $\mathcal{D}$ be a derivable net and T a transversal to it, which geometrically constructs $\mathcal{P}$ by the embedding process.*

Then there is a dual translation plane $\pi_{\mathcal{D}}^T$ constructed by the algebraic extension process. For any line of $\pi_{\mathcal{D}}^T - \mathcal{D}$, the spread in the associated 3-dimensional projective geometry obtained by the geometric embedding process produces a translation plane with spread $S(T)$ isomorphic to $\mathcal{P}$ and whose dual is isomorphic to $\pi_{\mathcal{D}}^T$.

Hence, all spreads are planar transversal spreads.

PROOF. We refer the reader to Johnson [114] for any background information not given explicitly. The reader might note that when a translation plane is defined from a spread in $PG(3, K)$, it is usually most convenient to consider the vector space as a left vector space and spread components left 2-dimensional vector spaces of the general form $x = 0$, $y = xM$, where M is a 2×2 matrix over K. Ultimately, we shall be constructing a derivable net from $K \oplus K$ considered as a right K-space. On the other hand, following the structure theory of the embedding of the derivable net into $PG(3, K)$ considered as a left vector space, we obtain a derivable net with components $x = 0, y = \alpha x$ as opposed to $x = 0, y = x\alpha$. Furthermore, we may define $x \diamond \alpha = \alpha x$, which forces the vector-space derivable net to be defined over K^{opp} as opposed to K. However, we shall see that the derivable net arising from a spread and, hence, a translation plane, is a right 2-dimensional vector space over K. To be clear, a derivable net written as $x = 0, y = x\alpha$ for α in K embeds within $PG(3, K)$ considering the associated vector space as a 'right' vector space. If we start with a 'left' vector space to facilitate the spread and the translation plane, we end up with a derivable net contained in a dual translation plane written as $x = 0, y = x\alpha$ with

α in K, which then embeds within $PG(3, K)$ as a 'right' vector space, which may be taken as a 'left' K^{opp} space $PG(3, K^{opp})$.

Given a derivable net $\mathcal{D}$, there is a skewfield K such that the set of points of the net is $K \oplus K \oplus K \oplus K$, and there is an associated 4-dimensional K-vector space V, which we fix as a left space, such that the corresponding 3-dimensional projective space Σ isomorphic to $PG(3, K)$ has a fixed line N generated as

$$\langle (1, 0, 0, 0), (0, 1, 0, 0) \rangle$$

and such that the points of $\mathcal{D}$ (d_1, d_2, d_3, d_4) correspond to

$$\langle (d_1, d_2, 1, 0), (d_3, d_4, 0, 1) \rangle.$$

In this context, the derivable net will have components written in the form $x = 0, y = \alpha x$ for all $\alpha \in K$. We note that the set of vectors of $y = \alpha x$ is not necessarily always a left K–subspace, although each point of the net may be considered a left K–subspace embedded in the associated projective space.

When there is a transversal T to $\mathcal{D}$, we may form the algebraic extension process, dualize, and construct a 'left' spread in $PG(3, K)$. We will be taking the spread in $PG(3, K)$, more properly a 'spread set' and forming the associated translation plane. We choose the spread set as follows: We choose a particular set of three lines and vectorially denote these by

$$x = 0, y = 0, y = x.$$

Then, any other line (including $y = 0$ and $y = x$) has the following form:

$$y = x \begin{bmatrix} g(t, u) & f(t, u) \\ t & u \end{bmatrix} \forall u, t \in K,$$

where g and f are functions from $K \times K$ to K, and x and y are denoted by row 2-vectors over K.

We define a multiplication

$$x \circ (t, u) = x \begin{bmatrix} g(t, u) & f(t, u) \\ t & u \end{bmatrix}.$$

Note that we assume when $t = 0$ and $u = 1$, we obtain $y = x$ so that $g(0, 1) = 1$ and $f(0, 1) = 0$.

To define an associated dual translation plane, we define

$$x * m = m \circ x.$$

However, when $x \in K$, we see that the coordinate structure for the dual translation plane contains K^{opp} instead of K and is a 'right' 2-dimensional K^{opp} vector space. Furthermore, we may take lines to have the following equations:

$$x = c, \; y = x * m + b$$
$$\forall c, m, b \; \in \; K^{opp} \oplus K^{opp}.$$

We note that allowing $(0, \alpha) = \alpha \in K$ we have a dual translation plane containing a vector-space derivable net defined by lines

$$x = c, y = x * \alpha + b$$
$$\forall \alpha \; \in \; K^{opp} \; and \; \forall b, c \in K^{opp} \oplus K^{opp},$$

which is isomorphic to the original net $\mathcal{D}$. Now let $\{e, 1\}$ be a right K^{opp}-basis for $K^{opp} \oplus K^{opp}$ so that a general element $m = \alpha * e + \beta = (\alpha, \beta)$. Since

$$x * m = (x * e) * \alpha + x * \beta,$$

we have the representation of the lines of the dual translation plane. Note that basically all that we have done is return to the 'right' vector-space derivable net over K^{opp}, from which we started. The transversal T is simply a line not in the derivable net.

Now we combine the two concepts and consider the spread $S(T)$ geometrically constructed and arising from a line of the form $y = x * (t, u) + (b_1, b_2)$ (i.e., the transversal T).

We note that

$$y = x * (t, u) + (b_1, b_2) = (t, u) \circ (x_1, x_2) + (b_1 b_2),$$

where

$$(t, u) \circ (x_1, x_2) = (t, u) \begin{bmatrix} g(x_1, x_2) & f(x_1, x_2) \\ x_1 & x_2 \end{bmatrix}$$
$$= (tg(x_1, x_2) + ux_1, tf(x_1, x_2) + ux_2),$$

is the following set of points:

$$\left\{ \begin{array}{c} (x_1, x_2, tg(x_1, x_2) + ux_1 + b_1, tf(x_1, x_2) + ux_2 + b_2); \\ b_1, b_2 \in K \end{array} \right\}.$$

Now we have a delicate issue. In order to consider this set of points as 'left' vectors so as to apply the appropriate embedding as 'left' 2-dimensional vector spaces or perhaps 'left lines,' we need to consider the vector subspace as a left space over K.

Now each of these points embeds as a 'left' line (a two-dimensional left K subspace) as

$$\langle (x_1, x_2, 1, 0), (tg(x_1, x_2) + ux_1 + b_1, tf(x_1, x_2) + ux_2 + b_2, 0, 1) \rangle,$$

which is

$$\alpha(x_1, x_2, 1, 0) + \beta(tg(x_1, x_2) + ux_1 + b_1, tf(x_1, x_2) + ux_2 + b_2, 0, 1),$$

and this, in turn, is

$$(\alpha x_1 + \beta(tg(x_1, x_2) + ux_1 + b_1),$$
$$\alpha x_2 + \beta(tf(x_1, x_2) + ux_2 + b_2), \alpha, \beta),$$

$\forall \alpha, \beta \in K$.

To reconstruct a spread, we choose to reconstruct a spread set, hence, letting

$$\widehat{x_2} = \alpha x_1 + \beta(tg(x_1, x_2) + ux_1 + b_1),$$
$$\widehat{x_1} = \alpha x_2 + \beta(tf(x_1, x_2) + ux_2 + b_2),$$

it follows that when there is an inverse, we have:

$$(\widehat{x_1}, \widehat{x_2}) \begin{bmatrix} x_2 & x_1 \\ tf(x_1, x_2) + ux_2 + b_2 & tg(x_2, x_2) + ux_1 + b_1) \end{bmatrix}^{-1}$$
$$= (\alpha, \beta).$$

When $x_1 = x_2 = 0$, then we obtain the subspace generated by

$$(\widehat{x_1}, \widehat{x_2}) = (b_2, b_1).$$

Now translate by adding $-(b_2, b_1)$. We note that, in this form, N has equation $y = 0$.

Now change the spread set by applying the mapping $(x, y) \longmapsto (y, x)$ so that now N has the form $x = 0$, and, generally, we have the spread represented as

$$x = 0, y = 0, y = x \begin{bmatrix} v & s \\ tf(s, v) + uv & tg(s, v) + us \end{bmatrix}$$
$$\forall v, s \in K.$$

Now change bases by

$$\begin{bmatrix} 0 & 1 & 0 & 0 \\ 1 & 0 & 0 & 0 \\ 0 & 0 & 0 & 1 \\ 0 & 0 & 1 & 0 \end{bmatrix}$$

to obtain the form of the spread as:

$$x = 0, y = 0, y = x \begin{bmatrix} tg(s, v) + us & tf(s, v) + uv \\ s & v \end{bmatrix} \forall v, s \in K.$$

For fixed elements $t \neq 0$ (as we must have a proper transversal to the derivable net) and u in K, we have

$$
\left[
\begin{array}{ccc}
tg\,(s,v) + us & tf(s,v) + uv & v \\
s & v &
\end{array}
\right]
$$

$$
=
\left[
\begin{array}{cc}
t & u \\
0 & 1
\end{array}
\right]
\left[
\begin{array}{cc}
g(s,v) & f(s,v) \\
s & v
\end{array}
\right],
$$

a basis change again by

$$
\left[
\begin{array}{cc}
A & 0_2 \\
0_2 & I_2
\end{array}
\right],
$$

where $A = \left[\begin{array}{cc} t & u \\ 0 & 1 \end{array} \right]$ transforms the spread into the form:

$$
x = 0, y = x \left[
\begin{array}{cc}
g(s,v) & f(s,v) \\
s & v
\end{array}
\right] \forall s, v \in K.
$$

Hence, the geometric extension process produces (by dualization) the original spread constructed from the algebraic extension process. □

3.1. Planar Transversal-Spreads and Dual Spreads.

We have not yet dealt with the possibility that a planar transversal-spread may not actually arise from a derivable affine plane, that it may be possible that the spread is not a dual spread.

In Johnson [**116**], similar constructions to the following are given and the reader is referred to this article for additional details.

THEOREM 145. *Let $\mathcal{D}$ be a derivable net and let Σ be isomorphic to $PG(3, K)$ and correspond to $\mathcal{D}$ with special line N. If K is infinite then there exists a dual translation plane π extending $\mathcal{D}$ and a line T of $\pi - \mathcal{D}$ such that $S(T)$ is a spread of $PG(3, K)$, which is not a dual spread. In particular, if K is a field then $S(T)$ is non-Pappian.*

There exist planar transversal-spreads, which are not dual spreads.

PROOF. As noted above, if K is infinite, we may embed $\mathcal{D}$ into a non-derivable dual translation plane. Hence, there exists a line T such that there is some Baer subplane that does not intersect T in an affine point. Therefore, $S(T)$ is not a dual spread. If K is a field then any Pappian spread in $PG(3, K)$ is a dual spread (see, e.g., Johnson [**114**]). Thus, $S(T)$ is non-Pappian. □

4. Semifield Extension-Nets

Suppose that $\mathcal{D}$ is a derivable net, and there exists a transversal T and construct the spread $S(T)$. In Knarr [**163**], the question was raised when $S(T)$ is Pappian or what happens when $\mathcal{D}$ is contained in a translation plane. By the previous sections, we may apply the algebraic extension process to consider such questions. In particular, there are lines $x = c$, $y = f(x)\alpha + x\beta + b$, which define any affine plane containing the derivable net $\mathcal{D}$. Hence, if $\mathcal{D}$ is contained in a translation plane, then the dual translation plane containing $\mathcal{D}$ is a translation plane, which implies that $S(T)$ is a semifield spread. Knarr observes this fact by noting that the 'translation' collineation group of the derivable net would then act on the net extended by the transversal implying a collineation group fixing a component N and transitive on the remaining components of the spread (points of T). Hence, we obtain:

THEOREM 146. *(see Knarr [**163**], also see Johnson [**135**]) Let $\mathcal{D}$ be a derivable net and let T be a transversal. If the extension net $\langle \mathcal{D} \cup \{T\} \rangle$ defined by $\mathcal{D} \cup \{T\}$ is a translation net, then the spread $S(T)$ defines a semifield plane.*

Furthermore, all semifield spreads in $PG(3, K)$, for K a skewfield, are 'semifield planar transversal-spreads' (arising from semifield planes).

PROOF. Apply the main result of the section on planar transversal-spreads. $\qquad\square$

REMARK 47. *Let $\mathcal{P}$ be any non-Desarguesian semifield spread in $PG(3, K)$, for K a skewfield. If we choose the axis of the affine elation group to be N and view the spread as a transversal to a derivable net with the embedding in $PG(3, K) - N$, the corresponding dual translation plane will be a semifield plane. On the other hand, if any other line of $\mathcal{P}$ is chosen as N in the embedding, the affine dual translation plane will not be an affine semifield plane. So, a semifield spread in $PG(3, K)$ could arise as a planar transversal-spread without the affine plane containing the derivable net being an affine semifield plane.*

With the above remark in mind, we now examine the semifield spreads that can be obtained when $\langle \mathcal{D} \cup \{T\} \rangle$ is a translation extension-net (an algebraic extension is a dual translation plane and a translation plane). The affine plane containing the derivable net will correspond to the dual translation plane side where the components are left subspaces over the skewfield K^{opp}, provided the geometric embedding is in the left projective space $PG(3, K)$.

We choose coordinates so that here is a vector space V (over a prime field $\mathcal{P}$) of the form $W \oplus W$ such that points are the vectors (x, y) for $x, y \in W$, and we may choose a basis so that $x = 0, y = 0, y = x$ belong to the derivable net $\mathcal{D}$. We want to consider the derivable net as a right vector-space net over a skewfield at the same time we are considering the vector space and the components of the derivable net as left spaces over the same skewfield. Furthermore, there is a skewfield K^{opp} such that $W = K^{opp} \oplus K^{opp}$ as a left K^{opp}-vector space and components of $\mathcal{D}$ may be represented as follows:

$$x = 0, y = x \begin{bmatrix} B & 0 \\ 0 & B \end{bmatrix}; B \in K^{opp}.$$

We again note that the components of $\mathcal{D}$ are not necessarily all right K^{opp}-subspaces, although we say that $\mathcal{D}$ is a 'right' vector-space net over K^{opp}. We note that, following the ideas in the section on algebraic and geometric extensions, we are working in the dual translation plane side, which contains the derivable net. The translation plane obtained by dualization has its spread in $PG(3, K)$. Recall that this means that the so-called right-nucleus of the semifield in question is K^{opp}.

It also follows that any translation net has components of the general form $y = xT$, where T is a $\mathcal{P}$-linear bijection of W. If W is decomposed as $K^{opp} \oplus K^{opp}$ over the prime field $\mathcal{P}$, choose any basis $\mathcal{B}$ for K^{opp} over $\mathcal{P}$. Then, we may regard V as (x_1, x_2, y_1, y_2), where x_i, y_i are in K for $i = 1, 2$ and also may be represented as vectors over $\mathcal{B}$. That is, for example, $x_j = (x_{j,i}; i \in \lambda)$, for $j = 1, 2$, with respect to $\mathcal{B}$ for $x_{j,i} \in \mathcal{P}$, for some index set λ. With this choice of basis, we may represent T as follows:

$$T = \left(y = x \begin{bmatrix} T_1 & T_2 \\ T_3 & T_4 \end{bmatrix} \right)$$

where T_i are linear transformations over $\mathcal{P}$ represented in the basis $\mathcal{B}$. Note that we are not trying to claim that the $T_i's$ are K^{opp}-linear transformations, merely $\mathcal{P}$-linear.

The action is then

$$(y_1, y_2) = (x_1, x_2) \begin{bmatrix} T_1 & T_2 \\ T_3 & T_4 \end{bmatrix} = (x_1 T_1 + x_2 T_3, x_1 T_2 + x_2 T_4),$$

where the x_i and y_j terms are considered as $\mathcal{P}$-vectors.

Hence, the $T_i's$ are merely additive mappings on K^{opp} but not necessarily K^{opp}-linear.

REMARK 48. *It should be kept in mind that a derivable net may always be considered algebraically a pseudo-regulus net with spread in*

$PG(3, K^{opp})$, *when the geometric embedding is in* $PG(3, K)$. *When* K
*is a field, this is not to say that these two projective spaces are the same
as* $\mathcal{D}$ *can be a regulus in a 3-dimensional projective space while being
embedded in another, and both projective spaces are isomorphic.*

Before we state our theorem, we remind the reader of the definition
of a 'skew-Desarguesian plane.'

DEFINITION 97. *Let* K *be a skewfield and let* V *denote a 4-dimensional
left* K*-vector space.*
   (1) Then the set of left 2-dimensional K*–subspaces*

$$\left\{ x = 0, y = 0, y = x \begin{bmatrix} u + \rho t & \gamma t \\ t & u \end{bmatrix} \right\} ; u, t \in K$$

is a spread in $PG(3, K)$ *if and only if* $z^2 + z\rho - \gamma \neq 0$ *for all* $z \in K$,
where x *and* y *are 2-vectors.*
   (2) The translation plane $\pi_{\rho,\gamma}$ *corresponding to a spread of type (1)
is said to be a 'skew-Desarguesian plane.' The plane is Desarguesian
if and only if both* ρ *and* γ *are in the center of* K. *For more details,
the reader is directed to Theorem 26.2 of* [**114**].

THEOREM 147. *If* $\mathcal{D}$ *is a derivable net and* $\langle \mathcal{D} \cup \{T\} \rangle$ *is a translation
net regarded as a left vector space net over the associated prime field* $\mathcal{P}$,
and $\mathcal{D}$ *regarded as a right vector-space net over* K^{opp}, *then the geometric
embedding of* $\mathcal{D}$ *into* Σ *isomorphic to* $PG(3, K)$ *is considered as a 'left'
space embedding.*
   Representing $\mathcal{D}$ *as*

$$x = 0, y = x \begin{bmatrix} B & 0 \\ 0 & B \end{bmatrix} ; B \in K^{opp},$$

we may represent T *as*

$$\left(y = x \begin{bmatrix} T_1 & T_2 \\ T_3 & T_4 \end{bmatrix} \right)$$

where the T_i *are additive mappings of* K^{opp} *and* $\mathcal{P}$*-linear transformations.*
   (1) The line $y = x \begin{bmatrix} T_1 & T_2 \\ T_3 & T_4 \end{bmatrix}$ *determines a semifield spread admitting an affine homology group with axis* $y = 0$ *and coaxis* $x = 0$
isomorphic to $K^{opp} - \{0\}$ *(the dual semifield plane has its spread in*
$PG(3, K)$ *and is* $S(T)$*).*

The semifield spread has the following form:

$$x = 0, y = x \left(\begin{bmatrix} T_1 & T_2 \\ T_3 & T_4 \end{bmatrix} \begin{bmatrix} A & 0 \\ 0 & A \end{bmatrix} + \begin{bmatrix} B & 0 \\ 0 & B \end{bmatrix} \right)$$

$\forall A, B \in K^{opp}.$

(2) The semifield spread with spread in $PG(3, K)$ (the $S(T)$) has the following form (where here the T_i's are considered additive mappings of K):

$$x = 0, y = x \begin{bmatrix} sT_1 + vT_3 & sT_2 + vT_4 \\ s & v \end{bmatrix} \forall v, s \in K.$$

(3) The semifield spread in $PG(3, K)$ is a skew-Desarguesian spread if and only if the $T_i's$ are all K-linear transformations (i.e., multiplication by elements of K) if and only if $\mathcal{D}^{+T}$ is a partial spread in $PG(3, K^{opp})$; considering $\mathcal{D}$ as a pseudo-regulus in $PG(3, K^{opp})$, T is a subspace in the projective spread $PG(3, K^{opp})$.

(4) (see Knarr [163]) If K is a field then the semifield spread in $PG(3, K)$ is Pappian if and only if the $T_i's$ are all K-linear transformations (i.e., multiplication by elements of K) if and only if $\langle \mathcal{D} \cup \{T\} \rangle$ is a partial spread in $PG(3, K)$; T is a subspace in the projective spread $PG(3, K)$ wherein $\mathcal{D}$ is considered a regulus.

PROOF. Although some of the following has been previously presented in the section on algebraic and geometric extension, we revisit these ideas here. Part (1) follows immediately from the algebraic extension process considering T as a transversal function.

We consider a point $(x_1, x_2, x_1T_1 + x_2T_3, x_1T_2 + x_2T_4)$. The reader should consult the Subplanes text [114] for additional details supporting the following coordinate proof. Now we may represent N by

$$\langle (1, 0, 0, 0), (0, 1, 0, 0) \rangle,$$

the zero vector

$$(0, 0, 0, 0) \ by \ \langle (0, 0, 1, 0), (0, 0, 0, 1) \rangle$$

and a general point

$$(d_1, d_2, d_3, d_4) \ by \ \langle (d_1, d_2, 1, 0), (d_3, d_4, 0, 1) \rangle,$$

where the 2-dimensional K–subspaces are considered right spaces and lines in $PG(3, K)$.

Hence, the lines associated with the net-points

$$(x_1, x_2, x_1T_1 + x_2T_3, x_1T_2 + x_2T_4)$$

are

$$\langle (x_1, x_2, 1, 0), (x_1T_1 + x_2T_3, x_1T_2 + x_2T_4, 0, 1) \rangle$$
$$= \alpha(x_1, x_2, 1, 0) + \beta(x_1T_1 + x_2T_3, x_1T_2 + x_2T_4, 0, 1),$$

for all $\alpha, \beta \in K$.

Let

$$x_2^* = (\alpha x_1 + \beta(x_1T_1 + x_2T_3)), \quad and \quad x_1^* = (\alpha x_2 + \beta(x_1T_2 + x_2T_4)).$$

Then

$$(x_1^*, x_2^*) \begin{bmatrix} x_2 & x_1 \\ x_1T_2 + x_2T_4 & x_1T_1 + x_2T_3 \end{bmatrix}^{-1} = (\alpha, \beta).$$

Now change bases by interchanging $x = 0$ and $y = 0$ to obtain the spread as

$$x = 0, y = x \begin{bmatrix} v & s \\ sT_2 + vT_4 & sT_1 + vT_3 \end{bmatrix} \forall v, s \in K.$$

Change bases by

$$\begin{bmatrix} 0 & 1 & 0 & 0 \\ 1 & 0 & 0 & 0 \\ 0 & 0 & 0 & 1 \\ 0 & 0 & 1 & 0 \end{bmatrix}$$

to change the form into:

$$x = 0, y = x \begin{bmatrix} sT_1 + vT_3 & sT_2 + vT_4 \\ s & v \end{bmatrix} \forall v, s \in K.$$

This proves (2). The proofs to (3) and (4) and then immediate. $\square$

We noted in part (4) that the associated spreads in $PG(3, K)$ are Pappian if a derivable net is a regulus net in $PG(3, K)$ and the transversal is a subspace within the same $PG(3, K)$. We might inquire as to the nature of the semifield spreads if we assume initially that the transversal is a subspace in $PG(3, K)$, but the derivable net is not necessarily a K-regulus.

We recall from our results on Baer groups that and derivable nets in $PG(3, K)$, for K a field that a representation may be taken so that a derivable net $\mathcal{D}$ with partial spread in $PG(3, K)$, for K a field, may be represented in the following form:

$$x = 0, y = x \begin{bmatrix} u & A(u) \\ 0 & u^\sigma \end{bmatrix} \forall u \in K$$

and where σ is an automorphism of K, and where x and y are 2-vectors over K and A is a function on K such that

$$\begin{bmatrix} u & A(u) \\ 0 & u^\sigma \end{bmatrix} \forall u \in K$$

is a field isomorphic to K. We also recall that $A \equiv 0$ in the finite case, or when there are at least two Baer subplanes incident with the zero vector that are K–subspaces. When there is exactly one K–subspace Baer subplane, the characteristic is two, $\sigma = 1$ and $A(u) = Wu + uW$ for some linear transformation W of K over the prime field.

First, consider the situation when there are two Baer subplanes that are K–subspaces so that A is identically zero. Suppose that we have

$$y = x \begin{bmatrix} a & b \\ c & d \end{bmatrix}$$

a transversal to the derivable net for a, b, c, d in K. We need to re-coordinatize so as to realize the derivable net as a regulus in an associated projective space.

We consider the mapping:

$$\tau : (x_1, x_2, y_1, y_2) \longmapsto (x_1, x_2^{\sigma^{-1}}, y_1, y_2^{\sigma^{-1}}).$$

It follows that τ maps

$$(x_1, x_2, x_1 u, x_2 u^\sigma) \longmapsto (x_1, x_2^{\sigma^{-1}}, x_1 u, x_2^{\sigma^{-1}} u),$$

which implies that

$$x = 0, y = x \begin{bmatrix} u & 0 \\ 0 & u \end{bmatrix} \forall u \in K$$

in the associated projective space. Now we consider the τ-image of

$$y = x \begin{bmatrix} a & b \\ c & d \end{bmatrix}.$$

As a linear transformation over the prime field, consider $x^{\sigma^{-1}} = xM$. Then, the image of

$$y = x \begin{bmatrix} a & b \\ c & d \end{bmatrix}$$

can be written in the following form:

$$y = x \begin{bmatrix} a & Mb^{\sigma^{-1}} \\ M^{-1}c & d^{\sigma^{-1}} \end{bmatrix} = \begin{bmatrix} T_1 & T_2 \\ T_3 & T_4 \end{bmatrix},$$

in the notation of the previous section, and where $x = (x_1, x_2^{\sigma^{-1}} = x_2 M)$ and $y = (y_1, y_2^{\sigma^{-1}} = y_2 M)$. Hence, we obtain the semifield spread in $PG(3, K)$ in the form

$$x = 0, y = x \begin{bmatrix} sa + v^{\sigma^{-1}} c & s^{\sigma^{-1}} b^{\sigma^{-1}} + v d^{\sigma^{-1}} \\ s & v \end{bmatrix} \forall v, s \in K.$$

Now change bases by

$$\begin{bmatrix} c & b^{\sigma^{-1}} & 0 & 0 \\ 0 & 1 & 0 & 0 \\ 0 & 0 & 1 & 0 \\ 0 & 0 & 0 & 1 \end{bmatrix}$$

to transform the spread set into the form:

$$x = 0, y = x \begin{bmatrix} v^{\sigma} + s(ac^{-1} + c^{-1} d^{\sigma^{-1}}) & s^{\sigma^{-1}} c^{-1} b^{\sigma^{-1}} \\ s & v \end{bmatrix}.$$

Hence, we obtain

THEOREM 148. *A derivable net $\mathcal{D}$ with transversal extension T giving a partial spread $\langle \mathcal{D} \cup \{T\} \rangle$ that is in $PG(3, K)$, for K a field, and such that there are at least two Baer subplanes that are K–subspaces constructs a semifield spread in $PG(3, K)$ of the following form:*

$$x = 0, y = x \begin{bmatrix} v^{\sigma} + sk & s^{\sigma^{-1}} l \\ s & v \end{bmatrix} \forall v, s \in K,$$

for σ an automorphism of K, and constants $k, l \in K \in K$.

REMARK 49. *The spreads mentioned above are considered in Johnson [118] and are generalization of spreads originally defined by Knuth and hence, perhaps, called should be called 'generalized Knuth spreads.' The reader is also referred to Chapter 20 on σ-flokki and (generalized) Hughes-Kleinfeld semifield planes.*

We now consider the possibility that the function A is not identically zero, for all $u, v \in K$. With $A(u) = Wu + uW$, a basis change by

$$\begin{bmatrix} I & W & 0 & 0 \\ 0 & I & 0 & 0 \\ 0 & 0 & I & W \\ 0 & 0 & 0 & I \end{bmatrix}$$

will change the form of the derivable net into a regulus in $PG(3, K)$. We note that the regulus will have the form:

$$x = 0, y = x \begin{bmatrix} u & 0 \\ 0 & u \end{bmatrix} \forall u \in K.$$

Moreover, a transversal in $PG(3, K)$ of the net D (in the original form) of the form

$$y = x \begin{bmatrix} a & b \\ c & d \end{bmatrix}$$

for a, b, c, d becomes a transversal to the standard regulus (note that the transversal is no longer then in $PG(3, K)$) and has the form

$$y = x \begin{bmatrix} a + Wc & (a + Wc)W + b + Wd \\ c & cW + d \end{bmatrix}.$$

Hence, we obtain in $PG(3, K)$, the spread:

$$x = 0, y = x \begin{bmatrix} sT_1 + vT_3 & sT_2 + vT_4 \\ s & v \end{bmatrix} \forall v, s \in K,$$

where

$$a + Wc = T_1, (a + Wc)W + b + Wd = T_2, c = T_3$$

and

$$cW + d = T_4.$$

It is important to note that the vector space is now of the form

$$(x_1, x_1 W + x_2, y_1, y_1 W + y_2),$$

which we identify with (x_1, x_2, y_1, y_2). With this identification, it follows that T_i's are additive mappings of K.

Hence, we obtain

THEOREM 149. *Let a derivable net $\mathcal{D}$ with transversal extension T giving a partial spread $\langle \mathcal{D} \cup \{T\} \rangle$ that is in $PG(3, K)$, for K a field, such that there is exactly one Baer subplane that is a K–subspace. Then there is an associated semifield spread in $PG(3, K)$ of the following form:*

$$x = 0,$$

$$y = x \begin{bmatrix} s(a + Wc) + vc & s((a + Wc)W + b + Wd) + v(cW + d) \\ s & v \end{bmatrix}$$

$$\forall v, s \in K,$$

where W is some prime field linear transformation of K, a, b, c, d constants in K.

CHAPTER 24

Partially Flag-Transitive Affine Planes

Now the question is how do we use the idea of a transversal to a derivable net to create some theory of interest? One way might be to assume that there is a derivable affine plane that admits a transitivity condition on the components that do not belong to the derivable net. We choose here the concept of a 'partially flag-transitive affine plane.' Recalling that a 'flag' is an incident point-line pair, we consider the flags that are on lines that do not belong to the derivable net and insist that there is a collineation group of the affine plane that is transitive on these flags.

We have shown that derivable nets that admit a transversal can be embedded either in dual translation planes or translation planes. However, a derivable affine plane may not be a dual translation plane or a translation plane but still we may use the extension theory to determine properties about the derivable affine plane, such as orders of groups and general group actions. This means, for example, that incompatibility results on Baer p-groups and elations in order q^2 translation planes might be able to be applied to p-groups that act on derivable affine planes.

DEFINITION 98. *A derivable affine plane that admits a collineation group leaving the derivable net $\mathcal{D}$ invariant and acting transitively on the flags on lines not in $\mathcal{D}$ shall be said to be 'partially flag-transitive.'*

So, what are some examples of partially flag-transitive affine planes? Suppose we take a translation plane π with spread in $PG(3, q)$ and consider any affine restriction to the dual of the projective extension of π. Then, it turns out that this dual translation plane is derivable, as we have seen in the chapter on transversal spreads. Also, the derived plane is a 'semi-translation plane.'

DEFINITION 99. *An affine plane is said to be a 'semi-translation plane' if and only if the plane admits a group of translations such that each orbit is a Baer subplane.*

We are not claiming that every semi-translation plane may be derived from a dual translation plane, but if one thinks of what would

345

happen if this were the case, the existence of a group of translation might become apparent. Specifically, a Baer subplane would then become a line of the original derivable net N. The orbits that we are interested in are those that correspond to lines on N that are in the parallel class (∞), where the dual to this point is the line at infinity of the associated translation plane. Since the translation group with center (∞) leaves the derivable net invariant and acts transitively on the affine points of the lines of this parallel class, then its group inherits as a collineation group of the derived plane and the orbits are Baer subplanes. It only then remains to check that this group becomes a translation group of the derived plane. So, we have shown that any plane derived from a dual translation plane using a derived net containing (∞) is a semi-translation plane. It has been an open question for about 50 years whether every semi-translation plane is derivable. There are semi-translation planes that do not derive to dual translation planes (the Hughes planes).

So, dual translation planes and derivable semi-translation planes might be candidates as examples of partially flag-transitive affine planes. Also, such partially flag-transitive affine planes might have flag-transitive groups that are somewhat connected to the translation plane case. We begin with the goal of making such a connection, and we begin with how this would work with a translation plane, since extensions of derivable nets can either give rise to translation planes or their duals.

Therefore, assume that we have a derivable net and a vector space transversal. We represent the derivable net as

$$x = 0, y = x \begin{bmatrix} B & 0 \\ 0 & B \end{bmatrix} \forall B \in K^{opp},$$

where $K^{opp} = F$ is a skewfield, and we represent the transversal in the form

$$\left(y = x \begin{bmatrix} T_1 & T_2 \\ T_3 & T_4 \end{bmatrix} \right) = xT,$$

where the $T_i's$ are prime field linear transformations and additive F-mappings.

First let F be isomorphic to $GF(q)$ for $q = p^r$, p a prime. Let $H_{p,f,g,u}$ be a group of order $q - 1$, whose elements are defined as follows:

$$
\tau_u = \begin{bmatrix} p(u) & 0 & 0 & 0 \\ 0 & f(u) & 0 & 0 \\ 0 & 0 & p(u)\lambda(u) & 0 \\ 0 & 0 & 0 & f(u)\lambda(u) \end{bmatrix},
$$

for $u \in$

and p, f, λ are functions on F. We require that the derivable net is left invariant under $\langle \tau_u; eF - \{0\} \rangle$. For this, we must have

$$
\begin{bmatrix} p(u)^{-1}p(u)\lambda(u) = \lambda(u) & 0 \\ 0 & f(u)^{-1}f(u)\lambda(u) = \lambda(u) \end{bmatrix}
$$

for some function v of u.

We consider situations under which

$$
x = 0, y = x \begin{bmatrix} p(u)^{-1} & 0 \\ 0 & f(u)^{-1} \end{bmatrix} T \begin{bmatrix} p(u)\lambda(u) & 0 \\ 0 & f(u)\lambda(u) \end{bmatrix} + wI
$$

for all $u \neq 0, w \in F$ defines a spread. We note that there is an associated elation group E of order q and the spread fixes $x = 0$.

DEFINITION 100. *When the above set defines a spread, we denote the spread by $\pi_{T,H_{p,f,p\lambda,f\lambda}}$ and call the spread a*

$$
`(T, EH_{p,f,p\lambda,f\lambda}) - spread.`
$$

More generally, it might be possible to have a group containing an elation group E of order q and a group H such that EH acts transitively on the components of the spread not in the derivable net but H may not be diagonal. In the more general case, we refer to the spread as a 'partially transitive elation group spread.'

REMARK 50. *Any $(T, H_{p,f,p\lambda,f\lambda})$-spread is derivable and the derived plane admits a collineation group fixing the spread and acting transitively on the components not in the derivable net. The elation group is turned into a Baer group B and the group $H_{p,f,p\lambda,f\lambda}$ is turned into the group $H_{p,p\lambda,f,f\lambda}$.*

We call such a spread a '$(T^*, BH_{p,p\lambda,f,f\lambda}) - spread.$' Also, more generally, if we have a partially transitive elation group spread, it derives to a so-called partially transitive Baer group spread.

PROOF. Change bases relative to F by the mapping that takes (x_1, x_2, y_1, y_2) to (x_1, y_1, x_2, y_2). Then, the group $H_{p,f,g,h}$ is changed to the group $H_{p,g,f,h}$. □

We now consider possible examples.

EXAMPLE 3. *(1) Consider any semifield plane of order q^2, whose semifield is of dimension two over its middle nucleus. This plane is a $(T, EH_{\lambda^{-1},\lambda^{-1},1,1})$-spread, which derives to a $(T^*, BH_{\lambda^{-1},1,\lambda^{-1},1})$-spread, which is also known as a 'generalized Hall spread' (i.e., of type 1).*

Note that if $y = xT$ is a vector-space transversal to a finite derivable net then we may realize the derivable net as a left vector space net and automatically use the 'left' extension process to construct a dual translation plane, which then becomes a semifield plane with spread:

$$x = 0, y = x(\alpha T + \beta I), \forall \alpha, \beta \in F.$$

(2) Consider a semifield plane of order q^2 whose semifield is of dimension two over its right nucleus. This plane provides a $(T, EH_{1,1,\lambda,\lambda})$-spread, which derives a $(T^*, BH_{1,\lambda,1,\lambda})$-spread that is also known as a 'generalized Hall spread of type 2.'

This is merely the situation with which we began, realizing the derivable net as a right vector space and constructing the dual translation plane, which is then a semifield plane with spread:

$$x = 0, y = x(T\alpha + \beta I), \forall \alpha, \beta \in F.$$

(3) Other known examples all correspond to situations where the $y = xT$ is a line in the projective space wherein the derivable net is a regulus. The partially transitive elation spreads correspond to flocks of quadratic cones. For example, in the odd order case, we consider the Kantor-Knuth semifield spread:

$$x = 0, y = x \begin{bmatrix} u & \gamma t^\sigma \\ t & u \end{bmatrix}; \forall u, t \in GF(q),$$

where γ is a nonsquare, q is odd, and σ is an automorphism of $GF(q)$.

We note that the elements of the elation group E have the following form:

$$\begin{bmatrix} 1 & 0 & u & 0 \\ 0 & 1 & 0 & u \\ 0 & 0 & 1 & 0 \\ 0 & 0 & 0 & 1 \end{bmatrix}; \forall u \in GF(q).$$

Furthermore, the group is $H_{1,f,\lambda,f\lambda}$, where

$$f(u) = u^\sigma, \lambda(u) = u^{\sigma+1}.$$

We now make the derivation of a dual translation plane more specific. The following discussion stems from similar ideas in the author's work [109]. We begin by considering arbitrary translation planes that are generalizations of semifield planes.

THEOREM 150. *Let π be any translation plane of order q^2 with spread in $PG(3,q)$. Assume that there exists a collineation group G in the linear translation complement of order q^2 that fixes a component and acts transitively on the remaining components of π.*

Then the dual translation plane is a partially flag-transitive derivable affine plane admitting a collineation group of order $q^5(q-1)$.

The dual translation plane is a translation plane (and hence a semifield plane) if and only if the group of order q^2 mentioned above is an elation group of the associated translation plane.

We shall give the proof as a series of lemmas. The first lemma should now be essentially immediate.

LEMMA 40. *Let π be a translation plane with spread in $PG(3,q)$. Choose any component $x = 0$, and let (∞) denote the parallel class containing the component. In the dual translation plane such that (∞) will become our line at infinity, take any 1-dimensional $GF(q)$-subspace X on any component $y = 0$. Then the lines $x(\infty)$ union (∞), for all x in X, generate a derivable net $\mathcal{D}^*$ in the dual translation plane obtained by taking (∞) as the line at infinity in the dual plane.*

LEMMA 41. *The dual translation plane of the previous lemma then admits a collineation group W consisting of a translation group T_∞ with center (∞) of order q^2 and an affine elation group E of order q. The kernel homology group K^* of order $q-1$ of the original translation plane acts on the dual translation plane so that $W = T_\infty E K^*$ has order $q^3(q-1)$. W leaves the derivable net $\mathcal{D}^*$ and leaves invariant an infinite point $(\infty)^*$.*

Furthermore, K^* fixes exactly two components of the derivable net and is transitive on the remaining components. E is normal in K^*E and this latter group has orbits of length q and $q(q-1)$ on the dual translation plane.

PROOF. Just reread the construction given in Lemma 40. The reader should verify why E becomes an affine elation group of the dual translation plane that leaves the derivable invariant. The rest of the proof is also left to the reader to complete. $\square$

LEMMA 42. *On the dual translation plane, there is a collineation group of order $q^5(q-1)$, which fixes the derivable net and acts partially flag-transitively.*

PROOF. We recall that the group $S = K^*E$ of order $q(q-1)$ mentioned previously is transitive on the parallel classes not in the derivable net. The translation group with center (∞) becomes a translation group of the dual translation plane with center $(\infty)^*$ and this group acts transitively on the affine lines of any parallel class not equal to $(\infty)^*$. Take any point P of the translation plane such that $P(\infty)$ is not a line of $(\infty)X$. Then, using the translation group, there is a group conjugate to S by a translation, which acts transitively on the lines incident with P other than $P(\infty)$. Note that the linear group of order q^2 acts on the dual translation plane since it permutes the points and lines and leaves invariant the line at infinity of the dual translation plane. □

The completes the proof of the theorem. The proof of the following corollary is now immediate.

COROLLARY 34. *Any semi-translation plane obtained by the derivation of a partially flag-transitive dual translation plane is also partially flag-transitive. The dual translation plane is of 'elation' type, whereas the semi-translation plane is of 'Baer' type (the dual translation plane admits an elation group of order q and the semi-translation plane admits a Baer group of order q).*

1. Derivable Affine Planes with Nice Groups

In this section, we give a complete classification of finite partially flag-transitive affine planes. The main ideas show that such planes admit groups of a certain required order. We may obtain some insight on what actually occurs, by merely assuming that we have a derivable affine plane admitting groups of such orders and develop the theory accordingly. This will give an indication of the types of planes that will show up under the classification of partially flag-transitive planes. First, we make some comments on our group assumptions.

Assume that π is a finite partially flag-transitive affine plane. The derivable net $\mathcal{D}$ is combinatorially equivalent to a projective space $PG(3, K)$, where K is isomorphic to $GF(q)$, relative to a fixed line N of $PG(2, K)$. Furthermore, the full collineation group of the net D is $P\Gamma L(4, K)_N$. Assume now that the given collineation group G of π is linear; i.e., in $PGL(4, K)_N$. It follows that the linear subgroup that fixes an affine point and the derivable net (which is now a regulus net) is a subgroup of $GL(2, q)GL(2, q)$, where the product is a central product with common group the center of order $q - 1$. In all of what follows, we shall assume that our groups acting on affine planes containing a derivable net are linear groups. Finally, we define the type of plane that will show up in our classification result.

DEFINITION 101. *A 'nonstrict semi-translation plane' is an affine semi-translation plane of order q^2 that admits a translation group of order $> q^2$.*

We note that dual translation planes or translation planes could be nonstrict semi-translation planes. We begin with a general result on elation groups in derivable planes. Note the special significance that this result has for conical flock planes or for α-flokki planes.

THEOREM 151. *Let π be a finite derivable affine plane of order q^2 admitting an elation group H of order q leaving invariant a derivable net. Then the H-orbits of infinite points union the center of H define a set of q derivable nets of π.*

PROOF. Let $\mathcal{D}$ denote the derivable net in question and choose (∞) to be the center of H. Let T be any line of $\pi - \mathcal{D}$. Then there is a dual translation plane defined by the algebraic extension process admitting H as a collineation group. But it is now clear that any dual translation plane has the property stated in the theorem and one of the derivable nets is defined by the image of T under H union the center of H. But the orbits of H in the dual translation plane share at least the orbit of π containing T. Since this argument is valid for any such line, it follows that the affine plane is a union of derivable nets whose infinite points share (∞).   □

We shall give the main results on groups acting on derivable affine planes in pieces so that the reader can digest the material a bit better. The proofs are intricate and combinatorial. All of the theorems and lemmas assume that π is a derivable affine plane of order q^2 that admits a linear p-group of order q^5 if q is odd or $2q^5$ if q is even, where $q = p^r$, and p is a prime.

THEOREM 152. *Assume that a derivable affine plane π of order q^2 admits a linear p-group S that fixes the derivable net $\mathcal{D}$ of order q^5 if q is odd or $2q^5$ if q is even. Then π contains a group that acts transitively on the affine points.*

Furthermore, the group contains either an elation group of order q or a Baer group of order q with axis a subplane of the derivable net $\mathcal{D}$.

PROOF. Clearly, S must fix an infinite point (∞) of $\mathcal{D}$. Since S acts on the remaining $q^2 - q$ infinite points of $\pi - \mathcal{D}$, it follows that the S-orbits cannot all have lengths strictly larger than q. Hence, there exists an orbit Γ of length $\leq q$ and, for $\alpha \in \Gamma$, S_α has order divisible by q^4 or $2q^4$ if q is even. Then, for ℓ a line of α, $S_{\alpha,\ell}$ has order divisible by q^2 or $2q^2$. Then the spread $S(\ell)$ admits a linear group of order q^2 or $2q^2$

that fixes the line N using the associated embedding in $PG(3,q)$. Since the group is linear, we must have an elation group of order divisible by q with axis N of the translation plane associated with $S(\ell)$. When q is odd, this implies that there can be no Baer p-elements by Theorem 298, which implies that the group of order q^2 acts regularly on the components of $S(\ell)$ and this is the largest possible stabilizer. So, when q is odd, S has an orbit of length exactly q, S_α is transitive on the affine lines of α, and $S_{\alpha,\ell}$ is transitive on the points of ℓ. Thus, S_α acts transitively and regularly on the affine points. When q is even, assume that we have a linear group of at least order $2q^2$ acting on the translation plane associated with the spread $S(\ell)$. In Johnson [**132**], the action of linear groups of order q^2 acting on spreads in $PG(3,q)$ is analyzed. It is shown that the group is transitive or non-Abelian and in the latter case, if not transitive, then has two orbits of components of length $q^2/2$, and there is an elation group E of order q in the center of the group. Each of the orbits of length $q^2/2$ are $q/2$ orbits of length q of E. Choose a group of order $2q^2$ containing a given group of order q^2. Hence, the two orbits of length $q^2/2$ are inverted or both fixed. The center E is characteristic in the group of order q^2 and hence normal in the group of order $2q^2$. If the two orbits are not inverted, then there is a Baer group B of order 4 which fixes an E-orbit. We may choose the elements of B to have the following general form:

$$\begin{bmatrix} 1 & a & 0 & 0 \\ 0 & 1 & 0 & 0 \\ 0 & 0 & 1 & a \\ 0 & 0 & 0 & 1 \end{bmatrix} \begin{bmatrix} I & 0 \\ 0 & T \end{bmatrix},$$

where T is a 2×2 matrix over $GF(q)$. At the same time, we may choose the elements of group E to have the following form:

$$\begin{bmatrix} 1 & 0 & u & m(u) \\ 0 & 1 & 0 & u \\ 0 & 0 & 1 & 0 \\ 0 & 0 & 0 & 1 \end{bmatrix} \quad \forall u \in GF(q) \text{ and } m \text{ is a function on } GF(q).$$

Furthermore, the lines of a net defined by the E orbit of $y = 0$ and incident with the zero vector are of the form:

$$y = x \begin{bmatrix} u & m(u) \\ 0 & u \end{bmatrix} \quad \forall u \in GF(q).$$

Note that B leaves invariant $x = 0$ and $y = 0$ and must leave the orbit of E containing $y = 0$ invariant. Since B has order $2^s \geq 4$, it follows that B must fix another component of this E-orbit, which we may take as $y = x$, without loss of generality. This implies that $T = 0$

so that B commutes with E. But, by Theorem 299, B has order less than or equal 2. Hence, the group of order divisible by $2q^2$ is transitive on the components of $S(\ell)$ and the stabilizer of a second component other than N has order exactly 2. Thus, when q is even and there is a 2-group of order $2q^5$, there is a subgroup that is transitive on the affine points. So in either the odd or even cases, we obtain a group of order q^4 or $2q^4$, which acts transitively on the affine points of the derivable affine plane.   □

The following proof is left for the reader to complete as an exercise.

THEOREM 153. *If π is non-Desarguesian in the elation case above, then π admits a set of q derivable nets sharing the axis of the elation group of order q.*

Before the statement of our next theorem, we give two fundamental lemmas.

LEMMA 43. *Let S_p be a Sylow p-subgroup. Hence, S_p leaves invariant an infinite point of $\mathcal{D}$, say (∞). The stabilizer of a second point fixes all infinite points of $\mathcal{D}$.*

PROOF. The group G is a subgroup of $GL(2,q)GL(2,q)T$ acting on the derivable net $\mathcal{D}$. A Sylow p-group is a subgroup of a group S_p^+ of order q^6 consisting of a Sylow p-group of each of the two $GL(2,q)'s$ and T of order q^4. The subgroup of S_p^+ that fixes two infinite points has order q^5 and consists of a Baer p-group together with T and hence fixes all infinite points of $\mathcal{D}$.   □

LEMMA 44. *Let σ be an nontrivial elation with axis $x = 0$ in $\mathcal{D}$ and let τ be a nontrivial Baer collineation of $\mathcal{D}$, also fixing $x = 0$ such that $\sigma\tau$ is in S_p (acts as a collineation of π). Then $\sigma\tau$ is a Baer collineation of π such that $Fix\sigma\tau$ has q parallel classes exterior to $\mathcal{D}$.*

PROOF. Coordinatize so that

$$\sigma = \begin{bmatrix} 1 & 0 & u & 0 \\ 0 & 1 & 0 & u \\ 0 & 0 & 1 & 0 \\ 0 & 0 & 0 & 1 \end{bmatrix}$$

and

$$\tau = \begin{bmatrix} 1 & a & 0 & 0 \\ 0 & 1 & 0 & 0 \\ 0 & 0 & 1 & a \\ 0 & 0 & 0 & 1 \end{bmatrix},$$

for a and $u \in F \simeq GF(q)$.

Then $\sigma\tau$ fixes $\{(0, y_1 a u^{-1}, y_1, y_2); y_i \in F\}$ pointwise and fixes no component of the derivable net except $x = 0$. Moreover, σ and τ share fixed points on $x = 0$. Hence, $\sigma\tau$ fixes exactly q^2 points, fixes $x = 0$, fixes exactly q points on $x = 0$, and fixes any line of the plane that contains points of $Fix\sigma\tau$. Suppose that $\sigma\tau$ fixes lines $x = 0$ and ℓ concurrent with a particular affine point P. So, $\sigma\tau$ fixes a Baer subplane π_o incident with P of $\mathcal{D}$ and induces an elation on π_o. Let (α) denote the parallel class containing ℓ. Then, $\sigma\tau$ fixes each of the q affine lines of π_o incident with (∞), fixes (α) so fixes each of the q affine lines incident with (α), as $\sigma\tau$ induces an elation on π_o. It follows that $\sigma\tau$ fixes q points on each of the lines fixed by $\sigma\tau$ incident with (∞). This argument shows that the set of fixed points belongs to a Baer subplane. $\qquad\square$

THEOREM 154. *If the order of the stabilizer of a point H is at least $2q$ then the order is $2q$ and H is generated by an elation group of order q and a Baer involution with axis in $\mathcal{D}$ or by a Baer group of order q and an elation.*

PROOF. We make the following claims,.

(1) Assuming that S_p fixes (∞), the stabilizer of a line $x = 0$ in S_p has order either q^3 or $2q^3$.

(2) Further, the stabilizer H of an affine point has order at least q in the q^3 case or at least $2q$ in the $2q^3$ case above.

(3) H contains either an elation group of order q, or a Baer group of order q with axis a subplane of $\mathcal{D}$.

(4) If the order of H is at least $2q$, then the order is $2q$ and H is generated by an elation group of order q and a Baer involution with axis in $\mathcal{D}$ or by a Baer group of order q and an elation.

These four items once proved complete the theorem.

We give the proofs to (1) through (4) as follows:

The proof of (1) is clear since the group acts transitively on the points of ℓ and fixes (∞) also acts transitively on the affine lines of (∞). Since $x = 0$ has q^2 points, the stabilizer of one of these has order at least q or $2q$.

Now H must actually fix a line of some Baer subplane π_o pointwise and permute the parallel class of subplanes to which π_o belongs. Since this parallel class has q^2 total members, H must leave invariant another subplane π_1 disjoint to π_o (the parallel class to which π_o belongs is the set of all subplanes of $\mathcal{D}$ disjoint from π_o union π_o). From the structure of derivable nets, it follows that π_o and π_1 must share a parallel class of lines.

If the shared parallel class is not (∞), then the group H fixes a second infinite point so it fixes all infinite points of $\mathcal{D}$ and hence fixes π_o pointwise. However, in this case, the order of H must, in fact be q.

Hence, assume that π_o and π_1 share a parallel class of lines of (∞). It follows that the elements of H are in $GL(2, q)GL(2, q)$ so that each element is the product of an elation σ and a Baer p-element τ. However, this implies that either σ or τ is 1 or there is an 'external' Baer p-element by Lemma 44. If this is so, then $p = 2$ and the order of H is at least $2q$. Note that each external Baer subplane contributes exactly q components outside of the derivable net. If there is an overlap, we may assume the overlap occurs at least on ℓ so there is a group generated by Baer elements of order at least 4 and fixing ℓ. But this means that within the translation plane of the associated spread corresponding to ℓ, we have a group of order at least $4q^2$, a contradiction. Hence, there are at most $q - 1$ 'external' Baer involutions (axes external to the derivable net). Let the external Baer involutions be denoted by $\sigma_i\tau_i$, where the σ_i are elations (not necessarily distinct and possibly trivial) and the τ_i are Baer involutions (not necessarily distinct but nontrivial) for $i = 1, 2, .., b_o$, where $b_o \leq q - 1$ and let the elements of H be denoted by $\sigma_i\tau_i = \rho_i$ for $i = 1, 2, .., \geq 2q$. Note that if $\sigma_i\tau_i$ is a Baer involution and σ_k is an elation not equal to σ_i then $\sigma_k\sigma_i\tau_i$ is an external Baer involution. Since the plane can be derived, it follows that if τ_s is a Baer involution (internal) not equal to τ_i then $\sigma_i\tau_i\tau_s$ is also a Baer involution. So, consider a given external Baer involution $\sigma_1\tau_1$. Then there are at least q elements that are non-identity elations or non-identity internal Baer involutions in H, and at least $q - 2$ of these are neither equal to σ_1 nor τ_1. Hence, multiplication of a given $\sigma_1\tau_1$ by elements of H results in $q - 2$ distinct other external Baer involutions. That is, if $\sigma_k\sigma_1\tau_1$ is $\sigma_s\sigma_1\tau_1$ or $\sigma_1\tau_1\tau_j$, then either σ_k is σ_s or τ_j. So, there are exactly $q - 2$ distinct external Baer involutions, assuming that there is one. The remaining two non-identity elements of H then cannot produce external Baer involutions. Hence, it must be that the remaining two elements are either σ_1 or τ_1. So, there are exactly $q-1$ external Baer involutions, implying there are exactly $q-1$ non-identity elations or internal Baer involutions.

Assume that there are s $\sigma_j's$ and hence $q - s$ $\tau_k's$. Note that the $\sigma_j\tau_k$'s are $s(q - s)$ distinct external Baer involutions. Hence, we must have

$$s(q - s) \leq q - 1.$$

So, we have

$$(s - 1)q \leq s^2 - 1$$

implying that if $s-1$ is not zero then $q \leq s+1$. But $q-1 \leq s \leq q-1$, which shows that $s = q-1$.

Thus, $s = 1$ or $q-1$ so that there is either an elation group of order q in H or an internal Baer group of order q in H. This implies that there is either an elation group of order q and an extra internal Baer involution that generates H if there is an internal Baer group of order q and an extra elation that generates H.

If the order of H is q, then it follows that in all cases, we can have only elations types or only Baer p-element types for otherwise we generate external Baer p-elements and a larger p-group than possible. That is, either H is a Baer p-group of order q with fixed axis in the derivable net or H is an elation group of order q.

Thus, the result is completed unless the order of H is at least $2q$. In this case, the stabilizer of ℓ has order at least $2q^2$ and hence, exactly $2q^2$ by Johnson [**132**] (section 6). So, the order of H is either $2q$ or q, and our above argument completes the proof of the four main points (1) through (4), which completes the proof of our theorem. $\qquad\square$

THEOREM 155. π *is a nonstrict semi-translation plane of order* q^2 *admitting a translation group of order* $q^3 p^\gamma$.

PROOF. By derivation, we may assume that the group H contains an elation group E of order q, which then acts transitively on the infinite points of $\mathcal{D} - (\infty)$. Then, the group Z that fixes a second infinite point and hence fixes all infinite points has order q^4 or $2q^4$. If this group is not fixed-point-free, then there is a Baer p-element with axis in $\mathcal{D}$. We may set up our argument so that such an element is in H and commutes with H and hence generates with E exactly $q-1$ external Baer p-elements and $p = 2$. (The reader should verify that since the group acts transitively on the lines of (∞), the axis of E may be chosen to correspond to the fixed point of the element of Z in question.) Hence, for each affine point P, there is a unique internal Baer involution τ_P with axis in $\mathcal{D}$. Thus, there is a set of q^2 mutually disjoint Baer subplanes in an orbit under Z. Therefore, Z is transitive on the set of Baer subplanes of some parallel class (as a parallel class of the derived net). So, in any case, there are at least $2q^4 - q^4 = q^4$ elements of Z that are products of Baer p-elements and translations on the net $\mathcal{D}$, and which are fixed-point-free. We now require some information on how a group or set can be fixed-point-free. But, the Baer p-group part comes from a group of order q, and we have a group of order q^4 or $2q^4$, and hence, we have a fixed-point-free set of cardinality at least q^4.

Suppose we represent a p-group fixing all infinite points of $\mathcal{D}$ as a subgroup of

$$\left\langle \sigma_a = \begin{bmatrix} 1 & a & 0 & 0 \\ 0 & 1 & 0 & 0 \\ 0 & 0 & 1 & a \\ 0 & 0 & 0 & 1 \end{bmatrix} ; a \in F \right\rangle T,$$

where T has the following form acting on the points of the net

$$\mathcal{D} : \left\langle \begin{array}{c} \tau_{(c_1, c_2, c_3, c_4)} : (x_1, x_2, y_1, y_2) \\ \longmapsto (x_1 + c_1, x_2 + c_2, y_1 + c_3, y_2 + c_4); c_i \in F \end{array} \right\rangle .$$

Then the subset that acts transitively on the points of $x = 0$ contains elements $\sigma_a \tau_{(0,0,c_3,c_4)}$, for various values of a in F and, for all $c_3, c_4 \in F$. Similarly, the subset that acts transitively on the points of $y = 0$ contains elements $\sigma_b \tau_{(c_1,c_2,0,0)}$, for various values of b in F, and, for all $c_1, c_2 \in F$. To see this, we note that the stabilizer of $x = 0$ in Z contains a set of cardinality q^2 that acts transitively and fixed-point-free on the affine points of $x = 0$. Hence, $(0, 0, 0, 0)$ is fixed by any element of the form σ_a, so in order to obtain a transitive action, we require all translations with center (∞) acting on $\mathcal{D}$. The proof for the action on $y = 0$ is analogous and left to the reader to complete.

Furthermore, if there is a fixed-point-free transitive action on $x = 0$, then the group acting contains a translation group of order q with center (∞). That is, we merely note that $(0, 0, y_1, y_2) \longmapsto (0, 0, y_1 + c_3, y_1 a + y_2 + c_4)$, for various values of a and c_3, c_4. Since we require all $c_3's$ and $c_4's$ for the transitive action, assume that $c_3 = 0$ and $c_4 \neq 0$. Then, if a is non-zero, we have the fixed points $(0, 0, -y_2 c_4 a^{-1}, y_2)$, for all y_2. Hence, all such values a must be zero. If $c_3 = 0$, then c_4 must be nonzero since otherwise the collineation is a Baer p-element. Therefore, we obtain that an element $\sigma_a \tau_{(0,0,0,c_4)}$ forces a to be zero.

Since we may repeat the above argument for any infinite point, we have a translation group of order q with centers (∞) and (0). The plane of order q^2 admits a translation group of order q^2, at least one of whose orbits is a subplane π_o of order q. The group leaving the infinite points of $\mathcal{D}$ pointwise fixed is transitive on the affine points and is transitive on the set of subplanes of the parallel class containing π_o. The proof of this claim is as follows: We note that the elements of the group Z fixing the infinite points are products of Baer collineations fixing π_o pointwise by a translation. Hence, Z permutes the parallel class of Baer subplanes containing π_o. The images of π_o by elements of Z admit the same translation group of order q^2 as π_o. Hence, the plane is a semi-translation plane.

It remains to show that the plane is a nonstrict semi-translation plane. We claim that the plane π admits a translation group of order at least q^3. We note that the elements of Z are products of Baer p-elements and translations. For any Baer p-element σ_a, assume that there are fewer than q^3 translations τ such that $\sigma_a \tau$ is a collineation. Since the translations are formally normal in the group of the net, it follows that each element $\sigma_a \tau$ has a unique representation of this kind. Therefore, the group order is strictly less than qq^3. Hence, there exists a collineation $\sigma_a \tau$, for which there is a group S_a of at least q^3 translations of the net such that $\sigma_a \tau g$ is a collineation for all $g \in S_a$. Fix g and consider $\sigma_a \tau g(\sigma_a \tau h)^{-1}$. This is a collineation that is also a translation, and which is not 1 if and only if $g \neq h$. It now follows that there are at least $q^3 - 1$ non-identity collineations that are translations. The translation group has order p^β so $p^\beta \geq q^3 - 1$, which implies that there is a translation group of order at least q^3. This completes the proof of the theorem. $\square$

THEOREM 156. *The translation group contains a $((\infty), \ell_\infty)$-transitivity. Furthermore, the translation group has order $q^3 p^\gamma$ and all infinite points not equal to (∞) are centers for translation groups of order qp^γ.*

Either π is a translation plane or there is a unique $((\infty), \ell_\infty)$-transitivity. Either (∞) is invariant under the full group or π is Desarguesian.

PROOF. First, we claim that if (∞) is moved by a collineation of G then π is Desarguesian.

We have an elation group of order q with axis $x = 0$. If (∞) is moved, we may assume that we have an elation group with axis $y = 0$ by the existing transitivity. Hence, the group generated by the elations acts on a regulus net and thus generated a collineation group isomorphic to $SL(2, q)$. Since the group has orbits of lengths $q + 1$ and $q^2 - q$, the translation group of order $q^3 p^\gamma$ has a decomposition into groups with fixed centers of orders qp^α and p^β such that

$$(q + 1)(qp^\alpha - 1) + (q^2 - q)(p^\beta - 1) + 1 = q^3 p^\gamma.$$

This equation reduces to:

$$qp^\alpha + p^\alpha + qp^\beta - p^\beta - q = q^2 p^\gamma.$$

Hence, it follows that $p^\alpha - p^\beta$ is divisible by q. First, assume that $p^\alpha = p^\beta$. Then

$$2p^\alpha - 1 = qp^\gamma,$$

which is a contradiction. If $p^\alpha \neq p^\beta$, then $\min(\alpha, \beta) = r$ if $q = p^r$, which implies that $p^\alpha = q$. If $p^\alpha = q$, then there are at least $q + 1$ centers of translation group of order q^2 so that plane is a translation plane. Thus, the plane is a translation plane of order q^2 which admits a collineation group isomorphic to $SL(2, q)$ generated by elations. Therefore, in this setting the plane is Desarguesian by Theorem 285. We may assume that the group fixes (∞). Let the translation group with center (∞) have order qp^δ.

Then,

$$qp^\delta - 1 + q(qp^\alpha - 1) + (q^2 - q)(p^\beta - 1) + 1 = q^3 p^\gamma.$$

This equation is equivalent to:

$$p^\delta + qp^\alpha + qp^\beta - p^\beta - q = q^2 p^\gamma.$$

So that $p^\delta - p^\beta$ is divisible by q. If $p^\delta \neq p^\beta$, then q must divide p^δ so that we have a $((\infty), \ell_\infty)$-transitivity. Thus, we are finished or we obtain:

$$p^\alpha + p^\beta - 1 = qp^\gamma,$$

which is a contradiction.

If $p^\delta = q$, then $p^\beta = qp^\rho$.

$$p^\alpha + p^\beta - p^\rho = qp^\gamma.$$

For this equation, it clearly follows that $p^\alpha = p^\rho = p^\gamma$. This completes the proof. □

We then simply summarize the main theorems as follows.

THEOREM 157. *(1) If a derivable affine plane π of order q^2 with derivable net $\mathcal{D}$ admits a linear p-group of order q^5 if q is odd or $2q^5$ if q is even that π contains a group which acts transitively on the affine points.*

(2) Furthermore, the group contains either an elation group of order q or a Baer group of order q with axis a subplane of $\mathcal{D}$. If the order of the stabilizer of a point H is at least $2q$ then the order is $2q$, and H is generated by an elation group of order q and a Baer involution with axis in $\mathcal{D}$ or by a Baer group of order q and an elation.

(3) π is a nonstrict semi-translation plane of order q^2 admitting a translation group of order $q^3 p^\gamma$, where $q = p^r$, for p a prime. Furthermore, either π is a translation plane or there is a unique $((\infty), \ell_\infty)$-transitivity and the remaining infinite points are centers for translation subgroups of orders qp^γ.

(4) If π is non-Desarguesian in the elation case above, then π admits a set of q derivable nets sharing the axis of the elation group of order q.

1.1. Classification. In this subsection, we use the previous section to classify partially flag-transitive planes.

THEOREM 158. *A finite derivable partially flag-transitive affine plane of order q^2 with linear group is a nonstrict semi-translation plane with a translation group of order $q^3 p^\gamma$.*

The plane admits either an elation group or a Baer group of order q.

Furthermore, in the elation case, either the plane is Desarguesian or the plane admits a $((\infty), \ell_\infty)$-transitivity, the point (∞) is invariant and the infinite points not equal to (∞) are centers for translation subgroups of order $q p^\gamma$.

PROOF. So, again let G denote the full collineation group of the associated affine plane π, under the assumption that the group is 'linear' with respect to the derivable net, and let T denote the translation group with center (∞) of S_p. We note that T is normal in S_p. Let ℓ be any transversal line to the derivable net. Then there exists a collineation group G_ℓ that acts transitively on the points of ℓ. We first claim that the p-groups of G_ℓ have orders q^2, or $2q^2$ and q is even. To see this, we note that there is a spread in $PG(3, q)$, $S(\ell)$. Furthermore, there is a group that fixes a line N of the spread and acts transitively on the remaining lines of the spread. We have assumed that the group is a subgroup of $PGL(4, q)_N$. Now consider the associated translation plane and realize that the collineation group, as a translation complement, acting here is a subgroup of $GL(4, q)$. Since the group fixes a component and is linear, the group induced on that component is a subgroup of $GL(2, q)$. Since the group is transitive on $S(\ell) - N$, let the order of the p-group be $p^\alpha q^2$. Hence, the elation group with axis N as order at least $p^\alpha q$. On the other hand, since the group is linear and transitive on the components not equal to N, the stabilizer of a second line is a Baer group of order p^α. Under these circumstances, if $p^\alpha > 1$, then $p^\alpha = 2$, which completes the proof of the assertion. Now the theorem is completed provided we show that the order of a Sylow p-group of G is either q^5, or $2q^5$ and q is even. Now since the group G acts transitively on components of the derivable affine plane $\pi - \mathcal{D}$, it follows that S_p permutes $q(q-1)$ points on the line at infinity and hence there must be an orbit of S_p of length q as otherwise pq would divide $q(q-1)$. Since this is, in fact, an orbit, the stabilizer of an infinite point (β) has order $|S_p|/q$. Since the lines of (β) are also in an orbit then the stabilizer of a line ℓ has order $|S_p|/q^3$, which is either q^2 or $2q^2$. This completes the proof.   □

1.2. Groups of Order q^6. Throughout this subsection, we use the group established in the previous section and subsection.

Let $p^r = q$, for p a prime, and let π be a partially flag-transitive affine plane. Then, there must be a group of order divisible by $q^2(q^2)(q^2 - q)$ by the assumed transitive action. Hence, the p-groups have orders divisible by q^5 and note that the full linear p-group of the derivable net has order q^6. Any such p-group S_p must leave invariant an infinite point (∞) of the derivable net $\mathcal{D}$, as the derivable net consists of $q + 1$ parallel classes of lines.

We have seen that dual translation planes arising from translation planes with spreads in $PG(3, q)$ that admit a collineation group of order q^2 in the translation complement and transitive on the components other than a fixed component admit collineation groups of order $q^5(q - 1)$ fixing a derivable net $\mathcal{D}$. Since the plane is a dual translation plane, there is an elation group of order q^2. Hence, there is a collineation group of order $q^6(q-1)$. We may ask are partially flag-transitive planes admitting the larger group dual translation planes?

LEMMA 45. *If π is a non-Desarguesian partially flag-transitive affine plane of order q^2 with elation group of order q, then H is normal in the subgroup of the stabilizer of the derivable net $\mathcal{D}$ that fixes the axis of H.*

PROOF. We have seen that (∞) is invariant. Since H is the maximal elation group with axis $x = 0$ fixing the derivable net, it follows that the stabilizer of $x = 0$ must normalize H. □

Hence, we see that

THEOREM 159. *Let π be a non-Desarguesian partially flag-transitive affine plane of order q^2 with linear group and of elation type. Then the corresponding group G fixes one derivable net containing the axis of (∞) and acts transitively on the remaining $q - 1$ derivable nets sharing (∞).*

THEOREM 160. *Let π be a non-Desarguesian partially flag-transitive affine plane of order q^2 with linear group and of elation type. If $q = p^r$, assume that $(p, r) = 1$.*

(1) If π admits a p-group S of order q^6, or $2q^6$ if $q = 2$, then π admits a collineation group, which fixes an infinite point (∞), and is transitive on the remaining infinite points.

(2) Furthermore, either the plane is a dual translation plane or the full group acts two-transitively on a set of q derivable nets sharing the infinite point (∞).

PROOF. Let H^+ denote the full elation group with center (∞). Since H^+ is elementary Abelian, H^+ permutes the orbits of H on the line at infinity, each of which defines a derivable net. It follows that the full collineation group of π that fixes the axis of H^+ must normalize H^+. Let S be a p-group of order q^6. Then S must fix (∞). Within S, there is a subgroup of order at least q^4 that fixes $x = 0$, and there is a subgroup S^- of order at least q^2 that fixes a point 0 on $x = 0$. Suppose that some element g of S^- fixes a line ℓ incident with 0. So, g leaves invariant the H^+-orbit containing ℓ.

Now assume that there exists another elation group of order q, H_1 of H^+, of order q, whose orbits define derivable nets. Then, either H_1 is H or $H_1 \cap H = \langle 1 \rangle$. Hence, if H is characteristic in H^+ and is, hence, normal or $\langle H_1, H \rangle$ has order q^2, that is, either the plane is a dual translation plane or H is normal in the full collineation group of the affine plane.

If g fixes ℓ then g fixes two derivable nets, one of which does not contain ℓ. Hence, the stabilizer of ℓ may be regarded as a subgroup of $P\Gamma L(4, q)_{N^*}$ for some line N^*, as indicated in the embedding. However, if $(p, r) = 1$, then the group is linear with respect to this group. In this setting, we have seen that p can only be 2 and the order of the stabilizer can only be 2. So, we have a group that acts transitively on the infinite points not equal to (∞) and acts two-transitive on the derivable nets sharing (∞). This completes the proof of the theorem. □

COROLLARY 35. *Let π be a partially flag-transitive affine plane with linear group, which is a translation plane and assume that the order is even. Then π is either a $(T, EH_{p,f,p\lambda,f\lambda})$-plane or a $(T^*, BH_{p,p\lambda,f,f\lambda})$-plane.*

PROOF. We may assume that we have an elation group of order q and that the plane is a translation plane.

When q is even and we have a group of order divisible by $q - 1$ acting on the derivable net, then, since $(q-1, q+1) = 1$, it follows that any element of order dividing $q - 1$ must leave at least two Baer subplanes invariant. We now assume that the groups in question are in the translation complement. Now there is a subgroup of $GL(2, q)GL(2, q)$, which normalizes the elation group E of order q. Assuming that E is a subgroup of the first $GL(2, q)$, it follows that we have a subgroup H^* of the first $GL(2, q)$ of order $(q-1)^2$ that normalizes E so that our group is a subgroup of $H^*GL(2, q)$. We note that an element of H^* will fix either exactly two or all Baer subplanes incident with the zero vector of $\mathcal{D}$. Let g be an element of order dividing $q - 1$. Let $g = g^*g_*$, where g^* is in H^* and g_* is in the second $GL(2, q)$, and which is generated by

Baer collineations of the net. It also follows that g_* also fixes exactly two or all Baer subplanes of the net. Since g^* and g_* commute and $q - 1$ is odd, it follows that an element g will fix either exactly two or all Baer subplanes incident with the zero vector. Hence, we may assume that there is a set of at least $q - 1$ group elements that fix two Baer subplanes incident with the zero vector. With the appropriate coordinate change, we notice that these group elements will act as distinct diagonal elements. These group elements will generate a group of order dividing $(q - 1)^4$, which is the direct sum of four cyclic groups of order $q - 1$. Let $q - 1 = \Pi p_i^{\alpha_i}$ be the prime decomposition. Then there is a p-subgroup of order $p_i^{\beta_i}$ for $\beta_i \geq \alpha_i$. Then, there is a subgroup of order $p_i^{\alpha_i}$. Since the group is Abelian, it follows that there is a subgroup of order $q - 1$. Because the group leaves two Baer subplanes invariant, we clearly have a $H_{p,f,p\lambda,f\lambda}$ group, which then gives rise to a $(T, EH_{p,f,p\lambda,f\lambda})$-plane. □

The following corollary is our main result on derivable affine planes admitting groups of order q^6.

COROLLARY 36. *Let π be a derivable affine plane of order q^2 that admits a p-group of order q^6 if q is odd or $2q^6$ if q is even containing a linear subgroup of order q^5 or $2q^5$ leaving invariant the derivable net $\mathcal{D}$ invariant. Assume that when $q = p^r$ for p a prime, then $(r, p) = 1$ and assume that π admits an elation group H of order q.*

(1) Then either the plane is Desarguesian or the center of H is invariant.

(2) Then π admits a collineation group fixing an affine point of order q^2 or $2q^2$ that fixes an infinite point (∞) of $\mathcal{D}$ and acts transitively on the remaining infinite points.

(3) Either the plane is a dual translation plane or the group acts transitively on a set of q derivable nets sharing (∞).

CHAPTER 25

Special Topics on Parallelisms

In this chapter, we offer a few special instances of where parallelisms and/or ideas arising from derivable nets might appear. We begin by consideration of what be a geometric property of transversal spreads.

1. Transversal Spreads and Dualities of $PG(3, K)$

In the Subplanes text [**114**], the idea was furthered that a geometric understanding of the concept of derivation could be explained using a duality of the ambient $PG(3, K)$-space. We examine these ideas further.

Let $\mathcal{D}$ be a derivable net and T a transversal such that $\mathcal{D} \cup \{T\}$ is a derivable-extension. Let $\widehat{\mathcal{D}} \cup \{T\}$ denote the corresponding derivable-extension, where $\widehat{\mathcal{D}}$ is the derived net of $\mathcal{D}$. When is the spread $S(T)$ with respect to $\mathcal{D} \cup \{T\}$, also denoted by $S(T)_{\mathcal{D}+T}$, isomorphic to the spread $S(T)_{\widehat{\mathcal{D}}+T}$ with respect to $\widehat{\mathcal{D}} \cup \{T\}$?

First of all, we note that if the corresponding projective space is $PG(3, K)$ of $\mathcal{D}^{+T}$, then $S(T)_{\mathcal{D}+T}$ is in $PG(3, K)$, whereas $S(T)_{\widehat{\mathcal{D}}+T}$ is in $PG(3, K^{opp})$. However, the points and (Baer) subplanes of $PG(3, K)$ are subplanes and points of $PG(3, K^{opp})$, respectively. Hence, the question only makes genuine sense when K is a field.

THEOREM 161. *Let $\mathcal{D}$ be a derivable net embedded in $PG(3, K)$, where K is a field, and let T be a transversal to $\mathcal{D}$, which is also a transversal to the derived net $\widehat{\mathcal{D}}$. Then the spread $S(T)_{\mathcal{D}+T}$ corresponding to $\mathcal{D}^{+T}$ is isomorphic to the spread $S(T)_{\widehat{\mathcal{D}}+T}$ corresponding to $\widehat{\mathcal{D}}^{+T}$ is and only if there is a duality of $PG(3, K)$ that maps one spread to the other.*

2. Skew Parallelisms

In the previous chapters and sections on extension of a derivable net, we have established various spreads using transversals of the derivable net. If all of the transversals are contained in a given affine plane, the question is whether there is a way to use various sets of spreads to construct or reconstruct the affine plane containing a derivable net.

We introduce a close relative of parallelisms, the 'skew-parallelisms,' which will play an important role in understanding the set of transversals to derivable nets. The work presented in this chapter basically follows the author's work [**109**], somewhat modified.

DEFINITION 102. *A 'skew parallelism' of $PG(3, K) - N$ is a set of spreads each containing N, which forms a disjoint cover of the lines of $PG(3, K)$ skew to N.*

DEFINITION 103. *Let S be a skew parallelism of $PG(3, K) - N$ and let $\mathcal{P}$ be any spread containing N. We shall say that $\mathcal{P}$ is 'orthogonal' to S if and only if $\mathcal{P}$ intersects each spread of S in a unique line $\neq N$.*

A set of skew parallelisms of $PG(3, K) - N$ is 'orthogonal' if and only if each spread of any one skew parallelism is orthogonal to each of the remaining skew parallelisms.

A set $\mathcal{A}$ of skew parallelisms of $PG(3, K) - N$ is said to be 'planar' if and only if, given any two lines ℓ_1 and ℓ_2 of $PG(3, K)$ that are skew to N, there is a skew parallelism of $\mathcal{A}$ containing a spread sharing ℓ_1 and ℓ_2.

If the spreads of a set of skew parallelisms are all dual spreads, we shall say that the set is a 'derivable' set of skew parallelisms.

THEOREM 162. *(1) Given an orthogonal and planar set $\mathcal{A}$ of skew parallelisms of $PG(3, K) - N$, then there is a unique affine plane $\pi_{\mathcal{A}}$ containing a derivable net such that the set of transversals to the derivable net are the spreads of the set $\mathcal{A}$.*

(2) Conversely, any affine plane containing a derivable net corresponds to a uniquely defined orthogonal and planar set of skew parallelisms.

(3) The set of derivable affine planes is equivalent to the set of derivable, orthogonal, and planar sets of skew parallelisms.

PROOF. We have seen that any spread $\mathcal{P}$ containing N may be considered as generated from a transversal to a derivable net. It is easy to verify that a skew parallelism corresponds exactly to a parallel class external to the derivable net. Two skew parallelisms that are orthogonal then correspond to two distinct parallel classes of a net extension of a derivable net and a planar and orthogonal set of skew parallelisms is such that any two distinct points of the net as lines of the projective space are incident with exactly one spread of some skew parallelism; two distinct points are incident with a unique transversal to the derivable net. Hence, an affine plane is constructed from a planar and orthogonal set of skew parallelisms. In order that the affine plane actually be derivable, it follows that each transversal spread must actually be a dual spread. Hence, a derivable, planar, and orthogonal set of skew parallelisms constructs a derivable affine plane. This proves all parts of the theorem. $\qquad\square$

Now we ask the nature of a 'transitive' skew parallelism.

DEFINITION 104. *A skew parallelism of* $PG(3, K) - N$ *is 'transitive' if and only if there exists a subgroup of* $P\Gamma L(4, K)_N$ *that acts transitively on the spreads of the skew parallelism.*

A planar and orthogonal set of skew parallelisms is 'transitive' if and only if there exists a subgroup of $P\Gamma L(4, K)_N$ *that acts transitively on the set.*

We shall say that the set is 'line-transitive' if and only if the stabilizer of a spread is transitive on the lines not equal to N *of the spread, for each spread of the skew parallelism.*

REMARK 51. *Let* π^D *be a dual translation plane with transversal function* $f(x)$ *to a right vector-space derivable net so that lines have the equations:*

$$x = 0, y = f(x)\alpha + x\beta + b$$

for all $\alpha, \beta \in K$ *and for all* $b \in V$.

Then, there is a collineation group of π^D *that leaves invariant the derivable net and acts transitively on the lines not in the derivable net and of the form* $y = f(x)\alpha + x\beta + b$, *where* $\alpha \neq 0$.

The 'translation group' T *is transitive on the lines of each such parallel class and represented by the mappings:*

$$(x, y) \longmapsto (x, y + b) \ \forall b \in V.$$

The affine elation group E *is represented by mappings of the form*

$$(x, y) \longmapsto (x, x\beta + y) \ \forall \beta \in K$$

and the affine homology group H *is represented by mappings of the form:*

$$(x, y) \longmapsto (x, x\alpha) \ \forall \alpha \in K - \{0\}.$$

Notice that T *and* E *correspond to certain translation subgroups with fixed centers of the corresponding translation plane and the group* H *corresponds to the kernel homology group of the associated translation plane.*

(1) It also follows that any derivable affine plane coordinatized by a cartesian group will admit a group isomorphic to T *and hence corresponds to a transitive skew parallelism.*

(2) Any such dual translation plane will produce a transitive planar and orthogonal set of transitive skew parallelisms.

(3) Any semifield spread that contains a derivable net as above will admit a translation group with center $(f(x))$, which fixes $f(x)$ and acts transitively on the points of $f(x)$, which implies that the transversal-spread is transitive.

Hence, any semifield spread produces a line-transitive planar and orthogonal set of skew parallelisms.

Part 6

Constructions of Parallelisms

We begin by showing that infinite regular parallelisms are usually possible for a variety of fields K, using variations on ideas of Betten and Riesinger. We then give the Penttila and Williams construction of finite regular parallelisms. We also generalize Beutelspacher's construction of line-parallelisms and provide some applications. The parallelisms of Johnson, which admit a collineation group fixing one spread and transitive on the remaining spreads, are also constructed. The infinite regular parallelisms constructed here have a particularly nice geometric construction from a corresponding Pappian affine plane.

CHAPTER 26

Regular Parallelisms

In this chapter, we give a brief sketch of the constructions of regular parallelisms (the spreads are Pappian) in $PG(3, K)$, for K an infinite field. We also present the constructions due to Penttila and Williams of finite regular parallelisms. We begin with the infinite case.

1. Infinite Regular Parallelisms

We begin by trying to find regular, i.e., Pappian parallelisms over infinite fields. The idea is to try to create a set of mutually disjoint Pappian spreads (on components) from a given Pappian spread by multiple derivation. This will mean that each such Pappian spread will share exactly two components with the original Pappian spread Σ, which, in turn, partitions Σ. This can be done for any characteristic. For infinite regular parallelisms, the main ingredient that makes the construction work is that K admits a quadratic extension K^+ such that if σ is the involution in $Gal_K K^+$, then $K^{+(\sigma+1)}$ is a proper subset of K. There is a natural collineation group of the associated Pappian plane Σ_{K^+} coordinatized by K^+ that admits a collineation group transitive on a partition of the spread by pairs of components.

THEOREM 163. *Let K be a field that admits a quadratic extension K^+. Let σ denote the involution in $Gal_K K^+$. Assume that $K - K^{+(\sigma+1)}$ is not empty. Let Σ denote the Pappian affine plane coordinatized by K^+ and let $c \in K - K^{+(\sigma+1)}$.*
The group

$$G_c = \left\langle \begin{array}{c} \tau_m : (x, y) \rightarrow (x^\sigma, y^\sigma) \begin{bmatrix} m & cm^{-\sigma+1} \\ 1 & m \end{bmatrix} \\ \forall m \in K^+ - \{0\} \end{array} \right\rangle$$

acts on and is transitive on the set of pairs of components

$$\{\{x = 0, y = 0\}, \{y = xm, y = xcm^{-\sigma}\} ; m \in K^+ - \{0\}\}.$$

PROOF. To prove part (1), notice that $\{x = 0, y = 0\}$ is mapped to $\{y = xm, y = xcm^{-\sigma}\}$ under the group element τ_m. Also,

$$(x, xt) \;\; \rightarrow \;\; (x^\sigma, x^\sigma t^\sigma) \begin{bmatrix} m & cm^{-\sigma+1} \\ 1 & m \end{bmatrix} =$$
$$(x^\sigma(m + t^\sigma), x^\sigma(cm^{-\sigma+1} + t^\sigma m)).$$

Therefore,

$$\{y = x(-m^\sigma), y = x(-c(m^\sigma)^{-\sigma} = y = x(-cm^{-1})\}$$
$$\rightarrow \;\; \{x = 0, y = 0\}$$

under τ_m. Finally, we claim that $\{y = xt, y = xct^{-\sigma}\}$ maps to

$$y \;\; = \;\; x\frac{cm^{-\sigma+1} + t^\sigma m}{m + t^\sigma},$$
$$y \;\; = \;\; xc(\frac{cm^{-\sigma+1} + t^\sigma m}{m + t^\sigma})^{-\sigma},$$
$$for \;\; m + t^\sigma \;\; \neq \;\; 0.$$

To see this, we note that under the mapping τ_m, $m + t^\sigma \neq 0$, and $cm^{-\sigma+1} + t^\sigma m \neq 0$, we see that

$$y = xt \rightarrow y = x\frac{cm^{-\sigma+1} + t^\sigma m}{m + t^\sigma},$$

so

$$y = xct^{-\sigma} \rightarrow y = x\frac{cm^{-\sigma+1} + ct^{-1}m}{m + ct^{-1}},$$

since $c^\sigma = c$. It remains to show that

$$c\left(\frac{cm^{-\sigma+1} + t^\sigma m}{m + t^\sigma}\right)^{-\sigma} = \frac{cm^{-\sigma+1} + ct^{-1}m}{m + ct^{-1}},$$

which follows by an easy calculation, which is left to the interested reader.

This completes the proof to the theorem. □

1.1. Multiple Derivation. Given any Pappian spread coordinatized by a field K^+ of degree 2 over K, we note that the subspread

$$A_\alpha = \{y = xm; m^{\sigma+1} = \alpha\}$$

is a derivable partial spread with derived spread

$$A_\alpha^\sigma = \{y = x^\sigma m; m^{\sigma+1} = \alpha\}.$$

Therefore,

$$\{\{x = 0, y = 0\} \cup_{\alpha \in K^{+(\sigma+1)}} A_\alpha\}$$

is a partition of the spread. It is also true that the spread obtained by the multiple derivation of A_α by A_α^σ is a Pappian spread.

DEFINITION 105. *The partition listed above is called a 'linear set' defined by the pair $\{x = 0, y = 0\}$. In general, the linear set defined by the pair of lines A, B gives rise to the Pappian spread $\Sigma_{A,B}$ with multiply derived Pappian spread $\Sigma^{\sigma}_{A,B}$.*

The general idea of a partition of the lines of $PG(3, K)$ by spreads, i.e., a 'parallelism' is to find a partition II of the Pappian spread Σ by pair of lines (components) $\{A, B\}$ such that

$$\cup_{\{A,B\}\in\mathrm{II}}\Sigma^{\sigma}_{A,B} = \mathcal{P}$$

is a regular parallelism.

DEFINITION 106. *A 'regular partial parallelism' is a set of regular spreads that are mutually disjoint on spread lines (components).*

In the following, we use the notation of the previous section in Theorem 163.

THEOREM 164. *Let $\Sigma_{A,B}$ denote the Pappian spread obtained by multiple derivation of the set of reguli in a linear set with carrying lines A, B of the Pappian spread Σ.*

(1) Then G_c acts transitively on

$$\mathcal{P}_c = \left\{\Sigma^{\sigma}_{x=0,\ y=0},\ \Sigma^{\sigma}_{y=xm,\ y=xcm^{-\sigma}}; m \in K^+ - \{0\}\right\}.$$

(2) $\mathcal{P}_c$ is a regular partial parallelism admitting a group acting transitive on the spreads.

PROOF. By Theorem 163, the group G_c permutes the partition

$$\left\{\{x = 0, y = 0\}, \{y = xm, y = xcm^{-\sigma}\}; m \in K^+ - \{0\}\right\}.$$

Since each pair defines a unique linear set and this linear set defines a unique multiply derived set, this clearly implies that G_c acts on and therefore acts transitively on $\mathcal{P}_c$.

To prove (2), we need to show that no two spreads have a component in common.

If two spreads have a component in common, then

$$\Sigma^{\sigma}_{x=0,y=0} \ and \ \Sigma^{\sigma}_{y=xm,\ y=xcm^{-\sigma}}$$

have a component in common by the transitivity of G_c. The components of $\Sigma^{\sigma}_{x=0,y=0}$ are

$$x = 0, y = 0, \ and \ y = x^{\sigma}z, \ where \ z \in K^+ - \{0\}.$$

The components of $\Sigma^{\sigma}_{y=xm,\ y=xcm^{-\sigma}}$ are

$$y = xm, y = xcm^{-\sigma}$$

and the images of components of $\Sigma_{x=0,y=0}^{\sigma}$, so the remaining components are the sets

$$\pi_{t,m} = \left\{ (x^{\sigma}m + xt^{\sigma}, x^{\sigma}cm^{-\sigma+1} + xt^{\sigma}m); x \in K^{+} \right\}.$$

Assume that $\pi_{t,m}$ is $y = x^{\sigma}z$. Then

$$(x^{\sigma}m + xt^{\sigma})^{\sigma}z = x^{\sigma}cm^{-\sigma+1} + xt^{\sigma}m, \ \forall \ x.$$

Hence, we must have $m^{\sigma}z = t^{\sigma}m$, and $tz = cm^{-\sigma+1}$, which implies that

$$\frac{m^{\sigma}}{t} = \frac{t^{\sigma}m}{cm^{-\sigma+1}} = \frac{t^{\sigma}m^{\sigma}}{c}.$$

Therefore,

$$c = t^{\sigma+1},$$

a contradiction. This proves (2).   □

2. The Set of 2-Secants

In Betten and Riesinger [9], the following important construction theorem is proved.

DEFINITION 107. *Let O be an elliptic quadric in $PG(3, K)$, where K is a field of characteristic not 2. A point Q of $PG(3, K)$ is an 'interior point' if and only if each line incident with Q intersects O in two distinct points.*

2.1. Betten and Riesinger 2-Secant Theorem.

THEOREM 165. *(Betten and Riesinger [9]) Let O be an elliptic quadric in $PG(3, K)$, where K is a field of characteristic not 2. Let S denote a set of lines, each of which intersects the elliptic quadric in two points (2-secants), such that the union of the lines of S is a partition of the non-interior points of $PG(3, K)$. Then there is an associated regular parallelism of $PG(3, K)$.*

The way the previous result works is as follows. Let Σ denote the Pappian spread associated with the elliptic quadric Ω in $PG(3, K)$ obtained by embedding the quadric in the Klein quadric in $PG(5, K)$ and using the Klein mapping from the Klein quadric to the associated $PG(3, K)$, which then maps the elliptic quadric bijectively to Σ. Choose any 2-secant M and let $M^{\perp}$ denote the line obtained as the image of M, using the induced polarity of $PG(3, K)$. The line $M^{\perp}$ is disjoint from Ω and taking the planes of $PG(3, K)$ containing $M^{\perp}$, a linear flock F_M of $\Omega - \{M \cap \Phi\}$ is constructed. The Thas-Walker construction (see Chapter 42) using F_M produces a Pappian spread Σ_M that contains the Klein images of $M \cap \Phi$ with Σ and is obtained from Σ by the multiple derivation of the linear André set of reguli defined by

$M \cap \Phi$. Thus, viewed in this manner, what is obtained is a regular parallelism of $PG(3, K)$ obtained as follows: Given a Pappian spread Σ,
there is a partition of the components of Σ into pairs $\{P, Q\}$ such that
multiple derivation of the linear André set defined by $\{P, Q\}$ is a partition of the 2-dimensional K–subspaces of the associated 4-dimensional
vector space V_4 over K.

Now we represent the field K^+ using a polynomial irreducible over
K, with matrix field

$$\left\{ \begin{bmatrix} u + tg & tf \\ t & u \end{bmatrix}; \atop u, t \in K; (u + tg)u + t^2 f = 0 \text{ iff } u = t = 0. \right\}.$$

Now consider the corresponding spread

$$\left\{ \begin{array}{l} x = 0, y = x \begin{bmatrix} u + tg & tf \\ t & u \end{bmatrix}; u, t \in K; \Delta_{u,t} = \\ (u + tg)u + t^2 f = 0 \text{ iff } u = t = 0 \end{array} \right\}.$$

Now represent the spread as an elliptic quadric of points on the Klein
quadric in $PG(5, K)$, where the points on the right in the following we
represent points on the quadric and 2-dimensional K–subspaces on the
left are the Pappian spread components.

$$x = 0 \rightarrow (0, 0, 0, 0, 0, 1),$$

$$y = x \begin{bmatrix} u + tg & tf \\ t & u \end{bmatrix} \rightarrow (1, u + tg, tf, t, u, \Delta_{u,t}).$$

Now consider a point $(d, 0, 0, 0, 0, 1)$ on the line (2-space)

$$\langle (0, 0, 0, 0, 0, 1), (1, 0, 0, 0, 0, 0) \rangle.$$

What we wish to show is that the line

$$\langle (d, 0, 0, 0, 0, 1), (1, u + tg, tf, t, u, \Delta_{u,t}) \rangle$$

intersects the Klein quadric in two distinct points; that this line is a 2-
secant, so that $(d, 0, 0, 0, 0, 1)$ is an interior point of the elliptic quadric
Ω associated with the Pappian spread. We shall use part (1) of the
following theorem in all of our constructions.

THEOREM 166. *(1)*

$$\langle (d, 0, 0, 0, 0, 1), (1, u + tg, tf, t, u, \Delta_{u,t}) \rangle$$

is a 2-secant if $-d\Delta_{u,t} \neq 1$.

(2) Hence, if $-d \notin K^{+(\sigma+1)}$, *then the set of all lines incident with*
$(d, 0, 0, 0, 0, 1)$ *is a set of 2-secants that forms an exact cover of the
point set of* $PG(3, K) - \{(d, 0, 0, 0, 0, 1)\}$. *Therefore, we obtain a regular
parallelism.*

(3) Choose any plane Π *of* $PG(3, K)$ *that contains the line*

$$\langle (0, 0, 0, 0, 0, 1), (1, 0, 0, 0, 0, 0) \rangle = L,$$

so

$$\Pi = \left\langle \begin{array}{c} (0, 0, 0, 0, 0, 1), (1, 0, 0, 0, 0, 0), \\ (1, u + tg, tf, t, u, \Delta_{u,t}) \end{array} \right\rangle =$$

$$\left\langle \begin{array}{c} (d, 0, 0, 0, 0, 1), (1, u + tg, tf, t, u, \Delta_{u,t}), \\ (1, 0, 0, 0, 0, 0) \end{array} \right\rangle$$

The points of the elliptic quadric on Π *are*

$$\left\{ \frac{-1}{d+v} m^{-\sigma}; d + v \neq 0 \ or \ -1; v \in K \right\}$$

and are paired incident with $(d, 0, 0, 0, 0, 1)$ *as follows:*

$$\left\{ \begin{array}{c} \frac{-1}{d+v} m^{-\sigma}, \frac{-1}{d+u} m^{-\sigma} \\ = \frac{-1}{d} \left(\frac{-1}{d+v} m^{-\sigma} \right)^{-\sigma}; u = \frac{-d}{d+v} m^{-(\sigma+1)} \end{array} \right\}.$$

Note that we choose $-d \neq m^{-(\sigma+1)}$, *which depends only on* $m \neq 0$ *of* K^+. *Then the corresponding pair is (defines) a 2-secant.*

PROOF. The point $\rho(d, 0, 0, 0, 0, 1) + (1, u + tg, tf, t, u, \Delta_{u,t}) = (\rho d + 1, u + tg, tf, t, u, \Delta_{u,t} + \rho)$ is a point on the Klein quadric corresponding to a component of the Pappian spread if and only if

$$(\rho d + 1)(\Delta_{u,t} + \rho) = \Delta_{u,t}.$$

This is true if and only if

$$(\rho d + 1)\rho = -\rho d \Delta_{u,t},$$

for $\rho \neq 0$, which is also equivalent to

$$(\rho d + 1) = -d \Delta_{u,t}.$$

What this means is that

$$\left(1, \frac{u + tg}{-d\Delta_{u,t}}, \frac{tf}{-d\Delta_{u,t}}, \frac{t}{-d\Delta_{u,t}}, \frac{u}{-d\Delta_{u,t}}, \frac{\Delta_{u,t}}{-d\Delta_{u,t}} \right)$$

is a point on the Klein quadric (elliptic quadric), which is distinct from $(1, u + tg, tf, t, u, \Delta_{u,t})$ if and only if

$$-d\Delta_{u,t} \neq 1.$$

This proves (1). Therefore, if

$$-d \notin K^{+(\sigma+1)}$$

we obtain a set of 2-secants with the property maintained in part (2).

$\square$

From the previous result, from the set of 2-secants, we obtain the following partition of the Pappian spread:

$$
\left\{
\begin{array}{c}
\{x = 0, y = 0\}, \\
\left\{ y = x \begin{bmatrix} u + tg & tf \\ t & u \end{bmatrix}, y = x \frac{1}{-d\Delta_{u,t}} \begin{bmatrix} u + tg & tf \\ t & u \end{bmatrix} \right\} \\
; u, t \in K, \text{ not both } 0
\end{array}
\right\}.
$$

Noting that

$$
\begin{bmatrix} u + tg & tf \\ t & u \end{bmatrix}^{\sigma+1} = \Delta_{u,t}, \text{ and letting } -d = c,
$$

we have part of the following theorem.

THEOREM 167. *(1)*

$$
\mathcal{P}_c = \left\{ \Sigma^{\sigma}_{x=0,\, y=0}, \ \Sigma^{\sigma}_{y=xm,\, y=xcm^{-\sigma}}; m \in K^+ - \{0\} \right\}
$$

corresponds to the set of 2-secants of the elliptic quadric embedded in the Klein quadric that share the point $(-c, 0, 0, 0, 0, 1)$.
(2) If the characteristic of K *is not 2, then*

$$
\mathcal{P}_c = \left\{ \Sigma^{\sigma}_{x=0,\, y=0}, \ \Sigma^{\sigma}_{y=xm,\, y=xcm^{-\sigma}}; m \in K^+ - \{0\} \right\}
$$

is a regular parallelism.

PROOF. We know that we have a regular partial parallelism. For fields not of characteristic 2, we see that we actually obtain a regular parallelism. The partial spreads A_α consist of components $y = xk$; $k^{\sigma+1} = \alpha$. Hence, i maps A_α onto $A_{\alpha^{-1}}$ and therefore preserves the linear set. Therefore, i is a collineation of $\Sigma^{\sigma}_{x=0,y=0}$ and maps $\Sigma^{\sigma}_{y=xm,y=xcm^{-\sigma}}$ to $\Sigma^{\sigma}_{y=xm^{-1},y=xcm^{\sigma}}$, and letting $m^{-1} = n$, then $cm^{\sigma} = cn^{-\sigma}$. We note that σ^* preserves all partial spreads A_α and maps $y = xt$ onto $y = xt^{\sigma}$. Clearly then σ^* is a collineation of the parallelism. $\square$

3. Isomorphisms of Transitive Parallelisms

Consider transitive partial parallelisms $\mathcal{P}_c$ and $\mathcal{P}_{c'}$ and assume they are isomorphic by f. We may assume that $\Sigma^{\sigma}_{x=0,y=0}$ is left invariant by f. Assume that $\{x = 0, y = 0\}$ is also left invariant by f. Then the linear set with respect to this pair is permuted, which implies that $\Sigma_{x=0,y=0}$ is also left invariant. We also assume that f fixes $x = 0$ and fixes $y = 0$ and that f is linear (the reader should verify why this is true). Hence, f necessarily has the form : $(x, y) \mapsto (xa, yb)$, for $a, b \in K^+ - \{0\}$. We then require that $\{m, cm^{-\sigma}\} \rightarrow \{ma^{-1}b, cm^{-\sigma}a^{-1}b\}$ must imply that $c'(ma^{-1}b)^{-\sigma} = cm^{-\sigma}a^{-1}b$, for all m. Therefore, $c' =$

$c(a^{-1}b)^{\sigma+1}$or $c'(cm^{-\sigma}a^{-1}b)^{-\sigma} = ma^{-1}b$, and the same condition holds. We have therefore proved the following theorem.

THEOREM 168. *If isomorphisms between regular partial parallelisms $\mathcal{P}_c$ and $\mathcal{P}_{c'}$ leave invariant*

$$\left\{\{x = 0, y = 0\}, \{y = xm, y = xcm^{-\sigma}\}; m \in K^+ - \{0\}\right\},$$

then $\mathcal{P}_c$ and $\mathcal{P}_{c'}$ are isomorphic if and only if $c' = cd^{1+\sigma}$, for some $d \in K^+ - \{0\}$.

Hence, the isomorphism classes are in $1 - 1$ correspondence with the non-identity elements of the quotient group $K^*/K^{+*(\sigma+1)}$.

PROOF. If the set $\{\{x = 0, y = 0\}, \{y = xm, y = xcm^{-\sigma}\}\}$ is invariant, then the argument given above shows without loss of generality that $x = 0$ and $y = 0$ are left invariant.   □

3.1. Examples. For c in $K^* - K^{+*(\sigma+1)}$, we note that $c^2 = c^{\sigma+1}$, so $K^*/K^{+*(\sigma+1)}$ is en elementary Abelian 2-group. In Betten and Riesinger [9], K is the field of real numbers, so there is a unique parallelism as $K^*/K^{+*(\sigma+1)} \simeq Z_2$. But our results here are also valid for any subfield of the reals. For example, take $K = Q_a$, the field of rationals and consider the field extension $\left\{\begin{bmatrix} u & -t \\ t & u \end{bmatrix}; u, t \in Q_a\right\}$. We note that $\{u^2 + t^2; u, t \in Q_a\} = K^{+(\sigma+1)}$. Let S denote the set of sums of the rational squares. So, we are considering Q_a/S. The reader is directed to the open problems Chapter 40 for a related problem.

3.2. The $SL(2, K)$-Partial Spreads. Now using the theory developed in Chapter 10, we have the following theorem.

THEOREM 169. *For every regular partial parallelism that admits a transitive group G, there is an associated partial spread in $PG(7, K)$ that admits a collineation group isomorphic to $SL(2, K)G$, that leaves invariant a K-regulus partial spread $\mathcal{R}$. The partial spread is a union of derivable partial spreads that mutually share $\mathcal{R}$, where $SL(2, K)$, generated by elations, leaves each derivable partial spread invariant and G acts transitively on the set of derivable partial spreads. When the characteristic of K is not 2, the partial spread is a spread in $PG(7, K)$ and an associated translation plane. The derivation of any derivable partial spread produces a translation plane admitting $SL(2, K)$, generated by Baer collineations.*

A 'generalized line star' is a set $\mathcal{S}$ of lines of $PG(3, K)$ that each intersect $\mathcal{Q}$ in exactly two points, such that the non-interior points of $\mathcal{Q}$ are covered by the set $\mathcal{S}$. Using the duality $\perp$ induced on $PG(3, K)$ by a quadric, every 2-secant ℓ maps to an exterior line $\ell^\perp$ to $\mathcal{Q}$. Now take the set of lines containing $\ell^\perp$, which defines a linear elliptic flock. Using the Klein mapping, we obtain a Pappian spread Σ corresponding to $\mathcal{Q}$ and the linear flock corresponds to the Pappian spread obtained from Σ by the replacement of a set of mutually disjoint reguli that covers Σ with the exception of two components. Furthermore, Betten and Riesinger show that generalized line stars produce regular parallelisms of $PG(3, K)$, which are coverings of the line set of $PG(3, K)$ by Pappian (regular) spreads. Now choose any conic $\mathcal{C}$ of intersection of $\mathcal{Q}$ and let $\pi_\mathcal{C}$ denote the corresponding plane containing $\mathcal{C}$ and form $\pi_\mathcal{C}^\perp = P_\mathcal{C}$. Since $P_\mathcal{C}$ is exterior to $\mathcal{Q}$, there is a unique 2-secant $\ell_{P_\mathcal{C}}$ containing $P_\mathcal{C}$ of the generalized line star. Then $\ell_{P_\mathcal{C}}^\perp \subset \pi_\mathcal{C}$ so that $\mathcal{C}$ corresponds to one of the lines of a spread of the parallelism. All of this is noted in Betten and Riesinger [9], but not using the language of parallelisms, assuming that the characteristic of K is not 2. Also, note that the existence of Pappian spreads in $PG(3, K)$ require the existence of a quadratic extensions F of K. We restate this theorem using our language.

THEOREM 170. *(Betten and Riesinger [9]) Let K be field of characteristic not 2 that admits a quadratic extension F. Every generalized line star of an elliptic quadric $\mathcal{Q}$ also produces a regular parallelism of $PG(3, K)$.*

REMARK 52. *There are a very large number of parallelisms over the field of real numbers, for which undoubtedly there are generalizations to various other fields, but which we have not tried to include here due to space requirements. The reader particularly interested in parallelisms in $PG(3, \mathcal{R})$, where $\mathcal{R}$ is the set of real numbers and or who is interested in topological spreads and parallelisms is particularly directed to the work of Betten and Riesinger [13], [10], [12], [11], [9] and the work by the author and Pomareda [152]. There are both regular and non-regular parallelisms constructed.*

4. Finite Regular Parallelisms

In the previous section, we have seen that there are a great variety of regular parallelisms of $PG(3, K)$, where K is an infinite field that admits a quadratic extension. When K is finite and isomorphic to $GF(q)$, the situation is much different.

Consider first a parallelism in $PG(3, 2)$. Since the associated spreads of order 4 are necessarily Desarguesian, any such parallelism is regular. Hence, there are associated translation planes of order 16, with spreads in $PG(7, 2)$, constructed as in the previous chapters. These translation planes are quite interesting in that they both admit $PSL(2, 7)$ as a collineation group. The two planes are transposes of each other called the Lorimer-Rahilly and Johnson-Walker planes of order 16. The planes admit a set of $1 + 2 + 2^2$ derivable nets of degree $1 + 4$ sharing a regulus of degree $1 + 2$. The group $PSL(2,7)$ acts doubly transitively on these derivable nets and induces a doubly transitive group on the 7 spreads of the two associated parallelisms. Since the process of taking the transpose of a translation plane corresponds to taking a duality of the associated projective space, it then follows that the two parallelisms in $PG(3,2)$ are dual to each other and are non-isomorphic.

Then Denniston [**43**] determined two regular parallelisms in $PG(3,8)$, again dual to each other. It was initially thought that only even order regular parallelisms were possible, however, Prince [**176**] using a computer, was able to determine all cyclic parallelisms in $PG(3,5)$ and found two regular parallelisms, again dual to each other. Penttila and Williams [**174**] extended all of these three types of regular parallelism to two infinite classes of regular parallelisms in $PG(3,q)$, where $q \equiv 2 \bmod 3$, where the two classes are dual to each other.

5. The Penttila-Williams Construction

The reader is also directed to the appendix Chapter 42 for background on hyperbolic quadrics in $PG(5,q)$ and the connection with translation planes of order q^2.

Let $GF(q^3) \oplus GF(q^3)$ be a 6-dimensional vector space V_6, where a quadric $\mathcal{Q}$ is defined as

$$\mathcal{Q}(x,y) = T(xy) = xy + (xy)^q + (xy)^{q^2},$$

for all $x, y \in GF(q^3)$. Let $\Sigma = \left\{ (y,z); y^{q^2} + z \in GF(q) \right\}$, for $y, z \in GF(q^3)$. Define a group action as follows: $h_u : (x,y) \to (ux, u^{-1}y)$, where u has order dividing $1 + q + q^2$. Let $G = \langle h_u; |u| \mid 1 + q + q^2 \rangle$, which clearly preserves the quadric. Then Penttila and Williams show that

$$\{(h_u\Sigma) \cap \mathcal{Q}\}, h_u \in G,$$

is a partition of the quadric $\mathcal{Q}$ by elliptic quadrics.

The main points of their ingenious proof are as follows:

(1) $\mathcal{Q}$ is a hyperbolic quadric provided $q \equiv 2 \bmod 3$ (this is, exactly when $(1 + q + q^2, q - 1) = (q - 1, 3) = 1$ and q is not of characteristic 3).

(2) Σ is a 4-dimensional vector space, upon which $\mathcal{Q}$ induces an elliptic quadric.

(3)

$$\{(h_u\Sigma) \cap \mathcal{Q}\}, h_u \in G,$$

is a partition of $\mathcal{Q}$.

Now, in the finite case, all hyperbolic quadrics in $PG(5, q)$ are equivalent so that by the Klein map, we have a partition of the lines of $PG(3, q)$ by a set of $1+q+q^2$ regular spreads, which then produces a regular parallelism. The dual parallelism turns out to be non-isomorphic to this parallelism, and it is also shown to correspond to the space $\Lambda = \{(y, z); y^q + z \in GF(q)\}$, for $y, z \in GF(q^3)$. We shall use the fact that $\mathcal{Q}$ induces an elliptic quadric on Σ if and only if $\Sigma^\perp$ is anisotropic. Define $\langle (x, y), (x^*, y^*) \rangle = T(x^*y + y^*x)$. Clearly, $\langle, \rangle$ defines a symmetric bilinear form on the vector space.

$$\mathcal{Q}((x, y) + (x^*, y^*)) = T((x + x^*, y + y^*)),$$

which is equal to

$$(x + x^*)(y + y^*) + ((x + x^*)(y + y^*))^q + ((x + x^*)(y + y^*))^{q^2}$$
$$= T(xy) + T(x^*y^*) + T(x^*y + y^*x)$$
$$= \mathcal{Q}(x, y) + \mathcal{Q}(x^*, y^*) + \langle (x, y), (x^*, y^*) \rangle.$$

Also,

$$\mathcal{Q}(\alpha(x, y)) = \alpha^2 \mathcal{Q}(x, y), \text{ for } \alpha \in GF(q).$$

Therefore, $\mathcal{Q}$ is a quadric. To see that it is non-degenerate, assume that for all (x, y) that there exists a vector (x^*, y^*) so that $\langle (x, y), (x^*, y^*) \rangle = T(x^*y + y^*x) = 0$. Suppose that $x = 1$ and $y = 0$, so that $T(y^*x) = 0 = (y^*x) + (y^*x)^q + (y^*x)^{q^2} = 0$, in x, which is identically zero and of degree $< q^3$. Hence, the coefficients are zero implying that $y^* = 0$. A symmetric argument shows that $x^* = 0$, showing that $\mathcal{Q}$ is a non-degenerate quadric.

We now give proofs to the above mentioned main points of the construction.

LEMMA 46. $\mathcal{Q}$ is a hyperbolic quadric provided $q \equiv 2 \bmod 3$.

PROOF. Note that $\{(x, 0); x \in GF(q^3)\}$ is a subspace of dimension 3, which is contained in the quadric. Hence, the Witt index 3 (maximum dimension of a subspace in the quadric), which implies that the quadric is hyperbolic. $\square$

LEMMA 47. Σ is a 4-dimensional vector space upon which $\mathcal{Q}$ induces an elliptic quadric.

PROOF. It suffices to show that $\Sigma^\perp$ is anisotropic. We claim that

$$\{(x, x^q); T(x) = 0\}$$

is in $\Sigma^{\perp}$.

$$\Big\langle (x, x^q), (y, z); y^{q^2} + z \in GF(q) \Big\rangle$$
$$= T(xz + x^q y)$$
$$= (xz) + (xz)^q + (xz)^{q^2} + (x^q y) + (x^q y)^q + (x^q y)^{q^2}$$
$$= \Big((xz) + (x^q y)^{q^2} \Big) + \big((xz)^q + (x^q y) \big) + \Big((xz)^{q^2} + (x^q y)^q \Big)$$
$$= \alpha T(x) = 0.$$

Since $\{(x, x^q); T(x) = 0\}$ is a subspace of dimension at least 2, it follows that $\Sigma^{\perp} = \{(x, x^q); T(x) = 0\}$. Let $w = (x, x^q)$ and $w^q = (x^q, x)$ such that $T(x) = 0$. Assume that $\beta w = w^q$, for $\beta \in GF(q)$, so that $\beta x = x^q$, so $\beta = x^{q-1}$, implying $x^{q^2} = \beta x^q = x^{2q-1}$, so that $x^{(q-1)^2} = 1$. But $(q-1, q^3 - 1) = (q-1)$, so that $x^{q-1} = 1$, and so $x \in GF(q)$. But then $0 = T(x) = 3x$, and since $q \equiv 2 \bmod 3$, then $x = 0$. Therefore, we may assume that $\Sigma^{\perp} = \langle w, w^q \rangle$, for some non-zero $w = (x, x^q)$, such that $T(x) = 0$.

Now suppose there is a vector $w = (x, x^q) \in \Sigma^{\perp}$ such that $\mathcal{Q}(w) = T(x^{q+1}) = 0$, and then $\mathcal{Q}(w^q) = T(x^{q(q+1)}) = 0$. Consider $\alpha, \beta \in GF(q)$ and a vector $\alpha w + \beta w^q$ of $\Sigma^{\perp}$. Then $\mathcal{Q}(\alpha w + \beta w^q) = \alpha^2 \mathcal{Q}(w) + \beta^2 \mathcal{Q}(w^q) + \alpha\beta \langle w, w^q \rangle$. It follows that

$$\langle w, w^q \rangle = T(x^{1+q^2}) + T(x^{2q}) = T(x^{1+q^2}) + T(x^2)$$
$$= T(x(x^{q^2} + x) = T(x(x + x^q + x^{q^2})) - T(x^{1+q})$$
$$= T(xT(x)) = T(x)T(x) = 0.$$

Hence, if there is a singular vector in $\Sigma^{\perp}$ then $\Sigma^{\perp}$ is totally singular.

If $\Sigma^{\perp}$ is totally singular, then for each x in $GF(q^3)$ so that $T(x) = 0$, then $\mathcal{Q}(x, x^q) = 0 = T(x^{q+1})$. Choose any element v of $GF(q^3)$ and notice that $T(T(v)) = 3T(v)$, which implies that $T(v - \frac{T(v)}{3}) = 0$, and thus shows that

$$(*) : T\big((v - \frac{T(v)}{3})^{q+1}\big) = 0, \text{ for all } v \in GF(q^2).$$

Now a short calculation, which the reader is encouraged to complete, shows that

$$(v - \frac{T(v)}{3})^{q+1} = \frac{1}{9}(-2v^2 + 5v^{q+1} - 2v^{2q} - v^{q^2+1} - v^{q^2+q} + v^{2q^2}).$$

Note that $T(v^2) = T(v^{2q}) = T(v^{2q^2})$ and $T(v^{q+1}) = T(v^{q^2+q}) = T(v^{1+q^2})$. Hence, $(*)$ shows that

$$(**) : \frac{1}{9}(-3T(v) - 3T(v^{q+1})) = -\frac{1}{3}(T(v) + T(v^{q+1})) = 0.$$

But, then
$$v + v^q + v^{q^2} + v^{q+1} + v^{q^2+q} + v^{1+q^2} = 0$$
for all $v \in GF(q^3)$. However, since $q^2 + q < q^3$, the coefficients of this polynomial in v must all be zero, a contradiction.

Hence, $\Sigma^\perp$ is anisotropic, which proves the lemma. □

LEMMA 48.
$$\mathcal{P}_{q^2} = \{(h_u\Sigma) \cap \mathcal{Q}\}, h_u \in G,$$
is a partition of $\mathcal{Q}$.

PROOF. Assume not! Then there is an element u of order dividing $1 + q + q^2$ such that $\Sigma \cap h_u\Sigma \cap \mathcal{Q}$ is non-empty. Note that

$$(h_u\Sigma)^\perp \cap \Sigma^\perp = \left\{(ux, \frac{x^q}{u}); T(x) = 0\right\} \cap \{(x, x^q); T(x) = 0\}$$
$$= \{(0,0)\},$$

since an intersection $(ux, x^q/u) = (x, x^q)$ if and only if $u^{1+q} = 1$, a contradiction as $(1 + q, 1 + q + q^2) = 1$, if u is not 1. This means that $(h_u\Sigma)^\perp \cup \Sigma^\perp$ has dimension 4, which implies that $((h_u\Sigma)^\perp \cup \Sigma^\perp)^\perp = \Sigma \cap h_u\Sigma$ has dimension 2.

We claim that
$$\left\langle (\frac{u^q}{u^q - u^{q^2+q}}, \frac{-u}{u - u^{q+1}} + 1), (\frac{u^q - 1}{u^q - u^{q^2+q}}, \frac{-u+1}{u - u^{q+1}} + 1) \right\rangle$$
$$= \Sigma \cap h_u\Sigma.$$

The two elements are easily seen to be linearly independent over $GF(q)$, assuming that u is not 1. To check that the two vectors are in $\Sigma \cap h_u\Sigma$, consider $(\frac{u^q}{u^q - u^{q^2+q}}, \frac{-u}{u - u^{q+1}} + 1)$. Then

$$\left(\frac{u^q}{u^q - u^{q^2+q}}\right)^{q^2} + \frac{-u}{u - u^{q+1}} + 1 = 1,$$

so that this vector is in Σ. Then
$$\left(\frac{u^q}{u^q - u^{q^2+q}}, \frac{-u}{u - u^{q+1}} + 1\right) = \left(u\left(\frac{u^{-1}}{1 - u^{q^2}}\right), \frac{1}{u}\left(\frac{-u^{q+1}}{1 - u^q}\right)\right)$$
and
$$\left(\frac{u^{-1}}{1 - u^{q^2}}\right)^{q^2} + \frac{-u^{q+1}}{1 - u^q} = 0,$$

showing the element is also in $h_u\Sigma$. Similarly, the second vector is in $\Sigma \cap h_u\Sigma$. The proof is complete if it can be shown that for $z = (\frac{u^q}{u^q - u^{q^2+q}}, \frac{-u}{u - u^{q+1}} + 1)$ and $w = (\frac{u^q - 1}{u^q - u^{q^2+q}}, \frac{-u+1}{u - u^{q+1}} + 1)$ then $\mathcal{Q}(\alpha z + \beta w) \neq 0$, for all $\alpha, \beta \in GF(q)$, not both zero.

Since

$$\mathcal{Q}(\alpha z + \beta w) = \alpha^2 \mathcal{Q}(z) + \beta^2 \mathcal{Q}(w) + \alpha\beta \langle z, w \rangle,$$

it remains to calculate $\mathcal{Q}(z)$, $\mathcal{Q}(w)$, and $\langle z, w \rangle$. Since

$$\mathcal{Q}(z) = T\left(-\frac{u^{2q+1}}{(u - u^{q+1})^{q+1}}\right),$$

the reader is challenged to complete this straightforward but lengthy calculation, which shows that $\mathcal{Q}(z) = 1$. Similarly,

$$\mathcal{Q}(w) = T\left(\frac{(u^q - 1)(1 - u^{q+1})}{(u - u^{q+1})^{q+1}}\right),$$

and a much shorter calculation than that as above determines this trace as 3, and similarly

$$\langle z, w \rangle = T\left(\frac{u^q(1 - u^{q+1})}{(u - u^{q+1})^{q+1}}\right) + T\left(\frac{-u^{q+1}(u^q - 1)}{(u - u^{q+1})^{q+1}}\right) = 3.$$

Therefore,

$$\begin{aligned}
\mathcal{Q}(\alpha z + \beta w) &= \alpha^2 \mathcal{Q}(z) + \beta^2 \mathcal{Q}(w) + \alpha\beta \langle z, w \rangle \\
&= \alpha^2 + 3\alpha\beta + 3\beta^2,
\end{aligned}$$

which is equal to 0 for $\beta = 0$ if and only if $\alpha^2 = 0$ and hence $\alpha = 0$. Therefore, assume that $\beta \neq 0$ so that $\alpha^2 + 3\alpha\beta + 3\beta^2 = 0$ if and only if $t^2 + 3t + 3 = 0$, for $t = \frac{\alpha}{\beta}$. If q is even, then the equation becomes $t^2 + t + 1$, which has trace over $GF(2)$ of $T_2(1) = r \bmod 2$, if $q = 2^r$, and since r is odd, and since $q \equiv 2 \bmod 3$, then there are no non-zero solutions in even order. If q is odd then the discriminant is -3. However, -3 is a non-square in $GF(q)$, where $q \equiv 2 \bmod 3$, and q is odd. This completes the proof. $\qquad\square$

Hence, using the fact that all finite hyperbolic quadrics are equivalent, the Klein mapping establishes a regular parallelism of $PG(3, q)$. Taking a polarity of $PG(3, q)$, there is an associated 'dual parallelism' obtained, where the corresponding spreads are the dual spreads of the original parallelism. Penttila-Williams show that these two parallelisms are never isomorphic by a detailed analysis of the collineation groups. The reader is directed to [174] for details on the group theory.

5.1. The New $SL(2, q) \times C_{1+q+q^2}$ Planes. We then have completed the proof of the Penttila-Williams construction of regular parallelisms. We note, of course, that there are then two infinite classes of translation planes of order q^4 with spreads in $PG(7, q)$, that are covered by $1+q+q^2$ derivable nets that share a fixed regulus net of degree $1 + q$, so that the planes admit $SL(2, q) \times C_{1+q+q^2}$, as a collineation group, where C_{1+q+q^2} is a cyclic group acting transitively on the set of $1 + q + q^2$ derivable nets. As mentioned previously, three of these classes of parallelisms in $PG(3, 2)$, $PG(3, 8)$, and $PG(3, 5)$ were known previously, the latter two being constructed by Denniston and Prince, respectively, and the first two parallelisms in $PG(3, 2)$ corresponding to the Lorimer-Rahilly and Johnson-Walker planes of order 16 admitting $PSL(2, q)$. The remaining parallelisms and their corresponding translation planes are then completely new.

CHAPTER 27

Beutelspacher's Construction of Line Parallelisms

In this chapter, we give the important construction of Beutelspacher of parallelisms of $PG(2^r - 1, q)$, and generalize the construction to arbitrary skewfields. This chapter is a modification of the author's work [**108**].

In 1974, Buetelspacher [**16**] gave a construction of parallelisms in $PG(2^r - 1, q)$, for any positive integer r and for any prime power $q = p^t > 2$. We offer a generalization of Buetelspacher's construction to arbitrary projective spaces over skewfields and construct a variety of new parallelisms.

We first somewhat generalize the notion of a geometric spread as used by Buetelspacher as follows. It will be assumed that all vector spaces are 'left' vector spaces and skewfield extensions are always considered to be on the left.

We recall from Chapter 3 the ideas of a Desarguesian line spread.

The following proposition is straightforward and left for the reader as an exercise.

PROPOSITION 6. *Let $K_1 \subseteq K_2$ be skewfields, K_2 is 2-dimensional vector space over K_1 and let V be a K_2-vector space. Let $\mathcal{R}$ denote the set of all 1-dimensional K_2-subspaces of V. Then, $\mathcal{R}$ is a Desarguesian line spread of $PG(V - 1, K_1)$.*

NOTATION 3. *In the following, we shall adopt the following notation: Let $\mathcal{R}$ denote a Desarguesian line spread in $PG(V - 1, K_1)$. If g and h are distinct elements of $\mathcal{R}$, let $\langle g, h \rangle$ denote the 2-dimensional K_2-vector subspace generated by g and h, which is also considered a 4-dimensional K_1-vector subspace and projectively as isomorphic to $PG(3, K_1)$.*

If L is a subfield containing K_1 and $z_1, z_2, .., z_s \in V$, the L-vector subspace generated by $\{z_i; i = 1, .., s\}$ shall be denoted by $\langle z_i; i = 1, .., n \rangle_L$.

The following propositions are analogous to similar ones in Buetelspacher [**16**].

PROPOSITION 7. *$\mathcal{R}$ as a set of 'points' and $\{\langle g, h \rangle_{K_2} ; g, \neq h\} \in \mathcal{R}$ as 'lines' is isomorphic to $PG(V - 1, K_2)$.*

*Let L be a line of $PG(V-1, K_1) - \mathcal{R}$. If L nontrivially inter-
sects $\langle g, h \rangle$ for $g, h \in \mathcal{R}$, then $L \subseteq \langle g, h \rangle_{K_2}$ and there exists a unique
$\langle g^*, h^* \rangle_{K_2}$, for $g^*, \neq h^*$ in $\mathcal{R}$ containing L.*

PROOF. Let $PQ = L$, where P and Q are points of $PG(V-1, K_1)$.
Since $\mathcal{R}$ is a spread, there exist unique lines g and h containing P
and Q, respectively, and $g \neq h$ since L is not in $\mathcal{R}$. Let $g = \langle e_g \rangle_{K_2}$
and $h = \langle e_h \rangle_{K_2}$, 1-dimensional K_2-subspaces. P in g means that $P = \langle a e_g \rangle_{K_1}$ where $a \in K_2$. Similarly, $Q = \langle b e_h \rangle_{K_1}$, where $b \in K_2$. Then
L, as a 2-dimensional K_1-subspace, is $\langle a e_g, b e_h \rangle_{K_1}$, which clearly is in
$\langle e_g, e_h \rangle_{K_2} = \langle g, h \rangle$.
    Let $L = P^* Q^* = PQ$, let $g^* = \langle e_{g^*} \rangle_{K_2}$, $h^* = \langle e_{h^*} \rangle_{K_2}$ contain
P^* and Q^*, respectively. Then $L = \langle a^* e_{g^*}, b^* e_{h^*} \rangle_{K_1}$, for some a^*, b^*
in K_2. Hence, $\langle a^* e_{g^*}, b^* e_{h^*} \rangle_{K_1} = \langle a e_g, b e_h \rangle_{K_1}$. This implies that $a^* e_{g^*}^*$
is in $\langle a e_g, b e_h \rangle_{K_1}$, implying that e_{g^*} is in $\langle e_g, e_h \rangle_{K_2}$. By symmetry,
$\langle g, h \rangle_{K_2} = \langle e_g, e_h \rangle_{K_2} = \langle e_{g^*}, e_{h^*} \rangle_{K_2} = \langle g^*, h^* \rangle_{K_2}$.   □

PROPOSITION 8. *$\mathcal{R} \mid \langle g, h \rangle_{K_1}$ is a line spread of $\langle g, h \rangle_{K_1}$ as a projective space isomorphic to $PG(3, K_1)$, for each 'line' $\langle g, h \rangle_{K_2}$ of $\mathcal{R}$.*

PROOF. When an element M of $\mathcal{R}$ nontrivially intersects $\langle g, h \rangle_{K_2}$,
it is contained in $\langle g, h \rangle_{K_2}$ since M is a 1-dimensional K_2-subspace and
$\langle g, h \rangle_{K_2}$ is a 2-dimensional K_2-subspace as well as a 4-dimensional K_1-
subspace. Since $\mathcal{R}$ is a spread, each point of $\langle g, h \rangle_{K_2}$ is covered uniquely
by some element N of $\mathcal{R}$.   □

1. Extension of Beutelspacher's Theorem

Since we are dealing with arbitrary skewfields, we begin with some
elementary properties of spreads and parallelisms.

PROPOSITION 9. *Let W be any vector space of finite dimension $d > 3$ over a skewfield L, let $\mathcal{S}_1$ and $\mathcal{S}_2$ be line spreads of $PG(W-1, L)$, and
let $\mathcal{P}_1$ and $\mathcal{P}_2$ be parallelisms of $PG(W-1, L)$. Then $\operatorname{card} \mathcal{S}_1 = \operatorname{card} \mathcal{S}_2$
and $\operatorname{card} \mathcal{P}_1 = \operatorname{card} \mathcal{P}_2$.*

PROOF. Since the proposition is trivial when L is finite, assume
that L is infinite. The proof of this result simply uses fundamental
properties of the cardinalities of infinite sets and is left to the interested
reader to complete.   □

THEOREM 171. *Let $K_1 \subseteq K_2$ be skewfields, K_2 is 2-dimensional
over K_1 and let V be a K_2-vector space. Let $\mathcal{R}$ denote the set of
all 1-dimensional K_2-subspaces; an associated Desarguesian spread of
$PG(V-1, K_1)$. Let $\mathcal{P}$ be a line parallelism of $PG(V-1, K_2)$ and for*

each $PG(3, K_1)$, $\langle g, h \rangle$, $g \neq h$ of $\mathcal{R}$, let $\mathcal{M}_{\langle g,h \rangle}$ denote a parallelism of $\langle g, h \rangle$ containing $\mathcal{R} \mid \langle g, h \rangle$.

More generally, index the spreads of the parallelism $\mathcal{P}$ by Ω, so that

$$\mathcal{P} = \cup \{\mathcal{M}^s \; ; \; s \in \Omega\}.$$

Let Λ be a set of cardinality the cardinality minus 1 of any parallelism of line spreads of any $PG(3, K_1)$ and let Δ be a set of cardinality the cardinality of lines of any spread of lines of $PG(V - 1, K_2)$.

For each spread $\mathcal{M}^s$ of $\mathcal{P}$ in $PG(V - 1, K_2)$;(the points are those of $\mathcal{R}$), for $s \in \Omega$, index the spread lines by $m^{s,k}$ by $k \in \Delta$.

Assume that each $m^{s,k}$ as a projective space isomorphic to $PG(3, K_1)$ admits a parallelism containing the parallelism induced by $\mathcal{R}$, and let $\mathcal{P}^{m^{s,k}}$ denote the set of spreads not equal to the induced spread. Denote by $S_\Lambda^{\mathcal{P}^{m^{s,k}}}$ the symmetric group on $\mathcal{P}^{m^{s,k}}$ permuting the spreads, a set of cardinality Λ, and let $m_t^{s,k}$, for $t \in \Lambda$, denote the spreads of $\mathcal{P}^{m^{s,k}}$.

For $i \in \Lambda$, denote by $\lambda^{\mathcal{P}^{m^{s,k}}}(i)$ the action of the associated permutation on the element i. Hence, we are considering a family of groups isomorphic to S_Λ and acting on Λ considered as $\mathcal{P}^{m^{s,k}}$.

(1) For each $m^{s,k}$ choose a permutation $\lambda^{\mathcal{P}^{m^{s,k}}} \in S_\Lambda^{\mathcal{P}^{m^{s,k}}}$. Denote by $m^{s,k}_{\lambda^{\mathcal{P}^{m^{s,k}}}(i)}$, the spread $m_j^{s,k}$ in $\mathcal{P}^{m^{s,k}}$ chosen by $j = \lambda^{\mathcal{P}^{m^{s,k}}}(i)$.

Then, for each $s \in \Omega$ and spread $\mathcal{M}^s$ of $\mathcal{P}$ in $PG(F - 1, K_2)$ for each $\lambda^{\mathcal{P}^{m^{s,k}}} \in S_\Lambda^{\mathcal{P}^{m^{s,k}}}$, and for each $i \in \Lambda$,

$$\Gamma^{\mathcal{M}^s}_{\{\lambda^{\mathcal{P}^{m^{s,k}}}(i)\}} = \left\{ m^{s,k}_{\lambda^{\mathcal{P}^{m^{s,k}}}(i)} ; m^{s,k} \in \mathcal{M}^s \right\}$$

is a line spread of $PG(V - 1, K_1)$.

(2)

$$\Gamma_{\{\lambda^{\mathcal{P}^{m^{s,k}}}\}} = \mathcal{R} \cup_{i \in \Lambda, s \in \Omega} \left\{ \Gamma^{\mathcal{M}^s}_{\{\lambda^{\mathcal{P}^{m^{s,k}}}(i)\}} \right\}$$

is a line parallelism of $PG(V - 1, K_1)$.

(3) From each parallelism of $PG(V - 1, K_2)$, and for each parallelism of each line as a projective space isomorphic to $PG(3, K_1)$ containing the induced spread of each spread not equal to $\mathcal{R}$, there is a corresponding line parallelism of $PG(V - 1, K_1)$. The cardinality of line parallelisms obtained is this way is

$$((card\ S_\Lambda)^{card\ \Delta})^{card\ \Omega}.$$

PROOF. To prove (1), we need to show that every point P of $PG(V - 1, K_1)$ is incident with a unique line of $\Gamma^{\mathcal{M}^s}_{\{\lambda^{\mathcal{P}^{m^{s,k}}}(i)\}}$. Every point P is incident with a unique line g_P of $\mathcal{R}$, which is, as a point, on a unique line m of $\mathcal{M}$. Now m is isomorphic to $PG(3, K_1)$ and contains g_P. Furthermore, m_i is a spread in m and each point P is incident with exactly one line of each spread m_i for each $i \in \lambda$. Hence, $\Gamma^{\mathcal{M}^s}_{\{\lambda^{\mathcal{P}m^{s,k}}(i)\}}$ is a line spread, which proves (1).

We note, by the index assumption, that no line of m_i is in $\mathcal{R}$.

To show that $\Gamma_{\{\lambda^{\mathcal{P}m^{s,k}}\}} = \mathcal{R} \cup_{i \in \Lambda, s \in \Omega} \left\{ \Gamma^{\mathcal{M}^s}_{\{\lambda^{\mathcal{P}m^{s,k}}(i)\}} \right\}$ is a line parallelism, we need to show that every line L that is not in $\mathcal{R}$ is in a unique spread of the form $\Gamma^{\mathcal{M}^s}_{\{\lambda^{\mathcal{P}m^{s,k}}(i)\}}$. We know that L is in a unique $\langle g, h \rangle_{K_2}$ for $g, \neq h$ of $\mathcal{R}$. Moreover, since $\langle g, h \rangle_{K_2}$ is a line of $\mathcal{R}$, and $\mathcal{P}$ is a parallelism of $\mathcal{R}$, where $\mathcal{R}$ is the set of points of $PG(V - 1, K_2)$, it follows that there is a unique spread $\mathcal{M}$ of $\mathcal{P}$ containing $\langle g, h \rangle_{K_2} = m$ as a line. But, m is isomorphic to $PG(3, K_1)$ and the assumed parallelism of m contains $\mathcal{R} \mid m$. Since L is a line not in $\mathcal{R}$, it follows that there is a unique spread m_j of m containing L. This completes the proof of part (2).

To obtain a line parallelism, we choose a fixed permutation from each symmetric group of each parallelism of $PG(3, K_1)$ arising from each line of each spread not equal to $\mathcal{R}$ of the parallelism $\mathcal{P}$. This proves (3).   □

The following corollary shows how parallelisms 'grow' using the construction.

COROLLARY 37. *In the finite case and K_1 isomorphic to $GF(q)$, $\Lambda = q + q^2$ (the number of spreads minus 1 in a parallelism of $PG(3, q)$) so that S_{q+q^2} has cardinality $((q + q^2)!)$. With K_2 isomorphic to $GF(q^2)$ and V a 2d-dimensional $GF(q^2)$-space, a line spread in $PG(d - 1, q^2)$ has $(q^{2d} - 1)/(q^4 - 1)$ lines and a parallelism has*

$$((q^{4d} - 1)(q^{4d} - q^2)) / ((q^4 - 1)(q^4 - q^2))$$

line spreads. Hence,

$$\text{card } \Delta = (q^{4d} - 1)/(q^4 - 1)$$

and

$$\text{card } \Omega = ((q^{4d} - 1)(q^{4d} - q^2))/((q^4 - 1)(q^4 - q^2)).$$

Then

(1) the number of possible line parallelisms of $PG(4d - 1, q)$ constructed as above is

$$(((q + q^2)!)^{\left(\frac{q^{4d}-1}{q^4-1}\right)\left(\frac{q^{2(2d-1)}-1}{q^2-1}\right)}.$$

(2) Assume that for $d = 2$, there is a parallelism of $PG(3, q^2)$ containing a regular spread $\mathcal{R}$ and there is a parallelism of $PG(3, q)$ containing a regular spread. Then there are

$$((q + q^2)!)^{\left(\frac{q^8-1}{q^4-1}\right)\left(\frac{q^{2(3)}-1}{q^2-1}\right)} = ((q + q^2)!)^{(q^4+1)(1+q^2+q^4)}.$$

parallelisms in $PG(2^3 - 1, q)$, each of which contains the regular spread induced from $\mathcal{R}$. Let $q = h^2$ and reapply the process constructing in $PG(2^4 - 1, \sqrt{q})$, then from each of

$$((q + q^2)!)^{\left(\frac{q^8-1}{q^4-1}\right)\left(\frac{q^{2(3)}-1}{q^2-1}\right)}$$

parallelisms of $PG(2^3 - 1, q)$, there are

$$((h + h^2)!)^{\left(\frac{h^{4d}-1}{h^4-1}\right)\left(\frac{h^{2(2d-1)}-1}{h^2-1}\right)} = ((\sqrt{q} + q)!)^{(q^2+1)(1+q+q^2)}$$

line parallelisms of $PG(2^4 - 1, \sqrt{q})$.
    Hence, there are

$$((\sqrt{q} + q)!)^{(q^2+1)(1+q+q^2)}{}^{((q+q^2)!)^{(q^4+1)(1+q^2+q^4)}}$$

possible line parallelisms of $PG(2^4 - 1, \sqrt{q})$.

2. Applications of Beutelspacher's Construction.

The previous section gives a construction of a large variety of line parallelisms in $PG(3, K_1)$, where K_1 is an arbitrary field that admits a quadratic field extension K_2. Furthermore, there are classification results using collineation group that may potentially be used in the extension of parallelisms in higher-dimensional projective spaces.

Using the extension theorem, we can construct line parallelisms in projective spaces as follows:

Assume that there is a set of fields $K_1 \subseteq K_2 \subseteq K_3 \subseteq ... \subseteq K_k \subseteq K_{k+1}$ such that K_{i+1} is a quadratic extension of K_i for $i = 1, 2, .., k$. To begin, assume that $k = 5$.

To construct a line parallelism of $PG(2^3 - 1, K_1)$, we consider K_4 as a K_2-vector space and take the Pappian spread of all 1-dimensional K_2-subspaces of K_4. This becomes a Pappian spread S_4 of $PG(2^3 - 1, K_1)$. Any two components g and h of S_4 generate a 2-dimensional K_2-subspace, which becomes a $PG(3, K_1)$ and S_4 induces a Pappian spread in $PG(3, K_1)$ coordinatizable by K_2. By the method of the previous section, we may first choose any line parallelism of such a $PG(3, K_1)$ that contains a Pappian spread coordinatizable by K_2 and consider that spread induced by S_4. We then may choose any line parallelism $\mathcal{P}_2$ of $PG(2^2 - 1, K_2)$. Since K_3 exists, our previous method allows the existence of a line parallelism here. Hence, we have obtained a variety of parallelisms in $PG(2^3 - 1, K_1)$ by the extension of Beutelspacher's construction.

Since $k = 5$ in our initial situation, we may also construct a variety of parallelisms in $PG(2^3 - 1, K_2)$. To continue, take K_5 as a K_2-vector space of dimension 2^3 and form the set of 1-dimensional K_2-subspaces of K_5, which then becomes a Pappian line spread S_5 of $PG(2^4 - 1, K_1)$. This Pappian line spread also induces in any associated $PG(3, K_1)$ determined by two components g and h the Pappian line spread coordinatizable by K_2. Another application of the extension theorem produces a line parallelism in $PG(2^4 - 1, K_1)$.

So, in the general case $z = 1, 2, .., k$, we may similarly construct line parallelisms in $PG(2^a - 1, K_1)$, for $a = 2, 3, .., k - 1$ and the reader is encouraged to complete this argument in the arbitrary and general case.

THEOREM 172. *(1) Let K_1 be a field and $K_1 \subseteq K_2 \subseteq K_3 \subseteq ... \subseteq K_k \subseteq K_{k+1}$ such that K_{i+1} is a quadratic field extension of K_i for $i = 1, 2, .., k$. Assume that V_z is a 2^z-dimensional K_z-vector space. Then, there exist line parallelisms in $PG(2^z - 1, K_1)$ for all $z = 2, .., k$.*

(2) If there is an infinite sequence of fields, each a quadratic extension field of the previous, then exist line parallelisms in $PG(2^z - 1, K_1)$ for all positive integers z.

(3) If there is another sequence $K_1 = K_1' \subseteq K_2' \subseteq K_3' \subseteq ... \subseteq K_k' \subseteq K_{k+1}'$, such that K_{i+1}' is a quadratic field extension of K_i', then, there exist another set of line parallelisms in $PG(2^z - 1, K_1)$, for all $z = 2, 3, .., k$.

If K_2 is not isomorphic to K_2' and the parallelisms are chosen so that there is a unique Pappian spread $\mathcal{R}$ and $\mathcal{R}'$ in each parallelism, respectively, using the corresponding sequence then none of these constructed line parallelisms can be isomorphic to their analogues of (1).

PROOF. Assume that there is an infinite sequence of fields. Whenever there is a construction of a line parallelism in $PG(2^a - 1, K_1)$, there is an analogous construction in $PG(2^a - 1, K_2)$. Then using the line parallelism in $PG(2^a - 1, K_2)$, together with a choice of parallelism for each associated $PG(3, K_1)$ containing a spread induced from a Pappian line spread of $PG(2^{a+1} - 1, K_1)$ taken as constructed from the 1-dimensional K_2-subspaces of K_{a+2}, there is a constructed line parallelism in $PG(2^{a+1} - 1, K_1)$.

If we have a parallelism of $PG(2^a - 1, K_1)$ containing $\mathcal{R}$ and a parallelism of $PG(2^a - 1, K_1)$ containing $\mathcal{R}'$ but all other spreads are non-Pappian, then an isomorphism from one to the other must map $\mathcal{R}$

onto $\mathcal{R}'$, implying that $PG(V - 1, K_2)$ is isomorphic to $PG(V - 1, K_2')$, for an appropriate vector space V, which, in turn, implies that K_2 is isomorphic to K_2'.   □

EXAMPLE 1. *Let K_1 be the field of rational numbers. Then there are $\varkappa_o$-distinct infinite sequences of fields each a quadratic extension of the previous field. Each such sequence provides a set of line parallelisms as above. Furthermore, there are choices of infinitely mutually non-isomorphic subfields K_2 quadratic over K_1. Any such sequence with the choice of exactly one Pappian spread within the parallelism produces an infinite number of mutually non-isomorphic line parallelisms.*

CHAPTER 28

Johnson Partial Parallelisms

In this chapter, we construct a variety of parallelisms in $PG(3, K)$, where K is an arbitrary field, finite or infinite. In the following, we shall let R be a regulus in $PG(3, K)$, where K is a field. This class also appears in the Subplane's Text [**114**]. Here we somewhat modify the construction for this text and is given here for convenience and completeness.

We first note the following fundamental property concerning Pappian spreads.

REMARK 53. *If K is any field that has a quadratic extension field K', then every element e of $K' - K$ defines an irreducible quadratic $x^2 - xf - g$ exactly when $e^2 = ef + g$.*

If K is not $GF(2)$, there exist two fields of matrices in $GL(2, K)$, which share exactly the matrix field $\left\{ \begin{bmatrix} u & 0 \\ 0 & u \end{bmatrix} ; u \in K \right\}$. Hence, there exist two distinct Pappian spreads containing a regulus.

PROOF. If K is not $GF(2)$, then there exist at least two elements e and e' in $K' - K$ that satisfy distinct irreducible quadratic equations over K. Let $e^2 = ef + g$ and $e'^2 = e'f' + g'$. Then the two matrix fields are

$$\left\{ \begin{bmatrix} u - tf & tg \\ t & u \end{bmatrix} ; u, t \in K \right\}$$

and

$$\left\{ \begin{bmatrix} u - tf' & tg' \\ t & u \end{bmatrix} ; u, t \in K \right\}.$$

These fields define Pappian spreads containing the regulus given by

$$\left\{ x = 0, y = x \begin{bmatrix} u & 0 \\ 0 & u \end{bmatrix} ; u \in K \right\}.$$

□

LEMMA 49. *Assume that there are two Pappian spreads Σ and Σ' that contain a regulus R.*

Let L be any line of R and let G denote the central collineation group of the Pappian affine plane associated with Σ, which fixes L pointwise.

Then G acts semi-regularly on the set of all lines of $PG(3, K)$ that are skew to L and not in Σ.

PROOF. We consider the associated Desarguesian affine plane $A(\Sigma)$. Let L' be any line of $PG(3, K)$ which is not in Σ and which is skew to L. Then L' is a 2-dimensional K-vector subspace. The intersection subspaces with L' and the spread lines of Σ define a spread for L'. Hence, L' becomes a Baer subplane of Σ incident with the zero vector, which is skew to L. Let g be in the central collineation group G of $A(\Sigma)$ with axis L. Then, $L'g \neq L'$ since otherwise, the center must be in the Baer subplane, whereas L' and L are disjoint. Hence, the group G is semi-regular. $\square$

LEMMA 50. *G_R fixes any Pappian spread Γ containing R and acts as a collineation group of the associated Pappian affine plane $A(\Gamma)$, which acts regularly on the partial spread $\Gamma - R$.*

PROOF. Any Pappian spread that contains R admits, as a collineation group, any group that fixes the opposite regulus linewise and which is generated by a central collineation group of the regulus net. But any central collineation of the ambient Pappian spread that fixes R and has center and axis in R projectively must induce a central collineation group of the net. Moreover, the subgroup G_R that leaves R invariant acts transitively on the components of the spread that are not in R. $\square$

LEMMA 51. *$\Sigma'g$ and $\Sigma'h$, for $g, h \in G$, share a line skew to L and not in Σ if and only if $\Sigma'g = \Sigma'h$.*

PROOF. Since G_R also is semi-regular on the images of L', it follows that the spread for $\Sigma' = R \cup \{L'g; g \in G_R\}$. Assume that $\Sigma'g$ and $\Sigma'h$ for $g, h \in G$ share a line skew to L and not in Σ. Then $\Sigma'gh^{-1}$ and Σ' share a line M' that is not in Σ and skew to L. But all lines not in R of Σ' are images of L' under the group G_R. Hence, $M' = L'w$, for some element $w \in G_R$. Moreover, M' in $\Sigma'gh^{-1}$ is the image under gh^{-1} of a line N' of Σ'. If N' is in R, then $N'gh^{-1}$ is a line of Σ, which implies that gh^{-1} is in G_R. If N' is not in R, then $M' = L'sgh^{-1}$, where s is in G_R and $N' = L's$. Thus, $L'sgh^{-1} = L'w$ but the group acts semi-regularly so this implies that $sgh^{-1} = w$. Since s and w are in G_R, this, in turn, implies that gh^{-1} is in G_R, which implies that $\Sigma'gh^{-1} = \Sigma'$ or equivalently that $\Sigma'g = \Sigma'h$. Hence, if two images share a line skew to L and not in Σ then they are equal. $\square$

THEOREM 173. *Let Σ be any Pappian spread in $PG(3, K)$ where K is a field. Let R be any regulus in the spread and L any line of R. Assume there exists a second Pappian spread Σ' in $PG(3, K)$ containing R (i.e., K not $GF(2)$).*

Let L' denote a line of Σ' that is not in the spread of Σ.

Let G denote the full central collineation group of Σ with axis L.

Then, $\{\Sigma'g; g \in G\}$ is a set of Desarguesian partial spreads that cover the lines that are skew to L and not in Σ.

PROOF. It remains to show that every line M', which is skew to L and not in Σ, is in the same G-orbit. Represent Σ in the form $x = 0, \; y = x \begin{bmatrix} u + \rho_1 t & \gamma_1 t \\ t & u \end{bmatrix}$, for all $u, t \in K$, where K is a field and $u^2 + u\rho_1 t - \gamma_1 t^2 = 0$ if and only if $u = t = 0$. Let M' be a line skew to L. This means that as a 2-dimensional K–subspace we may write M' in the form $y = xA$ for A any 2×2 K-matrix, $A = \begin{bmatrix} a & b \\ c & d \end{bmatrix}$. Note, for example, a basis for M' may always be taken in the form:

$$\{(1, 0, a, b), (0, 1, c, d)\}$$

so the M-space L' may be written as $y = x \begin{bmatrix} a & b \\ c & d \end{bmatrix}$. We consider the image of $y = x \begin{bmatrix} 0 & 1 \\ 0 & 0 \end{bmatrix}$ first under the affine homology group elements

$$(x, y) \longmapsto \left(x \begin{bmatrix} u + \rho_1 t & \gamma_1 t \\ t & u \end{bmatrix}, y \right)$$

for all u, t, $(u, t) \neq (0, 0)$ to obtain 2-subspaces of the following form:

$$y = x \begin{bmatrix} m + \rho_1 s & \gamma_1 s \\ s & m \end{bmatrix} \begin{bmatrix} 0 & 1 \\ 0 & 0 \end{bmatrix}$$

for all $m, s \in K - \{(0, 0)\}$. We then follow these images under the affine elation group elements:

$$(x, y) \longmapsto \left(x, x \begin{bmatrix} u + \rho_1 t & \gamma_1 t \\ t & u \end{bmatrix} + y \right)$$

for all $u, t \in K$. Hence, we need to show that

$$\begin{bmatrix} m + \rho_1 s & \gamma_1 s \\ s & m \end{bmatrix} \begin{bmatrix} 0 & 1 \\ 0 & 0 \end{bmatrix} + \begin{bmatrix} u + \rho_1 t & \gamma_1 t \\ t & u \end{bmatrix} = \begin{bmatrix} a & b \\ c & d \end{bmatrix}$$

has a unique solution (m, s, u, t), where $(m, s) \neq (0, 0)$ if $y = x \begin{bmatrix} a & b \\ c & d \end{bmatrix}$ is not a component of Σ. Thus, we obtain the equivalent system of equations:

$$u + \rho_1 t = a,$$
$$t = c,$$
$$m + \rho_1 s + \gamma_1 t = b,$$
$$s + u = d,$$

which clearly has a unique solution

$$(m, s, u, t) = (b - \rho_1(a - \rho_1 c) - \gamma_1 c, d - (a - \rho_1 c), a - \rho_1 c, c).$$

Now $(m, s) = (0, 0)$ if and only if $\begin{bmatrix} a & b \\ c & d \end{bmatrix} = \begin{bmatrix} u + \rho_1 t & \gamma_1 t \\ t & u \end{bmatrix}$ if and only if $M' \in \Sigma$. $\qquad\qquad\square$

THEOREM 174. *Let Σ be a Pappian spread in $PG(3, K)$, for K a field. Assume that there exists a regulus $\mathcal{R}$, which is contained in at least two distinct Pappian spreads Σ and Σ'. Let ℓ be a fixed component of Σ and let G denote the full group of central collineations of the affine translation plane $\mathcal{A}$ associated with Σ with axis ℓ.*

Consider the set of spreads $\{\Sigma'g; g \in G\}$ and form the Hall spreads $\overline{\Sigma'g}$ by derivation of each Rg.

(1) $\overline{\Sigma'g} = \overline{\Sigma'}g$; there is a group of Σ acting transitively on the set of Hall spreads.

(2) $\Sigma \cup \{\ \overline{\Sigma'g}\ ; g \in G\}$ is a parallelism consisting of one Pappian spread and the remaining spreads are Hall spreads.

PROOF. It is not difficult to show that the infinite Hall planes may be constructed by derivation of Pappian planes. It remains only to show that the lines are all uniquely covered by the indicated spreads. A line skew to L and in Σ is, of course, covered by Σ. The lines of $\Sigma'g$, which are not in Rg, are not in Σ and the set of such lines is covered by the set of lines of $\Sigma'g$'s that are not in any of the Rg's. A Baer subplane of Rg cannot be a Baer subplane of any Rh unless $Rg = Rh$ since a Baer subplane within a Desarguesian affine plane completely determines the regulus net containing it as a Baer subplane. Now take any line ℓ that does intersect $x = 0$. Then there exists a unique regulus net R_1 defined by ℓ that lies in $\mathcal{A}$. The question is whether any regulus net R_1 is the image of R under a collineation g on G. Take a regulus R_1 containing $x = 0$. Choose two distinct lines L_1 and L_2 of R_1 not equal to $x = 0$. Since the group of central collineations is doubly transitive on the components of Σ, it follows that there is a central collineation g with axis $x = 0$ that maps R onto R_1. Hence, we have a parallelism. $\qquad\square$

Using the previous construction, we may obtain another parallelism by the derivation of Σ and $\overline{\Sigma'}$.

THEOREM 175. *Under the assumptions of the previous theorem, let $\overline{\Sigma}$ denote the Hall spread obtained by the derivation of $\mathcal{R}$.*
Then $\overline{\Sigma} \cup \Sigma' \cup \{\overline{\Sigma'}g \text{ for } g \neq 1 \text{ of } G\}$ is a parallelism of $PG(3, K)$.

PROOF. Note that $\mathcal{R}$ is in Σ and Σ' so $\mathcal{R} \cup \mathcal{R}'$ is a set of lines of $PG(3, K)$, which is covered by $\Sigma \cup \overline{\Sigma'}$. Hence, this set is covered also by $\overline{\Sigma} \cup \Sigma'$. $\qquad\qquad\square$

Thus, there are parallelisms in $PG(3, K)$ of Johnson type, for every field $K \neq GF(2)$ that admits a quadratic extension.

1. Isomorphisms

In this section, we determine the possible isomorphisms of a parallelism constructed as in the previous section as well as a determination of the collineation group.

We begin with discussion of the collineation group of a Hall plane in both the finite and infinite settings.

THEOREM 176. *Let π be a Pappian plane with spread in $PG(3, K)$ for K a field. If $|K| > 3$, then the full collineation group of a Hall plane constructed from π by derivation of a regulus net R is the group inherited from π; leaves the regulus net R invariant.*

PROOF. Let σ be a collineation of the Hall plane π^* constructed from π by the derivation of R and let R^* denote the opposite regulus net. We note that if $\pi = R \cup M$ then $M\sigma - R^* \subseteq M$. Assume first that K is finite and isomorphic to $GF(q)$. Then $|M \cap M\sigma| \geq q^2 - q - (q+1)$, which is $> q + 1$ if and only if $q^2 > 3q + 2$ if and only if $q > 3$.

Then $|\pi\sigma \cap \pi| > q + 1$ so that $\pi\sigma$ and π are Desarguesian planes of order q^2 that share strictly more than $q + 1$ components. Since any two Desarguesian affine planes on the same points sharing at least three components contain a subnet coordinatized by a subfield of each coordinatizing field, it follows that $\pi = \pi\sigma$. That is, each collineation of π^* is a collineation of π and thus must leave R^* and hence R invariant.

Now assume that K is infinite.

Basically, it is possible to give the same proof as in the finite case without counting. For this, we appeal to the representation of the spread sets and leave the proof of the infinite case to the reader to complete. $\qquad\qquad\square$

COROLLARY 38. *Let π^* and ρ^* denote two Pappian affine planes with spreads in $PG(3, K)$, for K a field, and let π and ρ denote the* *associated Pappian planes that construct the indicated planes, respectively, by derivation of the regulus nets $\mathcal{R}_\pi$ and $\mathcal{R}_\rho$ with opposite regulus nets $\mathcal{R}_\pi^*$ and $\mathcal{R}_\rho^*$.*

If $|K| > 3$ and π^* is isomorphic to ρ^* by a mapping σ, then σ maps the regulus net $\mathcal{R}_\pi^*$ onto the regulus net $\mathcal{R}_\rho^*$.

PROOF. The full collineation group of ρ^* leaves the net $\mathcal{R}_\rho^*$ and admits a collineation group isomorphic to $GL(2, K)$ which acts transitively on the remaining components. Since there is a collineation group of ρ^* which acts transitively on $\rho^* - \mathcal{R}_\pi^* \sigma$, it follows that $\mathcal{R}_\rho^* = \mathcal{R}_\pi^* \sigma$. □

REMARK 54. *We note that we shall distinguish notationally between a spread $\mathcal{S}$ and the affine plane defined by the spread by using $\pi_\mathcal{S}$ to denote the plane. More generally, the translation net defined by a partial spread $\mathcal{Z}$ shall be denoted by $\pi_\mathcal{Z}$.*

THEOREM 177. *Let K be a field of cardinality > 3 and let Σ, Σ', and Σ'' denote Pappian spreads containing a regulus R, where Σ' and Σ'' are distinct from Σ, and let ℓ denote the axis of the central collineation group G of π_Σ.*

Let $\mathcal{P}_{\Sigma,\Sigma'} = \Sigma \cup \{ \overline{\Sigma' g} \; ; g \in G \}$ and $\mathcal{P}_{\Sigma,\Sigma''} = \Sigma \cup \{ \overline{\Sigma'' g} \; ; g \in G \}$ be parallelisms in $PG(3, K)$.

If σ is an isomorphism from $\mathcal{P}_{\Sigma,\Sigma'}$ onto $\mathcal{P}_{\Sigma,\Sigma''}$, then σ is a collineation of the Pappian plane π_Σ, which leaves invariant ℓ and may assumed to leave π_R invariant and map Σ' onto Σ''.

Furthermore, we may assume that σ leaves at least three parallel classes of π_R invariant, which implies that σ is an element of a group isomorphic to $\Gamma L(1, K^2)/GL(1, K)$, where K^2 denotes the quadratic extension of K corresponding to the Pappian plane π_Σ.

PROOF. An isomorphism σ, by definition, is a mapping from the spreads of $\mathcal{P}_{\Sigma,\Sigma'}$ onto the spreads of $\mathcal{P}_{\Sigma,\Sigma''}$ and since there is exactly one Pappian spread in both parallelisms, it follows that σ must map Σ onto Σ and hence induce a collineation of Σ. Since G acts transitively on the the spreads not equal to Σ of each parallel parallelism, we may assume that σ maps $\overline{\Sigma'}$ onto $\overline{\Sigma''}$. The collineation group of $\pi_{\overline{\Sigma''}}$ leaves $\pi_{R''}$ invariant and acts transitively on the components not in π_{R^*}, where R^* denotes the opposite regulus of R. Since the collineation group of $\pi_{\overline{\Sigma'}}$ leaves π_{R^*} invariant and acts transitive on the remaining components, it follows that $\pi_{\overline{\Sigma'}} \sigma$ leaves $\pi_{R^*} \sigma$ invariant and acts transitively on the remaining components. Hence, it follows that $\pi_{R^*} \sigma = \pi_{R^*}$, from which it follows that $R^* \sigma = R^*$ and hence that $R\sigma = R$.

Now assume that σ maps ℓ to ℓ_1. Assume that $\ell \neq \ell_1$. Let G^σ denote the central collineation group of π_Σ with axis ℓ_1. It follows that G and G^σ are collineation groups of the parallelism $\mathcal{P}_{\Sigma, \Sigma''}$.

Therefore, for every element g of G, there exists a corresponding element h_g of G^σ such that $\overline{\Sigma''}g = \overline{\Sigma''}h_g$. Note that the previous result on the collineation groups of Hall planes of orders >9 shows that two Hall planes are isomorphic (equal) if and only if the associated Pappian planes are isomorphic (equal).

Hence, it must be that $\pi_{\Sigma''}g = \pi_{\Sigma''}h_g$ and furthermore, it can only be that $Rg = Rh_g$. Thus, each regulus involved in the construction shares both components ℓ and ℓ_1, which is a contradiction. So, $\ell\sigma = \ell$. It remains to show that σ may be considered to leave at least three components of π_R invariant.

The subgroup of G, G_R, which leaves R invariant also acts 2-transitive on the components of π_R different from ℓ. Moreover, every collineation group that fixes each component of the opposite regulus net π_{R^*} acts as a collineation group of any Pappian plane that contains π_R. Therefore, it follows that G_R is a collineation group of each plane Σ' and Σ''. Hence, we may assume that σ fixes three components of π_R. Coordinatize π_Σ by K^2 and assume that σ fixes $x = 0, y = 0$, and $y = x$, with respect to some basis choice. Then, σ may be represented in the following form:

$$\sigma : (x, y) \longmapsto (x^\tau a, y^\tau a) \text{ for all } \tau \in \mathrm{Aut} K^2 \text{ and for all } a \in K^2 - \{0\}.$$

Therefore, σ is a group isomorphic to $\Gamma L(1, K^2)$. Note that all Pappian spreads are coordinatized by quadratic field extensions of K so that the mappings $(x, y) \longmapsto (x\alpha, y\alpha)$, for $\alpha \in K - \{0\}$ are collineations of each Pappian plane and hence are collineations of each corresponding Hall plane. Therefore, the group $GL(1, K)$ fixes each Hall and Pappian plane so that the permutation group induced is a subgroup of $\Gamma L(1, K^2)/GL(1, K)$.

This completes the proof of the theorem. $\qquad\square$

COROLLARY 39. *Let K be any field for which there exists a quadratic field extension K^2. Let $\mathcal{Q}$ denote the set of all quadratic field extensions of K.*

Then the group $\Gamma L(1, K^2)/GL(1, K)$ acts as a (not necessarily faithful) permutation group on $\mathcal{Q}$ and the orbits not equal to K^2 define the isomorphism classes of the parallelisms $\mathcal{P}_{\Sigma, \Sigma'}$, where Σ is the Pappian spread defined by K^2.

PROOF. Let ρ be an automorphism of K^2. Then $k^\rho \in K$ provided $k \in K$, which implies that the standard regulus net π_R is left invariant under the group $(x, y) \longmapsto (x^{\rho}, y^{\rho})$, for all elements $\rho \in Aut K^2$. Furthermore, the group elements $(x, y) \longmapsto (xa, ya)$ are kernel homology elements of the Pappian plane π_{Σ}, which implies, in particular, that such elements leave invariant the regulus net π_R. It then follows directly that any such mapping of $\Gamma L(1, K^2)$ must leave invariant the set of Pappian spreads containing the regulus R, when considered acting in the projective space and since the set of such Pappian spreads correspond bijectively to the set of quadratic field extensions of K, we see that $\Gamma L(1, K^2)$ acts as a permutation group on $\mathcal{Q}$. Furthermore, since $GL(1, K)$ leaves invariant such field extension, the group induced is a subgroup of $\Gamma L(1, K^2)/GL(1, K)$.

We have seen that the set of parallelisms containing a given Pappian spread Σ may be partitioned into isomorphism classes by the group $\Gamma L(1, K^2)$. This completes the proof of the corollary. $\qquad \square$

In the finite case and, for example, when K is the field of real numbers, all quadratic field extensions are isomorphic. Since any such Pappian spread may be embedded into $PG(3, K)$, we see that the isomorphisms may be taken within $\Gamma L(4, K)$. However, when there exist non-isomorphic quadratic extensions, we obtain other non-isomorphic parallelisms.

COROLLARY 40. *Let Σ and Δ denote Pappian spreads in $PG(3, K)$, for K a field of cardinality >3. Let R_{Σ} and R_{Δ} denote reguli in Σ and Δ, respectively. Let Σ' and Δ' denote Pappian spreads distinct from Σ and Δ, respectively, and containing R_{Σ} and R_{Δ}, respectively. Form the parallelisms $\mathcal{P}_{\Sigma, \Sigma'}$ and $\mathcal{P}_{\Delta, \Delta'}$.*

Then the two parallelisms are not isomorphic in any of the following situations:

(1) The field K_{Σ}^2 coordinatizing π_{Σ} is not isomorphic to the field K_{Δ}^2 coordinatizing π_{Δ},

(2) assuming Σ and Δ are isomorphic, the field $K_{\Sigma'}^2$ coordinatizing $\pi_{\Sigma'}$ is not isomorphic to the field $K_{\Delta'}^2$ coordinatizing $\pi_{\Delta'}$, or

(3) assuming Σ and Δ are isomorphic, identify Σ and Δ under the isomorphism, the field $K_{\Sigma'}^2$ coordinatizing $\pi_{\Sigma'}$ and the field $K_{\Delta'}^2$ coordinatizing $\pi_{\Delta'}$ are in distinct $\Gamma L(1, K^2)/GL(1, K)$ orbits.

PROOF. We have seen that there is exactly one Pappian spread in each parallelism. If the two parallelisms are isomorphic, then the two Pappian spreads must be isomorphic and hence the associated fields must be isomorphic. This proves (1).

If Σ and Δ are isomorphic, then by use of the central collineation groups, it follows that we may assume that Σ' maps onto Δ' (using the fact that if the associated Hall spreads map to each other then so do the corresponding Pappian spreads).

Hence, if Σ' maps to Δ', then the corresponding coordinatizing fields must be isomorphic. This proves (2) and we have seen previously that (3) is valid.   □

THEOREM 178. *Consider* $GL(1, K^2)$ *acting on* $\mathcal{Q}$. *If* $\sigma \in GL(1, K^2)$ *leaves an element of* $\mathcal{Q} - \{K^2\}$ *invariant, then* $\sigma^2 \in GL(1, K)$.

Hence, if K *is a characteristic two perfect field, then* $GL(1, K^2)$ *acts semi-regularly on* $\mathcal{Q} - \{K^2\}$.

PROOF. Acting on the associated Pappian spread, σ fixes all components. Assume that σ fixes a second quadratic extension or equivalently fixes a second Pappian spread. Since σ is then in $\Gamma L(2, L^2)$, for a quadratic extension L^2 of K and furthermore, the regulus R is in each such Pappian spread, it follows that σ has the following form acting on the second Pappian spread: $(x, y) \longmapsto (x^p b, x^p b)$, where the automorphism is the unique automorphism of $Gal_K L^2$ and juxtaposition denotes multiplication in L^2. Thus, σ^2 is in the kernel homology group of the second Pappian spread. In other words, σ^2 fixes all components of both Pappian spreads. However, any component of the second is a Baer subplane of the first Pappian spread so that σ^2 induces a kernel homology group on such Baer subplanes that are coordinatized naturally by K. That is, σ^2 is in $GL(1, K)$.

This implies that there exists an element $a \in K^2$ such that $a^2 \in K$. If the field is perfect and of characteristic two, then the mapping $x \longmapsto x^2$ is an automorphism, which implies that a is in K. This completes the proof.   □

Hence, we have immediately,

COROLLARY 41. *For finite parallelisms,* $GL(1, K^2)$, $K \simeq GF(q)$, *has orbits of lengths*

$$(q + 1)/2 \text{ or } (q + 1) \text{ if } q \text{ is odd and}$$
$$(q + 1) \text{ if } q \text{ is even.}$$

PROOF. For q odd, under $GL(1, K^2)$ there are exactly $a + b$ mutually non-isomorphic parallelisms, where $a + 2b = q - 2$, for a and b non-negative integers.   □

COROLLARY 42. *For* q *even, under* $GL(1, K^2)$, *there are exactly* $q/2 - 1$ *mutually non-isomorphic parallelisms.*

PROOF. Since any two quadratic field extensions of K are now isomorphic, and we know that there are exactly $q(q-1)/2$ Pappian spreads in $PG(3, K)$ containing a given regulus, the group $\Gamma L(1, K^2)$ permutes $q(q-1)/2 - 1 = ((q^2 - 1) - (q+1))/2$ spreads distinct from Σ.

We consider the orbits under $GL(1, K^2)$. If q is even, then the orbit lengths are all $q+1$ so that there are exactly $((q-1) - 1)/2 = q/2 - 1$ mutually non-isomorphic parallelisms.

If q is odd, then the orbit lengths are of length $q+1$ or $(q+1)/2$. Let a denote the number of orbits of length $(q+1)/2$ and b the number of orbits of length $q+1$. Then, $a(q+1)/2 + b(q+1) = ((q^2 - 1) - (q+1))/2$ so that $a + 2b = q - 2$.   □

For the next analysis, we shall require a fundamental result that gives an unusual construction of the $q(q-1)/2$ Pappian spreads containing a given regulus or equivalently the quadratic extensions of $GF(q)$.

THEOREM 179. *Let K be $GF(q)$ and K^2 be any quadratic extension of K. If K^2 is defined using the irreducible quadratic $x^2 + ax - b$, then let*

$$\left\{ \begin{bmatrix} u + at & bt \\ t & u \end{bmatrix} \forall u, t \in K \right\}$$

denote the associated 2×2 matrix field over K.

Let the spread Σ be denoted by

$$x = 0, y = x \begin{bmatrix} u + at & bt \\ t & u \end{bmatrix} \forall u, t \in K$$

and note that the regulus R is given by

$$x = 0, y = x \begin{bmatrix} u & 0 \\ 0 & u \end{bmatrix} \forall u \in K.$$

Then there is a Baer group B of π_R of order $q(q-1)$ such that the set of Pappian spreads containing R is given by ΣB.

By choice of coordinates, B has the following form:

$$\left\{ \begin{bmatrix} u & v & 0 & 0 \\ 0 & 1 & 0 & 0 \\ 0 & 0 & u & v \\ 0 & 0 & 0 & 1 \end{bmatrix} \forall v, u \neq 0 \text{ of } K \right\}.$$

Then, ΣB has the following form:

$$\left\{ \begin{aligned} & x = 0, \\ & y = x \left(\begin{bmatrix} u & v \\ 0 & 1 \end{bmatrix}^{-1} \begin{bmatrix} a & b \\ 1 & 0 \end{bmatrix} \begin{bmatrix} u & v \\ 0 & 1 \end{bmatrix} \right) \\ & \quad + w I_2 \end{aligned} \right\}$$

$\forall v, u \neq 0, w \in K.$

Furthermore, the associated quadratic extension fields are

$$\left\{ \begin{bmatrix} w + ((a - 2v)/u)t & (v(a - v) + b)/u^2)t \\ t & w \end{bmatrix} \right\}$$
$$\forall t, w \in K$$

$\forall v, u \neq 0 \in K.$

Note that there are exactly $q(q - 1)/2$ images, as there is always a unique non-trivial solution, namely, $(-1, a) = (u, v)$, to

$$(a - 2v)/u = a \text{ and } (v(a - v) + b)/u^2 = b.$$

Hence, if $x^2 + ax - b$ is irreducible, then so is $x^2 + ((a - 2v)/u)x - ((v(a - v) + b)/u^2)$ for all $v, u \neq 0$ of K.

PROOF. The group of a regulus contains a group isomorphic to $GL(2, q) \cdot GL(2, q)$, where the product is a central product and the intersection of the two groups is the scalar group of order $q - 1$. One of these groups is generated by central collineations, and one is generated by Baer groups. It follows that there is a Baer group of order $q(q - 1)$ acting. By appropriate choice of coordinates, we choose a Baer subplane to be given by $\{(0, x_2, 0, y_2); x_2, y_2 \in K\}$, when considering the planes in a 4-dimensional K-vector space. The form of the associated Baer group is now immediate.

To determine the images of the Pappian spreads, we may take the image set of $y = x \begin{bmatrix} a & b \\ 1 & 0 \end{bmatrix}$, which with R completely determines the image spread. We note that if an element of a Baer group leaves invariant a Pappian spread then the element must have order 2 since the element induces an element of $Gal_K K^2$, for some quadratic extension K^2 of K. Hence, each orbit has length $q(q - 1)/2$, which implies that the group B acts transitively on the set of Pappian spreads containing the regulus R.

To see that the fields are as maintained, note that

$$\left(\begin{bmatrix} u & v \\ 0 & 1 \end{bmatrix}^{-1} \begin{bmatrix} a & b \\ 1 & 0 \end{bmatrix} \begin{bmatrix} u & v \\ 0 & 1 \end{bmatrix} \right)$$
$$= \begin{bmatrix} a - v & (v(a - v) + b)/u \\ u & v \end{bmatrix}.$$

Now adding $\begin{bmatrix} v & 0 \\ 0 & v \end{bmatrix}$ and dividing by $\begin{bmatrix} u & 0 \\ 0 & u \end{bmatrix}$, we obtain:

$$\begin{bmatrix} (a - 2v)/u & (v(a - v) + b)/u^2 \\ 1 & 0 \end{bmatrix},$$

which defines the quadratic extension uniquely.

To see which elements fix which quadratic extensions, consider the system of equations:

$$(a - 2v)/u = a \text{ and } (v(a - v) + b)/u^2.$$

It is not difficult to verify the following:

For characteristic even or odd, the unique solution for (u, v) — $(-1, a)$. (Note when a is not zero, one obtains: $a^2(1 - u^2) = -4b(1 - u^2)$. But, $a^2 \neq -4b$ as the discriminant of $x^2 + ax - b$ is $a^2 + 4b$.) We note that the element $\begin{bmatrix} -1 & a \\ 0 & 1 \end{bmatrix}$ has order 2 when the order is odd or even). $\qquad \square$

COROLLARY 43. *When q is odd, let Σ be denoted by $\begin{bmatrix} 0 & \gamma \\ 1 & 0 \end{bmatrix}$, where γ is a non-square.*

There is an orbit length of $(q+1)/2$ of the Pappian spread given by $\begin{bmatrix} -2v/u & (\gamma - v^2)/u^2 \\ 1 & 0 \end{bmatrix}$ under $GL(1, K^2)$ if and only if

$$(-2v/u, (\gamma - v^2)/u^2) = (-2v/u, -\gamma).$$

Hence, the number of orbits of length $(q+1)/2$ is exactly the number of irreducible polynomials of the form

$$(x^2 + cx + \gamma)/((q+1)/2),$$

where γ is a fixed non-square in $GF(q)$.

PROOF. There is a element of $GL(1, K^2)$ that fixes the given Pappian spread if and only if σ^2 is in $GL(1, K)$. Hence, it follows that modulo $GL(1, K)$; we may assume that $\sigma = \begin{bmatrix} 0 & \gamma \\ 1 & 0 \end{bmatrix}$.

Therefore,

$$\begin{bmatrix} 0 & \gamma \\ 1 & 0 \end{bmatrix}^{-1} \begin{bmatrix} c & d \\ 1 & 0 \end{bmatrix} \begin{bmatrix} 0 & \gamma \\ 1 & 0 \end{bmatrix}$$

is

$$\begin{bmatrix} 0 & \gamma \\ d/\gamma & c \end{bmatrix},$$

which becomes

$$\begin{bmatrix} -c\gamma/d & \gamma^2/d \\ 1 & 0 \end{bmatrix}.$$

So, $(-c\gamma/d, \gamma^2/d) = (c, d)$ if and only if either $c = 0$ and $d = -\gamma$ (since it cannot be γ) or $c \neq 0$ and $d = -\gamma$. $\qquad \square$

COROLLARY 44. *The number of orbits of length $(q + 1)/2$ under $GL(1, K^2)$ is exactly one; call this orbit the 'short orbit.' Thus, the number of orbits of length $q + 1$ is $(q - 3)/2$. We call any such orbit a 'long orbit.' Hence, the number of mutually non-isomorphic parallelisms under $GL(1, K^2)$ is $(q - 1)/2$.*

PROOF. First assume that -1 is a non-square. Then, the polynomial $x^2 + c + \gamma$ is irreducible if and only if $c^2 - 4\gamma$ is a non-square. Choosing $\gamma = -1$, we have to ask how many $c's$ force the polynomial to be irreducible.

Hence, we are reduced to asking how many elements d have the property that $d^2 + 1$ is non-square. By Dickson [47] p. 48, Theorem 67, the number of nonzero elements d^2 is $(q+1)/4$ and hence the number of elements c is $(q + 1)/2$ as c is never zero in this case.

If -1 is a square, then $c^2 - 4\gamma$ is non-square if it has the form $-\gamma a^2$ for some a. This is true if and only if $\gamma(1 + a^2) = c^2$. The number of non-zero elements a^2 and hence the number of non-zero elements c^2 is the number of non-squares $1 + a^2$ which is, by Dickson [47], Theorem 67, $(q - 1)/4$ which gives $(q - 1)/2$ possible non-zero elements c. Since c can be zero in this situation, the number of irreducible polynomials is $(q - 1)/2 + 1 = (q + 1)/2$.

Since in each orbit of length $(q+1)/2$ of Pappian spreads, there are $(q + 1)/2$ Pappian spreads fixed by an element of $GL(1, K^2)$, we have that there must be exactly one orbit of length $(q + 1)/2$.

Hence, there are exactly $1 + d$ mutually non-isomorphic parallelisms, where $(q + 1)/2 + d(q + 1) = (q^2 - 1) - (q + 1))/2$, so it follows that $d = (q - 3)/2$, and therefore, we have $1 + (q - 3)/2 = (q - 1)/2$ mutually non-isomorphic parallelisms under $GL(1, K^2)$. □

REMARK 55. *We shall refer to the Pappian spreads and/or the fields coordinatizing the associated translation planes by (a, b) if and only if the associated irreducible polynomial is $x^2 + ax - b$.*

If $a = 0$ so that b is a non-square distinct from γ in K, we shall call the spreads 0-spreads and the associated fields, 0-fields.

THEOREM 180. *If q is odd, then the orbit of (a, b) under $GL(1, K^2)$ is*

(1)

$$\{(a_{u,t}, b_{u,t}); u, t \in K, \text{ for } u, t \text{ not both zero}\}$$
$$\text{where } a_{u,t} = a(u^2 + t^2\gamma) + 2(b - \gamma)ut)/\Delta,$$
$$b_{u,t} = (bu^2 - \gamma^2 t^2 + a\gamma ut)/\Delta),$$
$$\text{and } \Delta = u^2 - t^2 b - atu.$$

(2) In an orbit of a 0-spread $(0, b)$ of length $q + 1$ under $GL(1, K^2)$, there are exactly two 0-spreads, namely, $(0, b)$ and $(0, \gamma^2/b)$. Hence, the number of orbits of length $q + 1$ of 0-spreads is

$$(q - 3)/4 \text{ if } -1 \text{ is a nonsquare and}$$
$$(q - 5)/4 \text{ if } -1 \text{ is a square.}$$

(3) $\Gamma L(1, K^2)$ permutes the set of orbits of length $q + 1$ of 0-spreads. More generally, let $K[\theta] = K^2$, where $\theta^2 = \gamma$ and if τ denotes an automorphism of K^2, then

$$\tau : (a, b) \longmapsto (a^\tau/\rho, b^\tau/\rho^2), \text{ where } \theta^\tau = \rho\theta \text{ (thus } \rho^2 = \gamma^{\tau-1}).$$

Hence, a 0-orbit of $(0, b)$ is fixed by τ if and only if

$$b^{\tau-1} = \gamma^{\tau-1} \text{ or } b^{\tau+1} = \gamma^2\rho^2 = \gamma^{\tau+1}.$$

More generally, an orbit containing (a, b) is fixed by τ if and only if

$$(a^\tau/\rho, b^\tau/\rho^2) = (a_{u,t}, b_{u,t}) \text{ for some } u, t \in K.$$

PROOF. Form $\begin{bmatrix} u & \gamma t \\ t & u \end{bmatrix}^{-1} \begin{bmatrix} a & b \\ 1 & 0 \end{bmatrix} \begin{bmatrix} u & \gamma t \\ t & u \end{bmatrix} = \begin{bmatrix} a^* & b^* \\ c^* & d^* \end{bmatrix}.$

Then, form

$$\left\{ \begin{bmatrix} a^* & b^* \\ c^* & d^* \end{bmatrix} + \begin{bmatrix} -d^* & 0 \\ 0 & -d^* \end{bmatrix} \right\} \begin{bmatrix} c^{*-1} & 0 \\ 0 & c^{*-1} \end{bmatrix} \text{ to obtain}$$

$$\begin{bmatrix} a_{u,t} & b_{u,t} \\ 1 & 0 \end{bmatrix}, \text{ thus defining } (a_{u,t}, b_{u,t}).$$

The reader may easily verify that $a_{u,t}$ and $b_{u,t}$ are as maintained. This proves (1).

Now assume that $a = 0$. Then $a(u^2 + t^2\gamma) + 2(b - \gamma)ut$ the 'a'-entry is zero if and only if $ut = 0$ as $b \neq \gamma$. If $t = 0$, then $(0, b)$ is obtained and if $u = 0$, then $(0, \gamma^2/b)$ is obtained. Now $\gamma^2/b = b$ if and only if $b = \pm\gamma$, and if $b = -\gamma$, then $(0, -\gamma)$ is a field if and only if -1 is a square but then $(0, -\gamma)$ defines the unique orbit of length $(q + 1)/2$ of $GL(1, K^2)$.

If -1 is a non-square, there are therefore exactly

$$((q - 1)/2 - 1)/2 = (q - 3)/4$$

orbits of length $q + 1$ of 0-spreads and if -1 is a square, there are exactly

$$((q - 1)/2 - 2)/2 = (q - 5)/4$$

orbits of length $q + 1$ of 0-spreads. (Count the number of possible non-squares b defining orbits of $(0, b)$'s of length $q + 1$.) This proves (2).

Letting the $(0, \gamma)$ field be defined by $K^2 = K[\theta]$ such that $\theta^2 = \gamma$, if $x \longmapsto x^\tau$ defines an automorphism of K^2, then $\theta^\tau = \gamma^{(\tau-1)/2}\theta$.

Hence, if we represent vectors $(x_1\theta + x_2)$ as (x_1, x_2), it follows that the automorphism can be represented 4-dimensionally over K as follows:

$$(x_1, x_2, y_1, y_2) \longmapsto (x_1^\tau, x_2^\tau, y_1^\tau, y_2^\tau) \begin{bmatrix} \gamma^{(\tau-1)/2} & 0 & 0 & 0 \\ 0 & 1 & 0 & 0 \\ 0 & 0 & \gamma^{(\tau-1)/2} & 0 \\ 0 & 0 & 0 & 1 \end{bmatrix}.$$

From here, we assert that

$$(a, b) \longmapsto (a^\tau/\gamma^{(\tau-1)/2}, b^\tau/\gamma^{(\tau-1)})$$

defines the image of (a, b) under τ. To see this, we note that

$$\begin{bmatrix} a & b \\ 1 & 0 \end{bmatrix} \longmapsto \begin{bmatrix} \gamma^{(\tau-1)/2} & 0 \\ 0 & 1 \end{bmatrix}^{-1} \begin{bmatrix} a^\tau & b^\tau \\ 1 & 0 \end{bmatrix} \begin{bmatrix} \gamma^{(\tau-1)/2} & 0 \\ 0 & 1 \end{bmatrix}.$$

This gives

$$\begin{bmatrix} a^\tau & b^\tau/\gamma^{(\tau-1)/2} \\ \gamma^{(\tau-1)/2} & 0 \end{bmatrix},$$

from which the assertion follows.

Since $GL(1, K^2)$ is normal in $\Gamma L(1, K^2)$, the orbits of length $q + 1$ are permuted by $\Gamma L(1, K^2)$. An orbit containing $(0, b)$ is left invariant if and only if $b^\tau/\gamma^{(\tau-1)} = b$ or γ^2/b if and only if

$$b^{\tau-1} = \gamma^{\tau-1} \text{ or } b^{\tau+1} = \gamma^{\tau+1}.$$

If τ leaves invariant the orbit containing (a, b), then clearly each of the images of (a, b) is some $(a_{u,t}, b_{u,t})$.

This completes the proof of the theorem. $\square$

COROLLARY 45. *The automorphism defined by* $x \longmapsto x^q$ *fixes all* $GL(1, K^2)$ *orbits, when* q *is odd.*

PROOF. In this case, $\rho = -1$ and $c^\tau = c$ for all c in K. Hence, the orbit containing $(0, b)$ is mapped to $(0, b/(-1)^2)$. If $a \neq 0$, then we consider the possible solution to $(-a, b) = (a_{u,t}, b_{u,t})$. If $t = 0$, then $a_{u,0} = a$ and $b_{u,0} = b$ so this does not occur. Hence, assume that $t \neq 0$. Then, we need to verify the following equations: $(i) : -a = a(u^2 + t^2\gamma) + 2(b - \gamma)ut)/\Delta$ and $(ii) : b = (bu^2 - \gamma^2t^2 + a\gamma ut)/\Delta)$, where $\Delta = u^2 - t^2b - atu$.

Letting $s = u/t$, the first equation implies that $s = a/2$ or $s = (\gamma - b)/2$. The second equation implies $\gamma^2 - b^2 = as(\gamma + b)$. If $b = -\gamma$, then $(a, -\gamma)$ is in the orbit of length $(q + 1)/2$. Hence, $s = (\gamma - b)/a$ is the unique solution solving both (i) and (ii). Thus, this automorphism fixes all orbits under $GL(1, K^2)$. This completes the proof of the corollary.   □

THEOREM 181. *If $K \simeq GF(q)$, for q odd, then the number of mutually non-isomorphic parallelisms is at least $1 + [(q - 3)/2r]$, where $q = p^r$ for r a positive integer and p a prime.*

PROOF. We note that there is a unique orbit of length $(q + 1)/2$ so that $\Gamma L(1, K^2)$ must leave this orbit invariant. Furthermore, the automorphism of order 2 of K^2 fixes all $GL(1, K^2)$-orbits. There can be at least $1 + [(q - 3)/2r]$ mutually non-isomorphic parallelisms.   □

COROLLARY 46. *If $q = p$, an odd prime, then the number of mutually non-isomorphic parallelisms is exactly $(p - 1)/2$.*

1.1. The Full Collineation Group. In this subsection, the full collineation group of a Johnson parallelism may be determined.

THEOREM 182. *The full collineation group of $\mathcal{H}_\mathcal{P}$ of a parallelism of Johnson type $\{\Sigma, \Sigma'\}$ in $PG(3, q = p^r)$ contains the full central collineation group G of Σ of order $q^2(q^2 - 1)$ with a fixed axis and*

$$q^2(q^2 - 1) \mid |\mathcal{H}_\mathcal{P}| \mid q^2(q^2 - 1)2r, \text{ if } \Sigma' \text{ is in a long orbit,}$$
$$q^2(q^2 - 1) \mid |\mathcal{H}_\mathcal{P}| \mid q^2(q^2 - 1)4r, \text{ if } \Sigma' \text{ is in a short orbit.}$$

PROOF. The full collineation group of a parallelism of Johnson type is a subgroup of $\Gamma L(2, q^2)$ and depends on whether the second Desarguesian spread Σ' is in a short or long orbit. Indeed, since

$$|G_{\Sigma'}| = q(q - 1)|G_{x=0, y=0, y=x, R, \Sigma'}|,$$

where Σ' contains the regulus R, it follows the order of this last group divides $(q^2 - 1)2r$. The result now follows immediately.   □

1.2. Infinite Johnson Parallelisms. First, assume that we have an infinite field K admitting non-squares, and let γ be a non-square so that $K[\gamma] = K^2$ is a field coordinatizing a Pappian spread in $PG(3, K)$.

All of the previous results of finite fields of odd order basically apply in this setting. In particular, the mutually non-isomorphic parallelisms are the orbits under $\Gamma L(1, K^2)$.

Furthermore, we have seen that $\Gamma L(1, K^2)$ preserves the 0-orbits of $GL(1, K^2)$. However, since K is infinite, it is not clear what the orbits of $GL(1, K^2)$ are.

Never the less, we see that the corresponding orbits of elements $(a, -\gamma)$ must be permuted by $\Gamma L(1, K^2)$. The actuall number or cardinality of the set of such orbits depends on the field K.

It is still true that the orbit structure under of $GL(1, K^2)$ still has the same form mapping (a, b) onto $(a_{u,t}, b_{u,t})$. Furthermore, in each such orbit under $GL(1, K^2)$ of a 0-spread, for b not $-\gamma$ or γ, there are exactly two 0-spreads.

Considering the orbits of 0-spreads $(0, b)$ for b not $\pm\gamma$, then the number of such $GL(1, K^2)$-orbits is at least

$$(card\ K - 2)/2 = card\ K.$$

Note that if γ is a non-square then $a^2\gamma$ is also a non-square, so there are $card\ K$ non-squares.

Hence, we have

THEOREM 183. *Let K be an infinite field that admits a non-square. Let the automorphism group have cardinality $\mathcal{A}_o$.*

Then there are at least $card\ K/\mathcal{A}_o$ mutually non-isomorphic parallelisms.

COROLLARY 47. *Let K be any subfield of the reals. Then there are at least $card\ K$ mutually non-isomorphic parallelisms.*

In particular, if K is the field of real numbers, then there are $2^{\aleph_0}$ mutually non-isomorphic parallelisms.

2. The Derived Parallelisms

We now consider parallelisms of the second class of parallelism constructed. That is, let $\overline{\Sigma}$ denote the Hall spread obtained by the derivation of $\mathcal{R}$, and let $\mathcal{P}$ denote the previously constructed parallelism.

Then $\overline{\Sigma} \cup \Sigma' \cup \{\mathcal{P} - \{\Sigma, \overline{\Sigma'}\}\}$ is a parallelism of $PG(3, K)$. We shall call this a parallelism $\mathcal{P}^*$ 'derived' from $\mathcal{P}$.

We first note that any such parallelism $\mathcal{P}^*$ admits a collineation group isomorphic to $G_{\mathcal{R}}$.

THEOREM 184. *Any parallelism $\overline{\Sigma} \cup \Sigma' \cup \{\mathcal{P} - \{\Sigma, \overline{\Sigma'}\}\} = \mathcal{P}^*$ admits $G_{\mathcal{R}}$ as a collineation group that fixes $\overline{\Sigma}$ and Σ'.*

Furthermore, the full collineation group must leave $\overline{\Sigma}$ and Σ' invariant and permute the set of reguli $\{\mathcal{R}g; g \in G\}$.

PROOF. We note that $G_{\mathcal{R}}$ must fix Σ and $\overline{\Sigma'}$ and permute the remaining spreads of $\mathcal{P}$. It then follows that $G_{\mathcal{R}}$ fixes also $\overline{\Sigma}$ and Σ' and permutes $\{\mathcal{P} - \{\Sigma, \overline{\Sigma'}\}\}$, which implies that the group is a collineation group of the parallelism $\mathcal{P}^*$.

Clearly, Σ' is invariant under the full collineation group, as it is the only Pappian spread of the parallelism.

Assume that g is a collineation that maps $\overline{\Sigma}$ onto $\overline{\Sigma'}g$ for some $g \in G$. Then it follows from Corollary 38 that $\overline{\mathcal{R}}$ maps to $\overline{\mathcal{R}}g$ and hence that $\mathcal{R}$ maps to $\mathcal{R}g$. But since Σ' maps to Σ' it follows that Σ' contains $\mathcal{R}$ and $\mathcal{R}g$, both of which are reguli of Σ. Since $\Sigma \neq \Sigma'$, it follows that $\mathcal{R} = \mathcal{R}g$, which implies that $g \in G_{\mathcal{R}}$, a contradiction as then Σ' and $\overline{\Sigma'}$ would be in the parallelism. It similarly follows that the set of derived reguli are permuted so the original set of reguli is left invariant under the full collineation group.

Hence, $\overline{\Sigma}$ is left invariant by the full collineation group.   □

COROLLARY 48. *Assume that* $|K| > 3$.
Let Σ' *and* Σ'' *be Pappian spreads distinct from* Σ *and containing* $\mathcal{R}$.
Form the two parallelisms $\overline{\Sigma} \cup \Sigma' \cup \{\mathcal{P}-\{\Sigma, \overline{\Sigma'}\}\} = \mathcal{P}^{*}_{\Sigma'}$ *and* $\overline{\Sigma} \cup \Sigma'' \cup \{\mathcal{P}-\{\Sigma, \overline{\Sigma''}\}\} = \mathcal{P}^{*}_{\Sigma''}$.
Then any isomorphism from $\mathcal{P}^{*}_{\Sigma'}$ *onto* $\mathcal{P}^{*}_{\Sigma''}$ *must map* Σ' *onto* Σ'' *maps* Σ *onto* Σ, *permutes the set* $\{\mathcal{R}g \,;\, g \in G\}$ *and leaves the axis of* G *invariant.*

PROOF. The proof of the previous theorem shows that $\overline{\Sigma}$ is left invariant, which implies that Σ is also left invariant under any such isomorphism. Furthermore, it also follows from the previous theorem that the set of reguli with given axis ℓ are permuted by any isomorphism. Since these reguli are permuted and Σ is left invariant, it follows that the axis must also be left invariant under any isomorphism.   □

COROLLARY 49. *Two derived parallelisms are isomorphic if and only if the two original parallelisms are isomorphic.*
That is, $\mathcal{P}^{*}_{\Sigma'}$ *is isomorphic to* $\mathcal{P}^{*}_{\Sigma'}$ *if and only if the parallelism* $\mathcal{P}_{\Sigma'}$ *is isomorphic to* $\mathcal{P}_{\Sigma''}$.

PROOF. Let σ be an isomorphism from $\mathcal{P}^{*}_{\Sigma'}$ onto $\mathcal{P}^{*}_{\Sigma'}$. Since Σ' maps onto Σ'' and Σ onto Σ, and σ permutes the set $\{\mathcal{R}g \,;\, g \in G\}$ and leaves the axis of G invariant, it follows that σ induces an isomorphism mapping $\overline{\Sigma'}$ onto $\overline{\Sigma''}$ and maps Σ onto Σ and maps the set $\{\overline{\Sigma'}g \,;\, g \in G\}$ onto $\{\overline{\Sigma''}g \,;\, g \in G\}$. This proves that there is an induced isomorphism from $\mathcal{P}_{\Sigma'}$ onto $\mathcal{P}_{\Sigma''}$.   □

THEOREM 185. *A derived parallelism* $\mathcal{P}^{*}$ *cannot be isomorphic to* $\mathcal{P}$.

PROOF. If a derived parallelism is isomorphic to $\mathcal{P}$, then $\mathcal{P}^{*}$ admits a collineation group of the unique Pappian spread Σ' that acts transitively on the remaining spreads, which is contrary to the above theorem.   □

We then have directly the following theorem.

THEOREM 186. *Let $\mathcal{P}^*$ be a derived Johnson parallelism in $PG(3, q)$ for $q > 3$.*

(1) Then the full projective collineation group $H_{\mathcal{P}^*}$ of $\mathcal{P}^*$ is a subgroup of $\Gamma L(2, GF(q^2))$, which fixes a derivable net $\mathcal{R}$ and a component ℓ of $\mathcal{R}$. This group contains a central collineation group C of order $q(q-1)$ with axis ℓ that acts 2-transitively on the components of $\mathcal{R} - \ell$.

(2) $H_{\mathcal{P}^*}$ is a subgroup of $C \cdot GalGF(q^2)$.

Hence, we have the following conclusions:

CONCLUSION 1. *If q is odd p^r for p an odd prime, then there are at least $2(1+[(q-3)/2r]$ mutually non-isomorphic parallelisms in $PG(3, q)$ constructed as above. If q is an odd prim, then there are exactly $p - 1$ mutually non-isomorphic parallelisms in $PG(3, q)$ obtained by the group and derivation constructions.*

CONCLUSION 2. *If q is even, there are at least $2[(q/2-1)/2r]$ mutually non-isomorphic parallelisms obtained by the group and derivation constructions.*

CONCLUSION 3. *If K is the field of real numbers, then there are 2^{χ_0} mutually non-isomorphic group parallelisms and 2^{χ_0} mutually non-isomorphic derived parallelisms in $PG(3, K)$.*

2.1. Characterization. In this subsection, we ask if a 'full' central collineation group acting on a parallelism and fixing a spread characterizes the parallelism.

In this regard, the following structure theorem may be obtained.

THEOREM 187. *Let K be a skewfield and Σ a spread in $PG(3, K)$. Assume that there exists a partial parallelism $\mathcal{P}$ containing Σ that admits as a collineation group a central collineation group G of Σ with axis ℓ that acts 2-transitively on the remaining lines of Σ.*

Then
(1) Σ is Pappian,
(2) the spreads of $\mathcal{P} - \{\Sigma\}$ are Hall, and
(3) $\mathcal{P}$ is a Johnson parallelism.

PROOF. We recall that the central collineation group is transitive on the components of $\Sigma - \{\ell\}$, so that the associated translation plane is minimally a semifield plane. Moreover, the stabilizer of two components is also transitive, which implies that the translation plane is also a nearfield plane. This will force the spread Σ to be at least Desarguesian. To show that is Pappian, we see that the spread for Σ is covered by derivable nets, which are actually pseudo-regulus nets. That is, the components are left 2-dimensional K–subspaces and the Baer subplanes are right 2-dimensional K–subspaces. The central collineation group that fixes one of these pseudo-regulus nets must fix all Baer subplanes of the net (incident with the zero vector), and at least two of these Baer subplanes are both right and left 2-dimensional subspaces. There is a 'Baer group' that fixes one of the Baer subplanes and acts transitively on the remaining Baer subplanes incident with the zero vector of this net. This forces all Baer subplanes to be left 2-dimensional subspaces. What this means is that the pseudo-regulus net is a regulus net, implying that K is a field. From here, the reader should be able to complete the proof showing that the spreads other than Σ are Hall and that the parallelism is of Johnson type.   □

Part 7

Parallelism-Inducing Groups

In this part, we give a variety of techniques for the construction of parallelisms, all based on what are called 'parallelism-inducing groups.' The basic idea is to try to construct parallelisms that arise by finding a group that fixes a given spread and which acts transitively on the remaining spreads. The basic ideas work for both finite and infinite parallelisms. Once it is seen how to construct such 'deficiency one transitive partial parallelisms,' we see that the fixed spread is forced to be Desarguesian in the finite case. The process constructing the Johnson parallelisms used two distinct Pappian spreads that share exactly a regulus in $PG(3, K)$, and the group used is a central collineation group of one of these spreads thus permuting the remaining Pappian spreads to create a regulus. However, the spreads in the transitive orbit are not disjoint form the original Pappian spread but derivation of the reguli in the transitive group will make the corresponding Hall spreads disjoint on lines, thus creating a parallelism with one Pappian spread and the remaining spreads Hall. This can be generalized by taking a sequence of Pappian spreads sharing a regulus and using cosets of the group. This creates the idea of an m-parallelism. In the sequence of m Pappian spreads if there are n distinct spreads, this produces what is called an '(m, n)-parallelisms.' The results of this part are modifications of results of the author and R. Pomareda [153],[154][152],[150].

We study the type of collineation groups that can act in this transitive manner and analyze the groups in the finite case in basically three ways: First, as subgroups of $\Gamma L(2, q^2)$, or $GL(2, q)$ and as a nearfield group both in the infinite and infinite cases. We then ask what a 'minimal parallelism-inducing group' might look like and show the existence of such groups. It will turn out in the finite case that the Kantor-Knuth semifield planes will play a significant role in parallelism-inducing group. Similarly, as in the two Desarguesian spread situation mentioned above, the spreads involved in the parallelisms will be the Desarguesian spread and a derived Kantor-Knuth semifield spread.

One particularly nice feature of this method of parallelism-inducing groups is the ability to determine isomorphic parallelisms. It might be initially thought that two deficiency one transitive partial parallelism with all Hall planes are necessarily isomorphic if the same group is used. However, this is not the case as the isomorphism of the parallelisms generated depend strictly on the choice of the original two Desarguesian planes sharing a regulus.

We then consider two arbitrary spreads Σ and Σ', where Σ is Desarguesian in the finite case, where Σ' is an arbitrary spread sharing a regulus with Σ, but where there subgroup of the central collineation group of Σ that is transitive on the reguli, which suggests ideas of minimal parallelism-inducing groups.

Finally, in terms of our group actions, we consider what is called 'coset switching,' which generalizes the ideas of m-parallelisms and (m, n)-parallelisms. This may be accomplished in both the finite and infinite cases. In particular, we consider all of this over the field of real numbers.

Every finite parallelism in $PG(3, K)$ has a 'dual parallelism,' which is a parallelism in the finite case, obtained by taking the dual spreads to all of the spreads in the parallelism. For example, for the known regular parallelisms (see Chapter 26), the dual parallelism is not isomorphic to the original spread. It is a very intricate problem to determine whether a given dual parallelism in $PG(3, q)$ is isomorphic to its parallelism. In this part, all of the parallelisms originate from deficiency one transitive partial parallelisms. We end this part by showing (with the exception of two sporadic cases) that the dual of the parallelism arising from a deficiency one transitive partial parallelism can never be isomorphic to the original parallelism.

CHAPTER 29

Parallelism-Inducing Groups for Pappian Spreads

In the construction of the parallelisms of Johnson (see Chapter 28), for K a field that admits a quadratic extension, there are two Pappian spreads Σ and Σ' in $PG(3, K)$ that share a regulus R and the full central collineation group G with fixed axis ℓ of R of one of these spreads Σ. If Σ'^* denotes the associated Hall spread obtained by the derivation of R, then the parallelism is $\Sigma \cup \{\Sigma'^* g; g \in G\}$. Hence, the parallelism is essentially determined by the two spreads Σ, and Σ', and the group G. In a sense that we shall make clear below, the group G is 'parallelism-inducing.' The technique that we shall be discussing is suitable for when the base planes are Pappian, whereas, eventually we shall consider a generalization of the parallelism-inducing technique. When we consider 'elation-switching,' the generalization shall be essentially complete. Basically, the idea of the construction arises from the consideration of the proof that there is a cover of Baer subplanes of Σ disjoint from the axis of a central collineation group by images of another Pappian spread under the central group G. This group acts sharply doubly transitively on the components of $\Sigma - \{\text{axis}\}$ and acts sharply transitively on the Baer subplanes of Σ incident with the zero vector and which are disjoint from the axis of G.

DEFINITION 108. *Let K be a field. Let Σ and Σ' be any two distinct spreads in $PG(3, K)$ that share exactly a regulus R, and let ℓ be a line of R. Further assume that Σ is Pappian.*

Let G be a collineation group of the affine plane associated with Σ that leaves ℓ invariant and has the following properties:

(i) G is sharply 2-transitive on the set of components of Σ distinct from ℓ (condition 'sharp'),

(ii) G is regular on the set of Baer subplanes of the affine plane associated with Σ which are disjoint from ℓ (condition 'Baer'),

(iii) G_R fixes Σ' and acts regularly on the components of $\Sigma' - R$ (condition, 'fix'). (In the finite case, if G_R fixes Σ', then the group is regular on $\Sigma' - R$ by (ii).)

So, if G satisfies conditions sharp, Baer, and fix, the group is said to be 'Pappian parallelism-inducing' with respect to Σ and Σ'.

THEOREM 188. *Let G be a parallelism-inducing group with respect to Σ and Σ', spreads in $PG(3, K)$, for K a field that admits a quadratic field extension.*

Then $\Sigma \cup \{\Sigma'^*g; g \in G\}$ is a parallelism in $PG(3, K)$, where Σ'^* denotes the spread obtained by the derivation of R.

PROOF. We first consider the set of all 'lines' of $PG(3, K)$, which are not in Σ and which are disjoint from ℓ. These are Baer subplanes of the affine plane π_Σ associated with Σ. Since the group is regular on this set of lines (condition 'Baer'), it follows that this set is covered by $\cup_{g \in G} \Sigma'g$. If two spreads $\Sigma'g$ and $\Sigma'k$ share a component, then so do Σ' and $\Sigma'gk^{-1}$. Let $nh = m$, where $n, m \in \Sigma'$ and $h = gk^{-1}$.

If n and m are both not in R, then there exists an element in G_R say, j, such that $n = mj$. But then, $nh = mjh = m$ and m is a Baer subplane of Σ, as Σ and Σ' share exactly the derivable partial spread R. But G is regular on such Baer subplanes (again condition 'Baer') so, it follows that $jh = 1$, which implies that h is in G_R, so that $\Sigma' = \Sigma'h$ (condition 'fix'), which, in turn, implies that $\Sigma'^* = \Sigma'^*h$.

Now assume that n is in R but m is not in R. Then, $nh = m$ and G a collineation group of Σ implies that m is a common component of both Σ and Σ'. Hence, we have a unique cover by the spreads $\Sigma'g$, for $g \in G$, of all lines not in Σ and disjoint from ℓ.

Since the remaining lines intersect ℓ nontrivially, they are potentially Baer subplanes of a regulus net corresponding to Rg for some $g \in G$. Let π_o be such a Baer subplane and let $\ell, \mathcal{M}$, and $\mathcal{N}$ be three distinct components intersecting π_o. Then there exists a collineation g of G such that $\ell, \mathcal{M}g$, and $\mathcal{N}g$ are components of R. Hence, there is a unique regulus net of Σ containing $\pi_o g$, and therefore, there is a unique regulus net containing π_o. These regulus nets are in an orbit under G. Thus, if we derive these nets, we obtain a parallelism in $PG(3, K)$. □

Actually, properties sharp, Baer, and fix are almost sufficient to prove that Σ is necessarily Pappian. Also, note that it is not necessary to know that Σ is Pappian to prove that under the conditions of sharp, Baer, and fix, we obtain a parallelism.

THEOREM 189. *Let K be a field. Let G be group satisfying sharp, Baer and fix with respect to Σ and Σ' spreads in $PG(3, K)$. Let R denote the common regulus of Σ and Σ'.*

(1) If either K is finite or G contains a non-trivial affine elation then the full elation group E with axis ℓ is a subgroup, Σ' is a conical flock spread, and $G_{\Sigma'}$ acts regularly on the lines of Σ' not in the regulus R.

(2) Under the conditions of (1), Σ is Pappian.

PROOF. Because of the existence of reguli sharing ℓ, it follows that Σ is both a union of a set of reguli that mutually share ℓ (a 'conical spread') and Σ is a union of a set of reguli that mutually ℓ and another line ℓ_1 (a 'ruled spread'). But Since K is a field, we assert the only possibility is that Σ is Pappian. This is left to the reader to verify.

Since Σ is Pappian and the group is sharply 2-transitive on the components not equal to ℓ, the elation group E with axis ℓ must be normal in G, and since the group is 2-transitive, it follows that either E is trivial or transitive. The structure of the regulus nets implies that there is an elation group E^- of Σ that fixes each regulus and acts regularly on the components different from ℓ (see Johnson [114], Chapters 23 and 26). However, we don't know that E^- is a subgroup of G. Also, the stabilizer of R must act sharply 2-transitively on the components of R different from ℓ.

In the finite case, the group G is a subgroup of $\Gamma L(2, q^2)$ of order $q^2(q^2 - 1)$ and the p-subgroup of order q^2, where $p^s = q$ acts on ℓ as a subgroup of $\Gamma L(2, q)$. Let r_p denote the p-th part of s. Since $q^2 > r_p q$ for $q > 2$, it follows that there must be an elation with axis ℓ in G.

If there is an elation, then there is an elation group E acting regularly on the components not equal to ℓ. Hence, the stabilizer G_R acting on Σ'^* contains a Baer group that acts regularly on the Baer subplanes distinct from ℓ in the corresponding regulus net. By the Johnson, Payne-Thas Theorem, there is a partial conical flock of deficiency one, which, in this case, implies that the plane Σ' is a conical flock plane admitting a collineation group acting regularly on the components of $\Sigma' - R$. □

REMARK 56. *Assume that Σ and Σ' are both Pappian spreads in $PG(3, K)$. If $G_R S$, where S is the scalar group, contains the full central collineation group of Σ with axis ℓ that fixes R, then property (iii), condition 'fix' of Definition 108 is satisfied.*

DEFINITION 109. *Assume we have a subgroup G of a Pappian spread Σ and G satisfies conditions sharp and Baer. If Σ' is a Pappian spread containing a regulus R of Σ, and $(iii)'$ $G_R S$, where S is the scalar group, which contains the full central collineation group with axis ℓ of Σ that fixes the regulus R, we shall refer to this condition as 'central fix.'*

THEOREM 190. *Let K be a field. Let Σ and Σ' be distinct Pappian spreads in $PG(3, K)$ that share a regulus R. Then any collineation* *group G of Σ that fixes a line ℓ of R and has conditions 'sharp,' 'Baer,' and 'central fix' is a Pappian parallelism-inducing group.*

PROOF. It remains to show that G_R acts regularly on the lines of $\Sigma' - R$; condition fix is satisfied. Note every central collineation of Σ that fixes R and has axis in R also acts as a collineation of any Pappian spread that contains R. Since the actions of G_R and $G_R S$ on the components of Σ' are identical, it remains to show that the full central collineation group that fixes R has this property. However, this is built into the argument for the construction of Johnson parallelisms in Chapter 28.   □

DEFINITION 110. *Any collineation group G of a Pappian spread Σ with properties 'sharp,' 'Baer,' and 'fix,' which produces a parallelism relative to two Pappian spreads Σ and Σ' is said to be a 'parallelism-inducing group for Pappian spreads.'*

We shall see, in due course, that there are Pappian parallelism-inducing groups, which do not satisfy condition 'central fix.' This situation arises when there is a finite non-linear subgroup that produces a nearfield plane. For example, the elements of a group of a particular Desarguesian affine plane fixing a regulus could have the following form:

$$(x, y) \longmapsto (x^{\sigma_m} m^i, y^{\sigma_m} m^{i+1}) \text{ for all } m \in GF(q) - \{0\},$$

where is i a fixed integer, and σ_m is an automorphism depending on m. It will turn out that any second Desarguesian spread sharing the standard regulus will automatically admit the group

$$(x, y) \longmapsto (x m^i, y m^{i+1}) \text{ for all } m \in GF(q) - \{0\},$$

although this group does not necessarily act on the induced parallelism. However, the second Desarguesian spread does not necessarily admit the collineation corresponding to the Frobenius automorphism. When this does occur, we obtain a group, which is parallelism-inducing but does not have the property that $G_R S$ contains the full central collineation group with a fixed axis. Also, there are more general groups than the ones mentioned above, which may not satisfy such properties.

In general, we note that the absence of the full central collineation group does not obstruct the basic construction procedure if we assume that we have an elation in the group. In the finite case, condition 'sharp' implies that we have an associated nearfield group acting on the infinite points of Σ not equal to (∞). We consider what occurs if there is, in fact, an associated nearfield group.

DEFINITION 111. *Let a group G acting on a Pappian plane Σ act sharply 2-transitive on the set of components of Σ distinct from a fixed component ℓ and correspond to a nearfield $(K^2, +, *)$, and $Z * m = Z^{\sigma_m} m$, for a family of automorphisms of K^2, σ_m corresponding to m, and if $\{w * m + \epsilon n; m, n \in K\} \subseteq \{wm + \epsilon n; m, n \in K\}$, then $\{w * m + \epsilon n; m, n \in K\} = \{wm + \epsilon n; m, n \in K\}$, for $\epsilon = 0$ or 1 (the 'ϵ-condition').*

(i)$'$ In this situation, we shall say that the group satisfies the condition 'sharp nearfield.'

In particular, G is sharply 2-transitive on the set of components of Σ distinct from ℓ and corresponds to a nearfield group defined by

$$Z \longmapsto Z^{\sigma_m} m + b \,\forall b, m \neq 0 \text{ of } K^2 \text{ coordinatizing } \Sigma,$$

where σ_m is an automorphism that depends on m. (G is assumed to contain the full elation group E with axis ℓ.)

Note that in the finite case, since we have a nearfield, the ϵ-condition is automatically satisfied by cardinality.

(iii)$''$ Let Σ' be a Pappian spread containing the regulus R of Σ and assume that

$$\langle G_R, \{(x, y) \longmapsto (x^{\sigma_m}, y^{\sigma_m}); m \in K - \{0\}\}, S \rangle,$$

where S is the scalar group, contains the full central collineation group with axis ℓ of Σ, which fixes the regulus R. Furthermore, assume that Σ' admits

$$\{(x, y) \longmapsto (x^{\sigma_m}, y^{\sigma_m}); m \in K - \{0\}\}$$

as a collineation group. We shall call this condition 'nearfield fix.'

THEOREM 191. *Let K be a field. Let Σ and Σ' be distinct Pappian spreads in $PG(3, K)$ that share a regulus R. Then any collineation group G of Σ that fixes a line ℓ of R and has the conditions 'sharp nearfield,' 'Baer,' and 'nearfield fix' is parallelism-inducing for Pappian spreads Σ and Σ'.*

PROOF. We note that the assumptions, say, in particular, that we have an elation in G, and hence, we have the full elation group E with axis ℓ. Thus, in G_R, we have the regulus-inducing group E^-. So, we may assume that an element in G_R fixes $x = 0$ and $y = 0$ and has the following form $\sigma_m : (x, y) \longmapsto (x^{\sigma_m} a, y^{\sigma_m} b)$, for elements a, b in K^2. However, since $(y/x) \longmapsto (y/x)^{\sigma_m}(b/a)$, we may assume that $(b/a) = m$, since we are assuming we obtain a nearfield group acting on the line at infinity minus (∞). But, G_R is fixed and $y = x$ maps onto $y = x(b/a) = xm$. Hence, $m \in K$. So,

$$\sigma_m : (x, y) \longmapsto (x^{\sigma_m} a, y^{\sigma_m} am), \text{ for } m \in K - \{0\}.$$

If $\tau_m : (x, y) \longmapsto (x^{\sigma_m^{-1}}, y^{\sigma_m^{-1}})$, then

$$\tau_m \sigma_m : (x, y) \longmapsto (x a^{\sigma_m^{-1}}, y a^{\sigma_m^{-1}} m^{\sigma_m^{-1}}).$$

Since, we obtain in

$$\langle G_R, \{(x, y) \longmapsto (x^{\sigma_m}, y^{\sigma_m}); m \in K - \{0\}\}, S \rangle,$$

$$h_{m^{\sigma_m^{-1}}} : (x, y) \longmapsto (x, y m^{\sigma_m^{-1}}),$$

we then have in

$$\langle G_R, \{(x, y) \longmapsto (x^{\sigma_m}, y^{\sigma_m}); m \in K - \{0\}\}, S \rangle,$$

the kernel homology element of Σ:

$$k_a : (x, y) \longmapsto (x a^{\sigma_m^{-1}}, y a^{\sigma_m^{-1}}).$$

Since, we have a nearfield group induced, we have $\sigma_m \sigma_n = \sigma_{m^{\sigma_n} n}$. Now since, G_R is generated by E^- and the $\sigma_m's$, it follows that $h_{m^{\sigma_m^{-1}}}$ is in

$$\langle G_R, \{(x, y) \longmapsto (x^{\sigma_m}, y^{\sigma_m}); m \in K - \{0\}\}, S \rangle,$$

only if it is the product of element $\tau_m \sigma_m$ and an element of S. However, this implies that $a^{\sigma_m^{-1}}$ is in $K - \{0\}$, only if a itself is in $K - \{0\}$.

Hence, we obtain that

$$\sigma_m : (x, y) \longmapsto (x^{\sigma_m} a, y^{\sigma_m} am), \text{ for } m \in K - \{0\},$$

and a also in $K - \{0\}$. Now since G_R and $G_R S$ have the same action on the components of any spread in $PG(3, K)$, we may assume that $a = 1$.

Now we know that any Pappian spread Σ' containing R admits the full central collineation group fixing $x = 0$ pointwise and fixing R. If Σ' admits the collineations $(x, y) \longmapsto (x^{\sigma_m}, y^{\sigma_m})$ for all $m \in K - \{0\}$, then Σ' admits $E \langle \sigma_m; m \in K - \{0\} \rangle$ as a collineation group. We note that $E \langle \sigma_m; m \in K - \{0\} \rangle$ is sharply 2-transitive on the components of $R - \{x = 0\}$. If G_R is not $E \langle \sigma_m; m \in K - \{0\} \rangle$, then there is an element in G_R that fixes three components of Σ in R, a contradiction. Hence,

$$G_R = E \langle \sigma_m; m \in K - \{0\} \rangle$$

fixes Σ'. Moreover, it now follows that

$$\langle G_R, \{(x, y) \longmapsto (x^{\sigma_m}, y^{\sigma_m}); m \in K - \{0\}\}, S \rangle$$

is a collineation group of Σ' and the group of central collineations of Σ within

$$\langle G_R, \{(x, y) \longmapsto (x^{\sigma_m}, y^{\sigma_m}); m \in K - \{0\}\}, S \rangle$$

is regular on the components of $\Sigma' - R$. Let σ denote the involution in $Gal_K K^2$ and note that a component of $\Sigma' - R$ must have the form $y = x^\sigma n + xt$ for some $n \neq 0$ and $t \in K^2$. It then follows that

$$\Sigma' - R = \{y = (x^\sigma n + xt)\alpha + x\beta; \alpha \neq 0, \beta \in K\}.$$

But, G_R maps $y = x^\sigma n + xt$ onto

$$\{y = (x^\sigma n^{\sigma m} + xt^{\sigma m})m + x\delta; m \neq 0, \delta \in K\}.$$

Also, $\{n^{\sigma m} m = n * m; m \in K\} \subseteq \{n\alpha; \alpha \in K\}$ and

$$\{t^{\sigma m} m + \delta = t * m + \delta; m \neq, \delta \in K\} \subseteq \{t\alpha + \beta; \alpha \neq 0, \beta \in K\}.$$

Note that the ϵ-condition implies equality of the previous two inequalities.

Thus, G_R is sharply transitive on the components of $\Sigma' - R$. This proves the theorem. $\qquad\square$

DEFINITION 112. *A group G satisfying 'sharp nearfield,' 'Baer,' and 'nearfield fix' is a parallelism-inducing group for Pappian spreads Σ and Σ' as above. We shall refer to such groups as 'nearfield parallelism-inducing groups' for Pappian spreads.*

A special case of the above is when the nearfield is a field (i.e. K^2), so the conditions become 'sharp,' 'Baer,' and 'central fix.' In this case, we refer to the group as a 'linear parallelism-inducing group' for Pappian spreads.

CHAPTER 30

Linear and Nearfield Parallelism-Inducing Groups

In this chapter, we consider the possible groups of $\Gamma L(2, K^2)$ that are Pappian parallelism inducing, where K^2 denotes the quadratic field extension of K coordinatizing π_Σ. We begin with the linear groups in $GL(2, K^2)$. In this setting, we need to find subgroups that satisfy the conditions: Sharp, Baer, and central fix, and we begin with this last condition.

Let $K^2 - \{0\} = P_x \otimes P_y$ be a group direct product decomposition of the multiplicative group of K^2 that also decomposes $K - \{0\}$ as a group. Assume also that the decomposition is invariant under $Gal_K K^2$ (automatic if finite). Furthermore, we choose the notation that $K - \{0\} = P_x^- \otimes P_y^-$ for $P_x^- \subseteq P_x$ and $P_y^- \subseteq P_y$. We coordinatize a Pappian affine plane using K^2 and consider the following homology groups:

$$H^x : \left\langle \tau_h = \begin{bmatrix} 1 & 0 \\ 0 & h \end{bmatrix} ; h \in P_x \right\rangle$$

and

$$H^y : \left\langle \rho_k = \begin{bmatrix} k & 0 \\ 0 & 1 \end{bmatrix} ; k \in P_y \right\rangle .$$

Let

$$E = \left\langle \begin{bmatrix} 1 & a \\ 0 & 1 \end{bmatrix} ; a \in K^2 \right\rangle .$$

LEMMA 52. $EH^x H^y$ satisfies the central fix condition.

PROOF.

$$\langle (EH^x H^y)_R, S \rangle = \left\{ \begin{bmatrix} k\beta & \alpha \\ 0 & h\beta \end{bmatrix} ; k \in P_x^{-1}, h \in P_y^{-1}, \alpha, \beta \neq 0, \text{ in } K \right\} .$$

Letting $\beta = h^{-1}$, we have the set of elements

$$\left\{ \begin{bmatrix} kh^{-1} & \alpha \\ 0 & 1 \end{bmatrix} ; k \in P_x^{-1}, h \in P_y^{-1}, \alpha \text{ in } K \right\} .$$

Since we have a group partition, we obtain the full central collineation group with axis $x = 0$ that fixes R. $\square$

THEOREM 192. *The group EH^xH^y satisfies conditions 'sharp,' 'Baer,' and 'central fix.' Therefore, the group is a linear parallelism-inducing group for Pappian spreads.*

PROOF. Let σ denote the unique automorphism of K^2 that fixes K pointwise. Again, recall that any Baer subplane of π_Σ disjoint from $x = 0$ may be written in the form $y = x^\sigma a + xb$ for unique $a \neq 0, b \in K$.

We note that E is transitive on the components distinct from $x = 0$. Furthermore, $y = x$ maps to $y = xk^{-1}h$ under $\tau_h\rho_k$. Hence, the group is sharply doubly transitively on the components of π_Σ distinct from $x = 0$.

Furthermore, $y = x^\sigma$ maps under E as follows:
$(x, x^\sigma) \longmapsto (x, xa + x^\sigma)$ so that $y = x^\sigma$ maps to $y = x^\sigma + xa$, for all $a \in K^2$. Then $y = x^\sigma + xa$ maps as follows:
$(x, x^\sigma + xa) \longmapsto (xk, (x^\sigma + xa)h)$ so that $y = x^\sigma + xa$ maps onto $y = x^\sigma k^{-\sigma}h + xahk^{-1}$. Now since the decomposition $K^2 - \{0\} = P_x \otimes P_y$ is invariant under σ, it follows that $P_x = P_x^{-\sigma}$. Hence,

$$\{k^{-\sigma}h; k \in P_x \text{ and } h \in P_y\} = P_x \otimes P_y = K^2 - \{0\}.$$

Since a is arbitrary in K^2, it follows that $\{ahk^{-1}; a \in K^2\} = K^2$. Thus, $y = x^\sigma$ maps onto all possible elements of the form $y = x^\sigma a + xb$ for $a \neq 0$. So, the group EH^xH^y is transitive on the set of Baer subplanes of Σ disjoint from $x = 0$. Now let

$$y = x^\sigma k^{-\sigma}h + xahk^{-1} \text{ be equal } y = x^\sigma k^{*-\sigma}h^* + xa^*h^*k^{*-1}.$$

Then $k^{-\sigma}h = k^{*-\sigma}h^*$ and $ahk^{-1} = a^*h^*k^{*-1}$. Since we have a group partition, it follows that $k^{-\sigma} = k^{*-\sigma}$, implying that $k = k^*$ and $h = h^*$. This then implies that $a^* = a$. Hence, we have that EH^xH^y acts regularly on the Baer subplanes disjoint from $x = 0$. Therefore, the group EH^xH^y satisfies conditions sharp, Baer, and central fix implying by Theorem 69 that the group is parallelism-inducing. □

From the group H^xH^y given by

$$\left\langle \begin{bmatrix} h & 0 \\ 0 & k \end{bmatrix} ; h \in P_x, k \in P_y \right\rangle,$$

we construct the following group:

$$H^{j,i} = \left\langle \begin{bmatrix} h^j k^{i+1} & 0 \\ 0 & h^{j+1}k^i \end{bmatrix} ; h \in P_x, k \in P_y \right\rangle$$

where j, i are fixed integers.

LEMMA 53. *Use definitions given above, for $EH^{i,j}$.*
(1) The group $EH^{j,i}$ satisfies condition 'sharp.'

(2) Furthermore, the group $EH^{j,i}$ satisfies condition 'Baer' provided

$$\{h^{-\sigma j + j + 1}k^{-\sigma(i+1)+i}; h \in H_x, k \in H_y\} = K^2 - \{0\}, \text{ and}$$
$$h^{-\sigma j + j + 1} = 1 = k^{-(-\sigma(i+1)+i)} \text{ if and only if}$$
$$h = k = 1.$$

(3) The group $EH^{j,i}$ satisfies the condition 'central fix.'

PROOF. Note that $(h^j k^{i+1})^{-1}(h^{j+1}k^i) = hk^{-1}$. This implies that the group $H^{i,j}$ maps $y = x$ onto $y = xhk^{-1}$. Since E maps $y = 0$ onto $y = xa$ for all $a \in K^2$, it follows that we have condition 'sharp.' This proves (1)

We consider the image set of $y = x^\sigma$ under the group $EH^{i,j}$. It follows that

$$y = x^\sigma \longmapsto y = x^\sigma h^{-\sigma j + j + 1}k^{-\sigma(i+1)+i} + xa$$

for all $h \in P_x$, $k \in P_y$ and all $a \in K^2$. The conditions given guarantee that the image set contains all possible Baer subplanes. But since σ leaves the decomposition invariant, the mappings $Z \longmapsto Z^{-\sigma j + j + 1}$ map P_x into P_x and $W \longmapsto W^{-\sigma(i+1)+i}$ maps P_y into P_y. The condition given implies that we have surjective and injective mappings so that $EH^{i,j}$ is regular on the set of Baer subplanes of Σ that are disjoint from $x = 0$. Hence, we obtain the 'Baer' condition, so we have the proof of (2).

The group originated from a group partition of $K^2 - \{0\}$ that induces a group partition on $K - \{0\}$. So,

$$h^{-(j+1)}k^{-i}I_2 \begin{bmatrix} h^j k^{i+1} & 0 \\ 0 & h^{j+1}k^i \end{bmatrix} = \begin{bmatrix} h^{-1}k & 0 \\ 0 & 1 \end{bmatrix}.$$

Hence, we have the homology group $(x, y) \longmapsto (xu, y)$ for all $u \in K - \{0\}$ as a collineation group of the parallelism. Since, we also have E_R, it follows that the group $EH^{i,j}$ satisfies the condition central fix. This completes the proof of (3). $\square$

So, applying Theorem 69, we obtain the following result.

THEOREM 193. *The group $EH^{i,j}$ is a linear parallelism-inducing group for Pappian spreads, provided*

$$\{h^{-\sigma j + j + 1}k^{-\sigma(i+1)+i}; h \in H_x, k \in H_y\} = K^2 - \{0\}, \text{ and}$$
$$h^{-\sigma j + j + 1} = 1 = k^{-(-\sigma(i+1)+i)} \text{ if and only if}$$
$$h = k = 1.$$

COROLLARY 50. *In the finite case, $EH^{j,i}$ is a linear parallelism-inducing group for Desarguesian spreads if and only if*

$$(qj - j - 1, |P_x|) = 1 = (q(i + 1) - i, |P_y|).$$

In particular, we obtain a parallelism-inducing group for Desarguesian spreads if

$$(qj - j - 1, q^2 - 1) = 1 = (q(i + 1) - i, q^2 - 1).$$

PROOF. Since any group $GF(q^2) - \{0\}$ is cyclic, of order $q^2 - 1 = \Pi_{i=1}^{k} u_i^{\alpha i}$, where u_i are the distinct prime factors of $q^2 - 1$, for $i = 1, .., k$, it follows, by re-indexing, if necessary, that P_x is the set of elements of orders dividing $\Pi_{i=1}^{s} u_i^{\alpha i}$, and P_y is the set of elements of orders dividing $\Pi_{j=s+1}^{k} u_j^{\alpha j}$. Since $q - 1$ divides $q^2 - 1$, there is an induced group partition of $GF(q) - \{0\}$ so that P_x^{-1} is the set of elements of orders dividing $\Pi_{i=1}^{s} u_i^{\beta i}$ and P_y^{-1} is the set of elements dividing $\Pi_{j=s+1}^{k} u_j^{\beta j}$, by inserting $\beta_k = 0$, when u_k is not a divisor of $q - 1$. Since σ is the automorpism defined by $x \longmapsto x^q$, the partitions are invariant under σ.

We need to verify that we obtain the condition

$$\{h^{-\sigma j + j + 1} k^{-\sigma(i+1)+i}; h \in H_x, k \in H_y\} = K^2 - \{0\}, \text{ and}$$

$h^{-\sigma j + j + 1} = 1 = k^{-(-\sigma(i+1)+i)}$ if and only if $h = k = 1$, with $\sigma = q$. However, if

$$(qj - j - 1, |P_x|) = 1 = (q(i + 1) - i, |P_y|),$$

we obtain the required partition. Since surjective mappings are injective, in this case, we have proved the result.

We note that

$$(qj - j - 1, q^2 - 1) = 1 = (q(i + 1) - i, q^2 - 1)$$

implies the requirement.   □

In the finite case, we will show that all partitions may be obtained when $P_x = K^2 - \{0\}$ and $P_y = \langle 1 \rangle$ and the groups $H^{0,j}$. That is, the group

$$\left\langle \begin{bmatrix} m^j & 0 \\ 0 & m^{j+1} \end{bmatrix}; m \in K^2 - \{0\} \right\rangle.$$

In the general case, for arbitrary fields K^2, we obtain group partitions of $K^2 - \{0\}$ that are σ-invariant and induce group partitions on $K - \{0\}$. Hence, we obtain linear parallelism-inducing groups for Pappian spreads if and only if

$$\{h^{-\sigma j + j + 1}; h \in P_x\} = K^2 - \{0\}, \text{ and}$$

$h^{-\sigma j+j+1} = 1$ if and only if $h = 1$, if and only if the mapping on K^2
$Z \longmapsto Z^{-\sigma j+j+1}$ is bijective.

The next remark is left to the reader as an exercise.

REMARK 57. *In the finite case,*

$$Z \longmapsto Z^{-qj+j+1}$$

is bijective on $GF(q^2) - \{0\}$ if and only if

$$(2j + 1, q + 1) = 1.$$

1. Parallelism-Inducing Subgroups of $GL(2, q^2)$

Suppose that a Desarguesian parallelism-inducing group G is a subgroup of a Pappian affine plane with spread in $PG(3, q)$. Then G fixes
a component and has order $q^2(q^2 - 1)$. Assume that G is linear in
$GL(2, q^2)$. Then, it follows immediately that there must be a normal
subgroup E of G, which is the full elation group, with axis $x = 0$.
This group acts regularly on the components other than $x = 0$. Hence,
there is a subgroup of order $q^2 - 1$ that fixes a second component $y = 0$.
Since the group is linear, the second group is in the direct product of
two homology groups: one with axis $x = 0$ and one with axis $y = 0$.
Since the group is sharply transitive on the components not equal to
$x = 0$ or $y = 0$, given an element m of $GF(q^2) - \{0\}$, there exists a
unique element of the form $(x, y) \longmapsto (xa_m, ya_m m)$ mapping $y = x$
onto $y = xm$.

We know that the restriction of this group to either $x = 0$ or $y = 0$
is cyclic. If m has order $q^2 - 1$, then $a_m = m^i$ for some integer i. Hence,
the group is cyclic and generated by

$$\begin{bmatrix} m^i & 0 \\ 0 & m^{i+1} \end{bmatrix}.$$

Such a group maps $y = x^q$ onto $y = x^q m^{-qi+i+1}$. Therefore, we require
$(2i + 1, q + 1) = 1$.

Suppose that the order of m^i is t_i dividing $q^2 - 1$. Then, we obtain
a homology subgroup with axis $x = 0$ of order $(i, q^2 - 1)$, where $t_i =
(q^2 - 1)/(i, q^2 - 1)$, as we note that

$$(m^{i+1})^{t_i} = m^{(i+1)(q^2-1)/(i,q^2-1)} = m^{(q^2-1)/(i,q^2-1)} = m^{t_i}.$$

Similarly, if the order of m^{i+1} is $s_i = (q^2 - 1)/(i + 1, q^2 - 1)$, we
obtain a homology subgroup with axis $y = 0$ of order $(i+1, q^2 - 1)$. We
note that it is possible that $t_i = 1 = s_i$ so that there are no homology
groups within this group. However, note that if m is in $GF(q)$ then

multiplication by the kernel homology element $\begin{bmatrix} m^{-i} & 0 \\ 0 & m^{-i} \end{bmatrix}$, shows that we obtain the homology group with axis $x = 0$ of order $q-1$ that fixes the standard regulus. So, we see that we have the conditions of sharp, Baer and central fix so we obtain a linear parallelism-inducing group by theorem 69.

Hence, we obtain:

THEOREM 194. *(1) G is a finite linear Desarguesian parallelism-inducing group for Desarguesian spreads if and only if $G = EH^i$ where E is the full elation group with axis $x = 0$ and H^i is cyclic and generated by* $\begin{bmatrix} m^i & 0 \\ 0 & m^{i+1} \end{bmatrix}$, *where m is a generator of $GF(q^2) - \{0\}$ and $(2i + 1, q + 1) = 1$, for a fixed integer $i \le q^2 - 1$.*

(2) There is an affine homology group with axis $x = 0$ and co-axis $y = 0$ of order $(i + 1, q^2 - 1)$ and an affine homology group with axis $y = 0$ and co-axis $x = 0$ of order $(i, q^2 - 1)$. Furthermore, there are always homology groups of order $q - 1$ with either axis within GS.

DEFINITION 113. *We use the notation EH^i to denote the above parallelism-inducing linear group. Let $\mathcal{P}_i$ denote the corresponding parallelism using the same initial Desarguesian spread Σ and Desarguesian spreads Σ_i for $i = 1, 2$, respectively.*

The proof of the following is left to the reader as an exercise.

LEMMA 54. $\mathcal{P}_i \simeq \mathcal{P}_{i+(q+1)}$.

THEOREM 195. *If $\mathcal{P}_i$ is isomorphic to $\mathcal{P}_j$ then*

$$|(\langle H^i, H^j \rangle S)| \mid (2, q - 1)(q^2 - 1)(q - 1).$$

This is valid if and only if either
(i) $i \equiv j \mod (q + 1)$ when q is even and $(2i + 1, q + 1) = (2j + 1, q + 1) = 1$, or
(ii) q is odd and
(a) if one of the second Desarguesian spreads is left invariant under a kernel homology group of Σ of order $2(q - 1)$, then $i \equiv j \mod(q + 1)/2$,
(b) or if one of the second Desarguesian spreads is not left invariant by a kernel homology group of Σ of order $2(q - 1)$, then $i \equiv j \mod(q + 1)$ and $(2i + 1, q + 1) = (2j + 1, q + 1) = 1$.

PROOF. We may assume that an isomorphism ρ fixes the standard regulus R and fixes three components $x = 0, y = 0, y = x$. Let ρ have the form in Σ_1 of $(x, y) \longmapsto (x^\tau a, y^\tau a)$. Clearly, ρ normalizes both H^i

and H^j. Hence, the first assertion follows as previously. Note that we have seen that $\mathcal{P}_i \simeq \mathcal{P}_{i+(q+1)}$. Furthermore, $H^j S$ has index dividing 2 in $H^i H^j S$, where S is the kernel homology group of order $q - 1$. Since the orbit structures of $H^i S$ and H^i are identical, it follows either we have that $i \equiv j \bmod(q+1)/2$ and we have a group $H^i H^j S$ of order $2q^2(q^2 - 1)(q - 1)$, in which case the Desarguesian plane Σ' is fixed by a collineation group of order $2(q - 1)$, or, the group $H^i H^j S$ has order $q^2(q^2 - 1)(q - 1)$, and $i \equiv j \bmod(q+1)$. $\qquad \square$

2. The General Nearfield Parallelism-Inducing Group

We now return to the general cases and consider nearfield groups. For nearfield parallelism-inducing groups for Pappian spreads, we need a group nH_f with elements

$$\phi_m : (x, y) \longmapsto (x^{\sigma m} f(m), y^{\sigma m} f(m)m)$$

for all $m \in K^2 - \{0\}$, for some function $f : K^2 - \{0\} \longmapsto K^2 - \{0\}$, since the group EnH_f will act sharply transitively on the components of the Pappian spread not equal to $x = 0$. In order that we have a group, we must have the following condition holding:

$$f(m * n) = f(m)^{\sigma_n} f(n), \forall m, n \in K^2 - \{0\}.$$

To see this, note that $\phi_m \phi_n = \phi_{m \sigma_n n}$, since we have a nearfield. Then,

$$\phi_m \phi_m = \phi_{m \sigma_n n} : (x, y) \longmapsto (x^{\sigma m} f(m)^{\sigma_n} f(n), y^{\sigma m} f(m)^{\sigma_n} f(x)m^{\sigma_n}n).$$

We note that f is an homomorphism of $(K^2 - \{0\}, *)$ if and only if $\sigma_n = \sigma_{f(n)}$, for all $n \in K^2 - \{0\}$. In the previous cases, where $f(m) = m^i$, the condition is automatic.

Note that it is not necessarily the case that f restricted to $K - \{0\}$ is a function on $K - \{0\}$. When this is the case, we would obtain condition 'nearfield fix' analogously as previously, as follows: If $(x, y) \longmapsto (x^{\sigma m}, y^{\sigma m})$ leaves Σ' invariant, then the subgroup $\langle \phi_m; m \in K - \{0\} \rangle$, also leaves Σ' invariant.

$$\langle G_R, \{(x, y) \longmapsto (x^{\sigma m}, y^{\sigma m}); m \in K - \{0\}\}, S \rangle,$$

contains the group elements

$$(x, y) \longmapsto (x f(m)^{\sigma_m^{-1}}, x f(m)^{\sigma_m^{-1}} m^{\sigma m - 1}),$$

implying that we obtain the central collineations

$$(x, y) \longmapsto (x m^{-\sigma_m^{-1}}, y)$$

for all $m \in K - \{0\}$. Now conjugate by

$$(x, y) \longmapsto (x^{\sigma m} f(m), y^{\sigma m} f(m)m) :$$

$$(x, y) \longmapsto (x^{\sigma_m^{-1}} f(m)^{-\sigma_m^{-1}}, y^{\sigma_m^{-1}} (f(m)m)^{-\sigma_m^{-1}})$$
$$\longmapsto (x^{\sigma_m^{-1}} f(m)^{-\sigma_m^{-1}} m^{-\sigma_m^{-1}}, y^{\sigma_m^{-1}} (f(m)m)^{-\sigma_m^{-1}})$$
$$\longmapsto (xm^{-1}, y), \ \forall m \in K - \{0\}.$$

The image of $y = x^{\sigma}$ under ϕ_m is $y = x^{\sigma} f(m)^{1-\sigma} m$. Hence, we would require the mapping $m \longmapsto f(m)^{1-\sigma} m$ to be bijective. The analogous and more general theorem is as follows.

THEOREM 196. *Let Σ and Σ' be Pappian spreads, where K^2 coordinatizes the affine plane determined by Σ. If G is a nearfield parallelism-inducing group satisfying nearfield fix, then there is a function f on $K^2 - \{0\}$ whose restriction to $K - \{0\}$ is a function on $K - \{0\}$, satisfying $f(m^{\sigma_n} n) = f(m)^{\sigma_n} n$, for all $m, n \in K^2 - \{0\}$ and $G = EnH_f$.*

Then G is a nearfield parallelism-inducing group for the Pappian spreads Σ and Σ' if and only if for the corresponding nearfield $(K^2, +, *)$ and associated automorphisms, we have the following three conditions holding:

(1) if

$$\{w * m + \epsilon n; m, n \in K\} \subseteq \{wm + \epsilon n; m, n \in K\}$$

then

$$\{w * m + \epsilon n; m, n \in K\} = \{wm + \epsilon n; m, n \in K\},$$

for $\epsilon = 0$ or 1 (the ϵ-condition),

(2) $m \longmapsto f(m)^{1-\sigma} m$ is bijective on $K^2 - \{0\}$, and

(3) Σ' admits the corresponding collineations $(x, y) \longmapsto (x^{\sigma_m}, y^{\sigma_m})$.

3. Isomorphisms of Group-Induced Parallelisms

In Chapter 28, there was a considerable discussion on the set of isomorphisms of Johnson parallelisms. The ideas given in that chapter can be made more general and applicable for parallelism-inducing groups for Pappian spreads.

THEOREM 197. *Let K be a field of cardinality > 3. Let G_i be parallelism-inducing groups for the Pappian spreads $\{\Sigma, \Sigma_i\}$ for $i = 1, 2$ in $PG(3, K)$, where K^2 is a quadratic extension and let $\mathcal{P}_{\Sigma, \Sigma_i} = \mathcal{P}_i$ denote the corresponding parallelisms. Let ℓ denote the unique line of Σ fixed by G_i, $i = 1, 2$. Assume that Σ_i and Σ share a regulus R containing ℓ.*

(1) If ρ is an isomorphism from $\mathcal{P}_{\Sigma, \Sigma_1}$ onto $\mathcal{P}_{\Sigma, \Sigma_2}$, then ρ is a collineation of the Pappian plane π_{Σ}, which leaves ℓ invariant and may be assumed to leave R invariant, fix three components of R, and map Σ_1 onto Σ_2.

(2) Let G_i for $i = 1, 2$ be any two finite linear parallelism-inducing groups for Desarguesian spreads such that $G_{i,R}S$ contains the full central collineation group of Σ with fixed axis ℓ that leaves invariant the regulus R (G_i satisfies sharp, Baer, and central fix conditions), where S denotes the scalar homology group of order $q - 1$.

Then G_2S has index dividing $(2, q - 1)$ in $\langle G_1^\rho, G_2 \rangle S$. Furthermore, if the spread for Σ_2 is

$$\{x \;=\; 0, y = (x^q m + xn)\alpha + \beta;$$
$$\alpha, \beta \;\in\; GF(q) \text{ and } n \notin GF(q)\}$$

then $\langle G_1^\rho, G_2 \rangle S = G_2S$.

(3) For any parallelism-inducing group G, let $\mathcal{L}$ denote the subgroup of $GL(2, q^2)$ that can act on a parallelism given by a parallelism-inducing group. Then

$$|\mathcal{L}| \;\mid\; (2, q - 1)q^2(q^2 - 1)(q - 1).$$

Furthermore, if the spread for Σ_2 is

$$\{x = 0, y = (x^q m + xn)\alpha + \beta; \alpha, \beta \in GF(q)\}$$

and $n \notin GF(q)$ then

$$|\mathcal{L}| \;\mid\; q^2(q^2 - 1)(q - 1).$$

PROOF. (1) Recall that if an isomorphism turns ρ maps $\Sigma_1^* g_1$ to $\Sigma_2^* g_2$ then $\Sigma_1 g_1$ maps onto $\Sigma_2 g_2$ for $g_i \in G_i$. (If the associated Hall planes are isomorphic, then the corresponding Pappian planes are isomorphic.) Then $\Sigma \cap \Sigma_1 g_1$ is mapped to $\Sigma \cap \Sigma_2 g_2$; the reguli of Σ containing ℓ are permuted and hence ℓ is left invariant. Since G_2 is sharply 2-transitive on the components of Σ, it follows that we may assume that ρ fixes ℓ (chosen as $x = 0$) and two other components, say, $y = 0$ and $y = x$. Hence, we may assume that ρ fixes the regulus R containing these three components. Since there are exactly the Pappian spreads Σ_1 and Σ_2 that share R with Σ, it follows that we may assume that ρ maps Σ_1 onto Σ_2. This proves (1).

(2) If the assertion is not true, then there exists a linear collineation g in Σ acting on $\mathcal{P}_2$, which fixes at least three components of R and hence fixes R. Also, g is a collineation of the parallelism so must leave Σ_2 invariant (as there are exactly two Pappian spreads, Σ_1 and Σ_2, containing R) such that g^2 is not in S. The stabilizer of R in Σ that can act on Σ_2 and fix at least three components and be in $GL(2, q^2)$ has order dividing $2(q - 1)$. To see this, note that g must have the form

$(x, y) \longmapsto (xa, ya)$ and Σ_2 has the form

$$(y = x^q m + xn)\alpha + \beta I \ \forall \alpha, \beta \in K, for \ m \neq 0, n \in GF(q^2).$$

But, g maps $y = x^q m + xn$ onto $y = x^q m a^{1-q} + xn$ implying that $a^{(1-q)^2} = 1$. Therefore,

$$((1 - q)^2, q^2 - 1) = (1 - q, q + 1)(q - 1) = (2, q - 1)(q - 1).$$

Hence, the order of a divides $(2, q - 1)(q - 1)$. Moreover, if n is not in $GF(q)$, then we have

$$ma^{1-q}\alpha = m\alpha^* \ and \ n\alpha + \beta = n\alpha^* + \beta^*$$

for $\alpha, \beta, \alpha^*, \beta^* \in GF(q)$, so that $a^{1-q}\alpha = \alpha^* = \alpha$. So, the order of a is this case divides $q - 1$. This proves the theorem, as the proof of (2) implies (3).   □

4. Nearfield Parallelism-Inducing Groups

Previously, we have constructed the $EH^{i,j}$ parallelism-inducing groups for Pappian spreads. We have seen that the group $EH^{i,j}$ satisfies the central fix condition. We may also construct similar non-linear groups. The idea is to use the nearfield multiplication to construct a direct product of what would be homologies in the nearfield plane, with axes $x = 0$ or $y = 0$ but acting in the Desarguesian plane Σ. We first restrict the discussion to the groups $EH^{0,j}$ or $EH^{i,0}$. Referring to Theorem 191, we adopt the notation developed there.

We may form the corresponding non-linear groups nH_y^i and nH_y^j as follows:

If we have the group nH_y with elements $(x, y) \longmapsto (x^{\sigma^m}, y^{\sigma^m} m)$, we form the associated group nH_y^i with elements: $\phi_m : (x, y) \longmapsto (x^{\sigma^m} m^i, y^{\sigma^m} m^{i+1})$.

Similarly, considering the group nH_x with elements: $(x, y) \longmapsto (x^{\sigma^m} m, y^{\sigma^m})$, we form the associated group nH_x^j with elements $(x, y) \longmapsto (x^{\sigma^m} m^{j+1}, y^{\sigma^m} m^j)$.

The question is whether the indicated groups admit conditions 'sharp nearfield,' 'Baer,' and 'nearfield fix.'

Note that the group nH_y^i maps $y = x$ onto $y = xm$ for all $m \in K^2 - \{0\}$ and E maps $y = 0$ onto $y = xa$ for all $a \in K$. Hence, we have the condition 'sharp nearfield' provided if

$$\{w * m + \epsilon n; m, n \in K\} \subseteq \{wm + \epsilon n; m, n \in K\}$$

then

$$\{w * m + \epsilon n; m, n \in K\} = \{wm + \epsilon n; m, n \in K\},$$

for $\epsilon = 0$ or 1 (the ϵ-condition)

Let $m \in K - \{0\}$. Then, $y = xu$ maps onto $y = xu^{\sigma^m}m$ so that such elements leave R invariant. We need to show that G_R leaves invariant the second Pappian spread Σ' containing R. If $(x, y) \longmapsto (x^{\sigma m}, y^{\sigma m})$ leaves Σ' invariant then the subgroup $\langle \phi_m; m \in K - \{0\} \rangle$, also leaves Σ' invariant.

$$\langle G_R, \{(x, y) \longmapsto (x^{\sigma m}, y^{\sigma m}); m \in K - \{0\}\}, S \rangle$$

contains the group elements $(x, y) \longmapsto (xm^{i\sigma_m^{-1}}, xm^{(i+1)\sigma_m^{-1}})$, implying that we obtain the central collineations $(x, y) \longmapsto (xm^{-\sigma_m^{-1}}, y)$, for all $m \in K - \{0\}$. Now conjugate by

$$(x, y) \longmapsto (x^{\sigma m}m^i, y^{\sigma m}m^{i+1}):$$

$$
\begin{aligned}
(x, y) \quad &\longmapsto \quad (x^{\sigma_m^{-1}}m^{-i\sigma_m^{-1}}, y^{\sigma_m^{-1}}m^{-(i+1)\sigma_m^{-1}}) \\
&\longmapsto \quad ((x^{\sigma_m^{-1}}m^{-i\sigma_m^{-1}}m^{-\sigma_m^{-1}}, y^{\sigma_m^{-1}}m^{-(i+1)\sigma_m^{-1}}) \\
&\longmapsto \quad (xm^{-1}, y).
\end{aligned}
$$

Thus, we obtain the full central collineation group; we obtain condition 'nearfield fix.'

We consider the images of $y = x^\sigma$ under nH_y^i, and obtain: $y = x^\sigma m^{-\sigma i + i + 1}$. If the mapping $Z \longmapsto Z^{-\sigma i + i + 1}$ is bijective, we have the 'Baer' condition.

Hence, we obtain the following theorem.

THEOREM 198. *Let Σ and Σ' be Pappian spreads, where K^2 coordinatizes the affine plane determined by Σ, and let $G = EnH_y^i$.*

Then G is a nearfield parallelism-inducing group for the Pappian spreads Σ and Σ' if and only if for the corresponding nearfield $(K^2, +, *)$ and associated automorphisms, we have the following three conditions holding:

(1) if

$$\{w * m + \epsilon n; m, n \in K\} \subseteq \{wm + \epsilon n; m, n \in K\},$$

then

$$\{w * m + \epsilon n; m, n \in K\} = \{wm + \epsilon n; m, n \in K\},$$

for $\epsilon = 0$ or 1 (the ϵ-condition),
(2) $Z \longmapsto Z^{-\sigma i + i + 1}$ is bijective on K^2, and
(3) Σ' admits the corresponding collineations $(x, y) \longmapsto (x^{\sigma m}, y^{\sigma m})$.

The following corollary is now clear.

COROLLARY 51. *Let Σ and Σ' be Desarguesian spreads, where K^2 is isomorphic to $GF(q^2)$ and coordinatizes the affine plane determined by Σ, and let $G = EnH_y^i$.*

Then G is a nearfield parallelism-inducing group for the Pappian spreads Σ and Σ' provided $(2i + 1, q + 1) = 1$ and Σ' admits the corresponding collineations $(x, y) \longmapsto (x^{\sigma m}, y^{\sigma m})$.

COROLLARY 52. *Let Σ and Σ' be Desarguesian spreads, where K^2 is isomorphic to $GF(q^2)$ and coordinatizes the affine plane determined by Σ, and let $G = EnH_y^i$. Let the nearfield have Dickson pair (h, r), where $h^r = q^2$.*

(1) If r is an odd prime power suppose that $q^2 = p^{2rw}$, so that $h = p^{2w}$ and $q = p^{rw}$, and assume that $(\sqrt{h}, r)$ is a Dickson pair.

(2) If r is 2, assume that q is odd.

Then G is a nearfield parallelism-inducing group for the Pappian spreads Σ and Σ' provided $(2i + 1, q + 1) = 1$.

PROOF. (1) First assume that q is odd. The group F generated by $(x, y) \longmapsto (x^h, y^h)$ leaves invariant the orbit of length $(q + 1)/2$. Let s be a prime divisor of r. Then, s divides $\sqrt{h} - 1 = p^w - 1$. Also, $(p^w - 1, q + 1 = p^{wr} + 1) = (p^w - 1, p^{wr} - 1 + 2) = 2$. Since r is odd, s cannot divide $(q + 1)$ so it cannot divide $(q + 1)/2$. Since r is a prime power, the group F generated by $(x, y) \longmapsto (x^h, y^h)$ must leave a Desarguesian spread containing R invariant.

Now assume that q is even. Hence, the indicated group permutes the set of $q/2 - 1$ orbits of length $q + 1$. Since s divides $2^w - 1$ and $(2^w - 1, 2^{rw-1} - 1) = (2^{(w, rw-1)} - 1) = 1$, it follows that the group F must fix one of the orbits of length $q + 1$. Therefore, since $(2^w - 1, 2^{rw} + 1) = (2^w - 1, 2^{rw} - 1 + 2) = 1$, it follows that F must fix one of the spreads of the orbit.

The proof to (2) will be given more generally in the proof to the following example.   □

Also, in this section, we provide an example of a nearfield parallelism-inducing group, which is valid for certain infinite and finite fields.

EXAMPLE 5. *Let K be a field such that $K - \{0\}$ contains an index two subgroup of squares, and let K^2 be a quadratic extension of K such that $K^2 - \{0\}$ contains an index two subgroup of squares. Let σ denote the involution in $\mathrm{Gal}_K K^2$.*

Define $x^{\sigma m} m = x * m = x^{\sigma} m$ if m is non-square in K^2 and $x * m = xm$ if m is square in K.

(1) Then, $(K, +, *)$ is a nearfield.

(2) Furthermore, any group EnH_y^i is a nearfield parallelism-inducing group for Pappian spreads provided $Z \longmapsto Z^{-\sigma i + i + 1}$ is bijective on K^2.

(3) If $i = 0$ or -1, then EnH_y^i is a nearfield parallelism-inducing group.

PROOF. We need to verify the ϵ-condition: if

$$\{w * m + \epsilon n; m, n \in K\} \subseteq \{wm + \epsilon n; m, n \in K\},$$

then

$$\{w * m + \epsilon n; m, n \in K\} = \{wm + \epsilon n; m, n \in K\},$$

for $\epsilon = 0$ or 1. However, m is a square in K^2 since it is in K. Hence, we have equality.

If $i = 0$ or -1, we have the mapping $Z \longmapsto Z$ or $Z \longmapsto Z^\sigma$, respectively, so that the mapping is bijective on K^2.

It remains to show that we can find a Pappian spread Σ' containing R and admitting $(x, y) \longmapsto (x^\sigma, y^\sigma)$ as a collineation group. Choose the Baer subplane $y = x^\sigma m$ such that $m^{1+\sigma}$ is non-square. Choose a basis $\{1, e\}$ for K^2 such that $e^2 = \gamma$ is non-square in K. Note that

$$\{y = x^\sigma m\alpha + x\beta; \alpha, \beta \in K\}$$

is a quadratic field extension of K if and only if $x^\sigma m\alpha + x\beta$ is non-singular. If $x_o^\sigma m\alpha = -x_o\beta$ for $x_o \neq 0$, then $m^{\sigma+1} = \beta^{\sigma+1} = \beta^2$, a contradiction. Note that this then defines a Pappian spread containing R and distinct from Σ. The image of $y = x^\sigma m\alpha + x\beta$ under the group element $(x, y) \longmapsto (x^\sigma, y^\sigma)$ is $y = x^\sigma m^\sigma \alpha + x\beta$. But, since $m^\sigma = -m$, it follows that the image of $y = x^\sigma m\alpha + x\beta$ is $y = x^\sigma(-m\alpha) + x\beta$. Thus, we have the required Pappian spread. $\square$

5. Finite Regular Nearfield Groups

If we have a sharply 2-transitive group on q^2 points on the line at infinity of a Desarguesian affine plane, there is a corresponding regular nearfield group. That is, the group induced on the line at infinity is a regular nearfield group. We then are concerned with the possible preimage groups in $\Gamma L(2, q^2)$.

We see that any regular nearfield of order $h^r = q^2$ with kernel $GF(h)$ admits a sharply 2-transitive group of order $h^r(h^r - 1)$ acting in $\Gamma L(2, q^2)$ and so has the potential to be parallelism inducing.

However, we have noted previously that the most general group has the form EnH_f for some function f on $GF(q^2) - \{0\}$, such that $f(m^{\sigma_n}n) = f(m)^{\sigma_n}f(n)$. Since f is not necessarily a homomorphism, we need to assume that f induces a function on $GF(q) - \{0\}$ so that the condition nearfield fix is satisfied.

THEOREM 199. *(1) Any nearfield parallelism-inducing group for Desarguesian spreads has the form EnH_f^j, for some function f on $GF(q^2) - \{0\}$ satisfying $f(m^{\sigma_n}n) = f(m)^{\sigma_n}f(n)$ for all $n, m \in GF(q^2) -$*

$\{0\}$. *If the nearfield has kernel $GF(h)$ such that $h^r = q^2$ let $\sigma_m = h^{\lambda(m)}$.*
The group is, in fact, parallelism-inducing provided:
    (a) the function f induces a function on $GF(q) - \{0\}$,
    (b) the second Desarguesian spread admits

$$\left\langle (x,y) \longmapsto (x^{h^{\lambda(m)}}, y^{h^{\lambda(m)}}); m \in GF(q) - \{0\} \right\rangle$$

as a collineation group, and
    (c) $m \longmapsto f(m)^{1-q}m$ is an injective mapping on $GF(q^2) - \{0\}$.
    If particular, if $f(m) = m^i$, then the injective condition is equivalent
to $(2i + 1, q + 1) = 1$, where $h^r = q^2$.
    (2) In an EnH_y^j group, there are homology groups of orders $(i +$
$1, (q^2 - 1)/r)$ and $(i, (q^2 - 1)/r)$ with axes $x = 0$ (respectively, $y = 0$)
and coaxes $y = 0$ (respectively, $x = 0$).

PROOF. By Theorem 191, we need to verify that the conditions
nearfield sharp, Baer, and nearfield fix hold. We note that $y = x$ maps
onto $y = xm$ and $y = x^q$ maps onto $y = x^q f(m)^{1-q}m$, so that nearfield
sharp and Baer hold. The argument given in Theorem 191 shows that
nearfield fix holds provided the second Desarguesian spread admits the
collineation group

$$\left\langle (x,y) \longmapsto (x^{h^{\lambda(m)}}, y^{h^{\lambda(m)}}); m \in GF(q) - \{0\} \right\rangle.$$

Hence, we obtain a nearfield parallelism-inducing group.
    We note that $\lambda(m) = 0$ for a subgroup of order exactly $(q^2 - 1)/r$.
Moreover, the order of m^i is $((q^2 - 1)/r)/(i, (q^2 - 1)/r)$; we obtain a
homology group with axis $y = 0$ and coaxis $x = 0$ of order $(i, (q^2 -$
$1)/r)$, since if the order of m^i divides $((q^2 - 1)/r)/(i, (q^2 - 1)/r)$, then
$m^{i+1} = m$. Similarly, there is a homology group with axis $x = 0$
and coaxis $y = 0$ of order $(i + 1, (q^2 - 1)/r)$, since if the order of m
divides $(i + 1), (q^2 - 1)/r$, then $m^i = m^{i+1-1} = m^{-1}$ and there is a
corresponding homology group. This completes the proof.   □

    We now provide another isomorphism type theorem for finite nearfield
parallelism-inducing groups of type EnH_y^i.

THEOREM 200. *(1) There is a linear subgroup of nH_y^i corresponding*
to the Dickson pair (h, r) of order $(q^2 - 1)/r$.
    (2) Let nH_y^i and nH_y^j define isomorphic parallelisms corresponding
to Dickson pairs (h_1, r_1) and (h_2, r_2), respectively, and using the same
two Desarguesian spreads Σ_1 and Σ_2. There are linear groups of orders
$(q^2 - 1)/r_1$ and $(q^2 - 1)/r_2$ acting on either parallelism. Let $I_{i,j}$ denote
the order of the intersection of these two linear groups and let $s_{i,j} =$
$|S \cap H^i H^j|$.

Then, $(q^2 - 1)/I_{i,j}$ *divides* $2r_1r_2s_{i,j}$.

(3) Let EnH_y^i *and* EnH_y^j *define isomorphic parallelisms, with the second group corresponding to the Dickson pair* (h, r).

Then, $(q^2 - 1)/I_{1,j}$ *divides* $2r^2s_{1,j}$.

(4) In particular, when the Dickson pairs are the same $(q, 2)$, *we have* $(q^2 - 1)/I_{i,j}$ *divides* $8s_{i,j}$. *In this case,*

$$i \equiv j \ mod \ (q^2 - 1)/8s_{i,j}.$$

In particular,

$$i \equiv j \ mod(q + 1)/(8, q + 1).$$

(5) If one of the groups is linear and one has Dickson pair $(q, 2)$, *then*

$$i \equiv j \ mod \ (q + 1)/(4, q + 1).$$

PROOF. Since $\lambda(m) = 0$ for $(q^2 - 1)/r$ elements, we have the proof of part (1). Combining (1) and (2) proves part (3). Parts (4) and (5) are immediate consequences of the previous. □

REMARK 58. *Two nearfield groups with different kernels will, normally, induce non-isomorphic parallelisms.*

6. m-Parallelisms

In this section, we generalize the previous construction and instead of using two Pappian spreads and an associated group, we choose instead m such Pappian spreads. These ideas will lead us eventually to the concept of elation-switching. In the present setting, by a choice of cosets of a particular subgroup of G, we are able to construct a tremendous variety of parallelisms. The parallelisms that we obtain are called 'm-parallelisms' and, in the finite case, admit a central collineation group (of the original Desarguesian spread in $PG(3, q)$) of order $q^2(q^2 - 1)/m$. If m is not n, then an m-parallelism cannot be isomorphic to an n-parallelism.

In a sense, m-parallelisms are generated using a particular set of m Pappian spreads. If n of the m spreads are distinct, we call such parallelisms (m, n)-parallelisms. The original construction uses mappings from a particular second Pappian spread for the construction. Such spreads are subject to a choice of coset representations so further subclasses are obtained.

The construction of the Johnson parallelisms produced parallelisms in $PG(3, q)$ using a Desarguesian spread Σ_1 equipped with a central collineation group of Σ_1, G, with fixed axis ℓ, of order $q^2(q^2 - 1)$. It turns out that the set of Baer subplanes incident with the origin of Σ_1, which are disjoint from the axis ℓ are in a single orbit under G and the number of such Baer subplanes is exactly $q^2(q^2 - 1)$. That is, the group G is regular on this set of Baer subplanes.

The construction also depends on the choice of an initial regulus $\mathcal{R}$ within Σ_1 and containing ℓ. Choose a second Pappian spread Σ_2 containing $\mathcal{R}$ and let G denote the full central collineation group with axis ℓ of Σ_1. If s_2 is any line of $\Sigma_2 - \mathcal{R}$, then we note that $\Sigma_2 = s_2 G_{\mathcal{R}} \cup \mathcal{R}$. Let S be a normal subgroup of G containing $G_{\mathcal{R}}$, let $h \in S - \mathcal{R}$. We note that $\Sigma_2 S \cup \Sigma_2 h S$ is a partial parallelism. More important, if Σ_3 is any Pappian spread distinct from Σ_1 that contains $\mathcal{R}$, and $g \in S - G_{\mathcal{R}}$, then $\Sigma_3 = g s_2 G_{\mathcal{R}} \cup \mathcal{R}$. Note that this says that $\Sigma_3 S = \Sigma_2 S$ as a set, and furthermore, it is also true that $\Sigma_3 S \cup \Sigma_3 h S$ is a partial parallelism. Hence, it follows immediately that $\Sigma_2 S \cup \Sigma_3 h S$ is also a partial parallelism. These remarks basically prove the following lemma.

LEMMA 55. *Under the assumptions above, let S denote any normal subgroup of G that contains $G_{\mathcal{R}}$. Let Σ_2 and Σ_3 be Desarguesian spreads distinct from Σ_1 that contain $\mathcal{R}$. Assume that there is an element g of $S - G_{\mathcal{R}}$ that maps an element s_2 of $\Sigma_2 - \mathcal{R}$ onto an element s_3 of Σ_3.*

(1) Then s_3 is not in $\mathcal{R}$ and $\Sigma_3 - \mathcal{R} = s_2 g G_{\mathcal{R}} = s_3 G_{\mathcal{R}}$.

(2) Let $h \in G - S$, then $\Sigma_2 w$ and $\Sigma_3 h u$ share no line for all $w, u \in S$; $\Sigma_2 S \cup \Sigma_3 h S$ is a partial parallelism.

COROLLARY 53. *We denote the derived spreads of $\Sigma_i w$ by derivation of $\mathcal{R} w$ by $(\Sigma_i w)^* = \Sigma_i^* w$.*

(1) Then $\Sigma_1 \cup \Sigma_2^* w \cup \Sigma_3^* h k$, for h fixed in $G - S$ and for all $w, k \in S$ is a partial parallelism in $PG(3, q)$.

(2) If the order of the S is $q^2(q^2 - 1)/m$, where m divides $q + 1$, then there are $1 + 2(q(q+1)/m$ spreads in the partial parallelism. Note that we do not require Σ_2 and Σ_3 to be distinct.

COROLLARY 54. *Under the above assumptions, further assume that there are t Desarguesian spreads Σ_i for $i = 2, .., t+1$ distinct from Σ_1 and sharing $\mathcal{R}$ with the property that for each Σ_i $i > 2$, there is a line $s_{2,i}$ of $\Sigma_2 - \mathcal{R}$ and an element g_i of S such that $s_{2,i} g_i$ is a line of Σ_i.*

Assume that S is a normal subgroup of G. Let h_i, for $i = 2, 3, .., t+1$ belong to mutually distinct cosets of S.

(1) Then $\cup_{i=2}^{t+1} \Sigma_i h_i k_i$, for all $k_2, .., k_{t+1} \in S$ is a set of spreads $t|S|$ spreads that share no line of $PG(3, q)$ not in Σ_1 and disjoint from the axis ℓ of G (of S).

(2) $\Sigma_1 \cup_{i=2}^{t+1} \Sigma_i^* h_i k_i$, for all $k_2, .., k_{t+1} \in S$ is a partial parallelism in $PG(3, q)$ of $1 + t(q(q+1)/m$ spreads provided the order of S is $q^2(q^2 - 1)/m$ (we note below that any such group of this order is normal).

PROOF. Suppose that $\Sigma_i h_i k_i$ and $\Sigma_j h_j k_j$ share a component. Then $\Sigma_i h_i h_i k_j^{-1} h_j^{-1}$ and Σ_j also share a component t_j.

We know that there exist elements $s_{2,i}$ and $s_{2,j}$ of $\Sigma_2 - \mathcal{R}$, and elements g_i and g_j of S such that $s_{2,i} g_i$ and $s_{2,j} g_j$ are in $\Sigma_i - \mathcal{R}$ and $\Sigma_j - \mathcal{R}$ respectively. Let $\widetilde{g} = h_i h_i k_j^{-1} h_j^{-1}$. Let t_i be in Σ_i such that $t_i \widetilde{g} = t_j$. It is immediate that t_i and t_j cannot be in $\mathcal{R}$. Hence, there exist elements w_i and w_j of $G_\mathcal{R}$ such that $t_i = s_{2,i} g_i w_i$ and $t_j = s_{2,j} g_j w_j$.

Hence, we obtain: $s_{2,i} g_i w_i \widetilde{g} = s_{2,j} g_j w_j$.

Furthermore, since $s_{2,i}$ and $s_{2,j}$ are both in $\Sigma_2 - \mathcal{R}$, it follows that there is an element r of $G_\mathcal{R}$ such that $s_{2,i} = s_{2,j} r$. We, in turn, obtain $s_{2,j} r g_i w_i \widetilde{g} = s_{2,j} g_j w_j$.

Now since the group G acts regularly on Baer subplanes of Σ_1, which do not intersect the axis, it follows that $r g_i w_i \widetilde{g} = g_j w_j$ and thus $r g_i w_i h_i h_i k_j^{-1} h_j^{-1} = g_j w_j$.

Note that all group elements other than h_i and h_j^{-1} involved in the above expression are in S. But this says that h_i and h_j are in the same coset of S since S is a normal subgroup. Hence, this contradiction completes the proof of the corollary with the exception of the existence of a normal group of order $q^2(q^2 - 1)/m$ containing $G_\mathcal{R}$ provided m divides $q + 1$.

The full central collineation group $G = EH$, where E is the full elation group of order q^2 and H is a homology group of order $q^2 - 1$. Note that E is a normal subgroup and H is cyclic. Let H^- denote the unique cyclic subgroup of order $(q^2 - 1)/m$ provided m divides $q^2 - 1$.

Then we assert that EH^- is a normal subgroup of EH and if m divides $q + 1$, it contains $G_\mathcal{R}$. Let $g = eh \in EH = G$, where $e \in E$ and $h \in H$.

We recall that

$$h^{-1} e^{-1} E H^- eh = h^{-1} E H^- eh = E h^{-1} H^- eh \subseteq E h^{-1} H^- Eh,$$

which is $E h^{-1} E H^{-1} h = E h^{-1} H^- h = EH$. Hence, EH^- is normal in EH. Since E is in EH^-, then $E \cap G_\mathcal{R}$ is in EH^-. It remains to show that there is a subgroup of order $q - 1$ in $G_\mathcal{R} \cap EH^-$. However, H^- has order $(q^2 - 1)/m$ and is cyclic, so it contains a group of order $q - 1$ if and only if $q - 1$ divides $(q^2 - 1)/m$ if and only if m divides $q + 1$. $\square$

Hence, we obtain the following theorem.

THEOREM 201. *Let Σ_i, for $i = 1, 2, .., m + 1$, be Desarguesian spreads of $PG(3, q)$ containing a regulus $\mathcal{R}$ and assume that the spreads Σ_j for $j \neq 1$ are distinct from Σ_1.*

Let G denote the full central collineation group of Σ_1 with axis ℓ in $\mathcal{R}$ and assume that m divides $q + 1$.

Then there is a normal subgroup G^- of G of order $q^2(q^2 - 1)/m$ that contains $G_{\mathcal{R}}$.

Assume that for each Σ_i $i > 2$, there is a line $s_{2,i}$ of $\Sigma_2 - \mathcal{R}$ and an element g_i of $G^- - G_{\mathcal{R}}$ such that $s_{2,i}g_i$ is a line of Σ_i.

Choose any coset representative class $\{h_i; i = 2, .., m+1\}$ for G^- in G. Let Σ_i^* denote the spread obtained by the derivation of $\mathcal{R}$. Then

$$\Sigma_1 \cup_{i=2}^{m+1} \Sigma_i^* h_i k_i$$

for all $k_2, .., k_{t+1} \in G^-$ is a parallelism in $PG(3, q)$.

PROOF. We merely note that the number of spreads in the partial parallelism is $1 + m(q(q+1)/m) = 1 + q(q+1) = 1 + q + q^2$ so we obtain a parallelism. □

EXAMPLE 6. *In order to specify specific instances of the above theorem, assume that q is odd and assume that Σ_i are Desarguesian spreads for $i = 1, 2$ of the form $x = 0, y = x \begin{bmatrix} u & \gamma_i t \\ t & u \end{bmatrix}$ for all $u, t \in GF(q)$, where γ_i are non-squares in $GF(q)$ and $\gamma_1 \neq \gamma_2$. Let γ_3 be any non-square distinct from γ_1 and γ_2. Let $\theta = (\gamma_2 - \gamma_3)/(\gamma_3 - \gamma_1)$. Now consider the mapping of any group G^- of order $q^2(q^2 - 1)/m$ in E of the form*

$$\begin{bmatrix} 1 & 0 & 0 & \theta\gamma_1 \\ 0 & 1 & \theta & 0 \\ 0 & 0 & 1 & 0 \\ 0 & 0 & 0 & 1 \end{bmatrix}.$$

Then, $y = x \begin{bmatrix} 0 & \gamma_2 \\ 1 & 0 \end{bmatrix}$ maps onto $y = x \begin{bmatrix} 0 & \theta\gamma_1 + \gamma_2 \\ 1+\theta & 0 \end{bmatrix}$. Now it follows that

$$\begin{aligned} \gamma_3(1+\theta) &= \gamma_3(1 + ((\gamma_2 - \gamma_3)/(\gamma_3 - \gamma_1)) \\ &= ((\gamma_2 - \gamma_3)/(\gamma_3 - \gamma_1))\gamma_1 + \gamma_2 = \theta\gamma_1 + \gamma_2. \end{aligned}$$

Hence, we may apply the above theorem for any set of non-squares distinct from γ_1.

We note, however, that the above construction did not actually require finiteness. So, we obtain the more general result given as follows.

THEOREM 202. *Let Σ_i, for $i = 1, 2, .., m+1$, denote Pappian spreads in $PG(3, K)$, for K a field, containing the same regulus $\mathcal{R}$ and of the general form:*

$$\Sigma_i : x = 0, y = x \begin{bmatrix} u & \gamma_i t \\ t & u \end{bmatrix}, \forall u, t \in K,$$

for any finite set of distinct non-squares γ_i, $i = 1, 2, .., m + 1$.

Assume that there is an index m subgroup H^- of the homology group H of Σ_1 with axis $x = 0$. Let E denote the full elation group of Σ_1 with axis $x = 0$ and form EH^-, which is a normal subgroup of index m in EH. Further, assume that the full group $(EH)_R \subseteq EH^-$. In the finite case, this is accomplished if and only if m divides $q + 1$.

Let $H = \cup_{i=2}^{m+1} H^- g_i$, where $g_2 = 1$. Then $\cup_{i=2}^{m+1} \Sigma_i g_i h_i$, for all $h_2, .., h_{m+1} \in EH^-$ is a set of spreads, which covers all lines of $PG(3, K)$ which are disjoint from $x = 0$ and not in Σ_1.

PROOF. More generally, if Σ_i is given by

$$x = 0, y = x \begin{bmatrix} u + \rho_i t & \gamma_i t \\ t & u \end{bmatrix} \forall u, t \in K,$$

then the same elation mapping will work provided

$$(\gamma_2 - \gamma_i)s = (\gamma_i - \gamma_1)$$

and

$$(\rho_2 - \rho_i)s = (\rho_i - \rho_1)$$

has a unique solution for s. Hence, we have at least the solutions when either the $\rho_i = \rho_j$ for all i, j or when $\gamma_i = \gamma_j$ for all i, j. □

In particular, we may obtain examples of parallelisms in fields of any characteristic provided there is a quadratic extension superfield.

THEOREM 203. *Under the assumptions above,*

$$\Sigma_1 \cup_{i=2}^{m+1} \Sigma_i^* g_i h_i \; \forall h_2, .., h_{m+1} \in EH^-,$$

where Σ_i^* denotes the derived spread by deriving

$$\mathcal{R} g_i h_i \; \forall h_2, h_3, .., h_{m+1} \in EH^-$$

(i.e., $\mathcal{R}g$ for all $g \in EH$), is a parallelism of $PG(3, K)$.

PROOF. The only lines that are missing from the previous set and not in Σ_1 are the lines intersecting $x = 0$ non-trivially. Since these are the Baer subplanes of the regulus nets corresponding to $\mathcal{R}g$ for all $g \in EH$, we have all of the lines covered by $q^2 + q + 1$ spreads so we obtain a parallelism. □

THEOREM 204. *Assume that the set of $m+1$ Desarguesian spreads Σ_i are mutually distinct and $S = EH^-$ is a normal group of index m.*

Then the full central collineation group with axis $x = 0$ of the parallelisms constructed above is EH^-.

PROOF. Suppose that there is a central collineation $g \in EH - EH^-$ that acts on the constructed parallelism. Assume without loss of generality that $g = g_3$ in the context of the theorem. Then $\Sigma_3^* g_3 g_3^{-1} = \Sigma_3^*$ is a spread of the parallelism. However, so is Σ_2^* and both spreads cover the Baer subplanes of $\mathcal{R}$ and are distinct, which is a contradiction to the properties of a parallelism. ∎

DEFINITION 114. *Define a parallelism constructed from a group of index m, an 'm-parallelism.'*

We remark that conceivably different choices of coset representation sets determine non-isomorphic m-parallelisms.

Hence, with $\{g_i\}$ denoting a coset representation set, we denote the associated m-parallelism by '$(m, \{g_i\})$.'

COROLLARY 55. *An m-parallelism and an n-parallelism for $m \neq n$ are non-isomorphic.*

6.1. (m, n)-Parallelisms. Given an m-parallelism, we assume that there are at least $m - 1$ Pappian spreads containing a given regulus $\mathcal{R}$, which arise from a given Pappian spread by a mapping from a subgroup G^-. However, this is not necessary for the construction of a parallelism. Given a normal subgroup G^- of G containing $G_{\mathcal{R}}$, assume that we take m Pappian spreads distinct from Σ_1, but of these m, we assume that only n are distinct. Then, we still obtain a parallelism, but now it is not entirely clear what the full central collineation group is that acts on the parallelism. To be clear on this construction, first assume that m is finite and that we have n Pappian spreads distinct from Σ_1, say, Σ_i for $i = 2, 3, .., n + 1$. Assume further that we have i_j spreads equal to Σ_i for $\sum_{i=2}^{n+1} i_j = n$. Then we obtain the following parallelism: Let $\{g_i; i = 2, ..., n + 1\}$ be a coset representation set where $g_2 = 1$, then the parallelism is

$$\Sigma_1 \cup_{i=2}^{n+1} \sum_{j=1}^{i_j} \Sigma_i^* g_i h_i \ \forall h_i \in G^- \ \forall i = 2, .., n.$$

DEFINITION 115. *Any such parallelism constructed above shall be called an (m, n)-parallelism. Since $\{i_j\}$ forms a partition of n, the parallelism depends on the partition. Furthermore, the order is important in this case, so we consider that the partition is 'ordered.' Moreover, the parallelism may depend on the coset representation class $\{g_i\}$. When we want to be clear on the notation, we shall refer to the parallelism as a $(m, n, \{i_j\}, \{g_i\})$-parallelism. When $n = m$, we use simply the notation $(m, \{g_i\})$-parallelism.*

Furthermore, since each such parallelism depends on a choice of the initial Pappian spreads, the non-isomorphic parallelisms are potentially quite diverse.

CHAPTER 31

General Parallelism-Inducing Groups

In this chapter, a construction technique is introduced that allows the construction of transitive deficiency one partial parallelisms in a more general setting than the parallelism-inducing groups for Pappian spreads.

Let Σ be any Pappian spread in $PG(3, K)$, and let Σ' any spread, which shares a regulus R with Σ such that Σ' is derivable with respect to R. Assume that there exists a subgroup G^- of the central collineation group G with fixed axis L with the following properties:

(0) : Σ and Σ' share exactly R,

(i) : Every line skew to L, and not in Σ is in $\Sigma'G^-$,

(ii) : G^- is transitive on the reguli that share L, and

(iii) : a collineation g of G^- such that for $L' \in \Sigma'$, then $L'g \in \Sigma'$ implies that g is a collineation of Σ'.

Let $(Rg)^*$ denote the opposite regulus to Rg.

In this setting, G^- is said to be a 'general parallelism-inducing group.'

THEOREM 205. *Assume that G' is a general parallelism-inducing group for spreads Σ and Σ', where Σ is Pappian. Then*

$$\Sigma \cup \{(\Sigma'g - Rg) \cup (Rg)^* \ \forall g \in G^-$$

is a parallelism of $PG(3, K)$ consisting of one Pappian spread Σ and the remaining spreads derived Σ'-spreads.

PROOF. Assume that there exists two spreads $\Sigma'g$ and $\Sigma'h$ for $g, h \in G^-$ that share a common line. Then, $\Sigma'gh^{-1}$ and Σ' share a common line ℓ. But ℓ in Σ' and $\Sigma'gh^{-1}$ implies that ℓhg^{-1} is in Σ', which implies by (iii) that $\Sigma'hg^{-1} = \Sigma'$ if and only if $\Sigma'g = \Sigma'h$. Hence, every line skew to L and not in Σ is in some unique spread $\Sigma'g$ for some $g \in G$. Any other line M is either in Σ or non-trivially intersects Σ. In this case, M is a Baer subplane of the affine plane defined by Σ and hence lies in a unique regulus R_1.

The question becomes given any regulus R_1 of Σ containing the axis of G^-, does there exist a collineation of G^- that maps R onto R_1? However, this is guaranteed by (ii).

Hence, any line that is not skew to $x = 0$ lies in some unique regulus of Σ, which is then in some Rg for $g \in G^-$. Specifically, if a Baer subplane of the regulus net R and the regulus net Rg are equal, then $R = Rg$ (the defined regulus net of the associated affine plane defined by Σ is unique) so that by (iii), g leaves Σ' invariant. Hence, every line of $PG(3, K)$ is uniquely covered by a spread so that a parallelism is obtained. This completes the proof of the theorem. $\square$

THEOREM 206. *Assume that*

$$\Sigma \cup \{(\Sigma'g - Rg) \cup (Rg)^* \forall g \in G^-\}$$

is a parallelism. Then

$$\{\Sigma - R\} \cup R^* \cup \Sigma' \cup \{(\Sigma'g - Rg) \cup (Rg)^* \forall g \in G^- - \{1\}\}$$

is a parallelism. In this case, the spreads are Hall, Σ' (undetermined), and derived Σ' type spreads.

PROOF. The lines covered by $\Sigma \cup \{\Sigma' - R\} \cup R^*$ are the same as covered by $\{\Sigma - R\} \cup R^* \cup \Sigma'$. Since the remaining lines are covered as in the original parallelism, it follows that we have an associated parallelism. $\square$

REMARK 59. *(1) Note that in the finite case, there are exactly $\binom{q^2}{2} / \binom{q}{1} = q(q + 1)$ reguli that share $x = 0$. Hence, the order of the group G^- is divisible by $q(q + 1)$ and $G_{\Sigma'}^-$ is a collineation group of Σ'.*

(2) Let G^- be any central collineation group and G the full central collineation group with axis $x = 0$. Let G_R denote the subgroup, which leaves invariant the regulus R and acts sharply doubly transitively on the remaining components.

If $G^-/G^- \cap G_R \simeq G/G_R$ by the mapping that takes $gG^- \cap G_R$ to gG_R, then G^- acts transitively on the reguli of Σ that share $x = 0$.

Thus, the idea is to take G^- with the properties above and such that $G^- \cap G_R$ is as small as possible.

Assume that Σ is a Desarguesian affine plane of order $p^{2r} = q^2$ where p is odd. Let γ_1 denote a non-square in $GF(q)$. Represent Σ with components of the following form:

$$x = 0, y = x \begin{bmatrix} u & \gamma_1 t \\ t & u \end{bmatrix} \forall u, t \in GF(q).$$

Let Σ' have spread

$$x = 0, y = x \begin{bmatrix} u & \gamma_2 t^\sigma \\ t & u \end{bmatrix} \forall\, u, t \in GF(q)$$

and σ an automorphism of $GF(q)$, where γ_2 is a nonsquare in $GF(q)$ such that $\gamma_2 t^{\sigma-1} \neq \gamma_1$, for all nonzero $t \in GF(q)$.

For $q = p^r$, let $r = 2^b z$, where it is assumed that z is odd and > 1 and assume that 2^a is the largest 2-power dividing $q - 1$, written $2^a \parallel (q-1)$. In this setting, we consider those automorphisms σ defined as follows: $\sigma : x \longmapsto x^{p^{2^b s}}$, where s is any factor of z, including 1.

We assert that $(q - 1, 2^a(q + 1)) = 2^a$. To see this, note that

$$(q - 1, 2^a(q + 1)) = 2^a((q - 1)/2^a, (q + 1)),$$

which is $2^a(z, (q + 1)) = 2^a$ since $(z, q + 1) = 1$ as z divides $q - 1$ and is odd.

Furthermore, we assert that 2^a divides $p^{2^b s} - 1$ for any integer s. Since 2^a divides

$$q - 1 = p^{2^b z} - 1 = (p^{2^b} - 1)\left(\sum_{i=o}^{z-1} p^{2^b i}\right)$$

$$= (p^{2^b} - 1)\left(p^{2^a(z-1)} + \sum_{j=0}^{\frac{(z-3)}{2}} p^{2^{b+1}j}(1 + p^{2^b})\right),$$

and since $(1 + p^{2^b})$ is even and p is odd, it follows that 2^a divides $p^{2^b} - 1$, which, in turn, divides $p^{2^b s} - 1$ for any integer s.

Since $2^a \mid p^{2^b s} - 1$, we also shall use the notation that $2^a \mid (\sigma - 1)$.

Let E^+ denote the full elation group of Σ with axis $x = 0$, and let H denote the homology group with axis $x = 0$ and coaxis $y = 0$ of order $2^a(q + 1)$, where $2^a \parallel (q - 1)$, as above. We note that since Σ is a Desarguesian affine plane of order q^2, and $2^a(q + 1)$ divides $q^2 - 1$ if and only if 2^a divides $q - 1$, we have a cyclic homology group of order $2^a(q + 1)$ in the Desarguesian plane. We let $G^- = EH$.

Hence, the elements of E^+ have the following form

$$\begin{bmatrix} 1 & 0 & u & \gamma_1 t \\ 0 & 1 & t & u \\ 0 & 0 & 1 & 0 \\ 0 & 0 & 0 & 1 \end{bmatrix}$$

and the elements of H have the form:

$$\tau_{w,s} = \begin{bmatrix} w & \gamma_1 s & 0 & 0 \\ s & w & 0 & 0 \\ 0 & 0 & 1 & 0 \\ 0 & 0 & 0 & 1 \end{bmatrix},$$

where the upper 2×2 matrix is nonzero of order dividing $2^a(q+1)$.

Hence, $s = 0$ if and only if the element has order dividing $q - 1$ and $2^a(q+1)$. Since $(q-1, 2^a(q+1)) = 2^a$, it follows that the order of these elements divide both $q - 1$ and $2^a(q+1)$ if and only if $w^{2^a} = 1$.

Now note that when this occurs, then $\tau_{w,0}$ is a collineation of the Kantor-Knuth semifield plane listed above, since

$$y = x \begin{bmatrix} u & \gamma_2 t^\sigma \\ t & u \end{bmatrix} \longmapsto y = x \begin{bmatrix} w^{-1} & 0 \\ 0 & w^{-1} \end{bmatrix} \begin{bmatrix} u & \gamma_2 t^\sigma \\ t & u \end{bmatrix}$$

$$= \left(y = x \begin{bmatrix} uw^{-1} & \gamma_2 t^\sigma w^{-1} \\ tw^{-1} & uw^{-1} \end{bmatrix} \right) = \left(y = x \begin{bmatrix} uw^{-1} & \gamma_2 (tw^{-1})^\sigma \\ tw^{-1} & uw^{-1} \end{bmatrix} \right)$$

since $w^{-1} = (w^{-1})^\sigma$ if and only if $w^{\sigma - 1} = 1$, which is valid since 2^a divides $\sigma - 1$.

Note that Σ and Σ' share $x = 0, y = x \begin{bmatrix} u & 0 \\ 0 & u \end{bmatrix}$, but since $\gamma_2 t^{\sigma - 1} \neq \gamma_1$ for non-zero t in $GF(q)$, they share exactly these components.

Now assume that $\Sigma' g$ and Σ' share a line ℓ not in Σ.

$$\text{Let } g = \begin{bmatrix} w & \gamma_1 s & 0 & 0 \\ s & w & 0 & 0 \\ 0 & 0 & 1 & 0 \\ 0 & 0 & 0 & 1 \end{bmatrix} \begin{bmatrix} 1 & 0 & m & \gamma_1 r \\ 0 & 1 & r & m \\ 0 & 0 & 1 & 0 \\ 0 & 0 & 0 & 1 \end{bmatrix}, \text{ which maps the spread}$$

Σ':

$$x = 0, y = x \begin{bmatrix} u & \gamma_2 t^\sigma \\ t & u \end{bmatrix} \quad \forall \, u, t \in GF(q)$$

onto the spread $\Sigma' g$:

$$x = 0,$$

$$y = x \begin{bmatrix} w & \gamma_1 s \\ s & w \end{bmatrix}^{-1} \left(\begin{bmatrix} u & \gamma_2 t^\sigma \\ t & u \end{bmatrix} + \begin{bmatrix} m & \gamma_1 r \\ r & m \end{bmatrix} \right).$$

The component (line) ℓ is common to Σ' and $\Sigma' g$, which is not in Σ, if and only there exists a nonzero t^* and u^* in $GF(q)$ such that ℓ has the equation, for nonzero t^*:

$$y = x \begin{bmatrix} u^* & \gamma_2 t^{*\sigma} \\ t^* & u^* \end{bmatrix} \text{ is in } \Sigma' g \text{ of the above form.}$$

Hence, we have the following implications:

$$
\begin{bmatrix} u^* & \gamma_2 t^{*\sigma} \\ t^* & u^* \end{bmatrix}
$$

$$
= \begin{bmatrix} w & \gamma_1 s \\ s & w \end{bmatrix}^{-1} \begin{bmatrix} u+m & \gamma_2 t^\sigma + \gamma_1 r \\ t+r & u+m \end{bmatrix} \begin{bmatrix} w & \gamma_1 s \\ s & w \end{bmatrix} \begin{bmatrix} u^* & \gamma_2 t^{*\sigma} \\ t^* & u^* \end{bmatrix}
$$

$$
= \begin{bmatrix} u+m & \gamma_2 t^\sigma + \gamma_1 r \\ t+r & u+m \end{bmatrix}.
$$

Thus, we have:

$$
\begin{bmatrix} wu^* + \gamma_1 st^* & w\gamma_2 t^{*\sigma} + \gamma_1 su^* \\ su^* + wt^* & s\gamma_2 t^{*\sigma} + wu^* \end{bmatrix} = \begin{bmatrix} u+m & \gamma_2 t^\sigma + \gamma_1 r \\ t+r & u+m \end{bmatrix}.
$$

Hence, we must have the $(1,1)$ and $(2,2)$ entries on the left hand side matrix equal, which implies that $\gamma_1 st^* = s\gamma_2 t^{*\sigma}$. Therefore, we must have that $s = 0$. Then $w^{2^a} = 1$.

Thus, using the $(1,2)$-entries, we have

$$
\begin{aligned}
t + r &= wt^* \text{ and } \gamma_2 t^\sigma + \gamma_1 r = w\gamma_2 t^{*\sigma} \\
&= w\gamma_2 \left(\frac{t+r}{w}\right)^\sigma = w^{1-\sigma}\gamma_2 (t+r)^\sigma.
\end{aligned}
$$

Since $w^{1-\sigma} = 1$, it then follows that $\gamma_1 r = \gamma_2 r^\sigma$, which implies that $r = 0$. When $r = 0$ and $s = 0$, we have the group element

$$
\begin{bmatrix} w & 0 & 0 & 0 \\ 0 & w & 0 & 0 \\ 0 & 0 & 1 & 0 \\ 0 & 0 & 0 & 1 \end{bmatrix} \begin{bmatrix} 1 & 0 & m & 0 \\ 0 & 1 & 0 & m \\ 0 & 0 & 1 & 0 \\ 0 & 0 & 0 & 1 \end{bmatrix},
$$

and since both of the elements of the product are collineations of Σ', it follows that g is a collineation of Σ'.

In this case, we see that g must leave invariant Σ' and act as a collineation group.

Hence, if Σ' and $\Sigma'g$ share a common component not in Σ, then $\Sigma' = \Sigma'g$.

So, assume that $\Sigma'h$ and $\Sigma'j$ for $h, j \in E^+ H$, share a line M not in Σ.

Therefore, Σ' and $\Sigma'jh^{-1}$ for $jh^{-1} \in E^+ H$ share a line Mh^{-1} that is not in Σ, since $E^+ H$ is a collineation group of Σ. Thus, $\Sigma' = \Sigma'jh^{-1}$ if and only if $\Sigma'h = \Sigma'j$ if and only if $\Sigma'h$ and $\Sigma'j$ share a common line not in Σ.

Now we derive each plane $\Sigma'g$ by Rg, where R is the regulus

$$x = 0, y = x \begin{bmatrix} u & 0 \\ 0 & u \end{bmatrix} \ \forall \ u \in GF(q).$$

There are exactly $q(q+1)$ reguli in the spread for Σ, which share $x = 0$. Suppose some element in E^+H fixes one of these. Without loss of generality, we may assume that it is R. Then, clearly, $g \in E \langle \rho \rangle$, where E has elements

$$\begin{bmatrix} 1 & 0 & u & 0 \\ 0 & 1 & 0 & u \\ 0 & 0 & 1 & 0 \\ 0 & 0 & 0 & 1 \end{bmatrix} \text{ and } \rho = \begin{bmatrix} w & 0 & 0 & 0 \\ 0 & w & 0 & 0 \\ 0 & 0 & 1 & 0 \\ 0 & 0 & 0 & 1 \end{bmatrix}$$

for a primitive 2^a-element w. In other words, the stabilizer of R has order $2^a q$. Hence, the $q(q+1)$ reguli are in an orbit under E^+H.

Thus, we have $q^2 + q$ distinct spreads $\Sigma'g$ for $g \in E^+H$, and each such spread shares the regulus Rg with Σ. Since the reguli all share $x = 0$ and are in the spread for Σ, the regulus nets can share no common Baer subplane incident with the zero vector.

Hence, the $q(q+1)^2$ lines that lie in the opposite reguli lie exactly in the $q(q+1)$ derived spreads so that the derived spreads share no line that is not in Σ and share no line, since, if so, such a line would be a Baer subplane of some opposite reguli that lies in exactly one derived plane.

Therefore, we have a parallelism of $q^2 + q$ planes of derived Kantor-Knuth type and one Desarguesian plane.

THEOREM 207. *Let q be odd equal to $p^{2^b z}$, where z is an odd integer > 1. Assume that $2^a \parallel (q-1)$, then there exists a nonidentity automophism σ of $GF(q)$ such that $2^a \mid (\sigma - 1)$.*

Let γ_2 and γ_1 be non-squares of $GF(q)$ such that the equation $\gamma_2 t^\sigma = \gamma_1 t$ implies that $t = 0$.

(1) Then, there exists a parallelism $\mathcal{P}_{\gamma_2, \sigma}$ of derived Kantor-Knuth type with $q^2 + q$ derived Kantor-Knuth planes and one Desarguesian plane.

(2) The collineation group of this parallelism contains the central collineation group of the Desarguesian plane with fixed axis ℓ of order $q^2 2^a (q+1)$.

1. The Isomorphisms of Kantor-Knuth Type Parallelisms

The full collineation group of a Kantor-Knuth plane is completely determined in Johnson and Liu [140]. Furthermore, the collineation group of any derived semifield plane is known to be the group inherited from the semifield plane provided the order is not 9 or 16 (see Johnson [**126**]).

A Kantor-Knuth spread is covered by a set of q reguli that mutually share a line and it is known that the collineation group acts 2-transitively on this set of q reguli.

Hence, we may assume that any isomorphism between Kantor-Knuth spreads maps the standard regulus to the standard regulus, maps any given regulus of the first Kantor-Knuth spread to any given regulus of the second Kantor-Knuth spread, and fixes a component corresponding to the elation axis.

NOTATION 4. *We represent the Desarguesian spread by*

$$x = 0, y = x \begin{bmatrix} u & \gamma_0 t \\ t & u \end{bmatrix} \forall u, t \in GF(q)$$

and γ_o is a fixed non-square and the Kantor-Knuth semifield spreads π_1 and π_2 by

$$x = 0, y = x \begin{bmatrix} u & \gamma_i t^{\sigma_i} \\ t & u \end{bmatrix} \forall u, t \in GF(q),$$

for $i = 1, 2$ respectively, where γ_i are non-squares and σ_i are automorphisms of $GF(q)$.

THEOREM 208. *Two such parallelisms $\mathcal{P}_{\gamma_1, \sigma_1}$ and $\mathcal{P}_{\gamma_2, \sigma_2}$ are isomorphic if and only if one of the following two conditions hold:*

$$(\gamma_2, \sigma_2) = \left(\frac{\gamma_o^2}{\gamma_1^\rho}, \sigma_1^{-1} \right) \quad or \quad (\gamma_2, \sigma_2) = (\gamma_1^\rho, \sigma_1)$$

for some automorphism ρ of $GF(q)$.

PROOF. If the parallelisms are isomorphic by an element τ of $\Gamma L(4, q)$ then this element is a collineation of the unique Desarguesian spread Σ, which fixes a regulus R of Σ shared by the Kantor-Knuth semifield spreads π_1 and π_2 defined as follows:

$$x = 0, y = x \begin{bmatrix} u & \gamma_i t^{\sigma_i} \\ t & u \end{bmatrix} \forall u, t \in GF(q),$$

for $i = 1, 2$, respectively. An isomorphism τ must fix $x = 0$, where we are assuming that R is the standard regulus (net). Since we have an elation group acting transitively on the nonaxis lines (components) of the regulus, it follows that we may assume that τ fixes $x = 0, y = 0$, fixes the standard regulus net and maps the net when $t = 1$ in π_1 to the net for $t = 1$ in π_2.

Note that our construction requires that the Desarguesian spread be taken in a given manner. Hence, we let Σ have spread:

$$x = 0, y = x \begin{bmatrix} u & \gamma_0 t \\ t & u \end{bmatrix} \forall u, t \in GF(q)$$

and γ_o is a fixed nonsquare.

Let $A = \begin{bmatrix} v & \gamma_0 s \\ s & v \end{bmatrix}$ and consider that τ may be taken to have the following form:

$$\tau : (x_1, x_2, y_1, y_2) \longmapsto (x_1^\rho, x_2^\rho, y_1^\rho, y_2^\rho) \begin{bmatrix} A & 0 \\ 0 & A u_o I \end{bmatrix}.$$

Then, we must have: Let

$$B_{11} = (vu^\rho - \gamma_o s)v + s(v\gamma_1^\rho - \gamma_o su^\rho),$$

$$B_{1,2} = (vu^\rho - \gamma_o s)\gamma_o s + (v\gamma_1^\rho - \gamma_o su^{gr})v,$$

$$B_{2,1} = v(-su^\rho + v) + s(-s\gamma_1^\rho + vu^\rho),$$

$$B_{2,2} = \gamma_o s(-su^\rho + v) + v(-s\gamma_1^\rho + vu^\rho),$$

then

$$\begin{bmatrix} v & \gamma_0 s \\ s & v \end{bmatrix}^{-1} \begin{bmatrix} u^\rho & \gamma_1^\rho \\ 1 & u^\rho \end{bmatrix} \begin{bmatrix} v & \gamma_0 s \\ s & v \end{bmatrix} u_o I$$

$$= \begin{bmatrix} B_{11} & B_{1,2} \\ B_{2,1} & B_{2,2} \end{bmatrix},$$

which must have the general form $\begin{bmatrix} u_2 & \gamma_2^{\sigma_2} \\ 1 & u_2 \end{bmatrix}$ for all $u \in GF(q)$. Equating the $(1,1)$- and $(2,2)$-entries, we obtain $sv\gamma_1^\rho = -sv\gamma_1^\rho$. Hence, we must have either s or $v = 0$.

First, assume that $v = 0$. Since we have a 1 in the $(1,2)$-entry, we must have: $u_o = \frac{\gamma_o}{\gamma_1^\rho}$.

It then follows immediately that $\frac{\gamma_o^2}{\gamma_o^\rho} = \gamma_2$.

Since $y = x \begin{bmatrix} 0 & \gamma_1 t^{\sigma_1} \\ t & 0 \end{bmatrix}$ maps onto $y = x \begin{bmatrix} 0 & \frac{\gamma_o^2}{\gamma_o^\rho} t^\rho \\ t^\rho \sigma_1 & v \end{bmatrix}$, it then follows that $\sigma_2 = \sigma_1^{-1}$. Hence, in this case, we must have $(\gamma_2, \sigma_2) = (\frac{\gamma_o^2}{\gamma_o^\rho}, \sigma_1^{-1})$.

Now assume that $s = 0$. It then follows similarly that $\gamma_2 = \gamma_1^\rho$ and $\sigma_1 = \sigma_2$. $\square$

The above result allows the enumeration of the isomorphism classes albeit a bit cumbersome. For example, the following provides the idea.

COROLLARY 56. *Assume that $q \equiv -1 \bmod 4$. Let $q = p^{st}$, choose any proper automorphism $\sigma = p^s$ and non-squares γ_o and γ such that*

$$\gamma_o t^{\sigma - 1} \neq \gamma \ \forall t \in GF(q).$$

Then, there exist at least

$$\left[\frac{(q-1) - (\frac{(q-1)}{(p^s-1)} - 1)}{2st} \right]$$

mutually nonisomorphic parallelisms $\mathcal{P}_{\gamma,\sigma}$.

PROOF. We note that we obtain a parallelism as in our main theorem, since $2 \parallel (q-1)$ and $2 \mid (\sigma - 1)$ for any automorphism σ. There are $(q-1)/2$ non-square elements γ and there are $(q-1)/2(p^s - 1)$ nonsquare elements among the set of elements $\{t^{\sigma-1}; t \in GF(q) - \{0\}\}$. We may use the same automorphism and obtain at least the number of parallelisms as orbits under the Galois group. The number indicated in the result is the minimum number of such orbits.   □

Let $\mathcal{P}$ be any Kantor-Knuth type parallelism as above. Take the regulus R within the planes Σ and Σ'. Let R^* be the opposite regulus within the derived plane of Σ', Σ'^*. Now derive R within Σ and R^* within Σ'^* to produce the Hall plane Σ^* and the Kantor-Knuth plane Σ' leaving the remaining derived Kantor-Knuth planes.

THEOREM 209. *For each parallelism of type $\mathcal{P}_{\gamma,\sigma}$, there is a parallelism consisting of one Hall spread, one Kantor-Knuth semifield spread, and $q^2 + q - 1$ derived Kantor-Knuth spreads. We shall call such parallelisms the 'derived' parallelisms of the $\mathcal{P}_{\gamma,\sigma}$-parallelisms.*

We may discuss isomorphisms in this setting as well. We note that any isomorphism is a collineation of the Hall plane and for $q > 3$, we recall that the full collineation group of the Hall plane is a collineation group of the associated Desarguesian plane from which it was derived. Furthermore, the unique Kantor-Knuth semifield plane must map to the corresponding Kantor-Knuth semifield plane so we have the same situation as before.

THEOREM 210. *Given two parallelisms $\mathcal{P}_{\gamma_i,\sigma_i}$ for $i = 1, 2$. Then, the derived parallelisms are isomorphic if and only if the original parallelisms are isomorphic.*

2. Finite Regulus-Inducing Elation Groups

We are interested in the construction of parallelisms by groups G that fix a given spread Σ and act transitively on the remaining spreads.

All such groups will contain an elation group E^+ of order q^2 that fixes a component ℓ of Σ and acts transitively on the remaining components. In this section, we show that in order to construct such parallelisms, it is necessary that E^+ be partitioned into $q + 1$ 'regulus-inducing' subgroups of order q such that G acts transitively on these subgroups by conjugation.

In Chapter 38, we give a classification of all transitive deficiency one partial parallelisms in $PG(3,q)$. There is an extension of the partial parallelism by a spread Σ, but it is not known initially that Σ is or must be Desarguesian. In this section, the reader will note a certain similarity with the ideas is this later chapter, but here we always initially assume that Σ is a Desarguesian spread. In particular, the reader is directed to Theorem 261 for the statement of the main classification theorem and to Theorem 268, which gives an upper and lower bound for the order of the full collineation group of any transitive deficiency one partial parallelism in $PG(3,q)$. We tried to gather the main general results on 'transitivity' into one of the main parts of the text.

In the present section, there will be some repetition with the ideas in Chapter 38, but we decided to develop the material here in order to discuss the minimal parallelism-inducing groups.

THEOREM 211. *Let Σ be a Desarguesian spread in $PG(3,q)$, and let π_Σ denote the associated affine plane. If G is a collineation group of the translation complement of π_Σ that fixes a component ℓ of π_Σ, then the following are equivalent:*

(1) G is transitive on the $q(q+1)$ reguli of Σ that share ℓ and the p-elements in G are in $GL(2, q^2)$,

(2) There is an elation group E^+ of order q^2 in G, which is uniquely partitioned into $q+1$ subgroups E_i, $i = 1, 2, .., q+1$, of order q such that if m is any component of Σ distinct from ℓ, then $E_i m \cup \{\ell\}$ is a regulus and G is transitive on the $q+1$ elation groups.

PROOF. Let R be a regulus containing ℓ. Then there is a unique elation group E_R that has axis ℓ and acts regularly on the components of $R - \{\ell\}$. Furthermore, E_R fixes exactly q reguli of Σ. Thus, the $q(q+1)$ reguli are partitioned into $(q+1)$ sets of q reguli, each of which is fixed by a unique elation subgroup of order q. Assume that R_1 and R_2 are reguli fixed by the elation subgroups E_1 and E_2, respectively, E_1 and E_2 of order q. Without loss of generality, we may assume that R_1 and R_2 share ℓ and a second component m. If R_1 and R_2 are distinct, then they cannot share a third component. If $g \in E_1 \cap E_2$, then mg is common to R_1 and R_2 and g fixes R_1 and R_2. Hence, $g = 1$. We see then that the partition of E^+ into such 'regulus-inducing' elation groups is unique. So, any collineation group that permutes the reguli containing ℓ must also permute corresponding regulus-inducing elations. Hence, if R and R' are reguli sharing ℓ and E_R and $E_{R'}$ the corresponding regulus-inducing elation groups fixing R and R', respectively, then

$$R \overset{g}{\longmapsto} R' \iff E_R \overset{g}{\longmapsto} E_{R'}^g = E_{Rg}.$$

Thus, if G is transitive on the set of $q(q+1)$ reguli, consider the stabilizer of a regulus R, G_R. R determines a unique set of q reguli that mutually shares exactly the component ℓ with R. If R_1, R_2, R_3 are mutually distinct reguli of this set, assume that g_{23} maps R_2 into R_3. There is a unique elation group E_{R_1} of order q that fixes each of R_i, $i = 1, 2, 3$. Hence, g_{23} maps E_{R_1} onto $E_{R_1} = E_{R_2}$. In other words, g_{23} normalizes E_{R_1}. Thus, there is a set of elements that fix $\{R_i; i = 1, 2, , q\}$ and acts transitively on the set of q reguli fixed by E_{R_1}. So, this subgroup H is transitive on q elements, so q divides the order of H. Since any p-element of G is in $GL(2, q^2)$, all p-elements are of order p and are elations. Thus, there is a set of elations that generates an elation group $\widetilde{E}_1$ of order at least q. Assume that this group does not act transitively on the q reguli R_i, $i = 1, 2, .., q$. Then the order of p-subgroup must go up since the larger group is transitive. All of this implies that $\widetilde{E}_1$ must, in fact, act transitively on the q reguli R_i. Suppose that R_1 is fixed by the standard elation group E_1 with elements $\begin{bmatrix} 1 & u \\ 0 & 1 \end{bmatrix}$; $u \in GF(q)$. Then letting the associated Desarguesian spread be denoted by

$$x = 0, y = x \begin{bmatrix} u + t\alpha & t\beta \\ t & u \end{bmatrix}; u, t \in GF(q), \alpha, \beta \text{ constants},$$

with the elation axis being taken as $x = 0$, we see that there are elations g_t in $\widetilde{E}_1$, such that

$$g_t = \begin{bmatrix} I & M_t \\ 0 & I \end{bmatrix}, \text{ where } M_t = \begin{bmatrix} u_t + t\alpha & t\beta \\ t & u_t \end{bmatrix};$$

where u_t is a function of t. In this setting, R_1 is taken with components

$$x = 0, y = xu; u \in GF(q).$$

For a fixed element n in $GF(q^2)$, clearly $x = 0, y = xun; u \in GF(q)$ is a regulus with regulus-inducing elation group E_n, whose elements are

$$\begin{bmatrix} I & 0 \\ 0 & n \end{bmatrix}^{-1} \begin{bmatrix} I & M_t \\ 0 & I \end{bmatrix} \begin{bmatrix} I & 0 \\ 0 & n \end{bmatrix} = \begin{bmatrix} I & M_t n \\ 0 & I \end{bmatrix}.$$

Since this is valid for each n in $GF(q^2)$, it follows easily that the group generated by the $\widetilde{E}_n$'s is the full elation group of order q^2.

Now assume the that there is an elation group E^+ of order q^2 in G and G is transitive on the subgroups E_i of the partition of E^+ into $q + 1$ regulus-inducing elation groups of orders q. There are $q(q + 1)$ reguli sharing ℓ, and these are partitioned into $q + 1$ sets of q, each of which is fixed by some E_j. Since E^+ is transitive on each such set, it follows immediately that G is transitive on the reguli sharing ℓ. ☐

Since we are interested in using groups to construct parallelisms that admit a group fixing one spread and transitive on the other spreads, the question arises whether a group G that fixes one spread Σ and acts transitively on the remaining spreads of a parallelism must, in fact, be transitive on the reguli that share a component ℓ or, indeed, whether G need fix a component of Σ at all.

THEOREM 212. *Let $\mathcal{P}$ be a parallelism in $PG(3, q)$ and let Σ be a spread of $\mathcal{P}$. Let G be a collineation group of Σ that acts transitively on the remaining $q(q + 1)$ spreads of $\mathcal{P}$. Then Σ is Desarguesian, the remaining spreads are derived conical flock spreads, and G leaves invariant a component ℓ of Σ.*

Furthermore, G acts transitively on the reguli of Σ sharing ℓ.

PROOF. By the Classification Theorem for transitive deficiency one partial parallelisms of Chapter 38, it remains to show that G acts transitively on the reguli of Σ sharing ℓ. Also, we now know from the same theorem above that there is an elation group E^+ of order q^2 of Σ contained in G. Since there are $q(q + 1)$ spreads (translation planes), each translation plane $\pi_{\Sigma'}$ is fixed by an elation subgroup $E_{\Sigma'}$ of order q that acts as a Baer group on $\pi_{\Sigma'}$. The net of degree $q + 1$, which contains ℓ as a Baer subplane of $\pi_{\Sigma'}$, is a regulus net, by the Johnson, Payne-Thas Theorem. When we derive this net, we obtain a regulus R'^* containing ℓ as a component. However, R'^* need not necessarily be a regulus of Σ. Still $E_{\Sigma'}$ is a 'regulus-inducing' elation group, in that each component orbit of Σ union ℓ defines a regulus of Σ. Since we have E^+ acting transitively on these q reguli, this means that each such elation group $E_{\Sigma'}$ defines q reguli of Σ that are in an orbit under E^+ and so in an orbit under G. Even though R'^* need not be a regulus of Σ, all of the Baer subplanes incident with the zero vector (the opposite regulus R') are fixed by $E_{\Sigma'}$. Now $E_{\Sigma'}$ fixes a Baer subplane π_o of π_Σ if and only if the subplane non-trivially intersects ℓ and each element of $E_{\Sigma'}$ maps some infinite point of π_o back into an infinite point of π_o. Each such fixed Baer subplane defines a regulus of Σ sharing ℓ. Therefore, there are exactly $q(q + 1)$ Baer subplanes of Σ fixed by $E_{\Sigma'}$. Hence, R'^* accounts

for exactly $q + 1$ fixed Baer subplanes (whether they line up on a regulus net or not). These $q(q + 1)$ fixed Baer subplanes show up as lines
of spreads on the parallelism and since $E_{\Sigma'}$ is a collineation group of
the spread, they must fix each spread containing a fixed Baer subplane
and there can only be $q + 1$ (and must be exactly this number) of such
fixed Baer subplanes (lines of the spread) per fixed spread. Hence, it
follows that $E_{\Sigma'}$ fixes exactly q spreads. Since there are $q(q+1)$ distinct
spreads apart from Σ, it follows that there are exactly $q + 1$ mutually
disjoint regulus-inducing elation groups. It then follows easily that G
is transitive on the regulus-inducing elation groups. However, we have
shown previously that this is equivalent to G acting transitively on the
reguli of Σ sharing ℓ.   □

3. Finite General Parallelism-Inducing Groups

Previously, we mentioned that the sharply 2-transitive parallelism-
inducing groups are convenient for the analysis of parallelisms with one
Desarguesian spread and the remaining spreads Hall spreads. In this
section, we generalize these notions and consider subgroups that do not
act 2-transitively on the lines of Σ distinct from ℓ, but never the less
share the feature of essentially generating or inducing parallelisms, but
now such groups are generally applicable to any Desarguesian spread
Σ and any derived conical flock spread Σ'.

DEFINITION 116. *Let Σ and Σ' be two distinct disjoint spreads in
$PG(3, q)$. Assume that Σ is Desarguesian, and let G be a collineation
group of π_{Σ} such that all p-elements of G are in $GL(2, q^2)$, and which
is transitive on the reguli of Σ that share a line ℓ.*

*Then G is said to be 'parallelism-inducing relative to Σ'' if and
only if an element g of G that maps a line ℓ' of Σ' back into Σ' is a
collineation of Σ'.*

PROPOSITION 10. *If G is a parallelism-inducing group relative to
Σ', then Σ' admits a Baer group E of order q.*

PROOF. By the above theorem, G contains an elation group E^+
of order q^2. Each regulus-inducing Baer group of E^+ fixes $q(q + 1)$
Baer subplanes. Furthermore, any Baer subplane that intersects the
axis $x = 0$ of E^+ in a 1-dimensional $GF(q)$-subspace is fixed by some
regulus-inducing Baer group. Since Σ and Σ' are disjoint spreads, then
$x = 0$ becomes a Baer subplane of Σ', and there are exactly $q + 1$
components of Σ' that non-trivially intersect $x = 0$. This means that
each of these $q + 1$ components of Σ' is fixed by a regulus-inducing
elation group of order q. Under our assumptions, if a component M

of Σ' is fixed by an element g of G, then g becomes a collineation of Σ'. Hence, each of these $q + 1$ components of Σ' is fixed by a Baer group fixing $x = 0$ pointwise. Since each such component is fixed by a regulus-inducing elation group of order q, it follows that Σ' is fixed by a Baer group of order q, since Baer groups of order q are maximal. $\square$

PROPOSITION 11. *If G is a parallelism-inducing group relative to Σ', let E denote the unique Baer group of order q acting on Σ'. Then $G_{\Sigma'}$ normalizes E. Furthermore, if g is in $G_{\Sigma'}$ and g does not fix a regulus in Σ sharing $x = 0$, which is fixed by E, then no element ge for e in E fixes such a regulus in Σ.*

PROOF. G fixes $x = 0$ and Baer groups of order q are maximal. Hence $G_{\Sigma'}$ normalizes E. Hence, g permutes the q reguli fixed by E. Since E fixes each such regulus, g does not fix a regulus if and only if ge does not fix a regulus. $\square$

LEMMA 56. *$G_{\Sigma'}$ fixes a regulus R' of Σ' containing $x = 0$ as a Baer subplane and $G_{\Sigma'}$ fixes the derived regulus R'^* containing $x = 0$ as a component and also fixes a regulus $R_{\Sigma'}$ of Σ (R'^* and $R_{\Sigma'}$ may not be equal and R'^* may not be a regulus of Σ).*

PROOF. Let g be an element of G that fixes Σ'. We note that by the result of Johnson, Payne-Thas Theorem, the net of degree $q + 1$ containing $x = 0$ as a Baer subplane is a regulus net (components define a regulus) and furthermore, that a Baer group of order q produces a deficiency one partial flock of a quadratic cone, which is extendable if and only if the Baer subplane fixed pointwise by the Baer group lies in a derivable net. Hence, g fixes R' but the derived net R'^* may not correspond to a regulus in Σ. Suppose that whenever this occurs, g fixes a regulus of Σ'. Now suppose that g fixes a component M of Σ. Since g normalizes E, it follows that g fixes EM union ℓ, a regulus of Σ. So, assume that g fixes no regulus. E is transitive on the components of R'^*, not equal to $x = 0$. If any of these are components of Σ, then R'^* is a regulus of Σ fixed by all of $G_{\Sigma'}$. Hence, assume that each component of R'^* is a Baer subplane of Σ, necessarily disjoint from $x = 0$. Let N be a component of R'^*, not equal to $x = 0$, so is a Baer subplane of Σ. Thus, g is in $EG_{\Sigma',N}$. Assume that g does not fix a component of Σ and does not fix a regulus of Σ, fixed by E. Then ge does not fix a regulus of Σ fixed by E and hence does not fix a component of Σ. But we may assume then that ge, so also, without loss of generality, g fixes N. We may represent N by $y = x^q m + xn$, for $m \neq 0$ and g by

$$ (x, y) \longmapsto (x^\sigma, y^\sigma) \begin{bmatrix} a & b \\ 0 & d \end{bmatrix}, $$

for σ an automorphism of $GF(q^2)$ and a, b, d elements of $GF(q^2)$. Since N fixes $y = x^q m + xn$, then we obtain that $(x^\sigma a, x^\sigma b + (x^q m + xn)^\sigma d)$ is on N. This implies that

$$x^{\sigma q} a^q m + x^\sigma an = x^\sigma b + (x^q m + xn)^\sigma d.$$

By reducing modulo $2r$, where $q = p^r$, p a prime, it follows directly that

$$a^q m = m^\sigma d \text{ and } an = b + n^\sigma d.$$

Now consider the question of whether g can fix a component $y = xz$ of Σ. We have (x, xz) map to $(x^\sigma a, x^\sigma b + x^\sigma z^\sigma d)$ on $y = xz$ if and only if $az = b + z^\sigma d$. Thus, we see that g must, in fact, fix $y = xn$, a contradiction. Hence, we may assume that g does fix a regulus R_g of Σ. We need to check that R_g is not dependent upon g. Taking E as the standard regulus-inducing group, which we may do without loss of generality, we see that the components of R'^* are as follows:

$$y = x^q m + x(n + u), \forall u \in GF(q).$$

In general, we see that

$$(x^\sigma a, x^\sigma b + (x^q m + xn)^\sigma d)$$

is now on $y = x^q m + x(n + u_0)$ if and only if

$$a^q m = m^\sigma d \text{ and } a(n + u_0) = b + n^\sigma d.$$

But $y = xn$ will then map to $(x^\sigma a, x^\sigma b + x^\sigma n^\sigma d)$, which is on $y = x(n + u_0)$ if and only if $x(n + u_0) = b + n^\sigma d$.

This implies that $G_{\Sigma'}$ fixes the same regulus of Σ. This completes the proof of the lemma. $\square$

Our main construction device is contained in the following theorem.

THEOREM 213. *If Σ and Σ' are spreads in $PG(3, K)$, for K a field isomorphic to $GF(q)$, and G is a parallelism-inducing group relative to Σ', then*

$$\mathcal{P} = \{\Sigma\} \cup \{\Sigma'g; g \in G\}$$

is a parallelism.

PROOF. The previous lemma shows that $G_{\Sigma'}$ fixes a regulus $R_{\Sigma'}$ of Σ sharing $x = 0$. Since G is transitive on the reguli of Σ sharing $x = 0$, it follows that the number of images of Σ' is at least $q(q + 1)$. But all of these images are disjoint, since if two share a line l', we may assume that l' is in Σ' and there is an element g of G that maps l' into Σ'. Under our conditions, this means that g is a collineation group of Σ, so the two images are equal (or disjoint). Hence, by cardinality, we must have a parallelism of $1 + q(q + 1)$ spreads. $\square$

We have noted previously that the reguli in a given spread Σ' may not be the derived version of a regulus of Σ. We formalize this in the next definition.

DEFINITION 117. *A parallelism of the above type, where Σ and Σ'^* share a regulus, is said to be a 'parallelism of standard regulus type.' If Σ and Σ' share exactly $q + 1$ Baer subplanes invariant under E but these $q+1$ Baer subplanes do not fall into an E-invariant regulus of Σ, then the parallelism is said to be a 'parallelism of non-standard regulus type.'*

4. Minimal Linear Parallelism-Inducing Groups

By definition, a parallelism-inducing group G must contain an elation group E^+ of order q^2 and act transitively on the regulus-inducing elation subgroups of order q. The question is, how large must G actually be? A 'minimal parallelism-inducing group G' is a parallelism-inducing group of minimal order. In this section, we assume that G is contained in $GL(2, q^2)$ and call G a 'linear' parallelism-inducing group.

We recall that our fixed Desarguesian spread Σ has the following form:

$$x = 0, y = \begin{bmatrix} u + t\alpha & t\beta \\ t & u \end{bmatrix} ; u, t \in GF(q),$$

where $\{1, \theta\}$ is a basis for a quadratic extension $GF(q^2)$ of $GF(q)$, for $\theta^2 = \theta\alpha + \beta$. In this representation, ℓ is taken as $x = 0$, and the components of Σ are

$$x = 0, y = xm; m \in GF(q^2), \text{ as represented above.}$$

THEOREM 214. (1) *A minimal bound for a linear parallelism-inducing group is $q^2(q - 1)_2(q + 1)$, where $(q - 1)_2$ denotes the largest power of 2 that divides $(q - 1)$.*

(2) *A linear parallelism-inducing group of order $q^2(q - 1)_2(q + 1)$ may be written in the form E^+H, where H is cyclic or a direct product of cyclic groups.*

PROOF. Let G be a parallelism-inducing group. Since G must be transitive on $q(q+1)$ spreads of a parallelism, the order must be divisible by $q^2(q + 1)$, as G must also contain the elation group E^+ of order q^2. Since E^+ is transitive on the components of Σ not equal to ℓ, taken as $x = 0$, then $G = E^+G_{y=0}$. Let $H = G_{y=0}$. Since H is linear, each element of H has the following form: $\begin{bmatrix} a & 0 \\ 0 & b \end{bmatrix} ; a, b \in GF(q^2)$. Assume that the bound is exactly $q^2(q+1)$; that such a group could be transitive on $q(q+1)$ spreads. If q is odd, consider an element $g = \begin{bmatrix} a & 0 \\ 0 & b \end{bmatrix}$ of order 2. Then $a^2 = b^2 = 1$, so that both a and b are in $\{\pm 1\}$. If $a = b = -1$, this is a kernel homology group and hence fixes each 1-dimensional $GF(q)$-space and thus fixes each spread. So, assume, without loss of generality, that $a = 1$ and $b = -1$. Note that this is a homology group of Σ and fixes each Baer subplane π_o, defined by a 2-dimensional $GF(q)$-subspace, such that π_o intersects both $x = 0, y = 0$ non-trivially and two of whose infinite points are permuted by g. Thus, g must fix $(q+1)^2$ Baer subplanes, since each such involution will fix $q+1$ reguli that share two components $x = 0, y = 0$. Each such Baer subplane appears as a line of some spread of the parallelism, implying that g must fix spreads of the parallelism. Hence, the group H must have order at least $2(q+1)$. But, now assume that 2^{a^*} is the 2-power dividing $q - 1$, and consider a Sylow 2-subgroup S_2 of H. Note that H is a subgroup of the direct product of two cyclic groups, and so is Abelian. Hence, there is a unique Sylow 2-subgroup of H. Assume that S_2 has order 2^{b^*} strictly less than 2^{a^*+1}.

Now consider any element g as above, where a and b are both in $GF(q)$. Consider the elements $g_b = \begin{bmatrix} 1 & 0 \\ 0 & b \end{bmatrix}$ and $g_b = \begin{bmatrix} a & 0 \\ 0 & 1 \end{bmatrix}$ of $GL(2, q^2)$. We note that each such group fixes all of the Baer subplanes of the set of $q+1$ reguli of Σ that share $x = 0, y = 0$ and hence so does the product. If S_2 has order 2^b, then for any element $\begin{bmatrix} c & 0 \\ 0 & d \end{bmatrix}$ of H then $c^{2^b} = d^{2^b} = 1$. But since 2^{b^*} divides 2^{a^*}, it follows that c and d are in $GF(q)$ and hence the group elements must fix Baer subplanes. So, if 2^{b^*} is strictly less then 2^{a^*+1}, then G cannot be transitive on the spreads other than Σ of a parallelism. Therefore, a lower bound for the order of G is then $q^2(q-1)_2(q+1)$. This proves (1).

Now assume that H has order $(q-1)_2(q+1)$. Since H is a subgroup of the direct product of two cyclic groups, either H is cyclic or H is the direct product of two cyclic subgroups. Hence, part (2) is proved. □

5. Relative Linear Parallelism-Inducing Groups

In this chapter, we develop the possible minimal parallelism-inducing groups with respect to the Kantor-Knuth spreads.

THEOREM 215. *Let q be equal to p^r. Assume that we may choose an automorphism σ of $GF(q^2)$ such that $(q-1)_2$ divides $\sigma - 1$, considering σ as p^t for some t. (Note that this condition means that if $\sigma = p^t$ then $r/(t, r)$ is odd.) Let γ_2 and γ_1 be non-squares of $GF(q)$ such that the* *equation $\gamma_2 t^\sigma = \gamma_1 t$ implies that $t = 0$. Choose a Desarguesian plane Σ with the following spread:*

$$x = 0, y = x \begin{bmatrix} u & \gamma_1 t \\ t & u \end{bmatrix} ; u, t \in GF(q),$$

and let Σ' denote the spread derived from the following spread Σ'^* by derivation of the standard regulus net:

$$x = 0, y = x \begin{bmatrix} u & \gamma_2 t^\sigma \\ t & u \end{bmatrix} ; u, t \in GF(q).$$

Let $G_i = E^+ H$, where E^+ is the elation group of order q^2 of Σ with axis $x = 0$ and

$$H = \left\langle \begin{bmatrix} W & 0 \\ 0 & W^* \end{bmatrix} \right\rangle,$$

where

$$W = \begin{bmatrix} w & \gamma_1 s \\ s & w \end{bmatrix}.$$

Assume that the order of H is $(q-1)_2(q+1)$, and we have the following conditions:

(i) $\left\langle W^{-1}W^*; \begin{bmatrix} W & 0 \\ 0 & W^* \end{bmatrix} \in H \right\rangle$ has order $2^a(q+1)$ (this con-dition is also necessary to have a parallelism-inducing group).

(ii) H contains a subgroup of order $(q-1)_2$ acting as a collineation group of Σ', whose elements $\begin{bmatrix} W & 0 \\ 0 & W^* \end{bmatrix}$ are such that W and W^* are both scalar (note there cannot be any other elements of H whose elements are scalar matrices since all of these elements have order dividing $(q-1)$ and hence fix spreads of the parallelisms).

(iii) If WW^* is scalar then W and W^* are both scalar.

Then G is a minimal parallelism-inducing group relative to the Kantor-Knuth spread Σ'.

PROOF. We need to show that the group G_i is transitive on the reguli that share $x = 0$. We know that there is a unique partition of E into regulus-inducing subgroups E_c, where

$$E_c = \left\langle \begin{bmatrix} I & uc \\ 0 & I \end{bmatrix} ; u \in GF(q) \right\rangle,$$

and the set of $q+1$ regulus-inducing elation groups corresponds to the cosets of $cGF(q)^*$ of $GF(q^2)^*/GF(q)^*$.

We now consider the group H and conjugate E_1 to obtain

$$H_i^{-1} E_1 H_i = E_{W^{-1}W^*}.$$

Hence, we obtain that H_i is transitive on the groups E_c if and only if

$$\left\langle W^{-1}W^*; \begin{bmatrix} W & 0 \\ 0 & W^* \end{bmatrix} \in H \right\rangle$$

has order $2^a(q+1)$, since the order of W and W^* divides $2^a(q+1)$.

Thus, it remains to check that if a component ℓ of Σ' maps under an element eh_i in EH_i back into Σ', then eh_i is a collineation of Σ'.

Assume that ℓ is

$$y = x \begin{bmatrix} u & \gamma_2 t^\sigma \\ t & u \end{bmatrix},$$

$$e = \begin{bmatrix} 1 & 0 & m & \gamma_1 k \\ 0 & 1 & k & m \\ 0 & 0 & 1 & 0 \\ 0 & 0 & 0 & 1 \end{bmatrix},$$

and

$$h_i = \begin{bmatrix} w & \gamma_1 s & 0 & 0 \\ s & w & 0 & 0 \\ 0 & 0 & w^* & \gamma_1 s^* \\ 0 & 0 & s^* & w^* \end{bmatrix}.$$

Hence, if this component maps back into Σ', it follows that the $(1,1)$ and $(2,2)$-entries of the associated matrix are equal.

Therefore, the image matrix is as follows:

$$(*) \quad : \quad \begin{bmatrix} w & \gamma_1 s \\ s & w \end{bmatrix}^{-1} \left(\begin{bmatrix} m & \gamma_1 k \\ k & m \end{bmatrix} + \begin{bmatrix} u & \gamma_2 t^\sigma \\ t & u \end{bmatrix} \right).$$

$$\cdot \begin{bmatrix} w^* & \gamma_1 s^* \\ s^* & w^* \end{bmatrix}$$

$$= \begin{bmatrix} u^* & \gamma_2 t^{*\sigma} \\ t^* & u^* \end{bmatrix}.$$

Letting $\Delta = w^2 - \gamma_1 s^2$, we obtain a matrix with the $(1,1)$ and $(2,2)$-entries are as follows: the $(1,1)$-entry is

$$\frac{1}{\Delta}(w(m+u) - \gamma_1 s(k+t))w^* + (w(\gamma_1 k + \gamma_2 t^\sigma) - \gamma_1 s(m+u))s^*,$$

and the $(2,2)$-entry is

$$\frac{1}{\Delta}(-s(m+u) + w(k+t))\gamma_1 s^* + (-s(\gamma_1 k + \gamma_2 t^\sigma) + w(m+u))w^*.$$

Before we use the previous identity, assume that $t = 0 = t^*$. Then since the group G_i permutes the reguli of Σ sharing $x = 0$, it follows that the standard regulus R is left invariant by eh_i. Since the standard regulus is left invariant by the regulus-inducing elation group E_1, it follows that eh_i must normalize E_1, as E_1 is the only regulus-inducing subgroup of E that leaves R invariant. But, e centralizes E_1, implying that h_i normalizes E_1. But, as previously

$$H^{-1}E_1 H = E \begin{bmatrix} w & \gamma_1 s \\ s & w \end{bmatrix}^{-1} \begin{bmatrix} w^* & \gamma_1 s^* \\ s^* & w^* \end{bmatrix} = E_1.$$

This implies that the $(1,2)$-entry of $\begin{bmatrix} w & \gamma_1 s \\ s & w \end{bmatrix}$ is 0. Since the order divides $2^a(q+1)$, it follows that $s = 0$, which, in turn, implies that $s^* = 0$ and so then also $k = 0$. But, then, by our assumptions, the indicated element eh_i is a collineation of the derived plane of Σ' that fixes the regulus used in the construction on Σ', so that eh_i is a collineation of Σ'.

Hence, we have to assume that either t or t^* is nonzero.

Returning to equations giving $(1,1)$ and $(2,2)$, we see that this reduces to

$$s^* w(\gamma_2 t^\sigma - \gamma_1 t) = -sw^*(\gamma_2 t^\sigma - \gamma_1 t).$$

First assume that $t = 0$. Then note that all of the indicated 2×2 matrices on the left side of equation $(*)$ are in the field coordinatizing Σ and the right-hand side of the equation defines an element of the derived plane of Σ'. This implies that $t^* = 0$ and we have dealt with the case $t = 0 = t^*$ above.

Hence, we may assume that $t \neq 0$, so that we must have $\gamma_2 t^\sigma - \gamma_1 t \neq 0$, implying that

$$s^* w = -sw^*.$$

First assume that $s = 0$. Then $w \neq 0$ so that $s^* = 0$. Then

$$(*)' \quad : \quad \begin{bmatrix} w & 0 \\ 0 & w \end{bmatrix}^{-1} \left(\begin{bmatrix} m & \gamma_1 k \\ k & m \end{bmatrix} + \begin{bmatrix} u & \gamma_2 t^\sigma \\ t & u \end{bmatrix} \right) .$$

$$\begin{bmatrix} w^* & 0 \\ 0 & w^* \end{bmatrix}$$

$$= \begin{bmatrix} u^* & \gamma_2 t^{*\sigma} \\ t^* & u^* \end{bmatrix}.$$

Moreover, we know that such elements define collineations of Σ'^* and Σ', so this implies that

$$\gamma_2(t+k)^\sigma = \gamma_1 k + \gamma_2 t^\sigma \iff \gamma_k k^\sigma = \gamma_1 k \iff k = 0.$$

In this situation, we know that the indicated collineations are collineations of Σ'^*; this case is complete.

Thus, $s \neq 0$. Now assume that $w = 0$, implying that $w^* = 0$. Also,

$$\begin{bmatrix} 0 & \gamma_1 s \\ s & 0 \end{bmatrix}^{-1} \begin{bmatrix} 0 & \gamma_1 s^* \\ s^* & 0 \end{bmatrix} = \begin{bmatrix} s^*/s & 0 \\ 0 & s^*/s \end{bmatrix}.$$

But, by assumption (iii), noting that

$$\begin{bmatrix} 0 & \gamma_1 s \\ s & 0 \end{bmatrix} \begin{bmatrix} 0 & \gamma_1 s^* \\ s^* & 0 \end{bmatrix} = \begin{bmatrix} \gamma_1 s s^* & 0 \\ 0 & \gamma_1 s s^* \end{bmatrix},$$

implies that $\begin{bmatrix} 0 & \gamma_1 s \\ s & 0 \end{bmatrix}$ is scalar, a contradiction.

Therefore, we may assume that $w \neq 0$.

Thus, we obtain: $s^* = -sw^*/w$.

So,

$$\begin{bmatrix} w^* & \gamma_1 s^* \\ s^* & w^* \end{bmatrix} = w^* \begin{bmatrix} 1 & -\gamma_1 s/w \\ -s/w & 1 \end{bmatrix}.$$

Furthermore,

$$\begin{bmatrix} w & \gamma_1 s \\ s & w \end{bmatrix}^{-1} = \frac{w}{(w^2 - \gamma_1 s^2)} \begin{bmatrix} 1 & -\gamma_1 s/w \\ -s/w & 1 \end{bmatrix}.$$

All of this implies that

$$\begin{bmatrix} w & \gamma_1 s \\ s & w \end{bmatrix} \begin{bmatrix} w^* & \gamma_1 s^* \\ s^* & w^* \end{bmatrix} = \frac{w^*(w^2 - \gamma_1 s^2)}{w} \begin{bmatrix} 1 & 0 \\ 0 & 1 \end{bmatrix}.$$

This previous equation is impossible, unless we have $s = 0$ by assumption (iii), and we have considered this situation previously. Hence, the theorem is proved. $\qquad\square$

We now show that when $(q-1)/2$ is odd, we may improve the previous theorem by removing the condition (iii). In the proof of Theorem 268, it is shown that if G is any collineation group acting on a parallelism that fixes a Desarguesian spread Σ and acts transitively on the remaining spreads, then $G \cap$ (kernel homology group of π_Σ) has order dividing $(2, q-1)(q-1)$ and fixes each spread in the parallelism. Hence, in the minimal parallelism-inducing group situation, we would need to require that the only possible kernel homologies that occur in G are in the kernel homology group of order $q-1$. If we assume that H can only contain the kernel homology groups of order dividing $q-1$ at most, then for $(q-1)/2$ odd, we may remove condition (iii) in the previous theorem.

THEOREM 216. *Let q be odd equal to p^r, such that $(q-1)/2$ is odd. Choose any automorphism σ of $GF(q^2)$. Let γ_2 and γ_1 be non-squares* *of $GF(q)$ such that the equation $\gamma_2 t^\sigma = \gamma_1 t$ implies that $t = 0$. Choose a Desarguesian plane Σ with the following spread:*

$$x = 0, y = x \begin{bmatrix} u & \gamma_1 t \\ t & u \end{bmatrix}; u, t \in GF(q),$$

and let Σ' denote the spread derived from the following spread Σ'^* by derivation of the standard regulus net:

$$x = 0, y = x \begin{bmatrix} u & \gamma_2 t^\sigma \\ t & u \end{bmatrix}; u, t \in GF(q).$$

Let $G_i = E^+ H$, *where E^+ is the elation group of order q^2 of Σ with axis $x = 0$ and*

$$H = \left\langle \begin{bmatrix} W & 0 \\ 0 & W^* \end{bmatrix} \right\rangle,$$

where

$$W = \begin{bmatrix} w & \gamma_1 s \\ s & w \end{bmatrix}.$$

Assume that the order of H is $(q-1)_2(q+1)$, and we have the following conditions:

(i) $\left\langle W^{-1}W^*; \begin{bmatrix} W & 0 \\ 0 & W^* \end{bmatrix} \in H \right\rangle$ *has order $2^a(q+1)$ (this condition is also necessary to have a parallelism-inducing group).*

(ii) *H contains a subgroup of order $(q-1)_2$ acting as a collineation group of Σ', whose elements $\begin{bmatrix} W & 0 \\ 0 & W^* \end{bmatrix}$ are such that W and W^* are both scalar (note there cannot be any other elements of H whose elements are scalar matrices since all of these elements have order dividing $(q-1)$ and hence fix spreads of the parallelisms).*

(iii) *$H \cap$(kernel homology group of π_Σ) has order dividing $q-1$ and H contains either $\begin{bmatrix} I & 0 \\ 0 & -I \end{bmatrix}$ or $\begin{bmatrix} -I & 0 \\ 0 & I \end{bmatrix}$.*

Then G is a minimal parallelism-inducing group relative to the Kantor-Knuth spread Σ'.

PROOF. Referring back to the proof of the previous theorem, we need to argue those situations where we used condition (iii). There are basically two cases that occur: (1) When WW^* is scalar,

$$W = \begin{bmatrix} w & \gamma_1 s \\ s & w \end{bmatrix}, W^* = \begin{bmatrix} w^* & \gamma_1 s^* \\ s^* & w^* \end{bmatrix}, s^* w = -sw^*,$$

where all elements s, s^*, w, w^* are nonzero and (2) when $w = w^* = 0$.

Consider case (1). Since $(q-1)/2$ is odd, then the determinant of W and W^* are both ± 1. In the proof above, we have $WW^* =$

$(w^*(w^2 - \gamma_1 s^2)/w)I_2$. Let $\lambda = (w^*(w^2 - \gamma_1 s^2)/w)$. Then $\lambda^2 = \pm 1$, since both W and W^* are in the group of order $(q-1)_2(q+1)$. But, $\lambda^2 = \pm 1$, and λ in $GF(q)$, implies that $\lambda^2 = 1$, implying that $\lambda = \pm 1$.

Hence, since $(w^2 - \gamma_1 s^2) = \pm 1$, it follows that $w^* = \pm w$, so that $s^* = -(\pm s)$. Therefore,

$$W^* = \begin{bmatrix} \pm w & \gamma_1(-(\pm s)) \\ -(\pm s) & \pm w \end{bmatrix}.$$

Now note that

$$\rho = \begin{bmatrix} 1 & 0 & 0 & 0 \\ 0 & 1 & 0 & 0 \\ 0 & 0 & -1 & 0 \\ 0 & 0 & 0 & -1 \end{bmatrix}$$

is a collineation of $\pi_{\Sigma'}$ (the reader is invited to perform the calculation as an exercise). We note that this particular element may not actually be in H, but since our argument is symmetric with respect to W and W^*, we may make this assumption, without loss of generality. This means we may assume that $W^* = \begin{bmatrix} -w & \gamma_1 s \\ s & -w \end{bmatrix}$. Now reconsider the situation: $WW^* = \begin{bmatrix} -w^2 & 0 \\ 0 & -w^2 \end{bmatrix}$. Since $(-w^2)^2$ is ± 1, it must be 1, so that $-w^2$ is then ± 1, but this implies that $w^2 = 1$, since $(q-1)/2$ is odd. Also, we must have $(w^*(w^2 - \gamma_1 s^2)/w)I_2 = (-w^2)I_2 = -I_2$, and $(w^2 - \gamma_1 s^2) = \pm 1$. But, since $w^2 = 1$, and we are assuming that s is nonzero, we must have $\gamma_1 s^2 = 2$, and $w^2 - \gamma_1 s^2 = -1$. With w^* now equal $-w$, we have $(w^*(w^2 - \gamma_1 s^2)/w)I_2 = ((-w)(-1)/w)I_2 = -I_2$, a contradiction.

Now assume that $w^* = w = 0$.

Now we have $-\gamma_1 s^2 = \pm 1$, and since -1 is a non-square, as $(q-1)/2$ is odd, we have $\gamma_1 s^2 = -1$ and similarly $\gamma_1 s^{*2} = -1$, but this implies that $s^{*2} = s^2$. Hence, $s = \pm s^*$. Thus, the element of H is

$$h = \begin{bmatrix} 0 & \gamma_1 s & 0 & 0 \\ s & 0 & 0 & 0 \\ 0 & 0 & 0 & \pm \gamma_1 s \\ 0 & 0 & \pm s & 0 \end{bmatrix}.$$

Now since we have ρ as a collineation of $\pi_{\Sigma'}$, we may assume that $\pm 1 = 1$. But now h is a kernel homology of the Desarguesian plane π_Σ, implying that the order of this collineation must divide $(q-1)$. However, squaring and using the fact that $\gamma_1 s^2 = -1$, we see that the order of the collineation is 4, a contradiction as 4 does not divide $q - 1$. $\square$

Hence, we have basically completely classified all minimal parallelism-inducing groups relative to a Kantor-Knuth semifield plane.

6. Existence of Minimal Groups

Our examples of relative parallel-inducing groups include the cases when Σ' is Desarguesian or a derived Kantor-Knuth semifield plane. However, this is really no reason at all that there could not be groups relative to essentially any derived conical flock spread. In fact, when $(q-1)/2$ is odd, G is in $GL(2, q^2)$, and G does not contain the kernel homology group of the associated Desarguesian plane π_Σ, and acts on a parallelism, then basically G always contains a minimal parallelism-inducing group. Hence, even though a calculation might be quite difficult to accomplish, it is likely there are many new parallelisms involving derived conical flock spreads waiting to be found.

THEOREM 217. *Let $\mathcal{P}$ be a parallelism in $PG(3, q)$, let G be a collineation group in $\Gamma L(4, q)$ whose p-elements are in $GL(4, q)$, and assume that q is odd and $(q-1)/2$ is odd. If G fixes one spread Σ and is transitive on the remaining spreads, then Σ is Desarguesian. Let Σ' a spread of the parallelism not equal to Σ. Σ' is a derived conical flock spread whose derived plane has representation*

$$x = 0, y = 0, y = x \begin{bmatrix} u + g(t) & f(t) \\ t & u \end{bmatrix} ; u, t \in GF(q),$$

where g and f are functions on $GF(q)$.

If G is a subgroup of $GL(2, q^2)$, and G does not contain the kernel involutory homology, but Σ' is left invariant under an involution of G, then G contains a minimal linear parallelism-inducing group relative to Σ'.

PROOF. Under the conditions given, we may assume that

$$\sigma = \begin{bmatrix} 1 & 0 & 0 & 0 \\ 0 & 1 & 0 & 0 \\ 0 & 0 & -1 & 0 \\ 0 & 0 & 0 & -1 \end{bmatrix}$$

is a collineation of G. Since $(q-1)_2 = 2$, we need to find a group of order $2(q+1)$ containing σ that acts transitively on the spreads not equal to Σ of the parallelism. Assume that the order of G is $q^2(q+1)t$, where Theorem 268, the order of $G \cap GL(2, q^2)$ divides $q^2(q^2-1)(q-1)2$, and $G_{\Sigma'} \cap GL(2, q^2)$ has order dividing $q(q-1)^2 2$, where Σ' is a non-socle plane. Thus, the order of G divides $q^2(q^2-1)(q-1)4r$, where $q = p^r$, but since we are assuming that G is a subgroup of $GL(2, q^2)$, the order of G

divides $q^2(q+1)(q-1)^2 2$ and is divisible by $q^2(q+1)$. Since G contains a normal elation group E^+ of order q^2, then $G = E^+ G_{(y=0)}$, and recall that we may assume that G fixes the elation axis $x = 0$ of E^+. Since G is solvable, $G_{(y=0)}$ contains a subgroup of order $(q+1)_{2'}$, of odd order. Assume that 2^b is the order of a Sylow 2-subgroup of $G_{(y=0)}$ and let S_2 be a Sylow 2-subgroup of G of order 2^c, where $b \leq c \leq b + 2^8$. Note that the order of the stabilizer $G_{\Sigma'}$ subgroup divides $(q-1)^2 2$. Assume that $b = c$. Let σ be an involution in S_2 and since σ is in $GL(2, q^2)$, either σ is a kernel homology or an affine homology. In either case, it is easily seen, since $g(-t) = -g(t)$ and $f(-t) = -f(t)$, that σ leaves Σ' invariant. Elements of G of odd order dividing $(q-1)$ are either affine homologies or products of affine homologies. Such elements fix all Baer subplanes of the set of $(q+1)$ regulus nets of π_Σ sharing $x = 0$ and $y = 0$. Any such collineation of G must fix a Baer subplane, which is a line in some spread. Hence, in this case, G could not act transitively. Therefore, $c > b$. There are three possible involutions in G, the kernel involution, and two affine involutory homologies with axis $x = 0$ and coaxis $y = 0$ or with axis $y = 0$ and coaxis $x = 0$. Note, of course, that G contains either exactly one involution or contains all three as a Klein 4-group.

Case (1): G contains exactly one involution. Then a kernel involution fixes all Baer subplanes and an affine involution fixes $q(q+1)$ Baer subplanes (at least). In either case, since a Baer subplane becomes a line of some spread of the parallelism, it follows that any involution fixes a spread other than Σ.

Cases (2): G contains a Klein 4-group. Then the above note shows that the affine homologies and the kernel homology in the Klein 4-group will fix some Baer subplanes of π_Σ jointly. Therefore, the Klein 4-group will fix a spread.

Also, note in case (2), we have the kernel involution in G, so we must have case (1).

Note that $G_{(y=0)}$, since G is in $GL(2, q^2)$ is a subgroup of a direct product of two cyclic groups. First, assume Case Cyclic: S_2 is cyclic. Then take the unique subgroup S_2^- of order 2^{b+1}. Now let H^- denote the subgroup of order $(q+1)_{2'}$ of $G_{(y=0)}$ and consider $H = S_2^- H^-$. This group has order $(q-1)_2(q+1)$; it remains to show that $E^+ H$ is parallelism-inducing. If an element g of H fixes some spread Σ'', then we see that the order of g must divide 2^{b+1}. In this setting, we must have case (1) above, that G contains a unique involution. In the Sylow 2-subgroup S_2 of G, suppose that there is a cyclic group C_4 of order 4 fixing a spread Σ'. Then the order of S_2 is at least 2^{b+2}.

Hence, C_4 normalizes a regulus-inducing elation group $E_{\Sigma'}$, which, for purposes of argument, we may take as E_1 in the notation previously considered. Let the generator for C_4 be $\begin{bmatrix} c & 0 \\ 0 & d \end{bmatrix}$. Assume without loss of generality that c has order 4. Then $C_4^{-1}E_1C_4 = E_{c^{-1}d}$, implying that $c^{-1}d \in GF(q)$, say, α. Hence, $d = c\alpha$. Since the order of $c^{-1}d$ divides 4, it then divides 2 as $(q-1)/2$ is odd. Hence, $\alpha = \pm 1$. Therefore, the generator is $\begin{bmatrix} c & 0 \\ 0 & \pm c \end{bmatrix}$. Since S_2 is cyclic, let g be a generator for S_2 and note that g has order at least 2^{b+2}. However, note in this case that g^2 is the kernel involution, which is not allowed under our assumptions. Hence, no cyclic group of order 4 can leave invariant a spread. Therefore, $E^+H = S^-H^-$ is transitive on the non-socle spreads so that E^+H is a minimal parallelism-inducing group.

Case non-Cyclic: S_2 is not cyclic and hence is a direct product of two cyclic groups C_1 and C_2, so $S_2 = C_1 \times C_2$. In this case, there are involutions in both C_1 and C_2, generating a Klein 4-group, a contradiction as before. Hence, G contains a minimal parallelism-inducing group. □

When q is even, the problem is much easier in one sense, the existence of potential minimal parallelism-inducing groups is quite limited.

THEOREM 218. *Let $\mathcal{P}$ be a parallelism in $PG(3, q)$, where $q = 2^r$. Let G be a subgroup of $\Gamma L(4, q)$, such that all 2-elements are in $GL(4, q)$ and if $q = 8$, assume that G is in $GL(4, q)$. If G fixes one spread Σ and acts transitively on the remaining spreads of $\mathcal{P}$, then Σ is Desarguesian. If G is in $GL(2, q^2)$ acting on Σ, then G contains a minimal parallelism-inducing group relative to a spread Σ' of $\mathcal{P}$.*

PROOF. Just as before, since G is a subgroup of $GL(2, q^2)$, then the order of G divides $q^2(q+1)(q-1)^2 2$ and is divisible by $q^2(q+1)$. Since $(q+1)$ is odd and G is solvable, there is a subgroup H of order $q+1$. We claim that E^+H is a minimal parallelism-inducing group. The order is correct; we merely need to show that E^+H is transitive on $q(q+1)$ spreads. If not, there is an element of odd order dividing $(q+1)$ fixing a spread Σ', a contradiction to the fact that G is transitive. This completes the proof. □

Now when $(q-1)/2$ is even, the situation is less satisfactory as there may not actually be minimal parallelism-inducing groups. Consider the group G in $GL(2, q^2)$ and G of order dividing $q^2(q+1)(q-1)^2 2$. The previous argument when q is odd and the assumption that the kernel involution of Σ is not in G says that the Sylow 2-subgroup of G, S_2 is cyclic of order at least $(q-1)_2 2$. There is a unique subgroup S_2^- of this order. There is a subgroup H^- of order $(q+1)/2$, since G is solvable. Hence, consider $H = S_2^- H^-$, which has the correct order. We need to show that $E^+ H$ is transitive on the spreads not equal to Σ. Therefore, by order, we need only consider the Sylow 2-subgroup of order $(q-1)_2 2$ of H and determine the order of a subgroup fixing a spread other than Σ'. Notice that if a 2-group has order dividing $(q-1)_2$, then elements of the form $\begin{bmatrix} c & 0 \\ 0 & d \end{bmatrix}$ will force c and d to have order dividing $(q-1)$ so that c and d are necessarily in $GF(q)$ and the corresponding group elements fix spreads other than Σ. Consider $S_2 \cap G_{\Sigma'}$ and note that it has order exactly $|S_2|/2$, since S_2 cannot fix Σ' and any other elements of the form $\begin{bmatrix} c & 0 \\ 0 & d \end{bmatrix}$ that are in $GF(q)$ have order dividing $(q-1)_2$. Hence, the stabilizer of Σ' is S_2^2, the unique subgroup of index 2 in S_2. Since any proper subgroup of S_2^- is a subgroup of S_2^2, there cannot be a minimal parallelism-inducing group, unless G itself is a minimal parallelism-inducing group.

THEOREM 219. *Let $\mathcal{P}$ be a parallelism in $PG(3, q)$, and let G be a collineation group in $\Gamma L(4, q)$, whose p-elements are in $GL(4, q)$ and assume that q is odd and $(q-1)/2$ is even. If G fixes one spread and is transitive on the remaining spreads, then let Σ denote the G-fixed spread, which is Desarguesian, and let Σ' denote a derived conical flock spread.*

If G is a subgroup of $GL(2, q^2)$, and G does not contain the kernel involutory homology, but Σ' is left invariant under an involution of G, then G contains a minimal linear parallelism-inducing group relative to Σ' if and only if G is a minimal parallelism-inducing group relative to Σ' of $\mathcal{P}$.

The above results are important in that this says that when q is even or $(q-1)/2$ is odd then we may always assume that we have a minimal parallelism-inducing group to begin with and hope to determine the associated derived conical flock spreads that are relative to this group. When $(q-1)/2$ is even, initially one might also need to consider non-minimal groups in hopes of constructing parallelisms of this type.

7. Determination of the Minimal Groups

The determination of minimal parallelism-inducing groups $E^+ H$ then reduces to finding subgroups H of $GL(2, q^2)_{(x=0,y=0)}$ of order $(q-1)_2(q+1)$ that act transitively on the regulus-inducing elation subgroups of E^+. Of course, the actual determination of the groups depends on the representation of E^+, which, in turn, depends on the representation of the associated Desarguesian affine plane. Since we may choose any representation of the Desarguesian plane up to iso-morphism, we may fix a representation. When q is odd, it is easier to consider representations of the following form:

$$ x \ = \ 0, y = x \begin{bmatrix} u & \gamma_1 t \\ t & u \end{bmatrix}, $$

$$ \forall u, t \ \in \ GF(q), \gamma_1 \text{ a nonsquare in } GF(q), $$

and for q even, we take

$$ x \ = \ 0, y = x \begin{bmatrix} u+t & t\beta \\ t & u \end{bmatrix}, $$

$$ \forall u, t \ \in \ GF(q), x^2 + x + \beta \text{ irreducible over } GF(q). $$

Because of the requirement imposed by the theorems of the previous section, we always assume that a minimal group does not contain the kernel involution, when q is odd. When q is odd, this implies that the Sylow 2-subgroup of H is cyclic of order $(q-1)_2(q+1)_2$. We note that $O(H)$, the odd-order subgroup of H, is either cyclic or a direct product of two cyclic groups. We consider first the cyclic case.

7.1. When $O(H)$ is cyclic. Since S_2 is cyclic, when q is odd, and $S_2 = \langle 1 \rangle$ when q is even, we see that H is cyclic. Let g be a generator of H. Let ω be a primitive element of $GF(q^2)^*$ of order $(q-1)_2(q+1)$. Then, $g = \begin{bmatrix} \omega^i & 0 \\ 0 & \omega^j \end{bmatrix}$, for i and j nonnegative integers $< (q-1)_2(q+1)$. In this case, at least one of ω^i and ω^j has order $(q-1)_2(q+1)$. Since H is a subgroup of two cyclic subgroups of order (q^2-1) of π_Σ, the Desarguesian affine plane corresponding to Σ, the number of cyclic groups of order $(q-1)_2(q+1)$ merely depends on i and j. Assume first that the order of ω^i is $(q-1)_2(q+1)$. Then, in terms of the group, we may assume that $i = 1$. Then j could initially be any nonnegative integer $< (q-1)_2(q+1)$. However, if E_1 is the standard regulus-inducing elation group of order q, then we would require $E_{\omega^{j-1}}$ to generate all $q+1$ regulus-inducing elation groups of order q. This means that $\langle \omega^{j-1} \rangle$ produces a set of coset representatives for the group $GF(q^2)^*/GF(q)$. Hence, we require $(j-1, q+1) = 1$. Note that when q is odd this implies that j is even. Hence, if $\omega^{(q-1)_2(q+1)/2} = -1$, then $\omega^j = 1$, thus we have ensured our required condition that H does not contain the kernel homology involution. Now we show that case is precisely what occurs.

7.2. $O(H)$ is always cyclic. Under the assumptions above, q is odd, S_2, the unique Sylow 2-subgroup is cyclic and $O(H)$ is a direct product of cyclic groups $H_1 \times H_2$. Let $H_i = \langle g_i \rangle$, where $g_1 = \begin{bmatrix} a_1 & 0 \\ 0 & b_1 \end{bmatrix}$ and $g_2 = \begin{bmatrix} a_2 & 0 \\ 0 & b_2 \end{bmatrix}$. Since the Sylow 2-subgroup of H is cyclic, assume that the order of g_1 is $(q-1)_2(q+1)_2t_1$ and the order of $g_2 = t_2$, where t_1 and t_2 are both odd.

Assume that the prime decomposition for $q+1$ is $2^c \prod_{i=1}^k u_i^{\alpha_i}$, where u_i is an odd prime for $i = 1, 2, .., k$. Since $O(H)$ is not cyclic, t_1 and t_2 have non-trivial common factors $\prod_{i=1}^k u_i^{\beta_i}$, where β_i could be zero. Let $t_1 = \prod_{i=1}^k u_i^{\delta_i} = \prod_{i=1}^k u_i^{\beta_i} \prod_{i=1}^k u_i^{\delta_i - \beta_i}$ and $t_2 = \prod_{i=1}^k u_i^{\rho_i} = \prod_{i=1}^k u_i^{\beta_i} \prod_{i=1}^k u_i^{\rho_i - \beta_i}$. The group $H_1 \times H_2$ thus has order

$$(q-1)_2(q+1)_2 \prod_{i=1}^k u_i^{2\beta_i} \prod_{i=1}^k u_i^{\delta_i + \rho_i - 2\beta_i} = 2^c \prod_{i=1}^k u_i^{\alpha_i}.$$

Hence, the variety of different possible groups depends simply on the prime decomposition of $(q+1)_{2'}$. For example, we may assume that either a_1 or b_1 has order $(q-1)_2(q+1)_2t_1$ and the other element has order dividing this number. Similarly, either a_2 or b_2 has order t_2, where the other element has order dividing this number. Thus, there are four possibilities, one of which is as follows: $g_1 = \begin{bmatrix} \omega_1 & 0 \\ 0 & \omega_1^{j_1} \end{bmatrix}$, where ω_1 has order $(q-1)_2(q+1)_2t_1$ and j_1 is any even integer and say $g_2 = \begin{bmatrix} \omega_2 & 0 \\ 0 & \omega_2^{j_2} \end{bmatrix}$, also ω_2 has order t_2 and j_2 is any even integer. Consider now $\langle \omega_1^{j_1-1}, \omega_2^{j_2-1} \rangle$. Since this is a subgroup of $GF(q^2)^*$, it is cyclic. Consider an odd prime power $u_1^{\beta_1}$ dividing both t_1 and t_2. Since the group generated by g_1 and g_2 will have order divisible by exactly $u_1^{2\beta_1}$, we would require that we have a cyclic subgroup of order $u_1^{2\beta_1}$, which is impossible.

Hence, we see that there are never elements of common order, so that the case $O(H)$ non-cyclic never occurs. So, we have proved the following theorem.

THEOREM 220. *The minimal linear parallelism-inducing groups have the form $E^+ H_j$ or $E^+ H^j$, where*

$$H_j = \left\langle \begin{bmatrix} \omega & 0 \\ 0 & \omega^j \end{bmatrix}; \right. \left. \vphantom{\begin{bmatrix} \omega & 0 \\ 0 & \omega^j \end{bmatrix}} \right\rangle,$$
ω *has order* $(q-1)_2(q+1)$ *and* $(j-1,(q+1)) = 1$

or

$$H^j = \left\langle \begin{bmatrix} \omega^j & 0 \\ 0 & \omega \end{bmatrix}; \right. \left. \vphantom{\begin{bmatrix} \omega^j & 0 \\ 0 & \omega \end{bmatrix}} \right\rangle.$$
ω *has order* $(q-1)_2(q+1)$ *and* $(j-1,(q+1)) = 1$

Furthermore, we obtain:

THEOREM 221. *Assume that q is odd. Let $E^+ H$ be one of the groups $E^+ H_j$, or $E^+ H^k$, where $(j-1, q+1) = (k-1, q+1) = 1$, where j and k are odd.*

(1) *If $(q-1)/2$ is odd, then any group $E^+ H_j$ or $E^+ H^k$ is a parallelism-inducing group relative to any derived Kantor-Knuth semifield plane.*

(2) *If $(q-1)/2$ is even, then any group $E^+ H_j$ or $E^+ H^k$ such that $(j+1, q+1) = 1 = (k+1, q+1)$ is a parallelism-inducing group relative to any derived Kantor-Knuth semifield plane with associated automorphism σ such that $(q-1)_2$ divides $\sigma - 1$. That is, if $q = p^r$ and σ is p^t then $r/(t,r)$ is odd.*

PROOF. In general, from the theorems in the section on Kantor-Knuth semifield spreads, we would require $\langle w^{j-1} \rangle$ and $\langle w^{j+1} \rangle$ for w of order dividing $(q-1)_2(q+1)$ to each generate the group $GF(q^2)^*/GF(q)$. When $(q-1)/2$ is odd and j is even, our results guarantee these conditions. When $(q-1)/2$ is even, the stated conditions are sufficient as long as $(q-1)_2$ divides $(\sigma - 1)$. □

8. General Isomorphisms from Minimal Groups

Assume that we have a parallelism using a minimal parallelism-inducing group $E^+ H_j$ or $E^+ H^j$, such that $(j-1, q+1) = 1$ and denote the parallelism by $\mathcal{P}_j$ or $\mathcal{P}^j$, respectively. First, assume that $\mathcal{P}_j$ is isomorphic to $\mathcal{P}_k$, by a mapping g from $\mathcal{P}_j$ to $\mathcal{P}_k$, so that j and k are even integers such that $(j-1, q+1) = 1 = (k-1, q+1)$. Of course, we now have Σ, the Desarguesian spread common to both parallelisms, and since there is a unique Desarguesian spread in the parallelisms, it follows that g is a collineation of Σ. Moreover, since all collineation groups of both parallelisms fix the line ℓ by Theorem 267, and E^+, the elation group of order q^2 of π_Σ with axis ℓ, is transitive on the remaining components of π_Σ, we may assume that g has the following form: $g : (x, y) \longmapsto (x^\rho a, y^\rho b)$, where ρ is an automorphism of $GF(q^2)$

and a and b are in $GF(q^2)^*$. Then H_j^g is now a collineation group of
$\mathcal{P}_k$, as is H_k. Note that $\begin{bmatrix} \omega & 0 \\ 0 & \omega^j \end{bmatrix}^g = \begin{bmatrix} \omega^\rho & 0 \\ 0 & \omega^{j\rho} \end{bmatrix}$. Hence, we have a
linear collineation group of $\mathcal{P}_k$ of order $((q-1)_2(q+1))^2 / (|H_j^g \cap H_k|)$
(again the reader is directed to Theorem 268). The order of the linear
collineation group G of any such parallelism divides $q^2(q+1)(q-1)^2 2$.
Hence, $(q+1)/(2, q+1)$ divides $|H_j^g \cap H_k|$. In particular, this says when
q is even, the two groups H_j and H_k are equal, so we may assume that
$j = k$. So, assume that q is odd so that $(q+1)/2$ divides $|H_j^g \cap H_k|$.
Thus, we obtain $\omega^{2(q-1)2j} = \omega^{2(q-1)2k}$, which implies that

$$j \equiv k \mod (q+1)/2.$$

In the case when $|H_j^g \cap H_k| = (q+1)/2$, we have a generated group of
order $2(q-1)_2^2(q+1)$. In this case, the stabilizer of a spread Σ' not
equal to Σ has order $2(q-1)_2^2$. Hence, in the generated group G and
stabilizing Σ', we have a group of order 8. The Sylow 2-subgroup of
H_j^g is cyclic of order $(q-1)_2(q+1)_2$. So, we have elements $\begin{bmatrix} c & 0 \\ 0 & c^j \end{bmatrix}$
and $\begin{bmatrix} c & 0 \\ 0 & c^k \end{bmatrix}$, where the order of c is $(q-1)_2(q+1)_2$. Hence, we have
homologies $\begin{bmatrix} 1 & 0 \\ 0 & c^{k-j} \end{bmatrix}$. Furthermore, the existence of isomorphisms
produces central collineations of π_Σ acting on $\pi_{\Sigma'}$, which, in individual
cases, may be shown not to exist.

DEFINITION 118. *Let E^+H be a minimal parallelism-inducing group
acting on a parallelism and relative to Σ'. If Σ is the associated Desarguesian spread of the parallelisms, we shall use the notation $\mathcal{P}_{H,\Sigma,\Sigma'}$ to
denote this spread.*

THEOREM 222. *Assume that p is not 3 or 7. Then*
(1) *The parallelism $\mathcal{P}_{H_j,\Sigma,\Sigma'}$ and the parallelism $\mathcal{P}_{H^k,\Sigma,\Sigma'}$ are never
isomorphic.*
(2) *If q is even, then the parallelism $\mathcal{P}_{H_j,\Sigma,\Sigma'}$ and the parallelism
$\mathcal{P}_{H_k,\Sigma,\Sigma'}$ are never isomorphic, for j and k distinct. Hence, potentially,
there are $2\varphi(q+1)$ possible mutually non-isomorphic parallelisms using
minimal parallelism-inducing groups relative to Σ'.*
(3) *If q is odd and the parallelism $\mathcal{P}_{H_j,\Sigma,\Sigma'}$ is isomorphic to the
parallelism $\mathcal{P}_{H_k,\Sigma,\Sigma'}$, then $j \equiv k \mod (q+1)/2$ and there is an affine
homology group of order at least $2(q-1)_2$ of the Desarguesian affine
plane π_Σ that acts as a collineation of $\pi_{\Sigma'}$ as a Baer collineation group.*

Hence, there are at least $\varphi((q+1)/2)$ possible mutually non-isomorphic parallelisms using minimal parallelism-inducing groups.

PROOF. Consider part (1). If two such parallelisms are isomorphic then since Σ is the unique Desarguesian spread of both parallelisms, any isomorphism g from one parallelism to the other may be assumed to be a collineation of π_Σ, which fixes $x = 0$ (the full collineation group of such a parallelism leaves $x = 0$ invariant by Theorem 267. Moreover, both parallelisms admit the same group E^+, which acts transitively on the components of Σ apart from $x = 0$. Hence, we may assume that an isomorphism g has the following form:

$$g : (x, y) \longmapsto (x^\rho, y^\rho) \begin{bmatrix} a & 0 \\ 0 & b \end{bmatrix},$$

where ρ is an automorphism of $GF(q^2)$ and $a, b \in GF(q^2)^*$. Notice that H_j^g becomes a collineation group of $\mathcal{P}_{H^k,\Sigma,\Sigma'}$, assuming that g maps $\mathcal{P}_{H_j,\Sigma,\Sigma'}$ onto $\mathcal{P}_{H^k,\Sigma,\Sigma'}$. It follows immediately that $H_j^g = H_j$. Hence, $\mathcal{P}_{H^k,\Sigma,\Sigma'}$ admits a linear collineation group G of Σ of order $((q-1)_2(q+1))^2/J$, where, J is the order of the intersection of H_j and H^k. However, $\begin{bmatrix} \omega^i & 0 \\ 0 & \omega^{ij} \end{bmatrix} = \begin{bmatrix} \omega^{zk} & 0 \\ 0 & \omega^z \end{bmatrix}$, where the exponents are taken modulo $(q-1)_2(q+1) = N$, if and only if $z \equiv ij \bmod N$ and $i \equiv zk \bmod N$. Thus, $i \equiv ijk \bmod N$.

Let $j = 2j^*$ and $k = 2k^*$, then $i \equiv 4ij^*k^* \bmod N$. Hence, $i(4j^*k^* - 1) = tN$. It follows that $(q-1)_2(q+1)_2$ divides i. Hence, the order of the intersection can be at most $(q+1)_{2'}$. Thus, the order of G is divisible by $((q-1)_2(q+1))^2/(q+1)_{2'} = ((q-1)_2)^2(q+1)_2^2(q+1)_{2'} = ((q-1))_2^2(q+1)_2(q+1)$. However, by Theorem 268, the order of G must divide $2(q-1)^2(q+1)$, since G is a subgroup of $GL(2, q^2)$. Hence, $(q+1)_2 = 2$ or we are finished. Also, we are finished or the order of the intersection is exactly $(q+1)_{2'} = (q+1)/2$. This is the group generated by $\begin{bmatrix} \omega^{2(q-1)_2} & 0 \\ 0 & \omega^{2(q-1)_2 j} \end{bmatrix}$ and the group generated by $\begin{bmatrix} \omega^{2(q-1)_2 k} & 0 \\ 0 & \omega^{2(q-1)_2} \end{bmatrix}$. However, the $(1,1)$-entry for the first generator has odd order and the $(1,1)$-entry for the second generator has even order as k is even. Hence, these groups cannot be equal. This shows that these two parallelisms cannot be isomorphic. This proves (1). Now $\omega^{t(j-1)} = \omega^{s(j-1)}$ modulo $(GF(q^2))^*$ if and only if $\omega^{(t-s)(j-1)}$ is in $GF(q)^*$. We note that $\omega^{(q+1)(j-1)}$ is in $GF(q)$. So, ω has order $(q-1)_2(q+1)$ and so t, s vary from $1, \ldots, (q-1)_2(q+1)$. Suppose that $(j-1, q+1) = 1$ and j is even. Then $(j-1, (q-1)_2(q+1)) = 1$.

Therefore, the order of ω^{j-1} is actually $(q-1)_2(q+1)$. Hence, we may potentially take j to be any even integer from $1, 2, .., (q-1)_2(q+1)$, such that $(j-1)$ is relatively prime to $(q-1)_2(q+1)$, or equivalently that $(j-1, q+1) = 1$, so we may take any odd integer $j-1$ relatively prime to $q+1$. The previous argument of part (1) but using the parallelism $\mathcal{P}_{H_j, \Sigma, \Sigma'}$ and $\mathcal{P}_{H_k, \Sigma, \Sigma'}$ as done in the previous remarks shows that $j \equiv k$ mod $(q+1)/(2, q+1)$. Hence, in the even-order case, and j, k distinct integers such that $(j-1, q+1) = 1 = (k-1, q+1)$, the parallelisms cannot be isomorphic as the order of H_j and H_k is exactly $(q+1)$. Combining part (1) for q even and part (2), we see that for fixed Σ', the number of isomorphisms is exactly $2\varphi(q+1)$.

Now consider part (3). The above argument shows that if parallelisms $\mathcal{P}_{H_j, \Sigma, \Sigma'}$ and $\mathcal{P}_{H_k, \Sigma, \Sigma'}$ are isomorphic then $j \equiv k$ mod $(q+1)/2$. The number of integers between 1 and $(q-1)_2(q+1)$ that are congruent to each modulo $(q+1)/2$ is $2(q-1)_2/2$. Hence, if j and k are restricted to the even integers between 1 and $(q+1)/2$, then there are $\varphi((q+1)/2)/2$ possibilities. Combining this with part (1), we see we have at least $2\varphi((q+1)/2)/2 = \varphi((q+1)/2)$ mutually non-isomorphic parallelisms. $\qquad\square$

8.1. Non-standard Parallelisms. In our definition of minimal groups and our subsequent determination of the possible types, when q is odd, we always assumed that E^+H is such that H contains an involutory affine homology of the associated Desarguesian plane. Hence, the stabilizer of $\pi_{\Sigma'}$ will be fixed by a Baer involution g that fixes the pertinent regulus containing $x = 0$ considered as a subplane of $\pi_{\Sigma'}$. We note that if a Baer subplane of π_Σ is fixed by g then the Baer subplane must share the axis and coaxis of g. Hence, g fixes a unique subspace of dimension 2 over $GF(q)$ not equal to the component $x = 0$. Since the regulus-inducing elation group $E_{\Sigma'}$ fixing Σ' is transitive on the Baer subplanes not equal to $x = 0$ of the regulus net in question containing $x = 0$, it follows that any such parallelism obtained will always be a standard parallelism.

THEOREM 223. *If q is odd, then any minimal linear parallelism-inducing group relative to a spread Σ' always produces standard parallelisms.*

9. Isomorphic Kantor-Knuth Parallelisms

Previously, we have considered general isomorphisms of putative parallelisms using minimal parallelism-inducing groups. In this section, since we know when the parallelisms exist when we use Σ' as a derived

Kantor-Knuth semifield plane, including the Hall plane when $\sigma = 1$, we may more or less determine the isomorphisms in this setting. So, we assume that q is odd in this section. First, we recall that if Σ' is denoted by (γ_2, σ), then two Kantor-Knuth type semifield planes and hence two derived semifield planes (γ_2, σ) and (γ_2^*, σ^*) are isomorphic if and only if $\sigma^* = \sigma^{\pm 1}$. So, we may select any isomorphism of $GF(q)$, including the identity, and any non-square γ_2 with the property that $\gamma_1 t = \gamma_2 t^\upsilon$ implies that $t = 0$, where γ_1 is the non-square corresponding to the fixed Desarguesian spread Σ. For example, if we do select $\sigma = 1$, then any nonsquare $\gamma_2 \neq \gamma_1$ will suffice. Furthermore, we require $(q-1)_2$ to divide $\sigma - 1$, which is always true if $\sigma = 1$ or if $(q-1)/2$ is odd. Since $(q-1)/2$ odd is the easiest case to describe, we begin there.

If $(q-1)/2$ is odd, then we have shown that any abstract minimal parallelism-inducing group is actually a minimal parallelism-inducing group for any derived Kantor-Knuth semifield spread. Let $q = p^r$. If $r - 1$ is even, there are $1 + (r-1)/2$ mutually non-isomorphic derived Kantor-Knuth semifield spreads. If $r - 1$ is odd, then there are $2 + (r - 2)/2$ mutually non-isomorphic derived Kantor-Knuth semifield spreads. Hence, no $E^+ H_j$ or $E^+ H^k$ parallelisms could be isomorphic if they are relative to mutually non-isomorphic derived Kantor-Knuth semifield spreads. We note that even though (γ_2, σ) and (γ_2^*, σ) are isomorphic derived Kantor-Knuth semifield spreads, this does not say that the parallelisms that they induce are isomorphic, since any isomorphism must arise from a collineation of the associated Desarguesian spread, which is the unique Desarguesian spread in both parallelisms.

We see that if

$$x = 0, y = x \begin{bmatrix} u & \gamma_2 t^\sigma \\ t & u \end{bmatrix} ; u, t \in GF(q)$$

is the spread for $\Sigma_{\gamma_2}'^*$, the derived spread of Σ_{γ_2}', used in the construction, and

$$x = 0, y = x \begin{bmatrix} u & \gamma_2^* t^\sigma \\ t & u \end{bmatrix} ; u, t \in GF(q)$$

is the spread for $\Sigma_{\gamma_2^*}'^*$, the derived spread of $\Sigma_{\gamma_2^*}'$, used in the construction then

$$\theta = \begin{bmatrix} 1 & 0 & 0 & 0 \\ 0 & \gamma_2^*/\gamma_2 & 0 & 0 \\ 0 & 0 & 1 & 0 \\ 0 & 0 & 0 & \gamma_2^*/\gamma_2 \end{bmatrix}$$

is an isomorphism from Σ_{γ_2}' onto $\Sigma_{\gamma_2^*}'$. Any isomorphism between the derived spreads is an isomorphism between the spreads by a result of

Jha and Johnson [**86**]. Hence, if parallelisms using the same σ but different non-squares γ_2 and γ_2^* are isomorphic, then there is a collineation g of π_Σ that we may assume fixes $x = 0$ and $y = 0$ and has the form

$$g : (x, y) \longmapsto (x^\rho, y^\rho) \begin{bmatrix} a & 0 \\ 0 & b \end{bmatrix}.$$

If this mapping is considered to map the $(\gamma_2^*, \sigma)^*$ parallelism onto the $(\gamma_2, \sigma)^*$ parallelism, then $g\theta$ is a collineation of Σ'_{γ_2}, since again the full collineation group of Σ' is inherited from the collineation group of Σ'. Actually, we wish to take θg, which we may assume is a collineation of $\Sigma_{\gamma_2^*}$. Then let $\{\theta, 1\}$ be a basis for $GF(q^2)$ over $GF(q)$ so that $\theta^2 = \gamma$ and $\theta^q = -\theta$. Let $\theta^\rho = \theta\alpha_\rho + \beta_\rho$. Hence, $(z_1, z_2)^\rho = (\theta\alpha_\rho + \beta_\rho)z_1^\rho + z_2^\rho = \theta\alpha_\rho z_1^\rho + \beta_\rho z_1^\rho + z_2^\rho = (z_1^\rho, z_2^\rho) \begin{bmatrix} \alpha_\rho & \beta_\rho \\ 0 & 1 \end{bmatrix}$. Therefore, g has the following form within $\Gamma L(4, q)$:

$$g : (x_1, x_2, y_1, y_2) \longmapsto \left((x_1^\rho, x_2^\rho) \begin{bmatrix} \alpha_\rho & \beta_\rho \\ 0 & 1 \end{bmatrix}, (y_1^\rho, y_2^\rho) \begin{bmatrix} \alpha_\rho & \beta_\rho \\ 0 & 1 \end{bmatrix} \right) \begin{bmatrix} a & 0 \\ 0 & b \end{bmatrix},$$

where a and b are represented as 2×2 matrices over $GF(q)$.

$y = x \begin{bmatrix} u & \gamma_2^* t^\sigma \\ t & u \end{bmatrix}$ maps under g to

$$y = x \left(\begin{bmatrix} \alpha_\rho & \beta_\rho \\ 0 & 1 \end{bmatrix} a \right)^{-1} \begin{bmatrix} u^\rho & (\gamma_2^* t^\sigma)^\rho \\ t^\rho & u^\rho \end{bmatrix} \left(\begin{bmatrix} \alpha_\rho & \beta_\rho \\ 0 & 1 \end{bmatrix} b \right).$$

This subspace then maps under $g\theta$ to

$$y = x \begin{bmatrix} 1 & 0 \\ 0 & \gamma_2^*/\gamma_2 \end{bmatrix}^{-1} \left(\begin{bmatrix} \alpha_\rho & \beta_\rho \\ 0 & 1 \end{bmatrix} a \right)^{-1}$$
$$\cdot \begin{bmatrix} u^\rho & (\gamma_2^* t^\sigma)^\rho \\ t^\rho & u^\rho \end{bmatrix} \left(\begin{bmatrix} \alpha_\rho & \beta_\rho \\ 0 & 1 \end{bmatrix} b \right) \begin{bmatrix} 1 & 0 \\ 0 & \gamma_2^*/\gamma_2 \end{bmatrix},$$

and the element is then of the form $y = x \begin{bmatrix} u & \gamma_2^* t^\sigma \\ t & u \end{bmatrix}$.

We begin by considering the image of $y = x$ under $g\theta$. That is, taking $t = 0$ and $u = 1$, letting $a^{-1}b = c = \begin{bmatrix} w & \gamma_1 s_0 \\ s_0 & w \end{bmatrix}$, we obtain the following:

$$\begin{bmatrix} 1 & 0 \\ 0 & \gamma_2^*/\gamma_2 \end{bmatrix}^{-1} \begin{bmatrix} w & \gamma_1 s_0 \\ s_0 & w \end{bmatrix} \begin{bmatrix} 1 & 0 \\ 0 & \gamma_2^*/\gamma_2 \end{bmatrix} = \begin{bmatrix} w & \gamma_1 s_0 \gamma_2^*/\gamma_2 \\ s_0 \gamma_2/\gamma_2^* & w \end{bmatrix}$$
$$= \begin{bmatrix} u & \gamma_2^* t^\sigma \\ t & u \end{bmatrix}.$$

Hence, we must have

$$\gamma_2^*(s_0\gamma_2/\gamma_2^*)^\sigma = \gamma_1 s_0 \gamma_2^*/\gamma_2 \iff (s_0\gamma_2/\gamma_2^*)^\sigma = \gamma_1 s_0/\gamma_2,$$

for some s_0. For example, if $s_0 \neq 0$, then when $\sigma = 1$, we would have
the stipulation that

$$\gamma_1 \neq \gamma_2^2/\gamma_2^*.$$

Note this line of argument can be pushed further since we now
have a collineation of the associated plane. We are content to simply
establish the following theorem.

THEOREM 224. *Let $\mathcal{P}_{\gamma_2}$ and $\mathcal{P}_{\gamma_2^*}$ be parallelisms generated by the
parallelism-inducing group E^+H_j or E^+H^j relative to the derived Kantor-
Knuth semifield plane*

$$x = 0, y = x \begin{bmatrix} u & \gamma t^\sigma \\ t & u \end{bmatrix}; u, t \in GF(q)$$

and associated Desarguesian spread

$$x = 0, y = x \begin{bmatrix} u & \gamma_1 t \\ t & u \end{bmatrix}; u, t \in GF(q),$$

*where γ_1 and γ are non-squares such that $t\gamma_1 = \gamma_2 t^\sigma$ implies $t = 0$,
where $\gamma = \gamma_2$ or γ_2^*, respectively. Assume that $(s_0\gamma_2/\gamma_2^*)^\sigma \neq \gamma_1 s_0/\gamma_2$ for
all nonzero elements s_0 of $GF(q)$. Furthermore, assume that $(q-1)_2$
divides $\sigma - 1$. Then the two parallelisms are not isomorphic.*

*If $(q-1)/2$ is odd, then the number of mutually non-isomorphic
parallelisms obtained from minimal parallelism-inducing groups relative
to a derived Kantor-Knuth semifield plane is as follows: Assume that
the number of non-zero solutions to*

$$(s_0\gamma_2/\gamma_2^*)^\sigma - \gamma_1 s_0/\gamma_2 = 0$$

*is N_σ, as γ_2 and γ_2^* satisfy the property listed above for γ. There are
at least*

$$\sum_\sigma \varphi((q+1)/2)N_\sigma$$

mutually non-isomorphic parallelisms.

10. Even Order

So far our examples have been limited to odd-order cases. Our
arguments were made possible because of the representation of the
Kantor-Knuth semifield spread, so we also obtain odd-order examples
when Σ' is Desarguesian. When q is even, one could potentially try
essentially any even-order conical flock spread. However, there has
been no work done in this direction with one exception. In this section,

we give an idea of what can be done for at least for some groups when Σ' is Hall.

THEOREM 225. *Let $q = 2^r$ for r odd, and let Σ be a Desarguesian spread of the following form:*

$$x = 0, y = x \begin{bmatrix} u+t & t \\ t & u \end{bmatrix}; u, t \in GF(q),$$

and take a second Desarguesian spread Σ'^* of the form

$$x = 0, y = x \begin{bmatrix} u+\alpha t & t \\ t & u \end{bmatrix}; u, t \in GF(q),$$

where $\alpha \neq 1$.

Then the group $E^+ H^0$ is parallelism-inducing relative to the Hall plane Σ' (derived from Σ'^*).

COROLLARY 57. *Choose any irreducible polynomial $x^2 + x\alpha + \beta$ over $GF(q)$, for q even. Since β is a square, we may assume that there is a corresponding irreducible polynomial $x^2 + x\alpha^* + 1$. There are exactly $q/2$ irreducible polynomials of this form. Hence, there are at least $q/2 - 1$ distinct parallelisms using the same group $E^+ H^0$. Then there are at least $2(q/2 - 1) = q - 2$ distinct parallelisms using the groups $E^+ H_0$ and $E^+ H^0$.*

PROOF. Consider the group $E^+ H^{j=0}$. We need to show that this group is parallelism-inducing with respect to the Hall spread Σ'. Let l' be a component of Σ' and assume that an element g of $E^+ H^0$ maps l' into a component of Σ'. We note that $E^+ H^0$ is a collineation group of π_Σ and as such maps Baer subplanes that non-trivially intersect $x = 0$ back into such Baer subplanes. So, if l' non-trivially intersects $x = 0$ then $l'g$ non-trivially intersects $x = 0$. This would mean that g would leave invariant the net used in the derivation process to produce Σ' from Σ'^*. Thus, g would leave invariant this regulus net R' and therefore would leave invariant the net R :

$$x = 0, y = x \begin{bmatrix} u & 0 \\ 0 & u \end{bmatrix}; u \in GF(q).$$

But in $E^+ H^0$, then g must be in E_1, the regulus-inducing elation group, which is a collineation group of both Σ' and Σ'^*. Thus, we may assume that l' is a component of $\Sigma' \cap \Sigma'^*$ and by the same reasoning as above, we may assume that the image $l'g$ is a component of $\Sigma' \cap \Sigma'^*$. Then, there are an elation and an element of H^0, such that the following occurs:

$$\left(\begin{bmatrix} u + \alpha t & t \\ t & u \end{bmatrix} + \begin{bmatrix} m + k & k \\ k & m \end{bmatrix} \right)$$

$$= \begin{bmatrix} t^* \alpha & t^* \\ t^* & 0 \end{bmatrix} \begin{bmatrix} w + s & s \\ s & w \end{bmatrix}^{-1}$$

$$= \begin{bmatrix} t^* \alpha w + l^* s & l^* \alpha s + l^*(w + s) \\ t^* w & t^* s \end{bmatrix},$$

for $tt^* \neq 0$ and for u, t, m, k, w, s, t^* elements of $GF(q)$. Note that we may take the $(2,2)$ element zero in the first matrix of the right side in this equation by following g by an element e_1 of the regulus-inducing group E_1, if necessary, then ge_1 would enjoy the same conditions as g. We then obtain:

$$\left(\begin{bmatrix} u + \alpha t + m + k & t + k \\ t + k & u + m \end{bmatrix} \right) = \begin{bmatrix} t^* \alpha w + t^* s & t^* \alpha s + t^*(w + s) \\ t^* w & t^* s \end{bmatrix}.$$

This says that the $(1, 2)$ and $(2, 1)$ entries are equal in the matrix to the right. Therefore, $t^* w = t^* \alpha s + t^*(w + s)$, which leads to $t^* \alpha s = t^* s$, and since $t^* \neq 0$, we have $s = 0$. Since the order of $\begin{bmatrix} w + s & s \\ s & w \end{bmatrix}$ divides $q + 1$, since q is even, and $s = 0$, then $w = 1$. Thus, we obtain $\left(\begin{bmatrix} u + \alpha t + m + k & t + k \\ t + k & u + m \end{bmatrix} \right) = \begin{bmatrix} t^* \alpha & t^* \\ t^* & 0 \end{bmatrix}$. We arrive at the following conditions:

$$t^* = t + k, u = m \text{ and } u + m + \alpha t + k = t^* \alpha.$$

Combining these equations implies the following equation: $t\alpha + k = t^* \alpha = (t + k)\alpha$. However, this implies that $k = k\alpha$, so that $k = 0$. But, with $k = 0$, g becomes an element of the regulus-inducing elation group E_1, which is a collineation group of Σ', and this completes the proof of the theorem. $\qquad \square$

11. Derived Parallelisms

Since all of the parallelisms that we have constructed are standard, we may then choose any spread Σ' of the parallelism, i.e., a derived Kantor-Knuth semifield plane or a Hall plane, and note that the derived spread Σ'^* and Σ share a unique regulus R. If we derive Σ to Σ^*, we know that the following is a parallelism:

$$\Sigma^* \cup \Sigma'^* \cup \left\{ \Sigma'g; g \in E^+H - \{1\} \right\},$$

where E^+H is a minimal parallelism-inducing group. Hence, all of our previously constructed parallelisms produce derived parallelisms, where we further know that any two of these derived parallelisms are isomorphic if and only if the original parallelisms are isomorphic. We note that these derived parallelisms have one Hall spread, one Kantor-Knuth semifield spread and the remaining spreads are derived Kantor-Knuth semifield planes so clearly are not isomorphic to any of the parallelisms obtained using minimal parallel-inducing groups.

Part 8

Coset Switching

In this part, we formulate a generalization of certain of the constructions of parallelisms, in particular, the m-parallelisms are generalized. If we recall how m-parallelisms are constructed, we used m Desarguesian spreads that share a given regulus with an initial Desarguesian spread, called the 'socle.' We do not use a transitive group on the spreads but something close to it, and what actually occurs is that we are formulating what we call 'coset switching' to find the m Desarguesian spreads. In general, from a transitive deficiency one partial parallelisms, we develop a process of replacing sets of q spreads among the set of $q(q+1)$ spreads so as to create a number of new parallelisms.

This process could be considered analogous to net replacement in a Desarguesian affine plane of order q^2, where it is possible to consider a set of $q-1$ mutually disjoint reguli (disjoint on lines), where the reguli may be be chosen to be replaced or not by the associated opposite reguli). In that situation, there would be essentially $2^{(q-1)}$ possible replacements leading to distinct translation planes (of course, not all of which are mutually non-isomorphic).

In the construction technique given here, there are great variety of what we call 'E-switches' (or more simply 'switches'), whereby a given set of q-spreads of a parallelism is switched with another set of q-spreads, so as to create a related parallelism. In general, if there are t possible switches for each of $q+1$ sets of q-spreads, there are a possible t^{q+1} constructed parallelisms. In this setting, most of these constructed parallelisms will be new and most mutually non-isomorphic.

All of these ideas may also be considered more generally over infinite fields, and we construct a really vast number of new parallelisms over the real using these methods. In the finite case, always starting with a socle Desarguesian spread Σ_1, we could consider the Johnson parallelisms using a given second Desarguesian spread Σ_2, but here we wish to find spreads that 'switch' with Σ_2, in the manner mentioned above. In this setting, we prove that any other spread that can switch with a Desarguesian spread Σ_2 must be with Desarguesian or Kantor-Knuth. So, there are also a number of new parallelisms consisting of Desarguesian spreads and Kantor-Knuth spread. Since we may switch a Desarguesian spread or a Kantor-Knuth by other such spreads so as to obtain non-isomorphic parallelisms, this process produces also a tremendous number of new parallelisms. The theory given here and in the next chapter is modified from work of Diaz, Johnson, and Montinaro [46], [44].

CHAPTER 32

Finite E-Switching

1. E-and Desarguesian Switches

Let Σ_1 be a Desarguesian spread of odd order:

$$\Sigma_1 = \left\{ x = 0, y = x \begin{bmatrix} u + \rho_1 t & \gamma_1 t \\ t & u \end{bmatrix} ; u, t \in GF(q) \right\},$$

where $x^2 + \rho_1 x - \gamma_1$ is irreducible over $GF(q)$. Let the regulus R_0 be

$$R_0 = \left\{ x = 0, y = x \begin{bmatrix} u & 0 \\ 0 & u \end{bmatrix} ; u \in GF(q) \right\}.$$

Let E denote the elation group of order q^2 with axis $x = 0$ of the associated Desarguesian plane π_{Σ_1}. Note that

$$E = \left\langle \tau_{u,t} = \begin{bmatrix} 1 & 0 & u + \rho_1 t & \gamma_1 t \\ 0 & 1 & t & u \\ 0 & 0 & 1 & 0 \\ 0 & 0 & 0 & 1 \end{bmatrix} ; u, t \in GF(q) \right\rangle.$$

Let Γ_2 and Γ_3 denote two conical flock spreads of odd order distinct from Σ_1 in $PG(3, q)$ that share precisely R_0 with Σ_1, which may be represented as:

$$\Gamma_i = \left\{ x = 0, y = x \begin{bmatrix} u + g_i(t) & f_i(t) \\ t & u \end{bmatrix} ; u, t \in GF(q) \right\},$$

$$i = 2, 3$$

and assume that for any given t_0,

$$g_i(t_0) = \rho_1 t_0,$$
$$f_i(t_0) = \gamma_1 t_0$$

implies $t_0 = 0$ (this condition simply is that the given spreads share exactly R_0 with Σ_1). Note that either of these spreads could also be Desarguesian. Let E^- denote the subgroup of E that fixes R_0 and hence is a collineation group of Σ_1, and Γ_i, for $i = 1, 2$.

LEMMA 57. *(1) $\Gamma_2 E$ and $\Gamma_3 E$ both contain exactly q spreads.*

(2) If $\Gamma_2 E = \Gamma_3 E$ and $\Gamma_3 E$ consists of mutually disjoint spreads, then $\Gamma_2 E$ consists of mutually disjoint spreads.

PROOF. Since E^- fixes Γ_i, there are at most q image sets of Γ_i. For Δ either Γ_2 or γ_3, Δ is a conical flock plane and, as noted, has the general form

$$\Delta = \left\{ x = 0, y = x \begin{bmatrix} u + g(t) & f(t) \\ t & u \end{bmatrix} ; u, t \in GF(q) \right\},$$

where g and f are functions on $GF(q)$, such that $g(0) = 0 = f(0)$. Assume that an elation τ_{u^*,t^*} of E leaves Δ invariant. Then

$$y = x \left(\begin{bmatrix} u + g(t) & f(t) \\ t & u \end{bmatrix} + \begin{bmatrix} u^* + \rho_1 t^* & \gamma_1 t^* \\ t^* & u^* \end{bmatrix} \right) \in \Delta.$$

This means that

$$\begin{aligned} g(t + t^*) &= g(t) + \rho_1 t^*, \\ f(t + t^*) &= f(t) + \gamma_1 t^*, \end{aligned}$$

for all $t \in GF(q)$. Taking $t = 0$, our assumption that Δ shares exactly R_0 with Σ_1, implies that $t^* = 0$. Hence, there are precisely q mutually distinct images of Γ_i for $i = 2, 3$. Now assume that $\Gamma_3 E$ consists of q mutually disjoint (on lines) spreads. This means that there are exactly $(q^2 + 1)q$ lines of $\Gamma_3 E$ and each line is in exactly one spread $\Gamma_3 e$, for $e \in E$. There are also the same $(q^2 + 1)q$ lines of $\Gamma_2 E$ if $\Gamma_2 E = \Gamma_3 E$. Suppose that one of these lines ℓ lie in two distinct spreads $\Gamma_2 e_1$ and $\Gamma_2 e_2$, for $e_i \in E$, $i = 1, 2$.

Now ℓ lies in a unique spread $\Gamma_3 e_3$ for $e_3 \in E$. Since we are dealing with orbits under E, we may assume that $e_3 = 1$. Now $\ell \in \Gamma_2 e_1$ and $\Gamma_2 e_2$ means that $\ell \in \Gamma_2 e_1 (e_1^{-1} e_2)$ or rather that $\Gamma_3 (e_1^{-1} e_2)$ and Γ_3 both contain ℓ. Since $\Gamma_3 E$ consists of q mutually disjoint spreads, then $e_1^{-1} e \in E^-$. Let $e_1^{-1} e_2 = \alpha \in E^-$. Then $\Gamma_2 e_2 = \Gamma_2 e_1 \alpha = \Gamma_2 e_1$, since E is elementary Abelian and E^- leaves Γ_2 invariant. Hence, $\Gamma_2 e_1 = \Gamma_2 e_2$. This completes the proof. $\qquad\square$

DEFINITION 119. *If Γ_2 and Γ_3 are conical flock spreads (including the Desarguesian cases) that share exactly a regulus with an associated Desarguesian spread Σ_1 and E is an elation group of order q^2 of π_{Σ_1} that contains the regulus-inducing elation group of order q that defines the spreads Γ_i, for $i = 2, 3$, assume that at least one of $\Gamma_2 E$ and $\Gamma_3 E$ consist of q mutually disjoint spreads.*

If $\Gamma_2 E = \Gamma_3 E$, we shall say that $\Gamma_3 E$ 'E-switches' with $\Gamma_2 E$ (the definition is symmetric, so we may say, more simply, that $\Gamma_2 E$ and $\Gamma_3 E$ are 'switches' of each other).

REMARK 60. *If $\Gamma_3 E$ switches with $\Gamma_2 E$ and h is a collineation of Σ_1 that leaves invariant the axis $x = 0$ of E, then $\Gamma_3 h E$ switches with $\Gamma_2 h E$.*

PROOF. The proof is left to the reader to complete as an exercise.

$\square$

In the following, we consider the idea of switching in a very general manner. We note in the following there are exactly $(q + 1)$ regulus-inducing elation groups of order q in an elation group of order q^2.

In the following subsections, we consider finite E-switching, finite switching, and coset switching of Desarguesian spreads. We also are able to establish certain bounds for the numbers of Desarguesian t-spreads.

1.1. Switching Theorem for Parallelisms.

THEOREM 226. *Let Σ_1 be a Desarguesian spread in $PG(3, q)$ and let E denote the full elation group of the associated Desarguesian affine plane π_{Σ_1} with axis $x = 0$. Fix a second component $y = 0$ and let R_i, for $i = 1, 2, .., q+1$ denote the set of reguli of Σ_1 that share $x = 0, y = 0$ (a 'hyperbolic cover' of reguli). Let ρ_i and τ_i for $i = 1, 2, .., q + 1$ be conical flock spreads that share precisely R_i with Σ_1. Let ρ_i^* and τ_i^* denote the derived conical flock spreads obtained by the derivation of R_i. Assume that $\Sigma_1 \cup_{i=1}^{q+1} \rho_i^* E$ is a parallelism of $PG(3, q)$ and further assume that $\tau_i E = \rho_i E$, for each $i = 1, 2, .., q + 1$ (so necessarily $\tau_i E$ switches with $\rho_i E$).*

Then $\Sigma_1 \cup_{i=1}^{q+1} \tau_i^* E$ is a parallelism in $PG(3, q)$.

PROOF. By counting, we see that in

$$\Sigma_1 \cup_{i=1}^{q+1} \tau_i^* E,$$

there are $1 + q(q + 1)$ spreads. We need to show that these spreads are mutually disjoint. So take any line ℓ of $PG(3, q)$ that is not in Σ_1. This line lies in exactly one set $\rho_i^* E$. If ℓ is a Baer subplane (element of $R_i^* e$), then ℓ is also in $\gamma_i^* E$. So, assume that ℓ is not in one of the reguli of Σ_1 that share $x = 0$. Then ℓ is in some $\rho_i E$. Since $\tau_i E = \rho_i E$, it follows that every line of $PG(3, q)$ is in $\Sigma_1 \cup_{i=1}^{q+1} \tau_i^* E$, which completes the proof.

$\square$

2. The Switches of Desarguesian Spreads

In this section, we determine the possible switches $\Gamma_3 E$ of $\Gamma_2 E$, where Γ_2 is Desarguesian. The quite amazing result is that Γ_3 is either Desarguesian or derived Kantor-Knuth, which relies on the beautiful result of Thas that a flock of a quadratic cone whose planes share a point is either Kantor-Knuth of odd order, or Desarguesian (see Thas [**180**]).

Let

$$\Sigma_1 = \left\{ x = 0, y = x \begin{bmatrix} u + \rho_1 t & \gamma_1 t \\ t & u \end{bmatrix} ; u, t \in GF(q) \right\},$$

and

$$\Sigma_2 = \left\{ x = 0, y = x \begin{bmatrix} u + \rho_2 t & \gamma_2 t \\ t & u \end{bmatrix} ; u, t \in GF(q) \right\},$$

such that

$$(*) : \gamma_1 \neq \gamma_2 \text{ or if } \gamma_1 = \gamma_2 \text{ then } \rho_1 \neq \rho_2.$$

Then Σ_2 and Σ_1 are Desarguesian spreads that share precisely the regulus

$$R_0 = \left\{ x = 0, y = x \begin{bmatrix} u & 0 \\ 0 & u \end{bmatrix} ; u \in GF(q) \right\}.$$

Let

$$E = \left\langle \tau_{u,t} = \begin{bmatrix} 1 & 0 & u + \rho_1 t & \gamma_1 t \\ 0 & 1 & t & u \\ 0 & 0 & 1 & 0 \\ 0 & 0 & 0 & 1 \end{bmatrix} ; u, t \in GF(q) \right\rangle.$$

We know by Lemma 57 that $\Sigma_2 E$ consists of exactly q spreads and by the Johnson parallelisms, we know that they are mutually disjoint. We consider conditions that ensure that a conical flock spread Γ is such that ΓE switches with $\Sigma_2 E$.

THEOREM 227. *Assume the conditions and notation listed above. Let Γ be a conical flock spread that shares exactly R_0 with Σ_1. Specifically, let*

$$\Gamma = \left\{ x = 0, y = x \begin{bmatrix} u + g(t) & f(t) \\ t & u \end{bmatrix} ; u, t \in GF(q) \right\},$$

where for all non-zero t, $(g(t), f(t)) \neq (\rho_1 t, \gamma_1 t)$.

Then ΓE switches with $\Sigma_2 E$ if and only if one of the following cases occurs:

(i) $\gamma_2 \neq \gamma_1$ *and, we have*

$$g(t) = \rho_1 t + (\rho_2 - \rho_1)\frac{f(t) - \gamma_1 t}{\gamma_2 - \gamma_1}, \quad or$$

(ii) $\gamma_2 = \gamma_1$ *and* $f(t) = \gamma_1 t$

PROOF. We know that $\Sigma_2 E$ consists of q mutually disjoint spreads. Consider the following condition $(**)$, for each $t, u \in GF(q)$.

$$(**) \quad : \quad \begin{bmatrix} u + g(t) & f(t) \\ t & u \end{bmatrix}$$

$$= \begin{bmatrix} u + \rho_2 s & \gamma_2 s \\ s & u \end{bmatrix} + \begin{bmatrix} \rho_1(t-s) & \gamma_1(t-s) \\ t-s & 0 \end{bmatrix},$$

for s not zero, if t not zero, where $t - s$ not zero if t is not zero. Hence, in case $(**)$, we obtain

$$g(t) = \rho_1(t-s) + \rho_2 s$$

and

$$f(t) = \gamma_2 s + \gamma_1(t-s).$$

Note that these equations must be true for all t. First, assume that $\gamma_1 \neq \gamma_2$.

$f(t) - \gamma_1 t = 0$ if and only if $(\gamma_2 - \gamma_1)s = 0$, so that $s = 0$, but this situation cannot occur by assumption. Hence, $f(t) - \gamma_1 t$ is never zero for t non-zero. Thus,

$$s = \frac{f(t) - \gamma_1 t}{\gamma_2 - \gamma_1}.$$

Hence,

$$g(t) = \rho_1 t + (\rho_2 - \rho_1)\frac{f(t) - \gamma_1 t}{\gamma_2 - \gamma_1}.$$

The completes the proof of the theorem. $\square$

THEOREM 228. *With the same notation as above, if ΓE switches with $\Sigma_2 E$ then the plane π_Γ arising from the spread Γ is either Desarguesian or a Kantor-Knuth semifield plane. So, in particular, if q is even then Γ is a Desarguesian spread.*

PROOF. If we consider the corresponding flock of the quadratic cone with planes

$$\pi_t : \quad x_0 t - x_1 f(t) + x_2 g(t) + x_3 = 0.$$

From the previous theorem case (i), we have

$$g(t) = \rho_1 t + (\rho_2 - \rho_1)\frac{f(t) - \gamma_1 t}{\gamma_2 - \gamma_1}.$$

Then

$$g(t) = kt + zf(t),$$

for k and z constants in $GF(q)$. Therefore, we obtain:

$$\pi_t : \quad x_0 t - x_1 f(t) + x_2(kt + zf(t)) + x_3 = 0,$$

and we see that the planes contain the point $(-k, z, 1, 0)$, where the points of $PG(3, q)$ are taken homogeneously as (x_0, x_1, x_2, x_3).

In case (ii), when $\gamma_1 = \gamma_2$, we have the planes

$$\pi_t : \ x_0 t - x_1 \gamma_1 t + x_2 g(t) + x_3 = 0,$$

and the planes contain the point $(\gamma_1, 1, 0, 0)$.

Hence, in either case, the planes contain a common point. This means that $f(t) = \alpha t + \beta t^\sigma$, where σ is an automorphism by the theorem of Thas mentioned previously. This is equivalent, by coordinate change, to saying that Γ is a Kantor-Knuth conical flock spread. This completes the proof of the theorem. $\qquad \square$

2.1. General Switches. Now if one would try the same sort of game by trying to switch a (derived) conical flock spread, absolutely nothing is known.

A similar argument establishes the following switching result.

THEOREM 229. *Let*

$$\Gamma_i = \left\{ x = 0, y = x \begin{bmatrix} u + g_i(t) & f_i(t) \\ t & u \end{bmatrix} ; u, t \in GF(q) \right\}$$

be conical flock spreads, each of which share exactly R_0 with Σ_1. Assume $\Gamma_2 E$ is a set of q mutually disjoint spreads. Then $\Gamma_3 E$ switches with $\Gamma_2 E$ if and only if for each $t \in GF(q)$, there exists a bijective function $m : GF(q) \to GF(q)$ such that

$$\begin{aligned} g_3(t) - \rho_1 t &= g_2(m(t)) - \rho_1 m(t), \\ f_3(t) - \gamma_1 t &= f_2(m(t)) - \gamma_1 m(t). \end{aligned}$$

The reader is directed to Chapter 40 for the associated open problem.

3. Coset Switching again with Desarguesian Spreads

Now we know that switching with Desarguesian spreads can be accomplished only by Desarguesian or Kantor-Knuth spreads. However, not all Kantor-Knuth spreads are isomorphic, as it depends on the automorphism used in the spread. Also, even though all Desarguesian spreads are isomorphic, the form used will be important in the determination of various parallelisms and their isomorphisms.

Let σ_i be an automorphism of $GF(q)$. Let Σ_1 denote the Desarguesian spread:

$$\Sigma_1 = \left\{ x = 0, y = x \begin{bmatrix} u + \rho_1 t & \gamma_1 t \\ t & u \end{bmatrix} ; u, t \in GF(q) \right\}.$$

We shall be employing E-switches of Desarguesian spreads by spreads $\Sigma_i^{\alpha_i, \beta_i, \gamma_i, \sigma_i}$, where

$$\Sigma_i^{\alpha_i, \beta_i, \gamma_i, \sigma_i}$$

$$= \left\{ y = x \begin{bmatrix} x = 0, \\ u + g(t) = \rho_1 t + (\rho_2 - \rho_1)\frac{\gamma_i t^{\sigma_i} - \gamma_1 t}{\gamma_2 - \gamma_1} & \gamma_i t^{\sigma_i} \\ t & u \end{bmatrix} \right\},$$
$$; u, t \in GF(q)$$

Assume that

$$\gamma_j t^{\sigma_j} = \gamma_1 t, \text{ implies that } t = 0.$$

Also, we again note that when q is even, it can only be that $\sigma_j = 1$ since the spread is a semifield spread (see, e.g., Johnson [130]). Let F_{Σ_1} refer to the field coordinatizing Σ_1, namely

$$\left\langle \begin{bmatrix} u + \rho_1 t & \gamma_1 t \\ t & u \end{bmatrix}; u, t \in GF(q) \right\rangle.$$

Now we connect this with minimal Parallelisms.

We have noted that parallelisms may be obtained by choosing a Desarguesian Σ_2 sharing exactly a regulus R_0 with Σ_1 and using a group EH so that $\Sigma_1 \cup \Sigma_2^* EH$ becomes a parallelism, where Σ_2^* is the Hall spread obtained by derivation of R_0. We now detail at least some of the the particular groups that may be used. Recall that $(q-1)_2$ means the order of the 2-power that divides $q - 1$.

We recall the following groups:

$$H^j = \left\langle \begin{bmatrix} \omega^j & 0 \\ 0 & \omega \end{bmatrix}; \\ \omega \in F_{\Sigma_1} \backslash \{0\}; |\omega| = (q-1)_2(q+1) \text{ and } (j-1, q+1) = 1 \right\rangle$$

and

$$H_j = \left\langle \begin{bmatrix} \omega & 0 \\ 0 & \omega^j \end{bmatrix}; \\ \omega \in F_{\Sigma_1} \backslash \{0\}; |\omega| = (q-1)_2(q+1) \text{ and } (j-1, q+1) = 1 \right\rangle.$$

Let Σ_2 denote any Desarguesian plane

$$\Sigma_2 = \left\{ x = 0, y = x \begin{bmatrix} u + \rho_2 & \gamma_2 t \\ t & u \end{bmatrix}; u, t \in GF(q) \right\},$$

where q is odd, γ_2 is a non-square of $GF(q)$, not equal to γ_1 and $\rho_1 = \rho_2 = 0$. It turns out that we may now use $H_j E$ or $H^k E$ together with Σ_1 and Σ_2 to construct parallelisms in $PG(3, q)$.

THEOREM 230. (see Johnson and Pomareda [**151**] Theorems 9, 13, 15, 16) We now assume that q is odd and $\rho_1 = \rho_2 = 0$, and Σ_1 and Σ_2 are Desarguesian spreads. Let Σ_2^* denote the Hall plane obtained by the derivation Σ_2 by

$$R_0 = \left\{ x = 0, y = x \begin{bmatrix} u & 0 \\ 0 & u \end{bmatrix} ; u \in GF(q) \right\},$$

(1) If $(q-1)/2$ is odd, then

$$\mathcal{P}_j = \Sigma_1 \cup \Sigma_2^* H_j E$$

and

$$\mathcal{P}^k = \Sigma_1 \cup \Sigma_2^* H^k E$$

are parallelisms.

(2) If $(q-1)/2$ is even and $(j+1, q+1) = (k+1, q+1) = 1$, then $\mathcal{P}_j$ and $\mathcal{P}^k$ (as in (1)) are parallelisms.

(3) If q is not 3 or 7, then $\mathcal{P}_j$ is never isomorphic to $\mathcal{P}^k$.

(4) If q is odd and if $\mathcal{P}_j$ is isomorphic to $\mathcal{P}_k$, then $j \equiv k \bmod (q+1)/2$.

To see how this works, we offer some explanations. Let HE be one of the groups $H_j E$ or $H^k E$, subject to the restrictions given in the definitions of the groups. In order to apply Theorem 226, we write the parallelism $\Sigma_1 \cup_{i=1}^{q+1} \rho_i^* E$ as follows: Let $\Sigma_2^* H = \cup_{i=1}^{q+1} \rho_i^*$. We note that H has order $(q-1)_2(q+1)$, and the scalar group of order $(q-1)_2$ fixes Σ_2 and fixes the regulus net R_1. Hence, the number of distinct images of Σ_2^* under H is $q+1$. Now we want to choose a set of switches $\tau_i E$ for $\rho_i E$, for $i = 1, 2, .., q+1$. Note that there are exactly $q+1$ reguli R_i, $i = 1, 2, .., q+1$ of Σ_1 that mutually share two components $x = 0, y = 0$ and that the ρ_i, as defined, are Desarguesian spreads that contain R_i, for $i = 1, 2, .., q+1$ and are, of course, the set of images of Σ_2 under H. It follows that H is transitive on the reguli that share $x = 0, y = 0$. For each R_i, let h_i denote an element of H that maps R_1 to R_i and hence maps Σ_2 to $\Sigma_2 h_i$ and therefore maps Σ_2^* to $\Sigma_2^* h_i$.

We note that E must be normalized by H, since it is the full elation group of Σ_1 with axis $x = 0$. Therefore, if ΓE switches with $\Sigma_2 E$ then $\Gamma h E$ switches with $\Sigma_2 h E$, since $\Gamma h E = \Gamma E h = \Sigma_2 E h$ (see Remark 60).

Our main theorems concerning coset switching depend somewhat on whether q is odd or even due to the fact that Kantor-Knuth spreads can exist in the odd order case. In the odd order case, we also may some simplifications on the representations of the Desarguesian spreads.

3.1. Main Theorem on Coset Switching for Odd Order.

THEOREM 231. *Let*

$$\Sigma_1 = \left\{ x = 0, y = x \begin{bmatrix} u & \gamma_1 t \\ t & u \end{bmatrix} ; u, t \in GF(q) \right\},$$

$$\Sigma_2 = \left\{ x = 0, y = x \begin{bmatrix} u & \gamma_2 t \\ t & u \end{bmatrix} ; u, t \in GF(q) \right\},$$

where γ_1 and γ_2 are distinct non-squares, and let H any of the groups H^j or H_k. Let $\{h_i; i = 1, 2, .., q + 1\}$ be any coset representation set for H_{Σ_2}. For each h_i, choose either the spread Σ_2 or

$$\rho_i = \left\{ x = 0, y = x \begin{bmatrix} u & \eta_i t^{\sigma_i} \\ t & u \end{bmatrix} ; u, t \in GF(q) \right\},$$

where $\eta_i t^{\sigma_i} = \gamma_1 t$ implies $t = 0$ and σ_i is an automorphism of $GF(q)$, possibly 1.

(1) *Then $\rho_i h_i E$ switches with $\Sigma_2 h_i E$, for $i = 1, 2, .., q+1$ (the switch is trivial if $\rho_i = \Sigma_2$).*

(2) *Let ρ_i^* denote the spread derived from ρ_i by the replacement of R_1. Then*

$$\Sigma_1 \cup_{i=1}^{q+1} \rho_i^* h_i E$$

is a parallelism.

PROOF. Apply the previous analysis to Theorem 227. □

4. An Upper Bound

We have given a method by which there seem to be a tremendous number of new parallelisms constructed. In the odd order case, we would expect considerably more because we may use the Kantor-Knuth type spreads. Here we give a rough upper bound on the number of parallelisms obtained by our methods. Some of these parallelisms constructed will admit collineation groups HE that act transitively on the spreads not equal to Σ_1 of the parallelism and others with admit only the group E. We have mentioned Theorem 230, where all switches are trivial and note that there still are a large number of mutually non-isomorphic parallelisms, now depending solely on the group H chosen. If we choose another Desarguesian spread Σ_3 and switch $\Sigma_3 E$ with $\Sigma_2 E$, for all of the $q + 1$ switches, we also obtain HE as a group, but this parallelism may not be isomorphic to the one using Σ_2 as an isomorphism, from one to the other necessarily is a collineation of Σ_1. In this section, we basically are considering q odd.

REMARK 61. $\eta_i t^{\sigma_i} = \gamma_1 t$ implies $t = 0$ is the condition required. Let $\eta_i = \gamma_1 \tau_i$, where τ_i is a square in $GF(q)$. Let ω denote a primitive element of $GF(q)^*$ so the squares of $GF(q)$ are in $\langle \omega^2 \rangle$. Let $p^r = q$. We wish to choose τ_i not in $\left\langle \omega^{\frac{p^r-1}{p^{(z,r)}-1}} \right\rangle$. Thus, any element of $\langle \omega^2 \rangle - \left\langle \omega^{\frac{p^r-1}{p^{(z,r)}-1}} \right\rangle$, will suffice. Hence, there are $(q-1)/2 - (q-1)/(p^d-1)$ choices, where $d = (z,r)$. So, there are $(q-1)(p^d-1)-2)/2(p^d-1) =$

$$\frac{(q-1)}{2}\frac{(p^d-3)}{(p^d-1)}$$

possible choices for η_i where $Fix\sigma_i = GF(p^{d_i})$. Let r_{d_i} denote the number of distinct σ_i such that $Fix\sigma_i = GF(p^{d_i})$.

Therefore, the number of choices for each E-switch is

$$\sum_{d|r} \frac{(q-1)}{2}\frac{(p^d-3)}{(p^d-1)} r_d.$$

(1) Hence, the number of parallelisms constructible is this manner, from a given group H, and from a given coset representation is

$$\left(\sum_{d|r} \frac{(q-1)}{2}\frac{(p^d-3)}{(p^d-1)} r_d \right)^{q+1}.$$

(2) Let $\varphi(q+1)$ be the number of integers relatively prime to $q+1$. Then there are at least $2\varphi(q+1)$ groups H^j and H_k. Let H_{Σ_2} denote the stabilizer of Σ_2 in H. Note that it is not required that ρ_i^* admit any element of H_{Σ_2} as a collineation group to produce a parallelism. Since H is cyclic, let h generate H and h^j generate H_{Σ_2}. For h^z, for z between 1 and j, let r_z be any element of H_{Σ_2}. If $\{h^z r_z; z = 1, 2, .., j\}$ is a set of coset representatives then

$$\Sigma_1 \cup_{i=1}^{q+1} \rho_i^* h^i r_z E$$

is a parallelism. Therefore, there are at least

$$2\varphi(q+1) \left(\sum_{d|r} \frac{(q-1)}{2}\frac{(p^d-3)}{(p^d-1)} r_d \right)^{q+1}$$

distinct parallelisms constructible by E-switching (note also, we have not counted what differences might occur by varying the set of coset representatives).

PROOF. Choose a group, choose a set of E-switches. $\square$

REMARK 62. *The group $H_{\Sigma_2^*}$ can change the particular spreads that are used, but we do not make a separate count for these, as the changes are already previously considered in the sets of E-switches.*

4.1. Coset Switching for Even Order. We mentioned previously that the switches for $q = 2^r$, required r to be odd. Actually, the more general case is also valid, so we assume only that q is even in this subsection.

THEOREM 232. *Our notation for Σ_1 and Σ_2 is still assumed. Let Γ be a Desarguesian conical flock spread that shares exactly R_0 with Σ_1. Specifically, let*

$$\Gamma = \left\{ x = 0, y = x \begin{bmatrix} u + \alpha t & \beta t \\ t & u \end{bmatrix} ; u, t \in GF(q) \right\},$$

where one of the following conditions hold:
    (i) either $\beta \neq \gamma_1$ or
    (ii) or $\beta = \gamma_1 = \gamma_2$ and $\alpha \neq \rho_1$.
Then ΓE switches with $\Sigma_2 E$ if and only if
    $(i)'$: in case (i), we have

$$\alpha t = \rho_1 t + (\rho_2 - \rho_1) \frac{(\beta - \gamma_1)t}{\gamma_2 - \gamma_1},$$

    $(ii)'$: in case (ii), we have

$$\beta t = \gamma_1 t = \gamma_2 t.$$

4.2. Special Case. Although not as general, things become more transparent if we take $\gamma_1 = \gamma_2 = 1 = \beta$.
    Then for any spread of the form

$$\Gamma_\alpha = \left\{ x = 0, y = x \begin{bmatrix} u + \alpha t & t \\ t & u \end{bmatrix} ; u, t \in GF(q) \right\},$$

then $\Gamma_\alpha E$ switches with $\Sigma_2 E$, for α and ρ_2 not equal to ρ_1.

THEOREM 233. *Let q be even and Σ_1 be the Desarguesian spread*

$$\Sigma_1 = \left\{ x = 0, y = x \begin{bmatrix} u + \rho_1 t & t \\ t & u \end{bmatrix} ; u, t \in GF(q) \right\}.$$

Let Σ_2 be the Desarguesian spread

$$\Sigma_2 = \left\{ x = 0, y = x \begin{bmatrix} u + \rho_2 t & t \\ t & u \end{bmatrix} ; u, t \in GF(q) \right\}, \quad \rho_2 \neq \rho_1.$$

For a given group HE, let $\{h_i; i = 1, 2, .., q+1\}$ be a coset representation set for $H_{\Sigma_2^*}$. Assume that $\Sigma_1 \cup_{i=1}^{q+1} \Sigma_2^* h_i E$ is a parallelism, where Σ_2^* denotes the Hall spread derived from the spread Σ_2. Let Σ_{α_i} denote a Desarguesian spread

$$\Sigma_{\alpha_i} = \left\{ x = 0, y = x \begin{bmatrix} u + \alpha_i t & t \\ t & u \end{bmatrix} ; u, t \in GF(q) \right\}.$$

Then $\Sigma_{\alpha_i} h_i E$ switches with $\Sigma_2 h_i E$ so that we obtain a set of parallelisms

$$\Sigma_1 \cup_{i=1}^{q+1} \Sigma_{\alpha_i}^* h_i E.$$

5. Upper Bound, Even Order

We first note that at least for $H^0 E$ or $H_0 E$, we obtain a parallelism-inducing group, if q is even.

THEOREM 234. *For q even, consider the Desarguesian spread*

$$\Sigma_1 = \left\langle \begin{bmatrix} u + \rho_1 t & t \\ t & u \end{bmatrix} ; u, t \in GF(q) \right\rangle.$$

Then if we take a Desarguesian spread

$$\Sigma_2 = \left\langle \begin{bmatrix} u + \rho_2 t & t \\ t & u \end{bmatrix} ; u, t \in GF(q) \right\rangle, \text{ for } \rho_2 \neq \rho_1$$

and use the group $H^0 E$ or $H_0 E$, say, HE, then

$$\Sigma_1 \cup \Sigma_2^* HE,$$

is a parallelism.

PROOF. Assume first one considers the proof for the group H_0. The proof for H^0 will then follow by taking a duality of the projective space. This duality will change H_0 into H^0 by retaining E and an associated parallelism will be constructed. So, basically, we need only show that we obtain a mutually disjoint set of spreads. We leave the rest of the proof as an exercise for the reader, as it used ideas quite similar to those considered in the odd order cases. □

This will prove that following bound on the possible number of parallelisms produced by coset replacement in the even order case.

REMARK 63. *Now we note that there are $q/2 - 1$ possible elements α not equal to ρ_1 that produce Desarguesian spreads Γ such that ΓE switches with $\Sigma_2 E$. Hence, there are*

$$(q/2 - 1)^{q+1}$$

possible parallelisms constructed for each group H_0 and H^0. Therefore, by E-switching, we have constructed at least

$$2(q/2 - 1)^{q+1}$$

possible parallelisms.

5.1. The Number of Non-isomorphic Spreads within a Parallelism. In previous works on parallelisms, the number of mutually non-isomorphic spreads within a given parallelism is quite small; one if the spread is transitive, two if there is a transitive deficiency one partial parallelism. Here, we note that not all Kantor-Knuth spreads are isomorphic; in fact there are $[r/2]$ different isomorphism types. So, there could a Desarguesian spread, $[r/2]$ mutually non-isomorphic derived Kantor-Knuth spreads for $q = p^r$, p a prime, and a Hall spread. Therefore, this method produces parallelisms with possibly $2 + [r/2]$ mutually distinct non-isomorphic spreads, in the odd order case.

Choose any parallelism $\Sigma_1 \cup_{i=1}^{q+1} \rho_i^* h^i r_{z_i} E$ and choose any $\rho_i^* h^i r_{z_i} e_i$. This spread contains an opposite regulus D_i^* to a regulus D_i of Σ_1 sharing $x = 0$. Now we use D_i^* in Σ_1 and derive this spread producing $\Sigma_1^{D_i^*}$ and un-derive D_i^* (i.e., 'derive') to construct back D_i and the spread $\rho_i h^i r_{z_i} e$. Now the spread $\rho_i^* h^i r_{z_i}$ has exactly q images under E. Then

$$\mathcal{P}_{\rho_i^* h^i r_{z_i} e_i} = \Sigma_1^{D_i^*} \cup_{j=1, j \neq i}^{q+1} \rho_j^* h^j r_{z_j} E \cup \rho_i h^i r_{z_i} e \cup \left\{ \rho_i^* h^i r_{z_i} E - \rho_i^* h^i r_{z_i} e \right\}$$

is a parallelism, a 'derived-underive parallelism.'

If HE is a collineation group of the parallelism then the possible $q(q + 1)$ parallelisms $\mathcal{P}_{\rho_i^* h^i r_{z_i} e_i}$ are all isomorphic (see Diaz, Johnson, Montinaro [**46**]). However, if only E is a collineation group of the parallelism, there will be $q + 1$ mutually non-isomorphic parallelisms obtained in this way.

In this setting, we could obtain a Hall plane, a Kantor-Knuth semifield plane and $[r/2] - 1$ mutually non-isomorphic derived Kantor-Knuth planes.

5.2. Remarks on the Isomorphisms of Parallelisms. We have given a construction technique for a wide variety of parallelisms. What we have not done completely is to determine the collineation group of a parallelism $\Sigma_1 \cup_{i=1}^{q+1} \rho_i^* h_i E$ or to determine when two of our parallelisms are isomorphic. We offer a few remarks on how one would go about accomplishing this.

Let σ be an isomorphism from one of the parallelisms to another. Since we always have an invariant Desarguesian spread Σ_1, it follows that σ must be a collineation group of Σ_1 (or rather the Desarguesian affine plane corresponding to Σ_1).

Certainly, any collineation g of the parallelism is a collineation of Σ_1, and since E is a group of the parallelism, we may assume that g leaves $x = 0$ and $y = 0$ invariant. Since g then permutes the set of reguli sharing $x = 0$ and $y = 0$, it follows that any collineation will permute the $\rho_i^* h_i$. If g maps R_1 onto R_j, then gh_j^{-1} leaves R_1 invariant and is therefore a collineation of ρ_1^* and of ρ_1. Furthermore, g will also permute the $\rho_i h_i$. Let $\rho_i h_i g = \rho_{g^*(i)} h_{g^*(i)}$. We note that, as a set, $\{h_i; i = 1, 2, .., q + 1\}$ is sharply transitive on the set of reguli R_i that share $x = 0$ and $y = 0$. Anyway, it is possible that the full collineation group could contain HE or be simply E. So, in this manner, it is possible to sift the constructed parallelisms into isomorphism classes.

The results presented in this part generalize two classes of parallelisms constructed in prior work.

Previously, we have discussed the construction transitive deficiency one partial parallelisms admitting a group HE such that there is one Desarguesian spread Σ_1 and the remaining spreads are derived Kantor-Knuth spreads, where all corresponding Kantor-Knuth spreads are images under HE of a spread as follows:

$$\Sigma_2 = \left\langle \begin{bmatrix} u & \gamma t^\sigma \\ t & u \end{bmatrix} ; u, t \in GF(q) \right\rangle, \text{ for } \gamma t^\sigma = \gamma_1 t, \text{ if and only if } t = 0,$$

such that $(q - 1)_2$ divides $(\sigma - 1)$. This restriction on division was required to ensure that the associated parallelism $\Sigma_1 \cup \Sigma_2^* HE$ does, in fact, admit the group HG. However, the results presented in this part show that one may bypass this restriction as follows. Let $\{h_i; i = 1, 2, .., q + 1\}$ be a subset of H that acts transitively on the set of reguli of Σ that share $x = 0, y = 0$. Then, it follows that

$$\Sigma_1 \cup \Sigma_2^* HE = \Sigma_1 \cup_{i=1}^{q+1} \Sigma_2^* h_i E.$$

Furthermore, it is now possible to switch any $\Sigma_2^* h_i$, with any Kantor-Knuth spread that shares exactly a regulus R_0 with the Desarguesian spread Σ_1. Now replace $\Sigma_2^* h_i$ by any derived Kantor-Knuth or Hall spread $\rho_i^* h_i$, where ρ_i^* shares exactly R_0 with Σ_1. Then we obtain another parallelism that may not admit HE as a collineation group. For example, if we choose $\rho_i^* = \rho_0^*$, for all i and ρ_0 does not admit HG_{Σ_2}, then we new parallelism that contains one Desarguesian and the remaining spreads isomorphic Kantor-Knuth spreads, but the parallelism does not arise from a transitive deficiency one partial parallelism.

We have also seen the related parallelisms that might be considered the predecessors of the parallelisms constructed here, which are called 'm-parallelisms,' and related classes of the so-called (m, n)-parallelisms. We mention only one type of result on m-parallelisms. We recall that m-parallelisms and n-parallelisms for $m \neq n$ are necessarily non-isomorphic. One nice feature of m-parallelisms is for such parallelisms the full central collineation group of Σ_1 that acts on the parallelisms is precisely EH^-.

As an exercise, the reader could show that the set of that m-parallelisms form a subclass of the class of E-switchable parallelisms.

CHAPTER 33

Parallelisms over Ordered Fields

Here we first construct a variety of parallelisms over the reals that are of the deficiency one type and then show how to use coset switching to construct many interesting parallelisms.

As mentioned above, the application of this construction technique has been applied most successfully when the spreads other than the Pappian spread are derived conical flock spreads and when the group contains a large normal subgroup that is a central collineation group. (By 'conical flock spreads,' we intend to mean those spreads that correspond to flocks of quadratic cones.) For example, using Theorems 205 and 206.

So, it might be asked if the above constructions can be considered over infinite fields? Here we begin with the consideration of the question when the field is the field of real numbers, and we are able to show that there are a vast number of parallelisms, depending on the class of strictly increasing functions f on the reals that define a class of conical flock spreads. We point out that our construction process constructs not only parallelisms but (proper) maximal partial parallelisms and actually forms the first known classes of such objects.

So, we work over the field of real numbers $K = \mathcal{R}$, but many of our arguments will work for arbitrary ordered fields and we take up a more general analysis later in this chapter.

We consider a Pappian spread Σ_1 defined as follows:

$$x = 0, y = x \begin{bmatrix} u & -t \\ t & u \end{bmatrix} \forall u, t \in \mathcal{R}.$$

We let Σ_2 be a spread in $PG(3, \mathcal{R})$, defined by a function f:

$$x = 0, y = x \begin{bmatrix} u & -f(t) \\ t & u \end{bmatrix} \forall u, t \in \mathcal{R},$$

where f is a function such that $f(t) = t$ implies that $t = 0$ and $f(0) = 0$.

507

Thus, if a spread exists then the two spreads Σ_1 and Σ_2 share exactly the regulus $\mathcal{D}$ with partial spread:

$$x = 0, y = x \begin{bmatrix} u & 0 \\ 0 & u \end{bmatrix} \forall u \in \mathcal{R}.$$

LEMMA 58. *Let f be any continuous strictly increasing function on the field of real numbers such that $\lim_{x \longmapsto \pm\infty} f(t) = \pm\infty$.*
(1) Then Σ_2 is a spread.
(2) Let $G^- = EH^-$, where H^- denotes the homology group of Σ_1 (or rather of the associated affine plane), whose elements are given by

$$\left\langle \begin{bmatrix} u & -t & 0 & 0 \\ t & u & 0 & 0 \\ 0 & 0 & 1 & 0 \\ 0 & 0 & 0 & 1 \end{bmatrix} ; u^2 + t^2 = 1 \right\rangle.$$

and where E denotes the full elation group with axis $x = 0$.
(2) Then G^- is transitive on the set of reguli of Σ_1 that share $x = 0$.

PROOF. Part (1) is given in the remarks of Chapter 1, just preceding Theorem 116.

We consider part (2). Since E is transitive on the components of Σ_1 not equal to $x = 0$, then first assume that there is a regulus $\mathcal{D}_1$ that shares exactly $x = 0$ with $\mathcal{D}$. Then there is an elation subgroup $E_{\mathcal{D}_1}$ that acts transitively on the components not equal $x = 0$. It follows easily that $E_{\mathcal{D}_1} = E_{\mathcal{D}}$ and this group induces a partition of the components of Σ_1 into a unique set of reguli that mutually share $x = 0$. (In this context, the set of 'elation-base' reguli determine a flock of a quadratic cone in $PG(3, \mathcal{R})$.)

Since $E_{\mathcal{D}}$ is a normal subgroup of E, these elation-base reguli are permuted transitively by E.

Now assume that a regulus $\mathcal{D}_2$ shares two components with $\mathcal{D}$, which we may take without loss of generality to be $x = 0$ and $y = 0$. Now there is a unique set of reguli sharing $x = 0$ and $y = 0$, which cover the components of Σ_1. These reguli have the property that there is a collineation group H_1 of the full central collineation group with axis $x = 0$ and coaxis $y = 0$ such that H_1 acts transitively on the non-fixed components of each regulus. (Here the set of 'homology-base' reguli determine a flock of a hyperbolic quadric in $PG(3, \mathcal{R})$.)

The group H_1 has the following form:

$$\left\langle \begin{bmatrix} u & 0 & 0 & 0 \\ 0 & u & 0 & 0 \\ 0 & 0 & 1 & 0 \\ 0 & 0 & 0 & 1 \end{bmatrix} ; u \in \mathcal{R}-\{0\} \right\rangle .$$

We note that this is the form for the group because of the form of $\mathcal{D}$.

Hence, $\mathcal{D}_2$ has the following basic form:

$$x = 0, y = x \begin{bmatrix} w & -s \\ s & w \end{bmatrix} vI_2 \ \forall \ v \neq 0 \in \mathcal{R},$$

where w and s are fixed elements of $\mathcal{R}$.

It remains to show that a determinant 1 homology maps $\mathcal{D}$ onto $\mathcal{D}_2$. We note that, since we are dealing with reguli, if $y = x$ of $\mathcal{D}$ maps to some component of $\mathcal{D}_2$ then $\mathcal{D}$ must map to $\mathcal{D}_2$. Hence, equivalently, for a given w, s do there exist elements u and t such that $u^2 + t^2 = 1$ and a non-zero element v of $\mathcal{R}$ such that

$$\begin{bmatrix} w & -s \\ s & w \end{bmatrix} v^{-1}I_2 = \begin{bmatrix} u & -t \\ t & u \end{bmatrix}?$$

The determinant 1 group determines a circle of radius 1 and center $(0,0)$ in the real 2-dimensional plane. Since $(uv)^2 + (tv)^2 = v^2$ determines a circle of radius v in the real 2-dimensional plane, then any affine point (w, s) lies on one of these circles. Considering that the mapping $(x, y) \longmapsto (xv, yv)$, for v fixed and non-zero is a bijective mapping, it follows that if (w, s) is on the circle of radius v and center $(0,0)$ then $w^2 + s^2 = v^2$ if and only if $(w/v, s/v)$ is a point on the circle of radius 1 and equal to some (u, t) such that $u^2 + t^2 = 1$.

Hence, we must have

$$\begin{bmatrix} w & -s \\ s & w \end{bmatrix} v^{-1}I_2 = \begin{bmatrix} u & -t \\ t & u \end{bmatrix}.$$

This completes the proof of part (2). ☐

THEOREM 235. *Under the above assumptions, of Lemma 58, assume also that f is symmetric with respect to the origin in the real Euclidean 2-space and $f(t_o + r) = f(t_o) + r$ for some t_o and r in the reals implies that $r = 0$.*

Then $\Sigma_1 \cup \Sigma_2^* g$ for all $g \in G^-$ and where Σ_2^* denotes the derived spread of Σ_2 by derivation of $\mathcal{D}$, is a partial parallelism $\mathcal{P}_f$ in $PG(3, \mathcal{R})$.

PROOF. It suffices to show that the set of spreads $\cup \Sigma_2 g$ for all $g \in G^-$ covers uniquely a line of $PG(3, \mathcal{R})$ that does not lie in Σ_1 and, which is disjoint from $x = 0$ provided it covers it.

Assume that $\Sigma_2 g$ and $\Sigma_2 h$ share a component. Then so do $\Sigma_2 gh^{-1}$ and Σ_2 share a component M. Let $gh^{-1} = k$ and represent k as follows:

$$
k = \begin{bmatrix} 1 & 0 & m & -r \\ 0 & 1 & r & m \\ 0 & 0 & 1 & 0 \\ 0 & 0 & 0 & 1 \end{bmatrix} \begin{bmatrix} w & -s & 0 & 0 \\ s & w & 0 & 0 \\ 0 & 0 & 1 & 0 \\ 0 & 0 & 0 & 1 \end{bmatrix}
$$

such that $w^2 + s^2 = 1$.

Let M be $y = x \begin{bmatrix} u^* & -f(t^*) \\ t^* & u^* \end{bmatrix}$, and let the preimage of k be

$y = x \begin{bmatrix} u & -f(t) \\ t & u \end{bmatrix}$.

Hence, we must have

$$
\begin{bmatrix} w & -s \\ s & w \end{bmatrix} \begin{bmatrix} u^* & -f(t^*) \\ t^* & u^* \end{bmatrix}
$$

$$
= \begin{bmatrix} u & -f(t) \\ t & u \end{bmatrix} + \begin{bmatrix} m & -r \\ r & m \end{bmatrix}.
$$

Equating the $(1, 1)$ and $(2, 2)$ entries, we must have:

$$
wu^* - st = wu^* - sf(t^*).
$$

However, by our conditions on f, we must have that $s = 0$. Since $w^2 + s^2 = 1$, this implies that $w = \pm 1$. Note that since f is symmetric with respect to the origin then $-f(-t) = f(t)$ for all t in the reals. This means that the homology $(x, y) \longmapsto (-x, y)$ is a collineation of Σ_2. Hence, we may assume that $w = 1$.

Equating the $(1, 2)$ and $(2, 1)$ entries above, we obtain:

$$
\begin{aligned}
-f(t^*) &= -f(t) - r = f(-t) - r \text{ and} \\
t^* &= t + r.
\end{aligned}
$$

Therefore, we have

$$
f(t + r) = -(f(-t) - r) = f(t) + r.
$$

By our assumed condition, this implies that $r = 0$.

In this case, we see that k is a collineation of Σ_2, which completes the proof to our theorem. $\qquad \square$

The reader might check as an exercise that the follows functions give rise to examples satisfying the hypotheses of the previous theorem.

EXAMPLE 7. *For examples of continuous, strictly increasing functions f such that $f(t) = t$ implies that $t = 0$ and $f(t_o + r) = f(t_o) + r$ for some t_o implies $r = 0$, which are also onto functions, we consider the following set of examples:*

(1) Let $f(t) = t + a^t - 1$ for $a > 1$ if $t \geq 0$ and let
$f(t) = t - a^{-t} + 1$ if $t < 0$.

Note that we are basically 'defining' the function so that $-f(-t) = f(t)$.

(2) $f(t) = t + a^{g(t)} - 1$ for t positive and $f(t) = t - a^{-g(t)} + 1$ for t negative, provided we have the following conditions: $g(0) = 0$, $g(t)$ is differentiable and $g'(t) > 0$.

Now a slight additional assumption produces a parallelism.

THEOREM 236. *The above construction produces a parallelism if and only if $f(t) - t$ is surjective.*

PROOF. We have the group E as a collineation group of the partial parallelism. Any line disjoint from $x = 0$ has the form $y = x \begin{bmatrix} a & b \\ c & d \end{bmatrix}$.

A typical element of E has the following form:

$$\tau_{m,r} = \begin{bmatrix} 1 & 0 & m & -r \\ 0 & 1 & r & m \\ 0 & 0 & 1 & 0 \\ 0 & 0 & 0 & 1 \end{bmatrix}.$$

Therefore, the given line is covered if and only if an image is covered. If we let $r = b$ and $m = -d$, we see that it suffices to consider lines with $b = d = 0$.

Hence, we consider $y = x \begin{bmatrix} a & 0 \\ c & 0 \end{bmatrix}$. First, assume that $a = 0$. Since $t - f(t)$ is onto, there is an element t_o such that $c = t_o - f(t_o)$. Apply $\tau_{0,-f(t_o)}$ to $y = x \begin{bmatrix} 0 & -f(t_o) \\ t_o & 0 \end{bmatrix}$ to obtain $y = x \begin{bmatrix} 0 & 0 \\ t_o - f(t_o) = c & 0 \end{bmatrix}$.

Conversely, if the element $y = x \begin{bmatrix} 0 & 0 \\ c & 0 \end{bmatrix}$ is covered then $t - f(t)$ is forced to be onto.

Now assume that $a \neq 0$.

Consider the element

$$\sigma_{w,s,m,r} = \begin{bmatrix} w & -s & m & -r \\ s & w & r & m \\ 0 & 0 & 1 & 0 \\ 0 & 0 & 0 & 1 \end{bmatrix},$$

where $w^2 + s^2 = 1$. We consider the image of $y = x \begin{bmatrix} 0 & -f(t) \\ t & 0 \end{bmatrix}$ under

$\sigma_{w,s,m,r}$. Such an image will cover $y = x \begin{bmatrix} a & 0 \\ c & 0 \end{bmatrix}$ if and only if

$$\begin{bmatrix} w & -s \\ s & w \end{bmatrix} \begin{bmatrix} a & 0 \\ c & 0 \end{bmatrix}$$

$$= \begin{bmatrix} wa - sc & 0 \\ sa + wc & 0 \end{bmatrix}$$

$$= \begin{bmatrix} m & -f(t) - r \\ t - r & m \end{bmatrix}.$$

Hence, we must have

$$f(t) = r, m = 0, wa = sc, sa + wc = t - f(t).$$

Hence,

$$t - f(t) = s(a^2 + c^2)/a$$

and since $w^2 + s^2 = 1$, we have $s^2(c^2 + a^2)/a^2$ so that

$$s = \pm a\sqrt{1/(c^2 + a^2)}.$$

Thus, the requirement is that $t - f(t) = \pm\sqrt{a^2 + c^2}$, which is so if and only if $t - f(t)$ is surjective. □

EXAMPLE 8. *To see examples of functions f such that $f(t) - t$ is not surjective, we note that the projection of $y = -\tan t$ onto the lines $y = x$ or $y = -x$ is surjective. Thus, rotate $y = -\tan t$ thorough $\pi/4$ to find a continuous function on $(0, \infty)$, which is bounded between $y = x$ and $y = x + \pi/\sqrt{2}$. The function is continuous and strictly increasing and is bijective. Furthermore, since $y = x + \pi/\sqrt{2}$ is an asymptote and the function is concave down when x is positive, it follows that $f(t + r) = f(t) + r$ if and only if $r = 0$.*

REMARK 64. *In this case, $|f(t) - t| \leq \pi/\sqrt{2}$. Therefore, a partial parallelism which is not a parallelism is constructed that has the property that for each regulus R of Σ_1 (the Pappian spread) containing a fixed line ℓ, the opposite regulus R^* is in a unique spread of the parallelism.*

We assert that when the function f produces a partial parallelism, it must be a maximal partial parallelism.

To see this, suppose there is an another spread ρ that is not in the constructed partial parallelism $\mathcal{P}$. We have noted that none of the lines of ρ can intersect $x = 0$, the axis of the central collineation group G^-, since we have a covering of such lines by $\mathcal{P}$. However, this means that we have a spread ρ that covers the points of our unique Pappian spread Σ without intersecting the axis $x = 0$, a contradiction. Therefore, the partial parallelism is a maximal partial parallelism.

THEOREM 237. *When the function f produces a partial parallelism $\mathcal{P}$ and $f(t) - t$ is not an onto function then $\mathcal{P}$ is a proper maximal partial parallelism.*

COROLLARY 58. *If $\mathcal{P}$ is a proper maximal partial parallelism, then so is any derived partial parallelism $\mathcal{P}^*$.*

PROOF. All lines that non-trivially intersect the Baer subplane π_o corresponding to the axis of the central collineation group must be covered. So, any spread extra to the parallelism $\mathcal{P}^*$ must have its lines such that they are all disjoint from π_o, a contradiction. $\square$

CHAPTER 34

General Elation Switching

In this chapter, the ideas of finite coset switching are generalized for arbitrary fields K.

DEFINITION 120. *Let Σ_0 denote a Pappian spread in $PG(3, K)$, where K is a field:*

$$\left\{ x = 0, y = x \begin{bmatrix} u & \gamma_1 t \\ t & u \end{bmatrix} \forall u, t \in K \right\}, \text{ where } \gamma_1 \text{ is a non-square in } K.$$

Let E denote the full elation group of Σ_0 with axis $x = 0$:

$$E = \left\langle \begin{bmatrix} 1 & 0 & u & \gamma_1 t \\ 0 & 1 & t & u \\ 0 & 0 & 1 & 0 \\ 0 & 0 & 0 & 1 \end{bmatrix} ; u, t \in K \right\rangle.$$

Let Σ_2 and Σ_3 be distinct spreads of $PG(3, K)$ that share exactly the regulus

$$R = \left\{ x = 0, y = x \begin{bmatrix} u & 0 \\ 0 & u \end{bmatrix} \forall u \in K \right\}.$$

Assume the following two conditions:

(i)

$$\Sigma_2 E = \Sigma_3 E,$$

and

(ii) a line ℓ of $\Sigma_2 E$ is in a unique spread of $\Sigma_2 E$ if and only if ℓ is in a unique spread of $\Sigma_3 E$.

If the spreads Σ_2 and Σ_3 have properties (i) and (ii), we shall say that $\Sigma_2 E$ and $\Sigma_3 E$ are 'E-switches' of each other (or that $\Sigma_2 E$ has been 'switched' by $\Sigma_3 E$).

We have seen the restrictions in the finite case that force Σ_3 to be either Desarguesian or Kantor-Knuth. But now in the infinite case, there are a wide variety of spreads that have the necessary requirements. We begin with a necessary and sufficient condition on the associated functions to have spreads of this form. The following theorem should be very familiar from Chapter 16, as there is an associated flock of a quadratic cone.

THEOREM 238. *Let K be any field. Then*

$$\Sigma_f = \left\{ x = 0, y = x \begin{bmatrix} u & f(t) \\ t & u \end{bmatrix} \forall u, t \in K \right\},$$

where f is a function $K \to K$ such that $f(0) = 0$ is a spread if and only if for each $z \in K$, ρ_z is bijective where,

$$\rho_z(t) = f(t) - z^2 t.$$

Now we shall be interested in spreads of the above form that share precisely a regulus with a Pappian spread

$$\left\{ x = 0, y = x \begin{bmatrix} u & \gamma_1 t \\ t & u \end{bmatrix} \forall u, t \in K \right\},$$

where γ_1 is a non-square in K.

and such that the full elation subgroup of E that acts as a collineation group of the spread in question is

(i)

$$E^- = \left\langle \begin{bmatrix} 1 & 0 & u & 0 \\ 0 & 1 & 0 & u \\ 0 & 0 & 1 & 0 \\ 0 & 0 & 0 & 1 \end{bmatrix} ; u \in K \right\rangle,$$

where

(ii) the full elation of Σ_1 with axis $x = 0$ is

$$E = \left\langle \begin{bmatrix} 1 & 0 & u & \gamma_1 t \\ 0 & 1 & t & u \\ 0 & 0 & 1 & 0 \\ 0 & 0 & 0 & 1 \end{bmatrix} ; u, t \in K \right\rangle,$$

and

(iii) such that for any elation $e \in E - E^-$ then $\Sigma_{fe} \cap \Sigma_f$ is empty.

The following proposition is essentially immediate and is left to the read to verify.

PROPOSITION 12. *A spread*

$$\Sigma_f = \left\{ x = 0, y = x \begin{bmatrix} u & f(t) \\ t & u \end{bmatrix} \forall u, t \in K \right\},$$

shares exactly the regulus

$$R_0 = \left\{ x = 0, y = x \begin{bmatrix} u & 0 \\ 0 & u \end{bmatrix} \forall u \in K \right\}$$

with the Pappian spread Σ_1 if and only if

$$f(t) - \gamma_1 t = 0$$

implies $t = 0$.

DEFINITION 121. *If the function* g *defined by* $g(t) = f(t) - \gamma_1 t$ *is injective, we shall say that the function* f *has the 'regulus' property.*

If the full elation group of a spread Σ_f *of* E *is* E^-, *and for* $e \in E - E^-$, *then* $\Sigma_f e \cap \Sigma_f$ *is empty, we shall say that the spread has the 'regulus-inducing property.'*

For example, if $f(t) = \gamma_2 t$ and $\gamma_2 \neq \gamma_1$ then f will turn out to have the regulus-inducing property. For the automorphism type function $f(t) = \gamma_2 t^\sigma$ we need $\gamma_2 t_0^\sigma = \gamma_1 t_0$ for some t_0 if and only if $t = 0$.

More generally, we have the following description of spreads that have the regulus-inducing property.

PROPOSITION 13. *A spread*

$$\Sigma_f = \left\{ x = 0, y = x \begin{bmatrix} u & f(t) \\ t & u \end{bmatrix} \forall u, t \in K \right\}$$

has the regulus-inducing property if and only if for $t_0, r \in K$

$$f(t_0 + r) = f(t_0) + \gamma_1 r$$

implies $r = 0$.

PROOF. Let $e = \begin{bmatrix} 1 & 0 & u_0 & \gamma_1 r \\ 0 & 1 & r & u_0 \\ 0 & 0 & 1 & 0 \\ 0 & 0 & 0 & 1 \end{bmatrix}$ map $y = x \begin{bmatrix} u & f(t_0) \\ t_0 & u \end{bmatrix}$ to

$$y = x \left(\begin{bmatrix} u & f(t_0) \\ t_0 & u \end{bmatrix} + \begin{bmatrix} u_0 & \gamma_1 r \\ r & u_0 \end{bmatrix} \right).$$

This line is back in the spread if and only if

$$f(t_0 + r) = f(t_0) + \gamma_1 r.$$

Hence, we wish this never to hold for non-zero values r, so we require that $r = 0$ in this case. □

PROPOSITION 14. *The regulus property implies the regulus-inducing property.*

PROOF. Now let $f(t) = \gamma_1 t + g(t)$, so we have, by assumption, that g is injective. Then consider the equation

$$(***) \quad : \quad f(t_0 + r) = \gamma_1(t_0 + r) + g(t_0 + r)$$
$$= f(t_0) + \gamma_1 r = \gamma_1 t_0 + g(t_0) + \gamma_1 r.$$

So, for equation $(\ast\ast\ast)$ to imply that $r = 0$ is equivalent to the condition that

$$g(t_0 + r) = g(t_0), \text{ implies } r = 0.$$

Since g is assumed to be injective, this condition is automatically satisfied. So, the regulus property implies the regulus-inducing property. $\square$

COROLLARY 59. *The set of lines*

$$\Sigma_f = \left\{ x = 0, y = x \begin{bmatrix} u & f(t) \\ t & u \end{bmatrix} \forall u, t \in K \right\},$$

is a spread admitting the regulus property, and regulus-inducing property if and only

(i) $z \in K$, ρ_z is bijective where,

$$\rho_z(t) = f(t) - z^2 t$$

(ii) the function g such that $g(t) = f(t) - \gamma_1 t$ is injective.

These are the spreads that we shall use in the switching procedure, except that we shall further require that g is bijective.

DEFINITION 122. *Any spread admitting the properties (i), (ii) of the previous corollary with the extra condition that the function g is bijective shall be said to admit the 'switching property.'*

0.3. Main Theorem on General Elation Switching.

THEOREM 239. *Let Σ_0 denote a Pappian spread in $PG(3, K)$, where K is a field:*

$$\left\{ x = 0, y = x \begin{bmatrix} u & \gamma_1 t \\ t & u \end{bmatrix} \forall u, t \in K \right\},$$

where γ_1 is a non-square in K.

Let E denote the full elation group of Σ_0 with axis $x = 0$:

$$E = \left\langle \begin{bmatrix} 1 & 0 & u & \gamma_1 t \\ 0 & 1 & t & u \\ 0 & 0 & 1 & 0 \\ 0 & 0 & 0 & 1 \end{bmatrix} ; u, t \in K \right\rangle.$$

Assume that Σ_i is a spread in $PG(3, K)$ of the following form:

$$\Sigma_i = \left\{ x = 0, y = x \begin{bmatrix} u & f_i(t) \\ t & u \end{bmatrix} \forall u, t \in K \right\}, \text{ for } i = 2, 3,$$

where f_i is a function $K \to K$ and such that both spreads admit the switching property.

Then $\Sigma_2 E$ switches with $\Sigma_3 E$.

PROOF. Note that $\Sigma_2 E = \Sigma_3 E$ if and only if Σ_2 is in $\Sigma_3 E$ and Σ_3 is in $\Sigma_2 E$. Note that

$$y = x \begin{bmatrix} u & f_i(t) \\ t & u \end{bmatrix}$$

maps to

$$y = x \left\{ \begin{bmatrix} u & f_i(t) \\ t & u \end{bmatrix} + \begin{bmatrix} u^* & \gamma_1 t^* \\ t^* & u^* \end{bmatrix} \right\}, \forall u^*, t^* \in K$$

by E and note that E fixes $x = 0$ pointwise.

So consider, for $j \neq i$:

$$(*) \quad : \quad \begin{bmatrix} u & f_i(t) \\ t & u \end{bmatrix}$$

$$= \begin{bmatrix} u & f_j(s) \\ s & u \end{bmatrix} + \begin{bmatrix} 0 & \gamma_1(t-s) \\ (t-s) & 0 \end{bmatrix}.$$

Then

$$f_i(t) = f_j(s) + \gamma_1(t-s)$$

if and only if

$$f_j(t) - \gamma_1 t = f_j(s) - \gamma_1 s.$$

Therefore, given t in K, then there exists a unique s in K such that

$$f_j(t) - \gamma_1 t = f_i(s) - \gamma_1 s,$$

since ϕ_j and ϕ_i are both bijective. Hence, given u and t, there is a solution to $(*)$. Note that the argument is symmetric. Now suppose for u and t, there is another solution

$$(**) : \begin{bmatrix} u & f_i(t) \\ t & u \end{bmatrix} = \begin{bmatrix} u^* & f_j(s) \\ k & u^* \end{bmatrix} + \begin{bmatrix} w & \gamma_1 d \\ d & w \end{bmatrix},$$

it now follows easily that there is a unique solution to $(**)$, namely, the unique solution to $(*)$.

Since the argument is symmetric, we have $\Sigma_2 E = \Sigma_3 E$. This establishes condition (i) of Definition 120.

Now take an element $y = x \begin{bmatrix} u_0 & f_i(t_0) \\ t_0 & u_0 \end{bmatrix}$ and assume that there is an element e in E such that the image of this element is back in Σ_i. Then

$$\begin{bmatrix} u_0 & f_i(t_0) \\ t_0 & u_0 \end{bmatrix} + \begin{bmatrix} w & \gamma_1 r \\ r & w \end{bmatrix} = \begin{bmatrix} u_0 + w & f_i(t_0) + \gamma_1 r \\ t_0 + r & u_0 \end{bmatrix}.$$

But, this means that $f_i(t_0 + r) = f_i(t_0) + \gamma_1 r$, so that $r = 0$. Then this element e leaves Σ_{f_i} invariant. Again, since the argument is symmetric, we have that $\Sigma_2 E$ and $\Sigma_3 E$ are unions of disjoint spreads. Therefore, we have that $\Sigma_2 E$ switches with $\Sigma_3 E$. This completes the proof of the theorem. □

1. Deficiency One Transitive Groups

Let K be an ordered field such that all positive elements have square roots in K.

For example, if L is a subfield of the field of real numbers, then the numbers L^C constructible from L by straight-edge and compass is such an ordered field.

So, let K be such an ordered field and let Σ_1 denote the Pappian spread

$$\Sigma_1 = \left\{ x = 0, y = x \begin{bmatrix} u & \gamma_1 t \\ t & u \end{bmatrix} \forall u, t \in K \right\},$$

where γ_1 is a fixed negative element in K.

Let

$$H = \left\langle \begin{bmatrix} 1 & 0 \\ 0 & w \end{bmatrix}; w = \begin{bmatrix} u & \gamma_1 t \\ t & u \end{bmatrix}; u^2 - \gamma_1 t^2 = 1 \right\rangle$$

and let

$$\Sigma_2 = \left\{ x = 0, y = x \begin{bmatrix} u & \gamma_2 t \\ t & u \end{bmatrix} \forall u, t \in K \right\},$$

where γ_2 is a negative element in K, $\gamma_2 \neq \gamma_1$. Let R_0 denote the common regulus

$$\left\{ x = 0, y = x \begin{bmatrix} u & 0 \\ 0 & u \end{bmatrix} \forall u \in K \right\}.$$

THEOREM 240. $\Sigma_1 \cup \Sigma_2^* EH$ is a parallelism in $PG(3, K)$.

Furthermore, $E^- \left\langle \begin{bmatrix} 1 & 0 & 0 & 0 \\ 0 & 1 & 0 & 0 \\ 0 & 0 & -1 & 0 \\ 0 & 0 & 0 & -1 \end{bmatrix} \right\rangle$ is the subgroup of EH that leaves Σ_2 invariant, where

$$E^- = \left\langle \begin{bmatrix} 1 & 0 & u & 0 \\ 0 & 1 & 0 & u \\ 0 & 0 & 1 & 0 \\ 0 & 0 & 0 & 1 \end{bmatrix}; u \in K \right\rangle.$$

PROOF. First, we claim that EH is transitive on the set of reguli of Σ_1 that share $x = 0$. We note that E is transitive on the components of $\Sigma_1 - \{x = 0\}$. So the question then is H transitive on the reguli that share $x = 0$ and $y = 0$? Any such regulus has the following form:

$$
R_t = \left\{ x = 0, y = 0, \ y = x \begin{bmatrix} u & \gamma_1 t \\ t & u \end{bmatrix} ; u \in K - \{0\} \right\}.
$$

So, the question becomes can R_0 be mapped into R_t by an element of H? Hence, given $y = x \begin{bmatrix} 0 & \gamma_1 t \\ t & 0 \end{bmatrix}$, we need to find an element $y = x \begin{bmatrix} v & 0 \\ 0 & v \end{bmatrix}$, v not zero, such that

$$
\begin{bmatrix} v & 0 \\ 0 & v \end{bmatrix} w = \begin{bmatrix} u & \gamma_1 t \\ t & u \end{bmatrix},
$$

for some $u \in K$.

Note that since clearly the reguli sharing $x = 0, y = 0$ are permuted by H, it just takes one appropriate image to establish that R_0 is mapped onto R_t, as any three distinct components generate a unique regulus of Σ_1. Since w commutes with $\begin{bmatrix} v & 0 \\ 0 & v \end{bmatrix}$, we only need to show that for some element $\begin{bmatrix} u & \gamma_1 t \\ t & u \end{bmatrix}$, for t not zero, there exist elements $\begin{bmatrix} v & 0 \\ 0 & v \end{bmatrix}$ and $\begin{bmatrix} u^* & \gamma_1 t^* \\ t^* & u^* \end{bmatrix}$, where $u^{*2} - \gamma_1 t^{*2} = 1$ such that

$$
\begin{bmatrix} v & 0 \\ 0 & v \end{bmatrix} \begin{bmatrix} u^* & \gamma_1 t^* \\ t^* & u^* \end{bmatrix} = \begin{bmatrix} u & \gamma_1 t \\ t & u \end{bmatrix}.
$$

Let $u^2 - \gamma_1 t^2 = z$. Since $z > 0$, let $v = \sqrt{z}$, which exists by assumption. Now let $u^* = u/v$ and $t^* = t/v$. Then $(vu^*, vt^*) = (u, t)$ and $u^* - \gamma_1 t^* = (u/v)^2 - \gamma_1 (t/v)^2 = 1$. This establishes the transitivity.

This means that if Σ_2^* denotes the derived spread then $\Sigma_2^* E H$ will contain all Baer subplanes of Σ_1 that non-trivially intersect $x = 0$. Furthermore, if a Baer subplane of R_0 maps back into a Baer subplane of R_0 under an element g of EH then g leaves R_0^* invariant and hence leaves R_0 invariant. The subgroup of EH that leaves R_0 invariant is $E^- H^-$, where

$$
E^- = \left\langle \begin{bmatrix} 1 & 0 & u & 0 \\ 0 & 1 & 0 & u \\ 0 & 0 & 1 & 0 \\ 0 & 0 & 0 & 1 \end{bmatrix} ; u \in K \right\rangle,
$$

and

$$H^- = \left\langle \begin{bmatrix} 1 & 0 & 0 & 0 \\ 0 & 1 & 0 & 0 \\ 0 & 0 & -1 & 0 \\ 0 & 0 & 0 & -1 \end{bmatrix} \right\rangle.$$

These elements are collineations of Σ_2 and of Σ_2^*. It then follows that there can be no line that non-trivially intersects $x = 0$ that is in two distinct spreads.

The remaining 'lines' of $PG(3, K)$, apart from the components of Σ_1, are the Baer subplanes of Σ_1 that do not intersect $x = 0$. These have the basic form

$$y = x \begin{bmatrix} a & b \\ c & d \end{bmatrix} ; a, b, c, d \in K,$$

where if $a = d$, then $b \neq \gamma_1 c$. That is, either $a \neq d$ or $b \neq \gamma_1 c$. Such a line will lie in a unique spread of $\Sigma_2^* EH$ if and only if it lies in a unique spread of $\Sigma_2 EH$. Therefore, there must be an element of Σ_2, $y = x \begin{bmatrix} u & \gamma_2 t \\ t & u \end{bmatrix}$ and an element $\rho \in EH$ such that

$$\left(y = x \begin{bmatrix} a & b \\ c & d \end{bmatrix} \right) \rho = \left(y = x \begin{bmatrix} u & \gamma_2 t \\ t & u \end{bmatrix} \right).$$

Notice that we may apply an elation that adds $\begin{bmatrix} -u & -\gamma_1 t \\ -t & -u \end{bmatrix}$, as

$$\begin{bmatrix} u & \gamma_2 t \\ t & u \end{bmatrix} + \begin{bmatrix} -u & -\gamma_1 t \\ -t & -u \end{bmatrix} = \begin{bmatrix} 0 & (\gamma_2 - \gamma_1)t \\ 0 & 0 \end{bmatrix},$$

This means that $y = x \begin{bmatrix} 0 & b \\ 0 & 0 \end{bmatrix}$, for any non-zero b does lie in a unique spread of $\Sigma_2 EH$. Suppose we have

$$y = x \begin{bmatrix} 0 & b \\ 0 & 0 \end{bmatrix} \begin{bmatrix} w & \gamma_1 s \\ s & w \end{bmatrix} ; w^2 - \gamma_1 s^2 = 1$$

$$= x \begin{bmatrix} bs & bw \\ 0 & 0 \end{bmatrix}.$$

Choose any z and e in K, at least one non-zero. If $z^2 - \gamma_1 e^2 = m = p^2$, then $(z/p)^2 - \gamma_1(e/p)^2 = 1$. Then letting $b = p$, $w = z/p$ and $s = e/p$, we see that we obtain

$$y = x \begin{bmatrix} e & z \\ 0 & 0 \end{bmatrix},$$

for any e, z in K, not both zero, is in a unique spread of $\Sigma_2 E H_0$. Then adding $\begin{bmatrix} d & \gamma_1 c \\ c & d \end{bmatrix}$ (applying an elation), we obtain

$$y = x \begin{bmatrix} e+d & z+\gamma_1 c \\ c & d \end{bmatrix}.$$

Let $e + d = a$ and $z + \gamma_1 c = b$. Note that a, b, c, d are arbitrary, except that if $a = d$, then $e = 0$ so that z is not zero and hence $b \neq \gamma_1 c$. This shows that every line of $PG(3, K)$ is contained in a spread of $\Sigma_2 EH$. To ensure that no line is in two spreads of $\Sigma_2 EH$, we need only check that no line of $\Sigma_2 - R$ is in two spreads or equivalently that if

$$\left(y = x \begin{bmatrix} u & \gamma_2 t \\ t & u \end{bmatrix} \right) = \left(y = x \begin{bmatrix} u^* & \gamma_2 t^* \\ t^* & u^* \end{bmatrix} \right) \rho$$

for t non-zero and $\rho \in EH$ then $\Sigma_2 \rho = \Sigma_2$. Therefore, the question is whether there exist matrices $\begin{bmatrix} r & \gamma_1 s \\ s & r \end{bmatrix}$ and $\begin{bmatrix} w & \gamma_1 k \\ k & w \end{bmatrix}$, such that $w^2 - \gamma_1 k^2 = 1$ and

$$\begin{bmatrix} u & \gamma_2 t \\ t & u \end{bmatrix} \begin{bmatrix} w & \gamma_1 k \\ k & w \end{bmatrix}$$
$$= \left(\begin{bmatrix} u^* & \gamma_2 t^* \\ t^* & u^* \end{bmatrix} + \begin{bmatrix} r & \gamma_1 s \\ s & r \end{bmatrix} \right).$$

The elements that we are considering are now elements of Σ_1. The left-hand side is

$$\begin{bmatrix} uw + \gamma_2 tk & \gamma_1 uk + \gamma_2 tw \\ tw + uk & \gamma_1 tk + uw \end{bmatrix}$$

and note that the matrix on the right hand-side has equal $(1, 1)$ and $(2, 2)$-entries. Since $\gamma_1 \neq \gamma_2$, then it follows that $tk = 0$. Therefore, $k = 0$ then $w = \pm 1$. Now the only possible elements of E that map one element of the regulus R_0 back into an element of R_0 requires that $s = 0$. But now the element in question leaves Σ_2-invariant. This completes the proof of the theorem. $\qquad \square$

THEOREM 241. *Given any spread Σ_f, which is switchable (satisfies the switching property $f(t) - \gamma_1 t$ bijective and $f(t) - z^2 t$ bijective for all z).*

If Σ_2 is Pappian, then $\Sigma_f E$ switches with $\Sigma_2 E$.

PROOF. We need only check that $\gamma_2 t - \gamma_1 t$ and $\gamma_2 t - z^2 t = -(-\gamma_2 + z^2)t$ define bijective functions, which is clear since $\gamma_2 \neq \gamma_1$ and $(-\gamma_2 + z^2) > 0$. $\qquad \square$

2. Switching Theorem over Ordered Fields

We recall that our previous parallelism construction used the group

EH and two Pappian spreads Σ_1 and Σ_2, where E^- $\left\langle \begin{bmatrix} 1 & 0 & 0 & 0 \\ 0 & 1 & 0 & 0 \\ 0 & 0 & -1 & 0 \\ 0 & 0 & 0 & -1 \end{bmatrix} \right\rangle$

is the subgroup that leaves Σ_2 invariant. We note that EH is transitive on the reguli of Σ_1 that share $x = 0$. Hence, H is transitive on the reguli of Σ_1 that share $x = 0, y = 0$. Let $\{h_i; i \in \lambda\}$ be a coset representation set for

$$\left\langle \begin{bmatrix} 1 & 0 & 0 & 0 \\ 0 & 1 & 0 & 0 \\ 0 & 0 & -1 & 0 \\ 0 & 0 & 0 & -1 \end{bmatrix} \right\rangle$$

in H. Using this coset representation set, we may now give our main theorem on elation switching.

THEOREM 242. *Let K be any ordered field such the positive elements all have square roots. Let H denote the homology group with axis $y = 0$ and coaxis $x = 0$ of determinant 1, and let $\{h_i; i \in \lambda\}$ be a coset representation set for*

$$\left\langle \begin{bmatrix} 1 & 0 & 0 & 0 \\ 0 & 1 & 0 & 0 \\ 0 & 0 & -1 & 0 \\ 0 & 0 & 0 & -1 \end{bmatrix} \right\rangle$$

in H. For each $i \in \lambda$, choose any function f_i such that the functions $\rho_{i,z}$ and ϕ_i are bijective for each $z \in K$, where $f_i(0) = 0$, $\rho_{i,z}(t) = f_i(t) - z^2 t$, and $\phi_i(t) = f_i(t) - \gamma_1 t$.

Let Σ_{f_i} denote the following spread:

$$\Sigma_f = \left\{ x = 0, y = x \begin{bmatrix} u & f_i(t) \\ t & u \end{bmatrix} \forall u, t \in K \right\},$$

Then

$$\Sigma_1 \cup_{i \in \lambda} \Sigma_{f_i}^* E h_i$$

is a parallelism in $PG(3, K)$.

PROOF. We know that

$$\Sigma_1 \cup_{i \in \lambda} \Sigma_2^* E \left\langle \begin{bmatrix} 1 & 0 & 0 & 0 \\ 0 & 1 & 0 & 0 \\ 0 & 0 & -1 & 0 \\ 0 & 0 & 0 & -1 \end{bmatrix} \right\rangle h_i = \Sigma_1 \cup \Sigma_2^* EH$$

is a parallelism by Theorem 240 but then also

$$\Sigma_1 \cup_{i \in \lambda} \Sigma_2^* E h_i$$

is a parallelism. By Theorem 241, we know that $\Sigma_f E$ switches with $\Sigma_2 E$, where f is any of the functions f_i. Choose any line ℓ of $PG(3, K)$, then ℓ is in a unique spread of

$$\Sigma_1 \cup \Sigma_2^* EH = \Sigma_1 \cup_{i \in \lambda} \Sigma_2^* E h_i.$$

Either ℓ is a line of Σ_1 or there exists a unique h_j such that $\ell \in \Sigma_2^* E h_i$. Assume that ℓ non-trivially intersects $x = 0$. Then, ℓ is a Baer subplane of a Rg, for $g \in EH$. We note that $(Rg)^*$ for $g \in EH$ is also in $\Sigma_1 \cup_{i \in \lambda} \Sigma_{f_i}^* E h_i$. Hence, we may assume that $\ell \in (\Sigma_2^* - R) E h_i$. Assume that there is a line $m \in \Sigma_f E$ that does not intersect $x = 0$ that is in two spreads Σ_f and $\Sigma_f g$, for $g \in E$, then since $\Sigma_f E$ switches with $\Sigma_2 E$, we have a contradiction. Hence, every line of $PG(3, K)$ not in Σ_1 is in a unique spread of $\Sigma_2^* E h_i$ and therefore is in a unique spread of $\Sigma_{f_i}^* E h_i$. This completes the proof of the theorem. $\qquad \square$

2.1. Examples. Let F be any subfield of the field of real numbers and let $F^C = K$ denote the field of constructible numbers from F. If we let $\gamma_1 = -1$, choose any function f_i where $f_i(t) = \gamma_i t$, for $\gamma_i \neq -1$. Then it is clear that $\rho_{i,z}(t) = f_i(t) - z^2 t$, and $\phi_i(t) = f_i(t) - \gamma_1 t$ define bijective functions. Hence, we have the following result.

THEOREM 243. *Let $K = F^C$, a field of constructible numbers from a subfield F of the field of real numbers. Let Σ_1 denote the Pappian spread*

$$\Sigma_1 = \left\{ x = 0, y = x \begin{bmatrix} u & -t \\ t & u \end{bmatrix} \forall u, t \in K \right\},$$

and let

$$E = \left\langle \begin{bmatrix} 1 & 0 & u & \gamma_1 t \\ 0 & 1 & t & u \\ 0 & 0 & 1 & 0 \\ 0 & 0 & 0 & 1 \end{bmatrix} ; u, t \in K \right\rangle,$$

$$H = \left\langle \begin{bmatrix} 1 & 0 \\ 0 & w \end{bmatrix} ; w = \begin{bmatrix} u & \gamma_1 t \\ t & u \end{bmatrix} ; u^2 - \gamma_1 t^2 = 1 \right\rangle.$$

Let $\{h_i; i \in \lambda\}$ be a coset representation for

$$H^- = \left\langle \begin{bmatrix} 1 & 0 & 0 & 0 \\ 0 & 1 & 0 & 0 \\ 0 & 0 & -1 & 0 \\ 0 & 0 & 0 & -1 \end{bmatrix} \right\rangle in H.$$

For each $i \in \lambda$, choose a negative number γ_i in K such that $\gamma_i \neq -1$ and finally let

$$\Sigma_i = \left\{ x = 0, y = x \begin{bmatrix} u & \gamma_i t \\ t & u \end{bmatrix} \forall u, t \in K \right\}.$$

Then $\Sigma_1 \cup_{i \in \lambda} \Sigma_i^* E h_i$ is a parallelism in $PG(3, K)$.

REMARK 65. *Let σ_i be an automorphism of K. We consider functions f_i such that $f_i(t) = \gamma_i t^{\sigma_i}$, where γ_i is a negative number. In order to obtain parallelisms in a manner similar to that of the previous theorem, we need to check that $\rho_{i,z}(t) = \gamma_i t^{\sigma} - z^2 t$ and $\phi_i(t) = \gamma_i t^{\sigma} - \gamma_1 t$ define bijective functions. Note that first set of functions $\rho_{i,z}$ are always*

$$f_i(t) - z^2 t = \gamma_i t^{\sigma_i} - z^2 t = 0$$

if and only if $t = 0$ since $\gamma_i < 0$. Since the function is additive, we see $\rho_{i,z}$ is injective. In general, the surjectivity of $\rho_{i,z}$ is not always guaranteed.

2.2. Examples over the Reals. Let

$$f(t) = \left\{ \begin{array}{l} \gamma_1 t - a^t + 1, t \geq 0 \\ \gamma_1 t + b^{-t} - 1, t < 0 \end{array} \right\}, a, b \text{ both } > 1.$$

We see that $f(0) = 0$, f is continuous at all elements t of the reals and consider $f(t) - z^2 t$.

$$f(t) - z^2 t = \left\{ \begin{array}{l} \gamma_1 t - a^t + 1 - z^2 t, t \geq 0 \\ \gamma_1 t + b^{-t} - 1 - z^2 t, t < 0 \end{array} \right\}, a, b \text{ both } > 1.$$

Note that

$$\lim_{t \to \pm\infty} -f(t) = \pm\infty,$$

so that $f(t) - z^2 t$ is continuous and hence surjective. We note that the $\lim_{t \to 0} f(t) = 0$ and $f'(t)$ for t non-zero is

$$f'(t) - z^2 = \left\{ \begin{array}{l} \gamma_1 - a^t \ln a - z^2, t \geq 0 \\ -1 - b^{-t} \ln b - z^2, t < 0 \end{array} \right\}, a, b \text{ both } > 1,$$

which is never 0. Hence, $f(t) - z^2 t$ is bijective for each z. Now $f(t) - \gamma_1 t$

$$f(t) - \gamma_1 t = \left\{ \begin{array}{l} -a^t + 1, t \geq 0 \\ b^{-t} - 1, t < 0 \end{array} \right\}, a, b \text{ both } > 1,$$

and clearly this function is bijective. For example, assume that the function is not injective. Then the only questionable case is whether $b^{-t} - 1 = 1 - a^s$, for $t < 0$ and $s \geq 0$. But, then $a^s + b^{-t} = 2$ and both a and $b > 1$, then $a^s \geq 1$ and $b^{-t} > 1$, a contradiction.

The more general versions involving theorems reflecting the above set of examples is given in the following. The proofs are reasonably straightforward and are left for the reader to verify.

THEOREM 244. *Let r be a strictly increasing continuous real function of the positive real numbers, and let h be a strictly decreasing continuous real function on the negative real numbers. Choose any two real numbers a and $b > 1$ (possibly equal). Then a function f defined as follows is switchable.*

$$f(t) = \left\{ \begin{array}{l} \gamma_1 t - a^{r(t)} + 1, t \geq 0 \\ \gamma_1 t + b^{h(t)} - 1, t < 0 \end{array} \right\}, a, b \text{ both } > 1,$$

$$\lim_{t \to 0^+} a^{r(t)} = 1 = \lim_{t \to 0^-} b^{h(t)}.$$

THEOREM 245. *Under the above assumptions, any such function f may be used to construct E-switchable spreads $\Sigma_f E = \Sigma_2 E$, and thus*

$$\Sigma_1 \cup_{i \in \lambda} \Sigma_{f_i}^* E h_i$$

is a parallelism for any set of choices of functions f_i.

Finally, we may now construct parallelisms from any parallelism of the type $\Sigma_1 \cup_{i \in \lambda} \Sigma_{f_i}^* E h_i$, as follows: Choose an element eh_j of $E h_j$, for some $j \in \lambda$. There is a regulus R_{eh_j} of Σ_1 that is derived in $\Sigma_{f_j} eh_j$ to construct $\Sigma_{f_j}^* eh_j$. Derive R_{eh_j} in Σ_1 to construct the Hall plane $\Sigma_1^{R_{eh_j}^*}$ and underive $R_{eh_j}^*$ in $\Sigma_{f_j}^* eh_j$ to construct $\Sigma_{f_j} eh_j$.

THEOREM 246.

$$\Sigma_1^{R_{eh_j}^*} \cup_{i \in \lambda - \{j\}} \Sigma_{f_i}^* E h_i \cup_{g \in E - \{e\}} \Sigma_{f_j}^* E h_j \cup \Sigma_{f_j} eh_j$$

is a parallelism.

2.3. The Variety of Parallelisms. We note that although our original parallelism admits the group EH, the collineation group of certain of the constructed parallelisms can be made so that only E is a collineation group. Furthermore, certain of the derive-underive parallelisms can be found that do not admit a non-trivial collineation.

We note that to construct parallelisms over the reals using the switching technique, we have constructed $2^{\aleph_0}$ different functions f_i. Hence we have also constructed $2^{\aleph_0}$ distinct parallelisms of the type $\Sigma_1 \cup_{i \in \lambda} \Sigma_{f_i}^* E h_i$. Any isomorphism between two parallelisms $\Sigma_1 \cup_{i \in \lambda} \Sigma_{f_i}^* E h_i$ and $\Sigma_1 \cup_{i \in \lambda} \Sigma_{g_i}^* E h_i$ of this type necessarily is a collineation of Σ_1, the Pappian plane over the field of complex numbers (assuming that none of the derived conical flock spreads are Pappian). Since our parallelisms admit E, it follows that any isomorphism must be a collineation of Σ_1 that leaves $x = 0, y = 0$ invariant and must permute the set of reguli of Σ_1 sharing $x = 0$, $y = 0$ and so must permute the $\Sigma_{f_i}^* h_i$. Therefore, there is a collineation of Σ_1 that would map a function f_i to a function g_j. But this would mean that f_i and g_j are obtained as real linear combinations together with the automorphism of order 2. Clearly, it is easy to choose functions f_i and g_j that do not have this property and still produce parallelisms.

THEOREM 247. *(1) When K is the field of real numbers, there are $2^{\aleph_0}$ mutually non-isomorphic parallelisms of type $\Sigma_1 \cup_{i \in \lambda} \Sigma_{f_i}^* E h_i$.*

(2) Similarly, if any of derive-underive parallelisms are isomorphic, and if no derived conical flock spread can be a flock spread, any collineation would necessarily leave $\Sigma_{f_j} e h_j$ invariant and a conjugate would leave Σ_{f_j} invariant. Then there are $2^{\aleph_0}$ mutually non-isomorphic derive-underive parallelisms, none of which are isomorphic to any of the original parallelisms.

CHAPTER 35

Dual Parallelisms

Given any spread of $PG(3, K)$, take the duality of the projective space. What is obtained is called a 'dual spread,' and, as we have mentioned before, if the spread is represented by a set of matrices, the dual spread is represented by the set of transposes to these matrices. When we deal with K as a skewfield, everything becomes more complicated and we have been using the notation of $PG(3, K)_\mathcal{R}$ and $PG(3, K)_\mathcal{L}$ to represent the right and left projective spaces arising from a 4-dimensional vector space over K. For all of the known infinite classes of parallelisms of $PG(3, K)$, all spreads are either known to be dual spreads or suspected to be dual spreads. Now assume that we have a parallelism of $PG(3, K)_\mathcal{R}$, where the dual spreads (which, by the way, are in $PG(3, K)_\mathcal{L}$), also form a parallelism. The question is, are all of these dual parallelisms isomorphic to the original? Again, for all of the infinite classes of parallelisms, the answer is uniformly no!

DEFINITION 123. *Let K be a skewfield. Let S be a set of spreads that are also dual spreads and in $PG(3, K)_\mathcal{R}$ (or $PG(3, K)_\mathcal{L}$), which are mutually disjoint on lines. Then the set of dual spreads of S is said to be a 'partial dual parallelism.'*

REMARK 66. *Any partial dual parallelism in $PG(3, K)_\mathcal{R}$ is equivalent to a partial parallelism in $PG(3, K)_\mathcal{L}$ In particular, any finite parallelism produces a related 'dual' parallelism.*

THEOREM 248. *Let $\mathcal{P}_{-1}$ be any partial parallelism of deficiency one in $PG(3, K)_\mathcal{L}$ such that all spreads of $\mathcal{P}_{-1}$ are dual spreads.*

(1) Then there is a corresponding deficiency one partial parallelism $\mathcal{P}^*_{-1}$ in $PG(3, K)_\mathcal{R}$ called the 'dual' partial parallelism of deficiency one.

(2) Let $\mathcal{P}$ denote the unique extension of $\mathcal{P}_{-1}$ to a parallelism in $PG(3, K)_\mathcal{L}$ then there is a unique extension of $\mathcal{P}^*_{-1}$, say, $\mathcal{P}^*$, which is a parallelism in $PG(3, K)_\mathcal{R}$.

REMARK 67. *If $\mathcal{P}$ is a Pappian partial parallelism in $PG(3, K)$, then all spreads are dual spreads and there is a corresponding Desarguesian 'dual' partial parallelism.*

Hence, we obtain by an easy proof:

COROLLARY 60. *Let π be a translation plane corresponding to a deficiency one Pappian partial parallelism. Then, there is a corresponding translation plane π^* obtained from the dual partial parallelism.*

We have constructed a great variety of the partial parallelisms of deficiency one in the finite case. With every finite spread, the associated dual spread obtained by taking a polarity of the associated projective space is also a spread. Similarly, given a finite parallelism, there is an associated 'dual parallelism,' which is also a parallelism. The main fundamental question concerning parallelisms and their duals is how to distinguish between the two. That is, how does one find a class of parallelisms such that the dual parallelisms are always non-isomorphic to the original?

In this chapter, we intend to show that transitive deficiency one parallelisms also extend to parallelisms in $PG(3, q)$, whose dual parallelisms are never isomorphic to the original parallelism. This work follows the main results of the author [104], with appropriate changes for this text.

DEFINITION 124. *Let $\mathcal{P}$ be a parallelism admitting a Desarguesian spread Σ and a group G fixing a component $x = 0$. Let R be any regulus of Σ containing $x = 0$. If the opposite regulus R^* is a subspread of one of the spreads of $\mathcal{P}$ then we shall say that the parallelism is 'standard' and the associated group is 'standard.'*

THEOREM 249. *Let $\mathcal{P}$ be a standard parallelism in $PG(3, q)$, admitting a standard collineation group G that fixes one spread and is transitive on the remaining spreads. Let $\mathcal{P}_D$ be any 'derived' parallelism and let $\mathcal{P}_D^{\perp}$ be the corresponding dual spread.*
Then $\mathcal{P}_D$ is not isomorphic to $\mathcal{P}_D^{\perp}$.

We may assume that $q > 2$, since the two parallelisms in $PG(3, 2)$ are known to be non-isomorphic.

NOTATION 5. *We shall denote the translation plane associated with a spread S as π_S. The group G will be considered acting in the translation complement of some translation plane π_S.*

COROLLARY 61. *Let $\mathcal{P}$ be a parallelism admitting a standard group G in the translation complement of the socle Σ.*
(1) The stabilizer $G_{\pi_{\Sigma'}}$ of a non-socle plane $\pi_{\Sigma'}$ is a subgroup of the group that fixes a line of a regulus of Σ.
(2) This group $G_{\pi_{\Sigma'}}$ normalizes a regulus-inducing Baer group of order q.

(3) *The order of the full group of the parallelism G is as follows:*

$$q^2(q^2 - 1) \mid \ |G| \ \mid q^2(q^2 - 1)^2 2r,$$

where $q = p^r$.

PROOF. Exercise for the reader.   □

THEOREM 250. *Under the previous assumptions, let π_Σ be coordinatized by a field F containing K. Let F^* denote the kernel homology group of π_Σ of order $q^2 - 1$. Let Σ' be a non-socle spread and let $G_{\pi_{\Sigma'}}$ denote the collineation leaving Σ' invariant. Then*

$$\left| G_{\pi_{\Sigma'}} \cap F^* \right| \ \mid \ (2, q - 1)(q - 1).$$

PROOF. Assume there is an element g in the kernel homology group of Σ that fixes a non-socle plane Σ', whose order divides $q^2 - 1$. Initially, assume that the order of g is a prime power u^a. If u is odd, then g must fix at least two components of Σ'. Assume that g does not fix a component of Σ_2 then u divides $(q^2 - 1, q^2 + 1) = 2$. Hence, $u = 2$ or we are finished. But even when $u = 2$, there is a subgroup of $\langle g \rangle$ of order 2^{a-1} that fixes two components. Hence, we have that g or g^2 fixes two components of Σ', which are Baer subplanes of Σ that are fixed by g^i, for $i = 1$ or 2, assuming the order of g is a prime power. But, g^i then induces a kernel homology group on the Baer subplanes, implying that g^i is in K^*. Hence, we are finished or all odd prime power elements are in K^*. But, then g is in K^* or g^2 is in K^* when q is odd. This completes the proof of the lemma.   □

REMARK 68. *For the remainder of this section, we shall assume that G is a standard group acting on a Desarguesian spread Σ and transitive on the non-socle spreads of a parallelism $\mathcal{P}$. We note that G is then a subgroup of $\Gamma L(2, q^2)$ acting on π_Σ and acts in $\Gamma L(4, q)$ when consider acting on the parallelism. For notational purposes, Σ' shall always denote a non-socle spread.*

The following remark and two corollaries shall be left as (should be 'easy') exercises for the reader.

REMARK 69. *There are exactly $1 + q + q^2$ 2-dimensional K–subspaces that share a given 1-dimensional K–subspace X on the axis of the elation group. Each spread of the parallelism contains exactly one of these 2-dimensional K–subspaces. We call these 2-dimensional K–subspaces the "base" subspaces. Every base subspace is fixed by a subgroup of Σ of order $q(q - 1)^2 2r$.*

COROLLARY 62. *G is a solvable group.*

COROLLARY 63.

$$|G \cap GL(2, q^2)| \quad | \quad q^2(q^2 - 1)(q - 1)(2, q - 1) \text{ and}$$
$$|G_{\Sigma'} \cap GL(2, q^2)| \quad | \quad q(q - 1)^2(2, q - 1).$$

Thus,

$$|G| \mid q^2(q^2 - 1)(q - 1)2r(2, q - 1),$$

where $q = p^r$.

By Theorem 211 (part (2)), the elation group E may be partitioned into exactly $q + 1$ elation subgroups E_i, $i = 1, 2, .., q + 1$, of order q, each of which fixes exactly q spreads other than the socle spread. G acts transitively on the set of $q + 1$ elation subgroups.

COROLLARY 64. *If Δ is the set of $q+1$ sets of q spreads each fixed by an elation subgroup then the stabilizer of a set J of Δ normalizes a regulus-inducing elation subgroup.*

THEOREM 251. *Let g be any element of G such that $\langle g \rangle \cap GL(2, q^2)$ is not a kernel homology subgroup of Σ and assume that $p \nmid |\langle g \rangle|$.*
Then g fixes a unique component distinct from $x = 0$.

PROOF. Suppose that $g \in GL(2, q^2)$ fixes no component other than $x = 0$. Let g^j have order a prime power z^α, a divisor of the order of g. Then, z divides q^2, implying that $z = p$, a contradiction since $p \nmid |H|$. Hence, g^j fixes a component L of $\Sigma - \{x = 0\}$. More generally, If g does not a unique component distinct from $x = 0$, then clearly g^j is in the kernel homology group F^* of Σ. It follows then that either g, for some j fixes a unique component distinct from $x = 0$ or g is a kernel homology. Since $\langle g \rangle \cap GL(2, q^2)$ is not a kernel homology subgroup of Σ, we have completed the proof.   □

LEMMA 59. *$G_{\Sigma'}$ fixes a regulus net of Σ', on which $x = 0$ is a Baer subplane and there is a regulus-inducing elation group E' that acts on Σ' as a Baer group. E' acts transitively on the remaining subplanes that share the zero vector of the regulus net defined by $x = 0$ (as a Baer subplane).*
Hence, if π'_o is one of the q subplanes in a E'-orbit, then

$$|G_{\Sigma', \pi'_o}| = |G| / q^2(q + 1).$$

Furthermore,

$$|G_{L,E'}| = |G| / q^2(q + 1),$$

for any component L and regulus-inducing elation subgroup E'.
Furthermore, either $G_{\Sigma', \pi'_o} = G_{L', E'}$, for some unique component L' of Σ, or $G_{\Sigma', \pi'_o} \cap GL(2, q^2)$ is a subgroup of $(2, q - 1)K^*$, the kernel homology group of order $(2, q - 1)(q - 1)$.

PROOF. Each element g of $G_{\Sigma',\pi'_o} \cap GF(2,q^2)$ fixes a component L_g. We know that if g is a kernel homology then g is in $2K^*$ (the scalar group of order $(2,q-1)(q-1)$). By the above remarks, $G_{\Sigma',\pi'_o} \cap GF(2,q^2)$ either fixes a unique component L_g or is a subgroup of the $(2,q-1)K^*$ kernel homology group of order $(2,q-1)(q-1)$. In either case, there exists a component L' such that $G_{\Sigma',\pi'_o} \cap GL(2,q^2) \subseteq G_{L',E'} \cap GL(2,q^2)$, the latter group indicating the stabilizer of L' in the normalizer of E' in G.

We note that $|G_{E'}| = |G|/(q+1)$, so that $|G_{L',E'}| = |G|/q^2(q+1)$. Furthermore, $|G_{\Sigma'}| = |G|/q(q+1)$, so that $|G_{\Sigma',\pi'_o}| = |G|/q^2(q+1)$.

Thus, $|G_{L',E'}| = |G_{\Sigma',\pi'_o}|$. First, assume that $G_{\Sigma',\pi'_o} \cap GL(2,q^2)$ is not a kernel homology subgroup (necessarily in $2K^*$). Then, G_{Σ',π'_o} fixes a unique component L' and normalizes E'. Hence, $G_{\Sigma',\pi'_o} \subseteq G_{L',E'}$, implying equality by the above argument.

Now assume that $G_{\Sigma',\pi'_o} \cap GL(2,q^2)$ is the kernel homology group K^* of order $q-1$. Then the order of G_{Σ',π'_o} is $(q-1)z$, where z divides $2r$ if $q = p^r$. Therefore, the order of $G_{L,E}$ is also $(q-1)z$, for any component L and regulus-inducing elation subgroup E. □

LEMMA 60. $G \cap$ (Kernel Homology group of Σ) has order dividing $(2,q-1)(q-1)$. Furthermore, this group leaves invariant each spread of the parallelism.

PROOF. Let g be a kernel homology of Σ. Then, g normalizes each regulus-inducing elation subgroup (since it commutes with the linear subgroup). Each regulus-inducing group fixes exactly q non-socle spreads. Therefore, g permutes a set of q non-socle spreads. Let g^j have prime power order so divides q^2-1. Then, g^j fixes one of the q non-socle spreads so must in $2K^*$. Hence, every subgroup of g of odd prime power order is in K^*, implying that g is in K^* or g^2 is in K^*, from a previous lemma.

Now it also follows that E commutes with each regulus-inducing group E_i and so permutes the q spreads fixed by E_i. Then clearly, E is transitive on these spreads since the maximum Baer group fixing a spread has order q. Also, E permutes the spreads fixed by g^j, so g^j fixes each spread. It then follows that g fixes all spreads of the parallelism. □

1. The Isomorphisms of the Dual Parallelisms

LEMMA 61. *Represent $PG(3,q)$ as the lattice of subspaces of the 4-dimensional $GF(q)$-space V_4, with vectors as (x_1, x_2, y_1, y_2) and 3-dimensional subspaces written as* $\begin{bmatrix} a \\ b \\ c \\ d \end{bmatrix}$ *and a vector or 'point' incident with a 3-space or 'plane' exactly when*

$$ax_1 + bx_2 + cy_1 + dy_2 = 0.$$

Dualize by

$$(x_1, x_2, y_1, y_2) \longleftrightarrow \begin{bmatrix} a \\ b \\ c \\ d \end{bmatrix}.$$

(1) We note that $x = 0$ as $x_1 = x_2 = 0$ and $y = 0$ as $y_1 = y_2 = 0$ are interchanged by the duality. A line that may be written in the form $y = xM$, where M is a 2×2 $GF(q)$-matrix as a set of vectors becomes a line that is the intersection of a set of 3-dimensional $GF(q)$-subspaces (intersection of a set of planes). We shall use the notation $\perp$ to denote the images under the associated mapping.

We have that

$$(y = xM)^\perp = (y = x(-M^{-t}),$$

where M^t denotes the transpose of M.

(2)

   (a) If

$$\begin{bmatrix} I & A \\ 0 & I \end{bmatrix}$$

is a group element of a parallelism $\mathcal{P}$, then

$$\begin{bmatrix} I & 0 \\ -A^{-t} & I \end{bmatrix}$$

is a group element of the dual parallelism $\mathcal{P}^\perp$.

   (b) If

$$\begin{bmatrix} B & 0 \\ 0 & C \end{bmatrix}$$

is an element of the group of $\mathcal{P}$, then

$$\begin{bmatrix} B^{-t} & 0 \\ 0 & C^{-t} \end{bmatrix}$$

is an element of the group of $\mathcal{P}^{\perp}$.

PROOF. Let $x = (x_1, x_2)$ and $y = (y_1, y_2)$.

The vector (x, xM) maps under the duality to $\begin{bmatrix} x^t \\ M^t x^t \end{bmatrix}$.

Since

$$xx^t + x(-M^{-t})M^t x^t = 0,$$

it follows that

$$(y = xM)^{\perp} = (y = x(-M^{-t}),$$

which proves part (1).

The proof to part (2) involves merely a basis change and a similar argument and is left to the reader to complete as an exercise. □

NOTATION 6. *If $\mathcal{P}$ is a parallelism, then we denote the associated dual parallelism as $\mathcal{P}^{\perp}$. Hence, we may also use the notation $\mathcal{P}^{-t}$.*

The following lemma is now immediate.

LEMMA 62. *If $f : (x, y) \longmapsto (y, x)$, then $\mathcal{P}^{-t}$ maps to $\mathcal{P}^t$ and*
    (a) If

$$\begin{bmatrix} I & A \\ 0 & I \end{bmatrix}$$

is a group element of a parallelism $\mathcal{P}$, then

$$\begin{bmatrix} I & -A^{-t} \\ 0 & I \end{bmatrix}$$

is a group element of the dual parallelism $\mathcal{P}^t$.
    (b) If

$$\begin{bmatrix} B & 0 \\ 0 & C \end{bmatrix}$$

is an element of the group of $\mathcal{P}$ then

$$\begin{bmatrix} C^{-t} & 0 \\ 0 & B^{-t} \end{bmatrix}$$

is an element of the group of $\mathcal{P}^t$.

NOTATION 7. *We may assume that for any parallelism $\mathcal{P}$, then $\mathcal{P}^t$, is isomorphic to the dual parallelism $\mathcal{P}^{\perp} = \mathcal{P}^{-t}$. Hence, we may now refer to $\mathcal{P}^t$ as the dual parallelism of $\mathcal{P}$.*

We set things up so that the spread of Σ is represented by

LEMMA 63.

$$x = 0, y = x \begin{bmatrix} u + \rho t & t\gamma \\ t & u \end{bmatrix}; u, t \in GF(q),$$

where the indicated matrix set is a field of order q^2. Furthermore, the elation axis of E is denoted by $x = 0$ and we may take $\rho = 0$ if q is odd.

Then, again immediately, we obtain the next two lemmas.

LEMMA 64. *The dual parallelism consists of the dual spreads of all of the spreads of the original parallelism.*

(1) If a given spread is represented as a matrix spread set $x = 0, y = xM$, then the dual spread may be represented in the form $x = 0, y = xM^t$, where M^t is the transpose of the matrix M.

(2) The elation group E has the general form:

$$\left\langle \begin{bmatrix} I & U \\ 0 & I \end{bmatrix}; U = \begin{bmatrix} u + \rho t & t\gamma \\ t & u \end{bmatrix} \right\rangle,$$

considered as in the previous lemmas.

Then, an elation group in $\Sigma^\perp$, the dual spread of Σ, is given as

$$\left\langle \begin{bmatrix} I & U^t \\ 0 & I \end{bmatrix}; U = \begin{bmatrix} u + \rho t & t\gamma \\ t & u \end{bmatrix}^t \right\rangle.$$

(3) If a group of Σ has the form $\left\langle \begin{bmatrix} A & 0 \\ 0 & B \end{bmatrix} \right\rangle$, where A and B are non-zero elements of the field of order q^2, then in $\Sigma^\perp$, there is a group of the form $\left\langle \begin{bmatrix} B^t & 0 \\ 0 & A^t \end{bmatrix} \right\rangle.$

LEMMA 65. *Assume that a parallelism $\mathcal{P}$ is isomorphic to its dual parallelism $\mathcal{P}^t$, and σ is any element of $\Gamma L(4, q)$ then $\mathcal{P}$ is isomorphic to $\mathcal{P}^\perp \sigma$.*

LEMMA 66. *Let*

$$\sigma : (x, y) \longmapsto (x, y) \begin{bmatrix} 0 & 1 & 0 & 0 \\ 1 & 0 & 0 & 0 \\ 0 & 0 & 0 & 1 \\ 0 & 0 & 1 & 0 \end{bmatrix}.$$

Then (1) $\mathcal{P}^\perp \sigma$ contains Σ in the form given originally.

(2) $\mathcal{P}^\perp \sigma$ contains E, in the original form and if $\left\langle \begin{bmatrix} A & 0 \\ 0 & B \end{bmatrix} \right\rangle$ is in G acting on $\mathcal{P}$ then $\left\langle \begin{bmatrix} B & 0 \\ 0 & A \end{bmatrix} \right\rangle$ acts on $\mathcal{P}^\perp \sigma$.

PROOF. Consider the action of σ on the representation (the transposed spread; the dual spread of Σ)

$$x = 0, y = x \begin{bmatrix} u + \rho t & t \\ t\gamma & u \end{bmatrix} ; u, t \in GF(q) :$$

We note that σ fixes $x = 0$ and $y = 0$, and

$$y = x \begin{bmatrix} 0 & 1 \\ 1 & 0 \end{bmatrix} \begin{bmatrix} u + \rho t & t \\ t\gamma & u \end{bmatrix} \begin{bmatrix} 0 & 1 \\ 1 & 0 \end{bmatrix} =$$

$$\begin{bmatrix} u & t\gamma \\ t & u + \rho t \end{bmatrix}.$$

Hence, for q odd, $\rho = 0$, and we have verified that Σ is within $\mathcal{P}^\perp \sigma$. Assume that q is even and let $v = u + \rho t$ so that $u = v + \rho t$. That is,

$$\begin{bmatrix} u & t\gamma \\ t & u + \rho t \end{bmatrix} = \begin{bmatrix} v + \rho t & t\gamma \\ t & v \end{bmatrix}.$$

This proves (1).

To prove (2), we merely take the conjugate of the groups by σ. The part follows calculating

$$\left\langle \sigma^{-1} \begin{bmatrix} I & U^t \\ 0 & I \end{bmatrix} \sigma; U = \begin{bmatrix} u + \rho t & t\gamma \\ t & u \end{bmatrix} \right\rangle$$

and $\left\langle \sigma^{-1} \begin{bmatrix} B^t & 0 \\ 0 & A^t \end{bmatrix} \sigma \right\rangle$. The details are left for the reader to complete, noting that B, A are in $\left\{ \begin{bmatrix} u + \rho t & \gamma t \\ t & u \end{bmatrix} ; u, t \in GF(q) \right\}$. $\qquad\square$

LEMMA 67. *We assume that E_1 is*

$$E_1 = \left\langle \tau_u = \begin{bmatrix} 1 & 0 & u & 0 \\ 0 & 1 & 0 & u \\ 0 & 0 & 1 & 0 \\ 0 & 0 & 0 & 1 \end{bmatrix} ; u \in GF(q) \right\rangle,$$

the standard regulus-inducing group.

Then, there is a unique partition of E into regulus-inducing subgroups E_c, where

$$E_c = \left\langle \begin{bmatrix} I & uc \\ 0 & I \end{bmatrix} ; u \in GF(q) \right\rangle,$$

and the set of $q + 1$ regulus-inducing elation groups corresponds to the cosets $cGF(q)^*$ of $GF(q^2)^*/GF(q)^*$.

PROOF. Since the regulus-inducing groups are in an orbit under a collineation group of Σ, it follows that a regulus-inducing subgroup is exactly an image of E_1 under a collineation of Σ of the form $(x, y) \longmapsto (xa, yb)$, where the action is conjugation. Hence, E_1 maps onto

$$E_{a^{-1}b} = \left\langle \begin{bmatrix} I & ua^{-1}b \\ 0 & I \end{bmatrix} ; u \in GF(q) \right\rangle.$$

Therefore, the $(q+1)$ regulus-inducing elation subgroups then have the form

$$E_c = \left\langle \begin{bmatrix} I & uc \\ 0 & I \end{bmatrix} ; u \in GF(q) \right\rangle,$$

for $c \in GF(q^2)$. Moreover, two such elation groups E_c and E_d are identical if and only if $cd^{-1} \in GF(q)$. Hence, the $q+1$ regulus-inducing elation groups correspond to the $q + 1$ cosets of $GF(q^2)^*/GF(q)^*$. □

Assume that $\mathcal{P}$ is isomorphic to $\mathcal{P}^\perp$. Then $\mathcal{P}$ is isomorphic to $\mathcal{P}^\perp \sigma$ by some element ρ in $\Gamma L(4, q)$ mapping $\mathcal{P}$ onto $\mathcal{P}^\perp \sigma$. Since Σ is in both parallelisms and is the unique Desarguesian spread in each, it follows that Σ must be left invariant by ρ. That is, ρ is a collineation group of Σ. Since both parallelisms admit the collineation group E and the axis cannot be moved by an automorphism group of either parallelism, it follows that ρ must leave invariant the axis $x = 0$ of E. Since E has order q^2, it is transitive on the remaining components and we may assume that ρ fixes both $x = 0$ and $y = 0$. Represent ρ by $(x, y) \longmapsto (x^\tau c, y^\tau d)$, for c, d in $GF(q^2)$ associated with Σ and in the form defined in the previous lemmas, and where τ is an automorphism of $GF(q^2)$.

Define $g_{a,b} : (x, y) \longmapsto (xa, yb)$, and assume that $g_{a,b}$ acts on $\mathcal{P}$. Our previous section shows that $g_{b,a}$ acts on $\mathcal{P}^\perp \sigma$. Furthermore, we note that $E^\rho = E$, and $\langle g_{a,b} \rangle^\rho = \langle g_{a^\tau, b^\tau} \rangle$. Hence, we have in the group of $\mathcal{P}^\perp \sigma$, the groups $\langle g_{b,a} \rangle$ and $\langle g_{a^\tau, b^\tau} \rangle$. We note that $\langle g_{b,a} \rangle = \langle g_{b^\tau, a^\tau} \rangle$, since $(p^c, q^2 - 1) = 1$. Hence, we must have $\langle g_{b,a} \rangle$ and $\langle g_{a,b} \rangle$ as collineation groups of the parallelism and $g_{a,b} g_{b,a} : (x, y) \mapsto (x(ab), y(ab))$ is a collineation of the parallelism $\mathcal{P}^\perp \sigma$. However, since Σ has exactly the same form in both parallelisms, it follows that the above element is in the kernel homology group of Σ. But this means that $g_{a,b} g_{b,a} \in 2K^*$, which implies that the order of ab divides $(2, q-1)(q-1)$. Let $2K^* = \langle z \rangle$, z of order $2(q - 1)$ if q is odd and of order $q - 1$ if q is even.

Thus, we obtain a collineation

$$h_{a,\alpha} : (x, y) \longmapsto (xa, ya^{-1}\alpha), a \in GF(q^2) - \{0\},$$

where $\alpha \in 2K^*$ or K^*, respectively, as q is odd or even.

Then,

$$h_{a,\alpha}^{2(q-1)} : (x, y) \longmapsto (xa^{2(q-1)}, ya^{-2(q-1)}), a \in GF(q^2) - \{0\}.$$

We note that

$$h_{a,\alpha}^{2(q-1)} : (c, c^q) \longmapsto (ca^{2(q-1)}, c^q a^{-2(q-1)})$$

and $(ca^{q-1}, c^q a^{-(q-1)})$ is a point of $y = x^q$ since

$$(ca^{2(q-1)})^q = c^q a^{2(1-q)} = c^q a^{-2(q-1)}.$$

Hence, $h_{a,\alpha}^{2(q-1)}$ fixes a Baer subplane of Σ, implying that $h_{\alpha,\alpha}^{2(q-1)}$ also fixes a spread Σ' containing $y = x^q$ as a component.

We know that $G_{\Sigma'} \cap GL(2, q^2)$ divides $q(q-1)^2 2$. Therefore, $h_{a,\alpha}^{2(q-1)}$ has order dividing $(q-1)^2 2$.

First assume that $q^2 - 1$ has a p-primitive prime divisor u. Then u divides $q + 1$ and there is an element g of order u, which is necessarily in $GL(2, q^2)$, i.e. in $GL(4, q)$ and u odd. Since $h_{a,\alpha}^{2(q-1)}$ also has order u, then u is forced to divide $(q-1)^2 2$, a contradiction.

Hence, either $q^2 = 64$ or $q + 1 = p + 1 = 2^a$. First, assume the latter case.

In this situation, $(q-1)_2^2 2 = 8$. Now we have a subgroup of order divisible by $(q+1)(q-1)$ and either $(q-1)_2^2 = 4$ if the 2-order of the stabilizer of a spread, or we have $2K^*$ acting on the parallelism and fixing each spread. In this case, we have a subgroup of order divisible by $(q+1)(q-1)2$ acting on the parallelism.

Since the group is in $(\Gamma L(4, p) = GL(4, p)) \cap \Gamma L(2, p^2)$, it follows that the order of $G \cap GL(2, q^2)$ is divisible by $(q^2 - 1)/2^{(j+1) \bmod 2}$, when $2^j K^*$ acts on the parallelism, where $j = 0$ or 1.

We notice again that all elements of $G \cap GL(2, q^2)$ have the general form

$$g_{a,\alpha} : (x, y) \longmapsto (xa, ya^{-1}\alpha),$$

for a of order dividing $q^2 - 1$ and α of order dividing $2(q-1)$. Furthermore, we have the kernel homologies $2^j K^*$ acting, where $a^{2^j} \in K^*$ and the mapping is $(x, y) \longmapsto (xa, ya)$.

Then, we have a Sylow 2-subgroup S_2 of order divisible by $(q+1)$ or $(q+1)2$, respectively, as K^* or $2K^*$ acts. In S_2, it follows that all associated elements α, which may depend on the corresponding element a, have orders dividing 4, since $2(q-1)_2 = 4$.

If $2K^*$ acts, then we have a stabilizer subgroup S_2' of the Sylow 2-subgroup S_2 containing $2K^*$ of order $2(q-1)_2$, and it follows that $g_{a,\alpha}^4 : (x, y) \longmapsto (xa^4, ya^{-4})$ also fixes $y = x^q$. In this case, if $f \in S_2$

then $f^4 \in S_2'$. Notice that $S_2 \cap GL(2, q^2)$ is contained in

$$\left\langle (x, y) \longmapsto (xa^{(q-1)/2}, ya^{-(q-1)/2}), (x, y) \longmapsto (x, z^{(q-1)/2}); a \in GF(q^2)^* \right\rangle.$$

Since $(x, y) \longmapsto (xb, yb^{-1})$ fixes $y = x^q$ and hence fixes a spread Σ', it follows that in the possibly bigger group, the stabilizer of Σ' has index 4. Hence, S_2' has order at least $|S_2|/4$. Then this means that $(q+1)2/4 = (q+1)/2$ divides 8.

If $2K^*$ does not act, then $g_{a,\alpha}^2$ fixes $y = x^q$ and the same argument shows that $(q+1)/2$ divides 4.

Thus, in either case, $(q+1)$ divides 16, so that $p = 3$ or 7.

For $q = 8$, we have a subgroup of $GL(2, q^2)$ of order divisible by $63/(2, q-1)$, so of order 63. Take an element g of order 3. Then, $g^{(q-1)} = g^7$ has order 3 and must divide $(q-1)^2 = 7^2$, a contradiction.

Hence, the only special cases are $q = p = 3$ or 7. These two cases must be argued separately and the reader is directed to the authors article on this material. It is probably a stretch to suggest the 3 and 7 case as exercises for the reader, so instead we omit the rather lengthy argument.

Thus, we have the following theorem (again for $p = 3$ or 7, see Johnson [104]).

THEOREM 252. *Let $\mathcal{P}$ be a parallelism in $PG(3, q)$, $q = p^r$, admitting a standard automorphism group G that fixes one spread Σ and acts transitively on the remaining spreads (for example, assume the Sylow p-subgroups are linear). Let $\mathcal{P}^\perp$ denote the associated dual spread.*
Then $\mathcal{P}$ and $\mathcal{P}^\perp$ are never isomorphic.

1.1. The 'Derived' Parallelisms.
We know for our previous work in this text, that for any standard parallelism $\mathcal{P}$, the set of spreads $\Sigma^* \cup \Sigma'^* \cup (\mathcal{P} - \{\pm, \Sigma'\}$ is a parallelism, called a 'derived parallelism.'

THEOREM 253. *Let $\mathcal{P}$ be a standard parallelism in $PG(3, q), q > 3$, admitting a standard collineation group G that fixes one spread and is transitive on the remaining spreads. Let $\mathcal{P}_D$ be any 'derived' parallelism and let $\mathcal{P}_D^\perp$ be the corresponding dual spread.*
Then $\mathcal{P}_D$ is not isomorphic to $\mathcal{P}_D^\perp$.

PROOF. In $\mathcal{P}_D$, there is a unique conical flock spread Σ'^*, the remaining spreads are either Hall, i.e., Σ^*, or derived conical flock spreads $\Sigma'' \in \mathcal{P} - \{\Sigma, \Sigma'\}$. Suppose that τ is an isomorphism from $\mathcal{P}_D$ onto $\mathcal{P}_D^\perp$. Since we may assume that Σ is in both parallelism $\mathcal{P}$ and $\mathcal{P}^\perp$, in exactly the same form, we may assume that Σ^* is in both parallelisms, in exactly the same form. It follows that τ maps Σ^* onto Σ^*. Since $q > 3$, the full collineation group of Σ^* is the inherited group, i.e., leaves invariant the relevant derivable net. Hence, τ induces a mapping from Σ onto Σ. Furthermore, there is a unique conical flock spread Σ'^* in $\mathcal{P}_D$, which must map to the unique conical flock spread $\Sigma'^{*\perp}$ in $\mathcal{P}_D^\perp$. Since it is also true that for conical flock spreads, the inherited group is the full group, it follows that Σ' in $\mathcal{P}$ maps to $\Sigma'^{*\perp*}$ in $\mathcal{P}^\perp$. We note from Johnson [**133**] that the processes of derivation and transpose are commutative, in that the spread obtained from derivation then transpose is identical to the spread obtained from transpose and then derivation. Hence, $\Sigma'^{*\perp*} = \Sigma'^{**\perp} = \Sigma'^\perp$.

Therefore, we have an induced isomorphism from $\mathcal{P}$ to $\mathcal{P}^\perp$, a contradiction. $\qquad\qquad\qquad\qquad\qquad\qquad\qquad\qquad\qquad\qquad\qquad\qquad\qquad\square$

542

Part 9

Transitivity

In this part, we consider a variety of group theory results on t-parallelisms. We give the complete proof due to Biliotti, Jha, Johnson [22], of the classification of 'p-primitive parallelisms'; those parallelisms in $PG(3, q)$ that admit a collineation of order a p-primitive divisor of $\frac{q^3-1}{q-1}$, which leads to the classification of transitive parallelisms and then directly to the work of the author [112], where this is a complete determination of the doubly transitive parallelisms in $PG(3, q)$. We consider transitive t-parallelisms and show that this always implies that $t = 2$ (in the vector space version). We are able to give a completely new infinite family of examples, containing the Baker transitive parallelisms. There is considerable group theory involved in the transitive t-parallelism theorem, which is somewhat beyond the scope of the present text, but we do give a reasonable sketch of this result by listing many of theorems of which the proof depends in the appendix.

Furthermore, the classification of all finite deficiency one partial parallelisms in $PG(3, q)$ is proved, which was given first in Biliotti, Jha, Johnson [19], but with a particular and somewhat natural assumption on the collineation group and then more generally in Diaz, Johnson, Montinaro [45], where this assumption is removed.

Finally, we prove the classification of doubly transitive focal-spreads of Johnson and Montinaro [142] and show that this focal-spreads are k-cuts of Desarguesian spreads.

CHAPTER 36

p-Primitive Parallelisms

It is always of interest to determine the transitive groups that can act on a parallelism, and we consider such a study later on in this text. In this chapter, we consider parallelisms in $PG(3, q)$ that admit a collineation of order a p-primitive divisor of $q^3 - 1$, so we develop a more general study than merely the consideration of transitivity, noting that a group acting transitively on the parallelism has order divisible $\frac{q^3 - 1}{q - 1}$.

Certainly, the set of all transitive parallelisms is far from complete. For example, we do provide a new infinite class of transitive parallelisms in the next chapter, but there are also a large number of non-regular but transitive parallelisms in $PG(3, 5)$ because of Prince [**176**] (45 total parallelisms, of which 43 are not regular). Similarly, one might consider the classification of 2-transitive parallelisms, which we have noted has been resolved by the author, where is shown that only the two regular parallelisms in $PG(3, 2)$ are possible and these parallelisms admit a collineation group isomorphic to $PSL(2, 7)$ acting 2-transitively on the $1 + 2 + 2^2$ spreads. For work in $PG(3, q)$, it is possible to utilize the quite old work of Mitchell [**171**] 1911 and Hartley [**64**] 1928. Using this material, we are able to obtain strong restrictions on the nature of transitive parallelisms, which, in particular, provides an alternative proof for the classification of 2-transitive parallelisms, when $q \neq 4$ and furthermore shows that non-solvable groups admitting p-primitive elements have a very restrictive structure and involve only $PSL(2, 7)$ or A_7. In the transitive case, only $PSL(2, 7)$ can occur and $q = 2$. For more general work on doubly transitive t-parallelisms or transitive t-parallelisms, a considerable finite group theory is required. The work in this chapter follows the article by Biliotti, Jha, and Johnson [**22**], to a certain extent.

We first are reminded that a 'primitive group' is a transitive group that does not admit a non-trivial 'block' (there is not a non-trivial partition of the set upon which the group acts that permutes the elements of the partition). When the group is a linear group, the blocks tend to be vector subspaces that are permuted. In this chapter, primitive groups are discussed, but this is not where we start. We consider groups acting on parallelisms of $PG(3, q)$, especially when acting on the set of $1 + q + q^2 = (q^3 - 1)/(q - 1)$ spreads and ask what can be said of a collineation group that contains an element g, whose order is a (prime) p-primitive divisor of $(q^3 - 1)$? As we know a 'p-primitive divisor' of $p^k - 1$ is a divisor that does not divide $p^j - 1$, for $j < k$. Now there is one situation, namely, $4^3 - 1$, where prime p-primitive divisors do not exist, so our result will not apply for parallelisms of $PG(3, 4)$. It turns out that this assumption on p-primitivity is extremely powerful, and we can say a great deal about the possible parallelisms.

1. Biliotti-Jha-Johnson *p*-Primitive Theorem

The proof of the result describing the situation when there is a collineation acting on a parallelism of order a p-primitive divisor of $\frac{q^3-1}{q-1}$, while largely combinatorial, requires the group actions of $PSL(3, q)$ studied by Mitchell [171] and Hartley [64] almost a century ago, as well as the classification of primitive subgroups of $\Gamma L(4, q)$, which is due to a number of mathematicians and compiled in Kantor and Liebler [161] (the reader is also directed to the appendix Chapter 43, where the list of possible groups is also given). In the appendix, Chapter 43, we also provide a list of the possible subgroups or maximal subgroups of $PSL(3, q)$, as found in Mitchell and Hartley. At one point, we require some information from the Atlas of Simple Groups and at another point, some information on the nature of Baer groups. There is complete incompatibility between Baer p-groups and elations due to Foulser [53], when q is odd, where there is substantial incompatibility due to Jha and Johnson [96], for even order, where basically the existence of a Baer group of order $> \sqrt{q}$ in a translation plane of order q^2 forces any elation group to have order at most 2.

THEOREM 254. *Let $\mathcal{P}$ be a parallelism in $PG(3, q)$ and let G be a collineation group of $PG(3, q)$, which leaves $\mathcal{P}$ invariant and contains a collineation of order a p-primitive divisor u of $q^3 - 1$.*

Then, one of the two situations occurs:

(1) *G is a subgroup of $\Gamma L(1, q^3)/Z$, where Z denotes the scalar group of order $q - 1$, and fixes a plane and a point.*

(2) *$u = 7$ and one of the following subcases occurs:*

    (a) *G is reducible, fixes a plane or a point and either*

        (i) *G is isomorphic to A_7 and $q = 5^2$, or*

        (ii) *G is isomorphic to $PSL(2, 7)$ and $q = p$ for a prime $p \equiv 2, 4 \bmod 7$ or $q = p^2$ for a prime $p \equiv 3, 5 \bmod 7$.*

When $q = p = 2$, $\mathcal{P}$ is one of the two regular parallelisms in $PG(3, 2)$, or

(b) *G is primitive, G is isomorphic to $PSL(2, 7)$ or A_7 and $q = p$ for an odd prime $p \equiv 2$, $4 \bmod 7$ or $q = p^2$ for a prime $p \equiv 3$, $5 \bmod 7$.*

CONVENTION 3. *G denotes a collineation group of $PG(3, q)$ satisfying the assumptions of Theorem 283. When we need to consider the representation group of G in $\Gamma L(4, q)$, then we denote this group by G and we write $\bar{G}$, instead of G, to denote the corresponding group in $P\Gamma L(4, q)$.*

We start with a series of lemmas.

LEMMA 68. *Every plane of $PG(3, q)$ contains exactly one line from each spread of the parallelism.*

PROOF. Let π be any plane of $PG(3, q)$ and note that any spread is also a dual spread. Hence, π contains a line of any spread. However, π cannot contain two lines of the same spread as otherwise the lines would intersect. Since a parallelism is a covering of the lines of $PG(3, q)$, it follows that the lines of π belong one each to the spreads of the parallelism. □

LEMMA 69. *A parallelism cannot be left invariant by any non-trivial central collineation of $PG(3, q)$.*

PROOF. Let σ be a central collineation of $PG(3, q)$ and let π denote the plane pointwise fixed by σ. Let S be any spread of the parallelism and assume σ leaves invariant the parallelism. Since π contains a unique line of S, it follows that σ must leave S invariant since we have a parallelism. The remaining q^2 lines of S intersect π in the remaining q^2 points not on the line of π in S. Hence, it follows that σ fixes each line of S and hence induces a kernel homology on the plane π_S corresponding to S. Since the kernel homology group of π_S has order $q^2 - 1$ or $q - 1$ and σ has order dividing q or $q - 1$, we have that σ must be induced in π_S by multiplication by a scalar of $GF(q)$. Thus σ is trivial on $PG(3, q)$. □

LEMMA 70. *An element of $\Gamma L(4, q)$, $q = p^r$, of order a prime p-primitive divisor u of $q^3 - 1$ is in $SL(4, q)$ and projectively is in $PSL(4, q)$.*

PROOF. We note that since u and p are prime then u divides $p^{u-1} - 1$. Since u then divides $(p^{u-1} - 1, p^{3r} - 1) = p^{(u-1,3r)} - 1$ it follows that $(u - 1, 3r) = 3r$ so that $u \equiv 1 \bmod 3r$ and $(u, r) = 1$. Since

$\Gamma L(4,q)/GL(4,q)$ has order r, it follows that any such p-primitive element is in $GL(4,q)$. Now $GL(4,q)/SL(4,q)$ has order $q-1$ so that any p-primitive element is in $SL(4,q)$ and taken projectively is in $PSL(4,q)$. □

LEMMA 71. *Let g be a collineation of $PG(3,q)$ of order a p-primitive divisor u of q^3-1 leaving invariant a parallelism. Then g fixes a unique plane π and a unique point P exterior to π and acts semi-regularly on the spreads of the parallelism.*

PROOF. The number of points and planes of $PG(3,q)$ is $(q^4-1)/(q-1)$ and u is a prime dividing (q^3-1). If u divides $(q^4-1)/(q-1)$, then u divides $(q^4-1, q^3-1) = (q-1)$, a contradiction. Hence, g must fix a point P and a plane π. Note that we may assume that P is exterior to π since g is completely reducible. Now assume that g fixes two planes π and π'. Then g fixes the line of intersection l. Since l has $q+1$ points and $q+1$ divides q^2-1, g must fix at least three points on l and hence g fixes l pointwise by Lemma 70. Thus, by the p-primitivity assumption, g fixes both π and π' pointwise, a contradiction. The same argument shows that g cannot fix a spread of the parallelism for then it must fix at least three lines of the spread pointwise. □

LEMMA 72. *u can never be 3 or 5. If $u=7$, then $q=p$ for a prime $p \equiv 2, 4 \bmod 7$ or $q=p^2$ for a prime $p \equiv 3, 5 \bmod 7$.*

PROOF. Let $u=3$. If 3 divides $1+q+q^2 = 3+(q-1)+(q^2-1)$ then 3 divides $q-1$ as it always divides q^2-1. So, 3 is not a p-primitive divisor of q^3-1. Let $u=5$. Then since 5 divides q^4-1, for every $q=p^r$, $p \neq 5$ prime, we cannot have 5 as a p-primitive divisor of $1+q+q^2$. If $u=7$, then 7 divides p^6-1, for any prime $p \neq 7$. Since 7 must be a p-primitive divisor of $p^{3r}-1$, for $q=p^r$, then $3r \leq 6$, that is, $r=1$ or 2. Hence, in general $q=p$ or p^2. We must have $7 \mid q^2+q+1$. Suppose that $q=p$. Let $p=7f+i$, $0 \leq 1 \leq 6$. Then $7 \mid i^2+i+1$, which forces $i=2$ or 4. If $q=p^2$ then $7 \mid p^4+p^2+1$, but $7 \nmid p^3-1$, so that $7 \mid p^2-p+1$. In this case, $7 \mid i^2-i+1$, which implies that i is 3 or 5. □ □

PROPOSITION 15. *Assume that the group G fixes a plane π or a point P of $PG(3,q)$. Then the results of Theorem 254 hold.*

PROOF. Assume without loss of generality that the group G fixes the plane π. Regard G as a subgroup of $\Gamma L(4,q)$, $q=p^r$. Let $\overline{G}$ denote the action on $PG(3,q)$ of the group G and let $\overline{H} = \overline{G} \cap PSL(4,q)$. The action of $\overline{H}$ on π is faithful by Lemma 69. Let $H^* = \overline{H} \cap PSL(3,q)$ acting on π. We may assume that, in this case, all elements of order u, a p-primitive divisor of $q^3 - 1$, must sit in H^* by the same argument of Lemma 70. Suppose that $H^* \cong PSL(3, q)$. Then we have a linear group H_p^* of order q^3, which must fix a spread S and induce upon the corresponding translation plane π_S a linear group in the translation complement of π_S. Since this group also fixes a component l of π_S, the group induced on l is a subgroup of $GL(2, q)$ which has order q. Hence, there is an elation group with axis l of order q^2. Furthermore, there is a subgroup B^* of H_p^* of order q that fixes a second component. Since all subgroups are linear, we have a Baer group of order q. By the incompatibility results between Baer groups and elation groups of Theorem 299, it can only be that $q = 2$. Thus $H^* \cong PSL(3, 2) \cong PSL(2, 7)$. This case provides exactly the two regular parallelisms in $PG(3, 2)$. By the work of Mitchell and Hartley Chapter 2, the other possible groups are known. First, consider the case q even. A check of the possible maximal subgroups of $PSL(3, q)$, $q = 2^r$ shows that an element of order u may be contained only in a group of order dividing $3(q^2 + q + 1)/\nu$, where $\nu = (3, 2^r - 1)$. Now assume that q is odd. We have to check the list of Mitchell; again the reader is again directed to Chapter 2, where it is clear that the orders of all maximal subgroups do not have the appropriate divisor u, with the exception of cases 4, and 10 to 14 in that list. In the case 4, H^* is contained in a maximal subgroup of order $3(q^2 + q + 1)/\nu$, where $\nu = (3, p^r - 1)$ for $q = p^r$. In case 10, we have groups of orders 216, 72, and 36, which only have prime divisors 2 or 3, so there are no p-primitive elements. In case 11, we have a group isomorphic to $PSL(2, 7)$, so that $u = 7$ and $q = p$ for a prime $p \equiv 2$, 4 mod 7 or $q = p^2$ for a prime $p \equiv 3$, 5 mod 7 by Lemma 72. In cases 12 and 13, we have groups of order 360 or 720 which implies that $u = 3$ or 5. This cannot occur by Lemma 72. The last case is when the group is isomorphic to A_7 and $p = 5$ and r even. Since $n = 7$, this is only possible when $q = 5^2$.

Thus, we have only to consider the case when the order of H^* divides $3(1 + q + q^2)/\nu$. In this case, H^* has a normal u-group $\bar{N}$. However, N is normal in G and the centralizer of N is a field of order q^3 considered acting within $\Gamma L(3, q)$. It follows immediately that G must be a subgroup of $\Gamma L(1, q^3)$ since it normalizes the field centralizing N. $\qquad\square$

We need two other lemmas about the group G.

LEMMA 73. *If G does not fix any point or plane, then G is primitive in $PG(3, q)$.*

PROOF. Since G contains u-elements, then it cannot fix a line of $PG(3, q)$ by Lemma 71. Therefore, G is irreducible. If G leaves a pair of lines invariant, then a u-element in G fixes each line of this pair, which is impossible. If G leaves a tetrahedron invariant, then a u-element in G fixes each of the four vertices, in contrast with Lemma 71. Thus G is primitive in $PG(3, q)$.   □

LEMMA 74. *Suppose that G is irreducible, then $C_{PGL(4,q)}(\overline{G}) = \langle 1 \rangle$.*

PROOF. Let $h \in C = C_{PGL(4,q)}(\overline{G})$. Then h commutes with all elements of order u in G, so that any u-element $g \in G$ permutes the points and planes fixed by h. By Lemma 71 and by the irreducibility of G, we may assume there are at least two u-elements that fix both distinct planes and distinct points. Since any u-element fixes exactly one point and acts semiregular on the remaining points, it follows that there are at least eight distinct points that are fixed by h by Lemma 72. Suppose that g fixes the point P and the plane π. If a line l through P contains three points fixed by h, then h fixes l pointwise since $h \in PGL(4, q)$. Since g does not fix any line, there are at least 7 distinct lines pointwise fixed by h. If all these lines do not lie in the same plane, then $h = 1$. Otherwise, h fixes a plane π' pointwise and we must have $\pi = \pi'$, since g fixes π'. Now, there is in G a u-element moving π and centralizing h. Again, $h = 1$. Suppose no line through P contains three points fixed by h. Then h fixes at least 7 distinct points of π, so that either h fixes π pointwise or h fixes a line of π pointwise because $h \in PGL(4, q)$. The same argument as before shows that $h = 1$.   □

COROLLARY 65. *Suppose that G is irreducible and $G \cap PGL(4, q)$ contains a non-trivial normal subgroup N. Then $u \mid |Aut(N)|$.*

PROOF. Let U be the subgroup generated by the u-elements in G. Since U fixes at most one point and one plane and does not fix any line, then U must be irreducible as G is. Suppose that $u \nmid |Aut(N)|$. Then N centralizes U. By Lemma 74, $N = \langle 1 \rangle$.   □

We may now complete the proof of Theorem 254.

By Proposition 15 and by Lemma 73 we may assume that G is a primitive subgroup of $\Gamma L(4, q)$, $q = p^r$. $\bar{G}$ denotes the projective action of G on $PG(3, q)$. By Lemma 70, the elements of G of order a prime p-primitive divisor u of $q^3 - 1$ are in $SL(4, q)$.

From the list of primitive subgroups of $\Gamma L(4, q)$ (see Chapter 4), we simply consider the possible cases.

Case (a) $G \geq SL(4, q)$ cannot occur by Lemma 69, since $\bar{G}$ would contain central collineations.

Case (b) $G \leq \Gamma L(4, q')$, where $GF(q')$ a proper subfield of $GF(q)$. This cannot occur, since u is a p-primitive divisor of $q^3 - 1$ that forces $q' = q$.

Case (c) $G \leq \Gamma L(2, q^2)$ cannot occur, because of the existence of u-elements.

Case (d) G is a subgroup of $Z(G)H_1$, where H_1 is an extension of a special group of order 2^6 by S_5 or S_6 and q is odd. By Lemma 74, $\bar{G} \cap PGL(4, q) \leq \bar{H}_1$. The existence of u-elements in $\bar{G} \cap PGL(4, q)$ and Lemma 72 give $7 \mid |\bar{H}_1|$, a contradiction.

Case (e) $G^{(\infty)} \cong Sp(4, q)'$ or $SU(4, q^{1/2})$ cannot occur by Lemma 69, since $\bar{G}$ should contain central collineations.

Case (f) $G \leq \Gamma O^{\pm}(4, q)$ cannot occur because of the existence of u-elements (e.g., see [**74**], p. 247).

The remaining cases involve the last term $G^{(\infty)}$ of the derived series of G. So, we have a list of quasisimple groups $G^{(\infty)}$, which are clearly contained in $SL(4, q)$. The corresponding projective groups are non-abelian simple and belong to the following list.

$$
\begin{aligned}
\bar{G}^{(\infty)} &\cong PSL(2, q), A_5, A_6, A_7 \\
&\cong PSp(4, 3) \text{ for } q \equiv 1 \bmod 3, \\
&\cong PSL(2, 7) \text{ for } q^3 \equiv 1 \bmod 7, q \text{ odd} \\
&\cong PSL(3, 4) \text{ for } q \text{ a power of } 9 \\
&\cong Sz(q) \text{ for } p = 2.
\end{aligned}
$$

Since we know all of the outer automorphism groups of the simple groups in the list (e.g., see Atlas [**35**]) and u is a p-primitive divisor of $q^3 - 1$, not equal to 3 or 5, we may apply Corollary 65 using $\bar{G}^{(\infty)}$ as N. When $\bar{G}^{(\infty)} \cong PSL(2, q)$, $q = p^r$, Out $\bar{G}^{(\infty)}$ has order $(2, q - 1) \cdot r$, so that $u \mid |\text{Aut} \, \bar{G}^{(\infty)}| \Rightarrow u \mid |\bar{G}^{(\infty)}|$ by the same argument of Lemma 70, and this cannot occur by the p-primitivity of u. Similarly, we exclude the case $\bar{G}^{(\infty)} \cong Sz(q)$, $q = 2^r$, as Out $\bar{G}^{(\infty)}$ has order r. For the other cases it is easily to prove that, by question of order, the only possibilities are $u = 7$ and $\bar{G}^{(\infty)} \cong PSL(2, 7)$, A_7 or $PSL(3, 4)$.

The case $\bar{G}^{(\infty)} \cong A_7$ and $q = 2$ cannot occur as $PG(3, 2)$ has only two parallelisms both having the full group isomorphic to $PSL(2, 7)$. Again, we note that the case $q = 4$ is not considered as $2^6 - 1$ has no 2-primitive divisors.

When $\bar{G}^{(\infty)} \cong PSL(3, 4)$, then $q = 9$ by Lemma 72. Thus $\bar{G}^{(\infty)}$ acts on the 91 spreads of the parallelism. But $PSL(3, 4)$, regarded as a permutation group, has only one non-trivial representation of degree at most 91 and this representation has degree 40 (e.g., see [**48**]). Thus $\bar{G}^{(\infty)}$ must fix a spread S. Furthermore, $\bar{G}^{(\infty)}$ must fix some of the 82 lines of S, contrary to Lemma 71.

The use of Lemma 72 completes the proof.

2. Transitive Parallelisms

Although we shall provide a more general setting for transitive parallelisms when we consider transitive t-parallelisms, we begin the classifying of the groups arising from a transitive parallelism.

3. Biliotti, Jha, Johnson—Transitive Classification

THEOREM 255. *Let* $\mathcal{P}$ *be a transitive parallelism in* $PG(4, q)$, *where* $q \neq 4$.

Then either $\mathcal{P}$ *is one of the two regular parallelisms in* $PG(3, 2)$ *or the full collineation group of* $\mathcal{P}$ *is a subgroup of* $\Gamma L(1, q^3)/Z$, *where* Z *denotes the scalar group of order* $q - 1$.

PROOF. Suppose that G is a subgroup of $P\Gamma L(4, q)$ leaving $\mathcal{P}$ invariant and acting transitively on the spreads of $\mathcal{P}$. Then $1 + q + q^2 \mid |G|$. Since $q^3 - 1$ admits a p-primitive divisor u if $q \neq 4$ and $u \mid 1 + q + q^2$, we may use Theorem 254. Let $G^{(\infty)} \cong PSL(2, 7)$ or A_7. Then $q = p$ or p^2 for p a prime, so that $[G : PGL(4, q) \cap G] \leq 2$. Nevertheless, $C_{PGL(4,q)}(PGL(4, q) \cap G) = \langle 1 \rangle$ by Lemma 74 and hence $[PGL(4, q) \cap G : G^{(\infty)}] \leq 2$ as $\text{Out}\, G^{(\infty)}$ has order 2. Thus $[G : G^{(\infty)}] = 1$, 2, or 4. It follows that also $G^{(\infty)}$ must be transitive on the spreads of $\mathcal{P}$ as $1 + q + q^2$ is odd. Therefore, the order of $G^{(\infty)}$ must be divisible by $1 + q + q^2$, where $q = p$ for an odd prime $p \equiv 2$, 4 mod 7 or $q = p^2$, for a prime $p \equiv 3$, 5 mod 7. One can then merely run through the possibilities given in Theorem 254 to obtain a contradiction. $\qquad\square$

4. Johnson's Classification of 2-Transitive Parallelisms

In Johnson [112], the two transitive parallelisms were completely classified. The proof given in that paper uses the classification theorem of finite simple groups more for convenience than for actually requirement,, as the argument given is more about incompatibility of elations and Baer collineations and the degree of the permutation group action rather than the particular group. In the p-primitive element situation, we have argued using results about groups strictly within $\Gamma L(4, q)$ and we may easily obtain the same result except when $q = 4$, using the Biliotti-Jha-Johnson p-primitive Theorem.

THEOREM 256. *The 2-transitive parallelisms in* $PG(3, q)$, *for* $q \neq 4$, *are exactly the two regular parallelisms in* $PG(3, 2)$.

PROOF. If q is not 4, we are finished by Corollary 255 or $\Gamma L(1, q^3)/Z$ is divisible by $((q^3 - 1)/(q - 1))(q(q + 1))$, which implies that $q(q + 1)$ divides $3r$, where $q = p^r$, clearly an impossibility. $\Box$

CHAPTER 37

Transitive t-Parallelisms

We now consider the possibility of doubly t-parallelisms, as well as transitive parallelisms. If there are t-parallelisms, for $t > 2$, it might be possible to discover them using group theory, for example, perhaps there are doubly transitive or transitive t-parallelisms. However, Johnson and Montinaro [143], [144] first proved that doubly transitive t-parallelisms imply that $t = 2$, and more recently proved the same result for transitive parallelisms. The doubly-transitive theorem uses the classification theorem of doubly transitive groups, listed in the appendix. The very technical proof of the transitive theorem is sketched in the appendix by assembling the main building blocks of the proof. The reader is referred to Chapter 43 for the sketch of the proof.

1. Johnson-Montinaro t-Transitive Theorem

THEOREM 257. *Let $\mathcal{P}$ be a t-parallelism in $PG_n(q)$, $q = p^h$, p a prime, and let G be a collineation group of $PG(n, q)$, which leaves $\mathcal{P}$ invariant and acts transitively on it. Then $t = 2$. Furthermore, the group G fixes a point or a hyperplane of $PG(n, q)$, and one of the following occurs:*

(1) $Z_{\frac{q^n-1}{q-1}} \trianglelefteq G \leq Z_{\frac{q^n-1}{q-1}} \rtimes (Z_n.Z_h)$;
(2) $q = 2$, $\mathcal{P}$ *is one of the two regular parallelisms in $PG(3, 2)$, and $PSL(2, 7) \trianglelefteq G$.*

So, there is a outstanding problem whether t-parallelisms can exist for $t > 2$, but, in any case, they will be difficult to find without some use of group theory. However, since we do know what the group of a transitive 2-parallelism must look like, we may be able to determine new classes of transitive 2-parallelisms. In this next section, we give a new class of such parallelisms.

2. Transitive Parallelisms of $PGL(2r - 1, 2)$

In this section, we provide a general construction of an infinite class of transitive parallelisms in $PG(2r - 1, 2)$, which generalizes Baker's construction (see [2]). These parallelisms are due to the author and A.

Montinaro [**144**]. This chapter follows that article but is a somewhat more brief version.

Let $V = GF(2^{2m-1}) \oplus GF(2)$, considered as a $2m$-dimensional $GF(2)$-vector space, and let

$$(37.1) \qquad\qquad A = \{(a,0),(b,i),(a+b,i),(0,0)\},$$

where $a,b \in GF(2^{2m-1})^*$, $a \neq b$ if $i = 0$, and $i = 0$ or 1. Then A is a 2-dimensional $GF(2)$-space of V. Let $\mathcal{J}$ be the set of 2-dimensional $GF(2)$-spaces of V defined as in (37.1). *In the following, we shall say that A is a line of $PG_{2m-1}(2)$ meaning that A minus the zero vector is a 1-dimensional projective space of $PG_{2m-1}(2)$. If $A \in \mathcal{J}$, then A is said to be a line of $PG_{2m-1}(2)$ of type 0 or 1 according to whether $i = 0$ or 1, respectively.*

We also define $\sum^k A_0$ to be the sum of the k-powers of elements c in $GF(2^{2m-1})$, where $(c,0)$ is in A, and $\sum^k A_1$ is the sum of the k-powers of elements d in $GF(2^{2m-1})$, for $(d,1)$ in A, and $(\sum^1 A_1)^k$ to be the k-th power sum of the elements d in $GF(2^{2m-1})$, for $(d,1)$ in A.

For $\alpha \in GF(2^{2m-1})^*$, define

$$(37.2) \qquad \mathcal{A}_\alpha^k = \left\{ A \in \mathcal{J} : \sum^k A_0 + \sum^k A_1 + \left(\sum^1 A_1\right)^k = \alpha^k \right\}.$$

LEMMA 75. *$\alpha \neq 0$, and the following occurs:*

(1) *If A is a line of type 0 in $\mathcal{A}_\alpha^k$, then b and $a+b$ are the roots in $GF(2^{2m-1})$ of the equation*

$$(37.3) \qquad\qquad x^{2^t} + a^{2^t-1}x + \alpha^k/a = 0;$$

(2) *If A is a line of type 1 in $\mathcal{A}_\alpha^k$, then b and $a+b$ are the roots in $GF(2^{2m-1})$ of the equation*

$$(37.4) \qquad\qquad x^{2^t} + a^{2^t-1}x + a^{2^t} + \alpha^k/a = 0.$$

PROOF. Let A be a line of $PG_{2m-1}(2)$ lying in $\mathcal{A}_\alpha^k$. Then

$$A = \{(a,0),(b,i),(a+b,i),(0,0)\}.$$

Furthermore,

$$(37.5) \qquad \sum^k A_0 + \sum^k A_1 + \left(\sum^1 A_1\right)^k = \alpha^k,$$

where each term in the first part of the identity is above explained. If $i = 0$, then (37.5) becomes

$$(37.6) \qquad\qquad a^k + b^k + (a+b)^k = \alpha^k.$$

Since $k = 2^t + 1$, the equation (37.6) can be rewritten as follows

$$(37.7) \qquad a^{2^t}b + b^{2^t}a = \alpha^k.$$

From here, the proof is reasonably routine for this case and is left for the reader to complete.

Now, let $i = 1$. Then (37.5) becomes

$$(37.8) \qquad b^k + (a+b)^k = \alpha^k.$$

Again, the remainder of the proof is left for the reader to verify and to obtain that b and $a+b$ are the roots in $GF(2^{2m-1})$ of the equation (37.4). This completes the proof. $\qquad\square$

The next lemma provides a partition of $GF(2^{2m-1})$.

LEMMA 76. *Let $a \in GF(2^{2m-1})^*$ and $F_a = \{f_a(d); d \in GF(2^{2m-1})\}$, where $f_a(x) = x^{2^t} + a^{2^t-1}x$. Then*

$$GF(2^{2m-1}) = F_a \cup (F_a + a^{2^t}).$$

PROOF. Now let $f_a(x) = x^{2^t} + a^{2^t-1}x$ and

$$F_a = \{f_a(d); d \in GF(2^{2m-1})\}.$$

Since f_a is additive, we can compute the number of elements of F_a by determining the number of solutions of $x^{2^t} + a^{2^t-1}x = 0$. If $x \neq 0$, then $(x/a)^{2^t-1} = 1$, which implies $x = a$. Hence, there are exactly two solutions of $x^{2^t} + a^{2^t-1}x = 0$. Thus, F_a contains exactly $2^{2m-1}/2$ elements.

Now, assume that a^{2^k} is in F_a. Then $x^{2^t} + a^{2^t-1}x + a^{2^t} = 0$. Now dividing by a^{2^t}, we obtain the equation

$$(37.9) \qquad (x/a)^{2^t} + (x/a) + 1 = 0.$$

Let $z = x/a$. Then (37.9) becomes $z^{2^t} = z + 1$. By raising the previous equation to the 2^t-th power, we obtain $z^{2^{2t}} = z$, and this implies $z^{2^{2t}-1} = 1$. Therefore, $z = 1$, since $((2^t-1)(2^t+1), 2^{2m-1}-1) = 1$. This is a contradiction, since $z^{2^t} = z + 1$. Consequently, for each non-zero $a \in GF(2^{2m-1})^*$, the set $\{F_a, F_a + a^{2^t}\}$ is a partition of $GF(2^{2m-1})$. $\qquad\square$

Now, we are in position to prove the main result of this section.

Let ω be a primitive element of $GF(2^{2m-1})^*$, and let τ_ω be a transformation of $V_{2^{2m}}(2)$ which maps (e, i) onto (we, i). Since τ_ω fixes the zero vector, then τ_ω can be regarded as a collineation of $PG_{2^{2m}-1}(2)$. Set $H = \langle \tau_\omega : \omega \in GF(2^{2m-1})^* \rangle$, then $H \cong GF(2^{2m-1})^*$ and the following result is obtained.

THEOREM 258. *Under the previous assumptions, the following holds*

(1) *If* $k = 2^t + 1$ *and* $(t, 2m - 1) = 1$, *then* $\mathcal{A}_\alpha^k$ *is a spread of* $PG_{2^{2m-1}}(2)$;

(2) *The set*

$$\mathcal{P}_k = \cup_{a \in GF(2^{2m-1})^*} \mathcal{A}_\alpha^k$$

is a transitive parallelism of $PG_{2^{2m-1}}(2)$ *under the collineation group* H;

(3) $\mathcal{P}_{2^t+1} = \mathcal{P}_{2^{2m-1-t}+1}$.

PROOF. Let w be a non-zero vector of V. First, assume that $w = (a, 0)$, for $a \neq 0$. Then α^k/a is an element of $GF(2^{2m-1})^*$ and hence is either in F_a or $F_a + a^{2^t}$ by Lemma 76. If α^k/a is in F_a, since b and $a + b$ are the roots of (37.3) in $GF(2^{2m-1})$ by Lemma 75 (1), then $\{(0, 0), (a, 0), (b, 0), (a + b, 0)\}$ is the unique 2-space of $\mathcal{A}_\alpha^k$ containing w. If α^k/a is in $F_a + a^{2^t}$, since b and $a + b$ are the roots of (37.4) in $GF(2^{2m-1})$ by Lemma 75 (2), then $\{(0, 0), (a, 0), (b, 1), (a+b, 1)\}$ is the unique 2-space of $\mathcal{A}_\alpha^k$ containing w in this case.

Assume that $w = (b, 1)$. Again by Lemmas 75 and 76, it is easily seen that, there is a unique 2-space of $\mathcal{A}_\alpha^k$ containing w, and the proof of (1) is thus completed.

Assume that some 2-space

$$\{(0, 0), (a, 0), (b, i), (a + b, i)\}$$

lies in the spreads $\mathcal{A}_\beta^k$ and $\mathcal{A}_\alpha^k$. Then $\alpha^k = \beta^k$. This implies $(\alpha/\beta)^k = 1$ and hence $\alpha = \beta$, since $(k, 2^{2m-1} - 1) = 1$. Therefore, the number of such spreads of $PG_{2^{2m-1}}(2)$ is $2^{2m-1} - 1$, since $\mathcal{A}_\alpha^k$ is a spread of $PG_{2^{2m-1}}(2)$ for each $\alpha \in GF(2^{2m-1})^*$. As the number of lines in $PG_{2^{2m-1}}(2)$ is $(2^{2m-1} - 1)(2^{2m} - 1)/3$, then $\mathcal{P}_k$ is a partition of the lines in of $PG_{2^{2m-1}}(2)$ by spreads, that is, $\mathcal{P}_k$ is a parallelism of $PG_{2^{2m-1}}(2)$. Since τ_w maps $\mathcal{A}_\alpha^k$ onto $\mathcal{A}_{wa}^k$, then H preserves $\mathcal{P}_k$ and acts transitively on this one. Hence, $\mathcal{P}_k$ is a cyclic parallelism of $PG_{2^{2m-1}}(2)$. This completes the proof of (2).

Finally, let $A \in \mathcal{P}_{2^t+1}$, then

$$a^{2^t} b + b^{2^t} a + \delta a^k = \alpha^k,$$

for $\delta = 0$ or 1 and $\alpha \in GF(2^{2m-1})^*$. The previous identity is equivalent to

$$(a^{2^t} b + b^{2^t} a + \delta a^k)^{2^{2m-1-t}} = (\alpha^{2^t+1})^{2^{2m-1-t}},$$

being $(2^{2m-1-t} - 1, 2^{2m-1} - 1) = 1$, which in turn reduces to

$$ab^{2^{2m-1-t}} + ba^{2^{2m-1-t}} + \delta a^{2^{2m-1}+1} = \alpha^{2^{2m-1}+1},$$

implying that $A \in \mathcal{P}_{2^{2m-1-t}+1}$. Thus, we have proved that $\mathcal{P}_{2^t+1} = \mathcal{P}_{2^{2m-1-t}+1}$, which is the assertion (3), and which completes all parts of the theorem. □

As mentioned above, the previous construction provides a cyclic parallelism $\mathcal{P}_k$ that coincides for $k = 3$ with the cyclic parallelism discovered by Baker in [2]. Moreover, as we shall see in the next section, it provides exactly $\phi(2m-1)/2$ mutually non-isomorphic cyclic parallelisms in $PG_{2m-1}(2)$.

3. Isomorphisms of the Transitive Parallelisms

In this section, we consider the possible isomorphisms of the transitive parallelisms $\mathcal{P}_k$, for different values of k. Clearly, $\mathcal{P}_k$ cannot be isomorphic to $\mathcal{P}_{k'}$, for any field $GF(2^{2m-1})$, where $k = 2^t+1$, $k' = 2^{t'}+1$, and $(2^{2t} - 1, 2^{2m-1} - 1) = 1$ but $(2^{2t'} - 1, 2^{2m-1} - 1) \neq 1$. However, the main question is whether some $\mathcal{P}_k$ can be isomorphic to $\mathcal{P}_3$, the Baker transitive parallelism (see [2]).

Let $G \leq \Gamma L_{2m}(2)$ be the full group of a parallelism $\mathcal{P}_k$. Since $H \leq G$, where $H = \langle \tau_w : w \in GF(2^{2m-1})^* \rangle$, then G is transitive on $\mathcal{P}_k$ by Theorem 258. Then $H \trianglelefteq G \leq \Gamma L_1(2^{2m-1})$, for $m > 2$ from Theorem 257. Furthermore, G fixes the subspaces $\langle 0 \rangle \oplus GF(2)$ and $GF(2^{2m-1}) \oplus \langle 0 \rangle$.

Since the collineation $\gamma_\sigma : (a, i) \to (a^\sigma, i)$, where σ is an automorphism of $GF(2^{2m-1})$, maps $\mathcal{A}_\alpha^k$ onto $\mathcal{A}_{\alpha^\sigma}^k$, and hence maps $\mathcal{P}_k$ onto itself, we obtain the following:

THEOREM 259. *If $m > 2$, then $\Gamma L(1, 2^{2m-1})$ is the full collineation group of each parallelism $\mathcal{P}_k$.*

Let $\mathcal{P}_k$ and $\mathcal{P}_{k'}$ be two isomorphic parallelisms of $PG_n(q)$, and let $\varphi \in \Gamma L_{2m}(2)$ such that $\mathcal{P}_k \varphi = \mathcal{P}_{k'}$. Then $G_k^\varphi = G_{k'}$, where G_k and $G_{k'}$ denote the full groups of the parallelisms $\mathcal{P}_k$ and $\mathcal{P}_{k'}$, respectively. Since G_k and $G_{k'}$ leave invariant the subspaces $\langle 0 \rangle \oplus GF(2)$ and $GF(2^{2m-1}) \oplus \langle 0 \rangle$ in $V_{2m}(2)$, so does the element φ. Therefore, $\varphi(d, i) = (\varphi_1(d), \varphi_2(i))$, where φ_1 and φ_2 are collineations of $GF(2^{2m-1})$ and $GF(2)$, respectively. Actually, because of its linearity, φ_2 is the identity. Consequently, φ normalizes H and so is in $G_k \cong \Gamma L_1(2^{2m-1})$. Hence, φ is the identity, and we have proved the following:

THEOREM 260. *The parallelisms $\mathcal{P}_k$ and $\mathcal{P}_{k'}$ are isomorphic if and only if they are identical.*

COROLLARY 66. *The above construction provides exactly $\phi(2m - 1)/2$ mutually non-isomorphic transitive parallelisms in $PG_{2m-1}(2)$.*

PROOF. In order to prove the corollary, by Theorem 260, we just need to show that $\mathcal{P}_{2^{2m-1-t}+1}$ is the only parallelism that is identical to $\mathcal{P}_{2^t+1}$. Indeed, suppose that $\mathcal{P}_k = \mathcal{P}_{k'}$ for distinct $k = 2^t + 1$ and $k' = 2^{t'}+1$. Let L be a 1-subspace $\langle (a,0),(b,1)\rangle$ of $\mathcal{P}_k$. Since an isomorphism σ can only map L to L^σ, we may assume that the isomorphism is the identity mapping so that $(a+b)^k+b^k = a^k$, if and only if $(a+b)^{k'}+b^{k'} = a^{k'}$. In particular, if $a = 1$, then

$$b + (1 + b^k)^{1/k} = b + (1 + b^{k'})^{1/k'},$$

which is equivalent to

(37.10) $$(1 + b^k)^{k'} = (1 + b^{k'})^k.$$

Since there are 2^{2m-1} elements $(b,1)$ in $\cup \mathcal{A}_1^k = \cup \mathcal{A}_1^{k'}$, the previous equation is true for all elements b in $GF(2^{2m-1})$. By using (37.10), bearing in mind that $k - 1 = 2^t$ and $k' - 1 = 2^{t'}$, we obtain

(37.11) $$b^k + b^{k(k'-1)} + b^{k'} + b^{k'(k-1)} = 0,$$

for all elements b in $GF(2^{2m-1})$. Since $k \neq k'$ and both k and k' are less than 2^{2m-1}, by reducing all exponents modulo 2^{2m-1} in (37.11), we obtain that either $k \equiv_{2^{2m-1}-1} k(k'-1)$, or $k \equiv_{2^{2m-1}-1} k'(k-1)$. From here is fairly straightforward to verify that $k' = 2^{(2m-1)-t} + 1$ as $k = 2^t + 1$, and the proof is thus completed. $\qquad\square$

CHAPTER 38

Transitive Deficiency One

In the previous chapters, we have given a variety of constructions of the so-called 'transitive deficiency one partial parallelisms, which are partial parallelisms in $PG(3,q)$ of $q^2 + q$ spreads that admit a transitive group of the set of spreads. There is a unique extension to a parallelism, so the transitive group will act as collineation group of the extended plane (the 'socle' plane), perhaps not faithfully. An understanding of these parallelisms is of considerable importance and interest and is also a wonderful blend of geometric, combinatorial, and group theoretic ideas. If $q = p^r$, the transitive group G admits a Sylow p-group of order divisible by q. However, this group will be forced to 'grow' to order q^2 to accommodate the action on the spreads. When this occurs, it will turn out that every non-socle plane will be a translation plane with spread in $PG(3,q)$ admitting a Baer group. Hence, we may apply the Johnson, Payne-Thas theorem that states the the plane is a derived conical flock plane. This partial classification theorem is due to Biliotti, Jha, and Johnson [19] who proved the result under a certain restriction on the Sylow p-subgroups (that they are 'linear') and to Diaz, Johnson, and Montinaro [45], who removed this restriction. The proof given here combines ideas of both of these papers. The background material required for reading the following theorem is listed in the appendix Chapter 41.

1. BJJDJM-Classification Theorem

In this chapter with give a classification of transitive deficiency one partial flocks. This theorem is the cumulative work of Biliotti, Jha, and Johnson [19] and Diaz, Johnson and Montinaro [45], the so-called BJJDJM-Classification Theorem.

THEOREM 261. *Let $\mathcal{P}^-$ be a deficiency one partial parallelism in $PG(3,q)$ that admits a collineation group in $P\Gamma L(4,q)$ acting transitively on the spreads of the partial parallelism. Let $\mathcal{P}$ denote the unique extension of $\mathcal{P}^-$ to a parallelism. Let the fixed spread be denoted by Σ_0 (the 'socle'), and let the remaining $q^2 + q$ spreads of $\mathcal{P}^-$ be denoted by*

Σ_i, for $i = 1, 2, .., q^2 + q$. Let π_i denote the affine translation plane corresponding to Σ_i. Then Σ_0 is Desarguesian and Σ_i is a derived conical flock plane for $i = 1, 2, .., q^2 + q$.

The proof shall be given a series of lemmas.

In the following lemmas, we shall assume the hypothesis and notation of Theorem 261.

LEMMA 77. *Assume that $q^2 - 1$ has a p-primitive divisor u (an element that divides $q^2 - 1$ but does not divide $p^s - 1$, for $q = p^r$ and $s < 2r$). Let U be a Sylow u-subgroup of G. Then U is Abelian.*

PROOF. Let U be a Sylow u-group of G of order u^a. Since U has order u^a and u divides $q^2 - 1$, then U fixes at least two components of the socle, for if it fixes zero or one component, then every non-trivial orbit is divisible by u, forcing u to divide $q^2 + 1$ or q^2, where $q^2 + 1$ is the number of components of the affine socle plane. Let g_u be any element of U and assume that g_u fixes a non-zero point on a fixed component. Let X denote the $GF(p)$-subspace pointwise fixed by g_u. Then there is a Maschke complement C left invariant by g_u, on which g_u fixes no non-zero point. Hence, it follows that u divides $|C| - 1$, a contradiction unless C is trivial. This implies that g_u is an affine homology.

Thus, either there exists an element of U, which is an affine homology or U acts fixed-point-free on any fixed component. Let L and M be two components fixed by U and let $U_{[Z]}$ denote the subgroup of U fixing Z pointwise, where $Z = L$ or M. Then, since any homology group of odd prime power order is cyclic as such a group is a Frobenius complement, it follows that $U_{[L]}U_{[M]}$ is an Abelian subgroup. Hence, $U/U_{[L]}$ acts faithfully as a fixed-point-free subgroup acting on L. Since a fixed-point-free group is a Frobenius complement, then $U/U_{[L]}$ is cyclic. But, either any element of order u^β normalizing $U_{[L]}$ must centralize $U_{[L]}$ or u divides $u^t - 1$ for some t. Hence, it follows that U centralizes $U_{[L]}$, so it must be that U is Abelian. $\qquad\square$

LEMMA 78. *Any element g_u in $\Gamma L(4, p^r)$ of order a prime p-primitive divisor u is in $GL(4, p^r)$.*

PROOF. If u divides r, then consider the element

$$\theta : (x_1, x_2, y_1, y_2) \longmapsto (x_1^{p^{r/u}}, x_2^{p^{r/u}}, y_1^{p^{r/u}}, y_2^{p^{r/u}})$$

of order u. This element fixes each vector (x_1, x_2, y_1, y_2), for $x_i, y_i \in GF(p)$. The argument of the previous lemma then would state that θ must fix each element on both $x_1 = 0 = x_2$ and $y_1 = 0 = y_2$, a contradiction. $\qquad\square$

REMARK 70. *G acts as a collineation group on the socle plane Σ. Recall that when we say that G contains an elation it is meant that the collineation acts on translation plane corresponding to Σ as an elation.*

LEMMA 79. *Suppose that G contains an elation with axis L and there is a p-primitive divisor u of $q^2 - 1$. Let U denote a Sylow u-subgroup of the full group acting on the parallelism. If U leaves L invariant and does not centralize the elation subgroup with axis L, then the socle is a semifield plane.*

PROOF. Let S_p be a Sylow p-subgroup of G, so of order $\geq q$. Let E denote the elation subgroup. Since U then normalizes E, it follows that either U centralizes E or induces a non-trivial fixed-point-free group on E, implying that E has order q^2, that is, the socle is a semifield plane.   □

We also note that linear p-collineations are of three possible types. We have a Sylow p-group S of order divisible q in $\Gamma L((4, q)$. Then $S \cap GL(4, q)$ has order at least $q/(r, q) > 1$, where $q = p^r$, for p a prime, so there is always a linear p-collineation. We now regard the group G acting on the associated 4-dimensional vector space over $GF(q)$.

LEMMA 80. *Any linear collineation of order p in the translation complement of a translation plane of order q^2, $q = p^r$, p a prime, whose associated spread is in $PG(3, q)$ is one of the following three types:*
(1) elation,
(2) Baer p-collineation (fixes a Baer subplane pointwise), or
(4) quartic (an element of order p that fixes exactly a 1-dimensional subspace of V_4) that lies on a component (a 'line' of the associated spread).

PROOF. Simply use Theorem 294.   □

1.1. When *G* Admits an Elation.

PROPOSITION 16. *Let π be any translation plane of order q^2 with spread in $PG(3, q)$ that admits an affine elation σ in the translation complement. Then σ fixes exactly $q(q+1)$ Baer subplanes of π (incident with the zero vector), defined by 2-dimensional $GF(q)$-subspaces.*

PROOF. A central collineation of a projective plane will fix a subplane if and only if the center and axis of the collineation are in the subplane and there is some point of the subplane that is mapped back into the subplane under the collineation. (The reader might try to prove this as an exercise.) For affine elations σ of a translation plane π, it follows that any affine subplane must share the axis L (say, chosen as $x = 0$) and therefore share a 1-dimensional $GF(q)$-subspace. Choose a 1-dimensional subspace and coordinatize so that $(0,0,0,1)$ generates this 1-space. We may always represent a particular elation σ as follows: $\sigma : (x, y) \longmapsto (x, x + y)$, where x and y are 2-vectors $(x_1, x_2), (y_1, y_2)$, respectively, for $x_i, y_i \in GF(q)$. An invariant subplane is given by a 2-dimensional $GF(q)$-subspace that is not a component of π. Hence, if we choose a vector not on $x = 0$, say $(x_1^*, x_2^*, y_1^*, y_2^*)$, so x_1^* or x_2^* is not zero, in terms of generating a 2-dimensional $GF(q)$-subspace, we may assume that $y_2^* = 0$. Then

$$\sigma : (x_1^*, x_2^*, y_1^*, 0) \longmapsto (x_1^*, x_2^*, x_1^* + y_1^*, x_2^*).$$

This vector must be in $\langle (x_1^*, x_2^*, y_1^*, 0), (0,0,0,1) \rangle$, so that there exist $\alpha, \beta \in GF(q)$, where

$$(x_1^*, x_2^*, x_1^* + y_1^*, x_2^*) = \beta(x_1^*, x_2^*, y_1^*, 0) + \alpha(0,0,0,1).$$

Since at least one of x_1^* and x_2^* is non-zero, this implies that $\beta = 1$, $\alpha = x_2^*$, implying that $x_1^* = 0$. So, the invariant subspace is

$$\langle (0, x_2^*, y_1^*, 0), (0,0,0,1) \rangle$$

and since x_2^* is then non-zero, we may choose $x_2^* = 1$, without loss of generality. Hence, the subspace is

$$\langle (0, 1, y_1^*, 0), (0,0,0,1) \rangle.$$

Then for each element $y_1^* \in GF(q)$, we obtain a distinct 2-dimensional $GF(q)$-subspace that is invariant under σ. Hence, for each of the $q + 1$ 1-dimensional subspaces X of $x = 0$, there are exactly q Baer subplanes invariant under σ that contain X. Therefore, σ fixes exactly $q(q + 1)$ Baer subplanes of π. □

LEMMA 81. *Each elation leaves precisely q non-socle planes invariant.*

PROOF. Since every elation σ fixes precisely $q(q+1)$ Baer subplanes, all of which non-trivially intersect the axis L of σ, then each σ-invariant Baer subplane is a component in exactly one non-socle plane π_i. But then L is a Baer subplane of π_i and σ, as a Baer collineation of π_i, must fix exactly $q + 1$ components, the components that non-trivially intersect L as a Baer subplane. Hence, σ fixes exactly q non-socle planes. □

LEMMA 82. *If an elation exists in G, then there are at least $q + 1$ non-trivial elations.*

PROOF. Each elation σ fixes exactly q non-socle planes and G acts transitively on the non-socle planes. Hence, each non-socle plane is left invariant by some non-trivial elation of the socle plane. Therefore, there are at least $q+1$ non-trivial elations, if there is at least one. $\square$

LEMMA 83. *Every elation group of order p in a translation plane of order p^2 is regulus-inducing.*

PROOF. Consider the group generated by an element $\sigma : (x, y) \rightarrow (x, x+y)$, then the images union the axis of a component $y = xM$ are of the form:

$$x = 0, y = xiM; i \in GF(p),$$

which is clearly a regulus in $PG(3, p)$. $\square$

LEMMA 84. *If there is an elation in G and $q^2 = p^2$, for p a prime and $p + 1 = 2^a$, then the socle is Desarguesian*

PROOF. If $q = p$ and $p+1 = 2^a$ then, in this setting, since there are elations, and every elation group of order p is regulus-inducing, we have that Σ is a conical flock plane. Moreover, since any elation must also fix a Baer subplane, it follows that the order of the p-group is strictly larger than p, so it can only be that the order is p^2, since there can be no Baer p-elements by Theorem 298, as there are elations. Now we have a linear group of order p^2 with a regulus-inducing elation group contained in it. Let E denote the elation group with axis L. If E has order p^2, then Σ is a semifield plane, which is necessarily Desarguesian, since the order is a prime square. Hence, we may assume that E has order p and that a Sylow p-group fixes a 1-dimensional subspace X_L on L pointwise. If L is moved, then $SL(2, p)$ or $SL(2, 5)$ is generated by the elations. The first case implies that the plane π_0 is Desarguesian. In the latter case, there are exactly 10 elation axes and there is a group of order 12 that leaves L invariant and normalizes the Sylow 3-subgroup of order 9. This group leaves X_L invariant, and is transitive on the remaining 3 1-dimensional $GF(3)$-subspaces. Hence, we have a group of order 4 fixing X_L and a second 1-dimensional subspace on L. The group of order 4 must fix a second elation axis as well. The combinatorics of this situation imply that there is a Baer collineation, a contradiction.

Thus, L is invariant. If X_L is not invariant, then $SL(2, p)$ or $SL(2, 5)$, and $p = 3$. In this second situation, there are exactly 4 1-dimensional $GF(3)$-subspaces so the $SL(2, 5)$ case cannot occur. In the second case, then π_0 is Desarguesian, Hall, Hering or Walker of order 25 by Chapter 41, Theorem 285. But, since we have elations in the plane, only the Desarguesian plane survives. Therefore, assume that X_L is not invariant and the group induced on L is $SL(2,5)$.

Thus, X_L and L are both invariant under the full collineation group of the parallelism. The group induced on L is a subgroup of $\Gamma L(2,p) = GL(2,p)$, acting on L as a Desarguesian affine plane of order p. However, X_L is a fixed component and the stabilizer of a component of L has order dividing $p(p-1)^2$. But we have a group of order divisible by $p^2(p+1)$ fixing L and X_L. Hence, there must be an affine homology group with axis L of order at least $(p+1)/4$. An affine homology cannot commute with any elation of E, implying that $(p+1)/4$ must divide $(p-1)$, a contradiction unless $(p+1)/4$ is 1 or 2. Therefore, we have a contradiction unless possibly $p=3$. But then the plane must be Desarguesian or Hall, and the Hall plane does not contain affine elations.   □

LEMMA 85. *Assume that there is an element τ_u of G of order a p-primitive divisor of q^2-1. If U is a Sylow u-subgroup of G, then U is linear.*

PROOF. By assumption, there is a p-primitive prime divisor u of q^2-1, which must then divide $q+1$. This implies that there is an element of order u in G. Let U be a Sylow u-subgroup of G. From Lemma 77, U is Abelian and elements of order u are linear. We claim that U is linear. If not, let ρ be an element of order u^t, for $t > 1$, of U. We know that $(u,r) = 1$, so let ρ be given by

$$(x_1, x_2, y_1, y_2) \longmapsto (x_1^{p^z}, x_2^{p^z}, y_1^{p^z}, y_2^{p^z})\rho^*,$$

where ρ^* is in $GL(4,q)$. This means that the non-linear part of this equation will never drop off by taking powers of u. Hence, U is linear.   □

LEMMA 86. *If all elations of G have the same axis, then the order of the elation group is q^2. Furthermore, then the socle plane π_0 is a semifield conical flock plane and the non-socle planes are derived conical flock planes.*

PROOF. The order of the maximum elation subgroup that fixes a non-socle plane is an invariant. Let this order be p^t. Then there is an elation subgroup E_i of order p^t that fixes a non-socle plane π_i and also fixes a total of q such non-socle planes. Since G normalizes the full elation group E with axis L, it must permute the subgroups E_i and these elation subgroups are necessarily mutually disjoint. Then there is a partition of the full elation group E of order p^s by exactly $q+1$ mutually disjoint groups of order p^t. Note that $p^t \leq q$, since E_i acts on π_i as a Baer group and the maximum p-order of a Baer group is q. Therefore, $(q+1)(p^t-1) = p^s - 1$. Hence, we must have $qp^t + p^t - q = p^s$. Divide the equation by p^t, to obtain $q + 1 - q/p^t = p^{s-t}$. Since there are at least $q + 1$ non-trivial elations and these are assumed to have the same axis, then $s - t > 0$. If $p^t < q$, we have a contradiction, as p would divide all terms but the '1'-term. Hence, $p^t = q$, implying $p^{r-t} = q$, so that the order of the elation group is q^2. So, the non-socle planes admit a Baer group of order q and by the Theorem of Johnson, Payne-Thas, the non-socle planes are derived conical flock planes. But, then the groups E_i are what are called 'regulus-inducing' elation groups as acting on the socle plane π_0, so that π_0 is a semifield conical flock plane.   □

THEOREM 262. *Assume that q is not 8. If an elation exists and all elations have the same axis, then the socle plane π_0 is Desarguesian and the non-socle planes are derived conical flock planes.*

PROOF. It remains to show that π_0 is Desarguesian. Since all even order semifield conical flock planes are Desarguesian by a result of the author Theorem 296, we may assume that q is odd. In this setting, there is a prime p-primitive divisor of $q^2 - 1$, u and there is a collineation τ_u of order u, which is linear and furthermore the Sylow u-group U containing τ_u is also linear.

We wish to show that this plane is Desarguesian. If not, then the full collineation group of π_0 permutes a set of q regulus nets sharing the elation axis L (see, e.g., Johnson and Payne [147] or Johnson [124]). Hence, there is a group of order $q + 1$ that leaves $y = 0$ and $x = 0$ invariant and must therefore fix a regulus net R. It follows that τ_u must fix a third component, say, $y = x$ of R as u cannot divide $q - 1$. Therefore, by Theorem 286, there is an Ostrom phantom plane Π_0 consisting of the τ_u-fixed subspaces of dimension 2 over $GF(q)$ (of the associated 4-dimensional $GF(q)$-vector space). Since Π_0 is regular and contains $x = 0, y = 0, y = x$, it follows that Π_0 contains the regulus R in its spread. Since E_1 is regulus-inducing, it follows easily that E_1 is also a collineation group of Π_0 and τ is a kernel homology group of Π_0 (fixes all components of Π_0). However, then E_1 must commute with τ_u, a contradiction. Hence, $\pi_0 = \Pi_0$. This completes the proof in this case.   □

LEMMA 87. *If q is not 8, then either the socle plane π_0 is Desarguesian (for q not 8) or any Sylow u-subgroup U fixes exactly two components of the socle plane and these components are not elation axes.*

PROOF. Assume that U fixes at least three components of π_0. Then there is an associated Ostrom phantom Π_0 consisting of U-fixed 2-dimensional $GF(q)$-subspaces. Suppose there is a component J of $\Pi_0 - \pi_0$. Then J is a Baer subplane of π_0 and hence lies in a unique translation plane π_i, for i not 0, as a component. Hence, U fixes π_i. But this means that the order of G is divisible by $q(q+1)\,|U|$, a contradiction as u divides $q+1$ and U is a Sylow u-subgroup. Therefore, π_0 is Desarguesian. Moreover, U centralizes all elation groups. By Lemma 86, there is more than one elation axis.

So, assume that U fixes exactly two components of π_0. If either of these components L and M are elation axes, assume first that U centralizes the elation group E_L with axis L. Then since U fixes M and E_L does not, it follows that U must also leave invariant the image set of M under E_L. Hence, U does not centralize E. It follows that E has order q^2 and either the plane is Desarguesian or there is a unique elation axis. Theorem 262 then shows that plane is Desarguesian in this latter case. This completes the proof of the theorem. $\square$

So, if there exist elations in G, either we are finished or they cannot all have the same axis. The group generated by a set of elations in G with more than one axis is as follows: $SL(2, p^z)$, $SL(2, 5)$ and $p = 3$, $Sz(2^e)$, e odd and $p = 2$ or dihedral of order $2t$, for t odd and $p = 2$, by Theorems 290 and 291. We consider the dihedral case D_t first.

1.2. The Dihedral Case Cannot Occur. Note that in the dihedral case, there are exactly t elation axes, each of which is the axis for a unique elation of order 2. Since there are at least $q + 1$ elations, then $t \geq q+1$. Each elation fixes exactly q non-socle planes, and there is a partition of these into exactly $q + 1$ parts. If $t > q + 1$, then one of these parts admits two elations leaving invariant each non-socle plane of a set of q planes, a contradiction. Hence, $t = q + 1$. When q is not 8, we have 2-primitive elements of prime order. But, when $q = 8$, we now have a cyclic group of order $8 + 1 = 9$ acting transitively on the set of elation axes. So, we have a linear cyclic group of order 3^2, and 3^2 is a 8-primitive divisor of $8^2 - 1$. Let U denote a Sylow u-group, when q is not 8 or a 3-subgroup when $q = 8$. U is linear, when q is not 8, and possibly non-linear when $q = 8$, but has a linear subgroup of order divisible by 9. We note that U must fix at least two components of π_0, simply by order. Let N_E denote the net defined by the set of $q + 1$ elation axes. Let C_{q+1} denote the cyclic stem of order $q + 1$ of the dihedral group D_{q+1} generated by the elations.

We know by Theorem 297 that either the elation net N_E is a regulus net or the plane is 'twisted' through a Desarguesian plane (the order

$q = 8$ is not a special case, as may be seen using q-primitive divisors in place of p-primitive divisors, as there is a cyclic group of order 9, when $q = 8$). In either case, there are two Baer subplanes of an associated Desarguesian plane Π_0 that are fixed by D_{q+1} and which non-trivially intersect each elation axis of π_0. These two Baer subplanes of Π_0 are then not components of π_0, so that D_{q+1} fixes at least two Baer subplanes of π_0. Moreover, since D_{q+1} can only fix Baer subplanes that non-trivially intersect each Baer axis, then each fixed Baer subplane must lie within the elation net N_E. We know, from our work on Baer subplanes of a net, that if there are three Desarguesian planes of a net of degree $q + 1$ and degree q^2 then there are $q + 1$ and the net is a derivable net (regulus net in this case). Hence, there are either two or $q + 1$ fixed Baer subplanes of π_0 by D_{q+1}. However, since D_{q+1} is normal (as it is generated by the set of all elations in G), it follows that G permutes a set of at most $q+1$ fixed non-socle planes, a contradiction to transitivity. Hence, the dihedral case does not occur.

1.3. The Suzuki Group Case Does not Occur.
Let the group generated by the elations in G be isomorphic to $Sz(2^e)$, where e is odd. If q is not 8, then $Sz(2^e)$ is normalized by a 2-primitive collineation τ_u. In this setting, the socle plane cannot be Desarguesian. Hence, by Lemma 87, then any Sylow u-subgroup U fixes exactly two components of π_0, neither of which are elation axes. Therefore, U normalizes but does not centralize $Sz(2^e)$.

But τ_u permutes the set of $2^{2e} + 1$ elation axes and each elation group has order $2^e \leq q$. So, U does not centralize $Sz(2^e)$ then u divides $2^{2e} + 1$, so that u divides

$$(2^{2r} - 1, 2^{2e} - 1) = (2^{(2r,4e)} - 1) = 2^{2r} - 1,$$

so that $2r$ divides $4e$. Hence, r divides $2e$. If r is odd, then r divides e so that there is an elation subgroup of order at least q. Therefore, the order of a Sylow 2-group in $Sz(2^e)$ is at least q^2. But there are no Baer involutions in $Sz(2^e)$, so we must have $Sz(q)$ in this case. Now if τ_u centralizes $Sz(q)$, then τ_u is a kernel homology group of π_0, implying that the plane is Desarguesian, which cannot be the case. Hence, U normalizes but does not centralize $Sz(q)$, a contradiction to the order of the outer automorphism group of $Sz(q)$, since the order of the outer automorphism group is r, for $q = 2^r$, and u cannot divide r (see, e.g., the Atlas [**35**]), so that τ_u belongs to $Sz(q)$. But, the order of $Sz(q)$ is $q^2(q^2 + 1)(q - 1)$, and u therefore does not divide the order of $Sz(q)$.

Hence assume that r is even so that $r/2$ divides e, implying that there is an elation group of order at least $\sqrt{q}$. The elation group cannot be exactly $\sqrt{q}$ by Büttner [**29**], since there we would have a $Sz(\sqrt{q})$ acting on a translation plane of order q^2, which does not occur. Let the order of an elation group be $\sqrt{q}2^a$. We note by results of Hering [**67**] this will force π_0 to contain a Lüneburg-Tits subplane ρ_0 of order 2^{2e}, which is then of order $2^{2(r/2+a)} = 2^{2a}q$, which is strictly larger than the order of a Baer subplane. Therefore, this cannot occur.

Hence, if q is not 8, the Suzuki case does not occur. Therefore, assume $q = 8$ and assume that $Sz(2^e)$, for e odd acts. Since q is not a square and there is an invariant Lüneburg-Tits subplane of order 2^{2e} of π_0, and $2^e \geq 8$, we see that we obtain $Sz(8)$, forcing π_0 to be a Lüneburg-Tits plane. We have a 3-group U_3 of order at least 9, which normalizes $Sz(8)$. Since 3 does not not divide the order of $Sz(8)$, either U_3 centralizes $Sz(8)$ or 9 divides the order of the outer automorphism group of $Sz(8)$, a contradiction as the outer automorphism group has order 3. Hence, U_3 centralizes $Sz(8)$ and therefore fixes each elation axes and hence is a kernel homology group of π_0. But this means that π_0 is Desarguesian, a contradiction.

Therefore, the Suzuki group case cannot occur.

1.4. The $SL(2, p^t)$ Case Does Not Occur. When $SL(2, p^t)$ acts and is normalized by a Sylow U-subgroup, then either U centralizes $SL(2, p^t)$ or u divides $p^t + 1$, the number of elation axes. So, if q is not 8, then u dividing $p^t + 1$ implies that we have $SL(2, q)$ or $SL(2, q^2)$, both of which imply that the plane π_0 is Desarguesian, since the group is generated by affine elations.

In the case $SL(2, q)$, the set of $q + 1$ elation axes form a derivable net in π_0, and there is a set of $q + 1$ Baer subplanes each fixed by $SL(2, q)$. Since $SL(2, q)$ is normal in G, we see that G permutes a set of at most $q + 1$ non-socle planes containing the fixed Baer subplanes of $SL(2, q)$ as components.

So, again for q not 8, assume that τ_u centralizes $SL(2, p^t)$. Then τ_u fixes at least $p^t + 1$ components, so that there is an Ostrom phantom Π_0 admitting as a collineation group the group normalizing $\langle \tau_u \rangle$. Moreover, Lemma 87 shows that $\pi_0 = \Pi_0$. Similarly, U must centralize $SL(2, p^t)$ and centralizes τ_u. Therefore, $SL(2, p^t)U$ is a collineation group of Π_0, where U is a kernel homology group, as it is linear. Let L denote an elation axis. Now that π_0 is Desarguesian, we have a linear p-group S_p^- of order $> \sqrt{q}$ that fixes a 1-dimensional subspace X pointwise on L and hence must itself be an elation group acting on π_0, an subplane of order p^t. It follows easily that we must have $SL(2, q)$ or $SL(2, q^2)$ acting on π_0 and are subgroups of G. In the first case, $SL(2, q)$ defines a regulus net, containing Baer subplanes that are components in some non-socle planes. Since $SL(2, q)$ is normal in G, it follows that G permutes a set of at most $q + 1$ non-socle planes, a contradiction. If we have the group $SL(2, q^2)$, there is an elation group of order q^2, which means that there are Baer groups of order q acting on the non-socle planes, which basically is what we are trying to prove.

Hence, we are finished unless $q = 8$. When this occurs, and we have $SL(2, 2^t)$ generated by elations, as well as a 3-group U of order 9, which normalizes $SL(2, 2^t)$. So, we have $SL(2, 2^t)$, where the elation net contains a Desarguesian subplane of order 2^t. Therefore, we could have $SL(2, 4)$, $SL(2, 8)$, or $SL(2, 8^2)$. In the latter two cases, the plane π_0 is Desarguesian. Our arguments above show that the cases $SL(2, 8)$ or $SL(2, 8^2)$ establish the theorem. Hence, we are reduced to considering $SL(2, 4)$. Our group is in $\Gamma L(4, 8)$, so a linear 2-group of order $q/(3, 2) = q$. But, since we have an elation, this group (necessarily linear) has order at least 16, since every elation fixes a Baer subplane and hence fixes a non-socle subplane. Let L be an elation axis. A Sylow 2-subgroup (linear) fixes a 1-dimensional $GF(8)$-subspace X pointwise. The group induced on L has order at least 4 since the elations generate $SL(2, 4)$. If X is moved by the stabilizer of L, then we have $SL(2, 2^s)$ generated on L, where s divides 3 and $s \geq 2$. Therefore, $SL(2, 8)$ is generated on L, so that the plane π_0 is either Hall or Ott-Schaeffer, but neither of these planes admit affine elations. Hence, X is left invariant by the stabilizer of L. There are 5 elation axes so every 3-group S_3 (of order at least 9) fixes at least two of these axes and hence must stabilize some two corresponding 1-dimensional subspaces X_i, one on each of the two fixed elation axes L_i, $i = 1, 2$. Since the order of S_3 is at least 9, there must be a linear 3-element τ_3, which then fixes X_i, for $i = 1, 2$ pointwise (i.e. 3 does not divide 7), τ_3 is a linear Baer collineation of order 3. But linear Baer groups have order dividing $8(8 - 1)$, a contradiction by Theorem 298. This finishes the $SL(2, p^t)$-case.

1.5. The $SL(2, 5)$-Case Does not Occur. Here we must have $p = 3$ and there is a collineation τ_u, which permutes the 10 elation axes. If τ_u divides 10, then $\tau_u = 5$ and 5 divides $3^{2r} - 1$ but does not divide $3^e - 1$, for $e < 2r$. Since 5 divides $3^4 - 1$, this means that $4 \geq 2r$, so that $r = 2$ or 1 so that $q = 3$ or 9. Hence, $u = 5$. Also, U must normalize $SL(2, 5)$ and since the outer automorphism group of $SL(2, 5)$ has order 2, it follows that U centralizes $SL(2, 5)$, implying that U fixes all elation axes, contrary to Lemma 87.

THEOREM 263. *If G contains an elation, then the theorem is proved.*

COROLLARY 67. *q is not 8.*

PROOF. We need to show that there is an elation if q is 8. If $q = 8$, note that the Sylow 2-subgroups are in $\Gamma L(4, 2^3)$, so the Sylow 2-subgroups are linear. If there are no elations, then there is a Baer involutions in π_0. But this Baer involution fixes a non-socle plane π_1, which means that the Sylow 2-subgroup must have order at least $2 \cdot 8$. But such linear 2-group fixes a component L and fixes a 1-dimensional subspace X on L pointwise. Therefore, the group induced on L has order dividing 8, which implies that the group is not faithful, thus there is an elation in G. $\qquad \square$

1.6. If G contains a Baer p-Collineation, p-Primitive Case.
In the general case, we note the existence of an elation whenever, there is a linear p-group of order strictly larger than q.

LEMMA 88. *Any linear Sylow p-subgroup S_p of G that has order $p^a q$ contains an elation group of order p^a.*

PROOF. S_p is a subgroup of $GL(4, q)$ acting on the translation plane π_Σ corresponding to Σ. Therefore, S_p fixes a 1-dimensional K–subspace X pointwise. Let L denote the unique component containing X. The maximal order subgroup that can fix X pointwise and be in $GL(4, q)$ and act faithfully on L has order q. Thus, it follows that there must be a subgroup of order p^a that fixes L pointwise. $\qquad \square$

First, assume that we have a p-primitive collineation τ_u. Let U be a Sylow u-subgroup of G. We begin by establishing a fundamental lemma.

LEMMA 89. *U cannot fix a Baer subplane of π_0.*

PROOF. If U fixes a Baer subplane of π_0, then the order of G must be divisible by $q(q + 1) |U|$. However, u divides $q + 1$, a contradiction to the fact that U is a Sylow u-subgroup of G. $\qquad \square$

We let S_p^- denote the linear subgroup of S_p.

LEMMA 90. *If p is odd, then S_p^- cannot contain a Baer p-collineation.*

PROOF. If σ in S_p^- is Baer in π_0, let τ_u be a p-primitive collineation.
Let U be a Sylow u-subgroup of G. We see that if U leaves $Fix\sigma$, a Baer subplane of π_0 we obtain a contradiction to transitivity by the previous lemma.
Therefore, we must have that U does not leave $Fix\sigma$ invariant.
Assume first that p is odd. Then by Theorem 298, all p-elements lie on the same net of degree $q + 1$ or $p = 3$, and the group generated by the Baer collineations is either $SL(2, 5)$ or $SL(2, 3)$.

If $p = 3$ and $SL(2,5)$ is generated, there are exactly 10 Baer axes. This implies that $q = 9$. Hence, $u = 5$. Also, U must normalize $SL(2,5)$ and since the outer automorphism group of $SL(2,5)$ has order 2, it follows that U centralizes $SL(2,5)$, implying that U fixes all Baer axes, but U is forced to fix each Baer subplane, a contradiction by the previous lemma. If $p = 3$ and the group is $SL(2,3)$ then clearly U will centralize $SL(2,3)$, so that, in fact, U leaves invariant some $Fix\sigma$.

Now we must have a net N_σ of order $q+1$ containing all Baer axes. Since $Fix\sigma$ is Desarguesian, then this net N_σ contains at least two Baer subplanes admitting Baer groups, implying that there are at least three Baer subplanes, which implies by using Theorem 298, part (4), that there are exactly $q + 1$ Baer subplanes, so that the net containing the Baer p-elements is a regulus net. Furthermore, we must have $SL(2,q)$ generated by the Baer p-collineations. So, π_0 must be Hall, as the p-groups are generated by Baer p-collineations. Hence, we have a Baer group B of order q and this group fixes a plane π_j, j not 0, implying that the Sylow p-subgroup of G has order q^2 and admits a linear subgroup of order divisible by $q^2/(p^{2r}, r) > (\sqrt{q})^2$, so there is a linear p-group of order strictly larger than q. S_p^- then fixes a 1-dimensional $GF(q)$-subspace pointwise, which lines on exactly one component L of π_0. The group induced on S_p^- is an elementary Abelian subgroup of $GL(2,q)$ and hence has order dividing q. Therefore, there is an elation in S_p^-, a case that was considered previously, which completes the odd order case, when there is a Baer p-element. $\qquad \square$

LEMMA 91. *If $p = 2$, then S_2^- cannot contain a Baer involution.*

PROOF. So, assume that q is even. Let σ be a linear Baer involution and let the order of the Baer p-group B fixing $Fix\sigma$ pointwise be p^t, where $t \leq r$. Then B fixes a unique plane π_j that contains $Fix\sigma$ as a component, for j not 0. This also implies that the order of S_p^- is $> p^t\sqrt{q}$, since B is linear. We know that S_2^- must leave invariant some component L of the socle plane π_0. Furthermore, the group that S_2^- induces on L is faithful since there are no elations in G. Since S_2^- is linear then S_2^- induces a 2-group in $GL(2,q)$ acting on L. But any such 2-group is elementary Abelian, so S_2^- is elementary Abelian. Assume that B is in S_2^-, then since B fixes a 'unique' non-socle plane π_j that has $FixB$ as a component, it follows that S_2^- must fix π_j. But then the order of the Sylow 2-subgroup containing S_2^- must have order divisible by $q|S_2^-|$, which implies that the only possibility is that q divide (q, r), where $q = 2^r$, a contradiction. $\qquad \square$

Therefore, S_p^- has order $\leq q$ and does not contain a Baer p-element or an elation.

1.7. If G contains a Baer p-Collineation, No p-Primitive Case. Therefore, we consider the cases $q = 8$ or $q = p$ and $p + 1 = 2^a$, there are no elations and there is a Baer p-element. If q is not 8, then all p-elements are linear, so by Lemma 88, the Sylow p-group of G has order p. However, a Baer collineation σ fixing a Baer subplane that is a component of one of the spreads of the parallelism. Hence, σ fixes a spread, which implies that the Sylow p-subgroup in G is strictly larger than p, a contradiction. Hence, assume that $q = 8$ and there is a Baer involution and no elations, which is a contradiction by Corollary 67.

1.8. All Linear p-Elements Are Quartic. Hence, by the previous two subsections, we are finished if there are Baer p-elements or elations in S_p^-. Hence, all p-elements in S_p^- are quartic.

Let σ be a quartic element with axis L and center X. Case 1. τ_u leaves L invariant. If τ_u fixes X, then τ_u is an affine homology and fixes a coaxis M, which must be moved by σ, this then implies by Theorem 293 that in the generated group, there is an elation. Hence, τ_u cannot fix X if it fixes L. We note that S_p^- leaves L invariant and induces a faithful group on L, since there are no elations in S_p^-. Hence, S_p^- is elementary Abelian. But, then all p-elements fix X pointwise and we have a quartic group of order $> \sqrt{q}$. It follows easily, since X is moved that $SL(2, q)$ is generated on L and therefore the planes are known by Theorem 295.

But, by the structure of the planes with spreads in $PG(3, q)$ admitting $SL(2, q)$, we realize that the plane π_0 is Hall, for which the group $SL(2, q)$ is generated by Baer p-elements, a contradiction, as there are no linear Baer p-collineations.

Hence, we must have case 2, that τ_u moves L. Since there are no quartic elements if q is even, we may assume that q is odd. But, again we have by Theorem 295 that $SL(2, q)$ is generated and the planes are Hering or Walker of order 25.

If the plane is Hering or Ott-Schaeffer, $SL(2, q)$ is normal in the full collineation group of π_0, implying that $SL(2, q)$ is normal in G. Now let X_i, for $i = 1, 2, .., q + 1$ denote the set of axes for quartic groups (note that in then $\langle X_i, X_j \rangle$, for i not j is a Baer subplane of π_0, which lies in some unique plane π_j, for $j \neq 0$. There are exactly $(q + 1)q/2$ such Baer subplanes and as the set of quadric groups is $q + 1$, this implies that this set of Baer subplanes is permuted by the full group G. However, this means that G cannot act transitively on $q^2 + q$ planes.

Hence, the only case left is when $q = 5$ and the plane is one of the three Walker planes of order 25.

THEOREM 264. π_0 *cannot be a Walker plane of order* 25.

PROOF. Now assume that $q = 5$ and the plane is a Walker plane of order 25. We note that one of the three Walker planes of order 25 is actually a Hering plane. But, here we are considering the group action to be reducible but not completely reducible (there are two groups isomorphic to $SL(2,5)$ acting on the Hering plane/Walker plane of order 25). Hence, we have $SL(2,5)$ acting on Σ, where the 5-elements are quartic.

Then the group $SL(2,5)$ is reducible, and furthermore, we know from the structure of the three Walker planes of order 25 that there is a unique 2-dimensional subspace W that is $SL(2,5)$-invariant and W is a component L. Consider an element g of $SL(2,5)$ of order 3. Then it is known that there is an associated Desarguesian spread Δ such that g fixes each line of Δ (see the section in Lüneburg [167] on the three Walker planes of order 25). Moreover, the three Walker planes of order 25 share $5, 8$, or 11 lines of Δ. In any case, this means that g fixes any non-socle spread of the parallelism containing a line of $\Delta - \Sigma$. Since Σ is Walker, g fixes at least one non-socle spread. Hence, 3^2 divides the order of G, since G is transitive on $5(6)$ non-socle spreads. Let S_3 be a Sylow 3-subgroup. We know that the order of a Sylow 5-subgroup is exactly 5 and the Sylow 5-subgroups are quartic. If $SL(2,5)$ is not normal, then we are back to the Hering case. Hence, S_3 must normalize $SL(2,5)$ (the group generated by the 5-elements) so that there is an element g^* of order 3, which centralizes $SL(2,5)$. Note that the normalizer of $SL(2,5)$ is contained in $\Gamma L(2,5)$. Also, a group H_3 of order 9 is Abelian, so centralizes g, implying that H_3 is a collineation group of the Desarguesian plane whose spread is Δ. This implies that H_3 is a subgroup of $\Gamma L(2,5^2)$. Since g^* centralizes $SL(2,5)$, g^* fixes each 1-dimensional $GF(5)$-subspace and since 3 does not divide $5-1$, it follows that g^* is a homology with axis L and coaxis say M. Since L is $SL(2,5)$-invariant but M is not, this implies by André's theorem [1] that there is an elation in G with axis L. Hence, we have an elation group E of order 5, a contradiction since there are no elations in G. $\qquad\square$

This completes the proof of the theorem.

2. Deficiency One—The Spreads Are Isomorphic

When there is a doubly-transitive group acting on the parallelism, of course, the spreads are isomorphic and the stabilizer of a spread induces a group acting transitive on the remaining $q^2 + q$-spreads. This situation has been determined previous by the author.

THEOREM 265. *(Johnson [112]) Let $\mathcal{P}$ be a doubly transitive 1-parallelism of $PG(3, q)$. Then $q = 2$, $\mathcal{P}$ is one of the two regular parallelisms in $PG(3, 2)$ and the group is $PSL(2, 7) \leq G$.*

However, we may obtain this same conclusion with the Classification Theorem of transitive deficiency one partial parallelisms.

THEOREM 266. *Let $\mathcal{P}^-$ be a deficiency one parallelism in $PG(3, q)$ that admits a collineation group in $P\Gamma L(4, q)$ acting transitively on the spreads of the partial parallelism. Let $\mathcal{P}$ denote the unique extension of $\mathcal{P}^-$ to a parallelism. If the spreads of the parallelism are all isomorphic, then $q = 2$ and the parallelism is one of the regular parallelisms in $PG(3, 2)$.*

PROOF. We know that the socle plane π_0 is Desarguesian and the non-socle planes are derived conical flock planes. That is, each non-socle planes admits a Baer group of order q. If all of the planes are isomorphic, then we have a Desarguesian plane admitting a Baer group of order q. But such groups in Desarguesian planes have orders bounded by 2. Hence, $q = 2$ and the remaining part of the theorem is clear. □

3. The Full Group

We have shown that if $\mathcal{P}^-$ is a transitive deficiency one partial parallelism in $PG(3, q)$ and $\mathcal{P}$ is the unique extension then the spread of $\mathcal{P} - \mathcal{P}^-$ is Desarguesian and the other spreads are derived conical conical spreads. What we have not done is to complete the description of the full group of the parallelism $\mathcal{P}$. In determining the structure of the parallelism, we ultimately showed that there is a Baer group of order q fixing a given translation plane obtained from a given non-socle spread. This means that there must be an elation group E of order q^2 on the socle plane π_0. If E is not normal, then $SL(2, q^2)$ is generated by the set of elations of π_0. Moreover, since $SL(2, q)$ in $SL(2, q^2)$ fixes a regulus R of π_0, it follows that $SL(2, q)$ fixes exactly $q + 1$ Baer subplanes of π_0, which as a set form the opposite regulus R^* of R. All of the Baer subplanes must lie as components in the same non-socle plane π_1, and R^* a regulus of π_1, with Baer subplanes the $q + 1$

components of π_0 in R. Hence, each group $SL(2,q)$ fixes a unique non-socle plane. However, $SL(2,q^2)$ acts triply transitive on the $q(q^2+1)$ reguli of π_0, which means that there are at least two groups isomorphic to $SL(2,q)$ that fix a given non-socle plane. Let k denote the number of $SL(2,q)'s$ that fix a given non-socle plane. Then, by uniqueness, we must have $kq(q+1) = q(q^2+1)$, which implies that $q+1$ divides q^2-1, a contradiction. Therefore, we have the following theorem.

THEOREM 267. *The full collineation group of a transitive deficiency one partial parallelism in $PG(3,q)$ has a normal group E of order q^2. This group acts an an elation group of the associated socle plane given by the unique spread that extends the partial parallelism.*

We now consider the subgroup $G^- = G \cap GL(2,q^2)$ of a transitive deficiency one partial parallelism. We know that we have a normal elation group of order q^2. Hence, we may assume that any linear group, which does not contain a p-element fixes a second component. If the elation axis is $x = 0$, assume that second fixed component if $y = 0$. Thus, G^-/E is isomorphic to a subgroup of a direct product of two cyclic groups of order (q^2-1). We also know that E may be decomposed into $q+1$ regulus-inducing elation groups of order q, E_i, for $i = 1,2,..,q+1$. Furthermore, each elation subgroup E_i fixes exactly q non-socle planes. The stabilizer of a non-socle plane, say, π_1, admits one of the groups, say, E_1 as a normal subgroup. The full group of a derived conical flock plane is the inherited group for $q > 3$, by Jha and Johnson [85]. Therefore, consider the normalizer of E_1 in G^-. This group fixes a regulus R_1 of π_0 (the socle plane) and hence the possible order divides $(q-1)(q^2-1)$, where the (q^2-1) refers to the kernel subgroup K^+ of π_0 of the same order. Let K^- denote the subgroup of K^+ that acts on the parallelism. We note that E_1 fixes exactly q non-socle planes. Hence, we have a group of order divisible by $(q-1)|K^-|$ that permutes these q non-socle planes. Let τ be any collineation of K^- of prime power order v^α. Since τ must permute q^2-q components outside of the net that contains the Baer subplane fixed pointwise by E_1, suppose v^α does not divide $q-1$. Then some element of order v must fix a component. But, this component is a Baer subplane ρ_0 of π_0 and τ is a kernel homology group of π_0 and therefore induces a kernel homology on ρ_0, implying that τ divides $q-1$. Now assume that q is odd and τ has order 2^α and does not divide $q-1$. Then consider a Sylow 2-subgroup S_2 acting on the spread that leaves invariant a non-socle plane π_1. Then the corresponding parallelism-inducing elation group E_1 is necessarily normalized by S_2, which means that S_2 permutes the set of q^2-q components not intersecting the Baer subplane L fixed pointwise by E_1. We know that S_2 is a subgroup of order dividing $((q-1)(q^2-1))_2$. Indeed, considering the associated regulus as the standard regulus $x = 0, y = x\alpha; \alpha \in GF(q)$, of the socle plane, the elements of the normalizer of E_1 have the general form

$$\tau_{a,\alpha} : (x,y) \longmapsto (xa, ya\alpha), \text{ where } \alpha \in GF(q).$$

Notice that $\tau_{a,\alpha}^{q-1}$ is a kernel homology of π_0, and if any such kernel homology fixes a component of π_1–a Baer subplane of π_0 then this kernel element must have order dividing $q-1$. Suppose we try to determine a bound for the order of G^-. We claim that it is $q^2(q^2-1)(q-1)(2,q+1)$. In this case, we would have a stabilizer group of order $q(q-1)^2(2,q+1)$ containing the kernel group of order $q-1$. Moreover, in π_0, since the normalizer elements of E_1 have the above form and the group has order $(q-1)^2(2,q+1)$, it can only be that the elements a have orders dividing $2(q-1)$. Using the minimal bound, we have the following theorem.

THEOREM 268. *Let* $\mathcal{P}^-$ *be a transitive partial parallelism of deficiency one in* $PG(3,q)$. *Let* $\mathcal{F}$ *denote the full subgroup of* $GL(2,q^2)$ *that acts on* $\mathcal{P}^-$ *(that is,* $\mathcal{F}/K^*$, *for* K^* *the kernel subgroup of order* $q-1$*). Then*

$$q^2(q-1)_2(q+1) \mid |\mathcal{F}| \mid q^2(q^2-1)(2,q+1).$$

We have seen previously that the lower bound is taken on by a set of parallelisms, where the socle plane is Desarguesian and the remaining planes are derived Kantor-Knuth semifield flock planes. We now consider the upper bound. In this setting, it follows easily that we have a subgroup

$$H^- = \left\langle \begin{array}{c} (x,y) \longmapsto (x, y\alpha); \\ \text{for some elements } \alpha \in GF(q) \end{array} \right\rangle$$

of order at least $(q-1)/(2,q-1)$.

It is fairly direct to see that $E_1 H^-$ has component orbits of at least lengths $q(q-1)/(2,q-1)$. But, this also says that the $q-1$ orbits of E_1 of length q in π_1 are permuted into at most two orbits. For example, if there is one orbit, we notice that the indicated group H^- will induce a Baer group of order $q-1$ on π_1 and an affine homology group of the same order on the derived plane π_1^*, which is a conical flock plane. Hence, we have both a conical flock plane and a hyperbolic flock plane, implying that π_1^* is Desarguesian. But, even if q is odd and there is an affine homology group of order $(q-1)/2$, it is almost immediate that there is a Desarguesian partial spread that contains at least $1+(q-1)/2$ reguli that share a common component. But, it also now clear that the plane itself is Desarguesian. What this says is if the upper bound is taken on then the non-socle planes are Hall planes. It remains to show that there is a transitive partial parallelism whose group G has the upper bound as order.

Let π_0 denote the Desarguesian affine plane with spread

$$x = 0, y = xm; m \in GF(q^2)$$

and let R_0 denote the regulus net with in π_0 with partial spread

$$x = 0, y = x\alpha; \alpha \in GF(q^2).$$

The class of parallelisms of Johnson discussed in this text admit a collineation group G of order $q^2(q^2-1)$ consisting of central collineations with axis $x = 0$. Hence, the bound is taken on for q even. Let q be odd. Consider $y = x^q m$; m^{q+1} is a non-square in $GF(q)$. Consider the image of $y = x^q m$ under the group of order $q(q - 1)$ that fixes R_0. Note that $y = x^q m$ cannot intersect a subspace $x = 0$ or $y = x\alpha$, for $\alpha \in GF(q)$, since $\alpha^{q+1} = \alpha^2$ and m^{q+1} is non-square. The set of images of $y = x^q m$ under this subgroup is

$$y = x^q m\alpha + x\beta; \alpha \neq 0, \beta \in GF(q).$$

We claim that this set of images forms a partial spread. To see this note that for x non-zero if and only if $x^{q-1}(m(\alpha-1)) = \beta$. But since in the equation $(m(\alpha - 1))^{q+1} = \beta^{q+1}$, the left-hand side is a non-square and the right-hand side is a square then $\alpha = 1$ and $\beta = 0$.

We note that the spread Σ_1

$$x = 0, y = x^q m\alpha + x\beta; \alpha, \beta \in GF(q)$$

is Desarguesian. To see this, we claim that the elation group with elements $(x, y) \longmapsto (x + \delta y, y)$, for $\delta \in GF(q)$, is a collineation group of π_1. This element maps $y = x^q m$ to $y = x^q m + x(-m^{q+1}\delta)$. Since this element fixes R_0, it follows directly that we have an elation group of order q with axis $x = 0$ and an elation group of order q with axis $y = 0$. Hence, $SL(2, q)$ is generated by these elation groups, which implies that the plane is Desarguesian. If we derive this spread and take the image under the group G then the union of the set of images together with the spread for π_0, forms a parallelism. It remains to show that this parallelism admits a projective collineation group of order $q^2(q^2 - 1)(2, q + 1)$. Take the collineation $\tau : (x, y) \to (xa, ya^q)$, such that $a^q = -a$. We note that τ^2 is in the kernel group of order $q-1$, and therefore does not show up in the projective groups. Also, τ maps $y = x^q m\alpha + x\beta$ to $y = x^q m\alpha - x\beta$ and maps $y = x\delta$ to $y = -x\delta$, fixing $x = 0$ and $y = 0$. Therefore, is a collineation group of Σ_1 that fixes the regulus partial spread R_0. We have shown that τ is a collineation group of the derived plane π_1.

So, our parallelism is $\Sigma_0 \cup G\Sigma_1^*$, where Σ_1^* denotes the spread obtained by derivation of R_0 in Σ_1. Since τ normalizes G, and is a collineation group of both Σ_0 and Σ_1^*, then

$$\tau(\Sigma_0 \cup G\Sigma_1^*) = \Sigma_0 \cup G\Sigma_1^*.$$

Therefore, τ induces a collineation of order 2 induce on the parallelism. That is, there is a collineation group permuting the spreads of order $q^2(q^2 - 1)(2, q + 1)$.

We have therefore proved the following theorem.

THEOREM 269. *The upper bound of Theorem* 268 *is taken on by a transitive deficiency one partial parallelism if and only if the parallelism extending the partial parallelism has a Desarguesian socle and the remaining spreads are Hall.*

It should be pointed out that in the previous theorem the conclusion requires that the parallelism involved has a transitive deficiency one subpartial parallelism. A natural question is whether there are parallelisms that admit exactly one Desarguesian spread and the remaining spreads Hall that do not contain a transitive deficiency one sub-partial parallelism.

Consider again the parallelism $\Sigma_0 \cup G\Sigma_1^* = \mathcal{P}$ and choose any spread $g\Sigma_1^* = \Sigma_g^*$, derived this spread to produce the Desarguesian spread Σ_g and derive the Desarguesian spread Σ_0 to construct the Hall spread Σ_0^*. Then it is proved by the in the chapter on Johnson parallelisms that

$$\Sigma_0^* \cup \Sigma_g \cup \{h\Sigma_1^*; h \in G, \text{ and } h\Sigma_1^* \neq \Sigma_g^*\} = \mathcal{P}_g$$

is a parallelism, and we note that for $q > 2$, there is a unique Desarguesian spread Σ_g and the remaining spreads are Hall. Although these parallelisms have been known for awhile, there is nothing known regarding their isomorphisms or whether they could be isomorphic to $\Sigma_0 \cup G\Sigma_1^* = \mathcal{P}$.

THEOREM 270. *If $q > 9$ the parallelism*

$$\Sigma_0^* \cup \Sigma_g \cup \{h\Sigma_1^*; h \in G, \text{ and } h\Sigma_1^* \neq \Sigma_g^*\} = \mathcal{P}_g$$

of $PG(3, q)$ does not contain a transitive partial parallelism of deficiency one.

PROOF. If so, then there must be a group G_g fixing Σ_g and acting transitively on the remaining spreads. There are $q^2 + q$ remaining spreads of which $q^2 + q - 1$ are derived from Σ_0 by replacing reguli of Σ_0. Therefore, if $r \in G_g$, then $r\{h\Sigma_1^*; h \in G \text{ and } h\Sigma_1^* \neq \Sigma_g^*\} \cap \{h\Sigma_1^*; h \in G$

and $h\Sigma_1^* \neq \Sigma_g^*$} in at least $q^2 + q - 2$ reguli, say, by (re)indexing we have $\{\Sigma_i^*; i = 1, 2, .., q^2 - q - 2\}$. Hence, h maps Σ_i^* onto $\Sigma_{i(h)}^*$, for at least $q^2 + q - 2$. The full collineation group of a Hall plane of order > 9 fixes the regulus used in the replacement process (see, e.g., Jha and Johnson [**86**], [**88**]). This means that h is forced to map Σ_i to $\Sigma_{i(h)}$ and these are Desarguesian spreads that share a regulus with Σ_0. What this means is that h permutes a set of at least $q^2 + q - 2$ distinct reguli of Σ_0. Therefore, $h\Sigma_0 \cap \Sigma_0$ contains more than a regulus of components, implying that $h\Sigma_0 = \Sigma_0$, that is, h is a collineation of Σ_0. Hence, G_g is a collineation subgroup of Σ_0 and if Σ_0^* is mapped by h to Σ_k^*, it necessary means that Σ_0 is mapped to Σ_k, a contradiction. $\qquad\square$

COROLLARY 68. *If $q > 9$, the parallelism*

$$\Sigma_0^* \cup \Sigma_g \cup \{h\Sigma_1^*; h \in G, \text{ and } h\Sigma_1^* \neq \Sigma_g^*\} = \mathcal{P}_g$$

is not isomorphic to

$$\Sigma_0 \cup G\Sigma_1^* = \mathcal{P}.$$

We also note that there is a collineation z of G of Σ_0 that maps Σ_g to Σ_t and permutes the reguli containing the elation axis of the elation subgroup of G. This is true if and only if Σ_g^* maps to Σ_t^*, and note that z permutes the remaining reguli in question.

This proves the following theorem.

THEOREM 271. *The parallelism*

$$\Sigma_0^* \cup \Sigma_g \cup \{h\Sigma_1^*; h \in G, \text{ and } h\Sigma_1^* \neq \Sigma_g^*\} = \mathcal{P}_g$$

is isomorphic to the parallelism

$$\Sigma_0^* \cup \Sigma_t \cup \{h\Sigma_1^*; h \in G, \text{ and } h\Sigma_1^* \neq \Sigma_t^*\} = \mathcal{P}_t.$$

CHAPTER 39

Doubly Transitive Focal-Spreads

As pointed out in the chapters of focal-spreads, the only known focal-spreads are obtained from a construction of Beutelspacher [16]. In this chapter, the doubly transitive focal-spreads are completely determined as arising from Desarguesian spreads by k-cuts. The work in this chapter follows the article by the author and Montinaro [142], with appropriate modifications for this text.

Since any collineation group of a focal-spread necessarily leaves invariant the focus L, it then becomes a natural question if a doubly transitive group acting on an affine focal-plane then forces the focal-spread to have a similar structure as in the translation plane (i.e., t-spread) case. For example, is a doubly transitive affine focal-plane a k-cut of a semifield plane?

First, consider simply a group G acting on a focal-spread of type (t, k) over $GF(q)$ that admits a primitive group acting on the partial Sperner k-spread. Since the degree is q^t, by the O'Nan-Scott Theorem 7, either $G = TG_0$ and T is an elementary Abelian group of order q^t (T is the socle of G), or G is almost simple. The proof that we cannot obtain the almost simple case uses very technical aspects of Kleidman and Liebeck [161] and Guralnick's Theorem on simple groups of prime index, which provides us with a list of groups that can never apply in this situation. Although the proof is really beyond the scope of the text, we include it here but suggest that the reader have a copy of Kleidman and Liebeck in hand.

THEOREM 272. *If a group G is primitive on the partial Sperner k–spread of a focal-spread of type (t, k), then $G = TG_0$ and T is an elementary Abelian group of order q^t.*

PROOF. Assume not then G is almost simple. We assume that our group lies in $PG(t+k-1, q)$. Hence G fixes a $(t-1)$-subspace (projective dimension) and acts transitively on the q^t subspaces of projective dimension $k - 1$. Let $q^t = p^{rt}$, p prime. Let G_0 be the stabilizer of a K–subspace in G and let S be the socle of G (hence $S \trianglelefteq G \le Aut(S)$). Then either (i) $S \trianglelefteq G_0$ or (ii) $G = SG_0$ by [161].

We first consider case (ii), so let $S_0 = S \cap G_0$. As $G = SG_0$, then $[S : S_0] = [G : G_0] = q^t$. Then by [62], one of the following occurs:

(1) $S \cong A_{q^t}$ and $S_0 \cong A_{q^t-1}$;
(2) $S \cong PSL(d, u)$ and S_0 is the stabilizer of a line or a hyperplane in the natural representation. In particular, $[S : S_0] = \frac{u^d-1}{u-1} = q^t$ (clearly d must be prime);
(3) $S \cong PSL(2, 11)$ and $S_0 \cong A_5$;
(4) $S \cong M_{23}$ and $S_0 \cong M_{22}$;
(5) $S \cong M_{11}$ and $S_0 \cong M_{10}$;
(6) $S \cong PSU(4, 2) \cong PSp(4, 3)$ and H is the parabolic subgroup of index 27.

As $t > k \geq 1$, the cases (3)–(5) are ruled out. Assume that case (1) occurs. Then $S \cong A_{q^t}$. As $t > 1$ then since A_i is solvable for $i < 5$, we see that $q^t \geq 5$, but as $t > 1$, then $q^t \geq 8$. If $q^t \geq 9$, then $t + k \geq n - 2 \geq q^t$ by [161], Proposition 5.3.7. As $t > k$, then $2t \geq q^t$, which does not have a solution. So, $q^t = 8$ and we have again a contradiction by [161], Proposition 5.3.7. Hence, case (1) cannot occur.

Assume that case (2) occurs. Clearly, $u \neq t$. Then $t + k \geq u^{d-1} - 1$ by [161], Theorem 5.3.9, for $(d, u) \neq (3, 2), (3, 4)$. It is easily seen that, for our situation, it must be $(d, u) \neq (3, 2), (3, 4)$. Now

$$q^t = \frac{u^d - 1}{u - 1} = \frac{u}{u-1}(u^{d-1} - 1) + 1$$
$$\leq \frac{3}{2}(t + k) + 1 \leq 3t + 1,$$

which is impossible.

Finally, assume that the case (6) occurs. Then $S \cong PSp(4, 3)$ and $S_0 \cong 2^4 : A_5$ and $q^t = 27$. Hence $t = 3$ and $k = 1, 2$. Note that $S_0 \cong 2^4 : A_5$ is maximal in S. Moreover, the ambient space is either $PG(3, 3)$ (natural representation), or $PG(4, 3)$. We see that $PSp(4, 3)$ is a primitive rank 3 group on the 40 points, noting that $PG(3, 3)$ has exactly 40 points with (maximal) point-stabilizer not isomorphic to $S_0 \cong 2^4 : A_5$ (e.g., see the Atlas [35]).

So, it remains to investigate the action of $PSp(4, 3)$ in $PG(4, 3)$. Since $PSp(4, 3)$ fixes the focus, then it must lie in a maximal parabolic subgroup P_3 (3 in this case means the vector dimension) of $P\Gamma L(5, 3)$, which is impossible by [161], Proposition 4.1.17.

Now assume case (i). Then $q^t = [G : G_0] = [G/S : G_0/S]$. So $q^t \mid |OutS|$. Then S fixes each of the q^t projective $(k - 1)$-spaces, and fixes the focus. Therefore, S lies in a maximal parabolic subgroup of

$P\Gamma L(t+k, q)$. Thus, $S \leq PSL(t, q) \times PSL(k, q)$ by [161], Proposition 4.1.1, as S is nonabelian simple. As $t > k$, we may assume that $S \leq PSL(t, q)$. Recall that $p^{rt} = [G : G_0] \mid |OutS|$. Therefore, S cannot be alternating. Moreover, S cannot be sporadic by [161], Table 5.1.C. Hence, S is either classical or exceptional of Lie type.

Assume that S is in characteristic p. That is, we assume $S = S(p^f)$, $f \geq 1$. Then $p^{rt} \mid f$ by [161], Tables 5.1.A and 5.1.B. Hence a Sylow p-subgroup of S must have order at least $p^{p^{rt}}$. On the other hand, such a Sylow as must have order a divisor of $p^{rt(t-1)/2}$, as S lies in $PSL(t, q)$. So, $p^{p^{rt}} < p^{rt(t-1)/2} (\leq p^{rt(rt-1)/2})$, which is impossible in the admissible cases.

Assume that S is in characteristic $x \neq p$, so that $S = S(x^f)$, where $p^{rt} \mid f$. Now bearing in mind that $S \leq PSL(t, q)$, then the minimal modular representation in characteristic not p is at least t. But this degree is bounded below by a polynomial function of $F(x^f)$ with $p^{rt} \mid f$ by [161], Theorem 5.3.9 and Table 5.3.A. So, we have a contradiction by showing that $F(x^f) < t$ with $p^{rt} \mid f$ is impossible.   □

As mentioned previously, perhaps the most important open problem involving focal-spreads is whether every focal-spread of type (t, k) actually is a k-cut of some t-spread associated with a translation plane. In this text, we did note the following result:

THEOREM 273. *(Jha and Johnson* [77]*) Let F be a focal-spread of order q^t and type (t, k), which admits an affine homology group with coaxis the focus and axis a k-component that acts transitively on the remaining k-components of the partial Sperner k-spread. Then F may be constructed as a k-cut of a nearfield t-spread.*

In this chapter, we are able to take care of the doubly transitive case.

THEOREM 274. *Let π be a focal-spread of type (t, k) over $GF(q)$, where $q^t \neq 64$, that admits a collineation group G of π in $\Gamma L(t + k, q)$ that acts doubly transitive on the partial k-spread. Then there is a Desarguesian affine plane Π of order q^t such that π arises as a k-cut of the spread of Π.*

1. Johnson-Montinaro; Elementary Abelian Case

By Theorem 272, we may assume that G contains an elementary Abelian p-group T of order q^t, where $q = p^r$. Once this fact is known then the theorem may be given by more geometric methods. In fact, the following more general theorem is true:

THEOREM 275. *(1) If a focal-spread of type (t, k) over $GF(q)$, for $q = p^r$, p a prime, admits a non-trivial p-group P that is normalized by a group $\langle g \rangle$ of order a prime p-primitive divisor of $q^t - 1$ or*
(2) $q = p$ and $p + 1 = 2^a$,
then the focal-spread is a k-cut of a Desarguesian t-spread.

We first show that the elementary Abelian group T fixes the focus L pointwise. Again, we are assuming that $q^t \neq 64$. Also, if $q^t = p^2$, for p a prime, then $t = 2$ and $k = 1$ and the focal-spread will be shown to be a 1-cut of a 2-spread of any translation plane

We choose a focal-spread of type (t, k) and order q^t to be represented as

$$x = 0, y = 0, y = xB, \text{ where } B \in M,$$

where M is a set of $k \times t$ matrices of rank k such that the differences of any two distinct elements of M also have rank k, where x is a k-vector and y is a t-vector. Furthermore, we may assume that M contains $I_{k \times t}$, whose k rows are $[1, 0, 0, 0, .., 0]$, $[0, 1, 0, 0, .., 0]$, etc. where there are $t - 1$ zeros in each row and in row i, there is a 1 in the $(1, i)$-position, for $i = 1, 2, .., k$.

LEMMA 92. *If the focal-spread of type $(2, 1)$ of order p^2 is a k-cut of a 2-spread for a translation plane π of order p^2, then π is a k-cut of a Desarguesian plane that admits the group T as an elation group.*

PROOF. Assume that

$$x = 0, y = 0, y = xB^+, \text{ where } B \in M^+,$$

is a t-spread, where M^+ is a set of 2×2 non-singular matrices such that the differences of any two distinct elements of M^+ is also non-singular and where, in this context, x and y are 2-vectors. Assume also that for $x = (x_1, x_2)$ we consider the $2 + 1$ dimensional subspace

$$\left\{ \begin{array}{l} (x_1, 0, y_1, y_2), \ for \ all \\ x_1, y_i \in GF(q), and \ i = 1, 2 \end{array} \right\}$$

and choose $x = 0$ in the translation plane and the focus L of the focal-spread has equation $x = 0$, where now x is a 1-vector. Therefore, if $B^+ = \begin{bmatrix} B_{1 \times 2} \\ R_{1 \times 2} \end{bmatrix}$, then $(x_1, 0)B^+ = (x_1)B_{1 \times 2}$. Now since B^+ is a 2×2 matrix and M^+ is a 2-spread, we may represent B^+ in the form $\begin{bmatrix} u & t \\ g(t, u) & f(t, u) \end{bmatrix}$, where $u, t \in GF(p)$ and f and g functions from $GF(p) \times GF(p)$ to $GF(p)$, and where not both u and t are 0. Hence, $B_{1 \times 2} = [u, t]$. In other words, any focal-spread of type $(2, 1)$ is a k-cut of any 2-spread that we choose. So, choose a Desarguesian 2-spread π

and an elation group of order p^2 with axis $x = 0$. Any elation group fixes $\pi/(x = 0)$ pointwise, so will induce an elation group W on the focal-spread.

In other words, there is a spread S of 2-dimensional $GF(p)$-subspaces within a 4-dimensional vector space V_4 over $GF(p)$. We have a normal group T of order p^2 that leaves L invariant and is regular on the remaining p^2 components of dimension 1 over $GF(p)$. Any affine plane of order p^2 that admits an elation group W of order p^2 with axis L fixes $\pi/(x = 0)$ pointwise and hence is induced on the 'slice' that produces the associated focal-spread. This means that W is induced as a group of the focal-spread. If T does not fix L pointwise, then it fixes a 1-dimensional subspace Z pointwise and is transitive on the remaining p 1-dimensional $GF(p)$-subspaces. Furthermore, each such 1-space is fixed pointwise by a subgroup of order p. We know that the elements of W have the basic form $\begin{bmatrix} I_k & B \\ 0 & I_t \end{bmatrix}$, where B is a $k \times t$ matrix over $GF(q)$ and the components have form $y = xB$, for B in an additive group $\mathcal{B}$ of order p^2. Furthermore, the elements of T have the form $\begin{bmatrix} A & C \\ 0 & I_t \end{bmatrix}$ and the components have the form $y = xA^{-1}C$, where A is a $k \times k$ matrix and C is a $k \times t$ matrix. So, A is an element of $GF(p)$ and C is a 1×2 matrix. Since T is elementary Abelian, then $A^p = 1$, but since $A \in GF(p)$ then $A = 1$. Hence, $T = W$. $\square$

THEOREM 276. *T fixes the focus L pointwise.*

PROOF. T is the elementary Abelian p-group of order q^t that acts regularly on the partial Sperner k-spread. Therefore, it follows that the fixed point set of T must lie on L. Now suppose that for some element e of T, $Fix\,e$ is not the focus L. Since T is Abelian, then T leaves $Fix\,e$ invariant and fixes pointwise a subspace U, properly contained in L. We have two cases (a) $q^t - 1$ contains a p-primitive divisor and (b) $q^t = p^2$ and $p + 1 = 2^c$, which we have considered in the previous lemma.

Case (a). Since G is doubly transitive, the order of G is divisible by $q^t(q^t - 1)$. Assume that $g \in G_0$ is a element of order a prime p-primitive divisor of $q^t - 1$. Then there is a collineation g of G_0 that fixes U pointwise and letting N be a W-complement in L that is fixed by g, we see that g must fix L pointwise. But g leaves invariant a k-dimensional subspace and $t > k$. Therefore, g fixes the k-space pointwise. Hence, g is the identity, a contradiction. $\square$

We have seen the following lemma in a slightly different form from the basic theory of focal-spreads, we restate this here for emphasis.

LEMMA 93. *The elementary Abelian group T is an elation subgroup of the focal-space if and only if the elements of T have the form $\begin{bmatrix} I_k & B \\ 0 & I_t \end{bmatrix}$. Therefore, the focal-spread is additive if and only the T is an elation group.*

We shall now employ a result that we have seen in Chapter 2.

THEOREM 277. *If T is an elation group of the focal-spread and the focal-spread is a k-cut of a t-spread, then the focal-spread is a k-cut of a semifield t-spread.*

We may now assume that $q^t - 1$ admits a p-primitive divisor and that there is a collineation g of the focal-spread that fixes $x = 0$ and $y = 0$ of order a prime p-primitive divisor of $q^t - 1$. Clearly, g is linear and has the basic form $\begin{bmatrix} A & 0 \\ 0 & C \end{bmatrix}$, where A is a $k \times k$ matrix and C is a $t \times t$ matrix with elements in $GF(q)$. Since g restricted to $y = 0$, a k-space, it follows that g must fix $y = 0$ pointwise and hence $A = I_k$. Now $g = \begin{bmatrix} I_k & 0 \\ 0 & C \end{bmatrix}$. The associated parallel class of $x = 0$ consists of the elements $x = d$, for all k-vectors d. We note that (d, z) maps to (d, zC), for z a t-vector, so g fixes all elements of the parallel class containing $x = 0$ and hence is an affine homology.

We now connect these ideas with elation groups.

THEOREM 278. *If the group T that fixes L pointwise is an elation group so that the focal-spread is additive then the focal-spread is a k-cut of a Desarguesian spread.*

PROOF. We have a p-primitive divisor element g of the form $\begin{bmatrix} I & 0 \\ 0 & C \end{bmatrix}$, which means that we have a component $y = xI_{k \times t}C^j$, where, $0 \leq j \leq s$, where s is a p-primitive divisor. This means by additivity of the focal-spread that we have a component $y = x(\sum_{i=0}^{s} \alpha_i I_k C^j)$, for $\alpha_i \in GF(p)$. But note that the $GF(p)$-module generated by C, with elements $\sum_{i=0}^{s} \alpha_i C^j)$, is necessarily a field by Schur's Lemma. Since $\alpha_i I_{k \times t} = I_{k \times t} \alpha_i I_t$, we have a component $y = xI_{k \times t}D$, for all D in a field $GF(q^t)$. This means that we have a collineation group

$$\left\langle \begin{bmatrix} I & 0 \\ 0 & D \end{bmatrix} ; D \in GF(q^t) \right\rangle.$$

If we extend $y = xI_{k \times t}D$ to $y = xD$, we have then extended the focal-spread to a semifield plane that also admits the affine homology group of order $q^t - 1$ and hence the extended plane is Desarguesian. □

Finally, we show that there are, in fact, elations. Every element $g \in T$ has the form $\begin{bmatrix} A & B \\ 0 & I \end{bmatrix}$ so let h be an affine homology $\begin{bmatrix} I & 0 \\ 0 & D \end{bmatrix}$, for D a non-singular $t \times t$ matrix not equal to I_t, and form the commutator

$$g^{-1}h^{-1}gh = \begin{bmatrix} A^{-1} & -A^{-1}B \\ 0 & I \end{bmatrix} \begin{bmatrix} I & 0 \\ 0 & D^{-1} \end{bmatrix} .$$

$$\cdot \begin{bmatrix} A & B \\ 0 & I \end{bmatrix} \begin{bmatrix} I & 0 \\ 0 & D \end{bmatrix}$$

$$= \begin{bmatrix} A^{-1} & -A^{-1}B \\ 0 & I \end{bmatrix} \begin{bmatrix} A & BD \\ 0 & I \end{bmatrix}$$

$$= \begin{bmatrix} I & A^{-1}BD - A^{-1}B \\ 0 & I \end{bmatrix}$$

and

$$A^{-1}BD - A^{-1}B = A^{-1}B(D - I_t)$$

$z^\sigma = (z_1, .., z_s)^\sigma = (z_1^\sigma, z_2^\sigma, .., z_s^\sigma)$, is a $k \times t$ matrix. Since $D \neq I_t$, then $A^{-1}B(D - I_t)$ must have rank k and, in any case, there is an elation.

LEMMA 94. *T is an elation group.*

PROOF. Let E denote the subgroup of elations of T and note that any affine homology must normalize E since

$$\begin{bmatrix} I & 0 \\ 0 & D^{-1} \end{bmatrix} \begin{bmatrix} I & B \\ 0 & I \end{bmatrix} \begin{bmatrix} I & 0 \\ 0 & D \end{bmatrix}$$

$$= \begin{bmatrix} I & BD \\ 0 & I \end{bmatrix}$$

(see Lemma 93). Now E is an elementary Abelian p-group of order p^e and assume that $p^e < q^t$. Since we may assume that D has order a p-primitive divisor of $q^t - 1$, it follows that the homology must fix E elementwise, a contradiction, since the group of homologies acts semi-regularly on E. Therefore, $E = T$, so T is an elation group. □

Hence, combined with Theorem 278, the previous lemma shows that when there is a p-primitive divisor of $q^t - 1$, the theorem is finished. But, by Lemma 92, we may always make this assumption. This completes the proof of Theorem 275.

Part 10

Appendices

In this part, we list various results and background material required to read certain of the theory presented in this text. We also offer a chapter on open problems and directions for future research.

CHAPTER 40

Open Problems

PROBLEM 1. *Is there any translation plane of order q^2 and kernel containing $GF(q)$ whose associated spread cannot be embedded into a parallelism?*

PROBLEM 2. *Is it possible to have a parallelism where all associated translation planes are mutually non-isomorphic?*

PROBLEM 3. *Given any infinite skewfield K, is there always a parallelism in $PG(3, K)$? Is there always a regular parallelisms if K is a field?*

PROBLEM 4. *If a partial parallelism in $PG(3, K)$ is of deficiency one and all spreads of the parallelism are isomorphic, is there an extension to a parallelism and is the adjoined spread isomorphic to the others? (The answer is 'yes' and 'no' in that order but develop this idea in general.)*

The following problem is so obvious, why don't we call it 'The Obvious Problem P^3 (Pappian Partial Parallelism). Like the old refrain 'yes, this theorem is obvious, unfortunately the proof is false.'

PROBLEM 5. *Is there a partial parallelism of deficiency one where all spreads are Pappian but the parallelism extending the partial parallelism does not consist of Pappian spreads?*

In Chapter 12, the following theorem is proved, as a sort of addendum to the Jha-Johnson $SL(2, q) \times C$-theorem.

THEOREM 279. *Let π be a translation plane of order q^4 and kernel $GF(q)$ admitting a collineation group isomorphic to $SL(2, q)$, generated by Baer collineations, all of whose non-trivial orbits are of length $q^2 - q$. Let N denote the net of degree $q^2 + 1$ that is componentwise fixed by $SL(2, q)$.*

(1) There are at least $q + 1$ Baer subplanes of N that are incident with the zero vector and on any such Baer subplane π_0, considered as a 3-dimensional projective space over $GF(q)$, there is an induced Desarguesian partial parallelism of deficiency one.

(2) The Desarguesian partial parallelism of deficiency one may be extended to a Desarguesian partial parallelism if and only if the net N is derivable.

PROBLEM 6. *Solve the finite Obvious Problem P^3 directly using translation planes by proving that the net N in the above theorem is derivable.*

PROBLEM 7. *Give a suitable criterion for a parallelism and its dual parallelism to be isomorphic.*

PROBLEM 8. *Find an infinite class of parallelisms in $PG(3, q)$, each of which is isomorphic to its dual.*

PROBLEM 9. *Using isomorphism conditions of $PG(3, L)$, for various fields L, determine conditions that ensure that the parallelisms constructed in higher dimensional projective spaces using the generalized Beutelspacher construction are mutually non-isomorphic.*

PROBLEM 10. *Determine the possible conical flocks whose derived spreads may be embedded into a transitive deficiency one partial parallelism.*

PROBLEM 11. *Further develop the theory of α-flokki. In particular, the 2nd-cone problem is to determine which α-flokki produce flocks in both α-cones. For functions (g, f), show that if g is not zero then an associated α-flock of the first cone is never an α-flock of the second cone.*

PROBLEM 12. *Recall that an 'α-hyperbolic fibration' is a set Q of $q - 1$ α-derivable partial spreads and two carrying lines L and M such that the union $L \cup M \cup Q$ is a cover of the points of $PG(3, K)$. Show that the associated translation planes are somehow connected to α-flokki.*

1. Non-standard Groups and Non-standard Parallelisms

We have analyzed parallelisms of $PG(3, q)$ admitting a group G that fixes one spread and is transitive on the remaining spreads.

DEFINITION 125. *Let P be a parallelism admitting a Desarguesian spread Σ and a group G fixing a component $x = 0$. Let R be any regulus of Σ containing $x = 0$. If the opposite regulus R^* is a subspread of one of the spreads of P, then we shall say that the parallelism is 'standard.'*

If G is a standard group, then we have shown that the parallelism and its dual can never be isomorphic. However, it is an open question if there can be such transitive deficiency one partial parallelisms that do not admit standard groups. All of the known examples admit standard groups and, for example, if $(p, r) = 1$, all transitive groups are also standard.

PROBLEM 13. *Study deficiency one transitive partial parallelisms. Show that the spread extending the deficiency one partial parallelism to a parallelism is Desarguesian. Furthermore, show that the remaining spreads of the parallelism are derived conical flock spreads.*

Note that all of these would be true if it could be shown that any such transitive group is standard.

We previously mentioned 'standard' parallelisms. In this setting, we require that the socle plane have $q(q + 1)$ reguli sharing $x = 0$, and this will force the socle plane Σ to be Desarguesian. Furthermore, if one of the reguli R of Σ has its opposite regulus R^* as a subspread of a spread Σ' of the parallelism, then Σ' cannot contain another opposite regulus R_1^* to a regulus R_1 of Σ. To see this note that $x = 0$ is a Baer subplane of Σ' uniquely defining a regulus R', which we are requiring is an opposite regulus of a regulus of Σ. Hence, there are $q(q + 1)$ such reguli one each in the non-socle spreads of the parallelism.

If there is a parallelism admitting a Desarguesian socle plane Σ, where the remaining spreads are derived conical flock spreads, it is not clear that the corresponding conical flock spreads share reguli of Σ. If they do not, then the 'derived structures' of the previous section would not actually be parallelisms. That is, if the parallelism is not standard, this is a very wild situation.

1.1. Non-standard Parallelisms. When $q = 2$, there are exactly two parallelisms in $PG(3, 2)$, and these parallelisms admit $PSL(2, 7)$ as a collineation group acting two-transitively on the spreads. Hence, the stabilizer G of a spread Σ has order 24. So, there is a Sylow 2-subgroup of Σ that necessarily fixes a component. Since the Baer 2-groups in $\Gamma L(2, 4)$ have orders dividing 2, it follows that there is an elation group E of order 4. Hence, the group is standard. However, there are exactly $q(q - 1)/2$ Desarguesian spreads in $PG(3, q)$ containing a given regulus R. Hence, when $q = 2$, there is a unique such Desarguesian spread. Hence, the parallelisms, considered as extensions to transitive deficiency one partial parallelisms are non-standard but admit standard groups.

PROBLEM 14. *Show that any deficiency one transitive partial parallelism in $PG(3, q)$, for $q > 2$, lifts to a standard parallelism, or find a class of non-standard parallelisms.*

THEOREM 280. *Let P be a parallelism in $PG(2,q)$ with a standard group G, for $q > 2$.*

(1) If q is odd and P admits a group $2K^*$ of order $2(q-1)$ that is in the kernel homology group of the socle spread Σ, then P is a standard parallelism.

(2) Let Σ' be a non-socle spread of P. If Σ' is left invariant by an affine homology of Σ, then P is a standard parallelism.

More generally, if Σ is left invariant by a collineation of Σ that fixes exactly two components $x = 0, y = 0$ of Σ but fixes no 2-dimensional subspaces disjoint from $x = 0$ or $y = 0$, then P is a standard parallelism.

PROOF. Let E^- fix a spread non-socle spread Σ' and act as a Baer group of order q on Σ'. The net R' defined by the fixed point space of E^- is a regulus net. Since $2K^*$ acts on the parallelism and fixes each spread, and $FixE^-$ (i.e., $x = 0$) is left invariant under the full group of the parallelism, it follows that the q Baer subplanes incident with the zero vector of R' are permuted by $2K^*$. Since K^* fixes each such Baer subplane, it follows that either there is a fixed subplane under $2K^*$ or 2 divides q, a contradiction. But a fixed subplane that is fixed by $2K^*$ is a component of Σ. Since E^- acts regularly on these q subplanes, it follows that all subplanes of R' are components of Σ, implying that Σ' contains an opposite regulus to a regulus of Σ. This proves (1).

If Σ' is left invariant by an affine homology g, then g fixes exactly one 2-dimensional $GF(q)$-space that is disjoint from the axis of g, namely, the coaxis, a component of Σ. An affine homology g becomes a Baer p'-group, for $q = p^r$, which must fix a second Baer subplane of the net defined by $Fixg$. Hence, there is a second Baer subplane of R' that is a line of Σ and E^- is transitive on the set of Baer subplanes different from $x = 0$. Thus, the parallelism is standard, completing the proof of (2). $\qquad\square$

THEOREM 281. *Assume that P is a non-standard parallelism in $PG(3,q)$ admitting a standard group G.*

(1) If $q = 2^r$ is even, then the order of G/K^* divides $q^2(q+1)2r$. Hence, $G \cap GL(2,q^2)/K^*$ (the $GF(q)$-kernel) has order $q^2(q+1)$.

(2) If $q = p^r$ is odd, then the order of G/K^* divides $q^2(q+1)4r$. So, $G \cap GL(2,q^2)/K^*$ has order either $q^2(q+1)$ or $2q^2(q+1)$.

PROOF. Consider a non-socle spread Σ' and let g be a non-kernel collineation in $GL(2,q^2)$ and $G_{y=0}$. We may assume that g normalizes some regulus-inducing group, and so we may take the group to be the standard group E_1. Since g will have order dividing $q-1$ and permutes q spreads (the q spreads fixed by E_1), it follows that g will fix a spread

Σ', also fixed by E_1. Since we may assume that $q > 2$, it follows that the axis of the elation group, $x = 0$, is G-invariant. There are q remaining 2-dimensional $GF(q)$-subspaces that lie as Baer subplanes in the regulus net R' containing $x = 0$ as a Baer subplane. Let g have prime order. Since the order of g is prime and divides $q - 1$, it follows that g fixes one of these Baer subplanes. What we are trying to show is that this Baer subplane is actually $y = 0$, implying that the parallelism is standard. Since the parallelism is assumed to be non-standard, the fixed subplane must be a Baer subplane of Σ that is disjoint from $x = 0$ and hence has the form $y = x^q m + xn$, where $m, n \in GF(q^2)$ and $m \neq 0$.

Since g fixes $x = 0, y = 0$ and normalizes E_1, it follows that the form for g is $(x, y) \longmapsto (xb, xb\alpha)$, where $\alpha \in GF(q) - \{0\}$ and $b \in GF(q^2) - \{0\}$. If $\alpha = 1$, then g is a kernel homology, and if g is not in K^*, then the parallelism is standard from a previous result. Hence, $\alpha \neq 1$. Since g fixes $y = x^q m + xn$, we have the following two conditions:

$$b^q m = mb\alpha,$$
$$bn = nb\alpha.$$

If $n \neq 0$, then $\alpha = 1$. Hence, $n = 0$. Thus, $b^{q-1} = \alpha$, implying that $b^{(q-1)^2} = 1$. Since $(q^2 - 1, (q - 1)^2) = (2, q - 1)(q - 1)$, it follows that $\alpha = -1$ and q is odd, or we are finished.

But, then $g^2 : (x, y) \longmapsto (xb^2, yb^2)$ which is in K^*. This completes the proof. $\qquad\square$

PROBLEM 15. *The classification of transitive deficiency one partial parallelisms provides a set of $q^2 + q$ derived conical flock spreads. Derive each of these to construct a set of conical flock spreads. Assume that the associated parallelism is standard. Then the union of the replaced reguli form the socle spread. Each of the conical flock spreads produces a set of 2^{q-1} spreads, whose translation planes admit a cyclic homology group of order $q + 1$. Choose one such plane for each of the conical flock spreads. Develop theory to explain the nature of the set of $q^2 + q$ homology group spreads.*

PROBLEM 16. *If a vector space V admits a partition of type $(t_1, t_2, .., t_k)$, for $t_1 > t_2 > ... > t_k$, over $GF(q)$, of dimension $\sum_{i=1}^{k} \alpha_i t_i$, for α_i positive integer for $i = 1, 2, .., k$. Determine when the partition is a tower of general focal-spreads or find examples when the partition is not a tower of general focal-spreads.*

PROBLEM 17. *The matrix approach for focal-spreads is extremely powerful. Develop some generalizations using matrices of partitions of type (t, k, v), for $t > k > v$ by first creating a focal-spread of type*

$(t + k, v)$. *We have then a matrix representation:*

$$x = 0, y = xM, \ M \in \mathcal{M}$$

where x is a v-vector and y is a $t+k$-vector, $M's$ are $v \times (t+k)$ matrices of rank v whose differences are either 0 or of rank v. Now on the focal of dimension $t + k$, create a focal-spread of type (t, k) so we have as matrix representation

$$x_1 = 0, y_1 = x_2 N, \ N \in \mathcal{L}$$

where x_1 is a k-vector and y is a t vector. The $N's$ are $k \times t$ matrices of rank k, whose differences are either 0 or of rank k. Now see what the matrices are trying to tell in a converse setting.

PROBLEM 18. *Develop a theory of focal-spreads over infinite fields. Is the kernel a field (skewfield)? Generalize all of the questions posed in the finite case to the infinite case.*

PROBLEM 19. *Show that there are focal-spreads of type (t, k) that are not k-cuts.*

PROBLEM 20. *We have constructed double-spreads of type $(t, t - 1)$ of $st - 1$-dimensional vector spaces of $\frac{q^{t(s-1)} - 1}{q^t - 1}$ subspaces of dimension t and $q^{t(s-1)}$ subspaces of dimension $t - 1$. The double-spread of type is said to be a '$(t - 1)$-cut from a t-spread of a st-dimensional subspace.' Consider the natural extension problem: Must a $(t, t - 1)$-cut with this configuration arise from a t-spread of an st-dimensional vector space?*

PROBLEM 21. *Focal-spreads may or may not be at the heart of the theory of partitions of vector spaces. Show that the double-spreads of the previous open problem can never be considered as towers of focal-spreads.*

PROBLEM 22. *Assume that there are $(q^{st} - 1)/(q - 1)$ $st - 1$-dimensional subspaces each equipped with a double-spread of type $(t, t - 1)$ of $\frac{q^{t(s-1)} - 1}{q^t - 1}$ subspaces of dimension t and $q^{t(s-1)}$ subspaces of dimension $t - 1$, that are in a transitive orbit under a group G. Show that there is a vector space of dimension st such that each of the double-spreads are $t - 1$-cuts from hyperplanes.*

The ideas and theory of focal-spreads have produced the following insights:

- Construction of additive maximal partial spreads or new semi-field planes with predetermined exotic affine subplanes. Also, showing that there are proper additive maximal partial spreads will show that there are focal-spreads that cannot be obtained as k-cuts.

- Construction of translation planes of order p^t admitting affine subplanes of order p^k, where k does not divide t.
- New insights into the nature of subplanes of translation planes, most notably the somewhat difficult observation using isotopes of semifields that two known infinite families of (commutative) semifield planes due Knuth and to Kantor of orders 2^{5k} or 2^{7k}, where k is odd, admit subplanes of order 2^2.
- New developments of subgeometry partitions of finite projective spaces by a variety of different subgeometries; the first known examples of subgeometry partitions from a partition of a vector space that is not a t-spread.
- Constructions of associated 2-$(q^{k+1}, q, 1)$-designs, from focal-spreads of dimension $2k + 1$, with focus of dimension $k + 1$, and construction of associated double-spreads.
- Constructions of new triple-spreads obtained also from focal-spreads of type $(k + 1, k)$.
- The process of 'algebraic lifting' from spreads of $PG(3, q)$ to spreads of $PG(3, q^2)$ may be extended to focal-spreads and from here connections to subgeometry partitions become apparent as there is a clearly defined retraction group. The going-up process may be used in combination with lifting and twisting so that the groups obtained by standard algebraic lifting act on the focal-spreads and many of these admit retraction groups and hence new subgeometry partitions.
- The ideas of 'going up' also produce more diverse structures that might be called focal-spreads of type (t, t', k), where the vector space has dimension $t + k$ but the focus has dimension t' for $t' < t$.

Clearly, when dealing with partitions of vector spaces, there are a number of problems that emerge, and we point out these areas of general interest as follows:

PROBLEM 23. *Find a theoretically complete notion of the kernel of a partition of a vector space, particularly for focal-spreads, which is valid for either infinite or finite partitions.*

PROBLEM 24. *Determine when the middle and right nuclei of additive focal-spreads are fields.*

PROBLEM 25. *Use ideas from the theory of semifield planes to determine whether additive focal-spreads are always k-cuts.*

PROBLEM 26. *Determine whether additive focal-spreads obtained using the going up process are k-cuts.*

PROBLEM 27. *Analyze designs of type* $2 - (q^{k+1}, q, 1)$ *with the goal of reconstructing an associated focal-spread. That is, can every such design be embedded into a vector space of dimension* $2k+1$ *over* $GF(q)$?

PROBLEM 28. *Determine when two designs arising from different focal-spreads are isomorphic with no restrictions on the isomorphism mappings.*

PROBLEM 29. *Determine when double and triple spreads with appropriate parameters arise from focal-spreads.*

PROBLEM 30. *Study general subgeometry partitions with parameters equal to those that arise from focal-spreads of various types to determine when there is a 'lifting' back to a focal-spread.*

PROBLEM 31. *Study general partitions of vector spaces that admit retraction groups so as to construct subgeometry and quasi-subgeometry partitions.*

PROBLEM 32. *Determine whether the subgeometry partitions of a projective space* $PG(2t/w-1, q^w)$ *and a subgeometry partition of a projective space* $PG((t+k)/w-1, q^w)$ *are related geometrically, or rather, give a projective version of k-cuts that shows how the subgeometry partitions are related.*

We recall a result from the hyperplane constructions of subgeometry partitions to state an open problem

THEOREM 282. *Any k-cut focal-spread of type* $(k+1, k)$ *from a Desarguesian* $(k+1)$-*spread or from a nearfield plane of order* q^{k+1} *with kernel containing* $GF(q)$ *with* $q = h^2$ *produces a subgeometry partition of* $PG(2k-1, h^2)$ *by subgeometries isomorphic to (two)* $PG(k-1, h^2)$'s, $h+1$ $PG(k-1, h)$'s and $((h^{2(k+1)} - h^2)/(h+1))$ $PG(k-2, h)$'s, *using any hyperplane that shares a k-space with the focus.*

PROBLEM 33. *Can there be partitions of* $PG(2k-1, h^2)$ *with all four types of subgeometries? If so, there is a partition of a vector space with* $2 + h^2 - 1 = h^2 + 1$ K-*subspaces and a group of order* $h^2 - 1$ *acting on the remaining* $h^{2(k+1)} - h^2$ $k-1$-*dimensional subspaces with orbits of length 1 and* $h + 1$. *This means there are* $w(h+1)$ *orbits of length 1. Can these partitions arise as cuts from a planar* $2k + 1$-*spread?*

PROBLEM 34. *Find examples of translation planes of order* p^t *that admit affine subplanes of order* p^k, *where t does not divide t. Show, if possible, that commutative semifields satisfy the subsemifield dimension property (i.e, the dimension of a commutative subsemifield must divide the dimension of the commutative semifield).*

PROBLEM 35. *Given a t-spread, is there a subgeometry partition that gives rise to it (the interpretation of 'gives rise to' might be the issue)? Amazingly, if $t = 2$, the answer is 'yes.'*

PROBLEM 36. *Determine whether there are group theoretic characterizations of certain fundamental focal-spreads. For example, focal-spreads that admit groups fixing the focus and acting doubly transitive on the K–subspaces of the associated partial Sperner k-spread necessarily arise as k-cuts of Desarguesian spreads.*

PROBLEM 37. *Determine whether the 'transitive' problem is equivalent to the question of extensions of additive partial spreads.*

PROBLEM 38. *If the group fixing the focus and one K–subspace is transitive on the remaining k-components, is the focal-spread a k-cut? Recall in the section on Inherited Groups, it is shown that that any focal-spread admitting an affine homology group with this action must arise from a k-cut of a nearfield plane.*

PROBLEM 39. *Determine the collineation groups of a general focal-spread. If B is a focal-spread with group G and if B be extended to a spread set, when does G act as a collineation group of the spread set?*

PROBLEM 40. *Let P be a finite t-parallelism admitting a group fixing one spread Σ and acting transitively on the remaining spreads. Classify P. (The problem is solved for line parallelisms in $PG(3, q)$.)*

PROBLEM 41. *The following defines an additive maximal partial spread in $PG(3, q)$, where q is even and > 4:*

$$x = 0, y = \begin{bmatrix} u+t & tf+t^2 \\ t & u \end{bmatrix};$$

$u \in GF(q)$, $t \in GF(q)$ *of trace 0 over $GF(2)$, $tracef = 1$.*

The partial spread is maximal in $PG(3, q)$ since otherwise there would be a non-linear even order semifield flock, which does not occur.

Show that this family is maximal in $PG(4r - 1, 2)$, where $q = 2^r$. Using the companion focal-spread, a solution will provide the first non-extendable additive focal-spread.

PROBLEM 42. *Show that the going-up construction of focal-spreads gives examples of focal-spreads that cannot be extended or rather are not k-cuts.*

PROBLEM 43. *The construction process obtaining Sperner k-spreads from suitably many translation planes uses the idea of partitioning a vector space into its j-$(0$-sets$)$. There must be a theory of such Sperner k-spreads, which is somewhat analogous to the theory of translation planes of dimension $2k$ admitting k-spreads. Develop this theory.*

PROBLEM 44. *Let G be a parallelism-inducing group for spreads Σ and Σ'. Find examples where Σ' is not Pappian.*

PROBLEM 45. *Determine the number of mutually non-isomorphic parallelisms arising from linear parallelism-inducing groups in $PG(3,q)$. In particular, solve this problem when q is an odd prime.*

PROBLEM 46. *Determine the number of mutually non-isomorphic parallelisms arising from nearfield parallelism-inducing groups in $PG(3,q)$. In particular, consider this problem when q is an odd prime.*

PROBLEM 47. *Determine when two nearfield parallelism-inducing groups, whose nearfields have different kernels, can produce isomorphic parallelisms in $PG(3,q)$.*

We offer a few words on what makes the k-spreads constructed in this text 'new' in some sense. Given a rk-spread, it is trivial to obtain a k-spread simply by finding a k-spread of each component. We shall say that a k-spread is 'irreducible' if it cannot be obtained in this way. All of the k-spreads of tk-dimensional vector spaces that are constructed using the going up process are almost certainly irreducible. If a less restrictive definition of irreducible is taken, perhaps, say, that the k-spread cannot be obtained from any other vector space partition regardless of dimensions of the various subspaces in the partition then our k-spreads would not then be considered irreducible as for example, they could be obtained from certain $((t-1)k, (t-w)k, k)$-focal-spreads. Since the ideas of focal-spreads and their generalizations are important in various contexts, we will consider the connections with such partitions to our k-spreads to be of independent interest and allow the definition of an irreducible k-spread to be one that cannot be obtained from another z-spread by refining each component.

PROBLEM 48. *Give a suitable criterion to determine when a partition or k-spread is irreducible.*

PROBLEM 49. *In particular, find an oval-cone in $PG(3,K)$, for K finite or infinite that is not a quadratic cone that admits a parallelism.*

PROBLEM 50. *Determine the classes of all maximal partial Desarguesian t-parallelism, $t \neq 2$.*

Based on what we know of the internal structures of the known classes of translation planes and particularly what we know of semifield planes, our results show that either there are many classes of semifield planes left to be discovered that are quite different from the known families or there are great numbers of maximal additive partial spreads of very large deficiency. Since there are no non-semifield planes that are known to satisfy the subplane dimension problem in the negative, the same statement can be made for arbitrary translation planes.

To illustrate the complexity of the situation, recall again that there are semifield planes of order 2^5 that contain semifield subplanes of order 2^2, necessarily Desarguesian and there are semifields planes of orders 2^{5k} or 2^{7k}, for k odd, that admit Desarguesian subplanes of order 2^2. So take any semifield plane of order 2^5 and let c be any integer larger than 5 such that 5 does not divide c. Then either there is an additive partial spread which is a maximal partial spread of degree $\leq 1+2^{5+c-1}$ or there is a semifield plane of order 2^{5+c} that contains a semifield subplane of order 2^2 and of order 2^5. For example, if $c = 6$, either there is a semifield plane of order 2^{11} that contains subplanes of orders 2^2, 2^5 or there is an additive partial spread, which is maximal of degree $\leq 1+2^{10}$ and order 2^{11}. Now assume that we never obtain additive maximal partial spreads. Then choose any sequence of integers $2, 5, 11, i_4, i_5 .., i_n$ such that $i_{j+1} = i_j + t_j$, such that i_j does not divide t_j. Then there is a semifield plane of order 2^{i_n} admitting subplanes of orders 2^2, $2^5, 2^{11}, .., 2^{i_{n-1}}$. For example, take the sequence $2, 5, 11, 23, 47$, then there is an assumed semifield plane of order 2^{47} that contains semifield subplanes of orders 2^2, 2^5, 2^{11}, 2^{23}. Similar sequences are possible for semifields of order 2^{7k}, for k odd.

In general, we have shown that given any translation plane π_o of order p^d, we may find a partial spread of order p^{d+c} and degree $1 + p^d$ that contains a translation subplane of order p^d isomorphic to π_0. If this partial spread is not contained in a proper maximal partial spread, then there is a translation plane of order p^{c+d} that contains a translation subplane of order p^d. This seems improbable, assuming that d does not divide c. Thus, we would expect there to be a very large number of maximal partial spreads of large orders that may be generated in this manner.

PROBLEM 51. *Resolve the issue on additive maximal partial spreads or semifield planes admitting exotic subplanes.*

PROBLEM 52. *Do t-parallelisms exist when $t > 1$? If so, what could be their possible groups?*

PROBLEM 53. *Is there a way to use group theory to discover t-parallelisms?*

PROBLEM 54. *The answer to the last question requires a most sensitive use of the putative collineation groups, in that we have shown that none can be transitive (or doubly transitive).*

PROBLEM 55. *There are infinite classes of 1-parallelisms in $PG(3, q)$ that do not admit transitive groups but admit a collineation fixing one spread and transitive on the remaining spreads. Is such a construction possible for t-parallelisms, $t > 1$?*

PROBLEM 56. *There are partial t-parallelisms of $1 + h^{t+1} + h^{2(t+1)}$ t-spreads. If there are no t-parallelisms, what is the number of spreads in a maximal t-parallelism? Note that there are transitive partial t-parallelisms with the number of t-spreads as above. Does this make the associated t-spreads 1-spreads over a larger field?*

PROBLEM 57. *It is possible to consider the generalization of the problem on partially flag-transitive affine planes, this time with respect to a subplane covered net. Let π be a finite affine plane containing a subplane covered net S. If there exists a collineation group G which leaves S invariant and acts flag-transitively on the flags on lines not in S, can π be determined?*

PROBLEM 58. *Using the idea of a 'flock' of a sharply k-transitive set S_k to be a sharply 1-transitive subset and a 'parallelism' as a set of mutually disjoint flocks whose union is S_k, develop a theory and construct examples for arbitrary integers k.*

PROBLEM 59. *Develop the idea of a deficiency 1 partially sharp subset Λ of $P\Gamma L(n, K)$, where K is an arbitrary skewfield. The deficiency 1 definition could be that given any point P of $PG(n-1, K)$, there there is exactly one point in $PG(n - 1, K) - P\Lambda$. When $n = 2$, show that there is a translation plane admitting two Baer groups B_1 and B_2 whose component orbits are identical and that conversely such a translation plane constructs a partially sharp subset of some $P\Gamma L(2, K)$, where K is a skewfield.*

PROBLEM 60. *Find parallelisms of elliptic quadrics or hyperbolic quadrics in $PG(3, K)$, by maximal partial flocks.*

PROBLEM 61. *Extend the theory of flocks and parallelisms of quadratic cones to Laguerre planes and/or to α-flocks or to oval-flocks.*

PROBLEM 62. *Show that there are parallelisms of elliptic quadrics of characteristic 2.*

PROBLEM 63. *Betten and Riesinger have constructed a large variety of new parallelisms over the field of real numbers (see Betten and Riesinger* [13], [10], [12], [11], [9]). *Generalize any of this work using arbitrary fields K, essentially paying attention to the construction of a parallelism without worrying about topological aspects.*

PROBLEM 64. *Consider the isomorphism question for Johnson parallelisms. For c in $K^* - K^{+*(\sigma+1)}$, we note that $c^2 = c^{\sigma+1}$, so $K^*/K^{+*(\sigma+1)}$ is en elementary Abelian 2-group that corresponds to the isomorphism classes. For example, take $K = Q_a$, the field of rationals and consider the field extension $\left\{ \begin{bmatrix} u & -t \\ t & u \end{bmatrix}; u, t \in Q_a \right\}$. We note that*

$$\{u^2 + t^2; u, t \in Q_a\} = K^{+(\sigma+1)}.$$

So, we are considering $Q_a/(Sum\ of\ Rational\ Squares)$. Determine the cardinality of the set of isomorphism classes by determining this elementary Abelian 2-group.

PROBLEM 65. *We recall the general switching problem: Let*

$$\Gamma_i = \left\{ x = 0, y = x \begin{bmatrix} u + g_i(t) & f_i(t) \\ t & u \end{bmatrix}; u, t \in GF(q) \right\},$$

be conical flock spreads, each of which share exactly R_0 with Σ_1. Assume $\Gamma_2 E$ is a set of q mutually disjoint spreads. Then $\Gamma_3 E$ switches with $\Gamma_2 E$ if and only if for each $t \in GF(q)$, there exists a bijective function $m : GF(q) \to GF(q)$ such that

$$
\begin{aligned}
g_3(t) - \rho_1 t &= g_2(m(t)) - \rho_1 m(t), \\
f_3(t) - \gamma_1 t &= f_2(m(t)) - \gamma_1 m(t).
\end{aligned}
$$

Develop this idea: find conical flock spreads that have the switching requirement and construct perhaps new flocks of quadratic cones and new parallelisms.

PROBLEM 66. *In this setting, consider E-switching and necessarily all E-switches of Desarguesian spreads are also Desarguesian. However, it is not known whether all groups H^j or H_k may be used to construct parallelisms (see the chapter on coset switching, for odd order, for a definition of these groups). As mentioned, when $q = 2^r$, have shown that H^0 and H_0 may be used to construct parallelisms. Find Desarguesian planes for which any of the groups H^j or H_k can be used to construct new parallelisms.*

PROBLEM 67. *Are there parallelisms of quadratic cones or of hyperbolic quadrics that are not transitive?*

PROBLEM 68. *Find partially sharp subsets of q elements of $P\Gamma L(2,q)$. The associated translation planes of order q^2 admit two Baer groups. Do these exist?*

PROBLEM 69. *This problem refers to 'partially flag-transitive affine planes,' where the theory is developed for solvable groups. Are there 'nonsolvable' partially flag-transitive affine planes?*

REMARK 71. *The Hall and Desarguesian planes are the only translation planes of order q^2 that admit $SL(2,q)$ and are partially flag-transitive affine planes.*

PROOF. The translation planes admitting $SL(2,q)$ as a collineation group are determined in Foulser and Johnson [54] and [55]. The only derivable planes that admit a collineation group transitive on the components exterior to the derivable net are the Hall and Desarguesian. Any translation plane of this sort is partially flag-transitive.   □

PROBLEM 70. *In the classification theory of partial flag-transitive affine planes, a type of affine plane called a 'semi-translation plane' appears.   The only known semi-translation planes are either derived from dual translation planes or are the Hughes planes that derive the Ostrom-Rosati planes. Show that every semi-translation plane is derivable.   Suppose a semi-translation plane admits a Baer group of order q. Show that the plane is derivable. These problems on semi-translation planes have been open since about 1964.   Please solve these!*

PROBLEM 71. *Given any parallelism in $PG(3,q)$, show criteria for which the dual parallelism is not isomorphic to the original. Note that this is true for all transitive deficiency one partial parallelisms by work of the author.*

The only know parallelism where this is not true is in Prince's set of 45 transitive parallelisms in $PG(3,5)$ (see [176]). The set of these parallelisms contains the two regular parallelisms, which are known not be to isomorphic to each other. Find how many of the remaining 43 parallelisms are not isomorphic to their duals. Since there must be at least one, the question is what is the type of spread that allows that the corresponding parallelism is isomorphic to its dual?

If the above problem can be solved, use the ideas to find a infinite class of transitive parallelisms whose duals are not isomorphic to the original.

PROBLEM 72. *In this text, we have introduced the concept of an m-parallelism, which leads naturally to the so-called (m,n)-parallelism. All of these parallelisms depend on order and coset representations.*

Since $\{i_j\}$ forms a partition of n, the parallelism depends on the parti-
tion. Furthermore, the order is important in this case, so we consider
that the partition is 'ordered.' Moreover, the parallelism may depend
on the coset representation class $\{g_i\}$. When we want to be clear on
the notation we shall refer to the parallelism as a $(m, n, \{i_j\}, \{g_i\})$-
parallelism. When $n = m$, we use simply the notation of $(m, \{g_i\})$-
parallelism. Furthermore, since each such parallelism depends on a
choice of the initial Pappian spreads, the non-isomorphic parallelisms
are potentially quite diverse. Exploit this diversity and try to make
sense of how two m parallelisms (or, respectively, (m, n) parallelisms)
could be non-isomorphic.

The following problem involves the theorem of Biliotti-Jha-Johnson
on p-primitive collineations, which we restate.

THEOREM 283. *Let P be a parallelism in $PG(3, q)$ and let G be a*
collineation group of $PG(3, q)$, which leaves P invariant and contains
a collineation of order a p-primitive divisor u of $q^3 - 1$.
Then, one of the two situations occurs:

(1) *G is a subgroup of $\Gamma L(1, q^3)/Z$, where Z denotes the scalar*
group of order $q - 1$, and fixes a plane and a point.
(2) *$u = 7$ and one of the following subcases occurs:*
    (a) *G is reducible, fixes a plane or a point and either*
        (i) *G is isomorphic to A_7 and $q = 5^2$, or*
        (ii) *G is isomorphic to $PSL(2, 7)$ and $q = p$ for a prime*
           $p \equiv 2, 4 \bmod 7$ or $q = p^2$ for a prime $p \equiv 3, 5 \bmod$
           7.
           When $q = p = 2$, P is one of the two regular paral-
           lelisms in $PG(3, 2)$, or
    (b) *G is primitive, G is isomorphic to $PSL(2, 7)$ or A_7 and*
        $q = p$ for an odd prime $p \equiv 2, 4 \bmod 7$ or $q = p^2$ for a
        prime $p \equiv 3, 5 \bmod 7$.

PROBLEM 73. *When the group is transitive all of the subcases of*
part (2) cases do not occur other than $q = p = 2$, P is one of the two
regular parallelisms in $PG(3, 2)$. This is also true of the parallelisms of
Prince, which occur in case (1). The problem is to use combinatorial
and group theoretic arguments but only with the p-primitive assumption
to eliminate all other cases.
    For example, consider $u = 7$, G is isomorphic to A_7 acting on
$1 + 5^2 + 5^4$ spreads. Can this occur? If so, find an example.

PROBLEM 74. *Show that transitive parallelisms on $P(3, 4)$ cannot occur (this is the case not covered by the Biliotti-Jha-Johnson transitive theorem).*

CHAPTER 41

Geometry Background

In this chapter, we provide statements of results used in the proof of the following classification theorem, as well as in other results on translation planes and t-spreads.

THEOREM 284. *Let $\mathcal{P}^-$ be a deficiency one parallelism in $PG(3, q)$ that admits a collineation group in $P\Gamma L(4, q)$ acting transitively on the spreads of the partial parallelism. Let $\mathcal{P}$ denote the unique extension of $\mathcal{P}^-$ to a parallelism. Let the fixed spread be denoted by Σ_0 (the 'socle') and let the remaining $q^2 + q$ spreads of $\mathcal{P}^-$ be denoted by Σ_i, for $i = 1, 2, .., q^2 + q$. Let π_i denote the affine translation plane corresponding to Σ_i. Then Σ_0 is Desarguesian and Σ_i is a derived conical flock plane for $i = 1, 2, .., q^2 + q$.*

THEOREM 285. *(Biliotti, Jha, Johnson [17]) Let π denote a translation plane of order q^2. Assume that π admits a collineation group G that contains a normal subgroup N such that G/N is isomorphic to $PSL(2, q)$.*
   Then π is one of the following planes:
   (1) Desarguesian,
   (2) Hall,
   (3) Hering,
   (4) Ott-Schaeffer,
   (5) one of three planes of Walker of order 25, or
   (6) the Dempwolff plane of order 16.

THEOREM 286. *(Johnson [134], Theorem (2.3)) Let V be a vector space of dimension $2r$ over F isomorphic to $GF(p^t)$, p a prime, $q = p^t$. Let T be a linear transformation of V over F, which fixes three mutually disjoint r-dimensional subspaces. Assume that $|T|$ divides $q^r - 1$ but does not divide $LCM(q^s - 1)$; $s < r$, $s \mid r$. Then:*
   (1) all T-invariant r-dimensional subspaces are mutually disjoint and the set of all such subspaces defines a Desarguesian spread;
   (2) the normalizer of $\langle T \rangle$ in $GL(2r, q)$ is a collineation group of the Desarguesian plane Σ defined by the spread of (1);

(3) Σ may be thought of as a 2r-dimensional vector space over F; that is, the field defining Σ is an extension of F.

REMARK 72. *How the above theorem on $PSL(2, q)$ is the deficiency one Theorem is as follows: It is shown that there is a p-collineation τ of a translation plane π of order p^{2r} that fixes at least three components of π. Then, there is an associated Desarguesian affine plane Σ consisting of the τ-invariant subspaces of dimension r over F. In our situation, the vector space V will be 4-dimensional over $GF(q)$, forcing the associated Desarguesian affine plane to have its spread in $PG(3, q)$. Such a Desarguesian plane Σ is called an 'Ostrom phantom.'*

THEOREM 287. *(Gevaert and Johnson [59]) Let π be a translation plane of order q^2 with spread in $PG(3, q)$ that admits an affine elation group E of order q such that there is at least one orbit of components union the axis of E that is a regulus in $PG(3, q)$. Then π corresponds to a flock of a quadratic cone in $PG(3, q)$.*

In the case above, the elation group E is said to be 'regulus-inducing' as each orbit of a 2-dimensional $GF(q)$-vector space disjoint from the axis of E together with the axis will produce a regulus.

THEOREM 288. *(Johnson [124]) Let π be a translation plane of order q^2 with spread in $PG(3, q)$ admitting a Baer group B of order q. Then, the $q - 1$ component orbits union $FixB$ are reguli in $PG(3, q)$. Furthermore, there is a corresponding partial flock of a quadratic cone with $q - 1$ conics. The partial flock may be uniquely extended to a flock if and only if the net defined by $FixB$ is derivable.*

THEOREM 289. *(Payne and Thas [173]) Every partial flock of a quadratic cone of $q - 1$ conics in $PG(3, q)$ may be uniquely extended to a flock.*

THEOREM 290. *(Hering-Ostrom Theorem (see references in the Handbook [138]) Let π be a translation plane of order p^r, p a prime, and let E denote the collineation group generated by all elations in the translation complement of π. Then one of the following situations apply:*

(i) E is elementary Abelian,

(ii) E has order $2k$, where k is odd, and $p = 2$,

(iii) E is isomorphic to $SL(2, p^t)$, and the associated elation net is Desarguesian,

(iv) E is isomorphic to $SL(2, 5)$ and $p = 3$,

(v) E is isomorphic to $S_z(2^{2s+1})$ and $p = 2$.

THEOREM 291. *(Johnson and Ostrom [146] (3.4)) Let π be a translation plane with spread in $PG(3, q)$ of even order q^2. Let E denote a collineation group of the linear translation complement generated by affine elations. If E is solvable, then either E is an elementary Abelian group of elations all with the same axis, or E is dihedral of order $2k$, k odd and there are exactly k elation axes.*

THEOREM 292. *(Gleason [61]) Let G be a finite group operating on a set Ω and let p be a prime. If Ψ is a subset of Ω such that for every $\alpha \in \Psi$, there is a p-subgroup Π_α of G fixing α but no other point of Ω then Ψ is contained in an orbit.*

THEOREM 293. *(André [1]) If Π is a finite projective plane that has two homologies with the same axis ℓ and different centers then the group generated by the homologies contains an elation with axis ℓ.*

THEOREM 294. *(See Lüneburg [167] (49.4), (49.5)) Let τ be a linear mapping of order p of a vector space V of characteristic p and dimension 4 over $GF(p^r)$ and leaving invariant a spread.*

(1) Then the minimal polynomial of τ is $(x-1)^2$ or $(x-1)^4$. If the minimal polynomial is $(x-1)^4$ then $p \geq 5$, and τ is said to to be a 'quartic element.'

(2) The minimal polynomial of τ is $(x-1)^2$ if and only if τ is an affine elation (shear) or a Baer p-collineation (fixes a Baer subplane pointwise).

THEOREM 295. *(Biliotti, Jha, and Johnson [20]) Let π be a translation plane of order q^2, $q = p^r$, p odd, with spread in $PG(3, q)$. If π admits two mutually disjoint 'large' quartic p-groups (orders $> \sqrt{q}$) then the group generated is isomorphic to $SL(2, q)$ and the plane is Hering or the order is 25 and the plane is one of the Walker planes.*

THEOREM 296. *(Johnson [130]) Semifield flocks of quadratic cones in $PG(2, q^r)$ are linear (the corresponding semifield translation plane is Desarguesian).*

THEOREM 297. *(Johnson [134]) Let π be a translation plane with spread in $PG(3, q)$, for q even, admit a dihedral group D_{q+1} generated by elations with exactly $q + 1$ axes. Let C_{q+1} denote the cyclic stem of D_{q+1}.*

(1) If C_{q+1} leaves invariant two components of π then the net N_E of elations is a regulus net in $PG(3, q)$.

(2) If C_{q+1} does not leave invariant two components of π then there is an associated Desarguesian affine plane Σ such that C_{q+1} fixes exactly two components of Σ and these components non-trivially intersect each of the elation axes of π now realized as Baer subplanes of Σ.

THEOREM 298. *(Foulser* [53]*) Let π be a translation plane of odd order q^2.*

(1) Baer p-elements and elations cannot coexist.

(2) If $p > 3$ or $p = 3$ and the group generated by Baer 3-elements is $SL(2, 3^i)$, for $i > 1$, then the Baer subplanes fixed pointwise by Baer p-collineations in the translation complement all lie in the same net of degree $q + 1$.

(3) The group generated by Baer p-elements are those in the Hering-Ostrom theorem, except that Suzuki groups do not occur. In particular, if there are at least two Baer axes then the group generated is either $SL(2, 5)$ and $p = 3$ or $SL(2, p^t)$, for some integer t.

(4) If a net of degree $q + 1$ contains at least three Baer subplanes ρ_i, for $i = 1, 2, 3$, then the number of Baer subplanes of the net is $1 + |Ker\rho_1|$ (kernel of ρ_1).

THEOREM 299. *(Jha and Johnson* [96]*) Let π be a translation plane of even order and let E and B be elation and Baer 2-groups, respectively, in the translation complement. If $|B| > \sqrt{q}$, then $|E| \leq 2$.*

CHAPTER 42

The Klein Quadric

In this text, we shall use the device that connects translation planes of order q^2 with spreads in $PG(3, K)$, for K a field isomorphic to $GF(q)$, with ovoids of the hyperbolic quadric in $PG(5, K)$, the 'Klein Quadric.' Basically, the same theory is available for hyperbolic quadrics of standard type in $PG(3, K)$, where K is a field admitting a quadratic extension, and spreads in $PG(3, K)$. Let $(x_1, x_2, x_3, y_1, y_2, y_3)$, for $x_i, y_i \in K$,where $i = 1, 2, 3$, and K is a field admitting a quadratic extension. Let $x_1 y_3 - x_2 y_2 + x_3 y_1 = 0$ define a non-degenerate hyperbolic quadric Q in a 6-dimensional vector space V_6 over K (or in $PG(5, K)$). We note that $(1, a, b, c, d, \Delta)$, where $\Delta = ad - cb$, then is a point of Q, the other point (1-dimensional K–subspace) has the form $(0, 0, 0, 0, 0, 1)$. Let x and y be 2-vectors over K. We consider the 'Klein map' $\mathcal{K}$:

$$\mathcal{K} \; : \; (1, a, b, c, d, \Delta) \to y = x \begin{bmatrix} a & b \\ c & d \end{bmatrix},$$

$$\mathcal{K} \; : \; (0, 0, 0, 0, 0, 1) \to x = 0.$$

Then the image of Q under the Klein map is a bijection to the set of all 2-dimensional K–subspaces of a 4-dimensional K-space with vectors (x, y). Indeed, two mutually disjoint 2-dimensional K–subspaces map back to points on the quadric that are not incident to a line of the quadric. When K is isomorphic to $GF(q)$, an 'ovoid' of the quadric is a set of $q^2 + 1$ points no two of Q are incident with a line of the quadric. An analogous definition is available in the infinite case.

THEOREM 300. *(1) Ovoids of the hyperbolic quadric Q are equivalent to translation planes with spreads in $PG(3, K)$.*

(2) Ovoids that are elliptic quadrics are equivalent to Pappian Planes with spreads in $PG(3, K)$.

Furthermore, if V_6 is a 6-dimensional vector space, and $\mathcal{Q}$ is the associated non-degenerate hyperbolic quadric, then the associated bilinear form is given by the matrix

$$
M = \begin{bmatrix}
0 & 0 & 0 & 0 & 0 & 1 \\
0 & 0 & 0 & 0 & -1 & 0 \\
0 & 0 & 0 & 1 & 0 & 0 \\
0 & 0 & 1 & 0 & 0 & 0 \\
0 & -1 & 0 & 0 & 0 & 0 \\
1 & 0 & 0 & 0 & 0 & 0
\end{bmatrix}.
$$

If W is a subspace of V_6, then define $W^{\perp}$ to be the set of vectors v such that $wMv^T = 0$, for all $w \in W$. For the quadric $\mathcal{Q}$, any subspace W such that $W \cap \mathcal{Q} = \phi$, is said to an 'anisotropic' subspace. For example, take $\{(x_1, x_2, x_3, y_1, 0, 0)\} = W$, for certain elements x_i, y_1, $i = 1, 2, 3$. If W is a 2-dimensional subspace of V_6, then one of the following occur:

(1) $W \cap \mathcal{Q}$ is empty (so W is anisotropic), and $W^{\perp} \cap \mathcal{Q}$ is an elliptic quadric,

(2) $W \cap \mathcal{Q}$ is a point, and $W^{\perp} \cap \mathcal{Q}$ contains a 2-dimensional subspace,

(3) $W \cap \mathcal{Q}$ is a set of two points, and $W^{\perp} \cap \mathcal{Q}$ is a hyperbolic quadric,

or

(4) $W \cap \mathcal{Q} = W$, and $W^{\perp} \cap \mathcal{Q}$ contains a generator of Witt index 3.

The reader is directed to the appendices of the Handbook [**138**], for additional information and background on quadrics, and, of course, to Hirschfeld [**72**], [**71**].

1. The Thas-Walker Construction

The Thas-Walker construction of spreads from flocks of quadratic sets in $PG(3, K)$, where K admits a quadratic extension, is fundamental to our theory. This construction provides the following:

(1) Connects flocks of elliptic quadrics in $PG(3, K)$ by a set of mutually disjoint conics that cover all but two points of the quadric with translation planes with spreads in $PG(3, K)$, that are covered by mutually disjoint reguli together with two other components.

(2) Connects flocks of quadratic cones in $PG(3, K)$ by a set of mutually disjoint conics of intersection (plane intersections that do not contain the vertex of the cone) with translation planes with spreads in $PG(3, K)$, that are covered by reguli that mutually share exactly one component.

(3) Connects flocks of hyperbolic quadrics in $PG(3, K)$ by a set of mutually disjoint conics of intersection (plane intersections that do not

contain a line of the quadric) with translation planes with spreads in $PG(3, K)$, that are covered by reguli that mutually share two components.

The general construction is as follows: For each quadric set $\mathcal{E}$ (elliptic quadric, quadratic cone, hyperbolic quadric in $PG(3, K)$), embed $PG(3, K)$ into $PG(5, K)$ so that the quadric set $\mathcal{E}$ is a subset of the hyperbolic quadric $\mathcal{Q}$ in $PG(5, K)$. Let $\mathcal{F}$ be a 'flock' is defined in the above paragraph, as a set of planes of $PG(3, K)$. Let $\perp$ denote the polarity of $PG(5, K)$, induced by $\mathcal{Q}$, where

$$\begin{bmatrix} 0 & 0 & 0 & 0 & 0 & 1 \\ 0 & 0 & 0 & 0 & -1 & 0 \\ 0 & 0 & 0 & 1 & 0 & 0 \\ 0 & 0 & 1 & 0 & 0 & 0 \\ 0 & -1 & 0 & 0 & 0 & 0 \\ 1 & 0 & 0 & 0 & 0 & 0 \end{bmatrix}$$

defines the associated bilinear form.

Since $\pi \in \mathcal{F}$ is a 3-dimensional vector space, then $\pi^{\perp}$ is also a 3-dimensional vector space, and it will follow that $\mathcal{F}^{\perp} = \{\pi^{\perp}; \pi \in \mathcal{F}\}$ is an ovoid of $\mathcal{Q}$. Applying the Klein map associates the spreads in $PG(3, K)$ with the ovoids $\mathcal{F}^{\perp}$.

This bijection of

$$\mathcal{F} \iff \mathcal{F}^{\perp} \iff \mathcal{K}(\mathcal{F}^{\perp})$$

is called 'the Thas-Walker-Contruction.'

The reader is directed to Thas [180] for the basic theory in the finite case, which was done independently by Walker [181], and extended by Riesinger [177] in the infinite case.

CHAPTER 43

Major Theorems of Finite Groups

1. Subgroups of $PSL(2, q)$

We shall take the lists of certain subgroups of various classical groups from the article from O.H. King [162], also see [47].

The subgroups of $PSL(2, q)$, for $q = p^r$, p a prime, are as follows:

(a) a single class of $q + 1$ conjugate Abelian groups of order q;

(b) a single class of $q + 1$ conjugate cyclic groups of order d for each divisor d of $q - 1$ for q even and $(q - 1)/2$ for q odd;

(c) a single class of $q(q - 1)/2$ conjugate cyclic groups of order d for each divisor d of $q + 1$, for q even and $(q + 1)/2$, for q odd;

(d) for q odd, a single class of $q(q^2 - 1)/4d$ dihedral groups of order $2d$, for each divisor d of $(q - 10)/2$ with $(q - 1)/(2d)$ odd;

(e) for q odd, two classes each of $q(q^2 - 1)/(8d)$ dihedral groups of order $2d$ for each divisor $d > 2$ of $(q - 1)/2$ with $(q - 1)/(2d)$ even;

(f) for q even, a single class of $q(q^2 - 1)/(2d)$ dihedral groups of order $2d$, for each divisor d of $q - 1$;

(g) for q odd, a single class of $q(q^2 - 1)/(2d)$ dihedral groups of order $2d$, for each divisor d of $(q + 1)/2$ with $(q + 1)/(2d)$ odd;

(h) for q odd, two classes each of $q(q^2 - 1)/(2d)$ dihedral groups of order $2d$, for each divisor d of $(q + 1)/2$ with $(q + 1)/(2d)$ even;

(i) for q even, a single class of $q(q^2 - 1)/(2d)$ dihedral groups of order $2d$, for each divisor d of $(q + 1)$;

(j) a single class of $q(q^2 - 1)/24$ conjugate four-groups when $q \equiv \pm 3 \bmod 8$;

(k) two classes each of $q(q^2 - 1)/48$ conjugate four-groups when $q \equiv \pm 1 \bmod 8$;

(l) a number of classes of conjugate Abelian groups of order q_0, for each divisor q_0 of q;

(m) a number of classes of conjugate groups of order $q_0 d$, for each divisor q_0 of q and for certain d depending on q_0, all lying inside a group of order $q(q - 1)/2$ for q odd and $q(q - 1)$ for q even;

(n) two classes each of $[q(q^2 - 1)]/[2q_0(q_0^2 - 1)]$ groups $PSL(2, q_0)$, where q is an even power of q_0, for q odd;

(o) a single class of $[q(q^2-1)]/[q_0(q_0^2-1)]$ groups $PSL(2, q_0)$, where q is an odd power of q_0, for q odd;

(p) a single class of $[q(q^2-1)]/[q_0(q_0^2-1)]$ groups $PSL(2, q_0)$, where q is a power of q_0, for q even;

(q) two classes each of $[q(q^2-1)]/[2q_0(q_0^2-1)]$ groups $PGL(2, q_0)$, where q is a even power of q_0, for q odd;

(r) two classes each of $q(q^2-1)/48$ conjugate S_4, when $q \equiv \pm 1 \bmod 8$;

(s) two classes each of $q(q^2-1)/48$ conjugate A_4, when $q \equiv \pm 1 \bmod 8$;

(t) a single class of $q(q^2-1)/24$ conjugate A_4, when $q \equiv \pm 3 \bmod 8$;

(u) a single class of $q(q^2-1)/12$ conjugate A_4, when q is an even power of 2;

(v) two classes each of $q(q^2-1)/120$ conjugate A_5, when $q \equiv \pm 3 \bmod 10$.

2. The Lists of Mitchell and Hartley

2.1. Subgroups of PSL$(3, q)$(see [171], [64]). Mitchell q odd; The following is a list of subgroups of $PSL(3, q)$. A subgroup of $PSL(3, q)$ either fixes a point, a line, or a triangle (so is a subgroup of (a), (b), or (c) below), or is one of the groups in (d) through (k);

(a) the stabilizer of a point, having order $q^3(q+1)(q-1)^2/(3, q-1)$;

(b) the stabilizer of a line having order $q^3(q+1)(q-1)^2/(3, q-1)$;

(c) the stabilizer of a triangle, having order $6(q-1)^2/(3, q-1)$;

(d) the stabilizer of a triangle with coordinates in $GF(q^3)$ having order $3(q^2+q+1)/(3, q-1)$ ('imaginary triangle');

(e) the stabilizer of a conic, having order $q(q^2-1)$;

(f) $PSL(3, q_0)$, where q is a power of q_0;

(g) $PGL(3, q_0)$, where q is a power of q_0^3 and 3 divides q_0-1;

(h) $PSU(3, q_0^2)$, where q is a power of q_0^2;

(i) $PU(3, q_0^2)$, where q is a power of q_0^6 and 3 divides q_0-1;

(j) the Hessian groups of orders 216 (where 9 divides $q-1$ (isomorphic to $PU(3, 4)$)), 72 (isomorphic to $PSU(3, 4)$), and 36 (where 3 divides $q-1$, a subgroup of $PSU(3, 4)$);

(k) groups of order 168 (when -7 is a square in $GF(q)$ (isomorphic to $PSL(3, 2)$)), 360 (where 5 is a square in $GF(q)$ and there is a non-trivial cube root of unity) (isomorphic to A_6)), 720 (when q is an even power of 5 (isomorphic to $A_6.2$)), and 2520 (when q is an even power of 5 (isomorphic to A_7)).

Hartley q Even.
For even order, Harley's list is for the maximal subgroups.
Let q be even. The maximal subgroups of $PSL(3, q)$ are as follows:

(a) the stabilizers of a point, a line, a triangle or an imaginary triangle (coordinates in $GF(q^3)$), as for q odd, with the same orders;

(b) $PSL(3, q_0)$, where q is a prime power of q_0;

(c) $PGL(3, q_0)$, where $q = q_0^3$, and q_0 is a square;

(d) $PSU(3, q)$, where q is square;

(e) $PSU(3, q_0^2)$, where $q = q_0^6$, and q_0 is a non-square;

(f) groups of order 360, when $q = 4$ (isomorphic to A_6).

2.2. Subgroups of $PSU(3, q)$. The maximal subgroups of the projective special unitary group $PSU(3, q)$ (preserving a Hermitian form) are also given by Mitchell and Hartley. Let $\mathcal{H}(2, q)$ denote the Hermitian surface of absolute points.

Mitchell, q odd. Let q_0 be odd and $q = q_0^2$. The following is a list of the subgroups of $PSU(3, q)$, which either fix a point and a line, or a triangle (so is a subgroup of (a), (b), or (c) to follow), or is one of the groups listed in (d) through (i):

(a) the stabilizer of the center and axis of an elation (the stabilizer of a point on $\mathcal{H}(2, q)$ together with its polar line), of order $q_0^3(q_0 + 1)^2(q_0 - 1)/(3, q_0 + 1)$;

(b) the stabilizer of the center and axis of a homology (the stabilizer of a point not on $\mathcal{H}(2, q)$ together with its polar line), of order $q_0(q_0 + 1)^2(q_0 - 1)/(3, q_0 + 1)$;

(c) the stabilizer of a triangle, having order $6(q_0 - 1)^2/(3, q_0 + 1)$;

(d) the stabilizer of an imaginary triangle, of order $3(q_0^2 - q + 1)/(3, q_0 + 1)$;

(e) the stabilizer of a conic, of order $q_0(q_0 - 1)^2$;

(f) $PSU(3, q_1)$, where q is an odd power of q_1;

(g) $PSU(3, q_1)$, where q is an odd power of q_1^3 and 3 divides $\sqrt{q_1} + 1$;

(h) the Hessian groups of orders 216 (where 9 divides $q_0 + 1$ (isomorphic to $PU(3, 4)$)), 72 (isomorphic to $PSU(3, 4)$) and 36 (where 3 divides $q_0 + 1$, a subgroup of $PSU(3, 4)$);

(i) groups of order 168 (when -7 is a non-square in $GF(q_0)$ (isomorphic to $PSL(3, 2)$)), 360 (where 5 is a square in $GF(q_0)$ and there is not a non-trivial cube root of unity) (isomorphic to A_6)), 720 (when q_0 is an odd power of 5 (isomorphic to $A_6.2$)), and 2520 (when q_0 is an odd power of 5 (isomorphic to A_7)).

As with the subgroups of $PSL(3, q)$, for q even, only the maximal subgroups of $PSU(3, q)$ are given, for q_0 and $q = q_0^2$.

Hartley, q even. The maximal subgroups are as follows:

(a) the stabilizer of a point, line, a triangle or an imaginary triangle, as for q odd and with the same orders;

(b) $PSU(3, q_1)$, where q is an odd prime power of q_1;

(c) $PSU(3, q_1)$, where $q = q_1^3$, and $\sqrt{q_1}$ is non-square;
(d) groups of order 36 (when $q = 4$).

3. Finite Doubly Transitive Groups

We also list here the classification theorem of arbitrary finite doubly transitive groups $\bar{G}$.

Let v denote the degree of the permutation group.

The possibilities are as follows:

(A) $\bar{G}$ has a simple normal subgroup $\bar{H}$, and $\bar{H} \le \bar{G} \le \operatorname{Aut} \bar{H}$ where $\bar{H}$ and v are as follows:

(1) A_v, $v \ge 5$,
(2) $PSL(d, z)$, $d \ge 2$, $v = (z^d - 1)/(z - 1)$ and $(d, z) \ne (2, 2), (2, 3)$,
(3) $PSU(3, z)$, $v = z^3 + 1$, $z > 2$,
(4) $Sz(w)$, $v = w^2 + 1$, $w = 2^{2e+1} > 2$,
(5) $^2G_2(z)'$, $v = z^3 + 1$, $z = 3^{2e+1}$,
(6) $Sp(2n, 2)$, $n \ge 3$, $v = 2^{2n-1} \pm 2^{n-1}$,
(7) $PSL(2, 11)$, $v = 11$,
(8) Mathieu groups M_v, $v = 11, 12, 22, 23, 24$,
(9) M_{11}, $v = 12$,
(10) A_7, $v = 15$,
(11) HS (Higman-Sims group), $v = 176$,
(12) .3 (Conway's smallest group), $v = 276$;

(B) $\bar{G}$ has a regular normal subgroup $\bar{H}$ which is elementary Abelian of order $v = h^a$, where h is a prime. Identify $\bar{G}$ with a group of affine transformations $x \longmapsto x^g + c$ of $GF(h^a)$, where $g \in \bar{G}_0$. Then one of the following occurs:

(1) $\bar{G} \le A\Gamma L(1, v)$,
(2) $\bar{G}_0 \trianglerighteq SL(n, z)$, $z^n = h^a$,
(3) $\bar{G}_0 \trianglerighteq Sp(n, z)$, $z^n = h^a$,
(4) $\bar{G}_0 \trianglerighteq G_2(z)'$, $z^6 = h^a$, z even,
(5) $\bar{G}_0 \trianglerighteq A_6$ or A_7, $v = 2^4$,
(6) $\bar{G}_0 \trianglerighteq SL(2, 3)$ or $SL(2, 5)$, $v = h^2$, $h = 5, 11, 19, 23, 29$, or 59 or $v = 3^4$,
(7) $\bar{G}_0$ has a normal extraspecial subgroup E of order 2^5 and $\bar{G}_0/E$ is isomorphic to a subgroup of S_5, where $v = 3^4$,
(8) $\bar{G}_0 = SL(2, 13)$, $v = 3^6$.

4. Primitive Subgroups of $\Gamma L(4, q)$

THEOREM 301. (*Kantor and Liebler* [**159**, (5.1)])
Let H be a primitive subgroup of $\Gamma L(4, q)$ for $p^r = q$. Then one of the following holds:

(a) $H \geq SL(4, q)$,

(b) $H \leq \Gamma L(4, q')$ with $GF(q') \subset GF(q)$,

(c) $H \leq \Gamma L(2, q^2)$, with the latter group embedded naturally,

(d) $H \leq Z(H)H_1$, where H_1 is an extension of a special group of order 2^6 by S_5 or S_6 (here q is odd, H_1 induces a monomial subgroup of $O^+(6, q)$, and H_1 is uniquely determined up to $\Gamma L(4, q)$-conjugacy),

(e) $H^{(\infty)}$ is $S_p(4, q)'$ or $SU(3, q^{1/2})$,

(f) $H \leq \Gamma O^\pm(4, q)$,

(g) $H^{(\infty)}$ is $PSL(2, q)$ or $SL(2, q)$ (many classes),

(h) $H^{(\infty)}$ is A_5 (here H arises from the natural permutation representation of S_5 in $O(5, q)$ and $p \neq 2, 5$),

(i) $H^{(\infty)}$ is $2 \cdot A_5$, $2 \cdot A_6$ or $2 \cdot A_7$ (these arise from the natural permutation representation of S_7 in $O(7, q)$),

(j) $H^{(\infty)} = A_7$ and $p = 2$,

(k) $H^{(\infty)} = S_p(4, 3)$, and $q \equiv 1 \bmod 3$ (this arises from the natural representation of the Weyl group $W(E_6)$ in $O^+(6, q)$),

(l) $H^{(\infty)} = SL(2, 7)$, and $q^3 \equiv 1 \bmod 7$ (here $H^{(\infty)}$ lies in the group $2 \cdot A_7$),

(m) $H^{(\infty)} = 4 \cdot PSL(3, 4)$, and q is a power of 9,

(n) $H^{(\infty)} = Sz(q)$, and $p = 2$.

5. Aschbacher's Theorem

In this appendix, we list Aschbacher's theorem with a few remarks on the groups involved. We also give some results that rely on this theorem, and to the main theorem of Guralnick-Penttila-Praeger-Saxl [63]).

Let G_0 be a non-Abelian simple group and let $G_0 \trianglelefteq G \leq AutG_0$. In this case, G is said to be 'almost simple.' For example, if G_0 is isomorphic to $PSL(m, q)$ and G is a subgroup of $P\Gamma L(m, q) \leq Aut(PSL(m, q))$, the above situation would apply. Noting that a subgroup M of $\Gamma L(m, n)$ corresponds by $\overline{M} = M/(M \cap (scalars))$ to a subgroup of $P\Gamma L(m, q)$, then for a group K of $\Gamma = \Gamma L(m, q)$, we would have $G_0 \trianglelefteq \overline{K} \leq P\Gamma L(m, q)$. In the following, we shall be describing groups various subgroups of Γ or of $\overline{\Gamma}$. In particular, there are eight classes of such groups T either in Γ or $\overline{\Gamma}$, which we shall denote by $\mathcal{C}(T)$.

Aschbacher's theorem is:

Let Y_0 be one of the following subgroups of $GL_d(q)$: $SL_d(q)$, $Sp_d(q)$ (if d is even), $SU_d(\sqrt{q})$ (if q is a square), $\Omega_d^\varepsilon(q)$ ($\varepsilon = \pm$ is d is even, and

$\varepsilon = \circ$ if d is odd). Let $Y = GL_d(q) \cap Aut(Y_0) \circ Z$, where $Z \leq Z(GL_d(q))$. Then

THEOREM 302 (Aschbacher). *If $d \geq 2$ and $(d, q) \neq (2, 2)$ and $(2, 3)$ and let G be a subgroup of Y not containing Y_0, then either $G \in \cup_{i=1}^8 \mathcal{C}_i(Y)$ or $G \in \mathcal{S}(Y)$.*

In the setting used for transitive t-parallelisms, Aschbacher's theorem would read:

THEOREM 303. *(Ashbacher (restricted form) Let G be a subgroup of the classical group $\Gamma \cong \Gamma L(m, q)$. If $PSL(m, q) \not\leq \overline{G}$, then either $G \in \mathcal{C}(\Gamma) = \cup_{i=1}^8 \mathcal{C}_i(\Gamma)$, or $G \in \mathcal{S}$.*

5.1. The Groups $\mathbf{G} \in \cup_{i=1}^8 \mathbf{C}_i(\boldsymbol{\Gamma})$. In this setting, we assume that there is a group H such that $H_0 \trianglelefteq H \leq AutH_0$, where H_0 is a finite non-Abelian simple group. We assume that $\overline{G}$ is a subgroup of $AutH_0$ that does not contain H_0 so that $G \in \cup_{i=1}^8 \mathcal{C}_i(\Gamma)$, or $G \in \mathcal{S}$. Actually, in the major part of our discussion for the transitive t-parallelism problem, H_0 will be $PSL(k, q)$ and we may take $\overline{G}$ as a subgroup of $P\Gamma L(k, q) \subseteq AutH_0$. So, taking Γ as $\Gamma L(k, q)$, and allowing that G is in Γ, we offer an informal description of these preimage groups in $\cup_{i=1}^8 \mathcal{C}_i(\Gamma)$ taken basically from [**161**], p. 3 table 1.2A.

$\mathcal{C}_1(\Gamma)$: a stabilizer of a totally singular or non-singular subspace (a maximal parabolic group).

$\mathcal{C}_2(\Gamma)$: a stabilizer of a decomposition $V = \oplus_{i=1}^z V_i$, $\dim V_i = a$ (e.g., a wreath product $GL(a, q^b) \wr S_t$, $k = at$)

$\mathcal{C}_3(\Gamma)$: a stabilizer of an extension field of $F_q \cong GF(q)$ of prime index (e.g., group extensions $GL(a, q^b).b$, $k = ab$, b prime).

$\mathcal{C}_4(\Gamma)$: a stabilizer of a tensor product decomposition $V = V_1 \otimes V_2$ (e.g., central products $GL(a, q) \circ GL(b, q)$, $k = ab$).

$\mathcal{C}_5(\Gamma)$: a stabilizer of a subfield of F_q of prime index (e.g. $GL(k, q_0)$, $q = q_0^b$, b prime).

$\mathcal{C}_6(\Gamma)$: normalizers of symplectic-type r-groups (r-prime) in absolutely irreducible representations $(Z_{q-1} \circ R^{1+2a}).S_{p2a}(r)$, $k = r^2$).

$\mathcal{C}_7(\Gamma)$: a stabilizer of a z-tensor product $V = \otimes_{i=1}^z V_i$, $\dim V_i = a$ $(GL(a, q) \circ GL(a, q)... \circ GL(a, q)).S_z$, $k = a^z$ (where the central product is taken over z-productands).

$\mathcal{C}_8(\Gamma)$: a classical subgroup (e.g., $S_{pn}(q)$, k even, $O_k^\epsilon(a)$, q odd, $GU_k(q^{1/2})$, q a square).

Now to describe the class of groups of $\mathcal{S}$, we consider the definition given in Kleidman and Liebeck [**161**], p. 3.

5.2. The Subgroups of $\mathcal{S}$. Although we shall consider more general situations, assume that H_0 is $PSL(k, q)$, where we are adopting the notation of the previous subsection. In this setting, the subgroup $\overline{G}$ of $P\Gamma L(k, q)$ or G in $\Gamma L(k, q)$ lies in $\mathcal{S}$ if the following three conditions hold:

(a) The socle S of $\overline{G}$ is a non-Abelian simple group.

(b) If L is the full covering group of S, and if $\rho : L$ to $GL(V)$ is a representation of L such that $\rho(L)/\rho(L) \cap scalar\ subgroup$ is S, then ρ is absolutely irreducible.

(c) $\rho(L)$ cannot be realized over a proper subfield of F_q.

In a more general setting, when H_0 could be, say, $PSU_k(q)$, additional conditions are assumed. In general there are eight conditions to fully describe the subgroups of $\mathcal{S}$, for all possible situations.

Note also that we may consider G within $\Gamma L(k, q)$ to be in $\mathcal{S}$ to mean $\overline{G}$ is in $\mathcal{S}$.

REMARK 73. *Some of the subgroups will be doubly notated by P_i. This is also intended to indicate the stabilizer of a totally singular i-dimensional subspace. Each such group is a maximal parabolic subgroup (i.e., a maximal subgroup containing a Borel subgroup (see pp 179–181 [161] for further details)).*

In particular, the groups P_1 and P_n are then the stabilizers of a point or a hyperplane in $PG(n, q)$, respectively, assuming the original vector space is of dimension $n + 1$ over F_q.

REMARK 74. *If we assume that G is a subgroup of $\Gamma L(n + 1, q)$, then $\overline{G}$ acts on a projective space $PG(n, q)$ as the lattice of subspaces of a vector space V of dimension $n + 1$ over F_q. Again, recall that for the preimage groups E in $\Gamma L(n+1, q)$, we use $\overline{E} = E/(E \cap (scalars))$.*

Now for a group R of Γ, define $\mathcal{C}_i(R) = \mathcal{C}_i(\Gamma) \cap R$, and let $\mathcal{C}(R) = \cup_{i=1}^{8}\mathcal{C}_i(R)$. Then we let $\overline{\mathcal{C}_i(R)} = \mathcal{C}_i(\overline{R})$, so that $\overline{\mathcal{C}(R)} = \mathcal{C}(\overline{R})$.

6. Guralnick-Penttila-Praeger-Saxl Theorem

THEOREM 304. *(Guralnick-Penttila-Praeger-Saxl [63]) Let $G \leq GL(m, q)$ of order divisible by a primitive prime divisor of $q^e - 1$, where $\frac{1}{2}m < e \leq m$. Then G is classified.*

We sketch here the possible types. The classification is given in terms of nine possible types, listed as examples $(2.1) - (2.9)$.

It can be seen from the proof of Theorem 304 that, the *geometric classes* $\mathcal{C}_i(Y)$, where $1 \leq i \leq 8$, yield to the Examples 2.1–2.5, while the class $\mathcal{S}(Y)$ yields to the Examples 2.6–2.9. The reader is directed to the article for the details, but to give some idea of the results, we give a brief description of the groups involved in this theorem, these are as follows:

EXAMPLE 9. *Let $G^* = G \cap GL(n,q)$, for G considered within $\Gamma L(n,q)$. Let r be a p-primitive divisor of $q^n - 1$, that divides $|G|$. Then examples (2.1) through (2.9) of [63] for p-primitive divisors of $q^n - 1$ are as follows:*

(2.1) Classical examples. Assume also that r is a p-primitive divisor of $q_0^e - 1$, where $GF(q_0)$ is a subfield of $GF(q)$. There are the four basic types:

(a) $SL(d, q_0) \lhd G$, for $\frac{d}{2} < e \leq d$;

(b) $Sp(d, q_0) \lhd G$, for $\frac{d}{2} < e \leq d$, d even, e even;

(c) $SU(d, q_0^{1/2}) \lhd G$, for $\frac{d}{2} < e \leq d$, q_0 a square, e odd;

(d) $\Omega_d^{\in}(q_0) \lhd G$, $\in = \pm$, when d is even, $\in = \circ$, when d is odd, for $\frac{d}{2} < e \leq d$.

(2.2) Reducible examples. If U is a subspace or quotient space of dimension $m \geq e$, and G^U is the induced group on U of $GL(U)$, and G^U has the primitive prime divisor property;

(2.3) Imprimitive examples. Let $r = e + 1 \leq d$, the vector space $V = U_1 \oplus U_2 \oplus ... \oplus U_d$, where $\dim U_i = 1$, for $i = 1, 2, .., d$, G is a subgroup of $GL(1,q)$ wr S_d of $GL(d,q)$ that preserves the decomposition and G acts primitively on the set $\{U_1, U_2, .., U_d\}$. If $e \leq d - 4$, G induces A_d or S_d on this set.

(2.4) $A_c \unlhd \overline{G^*}$, where $n + 1 = c - 1$ or $n + 1 = c - 2$;

(2.5) $M_{11} \unlhd \overline{G^*} < PGL(11, 3)$;

(2.6) $M_{23} \unlhd \overline{G^*} < PGL(11, 2)$;

(2.7) $M_{24} \unlhd \overline{G^*} < PGL(11, 2)$;

(2.8) $J_3 \unlhd \overline{G^*} < PGL(9, 4)$;

(2.9) $PSL(2, w) \unlhd \overline{G}$, where $n + 1 = \frac{1}{2}(w + 1)$ and w is prime.

7. O'Nan-Scott Theorem

THEOREM 305. *If H is any proper subgroup of S_n other than A_n, then H is a subgroup of one or more of the following subgroups:*

1. An intransitive group $S_k \times S_m$, where $n = k + m$;

2. An imprimitive group S_k wr S_m, where $n = km$ (a wreath product);

3. A primitive wreath product, S_k wr S_m, where $n = km$;

4. An affine group $AGL(d,p) = pd : GL(d,p)$, where $n = pd$;

5. A group of shape $T^m.(Out\ (T) \times S_m)$, where T is a non-Abelian simple group, acting on the cosets of the 'diagonal' subgroup Aut $(T) \times S_m$, where $n = |T|^{m-1}$;

6. An almost simple group acting on the cosets of a maximal subgroup.

Finally, we mention:

8. Dye's Theorem

THEOREM 306. *(Dye [50]) If r is a proper divisor of d, then the stabilizer in $GL_d(q)$ of a r-spread of $V_d(q)$ is a subgroup of $GL_{d/r}(q^r) \langle \sigma \rangle$, where σ is a transformation of order r induced by an automorphism of $GF(q^r)$.*

9. Johnson-Montinaro t-Transitive Theorem

THEOREM 307. *Let $\mathcal{P}$ be a t-parallelism in $PG_n(q)$, $q = p^h$, p a prime, and let G be a collineation group of $PG(n,q)$, which leaves $\mathcal{P}$ invariant and acts transitively on it. Then $t = 1$ (the vector dimension is 2). Furthermore, the group G fixes a point or a hyperplane of $PG(n,q)$, and one of the following occurs:*

(1) *$Z_{\frac{q^n-1}{q-1}} \trianglelefteq G \leq Z_{\frac{q^n-1}{q-1}} \rtimes (Z_n.Z_h)$;*

(2) *$q = 2$, $\mathcal{P}$ is one of the two regular parallelisms in $PG(3,2)$, and $PSL(2,7) \trianglelefteq G$.*

The main outline of the proof is as follows:

To be clear, we recall some definitions. Here a t-parallelism is a partition of the t-dimensional subspaces of a finite vector space by a set of mutually disjoint t-spreads.

We note that if we use the vector space dimension s for a s-parallelism, the projective space version would be an $s-1$-parallelism. Here we use the projective space dimension t, so that the result will show that $t = 1$.

Assume that G is a subgroup of $P\Gamma L(n+1, q)$ acting transitively on a t-parallelism in $PG(n,q)$. We then first show that G does not contain the group $PSL(n+1, q)$. In this situation, we may rely on Aschbacher's Theorem, which describes the subgroups of $P\Gamma L(n+1, q)$ that do not contain $PSL(n+1, q)$ as related to various classes of naturally occurring groups. In particular, the maximal subgroups may be determined. In our case, the group G will contain an element of order a p-primitive divisor of $q^n - 1$ and we may use the Guralnick-Penttila-Praeger-Saxl Theorem 6 (types (2.4) through (2.9)) to eliminate possibilities that arise using Aschbacher's and Liebeck's theorems. We have listed most of the required technical results required for the reader to understand the proof given here. However, the interested reader is advised that there is also considerable reference to Kleidman and Liebeck [161], and other technical material. As it is not practical to list here all of the theory required, the reader is also directed to those articles for explicit statements on the results used.

We have listed some of the important results that are used in the proof of the transitive theorem. For example, Dye's Theorem 306 on the subgroups preserving an s-spread, is listed as there is also considerable use of this theorem.

There are some very strict Diophantine equations that arise when dealing with transitive t-spreads. Since, we are using the projective dimension, again a t-spread in $PG(n, q)$ occurs only if $t + 1 \mid n + 1$. Hence $n + 1 = (s + 1)(t + 1)$ for some positive integer s. Furthermore, if $\mathcal{W}$ is a t-spread of $PG((s + 1)(t + 1) - 1, q)$, then

$$|\mathcal{W}| = \frac{q^{(s+1)(t+1)} - 1}{q^{t+1} - 1}.$$

If $\mathcal{P}$ is a t-parallelism of $PG(n, q)$, it is easily seen that

$$|\mathcal{P}| = \prod_{j=1}^{t} \frac{q^{(s+1)(t+1)} - q^j}{q^{t+1} - q^j}.$$

DEFINITION 126. *Two t-parallelisms $\mathcal{P}_1$ and $\mathcal{P}_2$ in $PG_n(q)$ are isomorphic (or equivalent), if there is an element $\varphi \in P\Gamma L_{n+1}(q)$ such that $\mathcal{P}_1 \varphi = \mathcal{P}_2$.*

Now, consider a t-parallelism $\mathcal{P}$ of $PG_n(q)$ and a subgroup G of $P\Gamma L_{n+1}(q)$ acting transitively on $\mathcal{P}$. Denote by $G\hat{\ }$ the inverse image of G in $\Gamma L_{n+1}(q)$. So G is isomorphic to $G\hat{\ }$ reduced modulo $G\hat{\ } \cap Z(GL_{n+1}(q))$. Let $\mathcal{W}$ be any t-spread of the parallelism $\mathcal{P}$. Hence, $[G : G_\mathcal{W}] = |\mathcal{P}|$ as G is transitive on $\mathcal{P}$. Denote by $G\hat{}_\mathcal{W}$ the inverse image in $G\hat{\ }$ of $G_\mathcal{W}$. Since the center of $GL_{n+1}(q)$ fixes each t-subspace, then

$$|\mathcal{P}| = [G : G_\mathcal{W}] = [G\hat{\ } : G\hat{}_\mathcal{W}].$$

Now, let $X = G\hat{\ } \cap GL_{n+1}(q)$ and $X_\mathcal{W} = G\hat{}_\mathcal{W} \cap GL_{n+1}(q)$. Then $G\hat{\ }/X \leq \langle f \rangle$, where $q = p^f$. Moreover, $G\hat{}_\mathcal{W}/X_\mathcal{W}$ is isomorphic to a subgroup of $G\hat{\ }/X$, and

(43.1) $|\mathcal{P}| = [G\hat{\ } : G\hat{}_\mathcal{W}] = f_1 [X : X_\mathcal{W}],$

where f_1 is a divisor of f.

Note that, as G is transitive on $\mathcal{P}$, and

$$|\mathcal{P}| = \prod_{j=1}^{t} \frac{q^{(s+1)(t+1)} - q^j}{q^{t+1} - q^j}$$

then $p^{fn} - 1$ divides the order of G and, of course, divides the order of $G\hat{\ }$. Now we would like to use basic theory of p-primitive divisors of $p^z - 1$, which exist unless $p^z = 64$ or $z = 2$ and $p + 1 = 2^a$. Therefore, either we have $p^{fn} = 64$ or there exists a primitive prime divisor u of $p^{fn} - 1$ dividing the order of $G\hat{}$. However, since, $n + 1 = (s+1)(t+1)$, with $s, t \geq 1$, the 64-case cannot occur.

Therefore, the order of $G\hat{}$ is divisible by a p-primitive divisor u. It is not difficult to show that any group of order u is linear so u divides the order of X, where $X \leq GL_{n+1}(q)$. Now as we have a p-primitive divisor of $q^n - 1$ dividing the group in question, we may use Theorem 304 in order to determine the structure of X.

Finally, since $G_{\mathcal{W}}$ leaves invariant a t-spread in $PG_n(q)$, we may use Dye's Theorem 306. (Note that the terminology is mixed here, in Dye's theorem an r-spread refers to the vector space dimension), so that $G_{\mathcal{W}}$ is a subgroup of $GL_{\frac{n+1}{t+1}}(q^{t+1}) \langle \sigma \rangle$.)

Now the idea is to show that our group X in $GL(n + 1, q)$ cannot be one of the possible examples of Theorem 304, where the p-primitive element divides $q^n - 1$. Hence, since $(n + 1)/2 < n \leq n + 1$, we may apply Theorem 304. The reader should be alert to the fact that we shall be using this theorem twice for what seems like the same group X. What we actually do is to begin with X as a subgroup of $GL(n + 1, q)$ admitting a p-primitive divisor of $q^n - 1$ and basically rule out Example 2.1. We then realize X as a subgroup of $GL(n, q)$ and push the group through this sieve again to establish what shorts of subgroups X could possible contain.

Continuing our discussion, the remainder of the proof without the proofs are as follows:

LEMMA 95. *The group X is not listed in Example 2.1 of Theorem 304.*

PROOF. Assume that X is one of the groups listed in Example 2.1. Then one of the following occurs:

(1) $SL_{n+1}(q) \trianglelefteq X$;
(2) $SU_{n+1}(q) \trianglelefteq X$ for n even;
(3) $SU_{n+1}(\sqrt{q}) \trianglelefteq X$, for n odd and q square;
(4) $\Omega_{n+1}(q) \trianglelefteq X$ for n even and q odd;

       Then we use Dye's Theorem 306 on spreads and arguments are p-primitive divisors to rule out the various cases.

                                                            □

LEMMA 96. *The group X is isomorphic to a subgroup of $GL_n(q)$. Furthermore, X fixes either a 1-dimensional subspace or a hyperplane of $V_{n+1}(q)$.*

PROOF. For the p-primitive examples corresponding to the Examples 2.6-2.9 of Theorem 304 $n + 1$ is a primitive prime divisor of $q^n - 1$, contradicting the fact that $n + 1 = (t + 1)(s + 1)$, with $s, t \geq 1$. The remaining argument uses technical material principally from [**161**], Proposition 4.1.17.   □

Since $X = G^{\hat{}} \cap GL_{n+1}(q)$, where $G^{\hat{}}$ is the inverse image of G in $\Gamma L_{n+1}(q)$, it follows from the previous lemma that, the group G fixes a point or a hyperplane of $PG_n(q)$.

Since X is isomorphic to a subgroup of $GL_n(q)$ containing an element of order a primitive prime divisor of $q^n - 1$, we may apply Theorem 304 in order to obtain further information on the structure of X. Note that our previous lemma concerning Example 2.1 excluded the classical groups groups possible based on X as a subgroup of $GL(n + 1, q)$, where it would be asserted that X contains a particular group as a normal subgroup. When considering X as a subgroup of $GL(n, q)$, it then might be possible to X to contain an analogous group of different order. This is essentially what occurs.

LEMMA 97. *The group X belongs to either Example 2.1 or the Examples 2.4. (b) of Theorem 304*

PROOF. It is straightforward to see that the cases corresponding to the Examples 2.2, 2.3, 2.4.(a), and 2.5 are immediately ruled out. Furthermore, by Theorem 3.1 of [**8**] and by bearing in mind that $n+1$ is a composite number, the cases corresponding to the Examples 2.6–2.9 of Theorem 304 lead to the following admissible cases for X:

(1) $A_7 \trianglelefteq X \leq P\Gamma L_3(25)$;
(2) $Z_3.J_3 \trianglelefteq X < PGL(9, 4)$;
(3) $PSL_2(7) \trianglelefteq X \leq P\Gamma L_3(q)$, where $q = 2, 9$ or 25;
(4) $PSL_2(11) \trianglelefteq X \leq P\Gamma L_5(4)$;
(5) $PSL_2(17) \trianglelefteq X \leq PGL_8(2)$;
(6) $PSL_2(19) \trianglelefteq X \leq P\Gamma L_9(4)$;
(7) $PSL_2(41) \trianglelefteq X \leq PGL_{20}(2)$.

Nevertheless, by using [**35**], none of (2)-(7) satisfies the equation $d(X) = |\mathcal{P}|$, where $d(X)$ is a transitive permutation representation of X. Hence, these cases cannot occur, and the proof is thus completed.   □

LEMMA 98. $t = 1$, *and one of the following occurs:*

(I) $\mathcal{P}$ *is one of the two regular parallelisms in $PG_3(2)$ and $PSL_2(7) \trianglelefteq$ G*
(II) $X \leq GL_{n/b}(q^b) \cdot b$.

PROOF. By Lemma 97, the group X is listed within the Example 2.1, or within the Example 2.4(b)) of Theorem 304. In particular, one of the following occurs:

(1) $SL_n(q) \trianglelefteq X$;
(2) $Sp_n(q) \trianglelefteq X$;
(3) $SU_n(q) \trianglelefteq X$ for n even;
(4) $\Omega_n^-(q) \trianglelefteq X$ for n even;
(5) $X \leq GL_{n/b}(q^b) \cdot b$, $b > 1$, is an extension field example.

To see that the projective t-parallelism forces $t = 1$, we recall that

$$|\mathcal{P}| = \prod_{j=1}^{t} \frac{q^{n+1-j} - 1}{q^{t+1-j} - 1}.$$

Assume that $t \geq 2$. Then $n - 1 \geq 4$, and by arguing as in Lemma 95, we obtain $(p, f(n-1)) \neq (2, 6)$. Therefore, X is divisible by a primitive prime divisor of $q^{n-1} - 1$, and hence only the case where $SL_n(q) \trianglelefteq X$ is admissible. Then, by (43.1), the group X_W contains a Sylow p-subgroup of X, which has order $q^{n(n-1)/2}$. On the other hand, $X_W \leq GL_{\frac{n+1}{t+1}}(q^{t+1}).Z_{t+1}$ by Theorem 306. Hence, a Sylow p-subgroup of X_W must have order $q^{\frac{n+1}{t+1}(n-t)}$ at most. Consequently, $n(n-1)/2 \leq \frac{n+1}{t+1}(n-t)$, which is clearly impossible for $t \geq 2$.

The remaining groups are ruled using Dye's Theorem and technical results from Kleidman and Liebeck [**161**]. □

CHAPTER 44

The Diagram

Of course, the diagram is supposed to simultaneously represent flocks of hyperbolic quadrics as well as flocks of quadratic cones, and maybe parallelisms of these structures.

The very obvious clue to decoding the diagram is 'what does the red indicate?' Those of you who knew me back in the day recall when I had red hair as well as 'una barba rossa' will explain the red. But the red has a nice shape also, the hint being: What is 14? Unfortunately, there is a bit more—what do the yellow points above the 14 suggest (think noble thoughts)?

In any case, in spite of the tedious explanation of the diagram, I very much hope you enjoyed the book

—Cheers, Norm Johnson

Bibliography

[1] J. André, Über Perspektivitäten mit transitiver Translationsgruppe, Arch. Math. 6 (1954), 29–32.

[2] R.D. Baker, Partitioning the Planes of $AG_{2m}(2)$ into 2-Designs, Discrete Math. 15 (1976), 205–211.

[3] R.D. Baker, G.L. Ebert, and T. Penttila, Hyperbolic fibrations and q-clans, Des. Codes Cryptogr. 34 (2005), no. 2–3, 295–305.

[4] R.D. Baker and G.L. Ebert, A nonlinear flock in the Minkowski plane of order 11, Eighteenth Southeastern International Conference on Combinatorics, Graph Theory, and Computing (Boca Raton, Fl. 1989), Congr. Numer. 58 (1987), 75–81.

[5] L. Bader, Some new examples of flocks of $Q^+(3, q)$, Geom. Dedicata 27 (1988), no. 3, 371–375.

[6] L. Bader and G. Lunardon, On the flocks of $Q^+(3,q)$, Geom. Dedicata 29 (1989), no. 2, 177–183.

[7] S. Ball and M. Brown, The six semifield planes associated with a semifield flock. Adv. Math. 189 (2004), no. 1, 68–87.

[8] J. Bamberg, T. Penttila, Overgroups of cyclic Sylow subgroups of linear groups, Comm. Algebra 36 (2008), 2503–2543.

[9] D. Betten and R. Riesinger, Topological parallelisms of the real projective 3-space, Result. Math. 47 (2005), 226–241.

[10] D. Betten and R. Riesinger, Constructing topological parallelisms of $PG(3, R)$, via rotation of generalized line pencils, Adv. Geom. 8 (2008), 11–32.

[11] D. Betten and R. Riesinger, Generalized line stars and topological parallelisms of the real projective 3-space, J. Geom. 91 (2008), no. 1–2, 1–20.

[12] D. Betten and R. Riesinger, Hyperflock determining line sets and totally regular parallelisms of (to appear) Mh. Math.

[13] D. Betten and R. Riesinger, Parallelisms of $PG(3, R)$ composed to non-regular spreads (preprint).

[14] A. Beutelspacher, Partitions of finite vector spaces: an application of the Frobenius number in geometry, Arch. Math. (Basel) 31 (1978/79), no. 2, 202–208.

[15] A. Beutelspacher, Parallelismen in unendlichen projektiven Räumen endlicher Dimension, Geom. Dedicata 7 (1978), 499–506.

[16] A. Beutelspacher, Parallelisms in finite projective spaces, Geom. Dedicata, (1974) 35–40.

[17] M. Biliotti, V. Jha, and N.L. Johnson, Special linear group sections on translation planes. Bull. Belg. Math. Soc. Simon Stevin 13 (2006), no. 3, 401–433.

634

[18] M. Biliotti, V. Jha, and N.L. Johnson, Symplectic flock spreads in $PG(3, q)$, Note di Mat. 24, no. 1 (2005), 85–110.

[19] M. Biliotti, V. Jha, and N.L. Johnson, Classification of transitive deficiency one partial parallelisms, Bull. Belg. Math. Soc. Simon Stevin 12 (2005), 371–391.

[20] M. Biliotti, V. Jha, and N.L. Johnson, Large quartic groups on translation planes, I-Odd Order; A characterization of the Hering Planes, Note di Mat. 23 (2004), 151–166.

[21] M. Biliotti, V. Jha and N.L. Johnson, Foundations of Translation Planes, Monographs and Textbooks in Pure and Applied Mathematics, Vol. 243, Marcel Dekker, New York, Basel, 2001, xvi+542 pp.

[22] M. Biliotti, V. Jha, and N. L. Johnson, Transitive parallelisms, Results Math. 37 (2000), 308–314.

[23] M. Biliotti and N.L. Johnson, Bilinear flocks of quadratic cones, J. Geom. 64 (1999), no. 1–2, 16–50.

[24] M. Biliotti and N. L. Johnson, Maximal Baer groups in translation planes and compatibility with homology groups, Geom. Ded. 59, no. 1 (1996), 65–101.

[25] A. Bonisoli, On resolvable finite Minkowski planes, J. Geom. 36 (1989), no. 1–2, 1–7.

[26] A. Bruen, Spreads and a conjecture of Bruck and Bose, J. Algebra 23 (1972), 519–537.

[27] A. Bruen and J.A. Thas, Partial spreads, packings and Hermitian manifolds, Math. Z. 151 (1976), 201–214.

[28] R.P. Burn, Finite Bol loops, Math. Proc. Cambridge Philos. Soc. 84 (1978), no. 3, 377–385.

[29] W. Büttner, On translation planes containing $Sz(q)$ in their translation complement, Geom. Dedicata 11 (1981), no. 3, 315–327.

[30] W.E. Cherowitzo and N.L. Johnson, The doubly-transiitve α-flokki (preprint).

[31] W.E. Cherowitzo and N.L. Johnson, α-flokki, partial α-flokki, and Baer groups, (submitted).

[32] W.E. Cherowitzo and N.L. Johnson, Parallelisms of Quadric Sets, (submitted).

[33] W.E. Cherowitzo and N.L. Johnson, Net Replacement in the Hughes-Kleinfeld Semifield Planes (submitted).

[34] W.E. Cherowitzo, Flocks of Cones: Star Flocks, (submitted).

[35] J.H. Conway, R.T. Curtis, R.A. Parker, R.A. Wilson, An Atlas of Finite Groups. Maximal subgroups and ordinary characters for simple groups, with computational assistance from J.G. Thackrey, Oxford University Press, Eynsham, 1985, xxxiv + 252 pp.

[36] B.N. Cooperstein, Minimal degree for a permutation representation of a classical group, Israel J. Math. 30 (1978), 213–235.

[37] M. Cordero and R. F. Figueroa, Transitive autotopism groups and the generalized twisted field planes, Mostly finite geometries, Iowa City, IA, 1996, 191–196, Lecture Notes in Pure and Appl. Math., 190, Dekker, New York, 1997.

[38] P. Dembowski, Finite geometries, Springer-Verlag, Berlin, 1997.

[39] U. Dempwolff, A note on semifield planes admitting irreducible planar Baer collineations, Osaka J. Math. 45 (2008), no. 4, 895–908.

[40] U. Dempwolff, A characterization of the generalized twisted field planes, Arch. Math. (Basel) 50 (1988), no. 5, 477–480.

[41] R.H.F. Denniston, Cyclic packings of projective spaces, Atti Acc. Naz. Lincei, 8 (1976), 36–40.

[42] R.H.F. Denniston, Cyclic packings of the projective space of order 8. Atti Accad. Naz. Lincei Rend. Cl. Sci. Fis. Mat. Natur. (8) 54 (1973), 373–377 (1974).

[43] R.H.F. Denniston, Some packings of projective spaces, Atti Accad. Naz. Lincei Rend. Cl. Sci. Fis. Mat. Natur. (8) 52 (1972), 36–40.

[44] E. Diaz, N.L. Johnson, and A. Montinaro, Elation Switching in Real Parallelisms, Innovations in Incidence Geometries (to appear).

[45] E. Diaz, N.L. Johnson, and A. Montinaro, Transitive deficiency one partial parallelisms, Advances and Applications of Discrete Mathematics, 1, no. 1 (2008), 1–34.

[46] E. Diaz, N.L. Johnson, A. Montinaro, Coset switching in Parallelisms, Finite Fields and Applications, 14 (2008), 766–784.

[47] L.E. Dickson, Linear groups, with an Exposition of Finite Groups and Associative Algebras, Wiley, New York, 1962.

[48] J.D. Dixon and B. Mortimer, Permutation groups, Springer-Verlag, New York, Berlin, Heidelberg 1996.

[49] D. Draayer, N.L. Johnson, and R. Pomareda, Triangle Transitive Planes, Note di Mat. 26, no. 1 (2006), 29–53.

[50] R.H. Dye, Spreads and Classes of maximal subgroups of $GL_n(q)$, $SL_n(q)$, $PGL_n(q)$ and $PSL_n(q)$, Ann. Mat. Pura Appl. 58 (1991), 33–50.

[51] G.L. Ebert and K.E. Mellinger, Mixed partitions and related designs, Des. Codes Cryptogr. 44 (2007), no. 1–3, 15–23.

[52] D.A. Foulser, Planar collineations of order p in translation planes of order p^r, Geom. Dedicata 5 (1976), no. 3, 393–409.

[53] D.A. Foulser, Baer p-elements in translation planes, J. Algebra 31 (1974), 354–366.

[54] D.A. Foulser and N.L. Johnson, The translation planes of order q^2 that admit $SL(2, q)$ as a collineation group I. Even Order, J. Alg. 86 (1984), 385–406.

[55] D.A. Foulser and N.L. Johnson, The translation planes of order q^2 that admit $SL(2, q)$ as a collineation group II. Odd order, J. Geom. 18 (1983), 122–139.

[56] D.A. Foulser, N. L. Johnson and T.G. Ostrom, Characterization of the Desarguesian planes of order q^2 by $SL(2, q)$, Internat. J. Math. and Math. Sci. 6 (1983), 605–608.

[57] M.J. Ganley, Baer involutions in semifields of even order, Geom. Dedicata. 2 (1974), 499–508.

[58] M.J. Ganley and V. Jha, On translation planes with a 2-transitive orbit on the line at infinity. Arch. Math. (Basel) 47 (1986), no. 4, 379–384.

[59] H. Gevaert and N.L. Johnson, Flocks of quadratic cones, generalized quadrangles, and translation planes, Geom. Dedicata 27 (1988), 301–317.

[60] H. Gevaert, N.L. Johnson, and J.A. Thas, Spreads covered by reguli, Simon Stevin 62 (1988), 51–62.

[61] A.M. Gleason, Finite Fano planes, Amer. J. Math. 78 (1956), 797–807.

636

[62] R. Guralnick, Subgroups of prime power index in a simple group, J. Algebra 81 (1983), no. 2, 304–311.

[63] R. Guralnick, T. Penttila, C.E. Praeger, and J. Saxl, Linear groups with orders having certain large prime divisors, Proc. London Math. Soc. 78 (1999), 167–214.

[64] W. Hartley, Determination of the ternary collineation groups whose coefficients lie in the $GF(2^n)$, Ann. Math. 27 (1926), 140–158.

[65] O. Heden, The Frobenius number and partitions of a finite vector space. Arch. Math. (Basel) 42 (1984), no. 2, 185–192.

[66] G. Heimbeck, Translationsebenen der Ordnung 49 mit einer Quaternionengruppe von Dehnungen, J. Geom. 44 (1992), no. 1–2, 65–76.

[67] C. Hering, On the structure of finite collineation groups of projective planes, Abh. Math. Sem. Univ. Hamburg 49 (1979), 155–182.

[68] Y. Hiramine and N.L. Johnson, Regular partial conical flocks, Bull. Belg. Math. Soc 2 (1995), 419–433.

[69] Y. Hiramine, M. Matsumoto, and T. Oyama, On some extension of 1-spread sets, Osaka J. Math. 24 (1987), no. 1, 123–137.

[70] Y. Hiramine and N.L. Johnson, Regular partial conical flocks, Bull. Belg. Math. Soc. Simon Stevin 2 (1995), 419–433.

[71] J.W.P. Hirschfeld, Finite Projective Spaces of Three Dimensions, Oxford Univ. Press. 1985.

[72] J.W.P. Hirschfeld, Projective Geometries over finite fields, Oxford Univ. Press. 1979 (2nd Edition, 1998).

[73] J.W.P. Hirschfeld and J.A. Thas, General Galois Geometries, Oxford University Press, Oxford, 1991.

[74] B. Huppert, Endliche Gruppen I, Die Grundlehren der Mathematischen Wissenschaften, Band 134, Springer-Verlag, Berlin, Heidelberg, New York, 1967.

[75] D.R. Hughes and F.C. Piper, Projective Planes, Graduate Texts in Mathematics, 6, Springer-Verlag, New York, 1973.

[76] C. Jansen, K Lux, R. Parker, R. Wilson, An Atlas of Brauer Characters, Clarendon Press, Oxford, 1995.

[77] V. Jha and N.L. Johnson, Vector space partitions and designs; Part I; Basic Theory, Designs, Codes, Cryptography, Note di Mat. (to appear).

[78] V. Jha and N.L. Johnson, Vector space partitions and designs; Part II-Constructions, Note di Mat. (to appear).

[79] V. Jha and N.L. Johnson, The dimension of subplanes of translation planes, Bull. Belgian Math. Soc. (to appear).

[80] V. Jha and N.L. Johnson, Constructions of Sperner k-Spreads of dimension tk, Note di Mat. (to appear).

[81] V. Jha and N.L. Johnson, Spread-Theoretic Dual of a Semifield, Note di Mat. (to appear).

[82] V. Jha and N.L. Johnson, Subgeometry partitions from cyclic semifields, Note di Mat. 28, no. 1 (2008), 133–148.

[83] V. Jha and N.L. Johnson, Double-Baer groups, IIG (Gabor Korchmaros special volume), Innovations in Incidence Geometry, 6-7, 2007–2008, 227–248.

[84] V. Jha and N.L. Johnson, Algebraic and Geometric Lifting, Note di Mat., 27, n. 1 (2007), 85–101.

637

[85] V. Jha and N.L. Johnson, Ostrom-derivates, Innov. Incidence Geom. 1 (2005), 35–65.

[86] V. Jha and N.L. Johnson, The classification of spreads in $PG(3,q)$ admitting linear groups of order $q(q+1)$, I. Odd Order, J. Geom. 81 (2004/5), 46–80.

[87] V. Jha and N.L. Johnson, Nuclear fusion in finite semifield planes, Advances in Geom. 4 (2004), 413–432.

[88] V. Jha and N.L. Johnson, The classification of spreads in $PG(3,q)$ admitting linear groups of order $q(q+1)$, I. Even Order, Advances in Geometry (2003), S271–S313.

[89] V. Jha and N.L. Johnson, Lifting Quasifibrations II – Non-Normalizing Baer involutions, Note Mat. 20 (2000/01), no. 2, 51–68 (2002).

[90] V. Jha and Norman L. Johnson, Almost Desarguesian Maximal partial spreads. Designs, Codes, Cryptography 22 (2001), 283–304.

[91] V. Jha and N. L. Johnson, Conical, ruled, and deficiency one translation planes, Bull. Belgian Math. Soc. 6 (1999), 187–218.

[92] V. Jha and N.L. Johnson, Structure theory for point-Baer and line-Baer collineations in affine planes, Mostly Finite Geometries (Iowa City, IA, 1996), Lecture Notes in Pure and Appl. Math., 190, Dekker, New York, 1997, 235–273.

[93] V. Jha and N.L. Johnson, On regular r-packings, Note di Matematica, Vol. VI, (1986), 121–137.

[94] V. Jha and N.L., Regular parallelisms from translation planes, Discrete J. Math. 59 (1986), 91–97.

[95] V. Jha and N.L. Johnson, Derivable nets defined by central collineations, J. Combin. Inform. System Sci. 11 (1986), no. 2–4, 83–91.

[96] V. Jha and Norman L. Johnson, Coexistence of elations and large Baer groups in translation planes, J. London Math. Soc. (2)32 (1985), 297–304.

[97] N.L. Johnson, The non-existence of Desarguesian t-parallelisms, t an odd prime, Note di Mat. (to appear).

[98] N.L. Johnson, Extended André Sperner Spaces, Note di Mat. 28, no. 1 (2008), 149–170.

[99] N.L. Johnson, Constructions of Subgeometry Partitions, Bulletin Belgian Math. Soc. Bull. Belg. Math. Soc. Simon Stevin 15 (2008), 437–453.

[100] N.L. Johnson, m th-root subgeometry partitions. Des. Codes Cryptogr. 46 (2008), no. 2, 127–136.

[101] N.L. Johnson, Homology groups of translation planes and flocks of quadratic cones, II; j-Planes, Note di Mat. 28, no. 1 (2008), 85–101.

[102] N.L. Johnson, Bol Planes of orders 3^4 and 3^6. Note di Math. 26 (2006), 139–148.

[103] N.L. Johnson, Homology groups of translation planes and flocks of quadratic cones, I: The structure, Bull. Belg. Math. Soc. Simon Stevin 12 (2006), no. 5, 827–844.

[104] N.L. Johnson, Dual deficiency one transitive partial parallelisms, Note Mat. 23 (2004), 15–38.

[105] N.L. Johnson, Parallelisms of projective spaces, J. Geom. 76 (2003), 110–182.

[106] N.L. Johnson, Quasi-subgeometry partitions of projective spaces, Bull. Belg. Math. Soc. Simon Stevin 10 (2003), no. 2, 231–261.

638

[107] N.L. Johnson, Hyper-Reguli and non-André Quasi-subgeometry partitions of projective spaces, J. Geometry, 78 (2003), 59–82.

[108] N.L. Johnson, Buetelspacher's Parallelism Construction, Note di Mat., 21, (2002/2003), 57–69.

[109] N.L. Johnson, Transversal spreads, Bull. Belg. Math. Soc. 9 (2002), 109-142.

[110] N.L. Johnson, Dual parallelisms, Note Mat. 21 (2002), 137–150.

[111] Johnson, N.L., Some new classes of finite parallelisms, Note di Mat. 20 (2000/2001), 77–88.

[112] N. L. Johnson, Two-transitive parallelisms, Des. Codes Cryptogr. 22 (2001) 179–189.

[113] N.L. Johnson, Retracting spreads, Bull. Belg. Math. Soc. 8 (2001), 1–20.

[114] N.L. Johnson, Subplane Covered Nets, Pure and Applied Mathematics, Vol. 222 Marcel Dekker, 2000.

[115] N.L. Johnson, Infinite nests of reguli, Geom. Dedicata 70 (1998), 221–267.

[116] N.L. Johnson, Derivable nets can be embedded in nonderivable planes, Trends in Mathematics (1998), Birkhäuser Verlag Basel/Switzerland, 123–144.

[117] N.L. Johnson, Flocks of infinite hyperbolic quadrics, Journal of Algebraic Combinatorics, 6, no. 1 (1997), 27–51.

[118] N.L. Johnson, Lifting quasifibrations, Note di Mat. 16 (1996), 25–41.

[119] N.L. Johnson, Partially sharp subsets of $P\Gamma L(n, q)$, London Math. Soc. Lecture Notes #191. Finite Geometry and combinatorics, (1993), 217–232.

[120] N.L. Johnson, Translation planes covered by subplane covered nets, Simon Stevin, Vol. 66, no. 3-4, (1992), 221–239.

[121] N.L. Johnson, Ovoids and translation planes revisited, Geom. Dedicata 38 (1991), no. 1, 13–57.

[122] N.L. Johnson, Derivation by coordinates, Note di Mat. 10 (1990), 89–96.

[123] N.L. Johnson, Derivable nets and 3-dimensional projective spaces. II. The structure, Archiv d. Math. 55 (1990), 84–104.

[124] N.L. Johnson, Flocks and partial flocks of quadric sets, Finite Geometries and Combinatorial Designs (Lincoln, NE, 1987), Contemp. Math., vol. 111 Amer. Math. Soc., Providence, RI, 1990, no. 3–4, 199–215.

[125] N.L. Johnson, Translation planes admitting Baer groups and partial flocks of quadric sets, Simon Stevin 63 (1989), no. 3, 167–188.

[126] N.L. Johnson, The derivation of dual translation planes, J. Geom. 36 (1989), 63–90.

[127] N.L. Johnson, Semifield planes of characteristic p that admit p-primitive Baer collineations, Osaka J. Math. 26 (1989), 281–285.

[128] N.L. Johnson, Flocks of hyperbolic quadrics and translation planes admitting affine homologies, J. Geom. 34(1989), 50–73.

[129] N.L. Johnson, Derivable nets and 3-dimensional projective spaces, Abhandlungen d. Math. Sem. Hamburg, 58 (1988), 245–253.

[130] N.L. Johnson, Semifield flocks of quadratic cones, Simon Stevin, 61, no. 3–4 (1987), 313–326.

[131] N.L. Johnson, The maximal special linear groups which act on translation planes, Boll. Un. Mat. Ital. A (6) (1986), no. 3, 349–352.

[132] N.L. Johnson, Lezioni sui piani di traslazione, Quaderni d. Dipart. d. Mat. dell'Univ. di Lecce, Q. 3 (1986), 1–121.

[133] N.L. Johnson, A note on net replacement in transposed spreads, Bull. Canad. J. Math., 28 (4) (1985), 469–471.

[134] N.L. Johnson, Translation planes of order q^2 that admit $q+1$ elations, Geom. Ded. 15 (1984), 329–337.

[135] N.L. Johnson, A note on the construction of quasifields, Proc. Amer. Math. Soc. 29 (1971), 138–142.

[136] N.L. Johnson, Nonstrict semi-translation planes, Arch. Math. 20 (1969), 301–310.

[137] N.L. Johnson and M. Cordero, Transitive Subgeometry Partitions, Note di Mat. (to appear).

[138] N. L. Johnson, V. Jha, and M. Biliotti, Handbook of Finite Translation Planes. Pure and Applied Mathematics (Boca Raton), 289. Chapman & Hall/CRC, Boca Raton, FL, 2007, xxii+861 pp.

[139] N.L. Johnson and X. Liu, Flocks of quadratic and semi-elliptic cones, Mostly finite geometries (Iowa City, IA, 1996), Lecture Notes in Pure and Appl. Math., vol. 190, Dekker, New York, 1997, pp. 275–304.

[140] N.L. Johnson and X. Liu, The generalized Kantor-Knuth flocks, pp. 305–314 in N.L. Johnson, Ed., Mostly finite geometries, Lecture Notes in Pure and Appl. Math., Vol. 190, Marcel Dekker, New York–Basel–Hong Kong, 1997.

[141] N.L. Johnson and K. Mellinger, Multiple spread-retraction, Advances in Geom. 3 (2003), 263-0286.

[142] N.L. Johnson and A. Montinaro, The doubly-transitive focal-spreads (submitted).

[143] N.L. Johnson and A. Montinaro, Doubly transitive t-parallelisms, Results in Mathematics, 52 (2008), 75-089.

[144] N.L. Johnson and A. Montinaro, Transitive t-Parallelisms (submitted).

[145] N.L. Johnson and T.G. Ostrom, Direct products of affine partial linear spaces, Journal of Combinatorial Theory (A), 75, no. 1 (1996), 99–140.

[146] N.L. Johnson and T.G. Ostrom, Translation planes of characteristic two in which all involutions are Baer, J. Algebra (2) 54 (1978), 291–315.

[147] N.L. Johnson and S.E. Payne, Flocks of Laguerre planes and associated geometries, pp. 51–122 in N.L. Johnson, Ed., Mostly finite geometries, Lecture Notes in Pure and Appl. Math., Vol. 190, Marcel Dekker, New York–Basel–Hong Kong, 1997.

[148] N.L. Johnson and R. Pomareda, A maximal partial flock of deficiency one of the hyperbolic quadric in PG(3,9), Simon Stevin, vol. 64, no. 2(1990), 169–177.

[149] N.L. Johnson and R. Pomareda, Minimal parallelism-inducing groups, Aequationes Math. 20 (2007), 92–124.

[150] N.L. Johnson and R. Pomareda, Partial parallelisms with sharply two-transitive skew spreads, Ars Combinatorica 70 (2004), 275–287.

[151] N.L. Johnson and R. Pomareda, Parallelism-inducing groups, Aequationes Math. 65 (2003), 133–157.

[152] N.L. Johnson and R. Pomareda, Real Parallelisms, Note d. Mat. 21 (2002), 127–135.

[153] N.L. Johnson, and R. Pomareda, Transitive partial parallelisms of deficiency one, European J. Combinatorics, 232 (2002), 969–986.

[154] N.L. Johnson and R. Pomareda, m-Parallelisms, International J. Math. and Math. Sci. (2002), 167–176.

[155] N.L. Johnson, R. Pomareda, and F.W. Wilke, j-planes, J. Combin. Theory Ser. A 56 (1991), no. 2, 271–284.

[156] W.M. Kantor, Commutative semifields and symplectic spreads, J. Algebra 270 (2003), no. 1, 96–114.

[157] W.M. Kantor, Ovoids and translation planes, Canad. J. Math. 34 (1982), no. 5, 1195–1207.

[158] W. Kantor and T. Penttila, Flokki Planes and Cubic Polynomials, Note di Mat. (to appear).

[159] W.M. Kantor and R.A. Liebler, The Rank 3 permutation representations of the finite classical groups, Trans. Amer. Math. Soc. 271 (1982), 1–71.

[160] P. B. Kleidman, The low dimensional finite classical groups and their subgroups. Ph.D. Thesis.

[161] P. B. Kleidman and M. Liebeck, The subgroup structure of the finite classical groups, Cambridge University Press, Cambridge, 1990, Acad. Press, Boston, 1994.

[162] O.H. King, The subgroup structure of finite classical groups in terms of geometric configurations, Surveys in combinatorics 2005, 29–56, London Math. Soc. Lecture Note Ser., 327, Cambridge Univ. Press, Cambridge.

[163] N. Knarr, Derivable Affine Planes and Translation Planes, Bull. Belg. Math. Soc. 7 (2000), 61–71.

[164] D.E. Knuth, Finite semifields and projective planes, J. Algebra 2 (1965), 541–549.

[165] A class of projective planes, Trans. Amer. Math. Soc. 115 (1965), 182–217.

[166] G. Lunardon, On regular parallelisms in $PG(3,q)$, Discrete Math. 51 (1984), 229–235.

[167] H. Lüneburg, Translation Planes, Springer-Verlag, Berlin 1980.

[168] H. Lüneburg, Charakterisierungen der endlichen desarguesschenprojektiven Ebenen, Math. Z. 85 (1964), 419–450.

[169] B. Mwene, On the subgroups of $PSL_4(2^m)$, J. Algebra 41 (1976), 79–107.

[170] B. Mwene, On some subgroups of $PSL(4,q)$, q odd., Geom. Dedicata 12 (1982), 189–199.

[171] H.H. Mitchell, Determination of the ordinary and modular ternary linear groups, Trans. Amer. Math. Soc. 12 (1911), 207–242.

[172] T.G. Ostrom, Derivable Nets, Canad. Bull. Math. 8 (1965), 601–613.

[173] S.E. Payne and J.A. Thas, Conical flocks, partial flocks, derivation, and generalized quadrangles, Geom. Dedicata 38 (1991), 229–243.

[174] T. Penttila and B. Williams, Regular packings of $PG(3,q)$, European J. Combin. 19 (1998), no. 6, 713–720.

[175] T. Penttila and L. Storme, Monomial flocks and herds containing a monomial oval, J. Combin. Theory (Ser. A) 83 (1998), no. 1, 21–41.

[176] A. Prince, The cyclic parallelisms of $PG(3,5)$, European J. Combin. 19 (1998), no. 5, 613–616.

[177] R. Riesinger, Spreads admitting net generating regulizations, Geom. Ded. 62 (1996), 139–155.

[178] G. M. Seitz, Flag-transitive subgroups of Chevalley groups, Ann. of Math. 97 (1973) 27–56.

641

[179] L. Storme and J.A. Thas, k-arcs and partial flocks, Linear Algebra Appl. 226/228 (1995), 33–45.

[180] J.A. Thas, Generalized quadrangles and flocks of cones, European J. Combin. 8 (1987), 441–452.

[181] M. Walker, Spreads covered by derivable partial spreads, J. Comb. Theory (Series A), 38 (1985), 113–130.

Index